D1283048

The Biology of the Blastocyst

The Biology of the Blastocyst

EDITED BY
R. J. Blandau

The University of Chicago Press

Chicago and London

International Standard Book Number: 0-226-05670-8
Library of Congress Catalog Card Number: 70-128713
THE UNIVERSITY OF CHICAGO PRESS, CHICAGO 60637
The University of Chicago Press, Ltd., London

Printed in the United States of America

Contents

Preface

The purpose of this book is to present a modern synthesis of our knowledge of the intricacies of the biology of the mammalian egg during the preimplantation period. It concentrates particularly on the blastocyst, the final stage of development in the preimplantation period.

This monograph covers the deliberations of an international symposium, The Biology of the Blastocyst, held at Lake Wilderness, the postgraduate training center of the University of Washington. Here amid the magnificent trees and mountains of the Northwest forty scientists discussed at length the biology of the blastocyst and groped for a better understanding of the complex embryo-endometrial interrelationships that lead finally to implantation.

To those who took part it was a valuable experience. Conflicting views sent them back to the laboratory with renewed vigor to seek resolutions. The hope was expressed repeatedly that young and imaginative minds could be enticed to join in the exploration of this realm of biology and medicine.

In view of the rapidly expanding world population with its serious attendant problems, advance in basic knowledge concerning reproductive phenomena is vital to the discovery of new and more acceptable methods of regulating family size. The conference brought to bear on a pivotal segment of knowledge of the reproductive process the approaches and techniques from the disciplines of biology, chemistry, and physics. Whatever its imperfections and limitations, the resulting monograph attempts to portray the present-day views of this subject from these many vantage points; it bears witness to the belief that new knowledge and understanding follow such concerted attack.

RICHARD J. BLANDAU

Acknowledgments

This symposium was generously supported and sponsored by the Reproduction and Population Research Branch and Center for Population Research, National Institutes of Child Health and Human Development, and by the Population Council.

The editor wishes to thank Dr. Ruth Rumery, Mrs. Annie Kennedy, and Mrs. Lucille Holt, who assisted so effectively in the organization of the conference.

Thanks are due to Dr. Penelope Gaddum, Dr. Louise Odor, Mrs. Lynn Langley, Mrs. Gina Whest, Mrs. Briskai Rothenberger, and Mrs. Eunice Sing-Nyor Wang, who painstakingly assisted with the galley proofs and the preparation of the author and subject indexes, and to Mr. Roy Hayashi for his help in evaluating and arranging the illustrations.

The skillful editing of Mrs. Mary Adams is especially appreciated.

1

The Biology of the Blastocyst in Historical Perspective

Charles W. Bodemer

Department of Biomedical History
School of Medicine
University of Washington, Seattle

Modern knowledge of the mammalian blastocyst derives ultimately from the gradual accretion of information regarding mammalian reproduction, the structural basis of animal bodies, and the facts of comparative and experimental embryology. This essay discusses the growth of knowledge regarding early mammalian development and strives to illumine concepts of development and interpretations of the mammalian blastocyst prevailing during the period from classical antiquity to the late nineteenth century. It is necessarily of morphological orientation, since early mammalian development only recently has been investigated with the degree of technical and experimental finesse requisite to developmental physiology.

I. Concepts of Development during Antiquity and the Medieval Period

The generation and development of plants and animals has been a subject of study throughout the history of Western science. Before the seventeenth century, however, embryological knowledge was based primarily on the writings of Aristotle and Galen, both of whom lived during the period of Graeco-Roman antiquity. Since their viewpoints prevailed for more than a millennium, their concepts of embryonic development are representative of embryological thought during classical antiquity and the medieval period.

Aristotle's works dealing with generation were written during the latter part of the fourth century of the pre-Christian era. His primary embryological treatise is *De generatione animalium,* but discussions of generation and development occur in his other biological works, in particular, *De historia animalium* and *De partibus animalium.* Aristotle's concept of mammalian development is based upon the idea of the interaction of qualitatively different maternal and paternal elements. The male, according to the view set forth in *De generatione animalium,* contributes the semen; the female contributes the *catamenia.* Consistent with the Aristotelian adaptation of the doctrine of opposites, the male semen is hot and active, the *catamenia* are the equivalent of the menstrual blood secreted through very fine vessels into the uterine lumen. A comparable secretion is said to exist in all mammals.

The *catamenia* represent the material foundation of the embryo, which is activated and guided in its development by the action of the male semen. This is, of course, consonant with Aristotle's concept of the difference between male and female. "The female," he writes, "is a mutilated male, and the *catamenia* are semen, only not pure; for there is only one thing they have not in them, the principle of soul." The semen contributes nothing material to the embryo, but accounts for its power and movement. The *catamenia,* on the other hand, are the actual stuff of which the embryo is made:

> If, then, the male stands for the effective and active, and the female, considered as female, for the passive, it follows that what the female would contribute to the semen of the male would not be semen but material for the semen to work upon. This is just what we find to be the case, for the catamenia have in their nature an affinity to the primitive matter.

Aristotle's scheme of mammalian development obviously requires no ovum, since the *catamenia* are secreted directly into the uterine cavity and activation to development occurs in the uterus. Indeed, for Aristotle, "egg is the name given to that class of perfected fetation out of which the forming animal comes into being," and fetation is the first mixture of male and female. In rather physicochemical Aristotelian terms, then, the initial stage of mammalian development is a variety of coagulation:

> When the material secreted by the female in the uterus has been fixed by the semen of the male this acts in the same ways as rennet acts upon milk, for rennet is a kind of milk containing vital heat, which brings into one mass and fixes the similar material, and the relation of the semen to the catamenia is the same, milk and the catamenia being of the same nature—when, I say, the more solid part comes together, the liquid is separated off from it, and as the earthy parts solidify membranes form all round it; this is both a necessary result and for a final cause, the former because the surface of a mass must solidify on heating as well as on cooling, the latter because the foetus must not be in a liquid but be separated from it.

De generatione animalium is a truly monumental work, the more impressive in light of its date of composition. Perhaps the greatest embryological treatise dating from classical antiquity, it was a permeating influence for centuries, and its errors therefore had unfortunately enduring effects upon embryological thought. Aristotle's doctrine of semen and *catamenia* interacting *in utero* to form the fetus became in fact a potent deterrent to adequate understanding of mammalian development long after the same concept embedded in a treatise of lesser stature would have become a historical curiosity.

Galen, who flourished during the second century of the Christian era, was the last significant writer on the subject of generation and development during antiquity. His views on embryonic development are contained in *De semine, De foetuum formatione, De usu partium,* and *De naturalibus facultatibus.* Galen's works were not superseded in antiquity, and with those of Aristotle they became the main source of biomedical knowledge until the modern period. Aristotle was ignorant of the human ovaries, and his concept of *in utero* formation of the *catamenia* did not require female generative organs other than the uterus. The human ovaries were first described during the century after Aristotle's death by the Alexandrian anatomist Herophilus of Chalcedon. Considering the ovaries to be the equivalent of the testes, Herophilus designated them as the *testiculi foeminis,* a name they retained for centuries. Herophilus denies the uterus a role in formation of the maternal generative material and asserts, instead, that the ovaries secrete a female semen comparable with that of the male. This concept enjoyed considerable currency during antiquity, and exerted its effect upon such influential writers as Rufus, Soranus, and Galen.

Galen studied at Alexandria around A.D. 152, there imbibing the anatomical and physiological traditions of the Museum and later incorporating many Alexandrian concepts into his own theoretical constructs. The views of Herophilus had particular influence upon Galen's interpretations of mammalian development. Thus, Galen follows Herophilus in considering the ovaries as structural and functional homologues of the testes and positing that these *testiculi foeminis* are responsible for formation of the female contribution to embryogeny in the same way that the male testes are responsible for formation of semen in the male. Rejecting thus the Aristotelian concept of *in utero* formation of the female generative material, Galen conceives of its formation in the ovaries and its transport to the uterus through the oviduct, which "runs to the apex or cornu of the uterus on each side, and there deposits the semen."

Semen formation, according to Galen, occurs in a comparable manner in both sexes. But the female is endowed with less heat than the male and is imperfect; therefore, "of course, the female must have smaller, less perfect testes, and the semen generated in them must be scantier, colder, and wetter." Galen concurs with Aristotle that the semen originates in the blood and in the last analysis the aliment. Semen, he says, is "blood thoroughly concocted by the vessels containing it." Formation of the semen actually begins outside the ovary, in the highly coiled ovarian vessels; the final action in generation of the semen occurs within the ovary, where the blood moves sluggishly through the tortuous vessels. This relative hemostasis is the sine qua non of semen formation; indeed, writes Galen, "it is not at all surprising if a spermatic juice accumulates when the blood stagnates, so to speak, in these coils." Thus the semen, begun with a preliminary concoction in the ovarian vessels, is concocted thoroughly and completed in the ovary itself; and, not surprisingly in view of its hematogenic nature, it is highly charged with life-giving vital pneuma, that is, inhaled air transformed successively by the lungs and heart.

Both semina are discharged at coitus, and the oviducts transport the female semen to the lumen of the uterus. In the Galenic scheme the uterus may contribute nothing to embryogeny, but it is equipped with a most formidable power, or "faculty," by which it attracts an appropriate quality. The quality appropriate to the uterus is obviously the materials of generation, and the organ is so endowed that the oviduct possesses an attractive faculty for female semen, while the uterine cervix possesses an attractive faculty for male semen. As a result, upon their discharge the two semina are attracted to the uterine cavity, where they meet to form a conceptus highly charged with vital pneuma and innate heat.

In *De foetuum formatione,* Galen divides prenatal development into four stages: an unformed stage; a vascularized fetus without distinct liver, heart, and brain; a fetus with undemarcated external features in which these three organs are clearly defined; and a fetus possessing all its organs and capable of movement. The first stage is described as a conceptus unformed and, although mixed with blood, white like semen. Genesis, which embraces formation of the conceptus and the earliest developmental stages, says Galen,

> is not a simple activity of Nature, but is compounded of alteration and shaping. That is to say . . . the underlying substance from which the animal springs must be altered; and in order that the substance so altered may acquire its appropriate shape and positions, its cavities, outgrowths, attachments, and so forth, it has to undergo a shaping or formative process. . . . The seed having been cast into the womb or into the earth (for there is no difference), then, after a certain definite period, a great number of parts become constituted in the substance which is being generated; these differ as regards moisture, dryness, coldness and warmth, and in all the other qualities which naturally derive therefrom. . . . Now Nature constructs bone, cartilage, nerve, membrane, ligament, vein, and so forth, at the first stage of the animal's genesis, employing at this task a faculty which is, in general terms, generative and alterative, and, in more detail, warming, chilling, drying, or moistening; or such as spring from the blending of these, for example, the bone-producing, nerve-producing, and cartilage-producing faculties.

In Galen's embryological theory the alterative faculty transforms the primitive unformed raw material created by the union of male and female semina into the configurations represented by different tissues. Tissue integration and organogenesis is a function of the internal formative faculty, which is teleological, "doing everything for some purpose, so that there is nothing ineffective or superfluous, or capable of being better disposed." And thus his description of those first developmental stages which he designates as *geniture* includes nothing resembling the processes from cleavage through gastrulation.

The schemata of development elaborated by the writers of antiquity obviously offer little regarding the mammalian blastocyst. The concepts contained in the Hipprocratic treatise *De natura purei* and the embryological works of Aristotle and Galen all invoke a commingling and coagulation of fluids to produce a structureless conceptus and presuppose a variety of temperature-regulated process producing heterogeneity from homogeneity. At best, as with Aristotle, the mammalian ovum is considered no more than an oviform body found in the uterus sometime after conception. The antique concepts of development had therefore neither need nor place for the ovum and blastocyst.

The character flaw of embryology in antiquity emerges in the opening lines of Galen's *De foetuum formatione:*

> Philosophers have attempted to write on the formation of foetuses although they do not base what they have to say on dissection. And it is no wonder that they go wide of the truth. . . . Those who trust to their own conjectures, innocent of what is to be seen in dissection, are of course much more likely to miss the mark.

The sum of observations on embryos of various animals in antiquity is not negligible; yet the observations are fortuitous, discontinuous, and based entirely upon ocular inspection. The lack of a systematic *corpus* of embryological knowledge, technical insufficiencies, and the absence of conceptual necessity operated against even a theoretical appreciation of the blastocyst by the most outstanding embryologists of Graeco-Roman antiquity.

Galen's time was followed by more than a thousand years of stagnation and retrogression in many spheres of intellectual development and a disastrous decline in secular learning and culture in the Latin West. From Galen's day until the sixteenth century, well after the beginning of the Renaissance, embryological thought made no substantial advance. It is true that some patristic writings have embryological content, since both Neoplatonic and Christian thought embrace creation; and, although they did so without any apparent direct study, some church fathers included within their sphere of interest discussions of the inception of human life and ensoulment.

A more important influence upon the development of embryology was provided by the Arab commentators. Following Arab contact with the literature of the Greeks attendant upon the spread of Islam during the seventh century, a lively interest in Greek learning existed in the Moslem cities of Asia, Africa, and Spain. Averroës and Avicenna were the most important of the Arab commentators in the field of biology and medicine. Avicenna exerted the greatest influence upon embryology; his *Canon* was the mainstay of anatomical teaching in Western universities until the sixteenth century, in some until the eighteenth century. Avicenna's thoroughly Galenic treatment of the formation of the fetus tinged thought on the subject well into the seventeenth century.

From the eleventh through the thirteenth century men like Constantine the African, Gerard of Cremona, and Michael Scot translated into Latin various Greek works preserved in Arabic. In their attempt to synthesize pagan learning and Christian doctrine, the Scholastics produced great medieval encyclopedias, such as those of Vincent de Beauvais and Bartholomaeus Anglicus, which discuss matters of embryological interest compiled from Hippocrates, Aristotle, Galen, and others. In addition, *De animalibus* of Albertus Magnus deals with the subject at greater length, although it is primarily an exegesis of Aristotle's work on animals. These and other purely medieval works were eventually superseded by the great biological works of antiquity in their original and complete form as a result of the labors of humanistic scholars during the fourteenth, fifteenth, and sixteenth centuries. Thus the complete works of Aristotle were available in Greek in 1498, and the *opera omnia* of Galen were available in Greek in 1525, in Latin beginning in 1541.

II. Seventeenth- and Eighteenth-Century Concepts of Embryonic Development

Original investigation in biology and medicine began again in the mid-sixteenth century, as indicated, at least symbolically, by publication of Vesalius's *De humani corporis fabrica* in 1543. The hold of authority and scholastic modes of thought, however, were not easily discarded; and hence, despite the emergence of new attitudes and the intellectual liberation characterized as the Renaissance, antique embryological theories, especially those of Galen, continued to color, if not to dominate, the study of generation and development. This is apparent in the work of such an enlightened seventeenth-century investigator as William Harvey.

Harvey's *Exercitationes de generatione animalium* (1651) is often considered to have established firmly the concept that all life arises from ova, and it is assumed that the phrase emblazoned on the frontispiece, *Ex ovo omnia* (fig. 1), was thenceforth established as scientific "truth." This fact was in reality scarcely established, as Harvey himself was scarcely liberated from past authority. "So possessed is he by

Fig. 1. Frontispiece from *De generatione animalium* by William Harvey (1651). The egg from which Zeus is liberating various living creatures is inscribed *ex ovo omnia.*

scholastic ideas," wrote his translator, Willis (1847), "that he winds up some of his opinions upon animal reproduction by presenting them in the shape of logical syllogisms."

Harvey was unable to find semen in the uterus of the deer following coitus, and he was equally unable to locate "even a trace of the conception" within the uterus for some days *post coitum*. He therefore concludes that the female semen is a fiction, and he denies the necessity for any prepared maternal matter in the uterus in embryogenesis. He proposes instead that animals arise from an inherent primordium, the ovum, which is,

> the primordium vegetable or vegetable incipience, understanding by this a certain corporeal something having life in potentia; or a certain something existing *per se,* which is capable of changing into a vegetative form under the agency of an internal principle.

The confusion Harvey introduced into the subject of mammalian reproduction was unresolved even by the descriptions of follicles in mammalian ovaries by van Horne, de Graaf, and Swammerdam later in the seventeenth century. The difficulty of ascertaining the presence of semen or ova in the reproductive tract of female mammals following coitus supported the concept of fertilization accepted by Harvey. This, one of the more prevalent concepts for some decades, postulates that a seminal aura or effluvium emitted by the male semen is absorbed into the blood and transported therein to the ovary to make the ova fertile and able to engender an embryo.

In his *De mulierum organis generationi inservientibus* (1672) Regner de Graaf describes the ovarian follicles now bearing his name and also makes progressive observations upon the contents of the oviducts and uteri of rabbits at intervals from mating through term. Unable to locate anything in the female reproductive tract on the 1st and 2d days *post coitum,* de Graaf observed on the 3d day tiny spherical bodies in the oviduct, and on the 4th day he found slightly larger spherical bodies in the uterus. De Graaf thus observed and described the rabbit blastocysts and recognized them as the forerunners of the embryo. He assumed, however, that they represented discharged follicular contents traversing the oviducts as eggs en route to the uterus, where they would settle and hatch into embryos. Given the microscope and spared his premature death in 1673, de Graaf might well have discovered the actual mammalian ovum. Withal, his observations upon tubal ova and the early blastocyst provided the first proof that the embryo is formed before reaching the uterus.

De Graaf's observations upon the relation of the blastocyst to the ovarian contents and the onset of development in the oviducts clearly invalidated the hypothesis elaborated by Harvey. This hypothesis was contradicted also by Nuck's demonstration (1691) that dog embryos develop only above ligatures applied to the uterine horns within 3 days after coitus. And yet the idea of *in utero* coagulation of the embryo not only persisted, but received sufficient impetus from Albrecht von Haller in the mid-eighteenth century to inhibit acceptance of the conclusions inevitable from the investigations of de Graaf, Nuck, and others.

In 1752 Haller and his student, Kühlemann, undertook embryological observations upon sheep after the manner of Harvey and de Graaf. Their results were affected profoundly by the peculiarities of sheep development. Unlike the rabbit blastocyst, which is spherical and visible to the unaided eye, the sheep blastocyst, like that of the deer, is composed of elongated and collapsed strands resembling threads of mucus or the "white filaments like spider-webs" Harvey believed to be secreted by the uterus in embryogenesis. Not surprisingly, then, Haller's results were in agreement with those of Harvey, not de Graaf. Haller placed the substantial weight of his authority behind the concept, outlined in his *Elementa physiologiae* (1766), that something of undefined corporeality passes from the ovary to the uterus, there to coagulate into the ovum. Perhaps more than any other single factor, Haller's concept retarded progress toward a more accurate understanding of mammalian reproduction and contributed to the delay of more than 150 years between the discovery of the ovarian follicle and the true mammalian ovum.

William Cruikshank, who is noted in Boswell's *Life of Samuel Johnson* as the surgeon in attendance at the lexicographer's last illness, reopened investigation into mammalian reproduction in 1788. His study, published in 1797, describes stages in the development of the rabbit from the 1st day *post coitum* to term, with special emphasis upon the first 10 days after mating. In his careful, unembellished account, Cruikshank concludes that the rabbit ovum departs from the ovary and passes through the oviduct to reach the uterus on the 4th day. From his observations upon a total of 28 tubal ova he recognized that development occurs while the ovum

Fig. 2. Cruikshank's illustrations of 6-day rabbit embryos. The four ova on the left are "natural size"; those on the right are "magnified, in the simple microscope." (From Cruikshank 1797)

traverses the oviduct. Cruikshank clearly saw blastocysts (fig. 2), recording his observations thus:

> Opened a doe sixth day complete; found the ova loose in the uterus. . . . The ova were transparent and of different sizes; they were double, and contained each an internal vesicle, there was a spot on one side in most of them, which I conceived to be the intended point of adherence between them and the uterus; the internal vesicle was not equally in proportion to the external, but in some larger, in others less . . . just where one of these vesicles had become stationary a white vascular belt was beginning to form, and in the middle of this a cavity where the vesicle lay; the inner membrane I take to be the amnion, the outer chorion.

Cruikshank obviously did not conceive of the blastocyst, and the interpretive hazards associated with his analysis of the early embryo is perhaps best illustrated by his description of a 3-day ovum: "In all of them the chorion and amnion are even now distinct, and in some of them the allantois, as I suspect."

Cruikshank's publication appeared coincidently with an essay of contradictory content by John Haighton (1797). Cruikshank's observations confirmed those of de Graaf, and his demonstration of the relatively slow oviductal passage of the ovum explained that delay between release of the ovarian contents and the appearance of the embryo in the uterus which had so influenced Haller; Haighton's observations, on the other hand, were consistent with those of Haller, and led him to a comparable conclusion:

> The experiments I have made on this simple question do not allow me to incline to the side of DE GRAAF; for in the rabbit I have never found any thing in the uterus which had a regular circumscribed form earlier than the sixth day, and even then the substance was bounded by a covering so very tender, that it scarcely had firmness sufficient to support the figure. Before the sixth day, I have never seen any thing but irregular mucus-like masses in the uterus . . . on the tenth day, in the rabbit, an opaque spot is seen in this ovum, which increasing daily in its bulk, progressively manifests the formation of the foetus.

Thus countered, Cruikshank's study made so little impress upon embryological thought that as late as 1824 the Göttingen Academy of Sciences awarded its prize to an essay affirming the Harvey-Haller concept of *in utero* formation of the ovum. Notwithstanding, the second decade of the nineteenth century is a critical period in the history of embryology and, in particular, the subject of mammalian reproduction.

III. The Nineteenth Century: The Foundations of Mammalian Embryology

In the same year the Göttingen Academy did obeisance to past authority, Prévost and Dumas published an account of the early development of the dog and rabbit. They describe an 8-day dog conceptus (fig. 3), remarking an internal mass which they suspect engenders the embryo:

> Grossis trente fois et vus par transparence, ces ovules paraissent sous une forme ellipsoïde, et semblent composés d'une membrane d'enveloppe unique et mince, dans l'intérieur de laquelle est contenu un liquide transparent. A la partie supérieure de l'ovule on remarque une espèce d'écusson cotonneux, plus épais et marqué d'un grand nombre de petits mamelons. Vers l'une des extrémités de celui-ci, on observe une tache blanche, opaque, circulaire, qui ressemble beaucoup à une cicatricule. . . . Dans le premier état on ne peut encore y reconnaître le foetus; peut-être se trouve-t-il situé à l'intérieur de la tache blanche circulaire dont nous avons parlé.

This prescient description of the blastocyst is perhaps the best extant prior to the discovery of the mammalian ovum.

The true mammalian ovum was finally discovered in 1827 by Karl Ernst von Baer. In his *De ovi mammalium et hominis genesi* (1828) von Baer describes the ovum and the early stages in development, including the blastocyst, of various mammals (fig. 4). His account of blastocyst formation is daringly mechanistic for a period when many naturalists were in thrall to *Naturphilosophie* and Romantic transcendentalism:

> Making its way through the oviduct, the mammalian ovum undergoes almost no metamorphosis unless it imbibes the albuminoso-gelatinous mucus and thence grows a little so that the granules of the internal globule more and more form a periphery surrounding the fluid. . . . When the ovum has been transmitted to the uterus it grows more rapidly, imbibing a larger quantity of fluid and thence it becomes hollow and transparent. The granules more and more retreat to the periphery, and from their surface they excrete material from which a very thin membrane is formed, to the internal surface of which the granules adhere. This membrane is the vitelline membrane. . . . When first the centrifugal formative force has driven the granules to the periphery, it then impels the denser mass in each granule to the periphery, whereby each granule is transformed into a globule consisting of small granules around a transparent center. Therefore the meta-

morphosis of the ovum corresponds completely to the metamorphosis of the Graafian vesicle. . . . In the Graafian vesicle the proligerous stratum is present with its cumulus, to which the description of the blastoderm corresponds, for part of the granules in the ovum constitute a mound under the vitelline membrane, little by little changing into a disc.

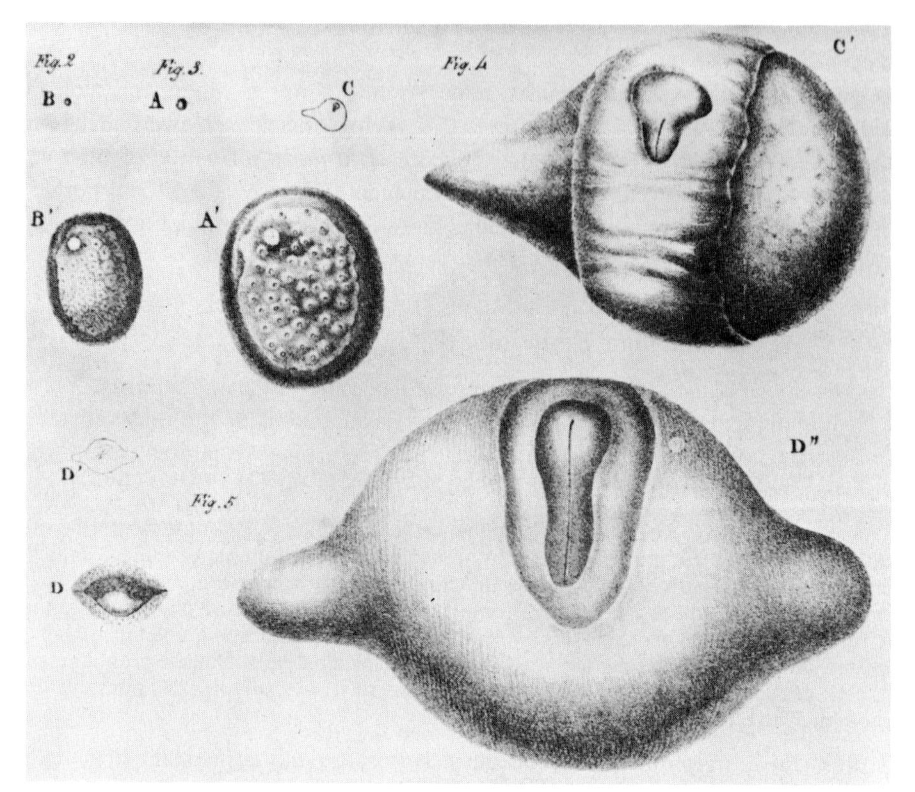

Fig. 3. Illustrations of early development of the dog from Prévost and Dumas (1824)

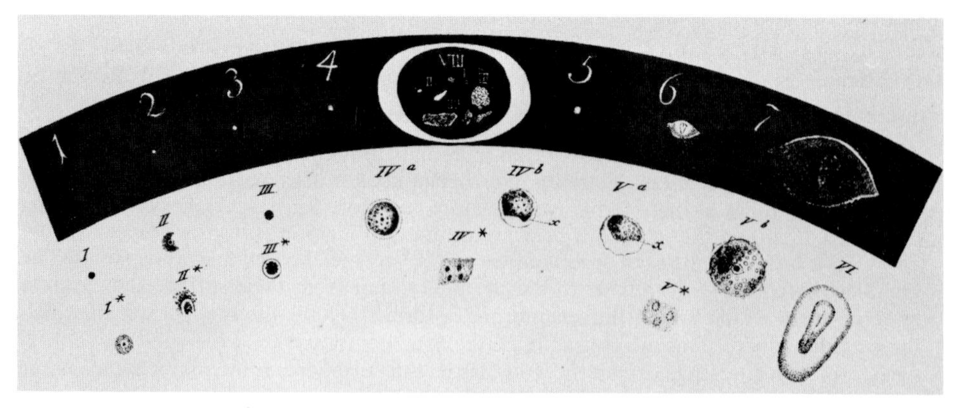

Fig. 4. Von Baer's illustrations of early developmental stages in the dog. (From von Baer 1828)

Laboring in the absence of the cell theory, it did not occur to von Baer that the mammalian ovum is a cell. Hence, even in his great *Entwickelungsgeschichte der Thiere* (1828, 1837) he never deals effectively with the initial stages of mammalian development and fails to comprehend the full meaning of the "hollow and transparent" structure he found in the uterus.

Uncertainty regarding the nature of early mammalian development remained well after the classical work of von Baer and Schwann's formulation of the cell theory. The vesicular character of embryos recovered from the uterus was no small barrier to understanding, and it figures prominently in the diverse attempts of the

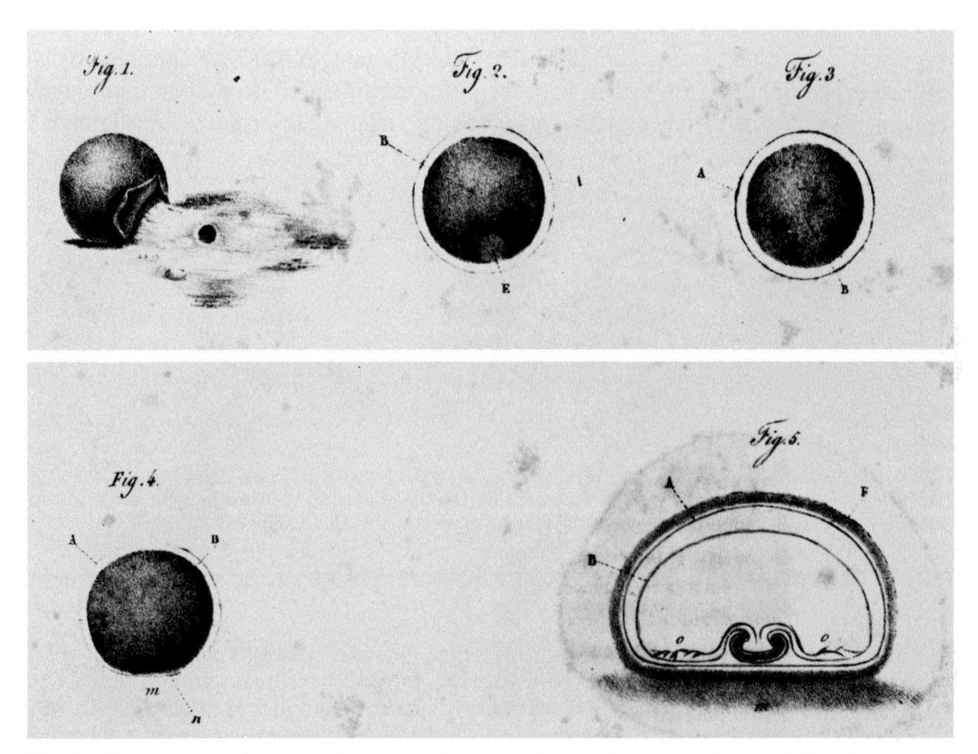

Fig. 5. Illustrations of progressive stages in the early development of the rabbit (from Coste 1834). The vitelline membrane is labeled *A*, the blastodermic vesicle *B*, and the embryonic spot *N*.

Romantic era to explain early mammalian development. Coste (1834, 1837), for example, describes the metamorphosis of the early embryo into a structure composed of two vesicles, one encased within the other (fig. 5). He designates the external vesicle the vitelline membrane; the internal vesicle, developed after fertilization, the blastodermic vesicle, "since it appears to have the same character as the ovum of birds, and also to result from the same causes." Once established, writes Coste, the *vésicule blastodermique* undergoes a new development:

> Alors on voit sur un point de la vésicule blastodermique, une tache circulaire, ayant l'aspect d'un nuage vague, et résultant d'un assemblage de granules qui se groupent peu à peu dans un ordre régulier. Cette tache que nous appellerons *embryonnaire,* parce que c'est elle qui constitue les premiers linéamens de l'embryon, placée dans l'épaisseur des

> parois même de la vésicule blastodermique, de circulaire qu'elle était d'abord, prend ensuite une forme elliptique dans laquelle on pourrait tracer deux foyers presque égaux, mais bien distincts l'un de l'autre. . . . Bientôt cette tache est assez dévelopée pour qu'il soit facile de distinguer quel sera la côté correspondant à la tête de l'embryon, et quel sera celui dans lequel se formera la queue.

Patently, Coste's *tache embryonnaire* is the early mammalian blastoderm, and the cloud of assembled granules probably corresponds with the inner cell mass, although his illustrations (fig. 6) scarcely illumine this point. Still, Coste is one of the first to approximate, however remotely, the true morphology of the mammalian blastocyst.

Comparative embryology enjoyed great vogue during the early nineteenth century, and consequently, the conditions of development in various animals, vertebrate and invertebrate, were often transferred indiscriminately to mammalian development. A relevant instance is the presence of yolk and its role in development.

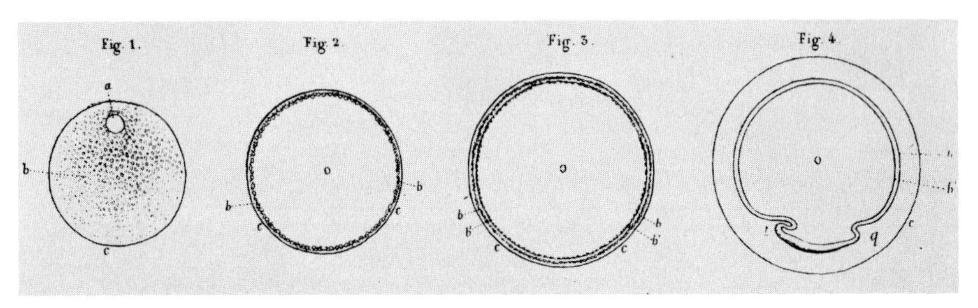

Fig. 6. Illustrations indicating the development of the mammalian blastocyst into a structure composed of the vitelline vesicle (*c*) surrounding the blastodermic vesicle (*b*). The latter subsequently becomes bilaminar and gives rise to the embryo. (From Coste 1837)

Rathke (1837), who is representative of the period, states a general position quite forcefully:

> The yolk of Crustacea—and the yolk of other animals—presents appearances which warrant us in concluding that at the time when the embryo forms and becomes developed, it [the yolk] is not merely a magazine of inert material, but rather leads a very powerful life, so that—at all events when the germ first arises—we might compare it to a particular organism, and subsequently to one of the organs of an animal being.

Many investigators thus embraced mammalian development and also applied to it Prévost's and Dumas finding (1824) that in the developing amphibian egg the yolk fragments into increasingly smaller "crystalline forms." Coste (1837) thus speaks of the space in the rabbit ovum once occupied by yolk, "the condensation of which has served to form the blastoderm," and T. Wharton Jones (1837) describes yolk grains in the ovum, "for the most part coherent," which subsequently form the vesicular blastoderm of the 7-day rabbit embryo. The contemporary comparative conceit led many others to assume the presence of an appreciable quantity of yolk in the mammalian ovum, an assumption seminal to discussions of early development suffused with comparisons and undisguised analogism.

Jones (1837) is the first investigator after Cruikshank to illustrate a developing mammalian embryo recovered from the oviduct. His illustrations are generally more informative than those of the English surgeon, as is his description of the 7-day rabbit embryo:

> No vitellary membrane was to be seen. The gelatinous-looking envelope constituted the only covering of the yelk, which now formed a vesicular blastoderma. The cavity of the gelatinous-looking envelope was much larger than the vesicular blastoderma. . . . The vesicular blastoderma was irregular on one side, that on which I supposed the embryo was about to be developed. It was beginning to present the separation into layers, and had the same peculiar friable globular structure as the blastoderma of the hen's egg.

Jones's work appeared the year before the first of three highly significant articles on the embryology of the rabbit published by Martin Barry (1838, 1839, 1840).

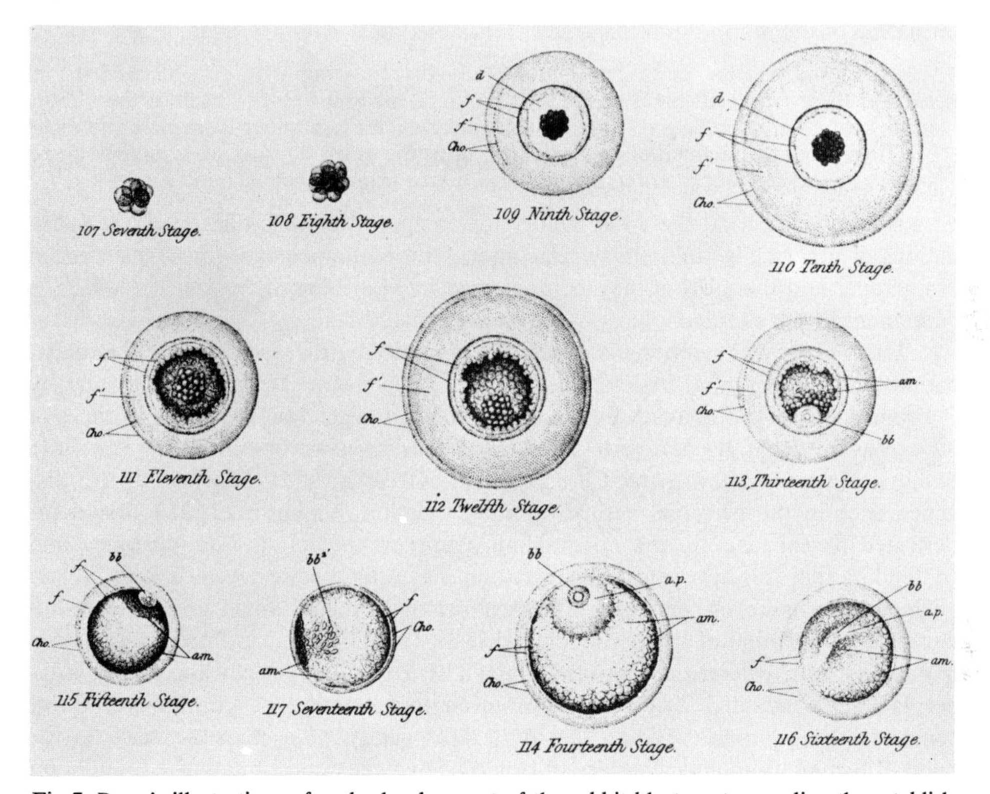

Fig 7. Barry's illustrations of early development of the rabbit blastocyst, revealing the establishment of the "mulberry-like structure" and the origin of the embryonic germ. (From Barry 1839)

Noting that the understanding of early mammalian development rested primarily on observations made upon the developing chick, Barry addresses himself to that "dark period of which very little is really known" between coitus and the appearance of the vertebrae. Studying more than 300 embryos recovered from the oviduct and uterus, Barry presents a rather detailed account of blastocyst formation (fig. 7):

> In the centre of the yelk, there arise several very large and exceedingly transparent vesicles. These disappear and are succeeded by a smaller and more numerous set. Several sets thus successively come into view . . . until a mulberry-like structure has been produced, which occupies the centre of the ovum. . . . In the uterus, a layer of vesicles of the

> same kind as those of the last and smallest set here mentioned, makes its appearance on the whole of the inner surface of the membrane which now invests the yelk. The mulberry-like structure then passes from the centre of the yelk to a certain part of that layer . . . and the interior of the mulberry-like structure is now seen to be occupied by a large vesicle containing a fluid and dark granules. In the centre of the fluid of this vesicle is a spherical body, composed of a substance having a finely granulous appearance, and containing a cavity filled with a colourless and pellucid fluid. This hollow spherical body seems to be the true germ. The vesicle containing it disappears, and in its place is seen an elliptical depression filled with a pellucid fluid. In the centre of this depression is the germ, still presenting the appearance of a hollow sphere.

In his third article (1840) Barry attempts to conjugate the cell concept and embryogenesis, interchanging facilely the terms "cell" and "vesicle," and describing formation of the morula in terms of cell reproduction:

> The germinal vesicle—or original parent cell—disappearing, twin cells succeed it . . . each of these twin cells gives origin to two others, making four . . . each of these four, in its turn a parent cell, gives origin to two, by which the number is increased to eight . . . this mode of augmentation continues, until the germ consists of a mulberry-like object, the cells of which are so numerous as not to admit of being counted. . . .

At a subsequent period the entire embryo is composed of cells filled with the foundations of other cells, and, Barry concludes, "the fundamental form in question in Mammalia, and therefore it may be presumed in Man himself . . . is that which is permanent in the simplest plants,—the single isolated cell."

Barry's essay at synthesis is affected profoundly by the contemporary preoccupation with the germinal vesicle, or nucleus, of the ovum. Described originally by Purkinje (1825) in the avian ovum, the germinal vesicle was identified in the ova of various invertebrates and vertebrates by various investigators, including von Baer (1828), Rathke (1829), and Carus (1831). Great mystery attached to the disappearance of the germinal vesicle after fecundation. Valentin (1835) first demonstrated its existence in the mammalian ovum, writing of its disappearance and contending that this apparent transformation into a fluid state revealed it to be "an analogue of semen, in particular a form of female semen." Some investigators believed disappearance of the germinal vesicle signaled little more than the initiation of development, whereas others (Bischoff 1839; Jones 1837; Schwann 1839; Wagner 1836*a*) contended its direct contribution to formation of the embryo. Barry considers the germinal vesicle "the mysterious centre of a nucleus which is the point of fecundation; and the place of origin of two cells constituting the foundation of the new being . . . which, having given origin to two cells disappears." From this concept of the ovular nucleus qua cell Barry concludes that "the embryo is the altered nucleus of a cell, and that the primitive trace of others appears to be no other than this same nucleus in a comparatively advanced stage." Proceeding, Barry states that the mulberrylike object derived ultimately from the germinal vesicle contains "a cell larger than the rest, elliptical in form, and having in its centre a thick-walled hollow sphere, which is the nucleus of this cell. . . . This nucleus is the rudimental embryo" (fig. 8). Barry's endeavor to relate mammalian embryogenesis to the cell theory within a year of its formulation, in ignorance of distinctions of nucleus and cytoplasm and the process of mitosis, is predictably imperfect. When finally the cell theory enabled virtually saltatorial advance in the analysis of development, knowledge of the structure and division of cells was sufficient to the task.

Unlike Botticelli's *Venus,* the cell doctrine did not emerge in attractively complete form. Indeed, considerable adscititious evidence and extensive revision were requisite to its final formulation. As late as 1861, Max Schultze lamented the "lack of agreement among histologists concerning what should be designated as a cell," and investigations of segmentation of the ova of invertebrates and oviparous vertebrates notwithstanding (Bergmann 1841; Kölliker 1844; Nägeli 1842; Reichert 1841, 1846; Vogt 1842), at mid-century this uncertainty was especially impedimental to elucidation of early mammalian development. The ambiguities occasioned by this incertitude are manifest in even the most ambitious of investigations, such as

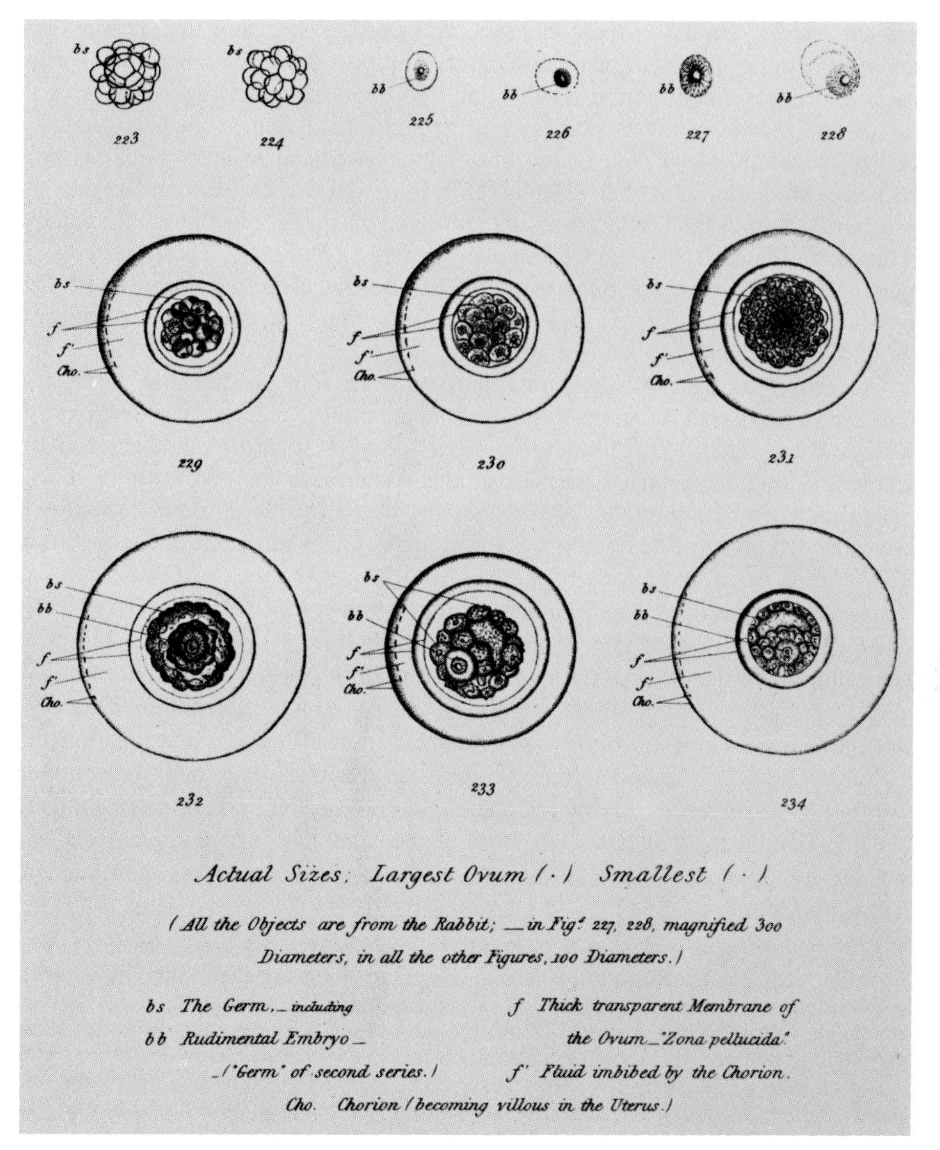

Fig. 8. Illustrations of blastocyst formation and development of the rabbit embryo. (From Barry 1840)

those of Theodor Bischoff. Bischoff's studies (1842, 1845, 1852, 1854) are thorough and based in observations upon numerous embryos. His important accounts of early mammalian development may thus exemplify the conceptual difficulties attending the neonatal period of the cell doctrine and the resultant transitional character of embryological thought.

Bischoff (1842) conceives of the mammalian ovum as consisting of the zona pellucida, the yolk, the germinal vesicle, and the germinal spot. The last three structures obviously correspond, respectively, with the cytoplasm, nucleus, and nucleolus of the ovum. Evaluating his observations within these structural parameters, Bischoff infers that the ovum is not a cell, but rather an organization surrounding the germinal vesicle. The latter, he concludes, is a primary cell, and the germinal spot enclosed therein is a nucleus of unique nature and fate. Released when the germinal vesicle undergoes dissolution at fertilization and dividing as the ovum traverses the oviduct, the germinal spot is said later to give rise to a central nucleus related to embryo formation. Movement of the blastocyst consequent to activity of oviductal cilia had been described earlier (Henle 1838; Barry 1839), and Bischoff argues that the rotatory motion thus induced assists the onset of changes in the yolk leading to its segmentation into increasingly smaller spheres (*Kugeln*). These spheres, he argues, are yolk granules grouped around the central clear nucleus derived from the liberated germinal spot. "It is clear," he writes, "that this process is not presented in any existing schemes of cell formation."

According to Bischoff's account, the ovum, its yolk divided and surrounded by an albuminous coat acquired during oviductal transit, reaches the uterus on the 4th day. There nucleated cells develop from the yolk spheres, establishing a thin membrane along the internal surface of the zona pellucida and forming thus a vesicle, named by Bischoff the *Keimblase,* or blastocyst. Soon after formation of the blastocyst a mass of distinctive cells aggregates on its inner surface, constituting the *Fruchthof,* or germinal area (fig. 9).

Bischoff's descriptions and numerous illustrations in his various treatises established securely the general idea of the blastocyst and the genetic relationship of the inner cell mass to the embryo. Despite his confusion regarding the cellularity of the ovum, a confusion scarcely unique to him, it may be argued that the modern study of mammalian development has a major foundation in his critical investigations.

Acceptance of Bischoff's interpretation of the blastocyst and its contained *Fruchthof* was favored greatly by his demonstration, on the basis of more than theoretical determinism, that the germ layer concept applies to mammalian development. He writes of his discovery of the laminated character of the germinal area in the rabbit blastocyst:

> Im Fruchthofe lagen Zellenkerne und Molecüle dicht gehäuft beieinander. . . . Als ich aber die Stelle des Fruchthofes genauer untersuchte, überzeugte ich mich, dass in demselben und etwas über ihn hinaus die Keimblase aus einer doppelten Lage bestand, indem sich hier an ihrer inneren Fläche eine sehr dünne Schichte von sehr zarten Zellen zu bilden, oder von ihr abzulösen begonnen hatte. Die Zellen dieser Schichte, welche es mir sowohl mit der Nadel von der äusseren zu trennen glückte, als sie sich auch von einem Ei in ihrer ganzen Ausdehnung von selbst abgelöset hatte, waren sehr blass, besassen noch alle ihre scharfen und bestimmten Contouren, zeigten alle einen Zellenkern und nur sehr wenige Molecüle als Zelleninhalt. Die bestimmte Ueberzeugung von diesem Verhältnisse machte mir keine kleine Mühe, da es äusserst schwierig war, die zarte und

> weiche Keimblase, die an jedem Instrumente hängen bleibt, so zu behandeln, dass ich die innere zarte Lage von ihr mit Bestimmtheit mit zwei feinen Nadeln unter der Loupe ablösete. Dennoch halte ich diese Entdeckung von der grössten Wichtigkeit, da sie in der Folge sich immer mehr bestätigend und für die weitere Entwicklung vom grössten Einflusse, zugleich eine wichtige Uebereinstimmung des Säugethiereies mit dem Vogeleie darthut.

At a time, then, when the germ layer concept was unrefined and knowledge of mammalian development was yet rudimentary, Bischoff records the existence of animal (ectoderm) and vegetative (endoderm) layers in the mammalian blastocyst, correctly equating them with the corresponding layers in the avian embryo. Bischoff's contri-

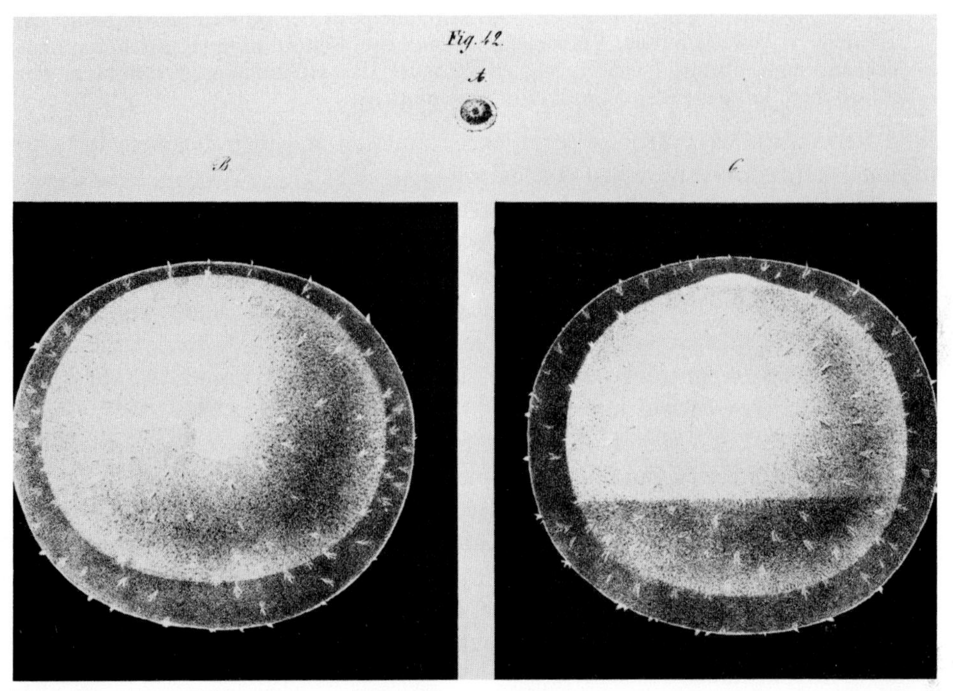

Fig. 9. The 7-day rabbit embryo as depicted by Bischoff, showing the embryonic area (*Fruchthof*) as a central light area at the dorsal surface of the blastocyst. The embryo is said to be bilaminar from this area to the beginning of the darkened area at the ventral aspect of the blastocyst. (From Bischoff 1842)

bution to the nineteenth century understanding of mammalian embryology was such that van Beneden (1875) would write later:

> La raison de cet abandon, dont les Mammifères ont été l'objet, se trouve avant tout, je crois, dans la perfection même des travaux de l'éminent embryogeniste de Munich. Plus je les ai étudiés, plus j'ai admiré comment, avec les moyens matériels dont on disposait alors à une époque ou l'histologie été à peine fondée, ou toute l'histoire des premières phénomènes du développement embryonnaire des Mammifères était à faire, Bischoff a pu pousser aussi loin ses recherches et arriver à des résultats aussi vrais et aussi complets.

Refinement of the germ layer concept of Pander and von Baer was very nearly completed at mid-century by Robert Remak. Remak interprets the germ layers as composed of cells which are derived from the single cell of the original ovum, and he demonstrates the specific histological future of each layer. He distinguishes two primary germ layers, the upper sensory layer (ectoderm) and a lower layer subdivided into the trophic layer (endoderm) and the motor-germinative layer (mesoderm). The chick embryo was Remak's primary object of study, but he records some vital information about mammalian development. The salient feature of Remak's observations upon the rabbit blastocyst (1855) is his description of the germ layers and their avian homologues:

> Es sind alsdann im Fruchthofe drei Blätter zu unterscheiden. Der obere zeigt die centrale schildförmige, aus kleinen Zellen bestehende Verdickung (den Bärschen Embryonalschild), von welcher sich alsbald die Medullarplatte abschnürt. Das mittlere hängt mit dem unteren noch innig zusammen und ist offenbar in der Ablösung von demselben begriffen. . . . Wie die fernere Entwickelung dieser drei Blätter zeigt, entspricht das obere durchaus dem oberen (sensoriellen) Keimblatte des Hühnchens, das mittlere dem motorischen und das untere dem Darmdrüsenblatte.

With Remak, the basic facts of germ layer formation and their relationship to the mammalian blastocyst were first elucidated.

Early in the nineteenth century, as before, embryos were usually studied *in toto,* sometimes after being spread on glass, under water, or following preservation in vinegar or alcohol. Freehand sections of embryos were prepared by some investigators, for example, Pander, von Baer, Rathke, and Reichert. About the time that Kupffer (1865) first prepared freehand serial sections, ca. 50 μ thick, of embryos hardened in potassium dichromate, Hensen (1863, 1866) introduced the *Querschnitter,* a mechanical aid to sectioning. It was Wilhelm His (1868, 1870), however, who developed a microtome truly eliminating the vagaries of freehand sectioning. He also introduced paraffin as an embedding medium, formulated a reliable procedure for fixation (osmium tetroxide), dehydration (alcohol), clearing (oil of lavender), and mounting (Canada balsam) prepared sections, and developed the wax-reconstruction technique for studying embryos in three dimensions. Staining techniques were not common until the last decades of the century, but the improved methods for preparation of embryos yielded a corresponding improvement in the acuity of investigations.

The innovations in embryological methodology, improved microscopes, and the advances in histology combined during the second third of the nineteenth century to orient investigators toward analysis of the internal organization of the embryo. This essentially technological proclivity blossomed into an intellectual environment conditioned profoundly by enthusiastic, if not abandoned, application of the evolutionary concept set forth by Darwin in 1859. This enthusiasm, buttressed by knowledge of the apparent commonality of various developmental stages, the universality of the germ layers, and the presence of gill slits in the embryos of higher vertebrates, engendered a form of transcendentalism according to which common descent replaced the archetype as the synthesizing factor. This new transcendentalism, culminating in Haeckel's (1872, 1891) concept of the gastraea and fervent oversystematization of all morphology through his law, *"die Ontogenie ist eine Recapitulation der Phylogenie,"* was not without its influence upon interpretations of the mammalian blastocyst.

The universality of the germ layers acquired new meaning following its integration with the doctrines of the universality of cells and the protoplasm. One symptom of the rampant transcendental overgeneralization was the emphasis on comparability of embryos of "higher" forms to adults of "lower" forms and the transference of this specious analogy to the germ layers. The ultimate generalization lay latent in Huxley's (1849) portentous description of the *Medusae* as constructed of two membranes "which appear to bear the same physiological relation to one another as do the serous and mucous layer of the germ." It was assured shortly after publication of *Origin of Species* when Kowalewski (1867) reported the development of invertebrate and vertebrate embryos alike from a bilaminar sac, the blastocyst in mammals:

> bei allen von mir erwähnten Embryonen geht die Bildung der beiden erwähnten Schichten oder Blätter (der äusseren und inneren) ganz auf dieselbe Weise vor sich. . . . Also wäre die erste Bildung des Embryos für alle diese verschiedenen Thiere ganz übereinstimmend; nur in den weiteren Veränderungen sehen wir die Unterschiede auftreten, welche jeden einzelnen Typus bezeichnen.

Various investigators (Kleinenberg 1872; Lankester 1873) devoted their efforts toward relating phylogeny and ontogeny on the basis of the compatibility of the germ layers. But it was Ernst Haeckel, possessed of a passion for methodical terminology and organization, a genius for expression with a penchant for magical metaphors, who seduced embryology with his gastraea theory, an exquisitely symmetrical scheme of ideas beautifully harmonizing fact with theory and absolutely devoid of scientific value. According to Haeckel's theory (1872, 1877) the gastraea, the parent of all biological forms, is a two-layered sac similar to the bilaminar stage in embryos. Haeckel designates this developmental stage, common to all animal groups from sponges to vertebrates, as the gastrula (fig. 10). In mammals, says Haeckel (1891), it is the blastocyst possessing two primary germ layers; and applying his Biogenetic Law, he concludes:

> Der Mensch und alle anderen Thiere, welche in ihrer ersten individuellen Entwickelungs-Periode eine zweiblätterige Bildungsstufe durchlaufen, müssen von einer uralten einfachen Stammform abstammen, deren ganzer Körperzeitlebens (wie bei den niedersten Pflanzentieren noch heute) nur aus zwei verschiedenen Zellenschichten oder Keimblättern bestanden hat.

Haeckel's influence upon embryology was immense, determining its thrust for decades. This is well illustrated by Balfour's (1885) declaration that comparative embryology aims "(1) to form a basis for Phylogeny, and (2) to form a basis for Organogeny or the origin and evolution of organs." The preoccupation with analogism and lines of descent resulting from the pervading influence of Haeckelian dogmas affected even the apostates. Thus Karl Reichert (1873), an able and experienced investigator of mammalian development, addresses himself lengthily to evaluation of "the meaning of the blastocyst in the developmental history of mammals and man, as well as in vertebrates generally." Reviewing conditions in numerous mammals, Reichert defines the blastocyst (*bläschenförmige Frucht*) as that structure existing during the period between the end of cleavage and the attachment of the embryo to the uterine wall (fig. 11). The blastocyst is not the embryo, he declaims, but only the developmental stage of a new creature, which is the embryo.

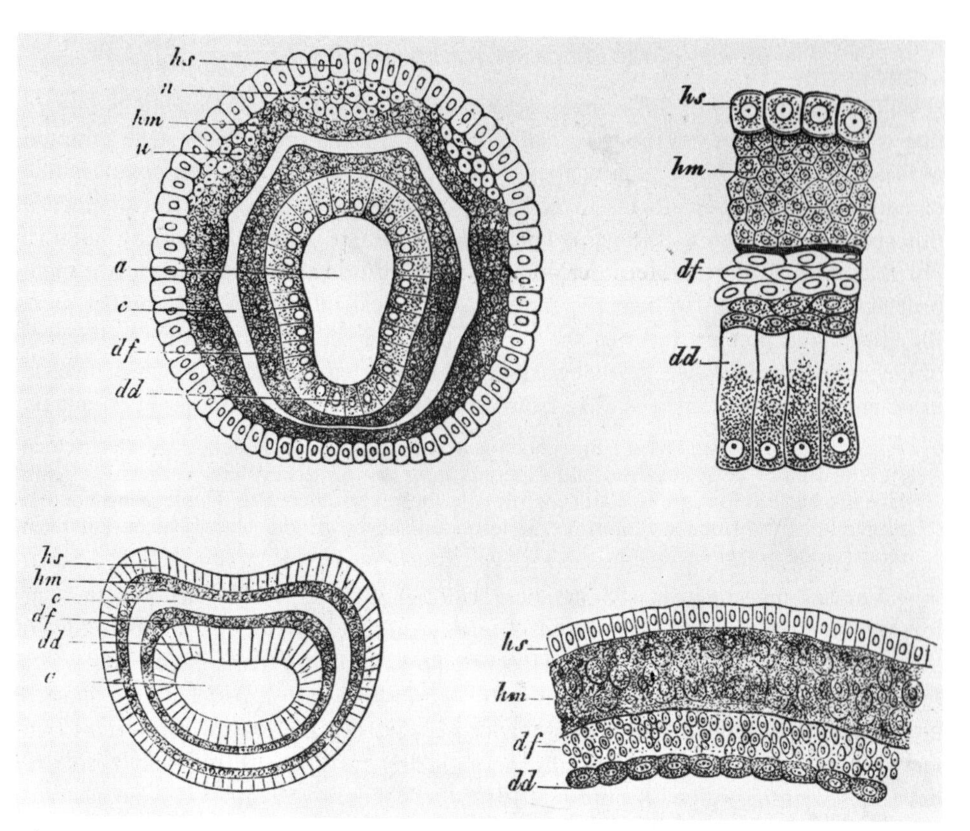

Fig. 10. Haeckel's comparison of the germ layers in earthworms and mammals in support of the universality of the germ layers and his gastraea theory. (From Haeckel 1891)

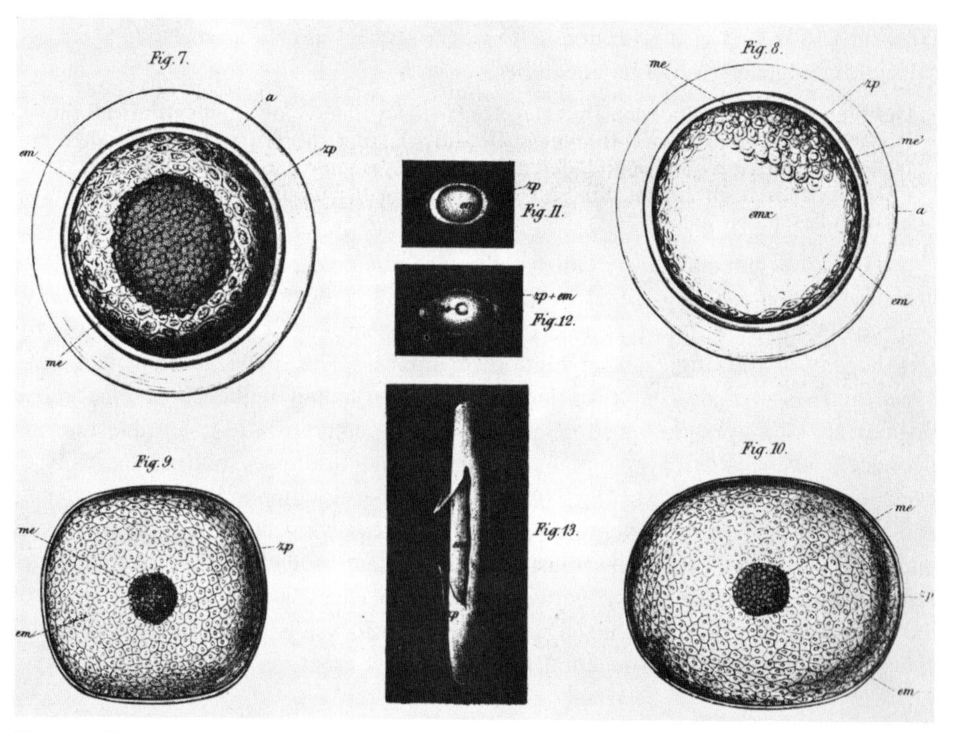

Fig. 11. Illustrations of several stages in development of the blastocyst of the rabbit and cat. (From Reichert 1873)

Noting that the blastocyst lies free in the uterine lumen for some time, Reichert concludes:

> Nach dem thatsächlichen Verhalten der bläschenförmigen Frucht und ihrer Umhüllungshaut während und vor der embyonalen Bildungsperiode unterliegt es keinem Zweifel, dass sie als eine selbstständige Bildungsphase in der Entwickelungsgeschichte der Säugethiere und des Menschen—als eine in den ersten Verkehr mit der Aussenwelt eintretende Larvenform derselben—zu betrachten ist. . . . In dieser Auffassung ist die bläschen-förmige Frucht, vornehmlich das Umhüllungshaut-Bläschen, eine Stammlarve; der am Embryonalfleck ausgebreitete Rest der Bildungsdotterzellen stellt ihren Germinations-und Vegetationspunkt, der Embryo die Knospe dar.

Haeckel's doctrines, steeped in romantic idealism, paradoxically emanating from objectively impersonal Darwinian doctrines, flourished with such strength that, directly or indirectly, the study of embryos was oriented toward what they might reveal of their ancestry. Thus, during the last third of the nineteenth century investigations of mammalian development were inordinately concerned with the origin of the germ layers and elucidation of embryonic homologies through comparison of mammalian development with that of other animals. This concern was compounded by Henle's (1876) description of the developmental history of the primitive streak in mammals, suggesting a possible genetic relationship of the primitive streak and the mesoderm. As a consequence of this overweening interest in the germ layers, heightened interest accrued to the blastocyst, in particular to the state of the germ layers at various stages of its development.

Van Beneden (1880) described the rabbit blastocyst as consisting of an outer layer of cells and an attached inner cell mass (fig. 12). In his account, the hypoblast arises by delamination from the inner mass, and the remaining portion of the mass retreats to the caudal end of the embryonic area to unite with the epiblast and form the mesoblast. Detecting a gap in the outer layer of the cleaved ovum, van Beneden ascribes it to involution of cells at that point and declares it the blastopore, homologous with the blastopore of other vertebrates. For van Beneden, then, the fully segmented ovum represents the gastrula stage in mammalian development. However, most investigators (Heape 1883; Hensen 1875; Kölliker 1882; Rauber 1875) described the embryonic area of the blastocyst prior to primitive streak formation as bilaminar and attributed the origin of the mesoblast to the primitive streak or to the primitive streak and the epiblast. The primitive streak now implicated in mesoderm formation in birds, reptiles, and mammals, the blastocyst was increasingly deemed a gastrula and the primitive streak an obliterated blastopore. Heape (1883), among others, used yet another argument to support the interpretation of the primitive streak as the homologue of the blastopore in lower vertebrates:

> It is very generally believed that mammals are descended from animals which possessed a large yolk sac, and it is stated that the blastodermic vesicle is a remnant of this sac. If this be true . . . the primitive streak of mammals is homologous with the same structure in birds, and the existence of such an arrangement, together with the presence of a complete neurenteric canal . . . in the mammal, is another instance of the morphological facts which led Balfour to conclude that the primitive streak was homologous with the true vertebrate blastopore.

This generalization, in the best tradition of the nineteenth century, proved a valuable stimulus and guide to further analysis of early mammalian development.

Scarcely more than a beginning; but by 1885 there existed the rudimentary facts and an emergent conceptual framework conducive to progressive elucidation of those crucial events occurring between the end of cleavage and establishment of the embryo. The blastocyst was recognized, in Kölliker's (1879) words, as "the primitive organ from which the development of mammals proceeds," and enough was known of its morphology to enable continued, more penetrating, analysis of its nature and activities. This analysis did continue, with the awesome awareness expressed so well by Remak in 1855 that "the entire primordium of the animal and its organized appendages consists of a single-layered vesicle which is scarcely distinguishable from an epithelial vesicle in the thyroid gland."

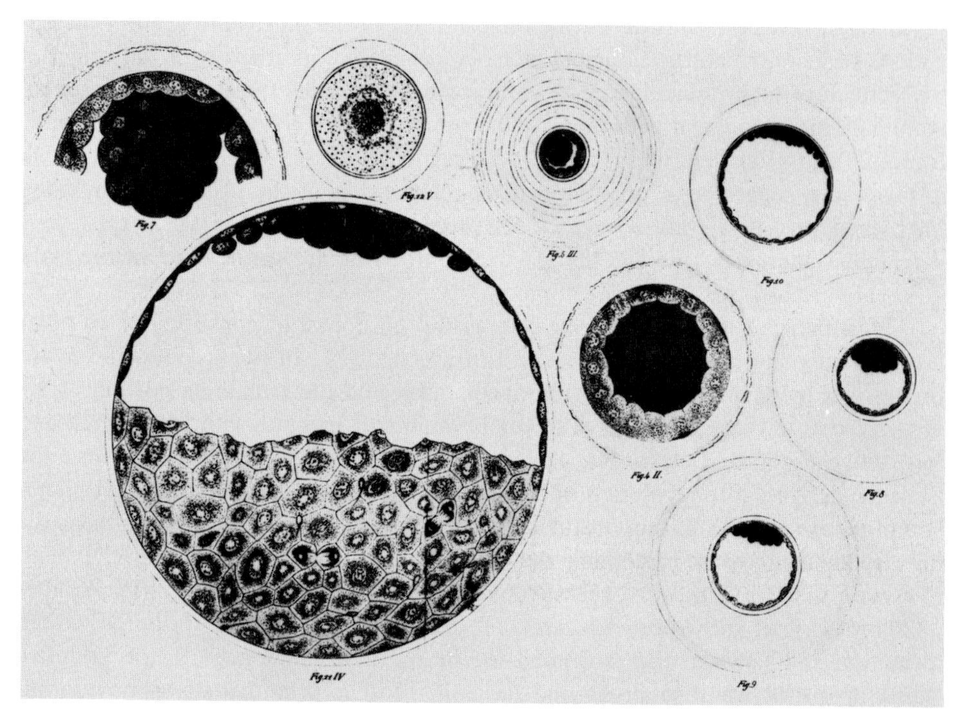

Fig. 12. Van Beneden's illustrations of various stages in development of the rabbit blastocyst. (From van Beneden 1880)

Acknowledgments

This research was supported in part by a grant from the National Library of Medicine.

References

Albertus Magnus. 1916–21. *De animalibus libri xxxi.* Münster.
Aristotle. 1912. *De partibus animalium,* trans. W. Ogle. Cambridge.

———. 1912. *De generatione animalium,* trans. and ed. A. Platt. Oxford.

———. 1965. *Historia animalium,* trans. A. L. Peck, Cambridge.

Avicenna. 1473. *Liber canonis.* Medionali.

Baer, K. E. von. 1827. *De ovi mammalium et hominis genesi.* Leipzig.

———. 1828, 1837. *Über Entwickelungsgeschichte der Thiere. Beobachtung und Reflexion.* Königsberg.

Balfour, F. M. 1885. *A treatise on comparative embryology,* 2d ed. London.

Barry, M. 1838. Researches in embryology: First series. *Phil Trans Roy Soc London,* part 1:301.

———. 1839. Researches in embryology: Second series. *Phil Trans Roy Soc London,* part 2:307.

———. 1840. Researches in embryology: Third series; A contribution to the physiology of cells. *Phil Trans Roy Soc London,* part 2:529.

Beneden, E. van. 1875. La maturation de l'oeuf, la fécondation, et les premières phases du développement embryonnaire des mammifères d'après de recherches faites chez le lapin. *Bull Acad Roy Belg* 60:1.

———. 1880. Recherches sur l'embryologie des mammifères: La formation des feuillets chez le lapin. *Arch d Biol* 1:137.

Bergmann, K. Die Zerklüftung und Zellenbildung im Froschdotter. *Arch Anat Phys u Wiss Med Jahrg* 1841:89.

Bischoff, T. L. W. 1839. Quoted in R. Wagner, *Lehrbuch der Physiologie: Erste Abtheilung, Physiologie der Zeugung und Entwickelung.* Leipzig.

———. 1842. *Entwicklungsgeschichte des Kaninchen-Eies.* Brunswick.

———. 1845. *Entwicklungsgeschichte des Hundeies.* Brunswick.

———. 1852. *Entwicklungsgeschichte des Meerschweinchens.* Giessen.

———. 1854. *Entwicklungsgeschichte des Rehes.* Giessen.

Boswell, J. 1791. *The life of Samuel Johnson, LL.D.* London.

Carus, K. G. 1831. *Erläuterungstafeln zur vergleichenden Anatomie.* Leipzig.

Coste, J. J. M. C. V. 1834. *Recherches sur la génération des Mammifères.* Paris.

———. 1837. *Embryogénie Comparée.* Paris.

Cruikshank, W. C. 1797. Experiments in which, on the third day after impregnation, the ova of rabbits were found in the Fallopian tubes; and on the fourth day after impregnation in the uterus itself; with the first appearances of the foetus. *Phil Trans Roy Soc London* 87:197.

Darwin, C. 1859. *The origin of species.* London.

Galen. 1821–33. *Opera omnia,* ed. C. G. Kühn. Leipzig.

Graaf, R. de. 1672. *De mulierum organis generationi inservientibus tractatus novus, demonstrans tam homines et animalia, caetera omnia quae vivipara dicuntur, haud minus quam ovipara, ab ovo originem ducere.* Leiden.

Haeckel, E. 1872. *Die Kalkschwämme: Eine Monographie.* Berlin.

———. 1877. *Studien zur Gastraea-theorie.* Jena.

———. 1891. *Anthropogenie oder Entwickelungsgeschichte des Menschen, Keimes und Stammes-Geschichte,* 4th ed. Leipzig.

Haighton, J. 1797. An experimental inquiry concerning animal impregnation. *Phil Trans Roy Soc London,* part 1:159.

Haller, A. von. 1746–52. *Disputationum anatomicarum selectarum.*

Haller, A. von. 1757–66. *Elementa physiologiae corporis humani.* Lausanne.
Harvey, W. 1651. *Exercitationes de generatione animalium.* London.
———. 1847. *Works,* trans. R. Willis. London.
Heape, W. 1883. The development of the mole (Talpa europa: The formation of the germinal layers, and early development of the medullary groove and notochord). *Quart J Micr Sci* 23:412.
Henle, F. G. J. 1838. Ueber die Ausbreitung des Epithelium im menschlichen Körper. *Arch Anat Phys Wiss Med* 2:103.
Hensen, V. 1863. Studien über das Gehörorgan der Decapoden. *Zeit f Wiss Zool* 13:319.
———. 1866. Ueber ein Instrument für mikroskopische Präparation. *Arch f mik Anat* 2:46.
———. 1875–76. Beobachtungen über die Befruchtung und Entwicklung des Kaninchens und Meerschweinchens. *Zeit Anat Entwick* 1:213.
Hippocrates. 1839–61. *Oeuvres complètes d'Hippocrate,* trans. E. Littré. Paris.
His, W. 1868. *Untersuchungen über die erste Anlage des Wirbelthierleibes: Die erste Entwickelung des Hühnchens im Ei.* Leipzig.
———. 1870. Beschreibung eines Mikrotoms. *Arch f mik Anat* 6:229.
Horne, J. van. 1668. *Suarum circa partes generationis in utroque sexu observationum prodromus.* Leiden.
Huxley, T. H. 1849. On the anatomy and the affinities of the family of the Medusae. *Phil Trans Roy Soc London* 139:413.
Jones, T. W. 1837. On the first changes in the ova of the mammifera in consequence of impregnation, and on the mode of origin of the chorion. *Phil Trans Roy Soc London,* part 1:339.
Kleinenberg, N. 1886. *Die Entstehung des Annelids aus der Larve von Lopadorhyncus: Nebst Bemerkungen über die Entwicklung anderer Polychaeten.* Leipzig.
Kölliker, A. 1844. *Entwicklungsgeschichte der Cephalopoden.* Zürich.
———. 1879. *Entwicklungsgeschichte des Menschen und der höheren Thiere: Zweite . . . Auflage.* Leipzig.
———. 1882. Die Entwicklung der Keimblätter des Kaninchens. In *Festschrift zur Feier des 300 jährigen Bestehens der Julius-Maximilians-Universität zu Würzburg.* Leipzig.
Kowalewski, A. 1867. Die Entwickelungsgeschichte des *Amphioxus lanceolatus. Mém de l'Acad de St Pétersbourg* 11:1.
Kupffer, C. W. von. 1865. Untersuchungen über die Entwicklung des Harn- und Geschlechtssystems: I. Die Entstehung der Niere bei Schaafembryonen. *Arch f mik Anat* 1:233.
Lankester, E. R. 1873. On the primitive cell-layers of the embryo as the basis of genealogical classification of animals, and on the origin of vascular and lymph systems. *Ann and Mag Nat Hist* 11:321.
Nägeli, C. 1842. *Zur Entwicklungsgeschichte des Pollens.* Zürich.
Nuck, A. 1691. *Adenographia curiosa et uteri foeminei anatome nova cum epistola ad amicum de inventis novis.* Leiden.
Prévost, J. L., and Dumas, J. B. A. 1824. De la génération dans les mammifères, et des premiers indices du développement de l'embryon. *Ann Sci Nat* 3:113.

Purkinje, J. E. 1825. *Symbolae ad ovi avium historiam ante incubationem. Vratislaviae. (Joan. Fried. Blumenbachio . . . semisaecularia . . . celebranti . . . gratulatur Ordo Medicorum Vratislaviensium.)*
Rathke, H. 1829. *Untersuchungen über die Bildung und Entwickelung des Fluss-Krebses.* Leipzig.
———. 1837. *Zur Morphologie, Reisebemerkungen aus Taurien.* Leipzig.
Rauber, A. 1875. Die erste Entwicklung des Kaninchens. *Sitz -ber d naturf Gesellsch Leipzig* 2:65.
Reichert, K. 1841. Ueber den Furchungsprocess der Batrachier-Eier. *Arch Anat Phys Wiss Med Jahrg* 1841:523.
———. 1846. Der Furchungsprozess und die sogenannte Zellenbildung um Inhaltsportionen. *Arch Anat Phys Wiss Med Jahrg* 1846:196.
———. 1873. Beschreibung einer frühzeitigen menschlichen Frucht im bläschenförmigen Bildungszustande ("sackförmiger Keim" von bär) nebst vergleichenden Untersuchungen über die bläschenförmigen Früchte der Säugethiere und des Menschen. *Phys Abhandl königl Akad Wiss Berlin* 1873:1.
Remak, R. 1855. *Untersuchungen über die Entwickelung der Wirbelthiere.* Berlin.
Schultze, M. 1861. Ueber Muskelkörperchen und das, was man eine Zelle zu nennen habe. *Arch Anat Phys Wiss Med Jahrg* 1861:1.
Schwann, T. 1839. *Mikroskopische Untersuchungen über die Uebereinstimmung in der Struktur und dem Wachsthum der Thiere und Pflanzen.* Berlin.
Swammerdam, J. 1672. *Miraculum naturae, sive uteri muliebris fabrica.* Leiden.
Valentin, G. 1835. *Handbuch der Entwickelungsgeschichte des Menschen mit vergleichender Rücksicht der Entwickelung der Säugethiere und Vögel.* Berlin.
Vesalius, A. 1543. *De humani corporis fabrica libri septem.* Basel.
Vogt, C. 1842. *Embryologie des Salmones.* Neuchatel.
Wagner, R. 1836*a*. *Lehrbuch der vergleichenden Anatomie.* Leipzig.
———. 1836*b*. *Prodromus historiae generationis.* Leipzig.

2

Comparative Embryology of Mammalian Blastocysts

Jack Davies/Hans Hesseldahl

Department of Anatomy
School of Medicine
Vanderbilt University
Nashville, Tennessee

A characteristic feature of mammalian development, interposed between the cleavage or segmentation stage of the fertilized egg and the stage of implantation, is the formation of a fluid-filled vesicle or blastocyst. Its structure is essentially simple and alike in all mammals, the differences being mainly in size and rate of development. The formation of the blastocyst by cavitation of an originally solid ball of cells or morula is of extreme importance, since it not only marks the beginning of a stage of enhanced metabolic activity but represents the stage in development at which the first organic contact between the embryonic and maternal tissues is achieved. The blastocyst is an epithelial structure surrounded by one or more noncellular layers, the zona pellucida in all species and an additional "mucin coat" in some forms such as the rabbit. The epithelial wall of the blastocyst is usually considered ectodermal and comprises the trophoblast or nutritive layer (Hubrecht). The trophoblast is continuous with a local or regional massing of epithelial cells at one pole of the blastocyst, the so-called inner cell mass, from which the embryo and its immediate membranes are developed (amnion, yolk sac, allantois). The wall of the blastocyst is originally unilaminar and is secondarily converted into a bilaminar structure by the growth of a layer of epithelial cells, the endoderm, emanating from the inner cell mass and growing partially or completely around the inner surface of the trophoblast. In this process, analogous in part to the process of gastrulation in submammalian forms, the cavity of the blastocyst or blastocoele is converted into an endodermally lined gastrocoele or primitive yolk sac. The blastocyst is usually presented to the uterine surface at implantation in this relatively simple bilaminar form.

In ungulates and carnivores, in which the preimplantation stage is normally prolonged, the blastocyst and inner cell mass show extensive differentiation at the time of implantation. In these instances the wall of the blastocyst is modified by the interposition of extraembryonic mesoderm, which may be vascular or avascular, and the functional maturation of the embryonic organs is relatively advanced.

In this chapter the general aspects of the development of the mammalian blastocyst are reviewed and some of the important biological problems associated with it are presented. In such studies the biology of the trophoblast must be considered as preeminent, since this epithelial layer is precociously developed in mammals and is at all times interposed between the embryo and the maternal tissues. It becomes at a later stage the epithelial component of the placenta, which may be defined as "an intimate apposition or fusion of the foetal organs to the maternal (or paternal) tissues for physiological exchange" (Mossman 1937). The complexities and species differences which mark the placental stages of mammals are in contrast to the relative simplicity of the blastocyst stage. The trophoblast may be of paramount importance also in interposing an immunological barrier between the mother and the fetus.

I. The Mammalian Zygote

At the time of fertilization the mammalian egg is 0.15–0.20 mm in diameter and is surrounded by a refractile "membrane" or zona pellucida about 0.019 mm in thickness. The mammalian egg is microlecithal in that it contains very little yolk material, in contrast to the macrolecithal egg of the bird. It shows no apparent polarity, that is, a differentiation into animal and vegetable poles, as observed in typical amphibian eggs. The first meiotic division, a reduction process resulting in halving of the nuclear DNA, occurs in the Graafian follicle during the process of maturation. The second meiotic division is begun just before ovulation but is not completed unless the egg is penetrated by a spermatozoon and fertilization occurs. The presence of two polar bodies is taken by Gregory (1930) as indicative of successful fertilization. The meiotic divisions are the occasions for the separation of nuclear material from the egg without a correspondingly large diminution in the volume of cytoplasmic material. The significance of polar body formation in its other biological aspects is unknown. Fertilization occurs in the ampullae of the oviducts. Penetration of the egg by the fertilizing spermatozoon is accompanied by changes in the plasma membrane of the egg, which does not allow additional spermatozoa to enter and so constitutes a "block to polyspermy" (Austin and Walton 1965). These surface changes in the egg, contingent on fertilization, are obvious in the large eggs of the sea urchin but are not clearly discerned in the mammalian egg. Molecular rearrangements in the plasma membrane, analogous to those involved in the antigen-antibody reactions, may occur, presenting formidable technical problems for successful analysis. The "block to polyspermy" may fail to develop in aged or damaged eggs. The resulting zygotes with an excess of DNA (polyploidy) usually fail to develop normally. Aberrations involving reductions or increases in the chromosomal material of an autosome or sex chromosome are determined also immediately after fertilization and so may affect all subsequent development.

Since the fertilizing spermatozoon must penetrate the accessory layer or layers enveloping the egg, their study is of the greatest significance. The zona pellucida, which is common to all mammalian eggs, is a highly refractile coat which stains intensely with stains for mucopolysaccharides such as the periodic acid-Schiff stain (formation of a purple Schiff base with active aldehyde groups released by oxidation). It also contains sialic acid (Soupart and Noyes 1964), which may serve to modify the surface charge on the egg and so affect sperm penetration. The zona may be removed by such proteolytic enzymes as pronase (Mintz 1962) and the enzyme neuraminidase, which selectively hydrolyzes sialic acid. The zona pellucida is laid down during the follicular development of the egg and appears to be a secretory product of the granulosa cells. It is penetrated by slender processes of the granulosa cells, which in some cases enter into intimate contact with processes of the egg; thickenings of the apposed plasma membranes may be observed at such points and may be sites of exchanges of vital materials between the egg and the follicular cells (Zamboni and Mastroianni 1966). The granulosa cells immediately around the egg (cumulus oophorus, discus proligerus) are shed with the egg at ovulation and are carried down the Fallopian tube into its ampullary part. They form a loose cluster of cells around the egg (corona radiata) which must be penetrated by the sperm before it can come into contact with the plasma membrane of the egg. The penetrating processes of the corona radiata are withdrawn gradually after the egg is shed from the follicle.

In some mammals, notably the rabbit, a second and much thicker coat is deposited on the surface of the zona pellucida, forming the so-called mucin coat (Bacsich and Hamilton 1954). This material, which gives the histochemical reactions for acid mucopolysaccharides, is secreted by the lining epithelial cells of the oviduct and is laid down incrementally (fig. 1). The mucin coat undoubtedly modifies the surface properties of the egg, morula, or blastocyst (e.g., stickiness) but it does not seem to pose a significant barrier to the penetration of soluble materials (Austin and Lovelock 1958).

II. Development of the Blastocyst in the Rabbit

The detailed development of the rabbit blastocyst has been studied by Gregory (1930) and serves as a useful descriptive prototype for blastocyst development in general.

Following fertilization the first cleavage division occurs within 24 hr. Note that all divisions of the egg after fertilization are mitotic and not meiotic and so involve the full normal or diploid number of chromosomes. The period preceding the first cleavage division is presumably characterized by: (1) a resting or presynthetic period (G_1), (2) a period of rapid DNA synthesis (S), and (3) a premitotic period (G_2). The speed or "tempo" of cleavage leading to the formation of a solid ball of cells or morula is estimated as follows (corrected to indicate hr after ovulation, since ovulation occurs about 10 hr after mating in the rabbit):

First cleavage spindle appears, 12 hr
2-cell stage, 11–21 hr
4-cell stage, 15–22 hr

8-cell stage, 22 hr
8–16-cell stage, 22–30 hr
16–32-cell stage, 31–45 hr
Morula stage, 31–65 hr
Differentiation into group of central cells and peripheral group (trophoblast) first observed, 17-cell stage
Cleft appears within morula (beginning of blastocyst stage), 65 hr
Blastocyst enters uterus about 70 hr (end of day 3)
Implantation begins, about 6 days.

The general form of the rabbit blastocyst at the 4th day is shown in figure 1.

Note that the trophoblast is unilaminar at the 4th day. The inner cell mass is undifferentiated and appears to be covered by a thin layer of trophoblastic cells (Rauber's layer). Mitoses are observed in the inner cell mass. The endoderm begins to delaminate from the under surface of the inner cell mass soon after the

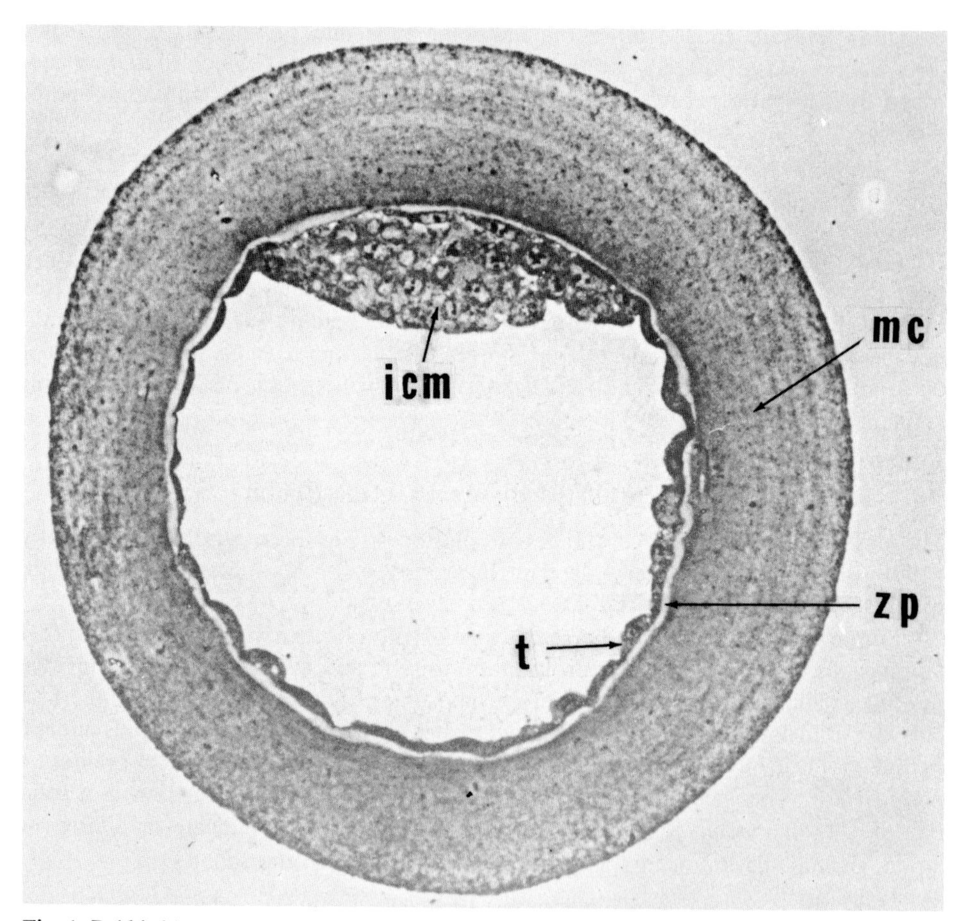

Fig. 1. Rabbit blastocyst at the 4th day: *icm*, inner cell mass; *mc*, mucin coat; *t*, trophoblast; *zp*, zona pellucida. Fixed in 8% osmic acid and phospate buffer; sectioned in araldite; stained with toluidine blue-borax; photographed under phase contrast. ×320. (Courtesy Dr. Pierre Soupart)

stage depicted in figure 1 and at the time of implantation (6 days) has spread to line the entire blastocyst. The expansion of the blastocyst is rapid after it has entered the uterus and increases in diameter from 1 or 2 mm to 4 mm at the time of implantation. The rabbit blastocyst is large, unlike that of most rodents. Its general form at implantation and its subsequent differentiation in the immediately postimplantation stages will be discussed latter.

III. Blastocyst Development in the Primate and the Human

Detailed information on the development of the macque is available (Lewis and Hartman 1941):

1-cell stage, 0–24 hr
2-cell stage, 24–36 hr
3- and 4-cell stage, 36–48 hr
5–8-cell stage, 48–72 hr
9–16-cell stage, 72–96 hr
Stage of free blastocyst, 96 hr to 9 ± 1 days

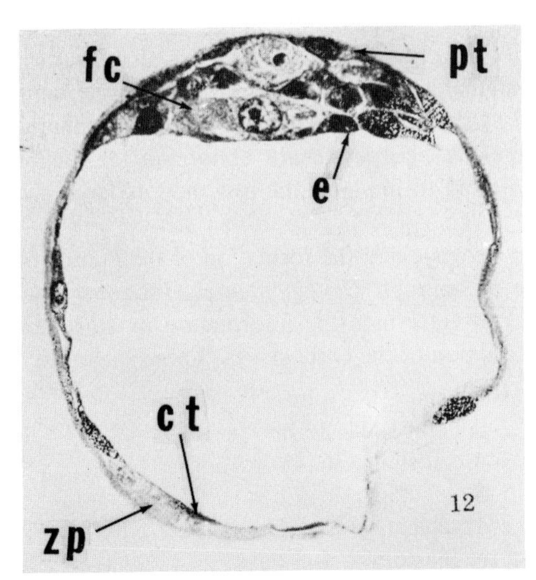

Fig. 2. Macaque blastocyst at the 8th day: *ct,* cavity wall trophoblast; *e,* primary endoderm; *fc,* formative (embryonic) cells; *pt,* polar trophoblast; *zp,* zona pellucida, ×450. (Courtesy C. H. Heuser and G. L. Streeter, *Contr Embryol* 29:17–55, 1941, plate 1, fig. 12)

The appearance of the blastocyst of the macaque at the 8th day is shown in figure 2.

The inner cell mass and trophoblast show identifiable differentiation at the 8th and 9th days, that is, just preceding implantation. The trophoblast or trophoblastic shell was differentiated by Hartman and Lewis into: (1) cavity wall trophoblast and (2) polar trophoblast (i.e., near the embryo or inner cell mass). The large clear cells of the inner cell mass were distinguished as formative or embryonic cells and the flattened epithelial cells apparently delaminating from the inner cell mass constituted the endoderm. The numbers of cells in these three categories at the 8th and 9th days in the macaque blastocyst were estimated as follows:

8 days:

cavity wall trophoblast, 58
polar trophoblast, 56
formative (embryonic) cells, 14
primary endoderm, 12
Total—140 cells

9 days:

cavity wall trophoblast, 224
polar trophoblast, 95 ± 10
formative (embryonic) cells, 32
primary endoderm, 24
Total—approximately 375 cells

This differentiation (on the basis of position) of the trophoblastic wall of the blastocyst has much merit, since there may be differential susceptibility of the trophoblast to toxic agents depending on its proximity to the inner cell mass. The relative counts for the different categories of cells are of interest also, and similar counts should be made in other mammalian species. That relatively large numbers of cells are not necessarily involved as shown in the elephant shrew Elephantulus (van Horst and Gilman 1944) where the blastocysts are abnormal: of the 60 or so blastocysts formed, only 1 is destined to implant; the rest may undergo swelling and hydropic degeneration.

Information on cleavage and the formation of the human blastocyst is less well documented. In the monograph *Developmental Horizons in Human Embryology* (1942, 1945, 1948) Streeter uses the information of Hartman and Lewis on the macaque for the description of the early stages. These stages are as follows:

Stage 1. 1-cell stage, 0–24 hr
Stage 2. 2-cell stage, 24–36 hr
3–4-cell stage, 36–48 hr
5–8-cell stage, 48–72 hr
9–16-cell stage (in uterus), 72–96 hr
Stage 3. Free blastocyst, 4–5 days
58-cell stage, 96 hr
107-cell stage, 108 hr
Stage 4. Implanting ovum, 6–9 days
Stage 5. Avillous ovum, 10–12 days
Stage 6. Primitive villi, 13–15 days

Hertig and Rock (1954) have made estimates of the number of cells comprising the human blastocyst at Streeter's stage 3:

58-cell stage:

trophoblastic cells, 53
formative (embryonic) cells, 5

107-cell stage:
- trophoblastic cells, 99
- formative (embryonic) cells, 4
- primary endoderm, 4

The precocity and rapid rate of division of the trophoblast as compared with the embryonic and endodermal cells is apparent here, as in the macaque. It is apparent also that since odd numbers of cells are regularly observed in the early cleavage stages, cleavage is asynchronous from the beginning or soon becomes so. The process of cleavage is accompanied also by changes in the size of the daughter cells (blastomeres) and in the nucleocytoplasmic ratios. The blastomeres become progressively smaller and there is no evidence of growth until late in the morula stage (Gregory 1930). It must be presumed, therefore, that the net synthesis of cytoplasmic protein is minimal in the cleavage and early morula stages. The source of nutritive materials is probably the stored yolk material. The nuclei of the blastomeres become progressively smaller also and the nucleocytoplasmic ratio is changed in favor of the nuclei. The role of the male and female pronuclei in subsequent development has not been determined in the mammalian zygote, mostly for technical reasons but also for lack of serious attempts to apply to mammalian development at any stage the technics, the intellectual concepts and biological questioning of the classical experimental embryologists of the Spemann tradition. Technics, such as the production of "hybrid merogonts" in which the male or female pronucleus has been removed (Balinsky 1965) or those used in the study of nuclear differentiation to remove nuclei from blastomeres (Briggs and King 1965), have not yet been applied to the problems of the mammalian zygote. Although the technics are formidable, they should not be impossible. The presence of cytoplasmic DNA in the zygote and the indirect evidence for "cytoplasmic inheritance" should serve to focus attention on the role of the cytoplasm in subsequent differentiation. Thus in the armadillo, which normally gives birth to identical quadruplets, there have been shown 2–140-fold variations in the observed values of twenty morphological and biochemical measurements in newborn monozygotic quadruplets, and these differences were attributed to segregation of cytoplasmic factors during formation of the quadruplets (Storrs and Williams 1968).

There are indications that the blastomeres in the rabbit are totipotent at least as far as the 2-cell stage (Seidel 1932) and possibly up to the 8-cell stage (Moore, Adams, and Rowson 1968). Beyond this point the small size of the blastomeres precluded more critical studies.

Earlier observations on the respiration of the egg in the cleavage, morula, and blastocyst stages (Boell and Nicholas [rat] 1948; Fridhandler, Hafez, and Pincus [rabbit] 1956) indicated that no significant consumption of oxygen occurred in the cleavage and morula stages. There was no glycolytic activity evident from the 1- to the 16-cell stage. Later morulae and blastocysts, but not freshly shed ova, showed glycolytic activity only in the presence of exogenous glucose. The energy requirements of the zygote were apparently met not from glucose or some glucose precursor (e.g., glycogen) but from some stored lipid material in the cytoplasm. The maturation of glycolytic mechanisms was interpreted as resulting from the maturation of appropriate enzyme systems. Transitory upsurges of oxygen consumption were observed in invertebrate eggs coincidental with the premitotic phases (G_2) when Cartesian diver technics were used (Scholander, Leivestad and Sundnes 1958).

IV. Cavitation and Expansion of the Blastocyst

The forces responsible for the movement of the cleaving zygotes and morulae down the oviducts are controlled by progesterone and estrogen in precise proportions and are mainly muscular. Excessive estrogen may lead to "tube locking of the ova" (Burdick, Whitney, and Pincus 1937). Failure of the blastocyst to arrive in the uterus at the correct time may lead to asynchrony between the blastocyst and the uterine mucosa. Under these circumstances the interaction between the trophoblast and the endometrium may be defective and implantation may be prevented. In addition to their effects on oviductal transport estrogen and progesterone may be involved also in the cavitation and expansion of the blastocyst. Cavitation appears to take place by an accumulation of fluid between the trophoblastic cells of the morula. Changes in the permeability of the trophoblastic cells are involved and may be dependent on estrogen or progesterone or both. Expansion of the blastocyst fails to occur in the absence of progesterone. The source of the water which is carried across the trophoblastic cell membranes, probably by active transport, must be the oviductal fluid. The formation of this fluid has been studied by Clewe and Mastroianni (1960) in the rabbit. Chemical studies are rare but have shown that active secretion may be involved and that the fluid is low in protein (Hamburger, Grossman, and Tregier 1955). The presence of proteins in the blastocyst fluid after the 7th day in the rabbit was first observed by Brambell (1954), who also noted its spontaneous clotting. These proteins were later shown (McCarthy and Kekwick 1949) to be identical with maternal plasma proteins, though present in smaller concentration. Lutwak-Mann (1954) has undertaken intensive studies of the composition of the blastocyst fluid in the rabbit with respect to simple sugars, electrolytes, and water-soluble vitamins. There is evidence of increased and selective permeability on the part of the blastocyst wall as development advances (see chapter 13). That the conversion of the unilaminar blastocyst into a bilaminar structure by growth of primary endoderm around it is responsible for this selective permeability of the blastocyst to soluble materials of maternal origin seems unlikely, since the endoderm does not completely line the blastocyst until the time of implantation (6–7 days in the rabbit), and in some forms never does (in the guinea pig). Of particular interest is the finding of relatively high concentrations of bicarbonate in the blastocyst fluid and of carbonic anhydrase activity, responsive to progesterone, in the endometrium of the rabbit. These observations raise the possibility that pumplike mechanisms may exist in the wall of the blastocyst similar to those in the proximal tubule of the nephron responsible for the exchange of sodium ions for hydrogen ions. Diamox, however, a carbonic anhydrase inhibitor, does not seem to have a deleterious effect on blastocyst development. The 6-day rabbit blastocyst has been shown (Fridhandler 1968) to incorporate nucleotides into DNA and also to be capable of synthesizing cholesterol and pregnenolone from acetate, the rate of incorporation being influenced by interstitial cell stimulating hormone (LH) and adrenocorticotrophic hormone (ACTH). The penetration of essentially unchanged maternal proteins into the blastocyst fluid is of theoretical interest in relation to the immunological problem of pregnancy. The exposure of the trophoblastic to maternal proteins at an early stage may possibly alter its immunological properties (e.g., antigenicity) at a later stage when they become an integral part of the placenta.

V. The Problem of Gastrulation

Occasion has been found to refer through this discussion to the spread of endoderm within the blastocyst. That this process represents at least a part of the gastrulation process common to all vertebrates seems likely (Balinsky 1965). Gastrulation throughout the vertebrate phylum is determined in large measure by the amount of yolk present in the egg. In the microlecithal egg of Amphioxus a primary gut cavity is produced by simple invagination, the invaginated layer comprising the endoderm. In the mesolecithal egg of the frog gastrulation takes place by epiboly in which there is an overgrowth of the cells of the animal (black) pole of the egg over the vegetable pole. In the macrolecithal egg of the birds a segmentation cavity appears beneath the embryonic disk and is converted to the gut cavity by the differentiation of endodermal cells in its wall (gastrulation by cavitation). In mammals the delamination of endoderm from the inner cell mass (gastrulation by delamination) occurs in all blastocysts so far described; the primary yolk sac so formed is, therefore, precocious but probably does not represent the definitive yolk sac. In man the primary yolk sac is secondarily converted into a definitive yolk sac, whereas in the macaque (Heuser and Streeter 1941) the definitive yolk sac appears to form by cavitation of the cells in relation to the under surface of the embryonic disk.

Regardless of how the definitive yolk sac is formed in mammals, it is evident that this process does not represent the essential and important part of the general phenomenon of gastrulation. It is of importance in converting the unilaminar blastocyst into a bilaminar structure, and the concomitant modifications in the permeability of the blastocyst must be considered, though they have not yet been interrelated. The more important part of gastrulation is the mass movement of cells around the dorsal rim of the blastopore or its homologous structure (Hensen's node) in mammals. The continuous migration of surface cells around the node to take up their position in the roof of the gut cavity or archenteron is perhaps the most important aspect of gastrulation in all higher vertebrates. By this means cells are brought into contact with the overlying ectoderm, and together with the mesoderm derived from them, comprise the chorda-mesoderm field which is the primary inducer of neural development. In considering homologies of the surface features and in constructing maps of prospective developmental potency of the gastrulae of lower vertebrates and mammals, it must be remembered that there is no homologue in submammalian forms for the trophoblast *stricto sensu.* The trophoblastic shell must be regarded as a peculiar mammalian structure developed solely in response to the demands of viviparity and prolonged gestation in the uterus.

VI. Implantation

Implantation is the adhesion or fixation of the blastocyst to the endometrial surface of the uterus. It is necessary to differentiate the site of primary adhesion from the site of definitive adhesion where the placenta is eventually developed. For example, in the rabbit (fig. 4) the site of primary adhesion is abembryonic, that is, at the pole opposite the inner cell mass; the placenta is developed later in relation to the embryonic pole.

For descriptive purposes it has been found useful to relate the site of primary adhesion to the side of the uterus in relation to or opposite to the attachment of the mesometrium (broad ligament in women). Implantation may thus be meso-

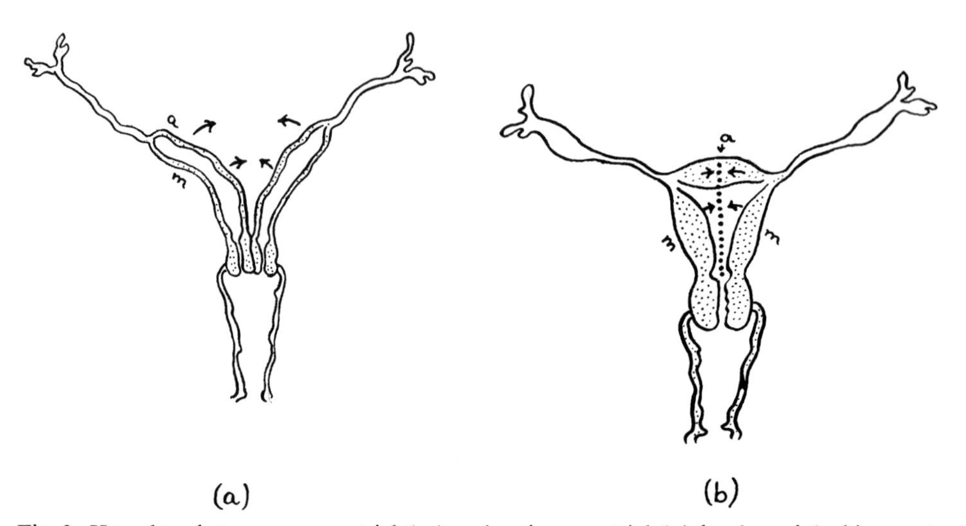

Fig. 3. Homology between mesometrial (*m*) and antimesometrial (*a*) borders of the bicornuate uterus of the rabbit (*a*) and the uterus simplex (*b*) of primates and man. Fusion of the two Müllerian ducts occurs along the axis of the dotted lines in (*b*) which, therefore, is homologous with the antimesometrial border (midventral and midsagittal lines).

metrial or antimesometrial. The homologous regions in the bicornuate uterus commonly found in most domestic and laboratory animals and the single uterus or uterus simplex of primates and man are illustrated in figure 3. The oviducts and uterus are formed in all eutherian mammals by the approximation of the paramesonephric or Müllerian ducts and, in some forms, their fusion. Fusion is absent in the rabbit, where there are two uterine horns and two cervices (note the vagina in mammals is probably of urogenital sinus origin). Fusion of the lower ends of the Müllerian ducts in the rat and guinea pig results in a single cervix but two uterine horns. In the uterus simplex characteristic of monkeys and man, there is extensive fusion above the level of the cervix producing a *corpus uteri.* The line of fusion of the Müllerian ducts in the uterus simplex (fig. 4) is indicated by dotted lines and corresponds morphologically to the antimesometrial side of the bicornuate uterus. Most implantation sites in women have been found along the midventral or middorsal line of the corpus uteri, so that it has been concluded that implantation in women is antimesometrial.

Primary implantation in the rat, guinea pig, and rabbit is antimesometrial. However, the definitive placenta is developed mesometrially in all these forms (see figs. 4, 9).

Implantation is defined also in terms of the degree of penetration of the uterine mucosa by the implanting blastocyst. Where the blastocyst remains within the uterine lumen but fails to sink beneath the surface (e.g., rabbit, fig. 4), the implantation is termed "superficial." This type of implantation is typical of large blastocysts (fig. 5).

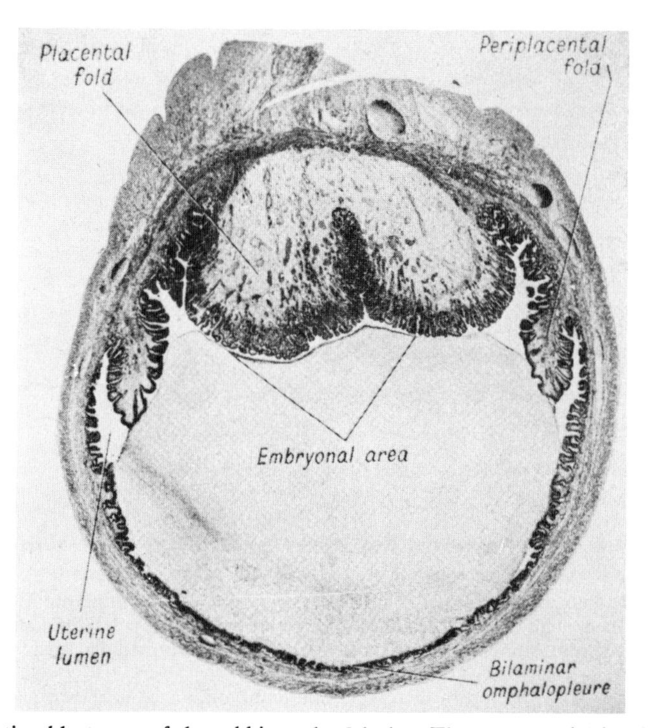

Fig. 4. Implanting blastocyst of the rabbit at the 8th day. The mesometrial border is at the top. Primary adhesion occurs between the bilaminar omphalopleure (trophoblast plus endoderm) and the uterine epithelium at the antimesometrial (obplacental) pole of the blastocysts. ×9. Primary adhesion occurs between the bilaminar omphalopleure (trophoblast plus endoderm) (Courtesy E. C. Amoroso, in *Marshall's Physiology of Reproduction,* vol. 2, 1964, figure 15.63)

Implantation is termed interstitial when the blastocyst sinks beneath the level of the endometrium and becomes enclosed in a recess of the uterine cavity or an eroded implantation cavity as in the guinea pig. This type of implantation is typical of small blastocysts, as in the rodents, and to this form of implantation the term nidation is frequently applied. Intermediate forms of implantation are found also. The relationship between the size of the blastocyst and the type of implantation in a group of mammals is evident in figure 5.

A further consequence of the type of implantation is the presence or absence of a decidua capsularis. No layer of covering decidua (decidua capsularis) is found where the implantation is superficial (e.g., rabbit, fig. 6). It is found wherever the blastocyst is significantly or completely submerged beneath the surface of the endometrium.

Not only is the primary attachment site of the blastocyst constant for a given species, but the inner cell mass also shows a constant orientation with respect to the uterine lumen and to the mesometrial or antimesometrial borders of the uterus

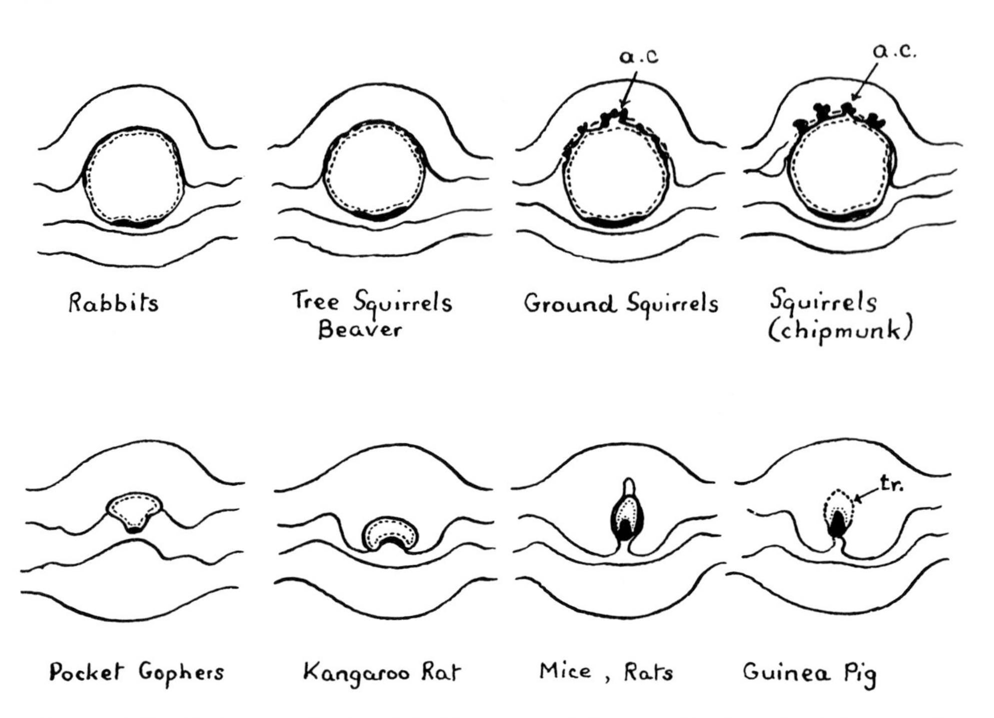

Fig. 5. Relationship between the size of the blastocyst and the type of implantation in lagomorphs (rabbits) and rodents. The largest blastocysts (e.g., rabbit) implant superficially. The smallest blastocysts (e.g., rat and guinea pig) implant interstitially. Intermediate types of implantation are shown.

The mesometrium is at the top in all cases. The embryonic area is shown as a thickening of the blastocyst inferiorly. The endoderm is indicated by stippling. Primary adhesion is shown abembryonically with the production of trophoblastic attachment cones in some rodents (*ac*). In the guinea pig the abembryonic trophoblast (*tr*) disintegrates early and the endoderm is never completed. (Redrawn from H. W. Mossman, *Contr Embryol,* 26:133–246, 1937, fig. 3, p. 144)

(see fig. 5). The spacing of blastocysts along the length of the uterus, especially in animals with large litters and bicornuate uteri, also presents many unsolved problems.

VII. Placentation: General Principles as Applied to the Mammalian Blastocyst

The detailed comparative aspects of placentation are beyond the scope of this chapter and may be found in the monographs of Amoroso (1964) and Mossman (1937). They are not wholly irrelevant here, however, since the general modifications in the wall of the blastocyst may profoundly modify its physiological function, especially with respect to the passage of materials across it. They become of even greater importance in those mammals, such as the ungulates and the carnivores, in which implantation is normally delayed (e.g., until the 17th day in the sheep), by which time the embryo and its ancillary membranes (yolk sac, amnion, allantois) are sufficiently differentiated to be functional. The manner in which the originally unilaminar wall of the blastocyst may be modified is illustrated in figure 6.

The first event is the spread of the endoderm around the inner surface of the trophoblastic shell forming a primitive yolk sac. This lining is complete in the rabbit but incomplete in the guinea pig, in which the abembryonic wall of the trophoblastic shell breaks down before the yolk sac is completed (fig. 5). The apposition of endoderm to trophoblast with no interposed mesoderm is termed a bilami-

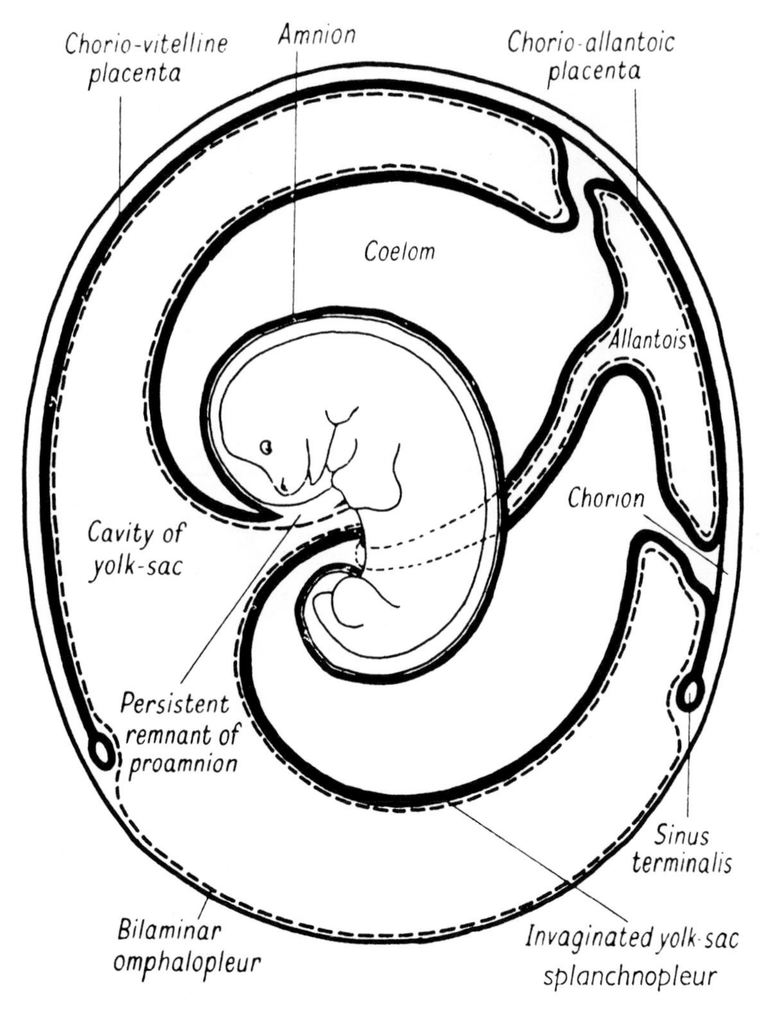

Fig. 6. Representation of placental modifications of the blastocyst wall based on the marsupial Perameles (redrawn from Hill). (Courtesy E. C. Amoroso in *Marshall's Physiology of Reproduction,* vol. 2, 1964, fig. 15.17)

nar omphalopleure (fig. 6). An example of this is the abembryonic pole of the blastocyst in the rabbit (fig. 4), where a bilaminar omphalopleure is formed and where the transitory primary fixation of the blastocyst develops; this area is often referred to as the obplacental area to distinguish it from the definitive placental area on the mesometrial side of the uterus (fig. 5). The bilaminar omphalopleure then undergoes conversion into a trilaminar omphalopleure following the appearance of extraembryonic mesoderm between the trophoblast and the endoderm. The extraembryonic mesoderm undergoes a varied degree of vascularization extending out from the area of the embryo. The limit of vascularization is usually marked by a circumferential vessel, the sinus terminalis. Inversion of the yolk sac or entypy (Duval) is a fundamental phenomenon observed typically in rodents and lagomorphs (rabbits, hares). In the rodents (fig. 7) it is early and precedes the appearance of extraembryonic mesoderm in relation to the trophoblast.

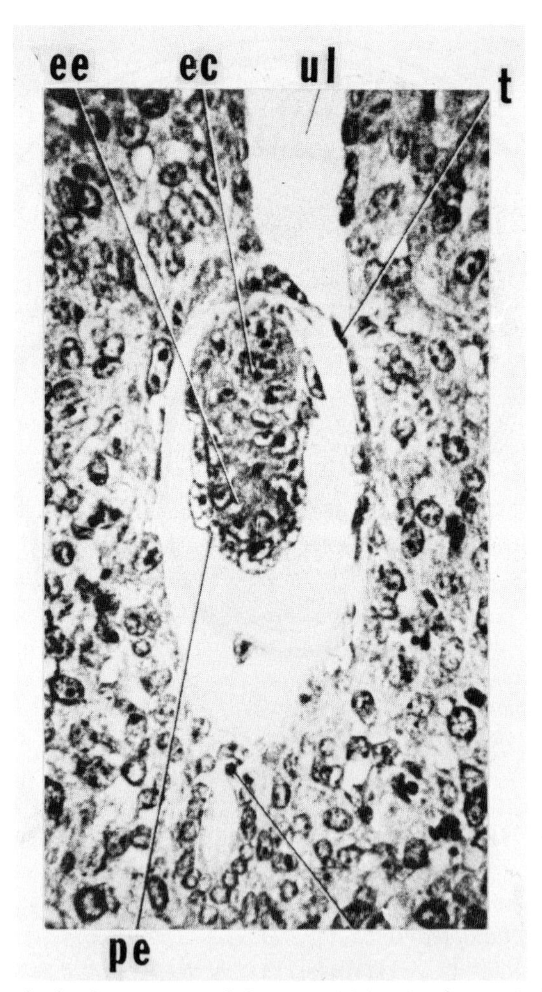

Fig. 7. Implantation site in the mouse at 6 days and 16 hr showing early inversion of the yolk sac. The mesometrial border is at the top: *ee*, embryonic ectoderm; *ec*, extraembryonic ectoderm; *pe*, proximal (visceral) endoderm; *t*, trophoblast; *ul*, uterine lumen. ×385. (Courtesy E. C. Amoroso in *Marshall's Physiology of Reproduction*, vol. 2, 1964, fig. 15.72)

It is associated with a precocious development of the inner cell mass which proliferates rapidly and becomes invaginated into the yolk sac cavity. The spread of the endoderm occurs pari passu and is completed in the mouse and rat, but fails to line the abembryonic wall of the chorionic sac in the guinea pig owing to its early breakdown. These events are illustrated in the mouse in figure 10.

In the large rabbit blastocyst the yolk sac is completed and lined by extraembryonic mesoderm before inversion occurs. Inversion occurs late (about the 12th day) around the axis of the sinus terminalis and results in the apposition of the inner (visceral) wall of the yolk sac, which is vascular, to the nonvascular outer (parietal) wall of the yolk sac. The latter, with its associated trophoblast, then disappears (about the 17th day in the rabbit), bringing the columnar absorptive cells of the visceral yolk sac into contact with the uterine epithelium and its secretions (fig. 9).

It is across this inverted vascular yolk sac that the transport of antibodies takes

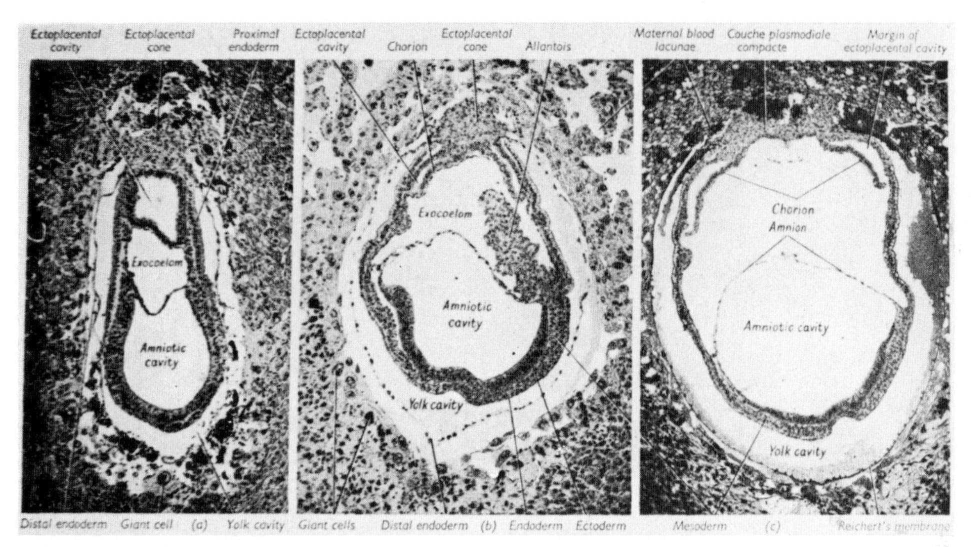

Fig. 8. Stages in the development of the mouse implantation site at (1) 7 days and 6 hr, (2) 7 days and 18 hr, and (3) 8 days. The definitive placenta is formed by the attachment of allantoic mesoderm to the thickened upper (mesometrial) pole of the blastocyst (ectoplacental cone). (Courtesy of E. C. Amoroso in *Marshall's Physiology of Reproduction,* vol. 2, 1964, fig. 15.73)

place (Brambell 1954). Similar conditions exist in the rodents following the disappearance of the parietal yolk sac and trophoblast about two-thirds of the way through gestation. Extensive areas of chorionic placentation (trophoblast plus avascular mesoderm) persist in some forms, for example, the rabbit, where they may have important physiological functions (Davies 1966).

The definitive placenta of mammals is developed by the apposition of the allantoic mesoderm to the chorion (see fig. 6). The allantoic diverticulum may persist as a hollow endodermal sac, often containing large amounts of fluid, as in the ungulates. It may be intermediate in size (cat, rabbit) or absent (guinea pig, rat). Whether or not a hollow endodermal allantois exists, the vascular mesoderm associated with it is believed to be the effective vascularizing agent of the definitive placenta in all eutherian mammals. The mesodermal stalk of the allantois is represented by the body stalk of the mammalian embryo; it is converted later into the umbilical cord following the extension of the amnion around it. Thus the definitive mammalian placenta is usually considered to be a chorioallantoic placenta.

The penetration of mesoderm into the trophoblast at the site of the chorioallantoic placenta carries embryonic or fetal vessels into intimate relationship with the maternal tissues. The trophoblast intervenes at all times, however, between the fetal and the maternal tissues and constitutes a "barrier" that selectively controls the flow of physiological materials across it. It may also be important in modifying the immunological factors tending toward the rejection of the placenta as a homograft or allograft. The trophoblast consists of a basal germinal layer of cytotrophoblast which undergoes mitotic division and gives rise to one or more layers of syncytiotrophoblast. Considerable syncytial transformation of the cytotrophoblast before its incorporation into the syncytiotrophoblast is now known to occur (Wynn 1967; Davies and Glasser 1967). The modified or differentiated cytotrophoblast then becomes part of the syncytial mass by disappearance of the plasma membranes. Desmosomes may persist for a considerable time. The cytotrophoblast seems to disappear later in gestation (by the 6th month in the human placenta), at which time the new formation of syncytiotrophoblast comes to an end. It may persist in certain areas of the human placenta, for example, the basal plate and septa placentae, but

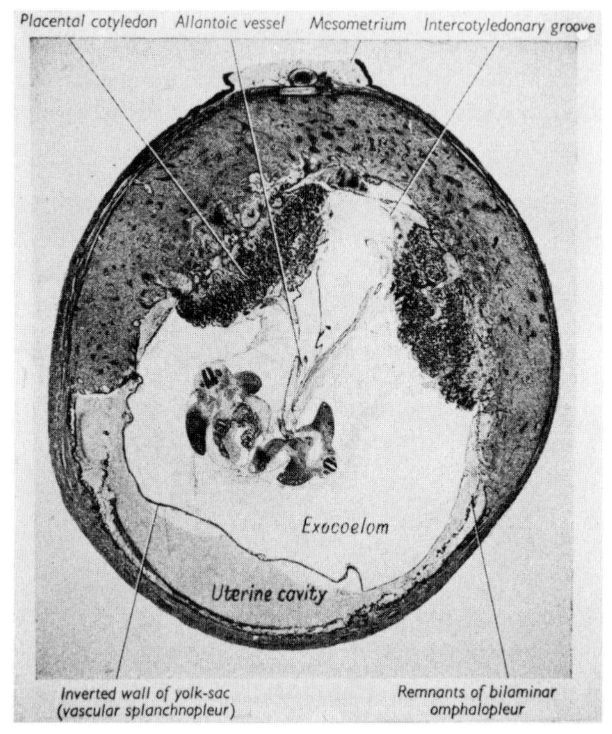

Fig. 9. Implantation site in the rabbit at the 12th day. The definitive placenta is developed as bilateral thickenings (cotyledons) of the embryonic (mesometrial) pole of the blastocyst. Inversion of the yolk sac by the embryo and its fluid-filled amnion has begun. ×5. (Courtesy E. C. Amoroso in *Marshall's Physiology of Reproduction*, vol. 2, 1964, fig. 15.68)

here it appears to be a specialized cell unlike the germinal cytotrophoblast or Langhans cell.

The intimacy of contact between the trophoblast and the uterine tissues in the chorioallantoic placenta has been made the basis of a histological classification of placental types in eutherian mammals (Grosser 1927):

1. epitheliochorial, an apposition of trophoblast to the uterine epithelium (odd-toed ungulates such as the horse and pig).
2. syndesmochorial, an apposition of trophoblast to the subepithelial stroma (even-toed ungulates such as the sheep and goat).
3. endotheliochorial, an apposition of trophoblast to the endothelial wall of the uterine blood vessels (e.g. carnivores).
4. hemochorial, a direct apposition of trophoblast to the maternal blood (e.g., rodents, primates, man).

Note that the layer of trophoblast in immediate contact with the maternal tissues is not invariably syncytial. No syncytiotrophoblast exists in the ungulates, for example, where the trophoblast is cellular. Even in the hemochorial placenta extensive modifications in the trophoblast bordering on the maternal blood spaces occur. Enders (1965) has distinguished a monochorial type (human, guinea pig), a dichorial type (rabbit), and a trichorial type (rat, mouse) in which the trophoblast is respectively single, double, or triple.

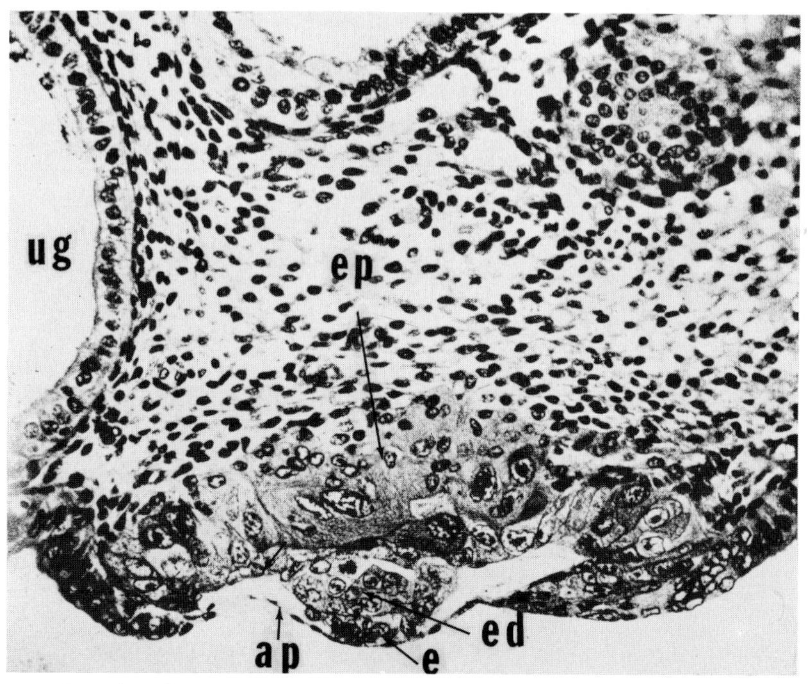

Fig. 10. Implantation site in the human female at the 7th day: *ap,* thin abembryonic pole of blastocyst; *ep,* thick embryonic pole with cytotrophoblast and syncytiotrophoblast; *ed,* embryonic disk; *e,* primary endoderm; *ug,* uterine gland. ×270. (Courtesy A. Hertig and J. Rock, *Contr Embryol,* 31:67–85, 1945, plate 1, fig. 3)

VIII. Placentation in Man

Implantation in man is interstitial and mesometrial (fig. 11).

No certain facts are known about the development of the primary yolk sac. In the 7½-day specimen (fig. 10) a cluster of cells ventral to the embryonic disk may represent the endoderm; the chorionic sac, however, is not lined to any significant extent by endoderm. The abembryonic pole of the trophoblastic shell is very thin. Amniogenesis appears to occur by cavitation within the trophoblast immediately dorsal to the embryonic disk. In the 12-day specimen, which is shown (fig. 11) to be superficially implanted in the decidua compacta of the endometrium, the primitive yolk sac is lined by flattened cells (Heuser's exocoelomic membrane), considered by some to be endodermal. There is a thick layer of sparsely cellular mesenchyme (primary mesoblast) between the yolk sac membrane and the trophoblast. The exocoelomic cavity is later developed within this layer, splitting off the yolk sac completely from the chorion (fig. 13).

The manner in which the definitive yolk sac (fig. 13) is formed is uncertain.

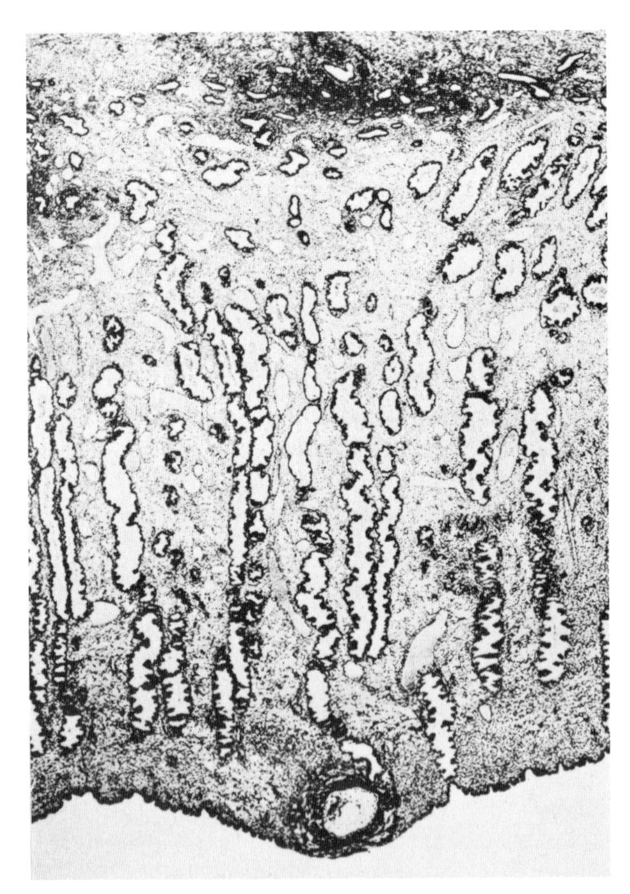

Fig. 11. Implantation site in the human female at the 11th day. The blastocyst is interstitially implanted but lies superficially in stratum compactum of the endometrium. ×30. (Courtesy A. Hertig and J. Rock, *Contr Embryol,* 29:127–56, 1951, plate 2, fig. 13)

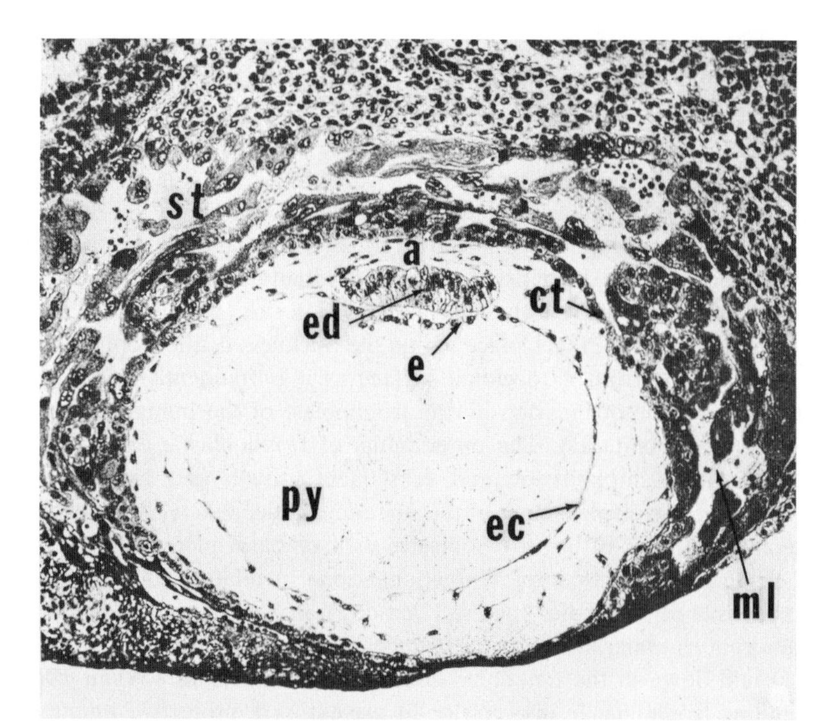

Fig. 12. Implantation site in the human female at the 12th day: *a,* amniotic cavity; *ct,* cytotrophoblast; *e,* endoderm; *ec,* exocoelomic membrane (Heuser); *ed,* embryonic disk; *ml,* maternal blood spaces or lacunae; *st,* syncytiotrophoblast; *py,* primary yolk sac. ×122. (Courtesy A. Hertig and J. Rock, *Contr Embryol,* 29:127–56, 1941, plate 6, fig. 23)

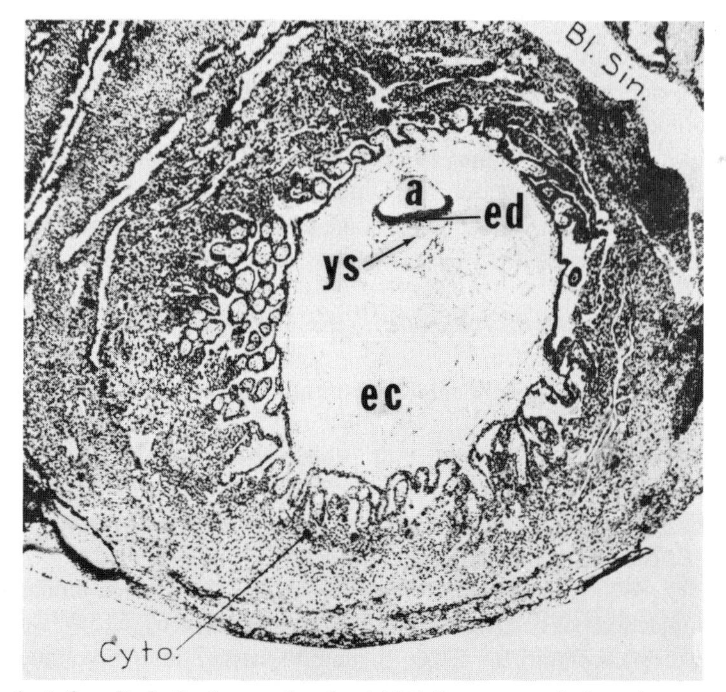

Fig. 13. Implantation site in the human female at 16½ days: *a,* amniotic cavity; *ec,* exocoelomic cavity; *ed,* embryonic disk; *ys,* definitive yolk sac. Chorionic villi are differentiated within the chorionic shell. "Cyto" marks the outer limit of the trophoblastic shell (basal plate) where cytotrophoblastic elements (cell columns) are abundant. ×428. (Courtesy C. H. Heuser, A. Hertig, and J. Rock, *Contr Embryol,* 31:87–99, 1945, plate 4, fig. 17)

There is some evidence that a constriction appears at the "waist" of the primary yolk sac followed by separation of the distal component (which becomes atrophic) from a proximal portion in relation to the embryo. The latter is the definitive yolk sac and is continuous with the embryonic gut.

The formation of lacunae containing maternal blood within the thick trophoblastic shell of syncytiotrophoblast is shown at the 12th day (fig. 12). The formation of chorionic villi by the extension of vascular allantoic mesoderm into the septa between the lacunae is shown in the later implantation site (fig. 13).

The formation of villi takes place within the thickness of the trophoblastic shell and not by sprouting from its decidual surface as it is frequently depicted. There is little if any evidence of invasion by the trophoblast of the human except during the few days of implantation. The outer limits of trophoblastic extension are established soon after implantation and subsequent growth and expansion of the chorionic sac involves displacement of the surrounding decidua without further invasion. The external limit of the trophoblastic shell or basal plate with the adjacent decidual elements forms a complex junctional zone in relation to which there are extensive deposits of "fibrinoid" material laid down in the later stages of pregnancy. Similar amorphous material, which is strongly acidophilic and periodic acid-Schiff positive, is laid down in the junctional zone of many placentas (Wynn 1967). Its role in limiting trophoblastic invasion or in presenting a protective immunological barrier between the maternal and fetal tissues has not been established.

References

Amoroso, E. C. 1964. In *Marshall's Physiology of Reproduction,* ed. A. S. Parkes, 2:127. London: Longman's Green and Co.

Austin, C. R., and Lovelock, J. E. 1958. Permeability of rabbit, rat, and hamster egg membranes. *Exp Cell Res* 15:260.

Austin, C. R., and Walton, Arthur. 1965. In *Marshall's Physiology of Reproduction,* ed. A. S. Parkes, 1:333. London: Longman's Green and Co.

Bacsich, P., and Hamilton, W. J. 1954. Some observations on vitally stained rabbit ova with special reference to their albuminous coat. *J Embryol Exp Morph* 2, part 1: 81.

Balinsky, B. I. 1965. *An Introduction to Embryology.* 2d ed. Philadelphia and London: W. B. Saunders Co.

Boell, E. J., and Nicholas, J. S. 1948. Respiratory metabolism of the mammalian egg. *J Exp Zool* 109:267.

Brambell, F. W. R. 1954. Transport of proteins across the fetal membranes. In *Cold Spring Harbor Symp Quant Biol* 19: *The Mammalian Fetus,* p. 71.

Briggs, T. J., and King, R. 1965. In *Molecular and Cellular Aspects of Development,* ed. E. Bell. New York: Harper and Row.

Burdick, H. O.; Whitney, R.; and Pincus, G. 1937. The fate of mouse ova tube-locked by injections of oestrogenic substances. *Anat Rec* 67:513.

Clewe, T., and Mastroianni, L. 1960. A method for continuous volumetric collection of oviduct secretions. *J Reprod Fertil* 1:146.

Davies, J. 1966. In *The Placenta and Fetal Membrane,* ed. C. A. Villee, p. 140. New York: Williams and Wilkins Co.

Davies, J., and Glasser, S. R. 1967. Observations on the human placenta two weeks after fetal death. *Amer J Obstet Gynec* 98:1111.

Enders, A. C. 1965. A comparative study of the fine structure of the trophoblast in several hemochorial placentae. *Amer J Anat* 116:29.

Fridhandler, L. 1968. In *Biology of Gestation,* ed. N. S. Assali, vol. 1, chap. 2. New York: Academic Press.

Fridhandler, L.; Hafez, E. S. E.; and Pincus, G. 1956. Respiratory metabolism of mammalian eggs. *Proc Soc Exp Biol Med* 92:127.

Gregory, P. W. 1930. The early embryology of the rabbit. *Carnegie Inst Wash Contrib Embryol* 21:141.

Grosser, O. 1927. In *Frühentwicklung, Eihautbildung und Placentation des Menschen und der Säugetiere,* p. 76. Munich: J. F. Bergmann.

Hamburger, F.; Grossman, M. S.; and Tregier, A. 1955. Experimental hydrouteri (hydrometra) in rodents and some factors determining their formation. *Proc Soc Exp Biol Med* 90:719.

Hertig, A. T., and Rock, J. 1941. Two human ova of the pre-villous stage, having an ovulation age of about eleven and twelve days, respectively. *Carnegie Inst Wash Contrib Embryol* 29:127.

———. 1945. Two human ova of the pre-villous stage, having a developmental age of about seven and nine days, respectively. *Carnegie Inst Wash Contrib Embryol* 31:67.

———. 1954. On the preimplantation stages of the human ovum: A description of four normal and four abnormal specimens ranging from the second to the fifth day of development. *Carnegie Inst Wash Contrib Embryol* 35:199.

Heuser, C. H.; Hertig, A. T.; and Rock, J. 1945. Two human embryos showing early stages of the definitive yolk sac. *Carnegie Inst Wash Contrib Embryol* 31:87.

Heuser, C. H., and Streeter, G. L. 1941. Development of the Macaque embryo. *Carnegie Inst Wash Contrib Embryol* 29:17.

Lewis, W. H., and Hartman, C. G. 1941. Tubal ova of the Rhesus monkey. *Carnegie Inst Wash Contrib Embryol* 29:9.

Lutwak-Mann, C. 1954. Some properties of the rabbit blastocyst. *J Embryol Exp Morph* 2:1.

McCarthy, E. F., and Kekwick, R. A. 1949. The passage into the embryonic yolk sac cavity of maternal plasma proteins in rabbits. *Addendum J Physiol* 108:184.

Mintz, B. 1962. Experimental study of the developing mammalian egg: Removal of the zona pellucida. *Science* 138:594.

Moore, N. W.; Adams, C. E.; and Rowson, L. E. A. 1969. Developmental potential of single blastomeres of the rabbit egg. *J Reprod Fertil* 17:527.

Mossman, H. W. 1937. Comparative morphogenesis of the fetal membranes and accessory uterine structures. *Carnegie Inst Wash Contrib Embryol* 26:133.

Scholander, P. F.; Leivestad, H.; and Sundnes, G. 1958. Cycling in the oxygen consumption of cleaving eggs. *Exp Cell Res* 15(3):505.

Seidel, F. 1932. Die Entwicklungsfähigkeiten isolierter Furchungszellen aus dem kaninchens Oryctolagus cuniculus. *Roux Arch* 152:43.

Soupart, P., and Noyes, R. W. 1964. Sialic acid as a component of the zona pellucida of the mammalian ovum. *J Reprod Fertil* 8:251.

Storrs, E. E., and Williams, R. J. 1968. A study of monozygous quadruplet armadillos in relation to mammalian inheritance. *Proc Nat Acad Sci USA* 60:910.

Streeter, G. L. 1942, 1945, 1948. Developmental horizons in human embryology. *Carnegie Inst Wash Contrib Embryol* 30:211, 31:27, 32:133.

Van Horst, C. J., and Gilman, J. 1944. On abnormal blastulas in Elephantulus. *Anat Rec* 90:101.

Wynn, R. M. 1967. Fetomaternal cellular relations in the human basal plate: An ultrastructural study of the placenta. *Amer J Obstet Gynec* 97:823.

Zamboni, L., and Mastroianni, L., Jr. 1966. Electron microscope studies on rabbit ova: 1. The follicular oocyte. *J Ultrastruct Res* 14:95.

3

Orientation and Site of Attachment of the Blastocyst: A Comparative Study

Harland W. Mossman

Department of Anatomy
University of Wisconsin, Madison

In most major taxonomic groups of eutherian memmals, available data show that at the time of first attachment of the blastocyst to the endometrium the inner cell mass or the embryonic disk has an almost constantly specific directional orientation to the uterus: it faces toward the mesometrium, away from it, or, in a few cases, to the side.

The direction in which the embryonic mass faces is correlated with much of the later development of the fetal membranes. Since the yolk sac extends in a direction opposite to that in which the disk faces, the various types of yolk sac placentas are necessarily opposite, although sometimes also somewhat lateral to the disk (fig. 1).

Because the developing body of the embryo tends to remain on the surface of the early yolk sac, the amnion and chorion and exocoelom at first form dorsal to the disk. Therefore when the allantois grows into the exocoelom it extends away from the yolk sac and dorsal to the amnion. Here, in most species, it makes its first contact with the chorion so that the first chorioallantoic placentation is established dorsal to the embryo, that is, on the side of the uterus toward which the disk faces (fig. 1). If the allantoic vesicle is absent or relatively small, as in the case in rabbits, rodents, and most insectivores and bats, this will also be the place of definitive attachment of the umbilical cord. However, if the allantoic vesicle expands at an early time to a large size and its stalk remains relatively short while the amnion expands rather slowly, the allantoic vessels will become widely separated and their attachments to the chorioallantois will move toward and eventually to the side opposite the first contact of the allantois with the chorion (fig. 1). This shift of allantoic vessel attachment to the chorioallantois is characteristic of carnivores and of other groups with expanded allantoic vesicles, very short cords, and diffuse or zonary placentas. A different situation occurs in artiodactyls. Here the rapidly expanding amnion brings the somewhat tardy allantoic outgrowth into line with the side of the yolk stalk and relatively small yolk sac before it can make contact with the chorion dorsal to the embryo (fig. 1). The amnion ensheathes the allantoic and yolk stalks rapidly to form a long umbilical cord, and the allantois is forced to make its first contact with the chorion ventral to the embryo alongside the small yolk sac.

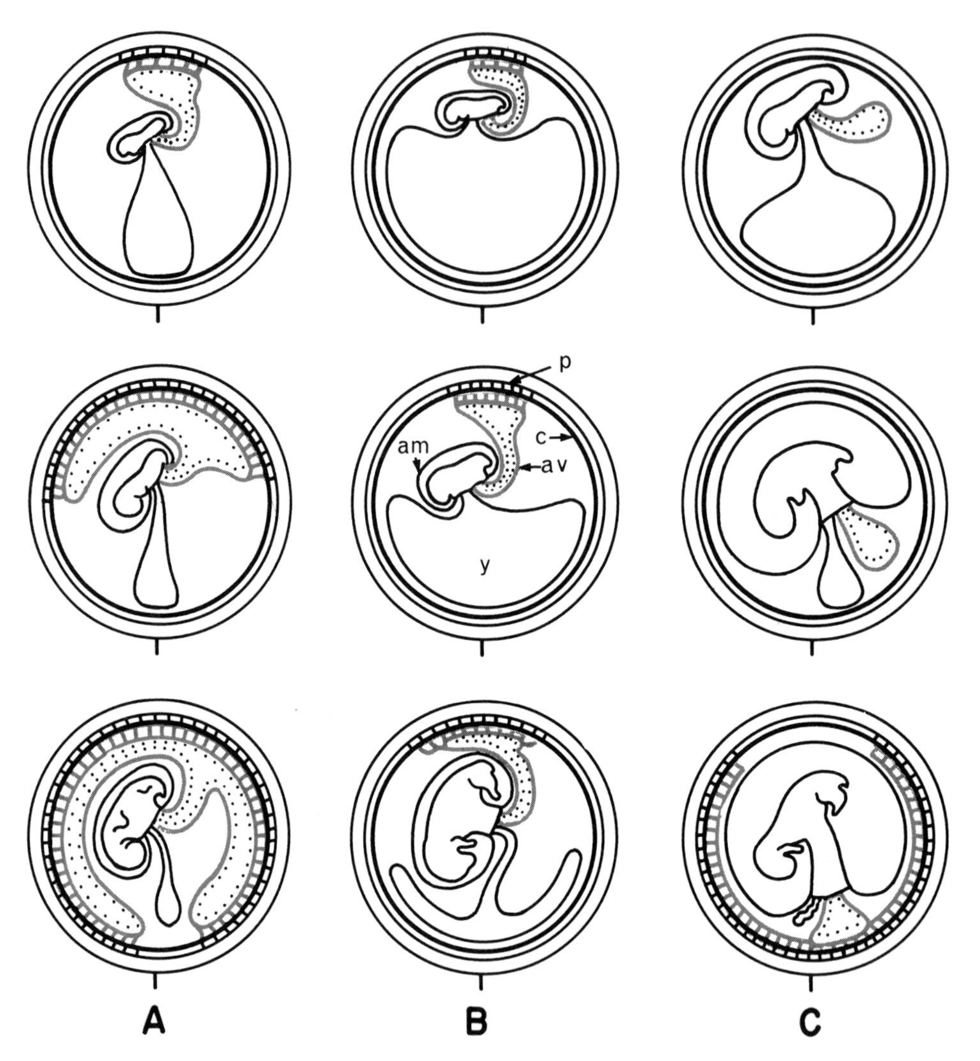

Fig. 1. Diagrams of the relation of the orientation of the embryonic disk to that of the yolk sac, first contact of the allantois with the chorion, and the definitive attachment of the allantoic vessels to the chorioallantoic placenta. *A,* as in Carnivora, where expansion of the allantoic vesicle moves the originally antimesometrial attachment of the allantoic vessels to the definitive mesometrial position. *B,* as in many insectivores, where the originally antimesometrial attachment of the allantois remains the definitive one, probably because the allantoic vesicle is small and the amnion is slow to expand. *C,* as in Artiodactyla, where rapid expansion of the amnion forces the allantois into line with the yolk stalk, so that its first contact with the chorion is mesometrial. *am,* amnion; *av,* allantoic vessels; *c,* chorion; *p,* chorioallantoic placenta; *y,* yolk sac cavity.

Although there is a large tubular expansion of the allantoic vesicle from this point toward each end of the chorion, the cord and allantoic vessel attachment remains fixed (fig. 1). Even though in this group allantoic vessels do supply almost the entire chorion and amnion, the allantoic vesicle does not fill the chorion as it does in carnivores.

Knowing these facts about later fetal membrane morphogenesis enables one to postulate with reasonable certainty the original orientation of the blastocyst. This

TABLE 1

SUMMARY OF ORIENTATION OF BLASTOCYST AND FETAL MEMBRANES IN CERTAIN MAMMALIAN TAXONOMIC GROUPS

Group	Genera[a]	Disk	First Troph. Attach.	First Allant. Attach.	Defin. Cord Attach.	Defin. Location Placenta	Vernacular Names
Tenrecoidea	4 (7)	anti	anti	anti	anti	anti	tenrecs
Erinaceoidea	1 (8)						hedgehogs
Soricoidea	8 (30)						shrews, moles
Anthropoidea	14 (21)	[b]					monkeys, apes, man
Bradypodidae	1 (1)	[b] (*i*)[c]					sloths
Dasypodidae	1 (8)	[b]					armadillos
Pholidota	1			?	meso	diffuse	scaly anteaters
Megachiroptera	2 (19)	meso	meso	meso	meso	meso	fruit bats
Macroscelidoidea	1 (4)						elephant shrews
Tarsiidae	1	anti	meso	meso	meso	meso	tarsiers
Chrysochloroidea	1 (3)	lat	lat	lat	lat	lat	gold moles
Tupaidae	5 (1)	anti	bilat	anti	bilat	bilat	tree shrews
Lagomorpha	4 (6)	meso	anti	meso	meso	meso	rabbits
Rodentia	37 (300)						rodents
Carnivora	11 (102)	anti (*i*)	central	anti	meso[d]	anti to zonary	carnivores
Hyacoidea	1 (2)	(*i*)				zonary	hyraxes
Artiodactyla	22 (63)	anti (*i*)	central	meso	meso	diffuse to coty	clovenhoofed animals

[a] Under genera, the number of genera on which the data are based is first. The number of genera in the same group about which nothing is known is in parentheses.

[b] These mammals have simplex uteri. The midsagittal plane of these is considered antimesometrial.

[c] (*i*) Based on indirect evidence, that is, the nature and orientation of the fetal membranes.

[d] The umbilical cord is very short in Carnivora, and the allantoic vessels radiate out from it in two leashes, one to each edge of the more or less zonary placenta (see fig. 1, *A*).

is especially useful in the case of large mammals, where it is impractical to section the uteri serially in order to find the disk of an implanting blastocyst; and also in rare forms where only later membrane stages may be available. In table 1 *i* indicates that the information is based on such indirect evidence.

Since blastocyst biology involves the phenomenon of "implantation," it is pertinent to consider here just what this term implies. Nidation (nesting), although more specific, is an adequate synonym for the process as it is usually thought of in man and rodents; but this connotation seems less appropriate when applied to the process in such uteri as those of the mare and cow, where the blastocyst must reach its definitive position and probably even its specific orientation several days before there is any anatomically demonstrable attachment to the endometrium. In these cases the blastocyst does not "nest"; it simply "perches"!

Once the blastocyst is on the nest site or "perch," the further "implantation" processes vary greatly among taxonomic groups. Whether the original position in the uterus is central or eccentric, attachment may involve no denudation of maternal tissue at all, denudation only opposite a certain area of the blastocyst, or both denudation and contact with endometrial stroma over the entire blastocyst surface. In other words, attachment may range from central to eccentric and from superficial to completely interstitial. Then too we speak of "delayed implantation" when the delay, in some species at least, is actually a delay in intrauterine transport and spacing.

In view of these anatomical considerations, "implantation," although useful as a general term, is nevertheless an inexact one. To avoid confusion it is time to recognize this and to avoid using it when the more exact terms—intrauterine transport, gestation site selection, blastocyst orientation, and blastocyst attachment—are needed.

I. Blastocyst Orientation

The constancy of a particular orientation of the blastocyst within a major taxonomic group, and of the correlated phenomena of first trophoblastic attachment, first contact of the allantois with the chorion, location of the placenta, and place of attachment to it of the cord, raises two basic questions: (1) Are there mechanisms within the blastocyst-uterine complex which assure this constancy? (2) Are there biological advantages in these specific orientations?

The first of these questions will certainly be considered in detail by other members of this conference; yet it may be well to introduce now some of the factors that may be associated with blastocyst spacing, site selection, orientation, and attachment. All uteri are muscular; some endometria are at least partially ciliated (rabbits and *Citellus*), and all undergo more or less hyperemia and edema during the implantation period. Since cilia are seldom present, muscular activity (Böving 1954, 1956), possibly abetted by the physical consistency of the endometrium, is probably the chief mechanism for the movement and the stopping of movement of blastocysts. The shape of the uterine lumen may also be important (fig. 2). In duplex, bicornuate, and bipartite uteri the cross-sectional shape of the lumen may be either more or less symmetrically radiate, laterally compressed, or T-shaped. Endometrial folds in these are primarily longitudinal, but often there are circular ones as well. Prominent semilunar folds, common in hystricomorph rodents, produce zigzag or spiral lumina. In most ruminants a combination of the longitudinal and circular fold pattern during development results in from 6 (Cervidae) to over 100 (Bovidae) prominent caruncles. Lumina of simplex uteri may be radiate in cross section, but more typically are dorsoventrally flattened. Their mucosae tend to be relatively smooth and free of large rugae.

The only easily observed special attachment areas of the endometrium are the protuberant areas called caruncles (implantation pads if only one or two are present). There may also be slight dilatations or "implantation chambers" as in many rodents, but these probably develop only at the time a blastocyst is present at the spot, or as the result of an artificial stimulus (Blandau, 1949 *a, b;* Bloch 1966; Fischer 1966; Orsini 1964, 1965). Implantation pads and caruncles are aglandular, but so are other regions in many uteri (fig. 2). In some species the mesometrial or antimesometrial region may be smooth and the rest of the mucosa rugose. Areas of relative vascular congestion are sometimes demonstrable microscopically. Again these seem to develop only in the immediate vicinity of a blastocyst even though it

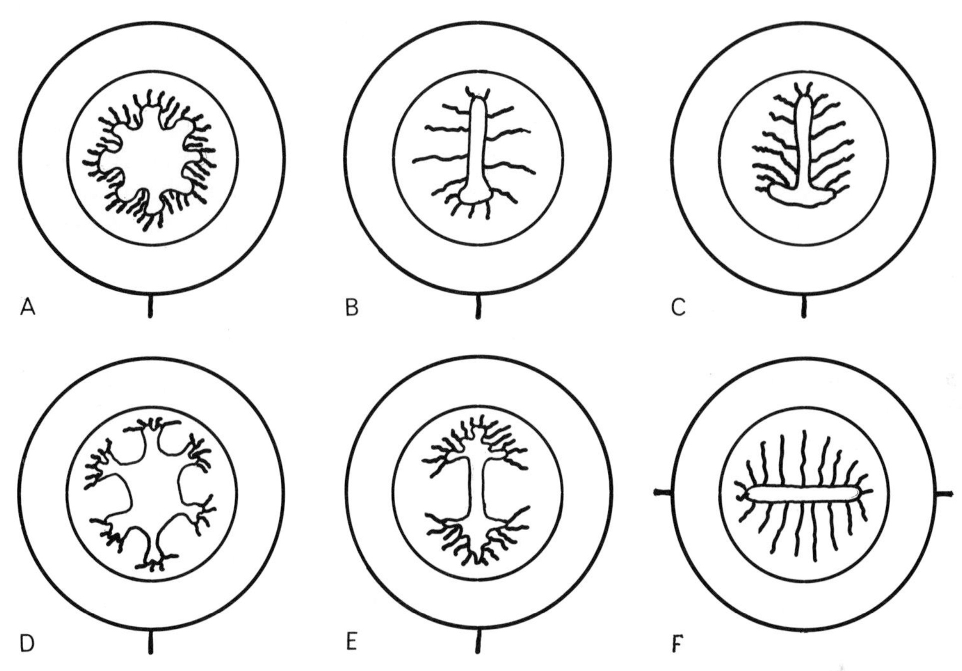

Fig. 2. Diagrams of common shapes of uterine lumina and the glandular distribution in the endometrium of eutherian mammals as seen in cross section. The mesometrium is below in each diagram. *A,* radiate (carnivores, pigs); *B,* laterally compressed (many rodents); *C,* T-shaped with aglandular mesometrial portion (shrews); *D,* radiate with many aglandular caruncles (most ruminants); *E,* with one pair of lateral aglandular pads or caruncles (tree shrews); *F,* dorsoventrally flattened (simplex uteri of monkeys and man).

is still unattached (Mossman and Weisfeldt 1939) or in response to artificial local stimuli such as those capable of producing deciduomata. Since the major vessels and nerves reach the uterus through the mesometrium, blastocysts are unavoidably oriented in one way or another to this region; however it is very unlikely that this region of gross entrance of vessels and nerves as such has any significant bearing on the intimate vascularization or innervation of any part of the endometrium. Large maternal placental vessels develop where the placenta develops, whether that is mesometrial, lateral, or antimesometrial.

There is of course good probability that morphologically unrecognizable but physiologically specialized regions for attachment do exist, at least in the endometria of some species (Horst 1950). Local physiological changes in vascular mechanisms, innervation, and secretory function could easily have escaped detection. Some such mechanism may account for the constancy of only right cornu pregnancies in certain African antelopes which ovulate as often from the left ovary as from the right (Buechner 1961; Kellas 1954; Mossman and Mossman 1962; Spinage 1969). Implantation phenomena also strongly suggest the existence af a uterine polarity, but there is little indication of its nature (Alden 1945). Wilson (1960) even showed that tumor fragments attach only at the antimesometrial side in pseudopregnant mouse uteri, the normal attachment site of mouse blastocysts.

Orientation factors in the blastocyst itself are also obscure. Certainly there is almost no possibility of locomotor capability. Cytoplasmic processes have been shown by Blandau (1949*a*) to penetrate the zona pellucida at attachment time, but there is no evidence that they are capable of moving the blastocyst. That the blastocyst might respond to favorable attachment areas by means of mechanisms involving differing degrees of stickiness has long been recognized (Böving 1954). Blastocyst secretions may play a part in this, and may even be correlated with the polarity of the blastocyst. Certainly different quantities, if not different qualities, of metabolites must be elaborated at the inner cell mass area from those produced by the thin trophoblast of the unilaminar portions. Kirby, Potts, and Wilson (1967) point to evidence that the inner cell mass of the mouse blastocyst may migrate actively inside its trophoblastic covering in order to orient properly to the region of first attachment. Certainly there is every reason to believe that early embryonic tissue cells are capable of migration and that they often do migrate in exact relation to a polarity of their environment. However Blandau (1949*a*) demonstrated that attachment cone cells differentiate at the abembryonic pole of the blastocyst of the guinea pig before attachment occurs. If this happens before orientation has taken place, there is no reason to assume migration of the inner cell mass. The concept of inner cell mass migration is an important one, for it implies a primary communication between the inner cell mass and the endometrium and places the trophoblast in a secondary role at this time. Therefore an effort should be made soon either to confirm or to invalidate this theory.

It is clear that the orientation of the inner cell mass or embryonic disk is a primary determinant of the position of the various fetal membrane elements, subject to modification by differences in relative growth rate and final size of the allantoic vesicle and amnion. Yet there are practically no recognized clues as to the functional advantages of the specific position of any of these organs. For instance, why should conditions be such as to bring about mesometrial chorioallantoic placentation in mice and antimesometrial placentation in shrews? This is fully as puzzling as its developmental background, the opposite orientation of the embryonic disks of these two families. Perhaps on the basis of the great variety of arrangements which have evolved, these differing orientations, although now established genetically, could be logically regarded as purely fortuitous. Yet to accept this hypothesis would certainly assure continued ignorance of the subject and delay discovery of possibly important principles of uterine-embryo interrelationships.

II. Evaluation of Present Information

Evaluation of our present knowledge of blastocyst orientation is now in order. Table 1 shows that almost nothing is known for certain about this subject in at least

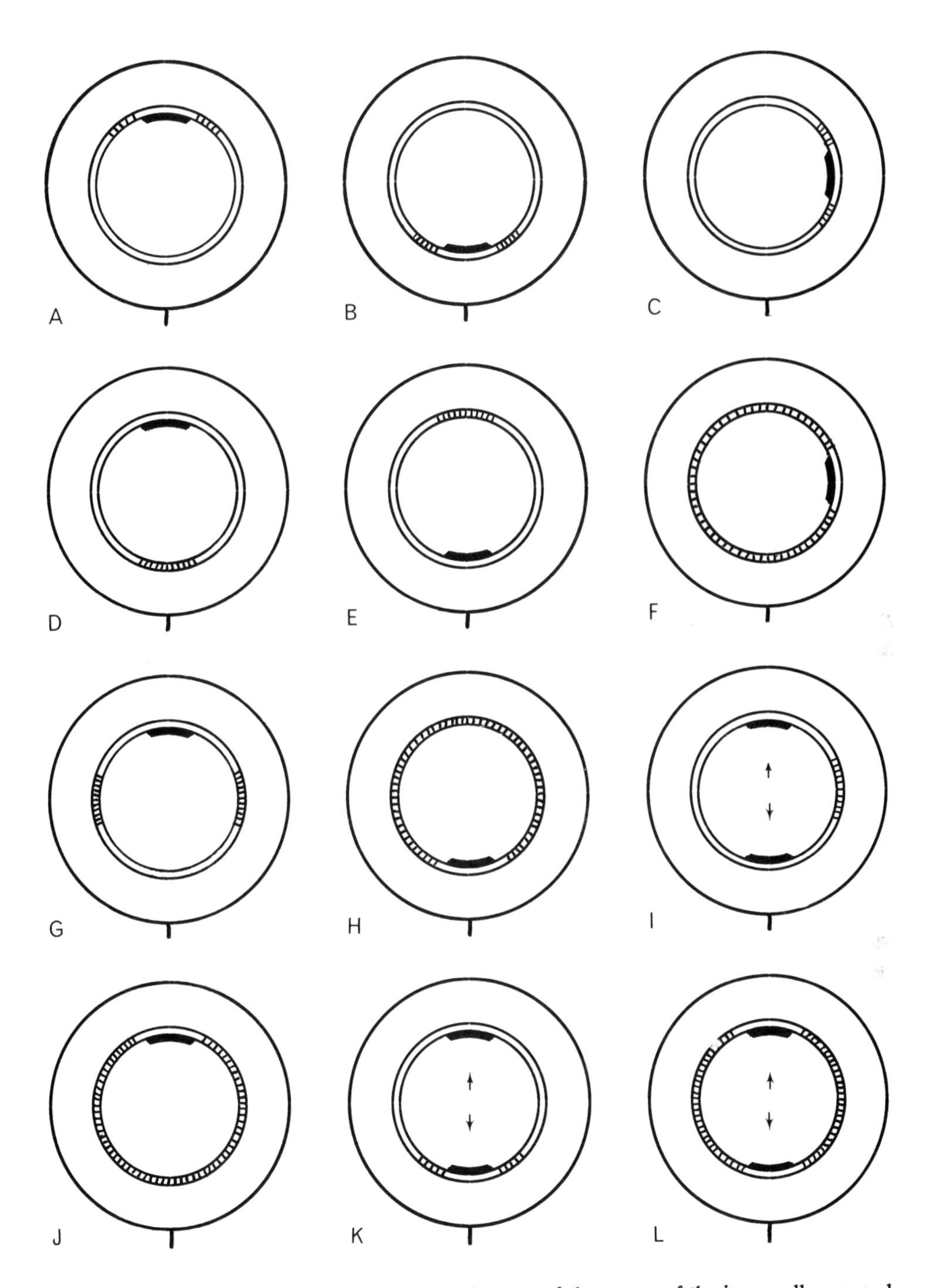

Fig. 3. Diagrams of the various orientations to the axes of the uterus of the inner cell mass and first trophoblastic attachment as described for eutherian mammals. The mesometrium is below in each diagram. *A*, shrews, monkeys, man; *B*, fruit bats; *C*, one bat (*Noctilio*); *D*, tarsiers; *E*, rabbits, rodents; *F*, some bats; *G*, tree shrews; *H*, some bats; *I*, gold moles; *J*, some bats; *K*, one galago; *L*, hyraxes. Several of these (*C, F, H, J, K,* and *L*) need confirmation (see table 1).

eight major taxonomic groups and that our knowledge of the other twelve is too often based on only a small number of the known genera in the group. It shows also that the consistency so typical of most of the major groups is absent in the Insectivora, where there are at least four different patterns. Taxonomists would not be surprised at this, because of the antiquity of this order and the great evolutionary radiation which has obviously occurred within it. The literature on the Microchiroptera ("insectivorous" bats) indicates wide variations similar to those in Insectivora. It is particularly confusing since it often indicates lack of consistency even within smaller subgroups, that is, superfamilies and families. Because of the constancy of pattern in other mammalian families and the obvious inadequacy of much of the literature on bat fetal membranes, it has seemed best not to conclude that Microchiroptera violate the general rules; instead, we should simply admit that there are insufficient reliable data on which to characterize them. They have therefore been omitted from table 1.

Figure 3 shows the great variety of combinations of disk and first attachment orientations that have been described. Some of the attachments shown are possibly wrong. For example, central attachment as described in carnivores and artiodactyls may very well be initiated antimesometrially or mesometrially. Indeed there may be different interpretations by different authors: I consider the first attachment in the rabbits and sciurids antimesometrial; others may consider it central.

III. Summary

The blastocysts of placental mammals orient in a variety of patterns in relation to the midsagittal plane of their uteri; yet the pattern is usually the same within any one major taxonomic group, that is, order or suborder. Insectivora is the only order in which several different patterns are known to occur. There are seven orders and two suborders of which too little is known to warrant an attempt at characterization. Information on most of the other groups is based on 10% or more of the genera included in the group and is consistent; therefore it is probable that it is essentially correct for the whole group.

The pattern of blastocyst orientation is correlated with the later development of the fetal membranes to such an extent that basic information on the latter for any taxon can be used to predict the former with reasonable accuracy.

Blastocyst orientation poses two basic questions for which we have no adequate answers. How is orientation brought about, and what is its biological significance to the species?

References

Alden, R. H. 1945. Implantation of the rat egg: I. Experimental alteration of uterine polarity. *J Exp Zool* 100:229.

Blandau, R. J. 1949*a*. Observations on implantation of the guinea pig ovum. *Anat Rec* 103:19.

———. 1949*b*. Embryo-endometrial interrelationship in the rat and guinea pig. *Anat Rec* 104:331.

Bloch, S. 1966. Beobachtungen zur Wechselwirkung zwischen Keim und Uterus bei Ovo-Implantation. *Acta Anat* (*Basel*) 65:594.

Böving, B. G. 1954. Blastocyst-uterine relationships. *Sympos Quant Biol* 19:9.

———. 1956. Rabbit blastocyst distribution. *Amer J Anat* 98:403.

Buechner, H. K. 1961. Unilateral implantation in the Uganda kob. *Nature* (*London*) 190:738.

Fischer, A. 1966. Die Abhängigkeit des Implantationsortes von Feinbau and Gefässanordnung im Uterus bei *Mesocricetus auratus* Waterhouse. *Z Anat Entwicklungsgesch* 125:189.

Horst, C. J. van der. 1950. The placentation of *Elephantulus*. *Trans Roy Soc S Afr* 32:435.

Kellas, L. 1954. Observations on the reproductive activities, measurements, and growth rate of the Dikdik (*Rhynchotragus kirkii thomasi* (Neumann). *Proc Zool Soc London* 124:751.

Kirby, D. R. S.; Potts, D. M.; and Wilson, I. B. 1967. On the orientation of the implanting blastocyst. *J Embryol Exp Morph* 17:527.

Mossman, A. S., and Mossman, H. W. 1962. Ovulation, implantation, and sex ratio in impala. *Science* 137:869.

Mossman, H. W., and Weisfeldt, L. A. 1939. The fetal membranes of a primitive rodent, the thirteen-striped ground squirrel. *Amer J Anat* 64:59.

Orsini, M. W. 1964. Implantation: A comparison of conditions in the pregnant and pseudopregnant hamster. *V Int Cong Anim Reprod A. I.* (*Trento*), p. 309.

———. 1965. In *Preimplantation stages of pregnancy,* ed. G. E. W. Wolstenholme and M. O'Connor, p. 162. London: J. and A. Churchill.

Spinage, C. A. 1969. Reproduction in the Uganda Defassa waterbuck, *Kobus defassa ugandae* Neumann. *J Reprod Fertil* 18:445.

Wilson, I. B. 1960. Implantation of tissue transplants in the uteri of pseudopregnant mice. *Nature* (*London*) 185:553.

4

Culture of Guinea Pig Blastocyst

Richard J. Blandau
Department of Biological Structure
School of Medicine
University of Washington, Seattle

Culture of guinea pig blastocysts is a promising technique for investigating many problems related to metabolism of the embryo, its escape from the zona pellucida, its orientation and attachment to endometrial tissue, and the origin and behavior of the syncytial trophoblast. Development of the blastocyst is unique in this species in that the zona pellucida is not shed until the embryo has made firm contact with the surface epithelium of the antimesometrium (Blandau 1949*a, b*). Since there is no evidence of delayed implantation in the guinea pig (Deanesly 1960), all phases of early development leading to complete interstitial implantation must proceed without interruption; otherwise the embryo perishes. Uterine attachment and interstitial implantation occur in Cavia during the latter part of the 6th day and early 7th day after fertilization. Several hours before the embryo attaches itself to the endometrium, the trophoblast cells of the abembryonal pole proliferate to form a multilayered attachment cone. The cells of the cone lining the zona pellucida send protoplasmic projections through the zona in the form of irregular, organelle-free pseudopodia. The details of this development, reported first in 1883 by Graf von Spee, were reinvestigated and amplified in 1957 by Blandau and Rumery, who noted that the zona pellucida in the region of the developing attachment cone thins out and ruptures, and the remaining zona then slips off the blastocyst, retaining its more or less round appearance. Attachment cones have been described in sectioned material in the blastocyst of the Sciuridae (Mossman 1937) and the macaque (Wislocki and Streeter 1938), but penetration of the zona pellucida by specialized cell processes of the trophoblast has been recorded *in vitro* only in the guinea pig.

Close examination of the appearance of the trophoblast in the youngest implanting human embryo (Hertig and Rock 1945) leads one to suspect that the human process of trophoblastic proliferation is similar to that found in the macaque and guinea pig. Hence the emphasis in this chapter is placed upon cultures of pre-implanting blastocysts and on implanted embryos of guinea pigs removed intact from the implantation chamber on the 8th day after fertilization.

I. Mating of the Animals and Recovery of the Embryos

During the late afternoon of each day all sexually mature females in the colony were examined to determine whether the vaginal membranes had ruptured, since this is indicative of the proestrous period. (Heat will occur in such animals within 24 to 48 hr.) Females with ruptured membranes were then placed with vigorous males and allowed to remain overnight. At 8:00 on the following morning the vaginal canal of each was examined for a vaginal plug or spermatozoa. The finding of either indicated mating and was recorded as day 0. The mated females were isolated from the males. If neither plugs nor spermatozoa were found the females were left with the males and examined daily until the vaginal membranes again sealed the vagina. To recover the blastocysts or implanted embryos the cornua of mated animals were removed under sterile conditions during either the late afternoon or evening of the 6th day or early morning of the 8th day of pregnancy. The females were killed by decapitation to assure exsanguination. Hank's solution was flushed gently through the cornua of animals killed on the 6th day. The washings were examined for blastocysts; when located they were transferred immediately to a large volume of the culture medium. Eight-day-old embryos were recovered by identifying the location of the decidual response and slitting the myometrium longitudinally along the antimesometrial border. A sharply pointed iridectomy scissors was slipped between the muscle and the decidua. (Cutting the muscle causes it to contract, exposing the whitish appearing, swollen decidua.) A sharply pointed watchmaker's forceps was used to gently separate the decidua in the midsagittal plane. (If this is done carefully the implantation chamber is opened and the cigar-shaped embryo is revealed.) Dissection of the walls of the chamber was continued until the attachment cone (Träger) of the embryo was identified. The Träger was grasped at its lower pole by watchmaker's forceps and lifted gently from its site of attachment. It, too, was transferred immediately to a large volume of culture medium.

II. Preparation of the Culture Medium

For the uninitiated, details of the method of tissue and organ culture as carried out in this laboratory may be reviewed in the book *Methods of Mammalian Embryology* edited by J. Daniels and published by W. H. Freeman Co., 1970.

The culture medium for the blastocysts and 8-day embryos consisted of 1/3 Locke-Lewis solution (NaCl, 8.5 gm; 0.42 gm; $NaHCO_3$, 0.2 gm; glucose, 0.2 gm; $CaCl_2 \cdot 2H_2O$, 0.25 gm dissolved in order in 1,000 ml triple glass-distilled water), 1/3 guinea pig serum, and 1/3 chick embryo extract. After this solution was well mixed 1.25 ml of 1% phenol red solution was added. The medium was sterilized by filtering through a 0.229 pore filter under light negative pressure. Its pH was adjusted to 7.2 or 7.5 by the addition of a few drops of either a 10% bicarbonate solution or 0.3N HCl.

The guinea pig serum was prepared by withdrawing blood by cardiac puncture from ether-anesthetized females after the animals had fasted for at least 12 hr. The tubed blood was then placed in the refrigerator to clot and the serum pipetted off, centrifuged, and filtered through several sets of millipore filters, the last of #0.22 pore size.

Chick embryo extract was prepared by mincing 10-day-old embryos in Locke-Lewis medium diluted to 25% and centrifuging. The supernate was stored frozen. Before use it was thawed and centrifuged twice at 3,000 RPM for 15 to 20 min. The supernate was filtered to remove bits of debris and other contaminants.

Attempts to use defined media, including 199 with addition of various concentrations of pyruvate, lactate, and albumen, did not have beneficial effects. The environmental requirement of the guinea pig blastocysts in artificial media needs to be explored more thoroughly.

III. Mounting of the Embryos

We have found Rose chambers ideal for the continuous observations of the guinea pig blastocysts and young implanted embryos in organ culture. The 6-day blastocysts or 8-day embryos were enclosed in the chamber filled with the culture medium. Since the attachment cone (Träger) is heavier than the inner cell mass, the embryos sank slowly through the medium when the chamber was inverted and came to rest on the coverglass. The embryos were maintained at 38° C and observed with the long working distance, dark contrast-medium phase objectives. The culture medium was replaced approximately every 3 days. The blastocysts and embryos growing in the Rose chambers were observed daily. The phase microscope was enclosed in a Plexiglas chamber in which the temperature was maintained at 38° C. We found a zirconium art lamp cooled by a water jacket the most satisfactory light source for the phase microscopy. A Zeiss green interference filter was used with the lamp and Polaroid filters were added to control the amount of illumination. Photographic records were made with Kodak plus-X film for still photography and Gevaert-type negative film #1-65 for cinematography.

IV. Observations of the Cultures

The appearance of a living 6½-day-old guinea pig blastocyst is portrayed in figure 1. The outer edge of the zona pellucida is somewhat irregular and, in comparison to mouse and rat blastocysts at this same stage, not particularly sticky. The trophoblast, composed of a single layer of cells, completely fills the perivitelline space. The plasma membranes of the trophoblast cells facing the blastocoelic cavity undulate almost continuously. After a variable length of time in culture the epithelial-like trophoblast cells of the abembryonal pole begin to proliferate actively. This process continues until the blastocyst has a bipolar appearance (fig. 2). One may then observe fine, clear protoplasmic projections tunneling through the zona pellucida. The projections apparently depolymerize the zonal materials, forming canaliculi which vary in shape and size (fig. 3).

Channelization continues until the zona of the abembryonal pole is riddled, appearing as full of holes as a sieve. The zona gradually thins out and, except for a few remaining fragments, disappears. When viewed with time-lapse cinematography the cytoplasmic projections display rhythmic pulsatile activity, each projection flowing out as an irregularly shaped bleb which may be withdrawn almost as rapidly as it is formed. Occasionally, clear fluid-filled vesicles break away from the cytoplasmic projections and disappear in the culture medium. The cytoplasmic projections originate only from the trophoblast cells lining the zona pellucida. When the embryo escapes from the zona *in vitro* these living cells continue to send out fingerlike projections for several days; the remaining cells of the attachment cone are transformed into the cyto- and syncytial trophoblast. The empty zona pellucida remains intact for as long as the culture is alive (fig. 4). After the blastocyst escapes from the zona pellucida it may retain its rounded vesicular shape for several days. During this time it expands and contracts rhythmically. This behavior appears to be related to the fluid secreted by the trophoblast cells into the blastocoelic cavity. At intervals some of the fluid escapes, causing the partial collapse of the blastocyst. The cycle is repeated many times.

There is no evidence in our cultures that the inner cell mass undergoes any kind of differentiation *in vitro*. The cells live for several days, when they undergo cytolysis and disappear. Within 24 hr the trophoblast cells have formed a mono-

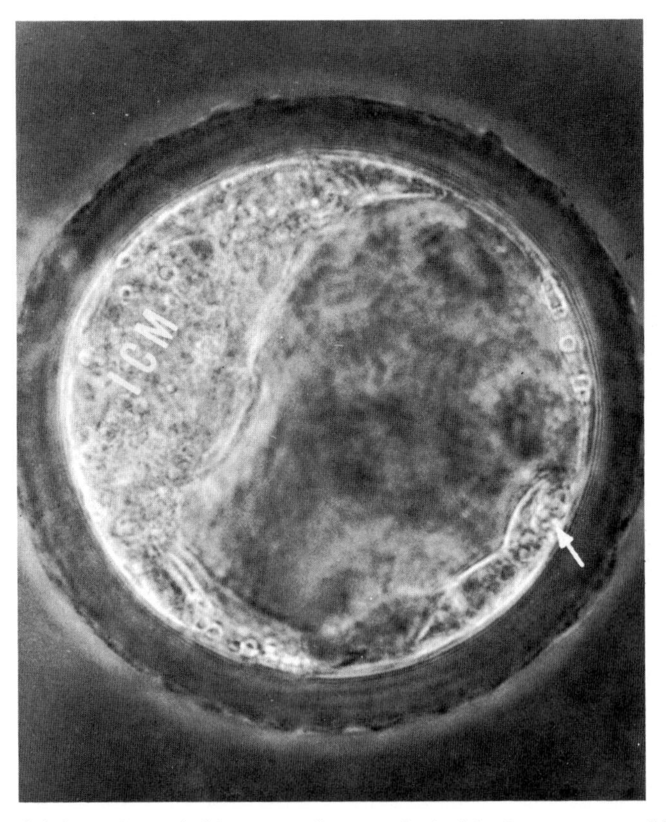

Fig. 1. A living 6½-day guinea pig blastocyst photographed with phase contrast objectives. Note single layer of trophoblast cells lining the zona pellucida and the irregular surface of the zona itself. ×550.

layer. The cells comprising the cytotrophoblast and syncytial trophoblast can be differentiated from one another readily. The remarkable feature of the cytotrophoblast is the large number of cells in various stages of mitotic divisions. Within 48 hr the syncytial trophoblast has spread out extensively on the coverglass. Its cytoplasm contains many nuclei varying in size and shape. The protoplasm of the syncytium is in constant motion flowing in different directions (figs. 5, 6). The leading edge

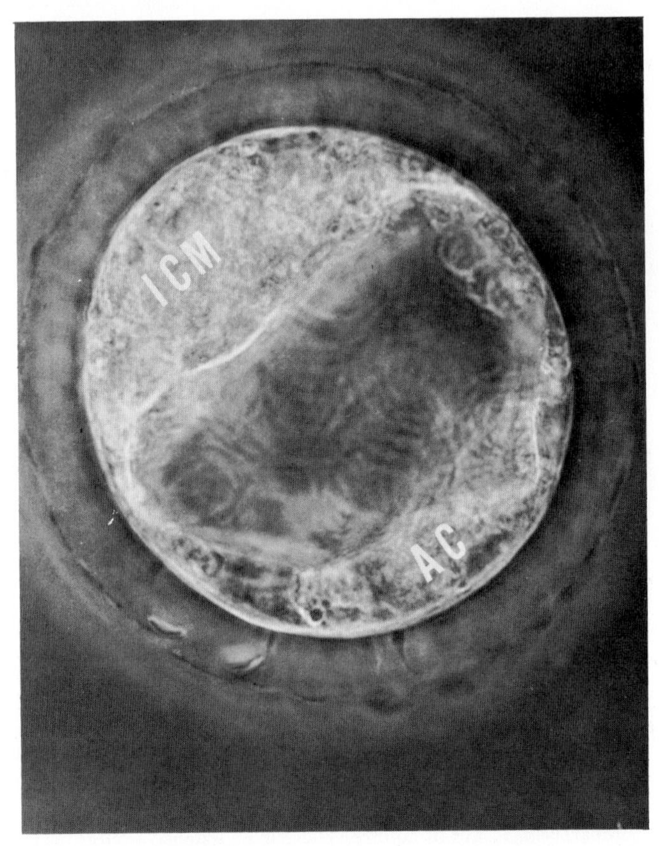

Fig. 2. A living 6½-day blastocyst in organ culture. The trophoblast cells of the abembryonal pole have proliferated to form a multilayered attachment cone. Several protoplasmic projections are tunneling through the zona pellucida (*arrows*). At least two sperm heads are embedded in the zona. ×500.

of the syncytium is often fluted and shows active pinocytosis. Nuclear divisions have never been seen in the syncytial trophoblast in any of our preparations.

Several studies in the rhesus monkey (Midgley et al., 1963) and in man (Tao and Hertig 1965) show clearly that the cytotrophoblastic nuclei labeled with tritiated thymidine migrate into the syncytial trophoblast. Initially such labeling could be found only in the nuclei of the cytotrophoblast; after 21 hr it appeared in the nuclei of the syncytial trophoblast as well.

After several weeks in culture the cytotrophoblast and particularly the syncytial trophoblast become filled with highly refractile granules (fig. 7). In many cultures bits of syncytial tissues move away from the main mass but retain connec-

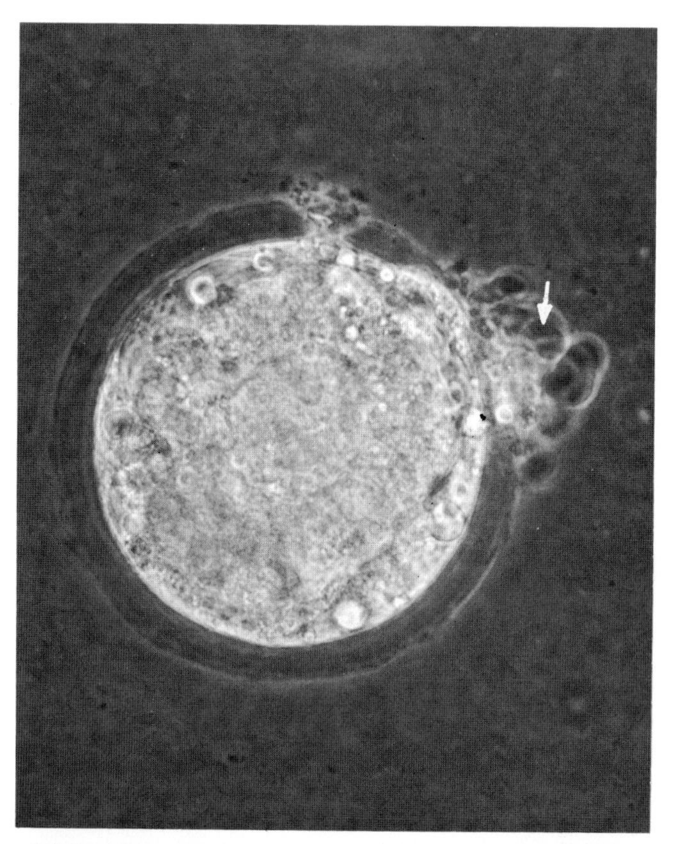

Fig. 3. A living 6½-day blastocyst showing the thinning of the zona pellucida as the attachment cone develops at the abembryonal pole.

Fig. 4. The extent of outgrowth of a guinea pig blastocyst which has been in culture for 48 hr. The embryo has escaped from the zona. Cyto- and syncytial trophoblast can be distinguished. Some of the cells continue to form cytoplasmic projections (*arrows*). Their behavior is similar to those which penetrate the zona pellucida originally. Note the three sperm heads embedded in the zona.

Fig. 5. The same preparation 5 days in culture. There is an extensive outgrowth of the syncytial trophoblast. The empty zona remains.

Fig. 6. Another blastocyst in whole organ culture for 5 days to show the variations in outgrowth. In this preparation the cytotrophoblast is more extensive than the syncytial trophoblast. There are some cells still enclosed by the zona pellucida.

Fig. 7. This preparation, the same as shown in figs. 4 and 5, has been in culture for 20 days. Note the vesicular organization involving certain of the syncytial cells. There is rather extensive cytolysis in the more peripherally located syncytium.

Figs. 8 and 9. Variety of patterns of outgrowths of syncytial trophoblast from the Träger region of 8-day guinea pig embryos. The plasma membranes are fluted and show active pinocytosis. The syncytium contains a variable number of pleomorphic nuclei; note the variations in their size.

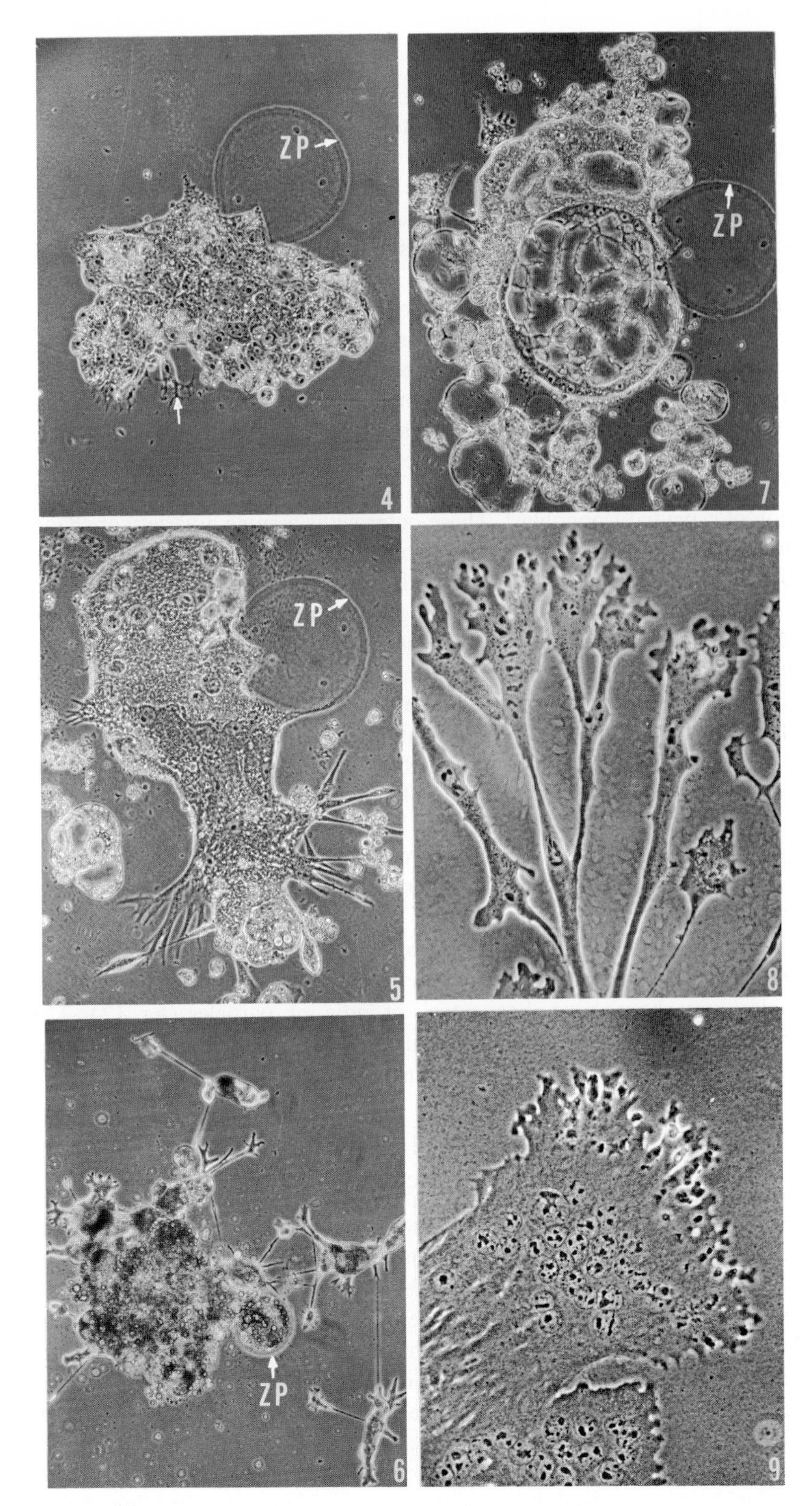
ZP
4
ZP
7
ZP
5
8
ZP
6
9

tions to one another by fine protoplasmic bridges. After 3 or more weeks in culture the trophoblast may become organized into vacuolated spheres which are lined by a single layer of cells (fig. 7). Soon the syncytium undergoes cytolysis, cytoplasmic vacuolization becomes extensive, and the vesicles fill with refractile granules. As mentioned earlier, in monolayer preparations (figs. 4, 5) the cytotrophoblast and syncytial trophoblast can be identified easily; it is much more difficult to observe the intermediate stages between the two. The loss of cell membranes from the cytotrophoblast cannot be clearly defined at the level of the light microscope but may be observed distinctly in electron micrographs (Pierce, Midgley, and Beals 1964). Of great interest is the remarkable change in behavior of the cytotrophoblast cells when transformed into the syncytium. Except for plasma membrane undulations the cells of the cytotrophoblast are relatively quiescent. Once they have been transformed into a syncytium, cytoplasmic flow becomes a constant feature.

The Träger of the 8-day guinea pig embryo in culture undergoes rapid proliferation of syncytial trophoblast. These outgrowths assume a variety of shapes and configurations (figs. 8, 9). Some are arranged in the form of lacework with protoplasm flowing backward and forward in all directions (fig. 10). The shapes of the nuclei may be distorted as they flow along the narrow channels. In some areas of the same culture the syncytial trophoblast contains hundreds of nuclei (fig. 11), but even with this impediment protoplasmic flow and movement are almost continuous. Since the 8-day embryo tends to settle within the Rose chambers, the Träger sticks to the glass while the rest of the embryo floats free. As the syncytial trophoblast begins to grow out, it attaches itself to the glass surface and spreads in all directions like the roots of a tree (fig. 13). A camera lucida representation of this outgrowth is shown in figure 14.

In cultures of the 8-day embryo there is a tendency for the yolk sac to swell. Often it attains a surprising size. The inner cell mass does not differentiate although the total number of cells does increase.

Occasionally when the 8-day embryo is cultured on gelatin membranes which have been prepared according to Ower's method (see chapter 12) the syncytial trophoblast may show active phagocytosis of india ink particles (fig. 12). The syncytium attached to the gelatin membrane depolymerizes the gelatin, freeing the india ink particles and incorporating them within vacuoles in the syncytial cytoplasm (fig. 12).

V. Orientation of the Blastocyst in Culture

As described by Mossman (chapter 3) the blastocyst of each species is oriented in a specific manner to the endometrium. Just how this embryo-endometrial relationship is established remains a question. Kirby, Potts, and Wilson (1967) (see Kirby, chapter 23) suggest that the surface of the mouse trophoblast has a uniform potential for attaching itself to the uterine epithelium and that after the loss of the zona pellucida attachment occurs at random. Kirby reasons that the inner cell mass moves independently inside the trophoblastic shell and assumes a position determined by some kind of gradient across the vertical axis of the cornua. In our observation of the guinea pig, formation of the attachment cone begins several hours before the zona pellucida is lost; indeed this activity is responsible for the shedding of the zona.

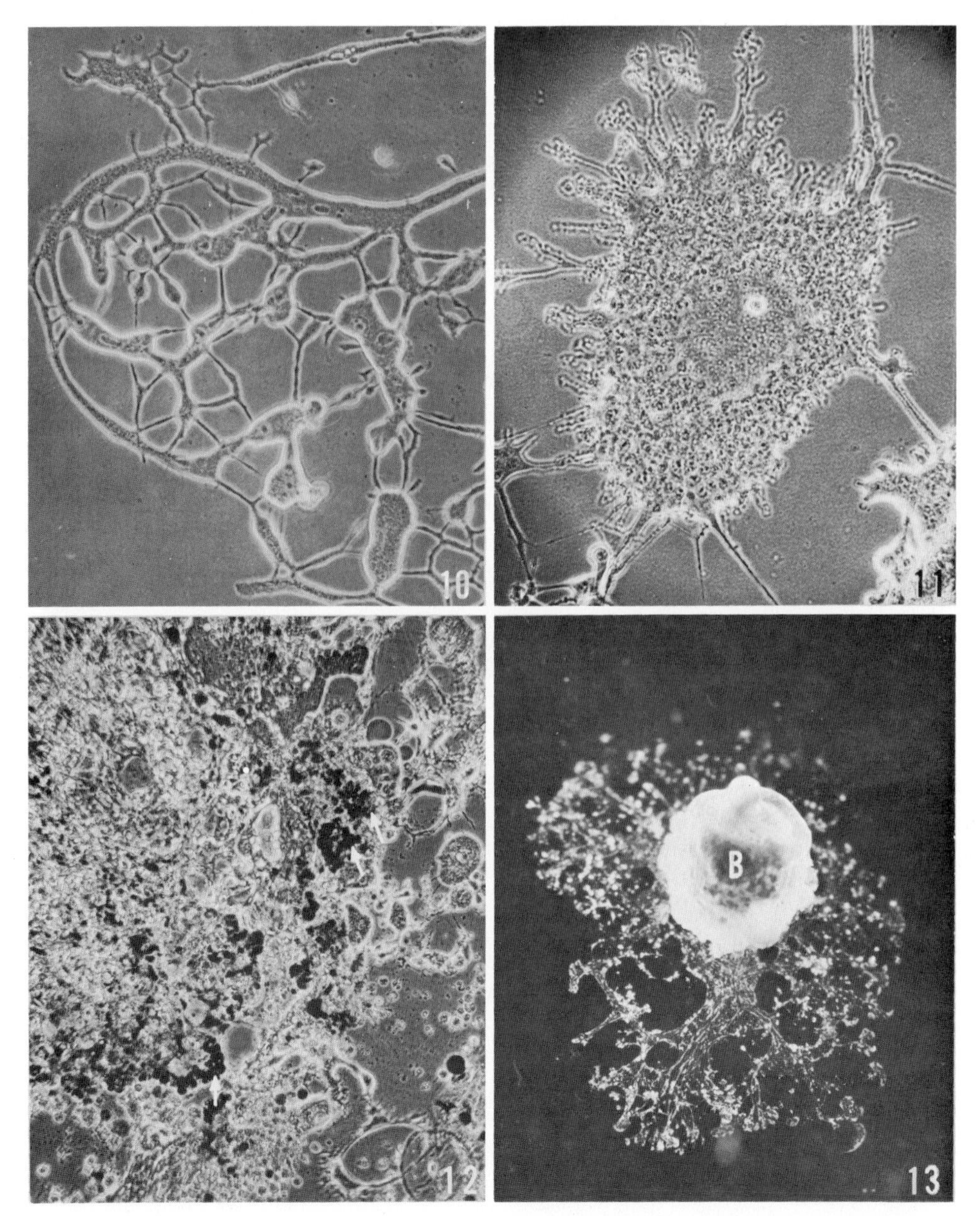

Figs. 10 and 11. The lacework pattern of outgrowth illustrated in fig. 10 is very common in younger preparations. Cytoplasm flows through all of the finer connections. Note the hundreds of nuclei in the syncytium of fig. 11. Movements of the cytoplasm in outgrowths of this kind are usually circumferential.

Fig. 12. The inclusion of india ink particles within the cytoplasm of the syncytium (*arrows*) shows the phagocytic potentiality of this tissue.

Fig. 13. Outgrowth of syncytial trophoblast from the area of the Träger of an 8-day embryo. The yolk sac has enlarged. The inner cell mass appears as a rounded nipple at the top.

A survey of many time-lapse sequences of intact blastocysts growing in organ culture reveals no movement of the inner cell mass whatsoever. The relationship between the inner cell mass and the attachment cone remains constant.

In one instance an abnormal attachment cone formed on the lateral side of the blastocyst wall rather than at the abembryonal pole. Although the attachment cone was somewhat smaller than normal, the sequence of its development was similar to that in normal blastocysts. There was no evidence of a shift in the position of the inner cell mass.

VI. Summary and Conclusions

The culture of unattached guinea pig blastocysts and recently implanted embryos is a valuable tool for the study of many problems in early development. These include:

1. The time, sequence, and nature of the development of the attachment cone
2. The role of the attachment cone in depolymerization of the zona pellucida, its orientation, and its attachment to the endometrium
3. The transformation of cytotrophoblast to syncytial trophoblast

Fig. 14. A drawing showing the extent of outgrowth of pure syncytium in an organ culture of an 8-day guinea pig embryo similar to that photographed in fig. 13.

4. The pinocytotic and phagocytic behavior and enzyme activity of the syncytial trophoblast
5. The role of hormones and drugs on the growth and differentiation of the syncytial trophoblast
6. Evaluation of the production of gonadotrophins and reproductive steroids
7. Metabolic requirements of the embryo during the preimplantation period.

Finally, the possibility of growing pure cultures of syncytial trophoblast may provide valuable tissue antigents for the study of various immunological phenomena related to implantation.

Acknowledgments

These observations could not have been accomplished without the technical skills of Mrs. Lynn Goldner and the photographic skills of Mr. Roy Hayashi.

This investigation was supported by research grants from the United States Department of Health, Education, and Welfare.

References

Blandau, R. J. 1949*a*. Observations on implantation of the guinea pig ovum. *Anat Rec* 103:19.

———. 1949*b*. Embryo-endometrial interrelationships in the rat and guinea pig. *Anat Rec* 104:331.

Blandau, R. J., and Rumery, R. E. 1957. The attachment cone of the guinea pig blastocyst as observed under time-lapse cinematography. *Fertil Steril* 8:570.

Deanesly, R. 1960. Implantation and early pregnancy in ovariectomized guinea pigs. *J Reprod Fertil* 1:242.

Hertig, A. T., and Rock, J. 1945. Two human ova of the pre-villous stage, having a developmental age of about seven and nine days respectively. *Contrib Embryol Carnegie Inst* 31:65.

Kirby, D. R. S.; Potts, D. M.; and Wilson, I. B. 1967. On the orientation of the implanting blastocyst. *J Embryol Exp Morph* 17:527.

Midgley, A. R.; Pierce, G. B.; Deneau, G. A.; and Gosling, J. R. S. 1963. Morphogenesis of syncytiotrophoblast *in vivo:* An autoradiographic demonstration. *Science* 141:349.

Mossman, H. W. 1937. Comparative morphogenesis of the fetal membranes and accessory uterine structures. *Contrib Embryol Carnegie Inst* 26:129.

Pierce, G. B.; Midgley, A. R.; and Beals, T. F. 1964. An ultrastructural study of differentiation and maturation of trophoblast of the monkey. *Lab Invest* 13:451.

Spee, Graf von. 1883. Beitrag zur Entwickelungsgeschichte der früheren Stadien der Mershweinchens bis zur Vollendung der Keimblase. *Archiv Anat Physiol* 7:44.

Tao, T. W., and Hertig, A. T. 1965. Viability and differentiation of human trophoblast in organ culture *Am J Anat* 116:315.

Wislocki, G. B., and Streeter, G. L. 1938. On the placentation of the macaque (*Macaca mulatta*) from the time of implantation until the formation of the definitive placenta. *Contrib Embryol Carnegie Inst* 27:1.

5

The Fine Structure of the Blastocyst

Allen C. Enders
Department of Anatomy
School of Medicine
Washington University
Saint Louis, Missouri

When Assheton wrote his description of the early development of the sheep in 1898, he stated that he had had the necessary "mehr Glück" that Bonnet (1884) had wished for in obtaining the necessary cleavage stages. At that time Assheton was personally familiar with or had written descriptions of the histology of the blastocysts of the mouse, sheep, pig, guinea pig, rabbit, bat, tree shrew, hedgehog, mole, and shrew. At the present time our knowledge of the reproductive cycles and of collection and preparative techniques has progressed to the point where obtaining comparable stages for electron microscopy requires more diligence than luck. Although eighteen species have been examined by electron microscopy, at this writing fewer than half this number have been examined in detail, and several of those first observed by light microscopists are notably absent from the list. On the other hand little improvement in detail of cytological observation of the blastocyst occurred from the time of van Beneden's (1880) study of the rabbit until the beginning of electron microscope studies.

Detailed description of all of the cytological information available from the recent material is inappropriate here and would include much information which we cannot interpret at this time for lack of knowledge. The material presented here will be restricted to some of the emerging generalizations and a few of the specific observations carrying pronounced functional implications. Quite possibly some of the generalizations may prove to be too broadly drawn and will need subsequent revision. It is hoped, however, that the summarizing of this information will encourage the continuation of these structural studies and make it easier to correlate the cytological information obtained with that from other procedures.

I. Materials and Methods

The material which forms the major basis of this report has been collected during the last decade in circumstances varying from field conditions to well-controlled laboratory conditions. Quite naturally the amount of information which can be obtained varies with the degree of tissue preservation. Unless otherwise noted all the blastocysts were fixed in 3% glutaraldehyde in 0.1 M phosphate buffer, postfixed in 2% osmium tetroxide following a buffer rinse, and embedded in one of the epoxy resins, usually Durcupan. Blastocysts collected in the field were usually kept in glutaraldehyde for several days until they could be postfixed and dehydrated in the laboratory. Blastocysts collected in the laboratory were fixed for 2 hr in glutaraldehyde.

TABLE 1

ANIMALS FROM WHICH BLASTOCYSTS HAVE BEEN COLLECTED

Family	Species	Results
Marsupialia	Red kangaroo (*Megaleia rufa*)	6 blastocysts during delayed implantation, 1 preimplantation (preservation poor)
Insectivora	Short-tailed shrew (*Blarina brevicauda*)	5 implanting blastocysts from 1 animal
Chiroptera	Little brown bat (*Myotis lucifugus*)	4 implanting blastocysts from 4 animals
	Funnel-eared bat (*Natalus stramineus*)	3 blastocysts from 3 animals
Edentata	Nine-banded armadillo (*Dasypus novemcinctus*)	18 blastocysts from both delayed implantation and early implantation
Carnivora	Ferret	Blastocysts from 10 animals, day 4 to day 12, including morulas to implanting blastocysts
	Mink	Blastocysts from 8 animals during delayed implantation
	Weasel (*Mustela erminea*)	Blastocysts from 3 animals late in the delayed implantation period
	Northern fur seal (*Callorhinus ursinus*)	8 blastocysts from 8 animals early in the delayed implantation period
Lagomorpha	Rabbit	18 animals, many blastocysts from morula through early implantation stages
	Swamp rabbit (*Sylvilagus aquaticus*)	2 blastocysts from 1 animal ($KMnO_4$-fixed only)
Rodentia	Laboratory rat	All stages from fertilization in large numbers
	Mouse	Blastocysts from 37 animals
	Gerbil	Blastocysts from 2 animals
	White-footed mouse (*Peromyscus maniculatus*)	Blastocysts from 2 wild-caught animals
	Guinea pig	Blastocysts from 8 animals

The animals from which blastocysts were obtained are given in table 1.

II. General Considerations

Many of the cytological features of the blastocyst are directly attributable to the peculiar nature of the ovum, whereas others reflect differentiation into trophoblast and embryonic tissues. Consequently it is necessary to examine the earlier stages of development as well as blastocyst formation.

The nuclear events associated with meiosis and fertilization are so dramatic that we tend to forget that ooplasm is also very specialized. The extent to which ova are highly specialized cells has been documented in recent years (Austin 1965; Enders and Schlafke 1965; Hadek 1965; Odor 1960; Szollosi 1965; Weakley 1966; Zamboni and Mastroianni 1966). The mitochondria are peculiar, usually round with septate cristae. The endoplasmic reticulum is largely agranular and often tubular or sacculate. The Golgi zones are distributed widely, with many vesicles and relatively short cisternae. Aggregations of storage materials are seen frequently that differ not only in amounts and in arrangement of macromolecules from species to species but also in class of compound.

In contrast to these features of the ovum embryonic cells are commonly characterized by an abundance of polyribosomes, elongate cisternae of granular endoplasmic reticulum, rod-shaped mitochondria with lamellar cristae, small juxtanuclear Golgi zones, and few inclusions other than glycogen. The extent to which the blastocyst reflects a differentiation away from the peculiar features of ooplasm to those of embryonic cells varies with the stage and the animal, but some general features are beginning to be discernible.

III. Nature of Cytoplasmic Inclusions

The inclusions found in blastocysts are of several types: those that are a conspicuous feature of the ovum and are present in reduced amounts in the morula and blastocyst stages, those largely restricted to the blastocyst stage, and those which though normal to the cells of adult organisms are exaggerated in some of the blastocysts. In addition bodies whose contents indicate that they may have been formed by degradative activities of the cell are found in most of the blastocysts.

A. *Inclusions Present in Ova*

Carnivores have a great deal of lipid in their ova, and exhibit a varying degree of diminution of this lipid during cleavage and blastocyst formation. Widespread distribution of lipid in the cat (Hill and Tribe 1924; Austin and Amoroso 1959), dog (van der Stricht 1923), bear (Wimsatt 1963), and ferret (Hamilton 1934) blastocysts has been described by other workers using light microscopy. In electron-microscope studies we have found lipid persisting into the blastocyst stage in the weasel, mink, ferret, and seal (figs. 16, 20). Lipid is the characteristic storage material in the ovarian and cleaving eggs in this order of mammals. The amount of lipid in other groups of mammals studied is relatively small. The mouse and rabbit have a few droplets of lipid in the ovum and relatively little in the blastocyst, whereas in the rat the reverse is true. In none of these species is lipid one of the more conspicuous features.

The predominant storage products of myomorph rodents are quite unlike any similar structures found in normal adult cells. As has been described by myself and others, these inclusions have a linear form in sections of mouse, rat, and hamster eggs, and are densely packed throughout most of the cytoplasm during cleavage stages (Enders and Schlafke 1965; Szollosi 1965; Weakley 1968). These structures have been referred to variously as plaques, fibrous elements, and cytoplasmic lamellae. In the hamster they take the form of a bilamellar disk (fig. 12), in the rat a unilamellar structure (fig. 6), and in the mouse groupings of individual linear elements (fig. 13). In all three they exhibit a subperiodicity.

The differential preservation of these structures by glutaraldehyde fixatives led us to the tentative conclusion that they were proteinaceous (Enders and Schlafke 1965). Weakley (1967), using 30% ethanol fixation followed by postfixation as well as glutaraldehyde, came to a similar conclusion.

Despite the unique appearance of these inclusions they have elicited little attention to date. No attempts have been made to use centrifugation to isolate them in one part of the ovum for identification or to fractionate the inclusions for chemical analysis. Calarco and Brown (1968) report that such inclusions are present, as would be anticipated, in the t^{12}/t^{12} morula as well as the normal mouse morula. In general we find that these elements are reduced in the blastocyst stage, but some of them persist until the time of implantation. When implantation is delayed, the elements may be completely eliminated before the time of implantation, although they are retained longer than in normal implantation. Young, Whicher, and Potts (1968) illustrate the presence of the lamellar elements in the hamster at the time of implantation. Clearly the time of disappearance of these special inclusions does not appear to be causally related to implantation.

Brinster (1967) has pointed out that from the fertilized ovum to the blastocyst stage in the mouse there is a 25% loss of protein. Although it is obviously risky to make quantitative estimates from distribution of materials in electron micrographs, it is clear that there is a greater than 25% reduction in concentration of fibrous elements. It is reasonable to suppose therefore that some of the constituents of these inclusions are reutilized by the blastocysts.

In the fertilized rabbit ovum there are mottled deposits within the largely agranular membranes of the endoplasmic reticulum (fig. 7). These structures have entirely disappeared by the time the rabbit blastocyst has commenced its swelling. Thus three different types of inclusions—lipid, intracellular lamellae, and intracisternal material—all exhibit the same pattern of diminution in three different orders of mammals.

Fig. 1. Ferret blastocyst (7 days *post coitum*) photographed after fixation. Note that the embryonic cell mass (*upper right*) is a ball of cells at this stage. The insert (*upper left*) is a mouse blastocyst at the same magnification. The ferret blastocyst enlarges further before implantation. ×91.

Fig. 2. Rat blastocyst from the period of delayed implantation. Note the large size of the cavity. ×259.

Fig. 3. Rat blastocyst flushed from the uterus on day 6 of pregnancy. Blastocysts have to be forcefully flushed from the implantation chamber at this stage. Note how elongate the blastocyst is. ×237.

Fig. 4. Mouse blastocyst on the afternoon of day 4, photographed with Nomarski optics. The zona pellucida is well defined, and the outline of some of the cells of the embryonic cell mass can be seen. ×387.

Fig. 5. Electron scan micrograph of three cells of a mouse blastocyst at implantation (day 5). The extreme irregularity of the surface is probably a shrinkage artifact, but the microvilli are present by other methods of preservation as well. ×6,984.

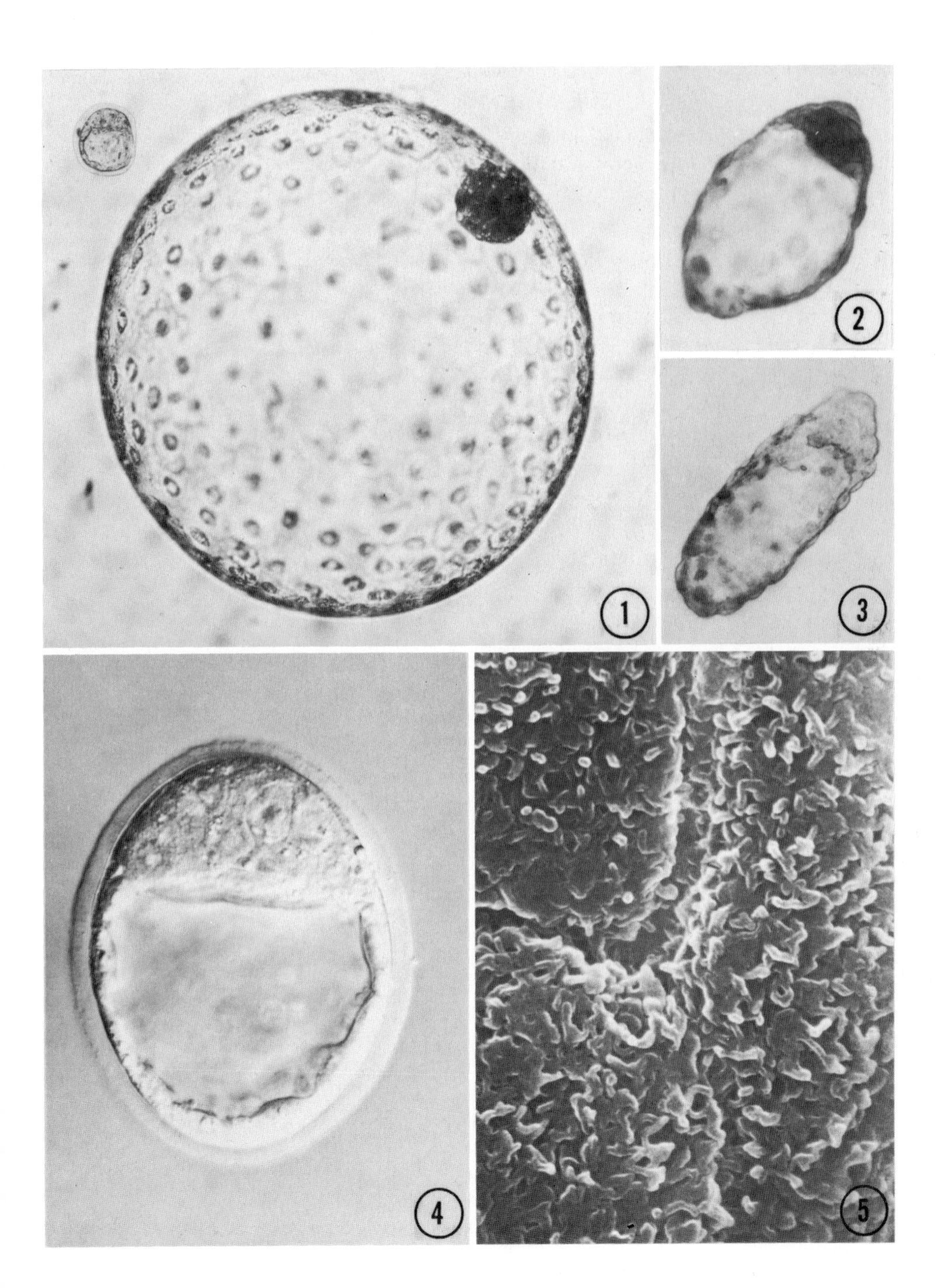
1
2
3
4
5

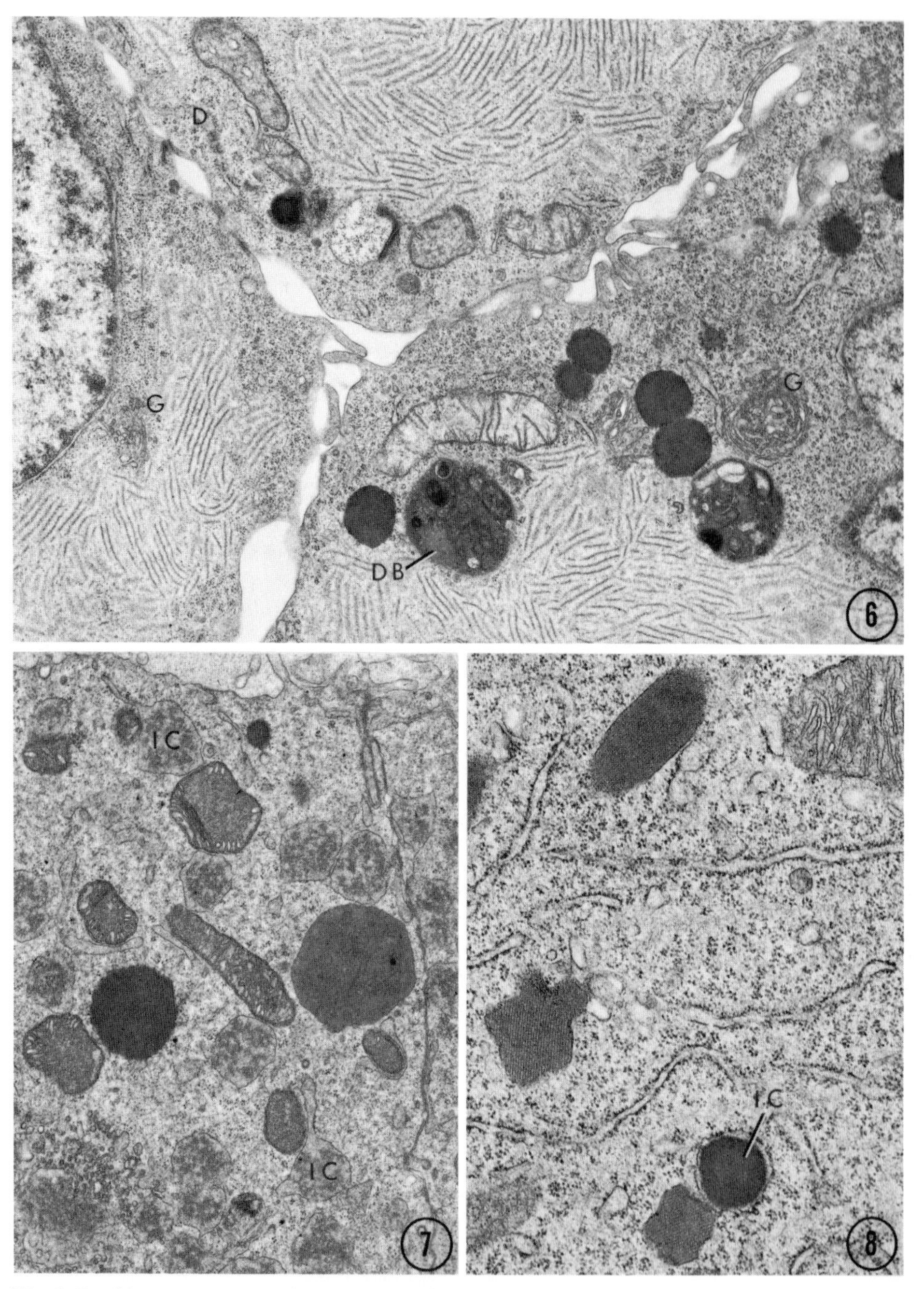

Fig. 6. **Rat** blastocyst (3 P.M. day 5). Parts of two embryonic cell mass cells and one trophoblast cell (*lower right*) are seen in this micrograph. Note that large areas of the cytoplasm are still occupied by lamellar elements, whereas other inclusions and the organelles are in the areas rich in polyribosomes. *D,* desmosome; *DB,* degradation body; *G,* Golgi. ×11,830.

Fig. 7. Rabbit morula (32 cells, 3 days *post coitum*). Note the mottled material in the cisternae of the endoplasmic reticulum (*IC*). Only a few ribosomes are associated with the membranes of the endoplasmic reticulum at this stage. ×15,934.

Fig. 8. Rabbit blastocyst (6 days *post coitum*) Note the dense intracisternal granule (*IC*), the crystalloids and the extensive granular endoplasmic reticulum. The cristae of the mitochondria are more lamellar at this stage. ×23,296.

B. *Inclusions Restricted to the Blastocyst Stage*

Inclusions restricted to the blastocyst stage are less common. In the mouse crystalloid inclusions appear in increasing numbers from the cleavage to blastocyst stages. These crystalloids, which form a conspicuous feature of the blastocyst, are usually juxtaposed to the endoplasmic reticulum (fig. 11). Often the major densities of the crystalloid are oriented at right angles to a smooth portion of otherwise granular endoplasmic reticulum. Why these inclusions should be elaborated in such numbers at this stage of development is not known.

A specific crystalloid is found in rabbit blastocysts, although it is not present in the cleavage stages (fig. 8). This crystalloid, which has been described by Hadek and Swift (1960) as well as van Beneden (1880), has been seen also in a few uterine epithelial cells. In addition unusually dense intracisternal granules are present within the endoplasmic reticulum of the rabbit trophoblast at the time of implantation (fig. 8). Recently Manes and Daniel (1969), using the amount of labeled protein as measure of synthesis, reported a 10-fold increase in protein synthesis between day 4 and day 6 in the rabbit blastocyst. It is precisely during this time period that the crystalloids accumulate and become such a conspicuous feature of the rabbit blastocyst. Autoradiographic data should show whether the majority of new protein is to be found in the crystalloids, intracisternal granules, or the less structured cytoplasm.

C. *Unusual Accumulations of Normal Cytoplasmic Constituents*

Fine cytoplasmic filaments of the type normally associated with desmosomes are not found in mammalian ova. However in the armadillo blastocyst (fig. 9), and to a lesser extent in the pig (Hall, Horne, and Perry 1965) and Mexican funnel-eared bat blastocysts, unusually large numbers of filaments are found in association with the desmosomes; they also ramify throughout the cytoplasm and extend completely around the nucleus from one end of the cell to the other.

Glycogen per se is not very abundant during the cleavage stages. Aggregates of glycogen particles accumulate in the blastocyst stage in a number of species and are particularly conspicuous in the armadillo (Enders 1962) (fig. 9). There is an increase in glycogen, especially in the trophoblast, at the time of implantation in the rat and mouse.

A wide variety of complex lipid-containing inclusions are found in most of the blastocysts. In some instances these complex granules are present from the time of fertilization and may well represent stored material. In other instances they increase in abundance during development and may be accumulation of material resulting from breakdown of intracellular or phagocytized material. In the rat and hamster we have found such bodies with partially disrupted fibrous lamellar structures (fig. 12). In the mouse, in contrast to the rat, the lamellar elements do not become increasingly segregated in the cytoplasm, and there are few structures which resemble the degradative bodies seen in the rat (fig. 6).

In several instances an introduced external marker has appeared in the degradation bodies in the rat. It is apparent therefore that both phagocytized material and products of autolysis can contribute to the formation of degradation bodies.

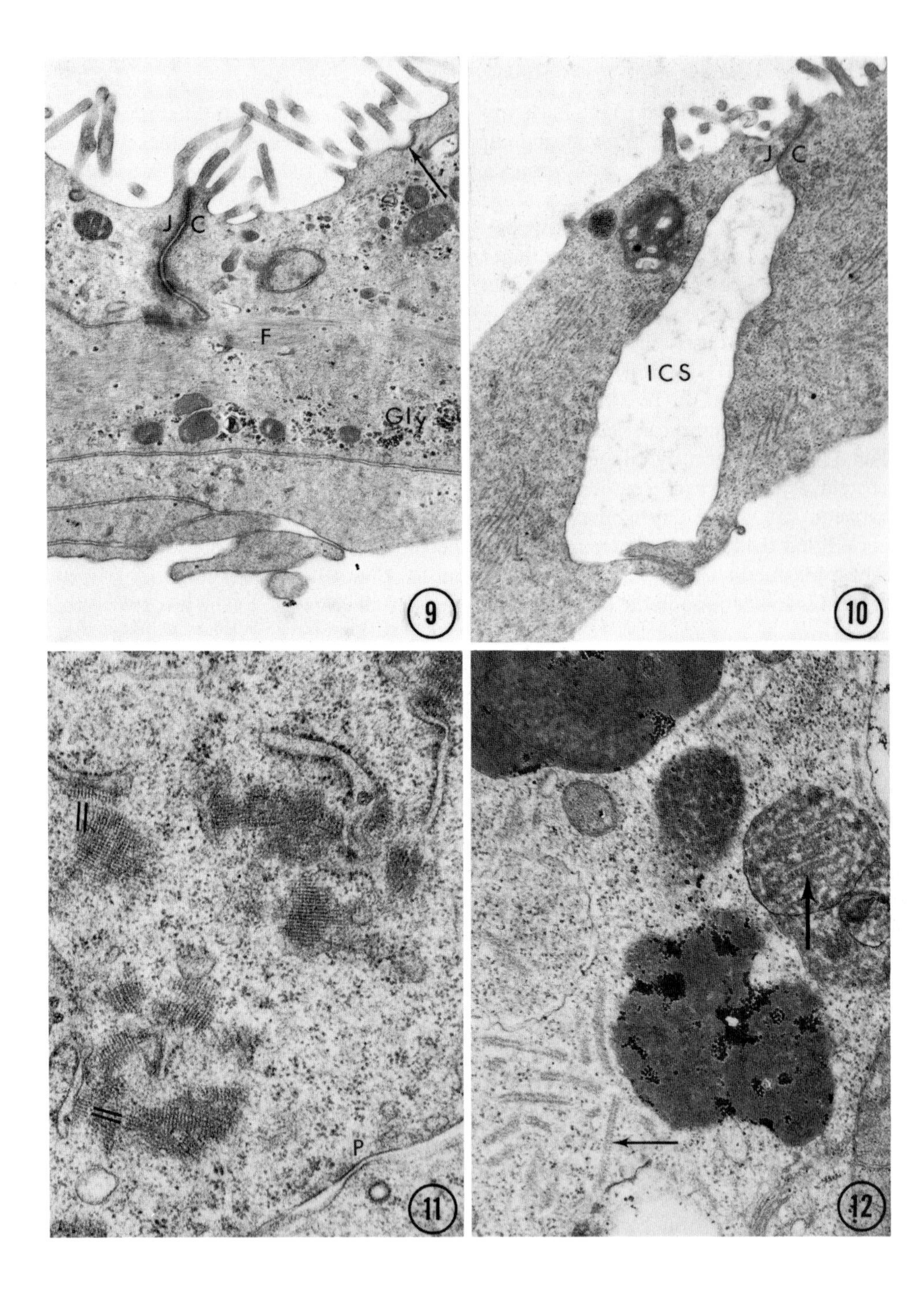
J C
F
Gly
9
J C
ICS
10
P
11
12

However the specific ways in which blastocysts segregate materials ordinarily expected to be subject to lysosomal digestion has not been studied and clearly deserves attention.

IV. General Changes in Cell Organization

In addition to changes occurring in the inclusions from ovum to blastocyst stage there are alterations in organelles and cell junctions (example, Schlafke and Enders 1967). Polyribosomes undergo tremendous increase, especially in the morula and early blastocyst stages. Granular endoplasmic reticulum appears in the later cleavage stages, its few rather extensive cisternae establishing the typical embryonic situation (compare figs. 7, 8). Golgi zones tend to assume the more common juxtanuclear position, and the number of cisternae increases. The species variations in the abundance of the Golgi (well developed in carnivores, poorly developed in myomorph rodents) are not understood at the moment. Mitochondria tend to assume typical rod form with lamellar cristae as opposed to the various bizarre round forms found in the ovum. Of particular interest is the history of cell junctions.

Changes in cell junctions in the rat appear to be typical; they have been most extensively followed in that animal (Schlafke and Enders 1967). The association of the first two cells formed by the first cleavage tends to be rather loose, with microvillous borders, no close apposition, and no well-formed cell junctions. Primitive junctions are found in the 4-cell stage; they are found abundantly in the 8-cell stage. These junctions are concentrated in a short region where apposed membranes are parallel to one another and show a slight increased density on their inner surface but are lacking the other characteristics of desmosomal structures such as tonofilaments, a central dense area, and more extensive inner cytoplasmic material. These early primitive desmosomes are the most common type of junction existing between cells during all the cleavage stages prior to formation of the blastocyst; they are found also between cells of the embryonic cell mass in the blastocyst stage (figs. 6, 11).

Fig. 9. Junction of two trophoblast cells of an armadillo blastocyst during the delayed implantation period. Numerous fine filaments (*F*) are present throughout the cytoplasm, and glycogen (*Gly*) is also present. Note the well-formed junctional complex (*JC*) with its region of close apposition. The arrow indicates a rough-surfaced micropinocytotic vesicle of a type found in all species. ×20,202.

Fig. 10. Rat blastocyst, illustrating an unusually large intercellular space (*ICS*) beneath the junctional complex (*JC*). Such spaces may be important in transport of fluid into the cavity of the blastocyst. Ferritin to which this blastocyst had been exposed *in vivo* may be seen in the zona pellucida in the upper left corner. ×22,386.

Fig. 11. Cytoplasm of a mouse blastocyst. A primitive junction (predesmosome) is seen at *P*. Note that in at least two places in the micrograph (*parallel lines*) crystalloids can be seen at right angles to a smooth portion of a membrane of the endoplasmic reticulum. Because of this association it is thought that the crystalloids are synthesized by the endoplasmic reticulum. ×43,680.

Fig. 12. Cytoplasm from a hamster blastocyst. Note the bilamellar inclusions (*small arrow*) and the presence of one of these within a degradation body (*large arrow*). Such complex inclusion bodies are probably concerned with utilization of stored material from the ooplasm. ×33,670.

The first more complex junctions, at the outer surface of some of the cells at the 8-cell stage, show extensive regions of closely apposed membranes and associated cytoplasmic density. They appear to be the precursor of the typical apical junctional complex, a region which acts as a barrier to diffusion between cells as well as a region of attachment (figs. 9, 15, 17).

Formation of continuous junctional complexes around the free borders of the trophoblast cells can be considered the necessary prerequisite to blastocyst formation, since this process converts what is essentially an intercellular space into a space surrounded by an epithelium. It is at this point that the blastocyst becomes an organism with an external and internal environment rather than just a collection of cells. Any secretion of fluid into a simple intercellular space could not contribute to blastocyst formation unless this space was sealed by junctional complexes. Whether any of these junctional complexes constitute true "tight junctions" permitting ionic flow between the trophoblast cells has not yet been determined. Such linkage would be expected in blastocysts such as that of the rabbit.

At the blastocyst stage desmosomes are frequently found on the lateral walls of the trophoblast cells. When the cells of the embryonic cell mass are still undifferentiated desmosomes are rare, but primitive desmosomes are common. As differentiation of these cells becomes more apparent, desmosomal complexes first seen between endodermal cells, then between cells of the epiblast, become abundant. Since the time of differentiation of endoderm prior to implantation depends largely on the length of the preimplantation period, desmosomal structures are more common in the embryonic cell mass of the free blastocyst of those species with a longer preimplantation period, such as the carnivores.

The lateral junctional complexes constitute a barrier to loss of fluid from the cavity of the blastocyst; they may also play a role in the active transport of fluid into this cavity. It has been suggested recently by Elfriede Gamow (personal communication) that in the case of the rabbit the transport of material may follow the three-compartment model of Curran and McIntosh (1962). This model has been interpreted in the case of the gall bladder (Kaye et al. 1966) to mean that there must be a lateral compartment between cells constituted by the space extending from the apical junctional complex to the simpler junction in proximity to the basement membrane (see fig. 10).

V. Differentiation of Endoderm

There is no cytological evidence of differentiation of separate groups of cells until after blastocyst formation (Enders and Schlafke 1965; Schlafke and Enders 1967). This cytological observation correlates well with the interesting results of Tarkowski (1965), Tarkowski and Wroblewska (1967), and Mintz (1968), whose experimental manipulations of blastomeres with regard to deletion and particularly to chimera formation (allophenic blastocysts) have indicated that the role of the individual blastomeres is not determined at an early stage. Unfortunately blastomeres of few species can be manipulated experimentally at present. In retrospect it is apparent that the early workers in mammalian embryology were unduly concerned with the germ layer problem (see arguments concerning epiblastic and hypoblastic

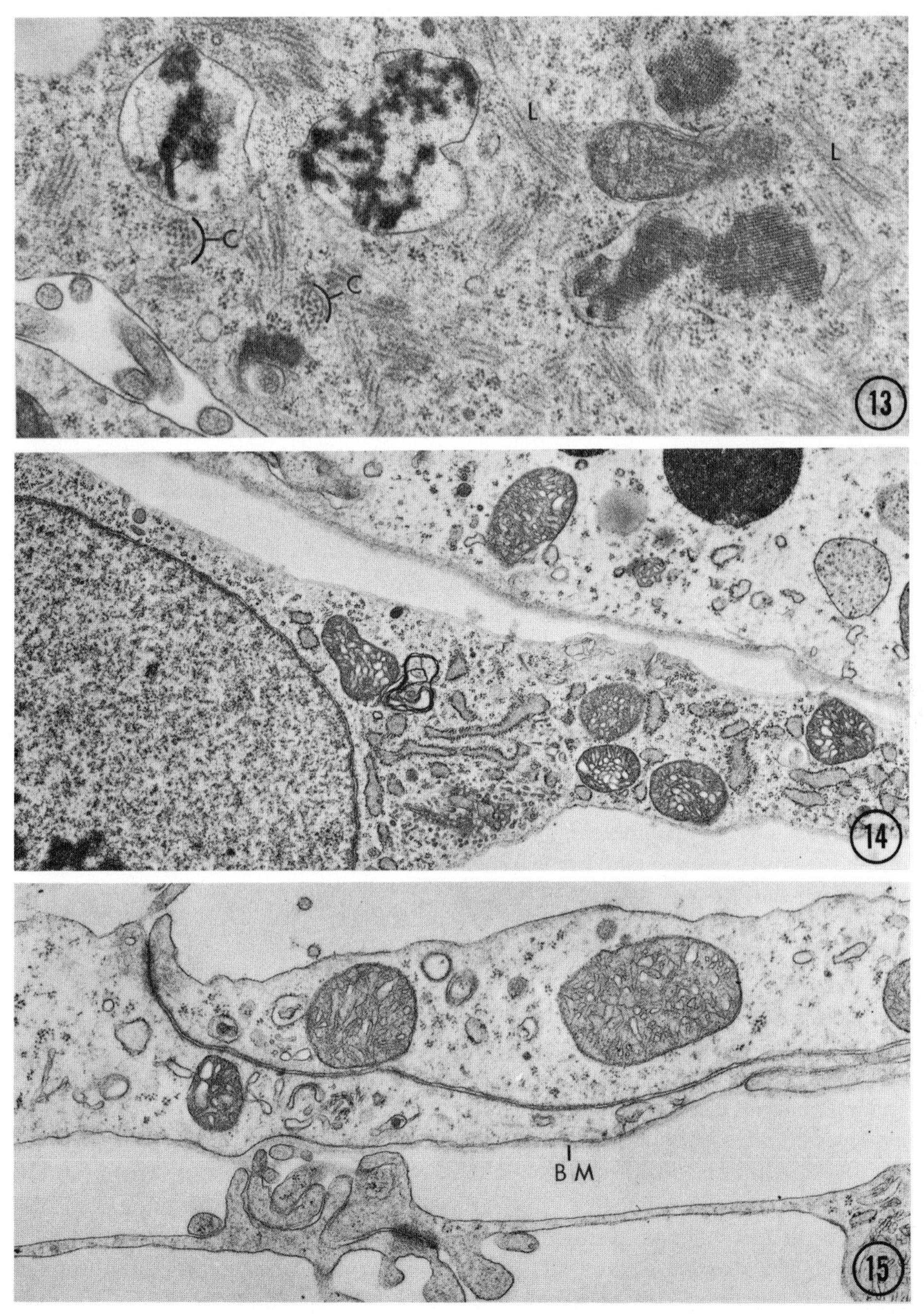

Fig. 13. Late day 4 mouse blastocyst. Both cross sections (*C*) and longitudinal sections (*L*) of the fibrous elements so common in the ovum and cleavage stages of the mouse can be seen in this micrograph. Unlike the rat there is no regional separation of the polyribosomes and fibrous elements. ×33,670.

Fig. 14. Trophoblast (*top*) and endoderm of the implanting shrew (*Blarina brevicauda*) blastocyst. Note the accumulation of dense material in the endoplasmic reticulum of the endoderm cell. ×11,830.

Fig. 15. Trophoblast and endoderm of an implanting shrew blastocyst. In this and many other species in which there is considerable swelling at implantation, portions of the endoderm cell become exceedingly thin. Note that although a basement membrane (*BM*) has formed under the trophoblast (*above*), there is no basement membrane associated with the endoderm. ×19,565.

origin of trophoblast, etc., in Hubrecht 1890, and Assheton 1898). Slight differences in size of blastomeres or intensity of staining were often taken to indicate divergence in developmental pathways of the early blastomeres by many authors in the nineteenth century and early twentieth century. However in all species examined recently by cytological methods the early blastomeres appear similar. We should expect, then, that the experimental results obtained from mouse blastomeres will be pertinent to the other species of mammals as well.

The endoderm cells are the first of the embryonic cell mass cells to show evidence of cytological differentiation. The cells of the embryonic cell mass adjacent to the cavity of the blastocyst become increasingly epithelial and develop apical junctional complexes and more extensive cisternae of the endoplasmic reticulum. Often these cisternae show an accumulation of dense intracisternal material (figs. 14, 27). The shape which the endodermal cells take is variable. In carnivores the sheet of cells is exceedingly thin, thinner than the width of a single mitochondrion; consequently it has areas which appear to be discontinuous under the light microscope. In these species and in the little brown bat and the shrew endoderm cells constitute a continuous sheet although not yet forming a yolk sac (fig. 15). In the myomorph rodents the early endodermal cells tend to be more cuboidal adjacent to the embryonic cell mass than elsewhere. Individual endodermal cells migrate along the abembryonic trophoblast.

The first basement membranes to appear in the blastocyst are flocculent extracellular accumulations (fig. 18). Basement membranes in blastocysts may occur beneath the trophoblast surrounding the epiblast and beneath the forming endodermal cells. In some instances there may be a basement membrane beneath the trophoblast cells while there is none underlying the adjacent endoderm. In all instances, however, the first basement membranes are formed prior to any mesoderm formation.

VI. External Coats of the Blastocyst

Under normal developmental conditions all blastocysts have at least one external coat when first formed. In the marsupials the coats include an egg shell membrane and an albuminous layer in addition to the zona pellucida (fig. 19). In rabbits coating material is deposited over the zona pellucida both while the ovum is in the oviduct and after its entry into the uterus. In the seal blastocysts we have examined there is in addition to the zona pellucida a narrow dense subzonal layer (figs. 20, 21).

There are common features of the zona pellucida in all species and some surprising differences. Almost all species have a more irregular outer portion of the zona, representing that area that was in close association with the granulosa cells, and a smoother inner region. In electron micrographs the zona is composed of fine granules or beaded strands of granules which are similar throughout its thickness (fig. 16).

The extent to which the zona pellucida thins during blastocyst expansion, its length of persistence, and mechanism of removal show distinct variations in different species. It is clear that the zona pellucida of the mouse and rat blastocyst is shed

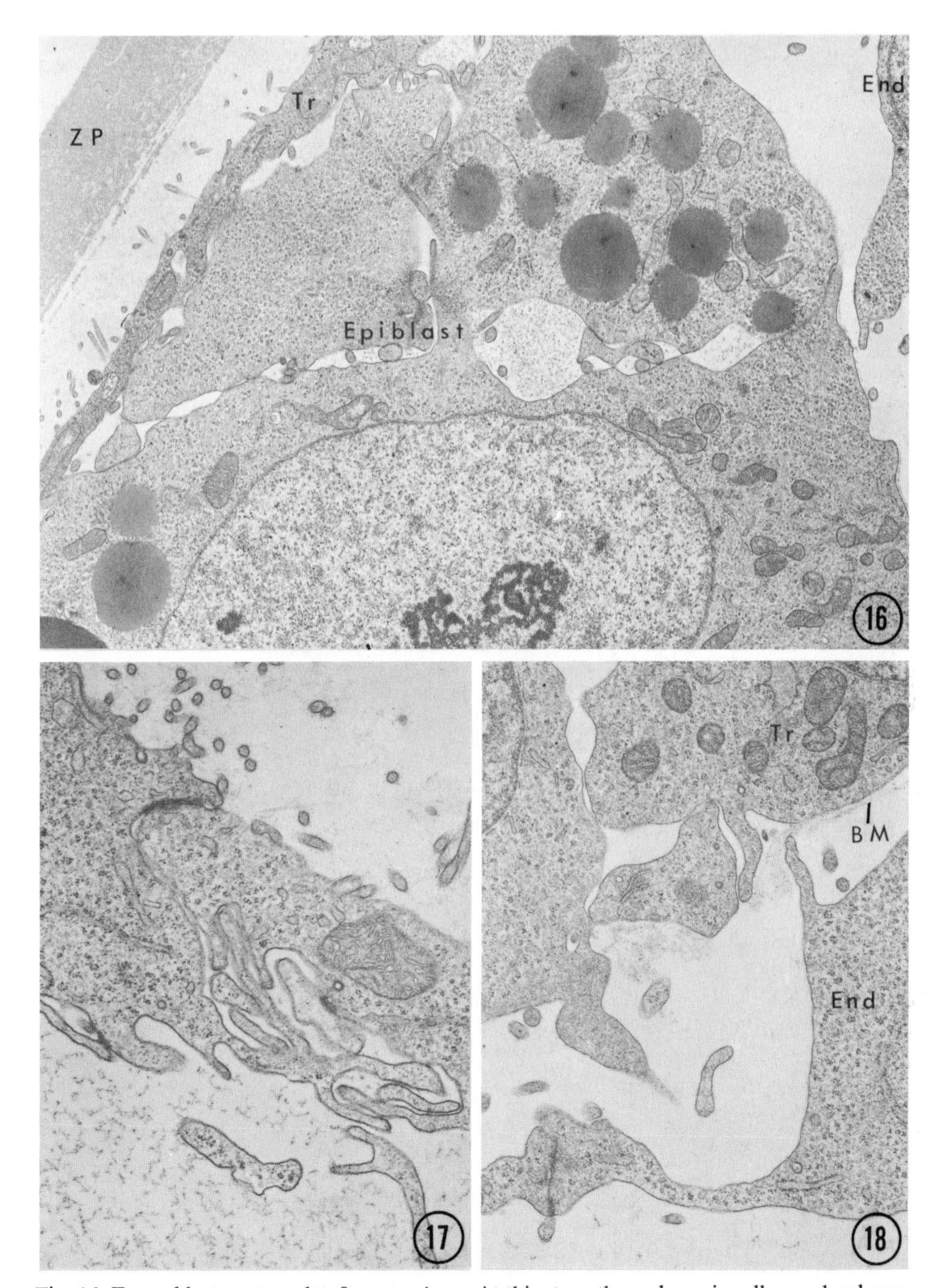

Fig. 16. Ferret blastocyst on day 9 *post coitum*. At this stage the embryonic cell mass has begun to expand and three cell types have begun to differentiate: a Rauber's layer of trophoblast (*Tr*), an endodermal layer (a small portion of the cell at the upper right, *End*) and a group of epiblast cells. Note that lipid is still present in some of the cells. Note also the strandlike arrangement of the zonal material (*ZP*). ×7,098.

Fig. 17. Trophoblast cells from a ferret blastocyst. Note that although there is some precipitate in the cavity of the blastocyst (*bottom*) there is no basement membrane at this stage. ×22,113.

Fig. 18. Trophoblast cell (*Tr*) and endoderm cell (*End*) from a mink blastocyst. A tenuous and incomplete basement membrane (*BM*) is beginning to form beneath the trophoblast cells but not as yet beneath the endoderm cells. ×14,560.

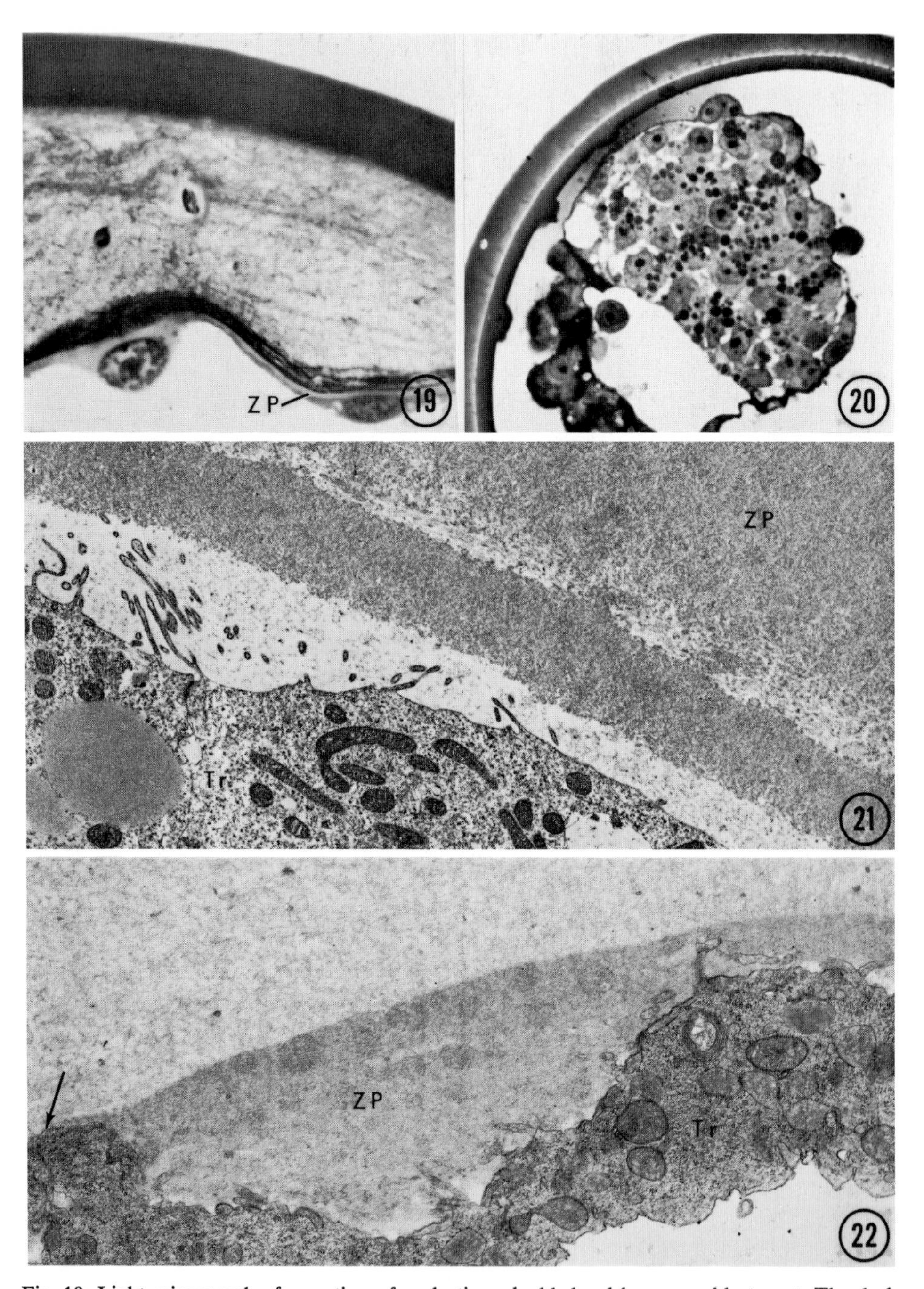

Fig. 19. Light micrograph of a section of a plastic-embedded red kangaroo blastocyst. The dark outer layer is the shell membrane. The albumin layer contains two sperm heads and is of irregular density. A thin pale zona pellucida (*ZP*) overlies the protoderm. ×510.

Fig. 20. Light micrograph of a section of a plastic-embedded fur seal blastocyst, showing the embryonic cell mass with many lipid droplets and a small portion of the collapsed blastocyst cavity. The distinct subzonal layer is not present before ovulation. The blebs on the inside of this layer are an indication that it was deposited from the interior. ×410.

Fig. 21. Electron micrograph of a fur seal blastocyst, showing a small portion of two trophoblast cells (*Tr*), the subzonal layer, and the inner part of the zona pellucida (*ZP*). ×7,553.

Fig. 22. Electron micrograph of a rabbit blastocyst. The zona pellucida (*ZP*) underlying the mucolemma has become irregular in thickness where it overlies bulges in the trophoblast cells. It is essentially discontinuous at the arrow at left. ×13,468.

during the preimplantation stage. Penetration and "hatching" of the zona of the mouse has been photographed *in vitro* (Cole 1967). The observation of rat and mouse blastocysts with a single point of penetration of the zona is common; individual processes that pass through the zona have been observed in electron micrographs of blastocysts of the former species (Schlafke and Enders 1967). Despite the evidence of active shedding of the zona, remnants of zonal material have occasionally been found in rat implantation sites. Multiple processes of the implantation cone penetrate the zona pellucida along much of its abembryonic extent in the guinea pig at the time of embryonic attachment (Blandau 1949).

In marsupials the extracellular membranes are retained until well after implantation. The albuminous layer diminishes in thickness during the preimplantation increase in size of the blastocyst. It is not known when the zona pellucida, which is difficult to see in ordinary preparations, is penetrated. The membrane found after rupture and displacement of the external coats is almost exclusively shell membrane (Enders and Enders 1969).

The zona pellucida of the rabbit blastocyst diminishes in thickness as the blastocyst expands. Unlike that in most other species this diminution is irregular; that is, it may be quite thin over projecting portions of trophoblast cells and thicker over other regions. By the 6th day it has been penetrated in many places (fig. 22) and by the time of implantation it is impossible to distinguish as a separate layer. The overlying layers of extracellular material do not completely disappear until well after the trophoblastic knobs have invaded the endometrium. However some vacuolation of this material may be seen prior to implantation, and there is appreciable thinning of the extracellular coat between trophoblastic knobs as well as at the sites of initial attachment (fig. 29).

The zona pellucida is surprisingly persistent in blastocysts of the fissiped carnivores. During swelling of the blastocyst the zona diminishes to approximately 1/5 of its original thickness while the surface area that it covers increases almost 50 times. Consequently there must be an increase in total volume occupied by zonal material. In the ferret we have calculated that there is a 4-fold increase in volume of the zona from the morula to the late preimplantation stage. There is no evidence of any addition of material to the zona during this period. The only difference in the appearance of the zonas from small and large blastocysts is the presence of circumferentially distributed clefts in the zonal material of the latter (fig. 16). Such clefts may represent a type of hydration phenomenon contributing to the increase in volume. When at implantation the zona pellucida finally ruptures, it is frequently seen to recoil between areas where the trophoblast penetrates the uterine epithelium.

A. *The Zona Pellucida as a Barrier to Large Molecules*

Austin and Lovelock (1958), in a pioneering series of experiments, determined the permeability *in vitro* of the zona pellucida of rat, hamster, and rabbit eggs to toluidine blue, digitonin, alcian blue, and heparin. Since heparin, with a molecular weight of 16,000, could not be demonstrated within any of these eggs and alcian blue, with a MW of 1,200, penetrated into all of them, these authors concluded that the zona pellucida of the eggs and mucin coat of the rabbit constituted a barrier to molecules with a MW of 16,000 or greater. Glass (1963), however, has demonstrated by use of immunofluorescent methods that serum proteins can pass into the egg and cleavage stages of the mouse *in vivo*. On the other hand Heyner, Brinster, and Palm (1969) found that anti-C3H serum did not cause degeneration of blastocysts of C3H mice if the zona pellucida was intact, but degeneration occurred rapidly in eggs lacking the zona.

To determine the extent to which the zona forms a barrier to large molecules we have exposed blastocysts *in vivo* to ferritin and thorotrast. Ferritin has a MW which varies with the iron content of between 480,000 and 1,000,000, and is approximately 110 Å in diameter. Thorotrast is a colloidal suspension of thorium dioxide in dextran, forming an aggregate particle of undetermined size but appreciably larger than ferritin. To our surprise both ferritin and thorotrast penetrated the zona pellucida of the rat readily on day 5 of pregnancy (figs. 23, 25). Ferritin, but not thorotrast, penetrates the expanded zona of the mink blastocyst during delayed implantation (figs. 24, 26). In the rabbit the smaller particles of thorotrast penetrate part way into the external coats within a few minutes, but it has not been determined whether or not these particles can eventually pass completely through.

Consequently in the blastocyst stage at least the zona pellucida is less of a barrier than anticipated from Austin and Lovelock's study. It would be useful to compare the permeability of the zona to these and other tracers in the early cleavage stages as well as the blastocyst stage.

VII. Micropinocytosis and Phagocytosis

Once material has passed through the zona pellucida, it must still penetrate into the ovum to be utilized. Although it is clear from the many *in vitro* studies with the mouse and rabbit that a wide variety of substances may get into the ovum at different stages, very little information is available on the mechanisms by which many of these substances enter the cytoplasm. Wales and Biggers's (1968) results with malic acid in the mouse 2- and 8-cell stages indicate the development of a transport mechanism in the later cleavage stages. Gwatkin (1966) has found that Mengo virus particles will get into mouse ova both with and without the zona. Since these particles normally have to be phagocytized by the host cell to penetrate it, his results are a strong indication of early phagocytic activity. The results of Glass and McClure (1965) with serum protein also illustrate the necessity of a mechanism for engulfing large molecules.

Micropinocytotic vesicles of the rough-surfaced variety can be found from fertilization on. Although these structures are not especially numerous, they are a common feature of all cells throughout the cleavage and early blastocyst stages. Other evidence of the transport of materials into the cytoplasm by surface activity has been found using mass tracers. Both ferritin (fig. 23) and thorotrast (fig. 25) are picked up readily by the rat blastocyst within 10 min of its deposition *in utero*. Ferritin can be demonstrated also in the coated vesicles of trophoblast cells of the mink blastocyst after introduction of this material into the uterine lumen (fig. 24).

Vacuoles containing cytoplasmic debris which may have been phagocytized are a common feature of late cleavage and blastocyst stages. The death of individ-

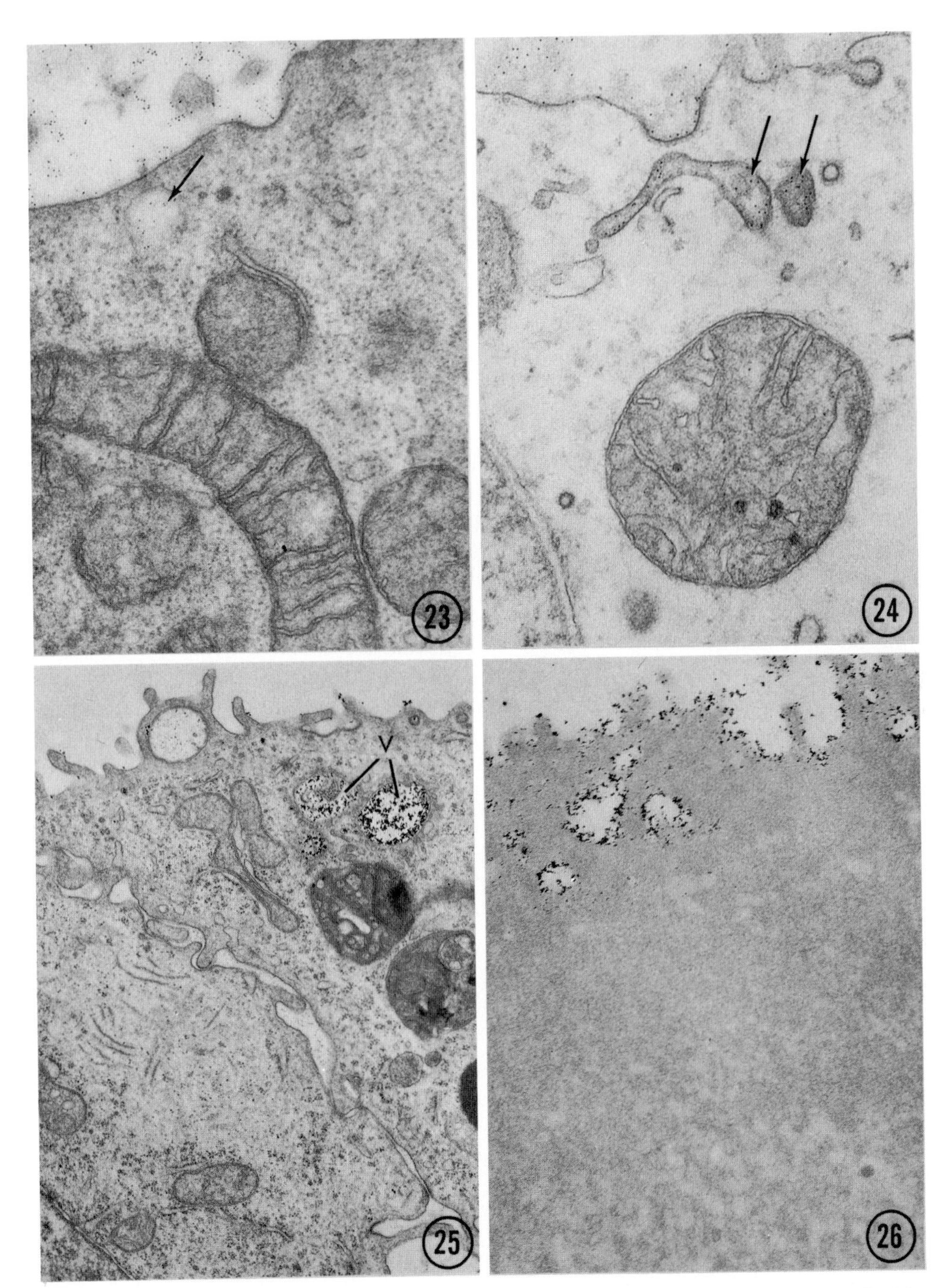

Fig. 23. Trophoblast cell from a rat blastocyst exposed to ferritin *in vivo*. Note that the ferritin has not only penetrated the zona pellucida but is in a micropinocytotic vesicle (*arrow*). A forming vesicle is seen immediately to the right. ×44,590.

Fig. 24. Trophoblast cell from a mink blastocyst fixed 10 min after the introduction of ferritin into the uterus. Ferritin has not only penetrated the zona pellucida but is seen associated with the fuzzy coat at the surface of the cell and in two vesicles beneath the surface (*arrows*). ×45,955.

Fig. 25. Trophoblast cells from a rat blastocyst fixed 10 min after thorotrast injection into the uterus. Thorotrast has been picked up by the trophoblast cells and is particularly clearly seen in the large vesicles (*V*) in the cell to the right. ×23,660.

Fig. 26. Zona pellucida of a mink blastocyst fixed 10 min after intrauterine injection of thorotrast. The thorotrast has penetrated into the indentation in the zona pellucida but has not passed further. ×21,840.

ual cells during late cleavage and blastocyst formation is apparently common, and cell fragments can be found both in trophoblast and in embryonic cell mass cells.

There is less evidence for deutoplasmolysis (yolk elimination) in material fixed for electron microscopy than might be expected from earlier descriptions using light microscopy. Much of the material that may result from disintegration of the polar bodies, from cell death, or from any fragmentation of blastomeres is apparently engulfed by the surviving cells. There is no evidence that cell death is important in formation of the cavity of the blastocyst.

VIII. Preimplantation Changes

A. *Shape of the Blastocyst*

The final shape achieved by the blastocyst prior to implantation shows a great deal of variation among species (figs. 1–4). There is even some difference in shape between the blastocyst at normal implantation and at implantation after lactational delay in the rat and mouse. The shape of the blastocyst may be a significant aspect of the blastocyst-endometrial relationship. The majority of blastocysts of the mouse and rat, for example, are larger at their abembryonic ends than at their embryonic ends and are flattened. This shape, plus the greater compressibility of the abembryonic end of the blastocyst, is probably of considerable significance in orienting the blastocyst at the implantation; yet the structural peculiarities of this stage have been ignored even in some of the most recent discussions (Kirby, Potts, and Wilson 1967). Blastocysts which undergo extensive swelling prior to implantation achieve apposition to the uterine epithelium by this change in shape. Despite their small size, murid blastocysts are closely apposed to the epithelium at the start of implantation (Sobotta 1903; Enders and Schlafke 1967; Reinius 1967; Potts 1968). Blastocysts, such as those of the armadillo and human, which apparently do not undergo preimplantation swelling and are situated in simplex uteri must have a different mechanism of apposition to establish contact with the epithelium of the endometrium prior to penetration.

Assheton (1895) pointed out that in the rabbit the differential compressibility of the embryonic cell mass might be responsible for its being oriented toward the mesometrially situated endometrial folds. His argument appears to have some validity, even though the blastocyst of the rabbit rapidly becomes spherical when removed from the uterus. At any rate the role that the shape of the blastocyst may have in orientation at the time of attachment deserves more study. It may prove to be an area of study amenable to the electron scan method (fig. 5).

In the apposition stage each trophoblast cell covers several uterine epithelial cells, a feature common to all species studied to date (Enders and Schlafke 1969). In material fixed *in situ* the shape of the trophoblast cells is reflected in deformation of the apposed surface of uterine luminal epithelial cells. There is relatively little deformation of the trophoblast cells.

B. *Syncytium Formation*

In some species, such as the rat, mouse, and little brown bat, in which there is little invasion of the endometrium by the trophoblast in the early stages of implantation the uterine epithelium is lost well before any formation of syncytial trophoblast occurs. In many of the more actively invading blastocysts syncytium formation occurs simultaneously with or just prior to implantation. Localized areas of syncytium are found in the implantation cone of the guinea pig, the trophoblastic knobs of the rabbit, and discoidal areas of the mustelid carnivore blastocyst at implantation. Isolated strips of cell membrane with conjoined ends and included desmosomes are found in the cytoplasm between nuclei in these syncytial regions (fig. 28). It should be emphasized that these strips of membrane indicate fusion of the membranes of adjacent cells, producing confluence of the cytoplasm rather than any breakdown of the individual cell membranes and consequent loss of the integrity of the cell surface. Regions of multiple folds of basal cell cytoplasm, membrane tubules, and unusual cytoplasmic densities are features associated with the process of syncytium formation in individual blastocysts but are not universally present.

C. *Trophoblast-Uterine Interaction*

That the trophoblast at implantation is capable of forming both desmosomes and apical junctional complexes with uterine epithelium is one of the more interesting phenomena emerging from its study (Enders and Schlafke 1969). Normally, junctional complexes are formed only between epithelial cells of similar origin within the same organism; when epithelia from different organisms are experimentally mixed, they tend to sort out by cell type rather than by organism or origin (Moscona 1957). Uterine epithelium in this sense does not "recognize" trophoblast as foreign. It should be noted that no protective extracellular coat or glycocalyx can be present between trophoblast and uterine epithelium at the apical junctional complex (fig. 30). It is interesting also that the junctional complex must be considered reversed as far as the trophoblast is concerned; that is, although its apical end is descending between epithelial cells, it precisely forms the counterpart to the apical junctional complex of the epithelial cells, reversing the usual order of close junction and desmosome.

The formation of cell junctions between trophoblast and uterine epithelium has now been documented for the rabbit, ferret, little brown bat, rat, and four-eyed opossum, and can be presumed to be of widespread occurrence (Enders and Schlafke 1969); but the significance of this phenomena has been only partially explored.

Acknowledgments

These studies have been supported by grants from the National Science Foundation.

I wish to express my sincere thanks to Dr. Joe Daniel, Dr. Hugh Tyndale-Biscoe, Dr. Robert K. Enders, Dr. Robert Sheldon, and Dr. William Wimsatt for their invaluable help in obtaining blastocysts from several species of non-laboratory animals.

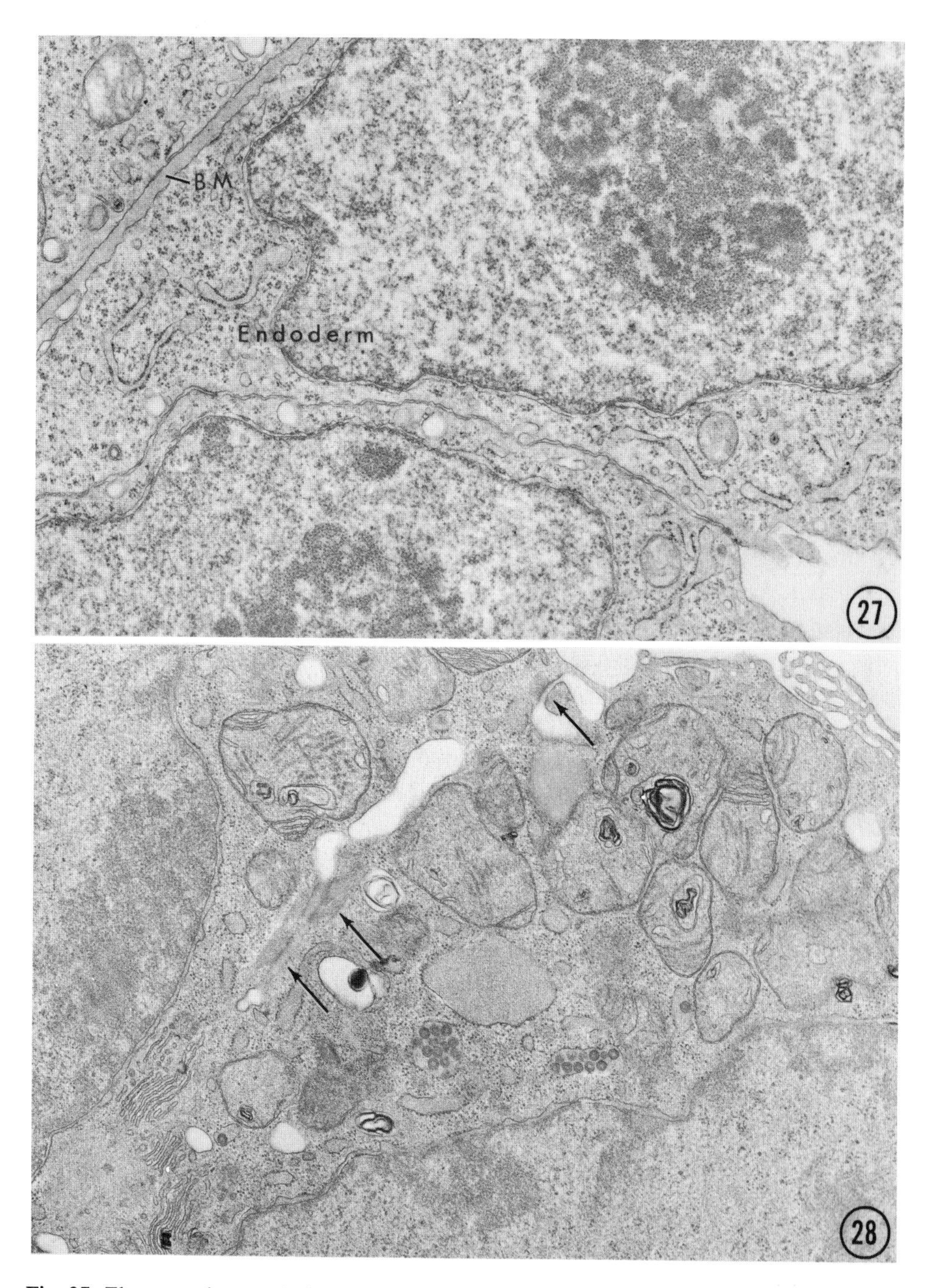

Fig. 27. Electron micrograph from a day 7 implanting rat blastocyst. Note the thick conjoint basement membrane (*BM*) between the embryonic cell mass (*upper left*) and the large differentiated endoderm cells. ×19,565.

Fig. 28. Electron micrograph of a guinea pig implantation cone. The cavity of the blastocyst is in the upper right. Note the lacework of cytoplasmic processes adjacent to the cavity of the blastocyst and the residual desmosomes (*arrows*) between the two nuclei in this newly syncytial tissue. The viruslike particles within the endoplasmic reticulum are commonly seen in guinea pig blastocysts but are also present in other stages. ×21,840.

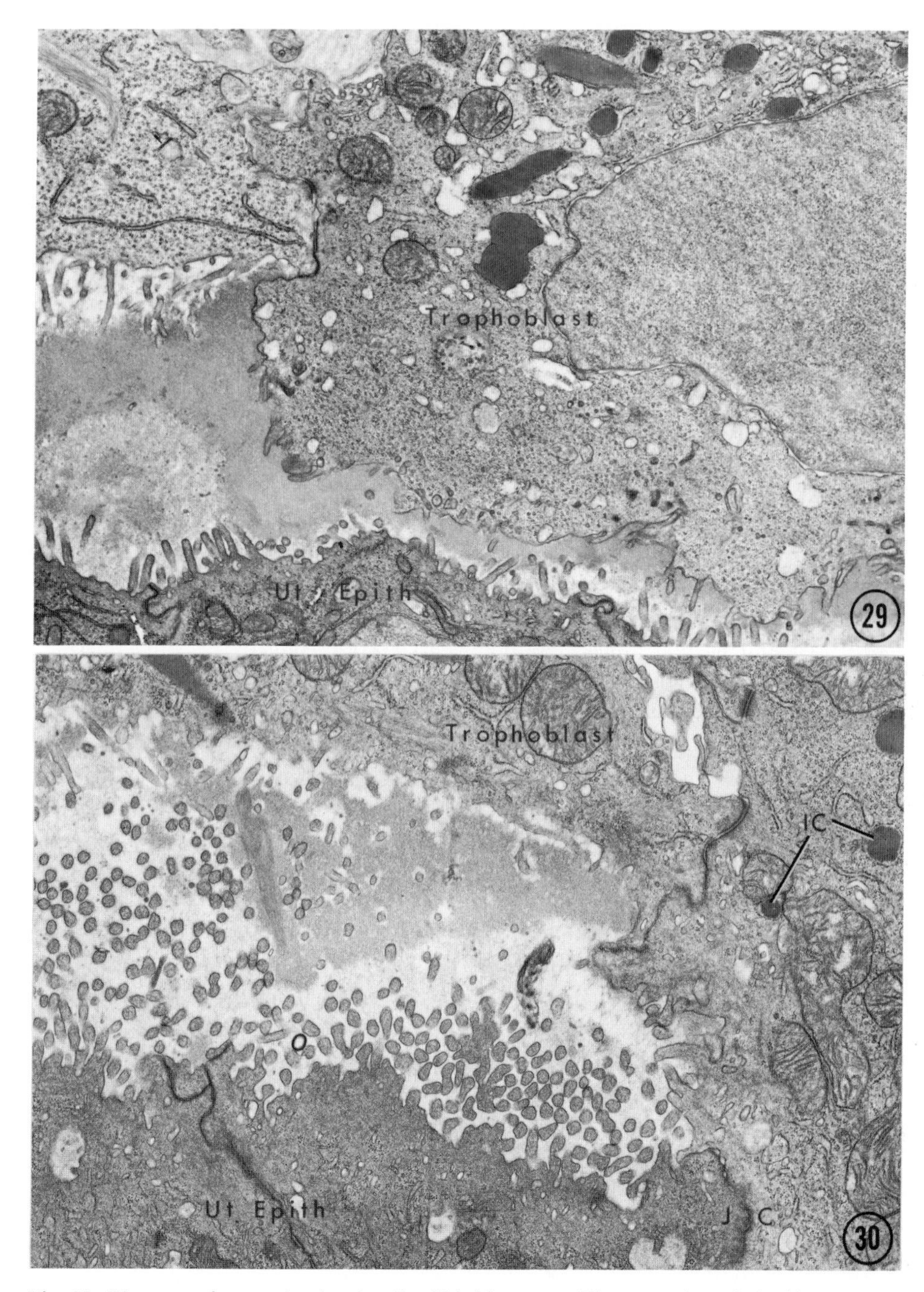

Fig. 29. Electron micrograph of a day 7 rabbit blastocyst. The expansion of the blastocyst has pressed the trophoblast close to the uterine epithelium. A small amount of extracellular material still remains between the trophoblast of a trophoblastic knob (*right*) and the uterus. ×10,556.

Fig. 30. Implanting rabbit blastocyst 7 days *post coitum*. At the margin of a site where the trophoblast has penetrated the uterine epithelium, the trophoblast shares a junctional complex (*JC*) with the uterine luminal epithelium. The trophoblast is readily identified by its cytological characteristics which include large mitochondria and intracisternal granules (*IC*). ×14,287.

References

Assheton, R. 1895. On the causes which lead to the attachment of the mammalian embryo to the walls of the uterus. *Quart J Micr Sci* 37:173

———. 1898. The segmentation of the ovum of the sheep, with observation on the hypothesis of a hypoblastic origin for the trophoblast. *Quart J Micr Sci* 41:205.

Austin, C. R. 1965. In *Preimplantation Stages of Pregnancy,* edited by G. E. W. Wolstenholme and M. O'Connor, p. 3. London: J. & A. Churchill.

Austin, C. R., and Amoroso, E. C. 1959. The mammalian egg. *Endeavour* 18:130.

Austin, C. R., and Lovelock, J. E. 1958. Permeability of rabbit, rat and hamster egg membranes. *Exp Cell Res* 15:260.

Blandau, R. J. 1949. Observations on implantation of the guinea pig ovum. *Anat Rec* 103:19.

Bonnet, R. 1884. Beiträge zur Embryologie der Wiederkäuer, gewonnen am Schafei. *Arch Anat Physio* 8:170.

Brinster, R. L. 1967. Protein content of the mouse embryo during the first five days of development. *J Reprod Fertil* 13:413.

Calarco, P. G., and Brown, E. H. 1968. Cytological and ultrastructural comparisons of t^{12}/t^{12} and normal mouse morulae. *J Exp Zool* 168:169.

Cole, R. J. 1967. Cinemicrographic observation on the trophoblast and zona pellucida of the mouse blastocyst. *J Embryol Exp Morph* 17:481.

Curran, P. F., and McIntosh, J. R. 1962. A model system for biological water transport. *Nature* (*London*) 193:347.

Enders, A. C. 1962. The structure of the armadillo blastocyst. *J Anat* 96:39.

Enders, A. C., and Enders, R. K. 1969. The placenta of the four-eyed opossum (*Philander opossum*). *Anat Rec* 165:431.

Enders, A. C., and Schlafke, S. J. 1965. In *Preimplantation Stages of Pregnancy,* ed. G. E. W. Wolstenholme and M. O'Connor, p. 29. London: J. & A. Churchill.

———. 1967. A morphological analysis of the early implantation stages in the rat. *J Anat* 120:185.

———. 1969. Cytological aspects of trophoblast-uterine interaction in early implantation. *Amer J Anat* 125:1.

Glass, L. E. 1963. Transfer of native and foreign serum antigens to oviducal mouse eggs. *Amer Zool* 3:135.

Glass, L. E., and McClure, T. R. 1965. In *Preimplantation Stages of Pregnancy,* ed. G. E. W. Wolstenholme and M. O'Connor, p. 294. London: J. & A Churchill.

Gwatkin, R. B. L. 1966. Effect of viruses on early mammalian development: III. Further studies concerning the interaction of Mengo encephalitis virus with mouse ova. *Fertil Steril* 17:411.

Hadek, R. 1965. The structure of the mammalian egg. *Int Rev Cytol* 18:29.

Hadek, R., and Swift, H. 1960. A crystalloid inclusion in the rabbit blastocyst. *J Biophys Biochem Cytol* 8:836.

Hall, F. J.; Horne, R. W.; and Perry, J. S. 1965. Electron microscope observations on the structure of cytoplasmic filaments in the pig blastocyst. *J Roy Micr Soc* 84:143.

Hamilton, W. J. 1934. The early stages in the development of the ferret: Fertilization to the formation of the prochordal plate. *Trans Roy Soc Edinburgh* 58:251.

Heyner, S.; Brinster, R. L.; and Palm, J. 1969. Effect of iso-antibody on pre-implantation mouse embryos. *Nature* (*London*) 222:783.

Hill, J. P., and Tribe, M. 1924. The early development of the cat (*Felis domestica*). *Quart J Micr Sci* 68:513.

Hubrecht, A. A. W. 1890. Studies in mammalian embryology: II. The development of the germinal layers of *Sorex vulgaris*. *Quart J. Micr Sci* 31:499.

Kaye, G. I.; Wheeler, H. O.; Whitlock, R. T.; and Lane, N. 1966. Fluid transport in the rabbit gallbladder. *J Cell Biol* 30:237.

Kirby, D. R. S.; Potts, D. M.; and Wilson, I. B. 1967. On the orientation of the implanting blastocyst. *J Embryol Exp Morph* 17:527.

Manes, C., and Daniel, J. C., Jr. 1969. Quantitative and qualitative aspects of protein synthesis in the preimplantation rabbit embryo. *Exp Cell Res* 55:261.

Mintz, B. 1968. Hermaphroditism, sex chromosomal mosaicism and germ cell selection in allophenic mice. *J Anim Sci* 27:51.

Moscona, A. 1957. The development *in vitro* of chimeric aggregates of dissociated embryonic chick and mouse cells. *Proc Nat Acad Sci USA* 43:184.

Odor, D. L. 1960. Electron microscopic studies on ovarian oocytes and unfertilized tubal ova in the rat. *J Biophys Biochem Cytol* 7:567.

Potts, D. M. 1968. The ultrastructure of implantation in the mouse. *J Anat* 103:77.

Reinius, S. 1967. Ultrastructure of blastocyst attachment in the mouse. *Z Zellforsch* 77:257.

Schlafke, S., and Enders, A. C. 1967. Cytological changes during cleavage and blastocyst formation in the rat. *J. Anat* 102:13.

Sobotta, J. 1903. Die Entwicklung des Eies der Maus vom Schlusse der Furchungsperiode bis zum Auftreten der Amniosfalten. *Archiv für Mikros Anat* 61:274.

Szollosi, D. 1965. Development of "yolky substance" in some rodent eggs. *Anat Rec* 151:424.

Tarkowski, A. K. 1965. In *Preimplantation Stages of Pregnancy,* ed. G. E. W. Wolstenholme and M. O'Connor, p. 183. London: J. & A. Churchchill.

Tarkowski, A. K., and Wroblewska, J. 1967. Development of blastomeres of mouse eggs isolated at the 4- and 8-cell stage. *J Embryol Exp Morph* 18:155.

Van Beneden, Edouard. 1880. Recherches sur l'embryologie des mammifères: La formation des feuillets chez le lapin. *Arch Biol* (*Liège*) 1:137.

Van der Stricht, O. 1923. The blastocyst of the dog. *J Anat* 58:52

Wales, R. G., and Biggers, J. D. 1968. The permeability of two- and eight-cell mouse embryos to L-malic acid. *J Reprod Fertil* 15:103.

Weakley, B. S. 1966. Electron microscopy of the oocyte and granulosa cells in the developing ovarian follicles of the golden hamster (*Mesocricetus auratus*). *J Anat* 100:503.

———. 1967. Investigations into the structure and fixation properties of cytoplasmic lamellae in the hamster oocyte. *Z Zellforsch* 81:91.

———. 1968. Comparison of cytoplasmic lamellae and membranous elements in the oocytes of five mammalian species. *Z Zellforsch* 85:109.

Wimsatt, W. A. 1963. Delayed implantation in the Ursidae, with particular reference to the black bear (*Ursus americanus* Pallus). In *Delayed Implantation,* ed. A. C. Enders, p. 49. Chicago: University of Chicago Press.

Young, M. P.; Whicher, J. T.; and Potts, D. M. 1968. The ultrastructure of implantation in the golden hamster (*Cricetus auratus*). *J. Embryol Exp Morph* 19: 341.

Zamboni, L., and Mastroianni, L. 1966. Electron microscopic studies on rabbit ova: II. The penetrated tubal ovum. *J Ultrastruct Res* 14:118.

6

Nucleoli and Ribonucleoprotein Particles in the Preimplantation Conceptus of the Rat and Mouse

Daniel Szollosi

Department of Biological Structure
School of Medicine
University of Washington, Seattle

The purpose of this paper is twofold: to explore the fine structure of nucleoli in the preimplantation stages of various mammals and to gain some understanding of their function. The distribution and arrangement of ribosomes during the same developmental period will also be examined, particularly in view of the more recent biochemical studies demonstrating that ribosomal ribonucleic acid (RNP) is the primary product of nucleoli (Brown and Gurdon 1964; Perry 1965).

Fertilized mammalian eggs differ from eggs of most other forms that have been studied in that several nucleoli are already formed in the pronuclei where these may undergo fusion at different developmental periods. These complex processes have been summarized by Austin (1961). Many nucleoli also appear subsequent to each cleavage division; with each cleavage their number decreases.

A reexamination of nucleoli and RNP particles is warranted because of some apparent contradictions: little or no RNA is detected in ovulated mammalian eggs (Alfert 1950; Flax 1953); nonetheless protein synthesis proceeds (Mintz 1964). Although detectable amounts of RNA (both nucleolar and cytoplasmic) have been found in ovarian oocytes, it appears to decrease prior to ovulation. However, little or no RNA has been detected in nucleoli of pronuclei by various cytochemical techniques (Austin and Braden 1953; Braden and Austin 1953). Nucleoli of the germinal vesicle break down prior to formation of the first polar body and no nucleoli or nucleolar remnants have been detected near the egg's chromosome set either during the first meiotic division or at time of ovulation when development is arrested at metaphase of the second meiotic division (fig. 1) (Odor and Renninger 1960; Zamboni and Mastroianni 1966 *a, b*). In freshly ovulated eggs ribosomes have not been demonstrated with conventional osmium

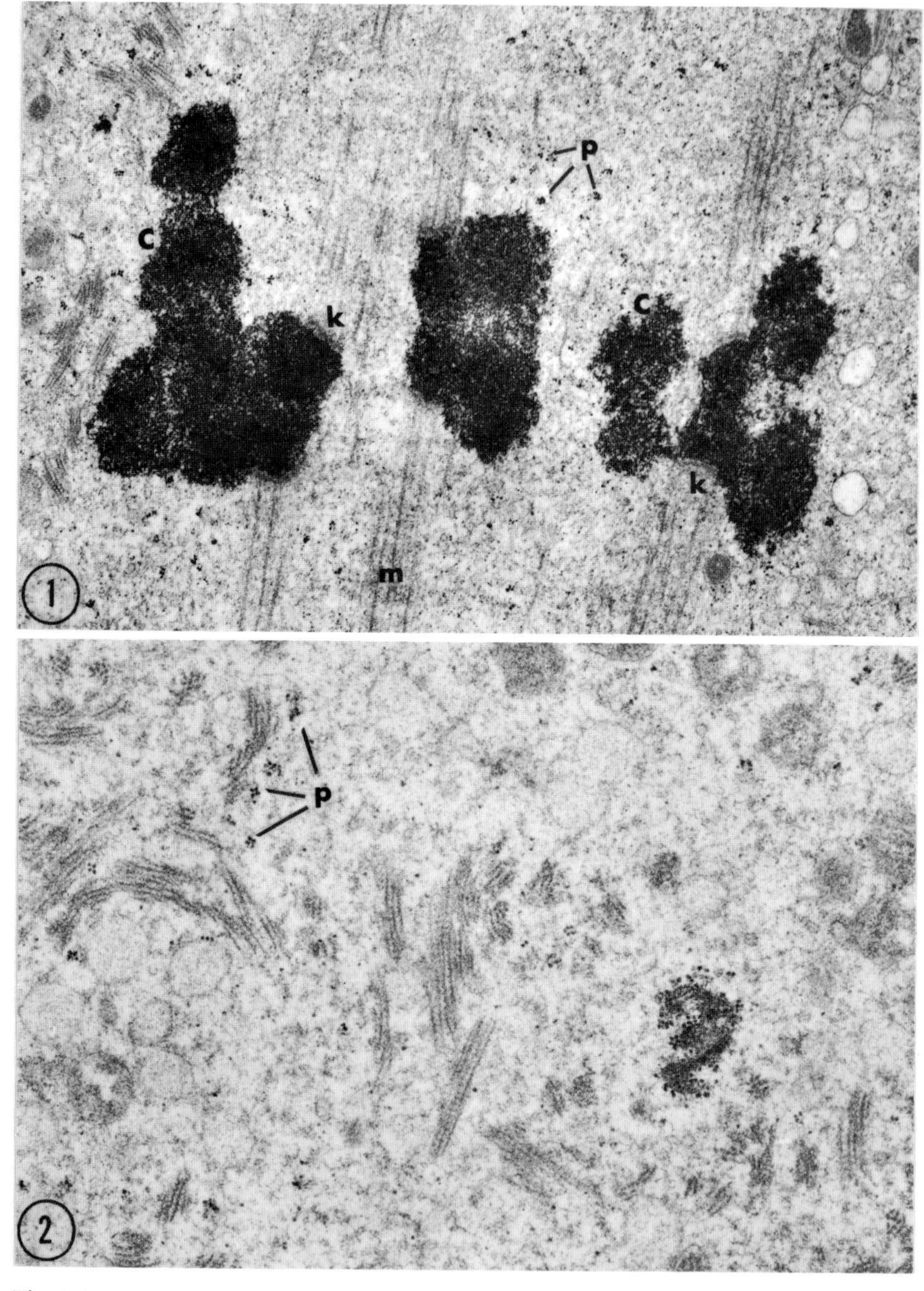

Fig. 1. Metaphase plate of the second meiotic division of an ovulated mouse oocyte. Chromosomes (*C*), kinetochores (*k*), microtubules (*m*), and several polysomal clusters (*p*) can be seen. ×15,925.

Fig. 2. Polysomes (*p*) and a polysome-filamentous complex are seen in the cytoplasm of a mouse oocyte. The filamentous elements with a prominent cross striation possibly represent yolk substances. ×28,392.

fixation procedures, nor have they been much more distinct after fixation in the more recently developed buffered glutaraldehyde solutions followed by postosmication (Schlafke and Enders 1967; Maraldi and Monesi 1968). These morphological observations have been puzzling because radioactively labeled amino acids seem to be incorporated into the egg cytoplasm prior to initiation of RNA synthesis (Mintz 1964). Lack of nucleolar RNA has raised the question whether nucleoli of cleavage stages can be considered "true" nucleoli, since RNA is generally considered an essential component of nucleoli.

A different type of ultrastructural image was presented when formaldehyde was incorporated into the primary fixative (Szollosi 1967) regardless of whether the latter was a solution of osmium tetroxide or glutaraldehyde (1% glutaraldehyde and 0.25% formaldehyde in 0.05 M phosphate or cacodylate buffer [Karnovsky 1965]). Clusters of ribosomes consisting of 4–6 or even more granules were preserved throughout the egg cytoplasm (fig. 2) with this modification. Studies of the region near the spindle apparatus (which usually is clear of cytoplasmic organelles other than microtubules) demonstrated the presence of large, helically arranged polysomes (fig. 3) or, on occasion, even a very large granular filamentous aggregate (figs. 2–4). These complexes are not limited to the proximity of the spindle. Admittedly it is difficult to identify cytoplasmic granules positively from ultrastructural features alone, but after a thorough examination of the size and arrangement of the granules and their staining characteristics, they are thought to be ribosomes.

Mammalian eggs are thus not in true contrast to the amphibian and sea urchin eggs which have been studied more fully by biochemical and ultrastructural methods (Brown and Gurdon 1966; Miller 1966; Tyler 1967; Anderson 1968). The differences are quantitative rather than qualitative.

Soon after formation of the pronuclei a number of small nucleoli (referred to as primary nucleoli) are produced amid the irregular chromatin masses, apparently coalescing when they contact each other. On further growth of the pronucleus more nucleoli are formed near the nuclear envelope (secondary nucleoli). When the synkaryon is formed the nucleoli slowly coalesce and disappear. Such nucleoli have been considered free of ribonucleic acid (for review see Austin 1961), but recently ribonuclease extractable basophilia has been demonstrated in them (Szollosi 1965).

All the above described nucleoli are composed of fine fibrils 40–50 Å thick which, upon electron-microscopic examination, can be likened to a ball of yarn (fig. 5). They lack prominent small granules, vacuoles, and amorphous components often described in a variety of vertebrate cells (Bernhard and Granboulan 1968; Hay 1968). The nucleolus-associated chromatin cannot be demonstrated in most cases. Occasionally a slightly thicker filamentous material can be discerned toward the periphery of nucleoli, sometimes in clusters, sometimes as a uniform peripheral ring (fig. 6). The more peripheral region may correspond to the nucleolus-associated chromatin. The Feulgen reaction has confirmed in some cases a positively reacting ring around the larger nucleoli (Alfert 1950; Austin 1961; Szollosi 1965). The conventional electron-microscopic methods are not suitable for making a clear distinction between various filamentous structures inside the nucleus.

After formation of secondary nucleoli a third group of dense, filamentous nuclear

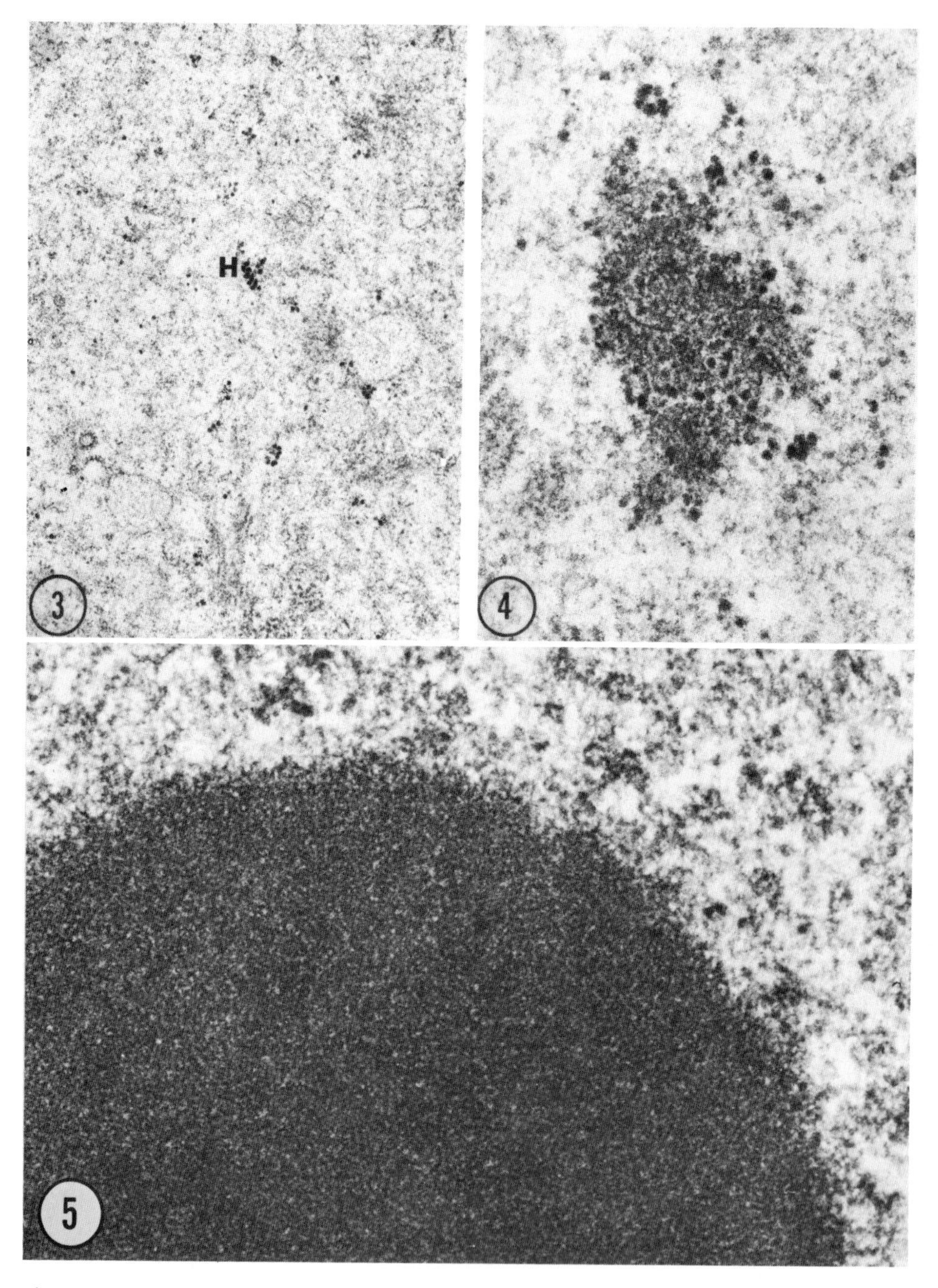

Fig. 3. A helically arranged polysome complex (*H*) in an unpenetrated tubal mouse oocyte. ×30,212.

Fig. 4. A polysome-filamentous complex in an unpenetrated mouse oocyte. ×75,530.

Fig. 5. Nucleolus of a mouse pronucleus is composed exclusively of coiled thin filaments measuring 40–50 Å. ×68,705.

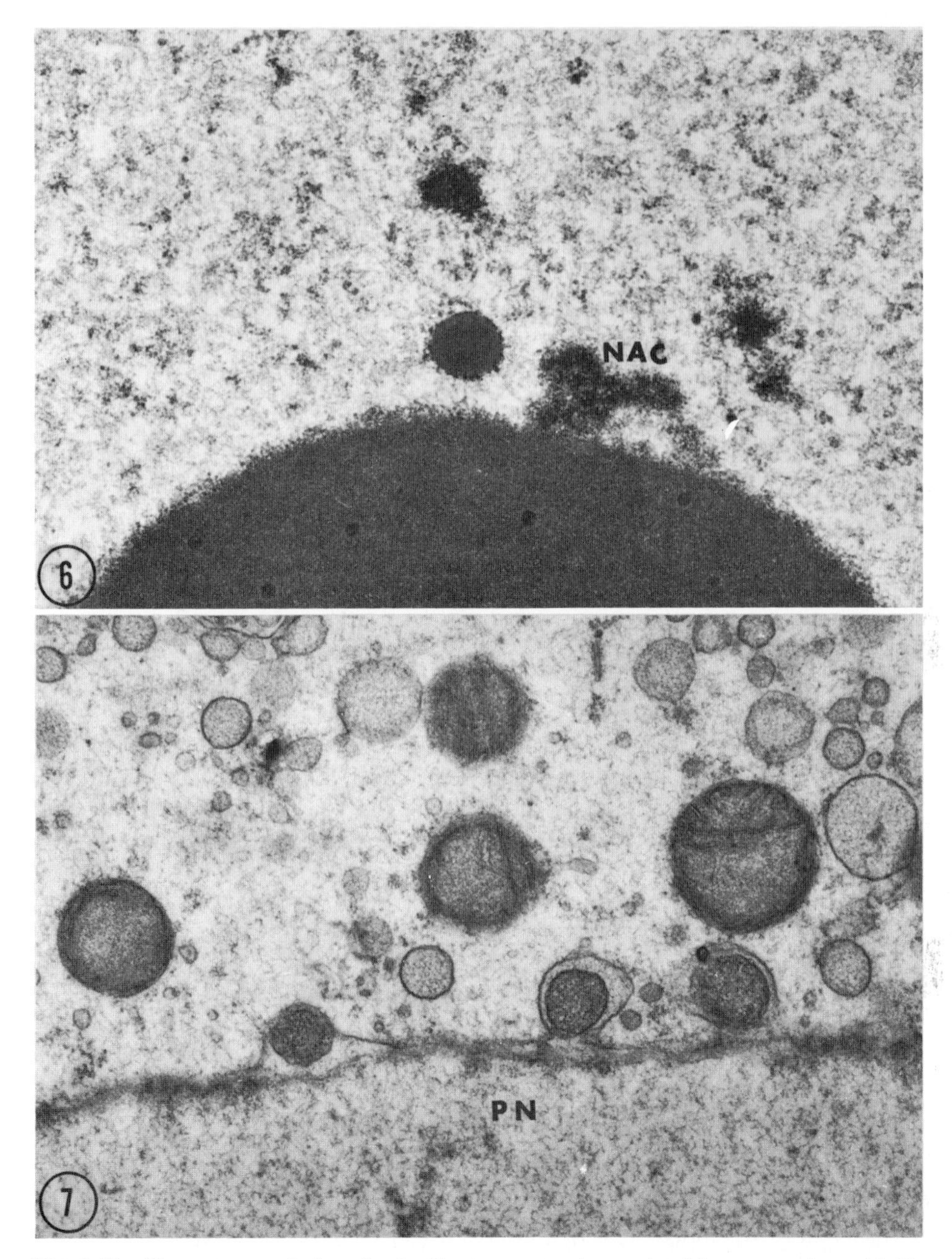

Fig. 6. The filamentous nucleolus of a 4-cell rat conceptus is associated in some regions with filaments of different density and diameter. These possibly represent the nucleolus-associated chromatin (*NAC*). ×27,573.

Fig. 7. Extrusion of three small filamentous "tertiary nucleoli" from pronuclei (*PN*) of the rat. ×34,125. By permission from Daniel Szollosi, "Extrusion of nucleoli from pronuclei of the rat," *J Cell Biol* 25:545, 1965.

bodies have been described as aggregating in distinct locations along the inner surface of the nuclear envelope in pronuclei (Szollosi 1965). These structures look morphologically identical to previously described nucleoli, and have identical histochemical reactions, but they are much smaller. (They are referred to as tertiary nucleoli.) They evaginate progressively from both pronuclei until they appear to be pinched off and set free into the cytoplasm surrounded by segments of both nuclear membranes (fig. 7). Although the original and most detailed studies were made on rat egg development, similar observations have been made on mouse, hamster, and rabbit eggs (Szollosi unpublished; Calarco and Brown 1969). Other investigators have observed the protrusion of small, spherical filamentous bodies from the surface of pronuclei but were reluctant to identify them as nucleoli; nor did they agree as to whether they were extruded from the pronuclei (Zamboni et al. 1966). The detailed time-course studies on rat eggs, however, leave little doubt on that matter. A gradual aggregation of the filamentous material along the inner surface of the nuclear envelope approximately 4½ hr after sperm penetration and their progressive evagination and pinching off could be followed. Parallel cytochemical studies indicated that all three classes of nucleoli contain some RNA, basic protein, and a polysaccharide other than glycogen, and should be considered nucleoli. The extrusion of such nucleoli can be observed for approximately 6 hr following its first detection, after which, several hours elapse prior to the first cleavage division.

Nucleoli of early 2- and 4-cell stages appear morphologically similar to those found in pronuclei. They are composed of tightly packed thin filaments. In the periphery of at least one of the nucleoli a network or reticulation (nucleolonema) forms shortly before the second cleavage. A few small granules later become visible among its meshes (fig. 8). In the 4-cell stage the peripheral network develops earlier in the cell cycle and the granules becomes more abundant (Hillman and Tasca 1968). Originally the network is composed only of thin, coiled filaments identical in size to those composing the bulk of nucleoli. Later, granules also appear. It seems that the coiled filaments of the network transform into the granules by condensation (which reportedly also occurs in other nucleoli [Bernhard 1966]), since a residual filamentous substructure can be demonstrated within the granules at higher magnifications (fig. 9).

With each subsequent cleavage fewer nucleoli are formed. In the 8-cell stage the nucleolar network develops early in the cell cycle in association with at least one of the nucleoli. However other nucleoli remain compactly filamentous. Granules appear only in association with that nucleolus which has developed the peripheral network (fig. 10). In the 8- and 16-cell stages the transformed nucleolus is morphologically very reminiscent of nucleoli of adult somatic cells having high metabolic activity, that is, those that are involved in RNA and protein synthesis (Calarco and Brown 1968). However nucleoli of cleavage stages are much larger (fig. 11).

The above developmental sequence of the peripheral nucleolar network and granules appears to follow a similar pattern in every cleavage cycle. The cell cycles differ only in the relative duration of each phase within the cycle. Nucleoli of pronuclei do not demonstrate any changes and are constituted of compacted filaments throughout their existence. The networks at the periphery of some nucleoli are

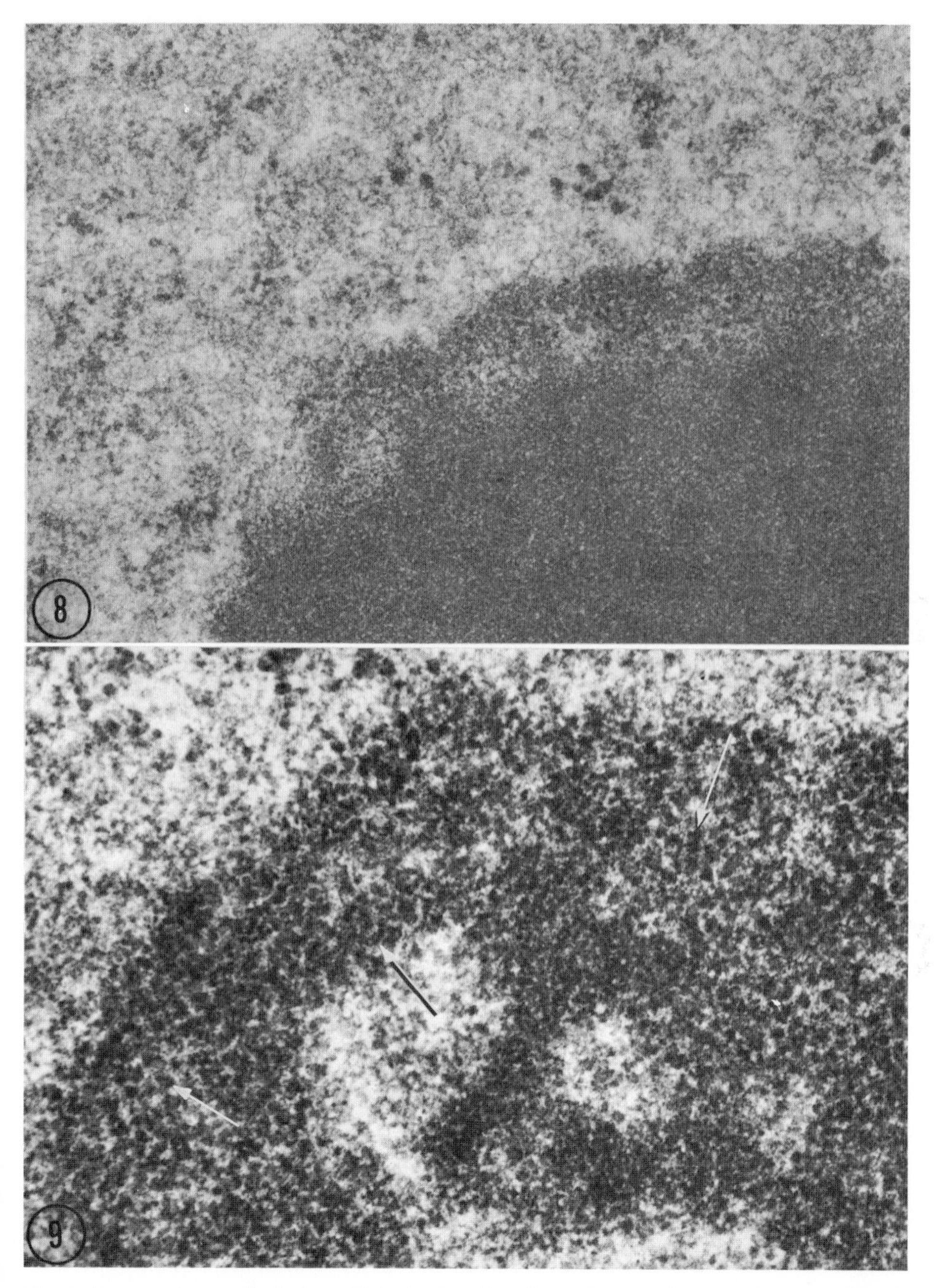

Fig. 8. A network or reticulation which developed at the periphery of a nucleolus preceding the second cleavage of a mouse egg. A few small granules are visible among the meshes. ×58,695.

Fig. 9. The small granules among the nucleolar meshes appear to be formed from a condensation or rearrangement of thin filament (*arrows*) (16-cell rat conceptus). ×81,263.

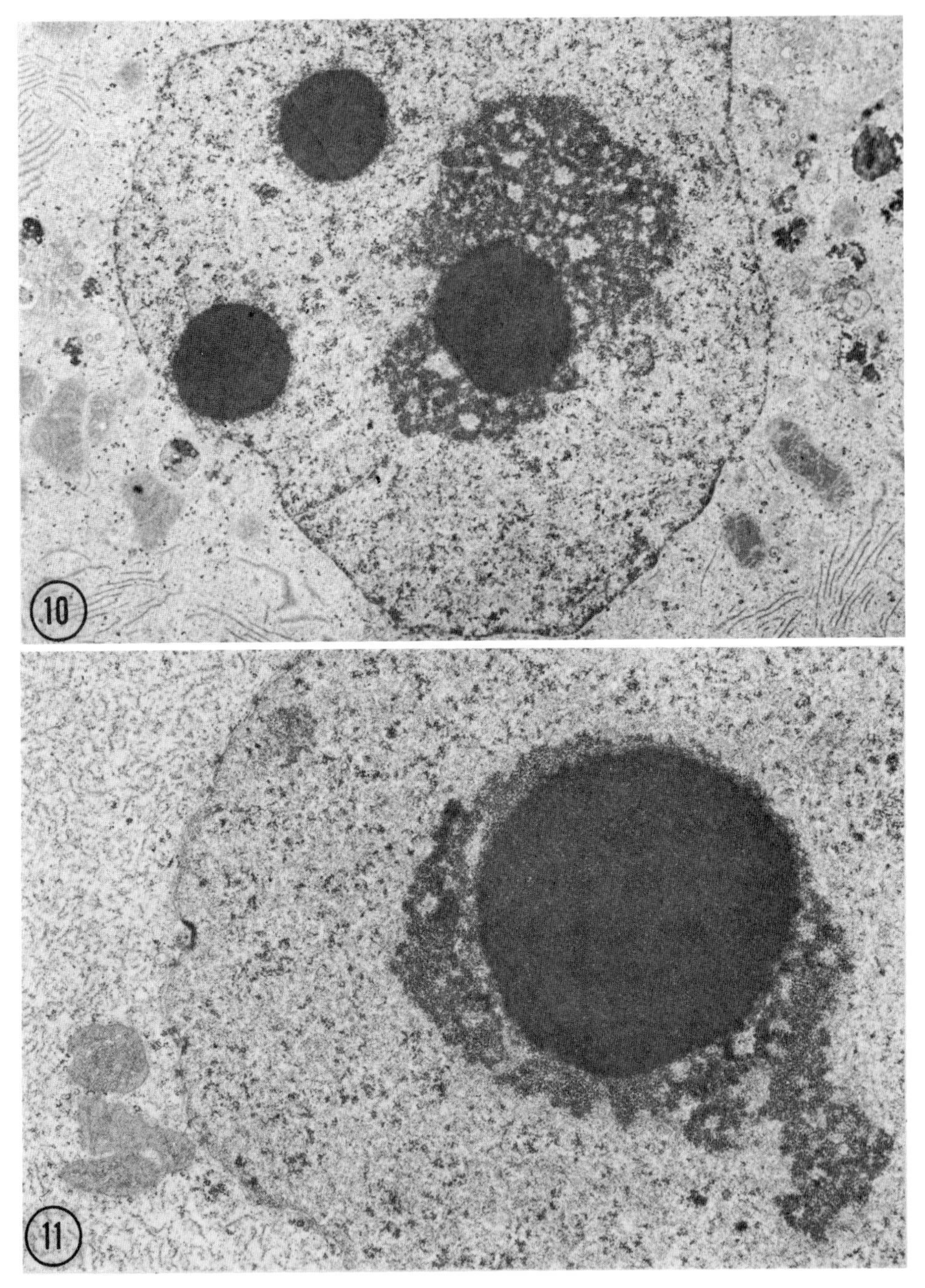

Fig. 10. The peripheral network developed around only one of the nucleoli at one time (8-cell rat conceptus). ×8,372.

Fig. 11. Among the meshes of the peripheral nucleolar network of a 16-cell rat conceptus many small granules appeared reminiscent of nucleoli of adult somatic cells involved actively in protein synthesis. ×12,285.

formed by groupings of thin filaments. Within this filamentous network some appear to condense and form small granules. With each cleavage the reticulated phase is more extensive. Appearance of granules is always preceded by a network formation.

Sequential events in cell activities are very difficult to assess by electron microscopy. In all the stages studied from 2-cell through the morula, usually only one nucleolus of the several per nucleus showed the morphological transformation at a given time. From these studies it cannot be determined whether each nucleolus will be transformed within a cell cycle. No adequate morphological nucleolar marker exists to resolve this question even if eggs are recovered and fixed at different time intervals. Since several nucleoli appear simultaneously at various locations in nuclei of blastomeres (or in pronuclei) and since each is associated with chromatin material, it can be assumed that:

1. Different chromosome segments and different cistrons are involved in their formation or;

2. The nucleolar cistron is duplicated and each nucleolus is associated with a copy of the same cistron (this latter case has been described in amphibian oocytes [Callan 1966]).

Autoradiographic experiments and biochemical determinations have demonstrated incorporation of ^{3}H-uridine into nucleoli possessing a peripheral network and granules in the 4- and 8-cell stages (Mintz 1964; Hillman and Tasca 1968). Most biochemical studies have disclosed gradually increasing levels of incorporation in the 4-cell, 8-cell, and morula stages with a rapid increase thereafter (Monesi and Salfi 1967; Ellem and Gwatkin 1968). Recently a low level of ^{3}H-uridine incorporation into a high molecular weight RNA has been seen as early as the 2-cell stage (Woodland and Graham 1969). Label was found in the 4S and rRNA fractions at the 4- and 8-cell stages of the conceptus whose syntheses were apparently activated in the same cell cycle.

Concomitant with the nucleolar morphological changes is an increase in the number of cytoplasmic polysomes. At the 4-cell stage such an increase is barely noticeable, but it is very distinct at the 8- and 16-cell stages (Maraldi and Monesi 1968). The majority of polysomes at first seem to be free in the cytoplasmic matrix, and cisternae of the rough endoplasmic reticulum are rarely found. But, starting with the 8-cell stage, membrane-associated polysomal complexes are more frequent; the outer component of the nuclear envelope is the first to become ribosome studded. The second membrane component which acquires polysomes is a small membrane saccule associated with nearly every mitochondrion (fig. 12) (Calarco and Brown 1969). These complexes of mitochondria and membrane saccules are first detectable at the time of ovulation when devoid of ribosomes and are present throughout the first cleavage stages. They might be remnants of more extensive mitochondria-ER complexes prominent during oogenesis (Szollosi 1969).

Recent studies on embryonic development of *Ascaris lumbricoides* (Kaulenas, Foor, and Fairbairn 1969) have demonstrated similar correlation between the appearance of a nucleolus with nucleolonema and rRNA synthesis at the 4-cell stage. Although the majority of newly synthesized RNA at this stage was distinctly heterogeneous, some minor components suggested that some rRNA has been synthesized and its synthesis continues at a higher rate up to blastulation. Nucleoli

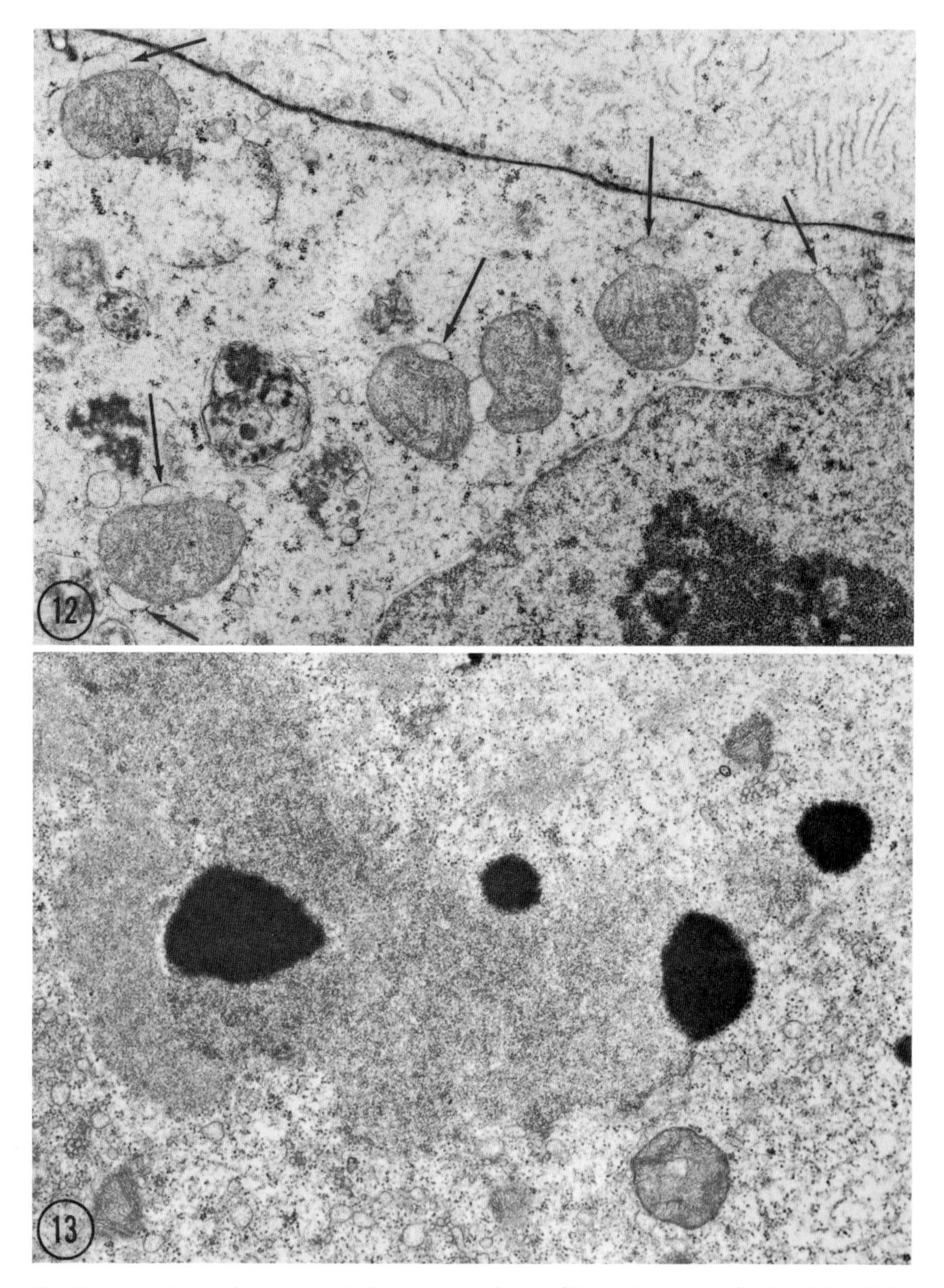

Fig. 12. A small membrane saccule became associated with nearly every mitochondrion. Polysomes were attached to them at the 8- or 16-cell stage. ×17,290.

Fig. 13. Compact filamentous nucleoli have appeared at various points in association with chromatin masses before reconstruction of the nuclear envelope (rat, second cleavage). ×15,015.

already appear in pronuclei similar to mammals, with poorly developed nucleolonema, and a brief rRNA synthesis apparently is limited exclusively to the male pronucleus. It is unclear, however, from the published material whether the granular nucleolar component is present in nucleoli at these early stages or whether it appears only when active rRNA synthesis is initiated at the 4-cell stage or later.

It is not known how ribosomal precursor particles are transferred from the nucleus to the cytoplasm. The passage of nucleolar fragments or other visible structures across nuclear pores or the nuclear envelope has not been observed. Cell division apparently plays an important role for communication between nucleus and cytoplasm in the rat egg. The nucleolus breaks down by prometaphase of the cell cycle and no evidence has been obtained in either rat or mouse cleavage stages for the retention of nucleolonema or other nucleolar components on the chromosomes (Estable and Sotelo 1955). Small filamentous nucleoli have been found associated with segments of some anaphase chromosomes at the second cleavage (fig. 13) and at several of the subsequent cleavages (Szollosi 1966, 1968). Along the chromosomes toward the spindle pole, short fragments of the reconstituting nuclear envelope have been detected also. Several of the small, filamentous nucleoli apparently lose their contact with the chromosomes and can be seen in the vicinity of the forming nucleus (fig. 14). Infrequently, similar structures have been observed even along the cell membrane. Structurally the extruded nucleoli are identical to the forming filamentous nucleolus in interphase of the same cell (fig. 15). Close scrutiny of blastomeres in division or shortly after cleavage has disclosed that several small nucleoli are set free during the second, third, fourth, and fifth cleavages. These extruded nucleoli are not surrounded by membranes, since they are formed and released prior to reconstruction of the nuclear envelope. After having reached the cytoplasm they undergo peripheral granular transformations similar to that displayed by nucleoli inside nuclei, but no peripheral network (nucleolonema) formation or reticulation is seen (figs. 16, 17). The process might be a rearrangement or condensation of coiled filaments similar to the formation of nucleolus-associated granules (Bernhard 1966). The granules display an initial higher concentration in certain well-defined cytoplasmic regions. Only secondarily do the granules attain uniform dispersion. Although the most likely interpretation is that the granules might be ribosomes, no positive identification is available. The individual granules demonstrate a filamentous substructure and, in sectioned material, lack the characteristic bipartite nature described for negatively stained isolated ribosomes from a variety of sources (Huxley and Zubay 1960).

Granule complexes of similar dimensions have been encountered also in a variety of invertebrate eggs. Several such complexes are scattered throughout the cytoplasm of the jellyfish egg, *Aequorea aequorea* (fig. 18) and eggs of the polychaete worm, *Armandia brevis.* These are present at the time of ovulation and their appearance has not been connected to either meiotic or mitotic divisions. In pronuclear stages of sea urchin eggs the "heavy bodies" (a granule complex surrounded by annulated membrane segments) possibly may be analogized to the granule complexes encountered in rat eggs.

In young rat and mouse blastocysts several filamentous nucleoli form among the chromatin material in early telophase (fig. 19). These must fuse to form one or two

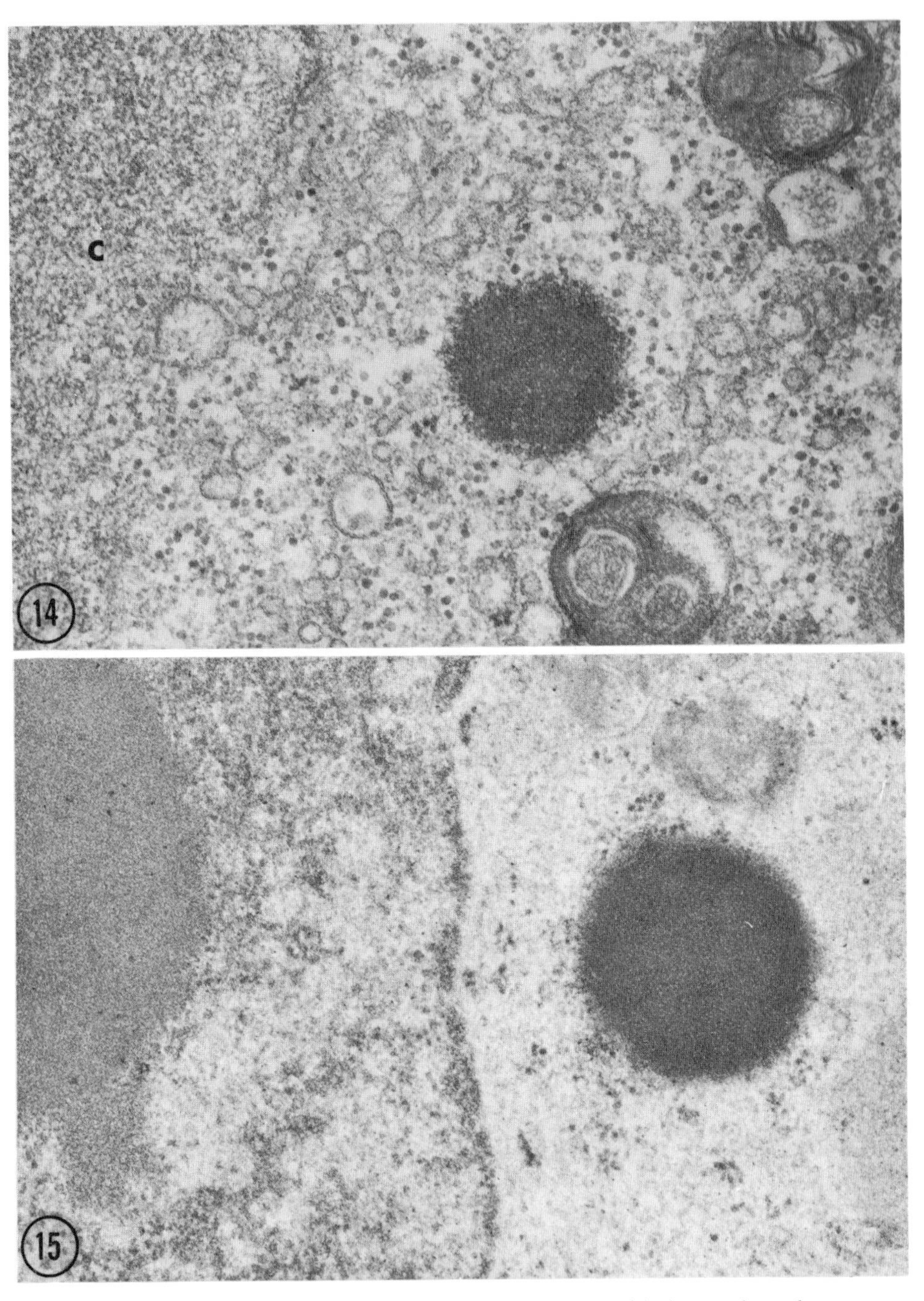

Fig. 14. Small filamentous nucleoli apparently lose their contact with the anaphase chromosomes (*C*) and are found freely in the cytoplasm (rat, second cleavage). ×25,207.

Fig. 15. "Nucleoli" in the cytoplasm are morphologically similar to that found in the nucleus. At the periphery of both structures small granules have appeared (16-cell rat conceptus). ×51,233.

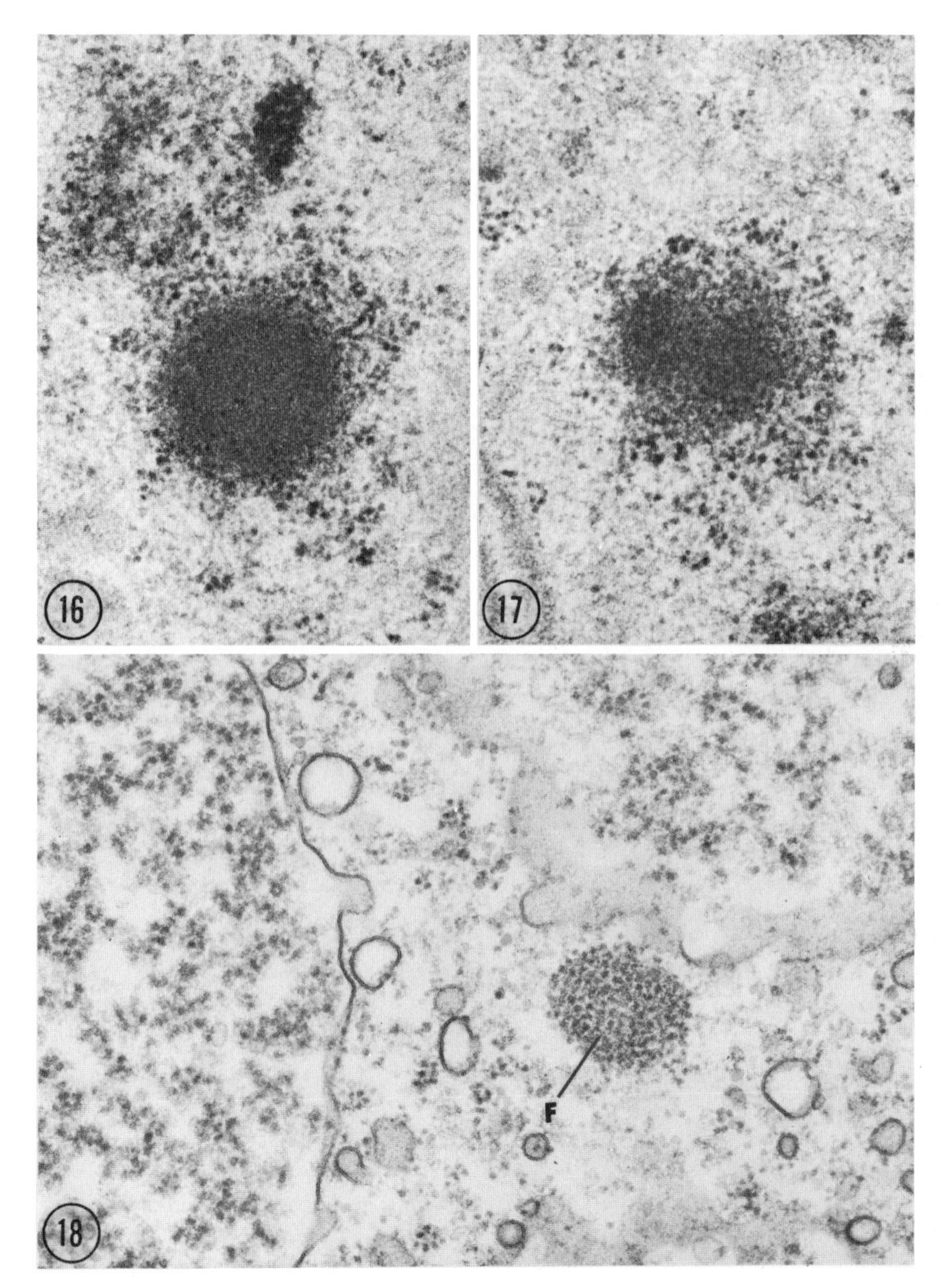

Figs. 16, 17. In the cytoplasm filamentous "nucleoli" transform peripherally into granular elements. ×54,145.

Fig. 18. Granule clusters can be detected in the cytoplasm of a jellyfish egg (*Aequorea aequorea*) which appear to have a filamentous substructure (*F*). ×50,414.

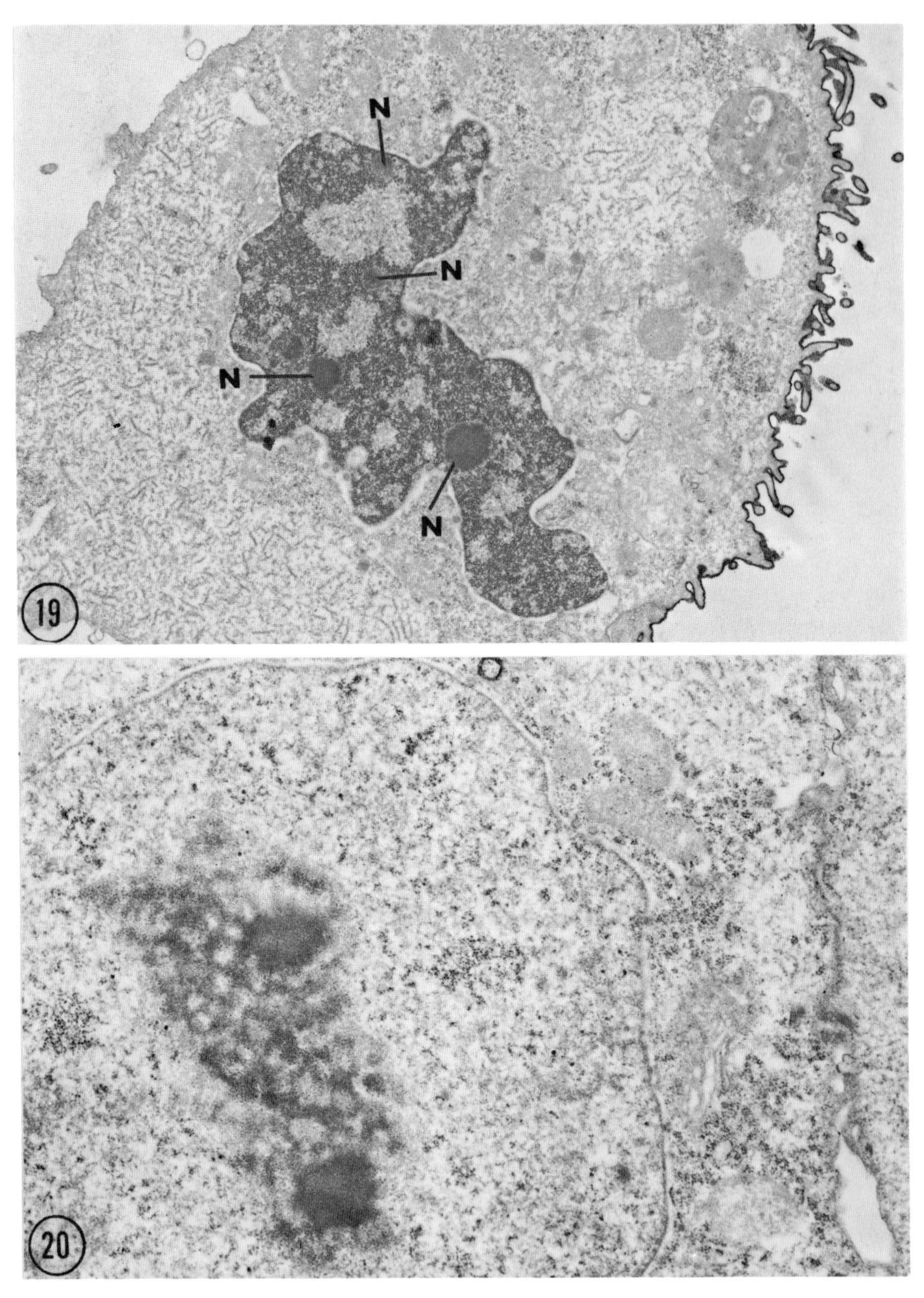

Fig. 19. Several small, compacted, filamentous nucleoli (*N*) develop among the chromatin substance of an early interphase cell of a rat blastocyst. ×9,555.

Fig. 20. Within a large reticulated nucleolus two filamentous structures can be found which may represent a partial nucleolar fusion (rat blastocyst). ×15,470.

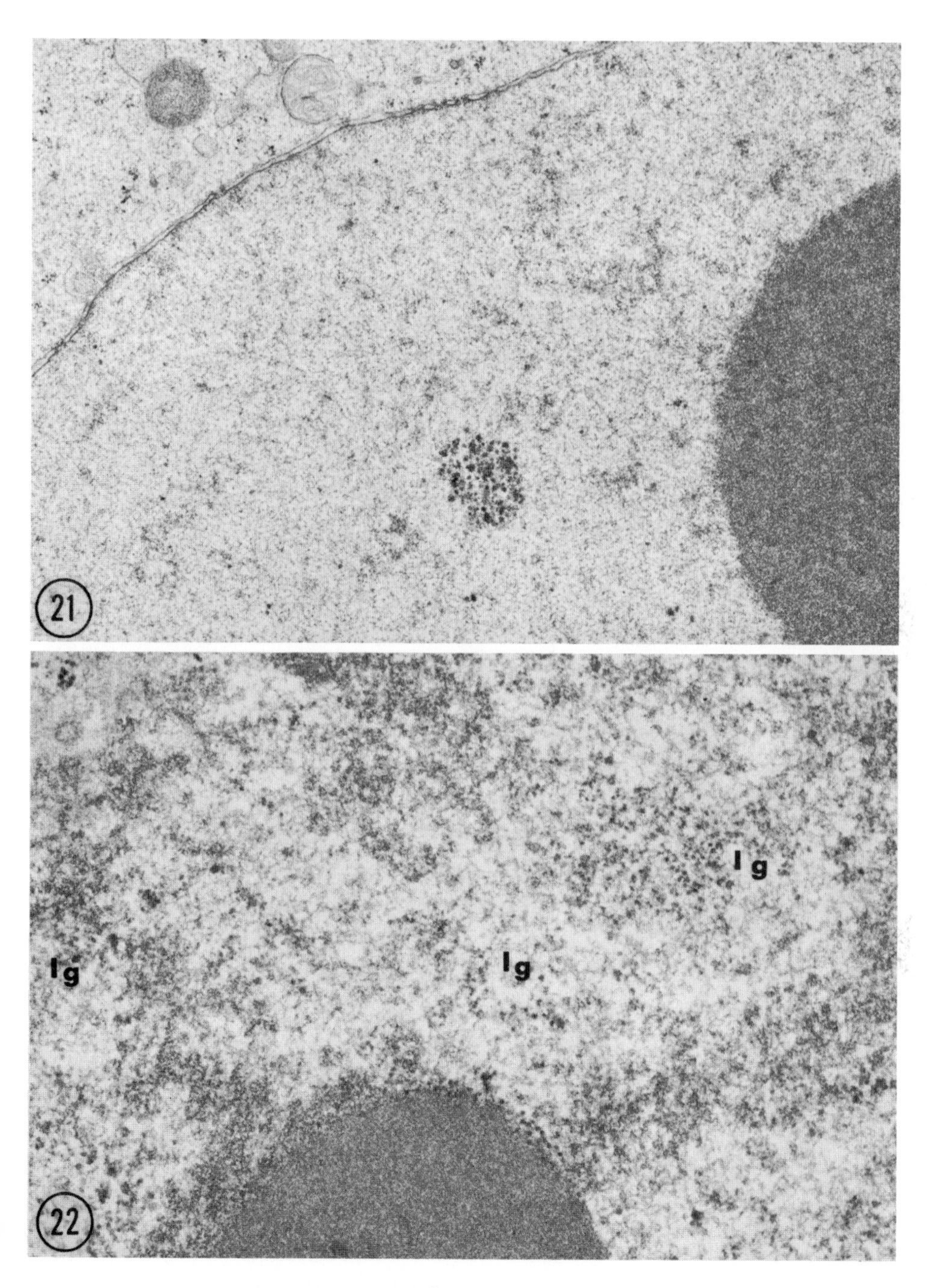

Fig. 21. Granules ranging from 200 to 500 Å can be found in the nucleoloplasm of pronuclei of a mouse egg. They are not evidently related to the nucleolus or to chromatin. ×20,020.

Fig. 22. Several interchromatin granule clusters (*IG*) are found in nuclei of a mouse blastocyst. ×32,305.

nucleoli later on as the chromatin expands. Frequently two filamentous core structures can be observed within a large reticulated nucleolus (fig. 20) (Calarco and Brown 1968). These possibly represent a partial nucleolar fusion. Superficially similar morphological transformations take place in every cell of the blastocyst, whether it belongs to the embryonic cell mass or to the trophoblast layer.

Frequently granules of irregular size range are found in nuclei. In pronuclei their diameter varies between 200 and 500 Å (fig. 21). In nuclei of blastomeres prior to the blastocyst stage they are slightly more uniform in dimension, measuring between 200 and 300 Å (fig. 22). The granules are not associated with chromosomal material and therefore the descriptive term "interchromatin granule" is

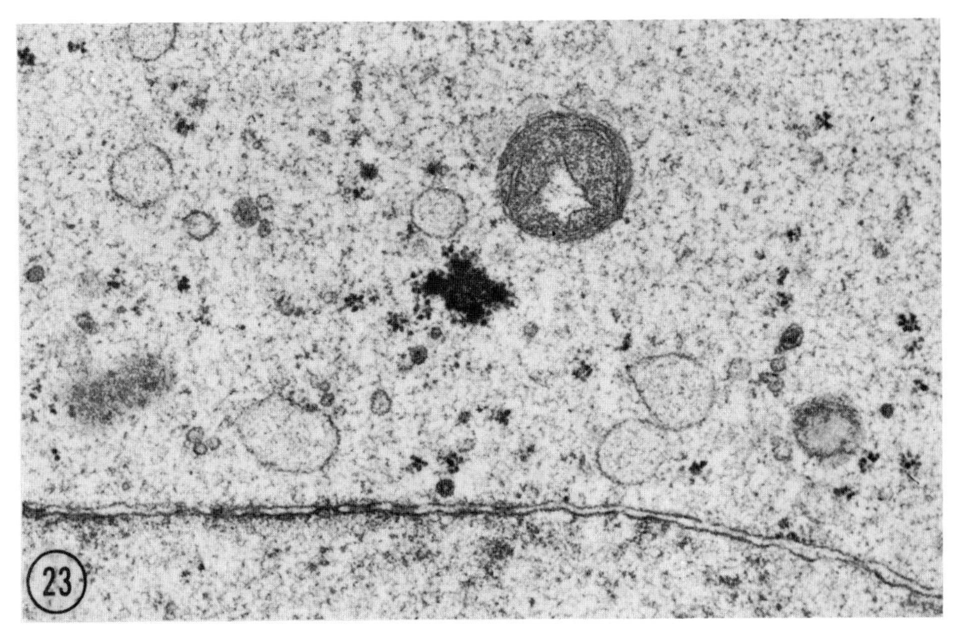

Fig. 23. Shortly after development of pronuclei in mouse eggs, granule clusters are found in the cytoplasm similar to those also found inside pronuclei. ×32,305.

applicable (Bernhard and Granboulan 1963). The term, as originally defined in the literature, seems to fit the granules well in cleavage nuclei, where several clusters of granules appear per section. The clusters are irregularly distributed between the chromatin strands. Occasionally a larger granule can be found among the more uniform granules, surrounded by a large, clear space. The larger granule population might be likened to the "perichromatin granules" (Swift 1962; Watson 1962). Neither the perichromatin nor the interchromatin granules are well characterized as yet and the similarity might be purely superficial. The irregular granules found in pronuclei do not fit either category. In earlier cytochemical analyses made at the electron microscope level of resolution (Watson 1962) it was suggested that the granule may be constituted of deoxyribonucleoproteins. It is true that cold perchloric acid or RN-ase extraction does not eliminate the granules; nor is their staining intensity diminished significantly with uranyl salts or with indium following similar extraction procedures. Even so the possibility cannot be excluded that the granules are ribonucleoprotein in nature but are somehow protected from extraction. No close morphological relationship exists between the granules and the nucleoli. It has been suggested that they may represent nucleolar RNA migrating to the cytoplasm (Smetana, Steel, and Bush 1963). Granule complexes occasionally encountered in the cytoplasm support this assertion (fig. 23).

Acknowledgments

This work was supported by United States Public Health Service grants HD 03752 and HE 02698 from the National Institutes of Health.

References

Alfert, M. 1950. A cytochemical study of oogenesis and cleavage in the mouse. *J Cell Comp Physiol* 36:381.

Anderson, E. 1968. Oocyte differentiation in the sea urchin, *Arbacia punctulata,* with particular reference to the origin of cortical granules and their participation in the cortical reaction. *J Cell Biol* 37:540.

Austin, C. R. 1961. *The mammalian egg.* Oxford: Blackwell Scientific Publications.

Austin, C. R., and Braden, A. W. H. 1953. The distribution of nucleic acids in rat eggs in fertilization and early segmentation: I. Studies on living eggs by ultraviolet microscopy. *Aust J Biol Sci* 6:324.

Bernhard, W. 1966. Ultrastructural aspects of the normal and pathological nucleolus in mammalian cells. *Nat Cancer Inst Monogr* 23:13.

Bernhard, W., and Granboulan, N. 1963. The fine structure of the cancer cell nucleus. *Exp Cell Res* 9:19.

———. 1968. Electron microscopy of the nucleolus in vertebrate cells. In *The nucleus,* ed. A. J. Dalton and F. Haguenau, vol. 3 of *Ultrastructure in biological systems.* New York: Academic Press.

Braden, A. W. H., and Austin, C. R. 1953. The distribution of nucleic acids in rat eggs in fertilization and early segmentation: II. Histochemical studies. *Aust J Biol Sci* 6:665.

Brown, D. D., and Gurdon, J. B. 1964. Absence of ribosomal RNA synthesis in the anucleolate mutant of *Xenopus laevis. Proc Natl Acad Sci USA* 51:139.

———. 1966. Size distribution and stablility of DNA-like RNA synthesized during development of anucleolate embryos of *Xenopus laevis. J Molec Biol* 19:399.

Calarco, P. G. and Brown, E. H. 1968. Cytological and ultrastructural comparison of t^{12}/t^{12} and normal mouse morulae. *J Exp Zool* 168:169.

———. 1969. An ultrastructural and cytological study of preimplantation development of the mouse. *J Exp Zool* 171:253.

Callan, H. G. 1966. Chromosomes and nucleoli of the axolotl, *Ambystoma mexicanum. J Cell Sci* 1:85.

Ellem, K. A. O., and Gwatkin, R. B. L. 1968. Patterns of nucleic acid synthesis in the early mouse embryo. *Develop Biol* 18:311.

Estable, C., and Sotelo, J. R. 1955. The behaviour of the nucleolonema during mitosis. In *Symposium on fine structure of cells.* VIII Int Cong Cell Biol (Leiden), p. 170. The Netherlands: Noordhoff, Groningen.

Flax, M. H. 1953. Ribose nucleic acid and protein during oogenesis and early embryonic development in the mouse. Ph.D. diss., Columbia University (quoted from Mintz 1964).

Hay, E. D. 1968. Structure and function of the nucleolus in developing eggs. In *The nucleus,* ed. A. J. Dalton and F. Haguenau, vol. 3 of *Ultrastructure in biological systems.* New York: Academic Press.

Hillman, N. W., and Tasca, R. J. 1968. Ultrastructural and biochemical studies of normal and mutant (t^{12}/t^{12}) mouse embryos. *Proc XII Int Cong Genetics* (Tokyo) abstract 7.5.1.

Huxley, H. E., and Zubay, G. 1960. Electron microscope observations on the structure of microsomal particles from *Escherichia coli. J Molec Biol* 2:10.

Karnovsky, M. J. 1965. A formaldehyde-glutaraldehyde fixative of high osmolarity for use in electron microscopy. *J Cell Biol* 27:137A.

Kaulenas, M. S.; Foor, W. E.; and Fairbairn, D. 1969. Ribosomal RNA synthesis during cleavage of *Ascaris lumbricoides* eggs. *Science* 163:1201.

Maraldi, N. M., and Monesi, V. 1968. Electron microscopic observations on the development of preimplantation mouse embryo. In *IV Europ conf electron microscopy,* ed. D. S. Bocciarelli, 2:321. Tipographia Poliglotta Vaticana.

Miller, O. L., Jr. 1966. Structure and composition of peripheral nucleoli of salamander oocytes. *Nat Cancer Inst Monograph* 23:53.

Mintz, B. 1964. Synthetic processes and early development in the mammalian egg. *J Exp Zool* 157:85.

Monesi, V., and Salfi, V. 1967. Macromolecular syntheses during early development in the mouse embryo. *Exp Cell Res* 46:632.

Odor, D. L., and Renninger, D. F. 1960. Polar body formation in the rat oocyte as observed with the electron microscope. *Anat Record* 137:13.

Perry, R. P. 1965. The nucleus and the synthesis of ribosomes. *Nat Cancer Inst Monogr* 18:325.

Schlafke, S., and Enders, A. C. 1967. Cytological changes during cleavage and blastocyst formation in the rat. *J Anat* 102:13.

Smetana, K., Steel, W. J.; and Bush, H. 1963. A nuclear ribonucleoprotein network. *Exp Cell Res* 31:198.

Swift, H. 1962. Nucleoprotein localization in electron micrographs: Metal binding and radioautography. In *Interpretation of ultrastructure,* ed. R. J. Harris, 1:213. New York: Academic Press.

Szollosi, D. 1965. Extrusion of nucleoli from pronuclei of the rat. *J Cell Biol* 25: 545.

———. 1966. Nucleolar transformation and ribosome development during embryonic development of the rat. *J Cell Biol* 31:115A.

———. 1967. Fixation procedures of embryonal tissues for electron microscopy. In *Methods in developmental biology,* ed. F. H. Wilt and N. K. Wessell. New York: Thomas Y. Crowell Company.

———. 1968. Nucleolar extrusion: A prerequisite for the formation of cytoplasmic polysomes in the rat conceptus. In *Ann Embryol Morph,* VI Int Cong Embryol (Paris).

———. 1969. Mitochondria-rough endoplasmic reticulum complexes in maturing oocytes and spermatocytes. *J Cell Biol* 43:143A.

Tyler, A. 1967. Masked messenger RNA and cytoplasmic DNA in relation to protein synthesis and processes of fertilization and determination in embryonic development. *Develop Biol* Suppl 1:170.

Watson, M. L. 1962. Observations on a granule associated with chromatin in the nuclei of cells of rat and mouse. *J Cell Biol* 13:162.

Woodland, H. R., and Graham, C. F. 1969. RNA synthesis during early development of the mouse. *Nature* (*London*) 221:327.

Zamboni, L., and Mastroianni, L., Jr. 1966*a*. Electron microscopic studies on rabbit ova: I. The follicular oocyte. *J Ultrastruct Res* 14:95.

———. 1966*b*. Electron microscopic studies on rabbit ova: II. The penetrated tubal ovum. *J Ultrastruct Res* 14:118.

Zamboni, L., Mishell, D. R., Jr.; Bell, J. H.; and Baca, M. 1966. Fine structure of the human ovum in the pronuclear stage. *J Cell Biol* 30:579.

7

Loss of Zona Pellucida and Prolonged Gestation in Delayed Implantation in Mice

Ruth E. Rumery/Richard J. Blandau
Department of Biological Structure
School of Medicine
University of Washington, Seattle

It is well established that if female mice are mated during the postpartum estrus and are allowed to suckle a specific number of young, implantation of the blastocysts is delayed and gestation prolonged (Enzmann, Saphir, and Pincus 1932; Brambell 1937; Krehbiel 1941; Bruce and East 1956).

In recent years emphasis has centered on experimentally delaying nidation by ovariectomizing or hypophysectomizing mated mice and subsequently treating them with progesterone and estrogen (Bloch 1958; Yoshinaga and Adams 1966; Brumley and De Feo 1964; Psychoyos 1966; Bindon and Lamond 1969). Even though such studies have contributed to the understanding of the interrelationships between blastocysts and the endometrium, there remains a hiatus in information as to when delayed blastocysts lose their zonae pellucidae and how long they remain free within the cornual lumina before implanting.

In normal mice implantation occurs between the 4th and 5th days *post coitum.* Blastocysts, free of zonae, may be recovered a number of hours prior to any decidual reactions (Finn and McLaren 1967; Orsini and McLaren 1967). In a preliminary report Rumery and Blandau (1966) stated that blastocysts from lactating mice retained their zonae at least 24 hr longer than did those from normal mice, and implantation was not completed until near the end of the 11th day after mating. Using the same approach McLaren (1967) confirmed these observations, noting that blastocysts did not shed their zonae until the second half of the 5th day. No endometrial reaction was observed even as late at the 8th day unless the litter size was reduced considerably.

The present investigation was undertaken to determine when blastocysts, under conditions of delayed implantation, lose their zonae and how long they remain within the cornual lumina before implanting in lactating mice nursing a standard number of 6 young. Data were obtained also on the time of parturition of these mice if gestation was not interrupted.

I. Materials and Methods

Virgin female mice of Swiss-Webster strain were placed with males and examined each morning for a copulation plug. When a plug was found, the female was considered to be in day zero of gestation. For this study the mated females were divided into control and experimental groups, and observations on the stage of development and condition of their ova were made in the following manner. Ova from the control group were examined during both the A.M. and the P.M. of days 3 through 5 of gestation. The mice in the experimental groups delivered their young and were remated during the postpartum estrus. Each female was allowed to retain and suckle just 6 of her offspring. Animals in the experimental group were killed each day from the 3d through the 15th day after mating and their ova recovered by the technique described below. Sixty-three mice were allowed to complete their second parturition; an accurate record was kept of the interval between the finding of the copulation plug and delivery.

All mice were killed by decapitation. Each cornu was removed to a separate Maximow slide where the ova were flushed out with Hanks's balanced salt solution through a 5 ml syringe fitted with a 25 gauge needle. The cornu was held in a vertical position over the well of the Maximow slide by grasping the mesosalpinx at its proximal end. The beveled tip of the hypodermic needle was inserted into the oviductal lumen and held in place with a fine forceps. Approximately 0.5 ml of Hanks's solution was flushed gently through the cornual lumen. To be certain that no ova were caught within the folds of the mucosa the organ was carefully stretched by grasping the mesosalpinx of the distal end, again with a fine forceps. The cornu was then flushed once more with 0.5 ml of Hanks's solution.

Ova usually were still in the oviducts in females killed on the 3d day following mating. In these females the tip of a fine glass pipette was inserted into the lumen of the fimbriated end of each oviduct and Hanks's solution flushed through by means of a simple device described earlier (Blandau 1961). Since mouse ova are fragile and easily ruptured, it is important that only gentle pressure be applied during the flushing process.

Although ova in the oviductal and cornual washings could be identified readily and counted under the dissecting microscope, they were examined at higher power with the phase microscope for more detailed observations. For this examination the ova were placed on a microscope slide in the center of a partially open Vaseline ring. When sufficient Hanks's solution was added to nearly fill the area, a round 18 mm coverglass was placed on the Vaseline ring and the small gap sealed with a drop of mineral oil.

Ova from each female were classified as blastocysts with zonae, blastocysts free of zonae, implanted blastocysts, or unfertilized ova. Phase photomicrography provided a permanent record of the appearance of the living ova. If the total number of blastocysts recovered from an animal was less than 6, both cornua were fixed in Bouin's fluid, sectioned serially, and stained with hematoxylin and eosin in order to locate any ova which might have been trapped in the endometrial crypts. The experimental females that delivered their litters were killed as soon as parturition was completed, and their cornua examined for resorption sites. All offspring were counted, weighed, and sexed.

II. Observations

A total number of 327 mice were used for the data reported here. From this number, 259 females were in the two delayed groups and 68 mice served as controls.

TABLE 1

OVA FROM CONTROL MICE

DAYS AFTER MATING	NUMBER OF MICE	NUMBER OF BLASTOCYSTS		UNFERTILIZED OVA	NUMBER OF IMPLANTED OVA	TOTAL NUMBER OF OVA
		With Z.P.	Without Z.P.			
3 days A.M.	6 ♀	18 morulae, 43 blast.	0	1	0	62
P.M.	9 ♀	4 morulae, 97 blast.	0	2	0	103
4 days A.M.	15 ♀	2 (1 ♀)	128	8	0	138
P.M.	12 ♀	0	105	0	0	105
	4 ♀	0	30	0	0	35 } 61
Eve	2 ♀	0	5	0	14	***26*** } ***61***
	1 ♀	0	0	0	12	
5 days A.M.	12 ♀	0	0	0	139	***139***
P.M.	3 ♀	0	0	0	28	***28***
Eve	4 ♀	0	0	0	48	***48***

NOTE: Italicized boldface numbers denote total number of implantation sites for each designated time interval.

A. *Ova from Control Mice*

All data relative to ova from control mice are shown in table 1 and figure 1.

The majority of ova recovered from the cornua of control mice killed on the morning of the 3d day after mating were either in the early blastocyst stage (43) or still in the morula stage (18). All ova appeared intact with distinct perivitelline spaces and no evidence of fractured zonae pellucidae (fig. 2).

By the afternoon of the 3d day blastocysts had begun to expand, usually obliterating the perivitelline spaces (fig. 3). The expanding blastocysts tended to thin out the zonal membranes. Even though most of the ova at this stage had developed into blastocysts, 4 were still in the morula stage. There was considerable variation in size and shape of the blastocysts recovered on the morning of the 4th day (figs. 4–6). The majority of them had elongated and appeared somewhat expanded. Some blastocysts had become 2 to 3 times larger than those seen on the previous afternoon; only 2 still possessed intact zonae (fig. 4). Although nearly all the blastocysts had shed their zonae pellucidae, no remnants of these were found in the washings even after a careful search. No gross decidual responses were observed in the cornua. Although the unfertilized ova were fragmenting, all had retained their zonae (fig. 7).

Little change was noted in the blastocysts recovered during the afternoon of the 4th day. Most of them had assumed an elongate and elliptical shape and all zonae had been shed. By this time the blastocysts had become somewhat sticky and

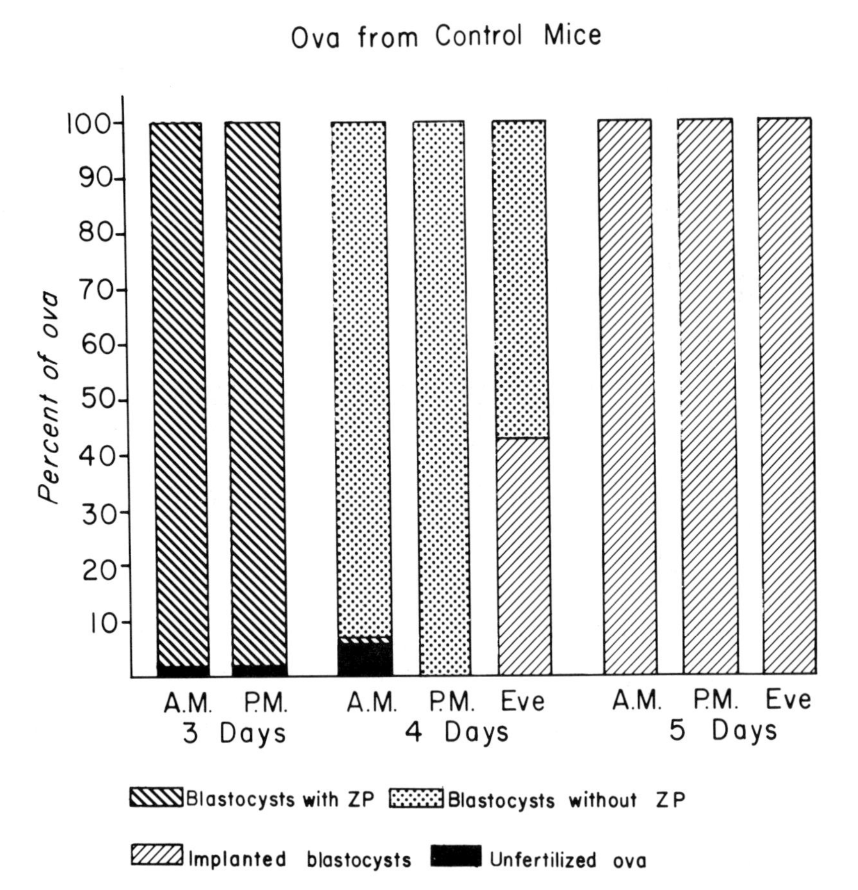

Fig. 1. Ova from control mice

tended to adhere to the glass surfaces causing some difficulty in their manipulation. No decidual responses were noted, nor were there any implantations.

During the evening of the 4th day, blastocysts from 3 females had begun to implant, even though 5 free, normal appearing blastocysts were flushed from the cornua of 2 of these mice, possibly from the decidual crypts. Cornua from the remaining females examined at this interval showed no gross decidual reactions and their blastocysts resembled those seen a few hours earlier.

No free blastocysts were recovered at any time on the 5th day. All ova either were in the process of implanting or had completed nidation.

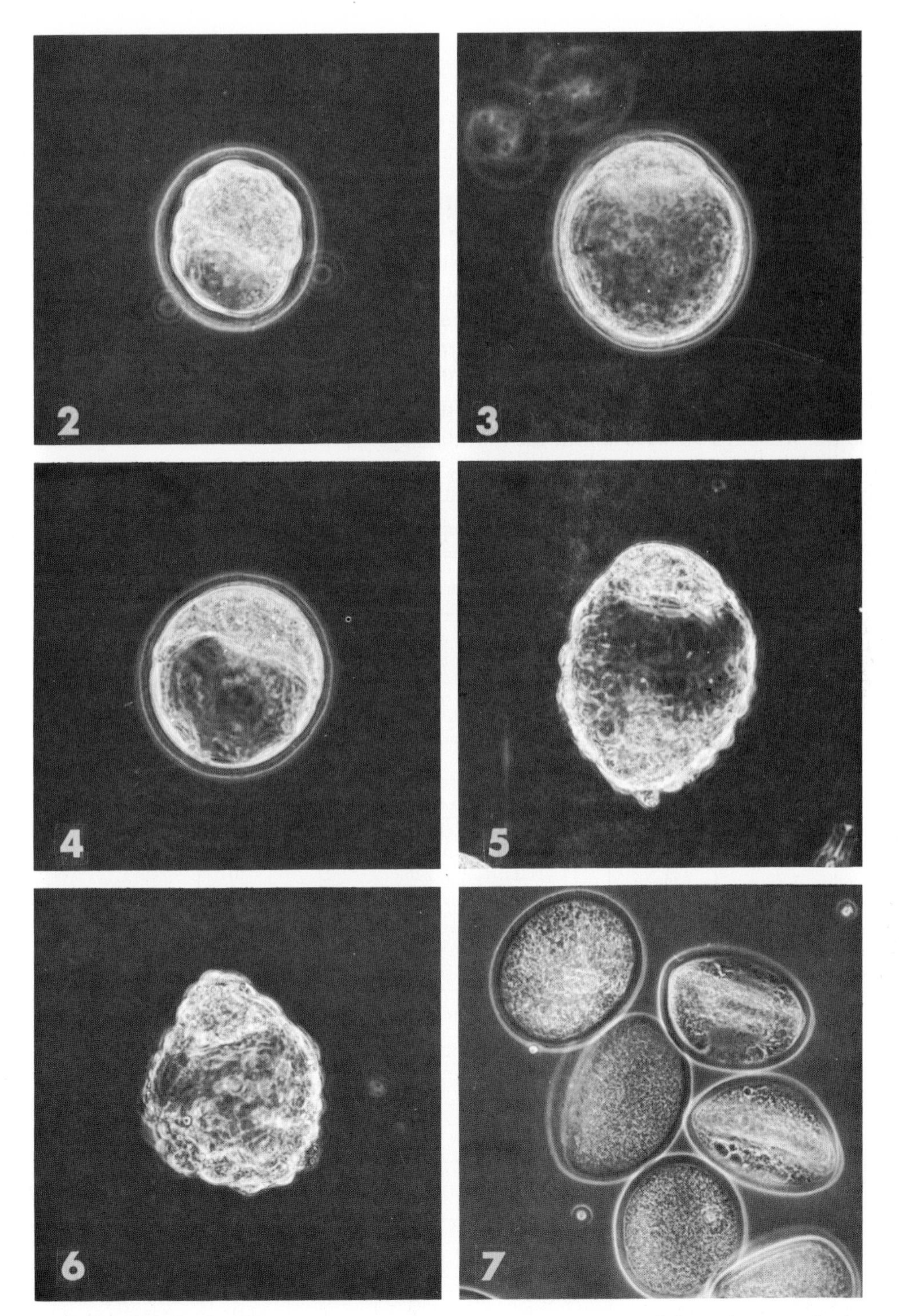

Figs. 2–7. All of these blastocysts and unfertilized ova were recovered from the control mice. All preparations were photographed in the living state. Dark medium-contrast phase objectives were used.

(2) Blastocyst recovered on the morning of the 3d day *post coitum*. Note the presence of the perivitelline space. ×270.

(3) Expanded blastocyst with thin zonal membrane as it appeared on the afternoon of the 3d day. ×270.

(4–6) Appearance of blastocysts on the morning of the 4th day. Blastocyst in fig. 4 still has its zona, while those in figs. 5 and 6 have shed their zonae. Already there is a significant variation in size and shape of blastocysts recovered at the same time interval. ×270.

(7) Unfertilized ova in cornual washings on the morning of the 4th day. Zonae are intact although the ova are degenerating. ×300.

B. *Ova from Experimental Mice*

Data on the delayed blastocysts are summarized in table 2 and in figure 8. Observations made on these blastocysts will be described under the following headings: days 3 through 6; and days 7 through 15.

1. *Days 3 through 6.* This is the period during which the blastocysts expanded and elongated, became more ovoid in shape and lost their zonae pellucidae. During this interval they became quite sticky and somewhat more difficult to handle.

All blastocysts recovered on the morning of the 3d day following mating were round in shape and of similar size, and all were enclosed in their zonae pellucidae. The perivitelline space was still visible in all ova. Of the 80 fertilized ova recovered in the washings 53 were still in the morula stage.

TABLE 2

OVA FROM DELAYED MICE

Days after Mating	Number of Mice	Number of Blastocysts		Unfertilized Ova	Implanted Ova	Total Number of Ova
		With Z.P.	Without Z.P.			
3 days	7 ♀	{53 morulae 27 blast.	0	7	0	87
4 days	19 ♀	163	31	47	0	241
5 days	11 ♀	20	107	0	0	127
6 days	26 ♀	4 (3 ♀)	261	20	0	285
7 days	{24 ♀ 1 ♀	1 (1 ♀) 0	226 0	22 0	0 14	249 ***14*** } 263
8 days	{16 ♀ 2 ♀ 5 ♀	0 0 0	146 3 0	8 0 0	0 20 71	157 ***91*** } 248
9 days	{8 ♀ 7 ♀	0 0	68 0	5 0	0 91	73 ***91*** } 164
10 days	{8 ♀ 8 ♀	0 0	84 0	1 0	0 98	85 ***98*** } 183
11 days	{2 ♀ 5 ♀	0 0	19 0	0 0	0 61	19 ***61*** } 80
12 days	{3 ♀ 6 ♀	0 0	22 0	4 0	0 84	26 ***84*** } 110
13 days	{1 ♀ 11 ♀	0 0	8 0	0 0	0 148	8 ***148*** } 156
14 days	{1 ♀ 15 ♀	0 0	0 0	2 0	9 202	2 ***211*** } 213
15 days	10 ♀	0	0	0	145	***145***

NOTE: Italicized boldface numbers denote total number of implantation sites for each daily interval.

Blastocysts recovered on the 4th day showed definite changes in both size and shape. Most of the remaining zonae had become thinned out and several were fractured (figs. 9–11). Sixteen percent of all ova had lost their zonae and one blastocyst was in the process of shedding its membrane.

Considerable variation in both size and shape of the blastocysts was evident on the 5th day (figs. 12–14). Although a few of them were round, the majority had expanded and appeared elliptical, and 84% had shed their zonae completely. Any zonae still intact appeared much thinner and occasionally wrinkled. Only 1 empty zona was found in the cornual washings.

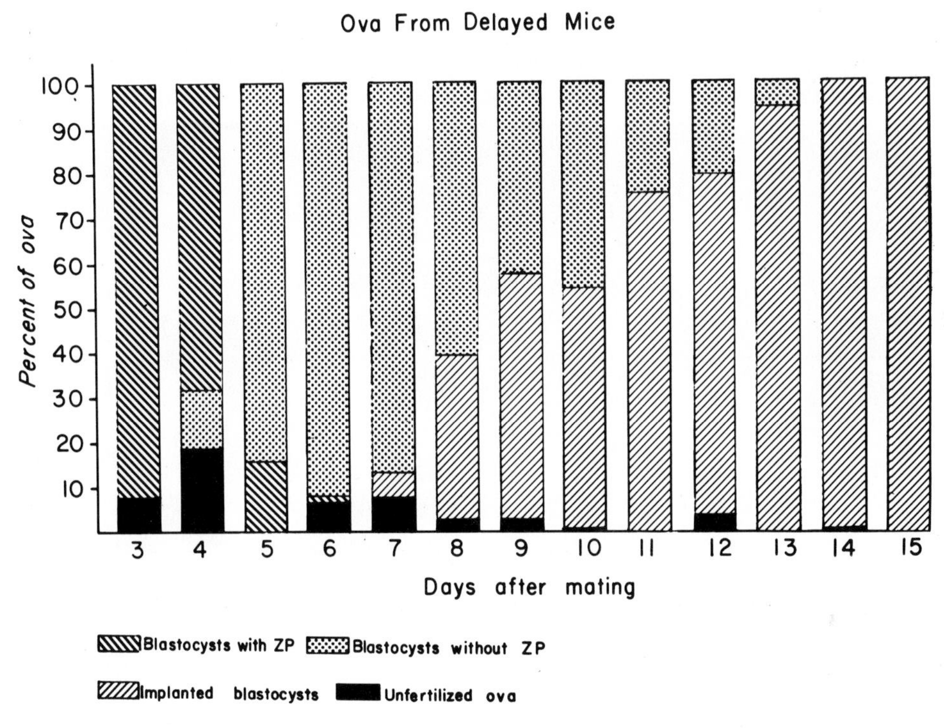

Fig. 8

On the 6th day zonae were still intact in only 4 of 265 blastocysts examined. These zonae were thin, granular in apparance, and so depolymerized that they could be removed from the ova quite readily. One empty zona was recovered (fig. 15). The appearance of these zonae-free blastocysts resembled those seen on the previous day (figs. 16 and 17). These, too, became increasingly sticky. There was no macroscopic evidence of decidual responses.

2. *Days 7 through 15.* There were 133 animals in this group. During this period almost all the blastocysts were implanting. The first gross decidual reactions were observed in the cornua of 1 female on the 7th day. During the next 6 days the proportion of implanting to free blastocysts gradually increased from 37% on the 8th day to 95% on the 13th day. The appearance of all ova recovered was the same as

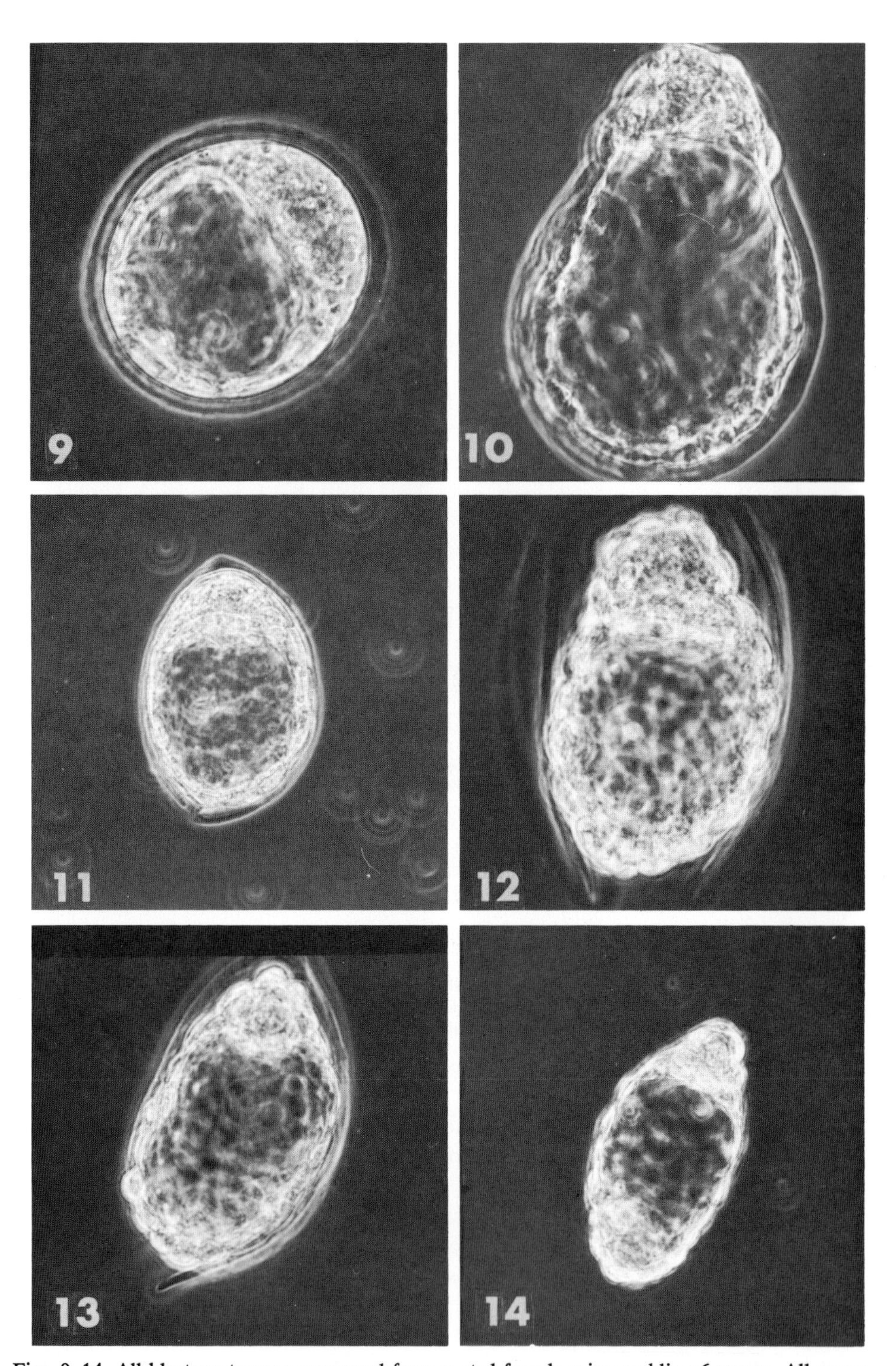

Figs. 9–14. All blastocysts were recovered from mated female mice suckling 6 young. All preparations were photographed in the living state. Dark medium-contrast phase objectives were used.

(9–11) Delayed blastocysts recovered from the cornual washings on the 4th day. All have intact zonal membranes. There is considerable variation in size and shape of these three blastocysts. In fig. 11 note the thinned-out zonal membrane which is already fractured. ×432, fig. 9; ×270, figs. 10 and 11.

(12, 13) The appearance of blastocysts at 5 days, still in the zonae, which are in the process of being shed. ×270 and ×540.

(14) This 5-day blastocyst had shed its zona and has attained the typical elliptical shape. ×270.

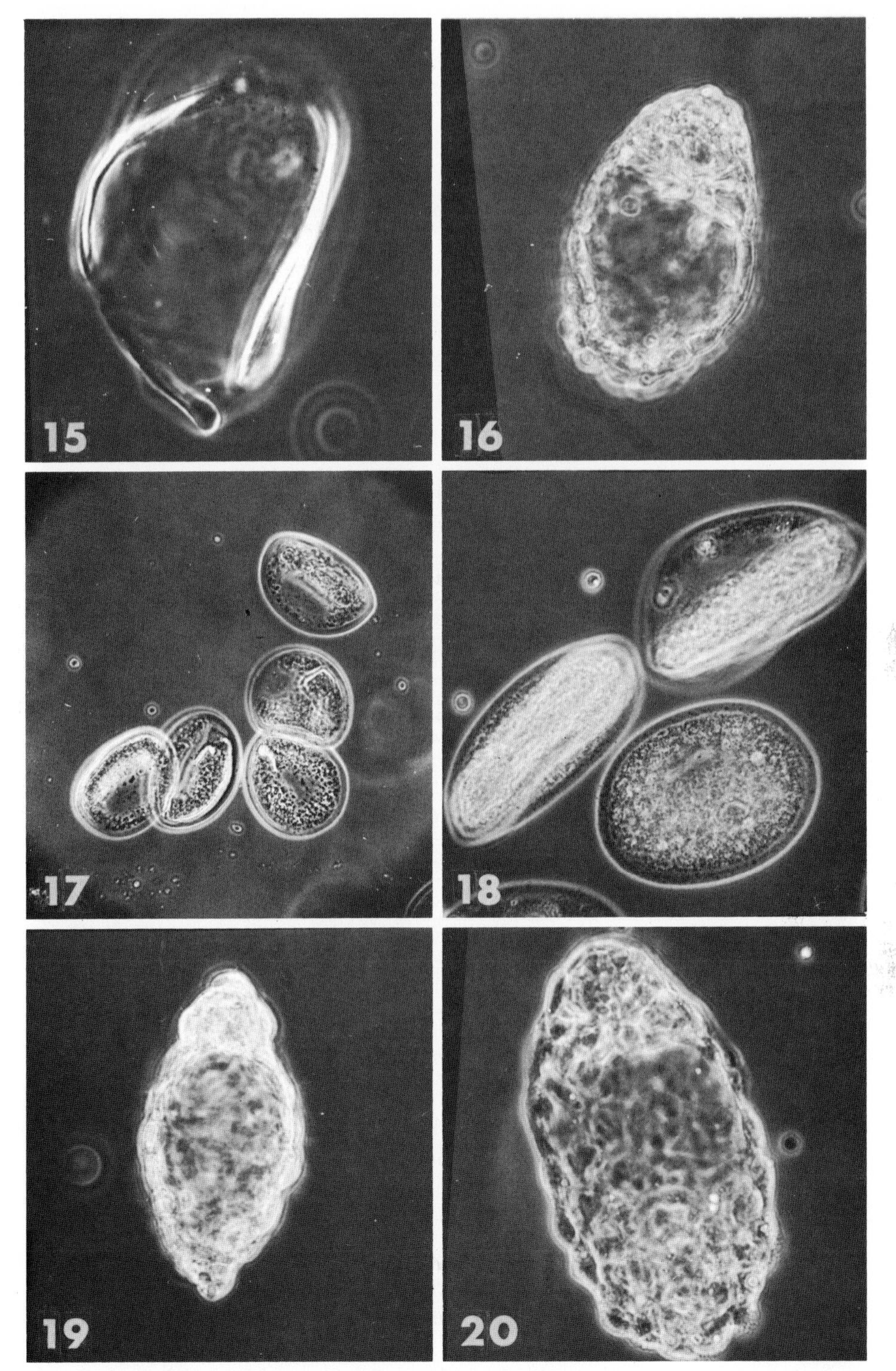

Figs. 15–20. Delayed blastocysts, one zona, and unfertilized ova recovered from the experimental mice. All preparations were photographed in the living state. Dark medium-contrast phase objectives were used.

(15) One of the rare empty zonae found in the cornual washings. This zona was recovered from a pregnant animal killed on the 6th day. ×432.

(16) Appearance of blastocyst, zona free, recovered on the 6th day. ×540.

(17) Unfertilized ova found also in the washings on the 6th day. All of the zonae are intact. ×173.

(18) Unfertilized ova recovered on the 9th day. As emphasized previously the ova have undergone cytolysis but the zonal membranes are of normal thickness. ×320.

(19) A small cigar-shaped free blastocyst recovered on the 9th day. ×540.

(20) This blastocyst was recovered on the 13th day, the last day any free blastocysts were found in the washings. This blastocyst is considerably expanded in comparison with the one in fig. 19. ×540.

that seen on the 7th day (figs 19 and 20). All cornua examined on the 15th day had well-developed implantation sites.

C. *Unfertilized Ova*

Unfertilized ova were found in the cornual washings of both control and experimental mice. After the morning of the 4th day no unfertilized ova were recovered from the controls. However they were present in the experimental females until the 15th day of gestation, with the greater number recovered during the first 7 days. Their appearance was always the same regardless of their length of stay within the cornua; they were granular and fragmenting, but the zonae were always intact (figs. 7, 17, and 18).

TABLE 3

GESTATION PERIODS AND LITTER SIZE OF MICE WITH DELAYED BLASTOCYSTS

Number of Mice That Delivered	Gestation Period (Days)	Average Litter Size
10 ♀	20 and 21	10
7 ♀	22	9
9 ♀	23	10
9 ♀	24	10
9 ♀	25	11
7 ♀	26	9
6 ♀	27	10
2 ♀	28	9
3 ♀	29	8
1 ♀	30	6

D. *Fixed and Stained Cornua from Experimental Mice*

At microscopic examination of the cornua which were fixed and sectioned serially very few ova were found. Uteri from 19 mice were examined, and only 7 ova were seen in the endometrial folds of 5 females.

E. *Mice Having Prolonged Gestations*

Mice with prolonged gestation periods, together with the average size of their litters, are shown in table 3. The 10 females which gave birth on the 20th and 21st days were not considered delayed in parturition since 5% of our stock mice delivered on the 20th day.

In 41 animals gestation varied from 22 to 26 days, while in 12 more females parturition was extended until the 27th to the 30th day. Neither the average litter size of 9 nor the average body weight of the young (1.69 gms) was significantly different from those of stock animals; the sex ratio of these newborn was normal. All of the young were free of any congenital abnormalities.

Resorptions were noted infrequently except in cornua from mice that delivered on the 27th to the 30th days. All of these animals had 1 or 2 resorptions except for 2 females with 3 and 4 resorptions. Most resorptions measured 3–4 mm; when sectioned, however, none of them contained any definitive embryonic tissue.

III. Discussion

The results of this investigation have shown that blastocysts from pregnant lactating mice with 6 sucklings retained their zonae pellucidae 24 to 48 hr longer than normal. After these blastocysts were freed from their zonae, they remained in the cornual lumina for varying intervals, implanting gradually over a 6-day period. Considerable variation in the length of gestation was observed also. In the majority of the experimental females parturition occurred at the 22d through the 27th days, but in 6 of the 63 mice delivery did not take place until the 30th day. This extended gestation was in sharp contrast to that of our stock mice, which for the most part deliver on the 19th day. Even though gestation was prolonged, the average litter size, body weight, and sex ratio of the newborn were not abnormal. An interesting feature of early development in mice and rats is the variablility in the growth rate of embryos within the same cornu. It appears that as gestation progresses, the smaller embryos are able to approximate the size of the larger ones so that at parturition there is no significant difference in body weight.

From the data presented here it is evident that the suckling stimulus provided by the 6 young was sufficient to delay gestation from 3 to 11 days beyond the usual 19-day period. The importance of litter size in affecting the extent of delay has been emphasized repeatedly. Enzmann, Saphir, and Pincus (1932) reported that each suckling delayed the ensuing parturition by 21 hr. In our experience it is very doubtful that the time of parturition can be quantitated so precisely. Brambell (1937) noted that gestation was not prolonged in female mice unless they retained at least 3 sucklings. Under these conditions the development of the corpora lutea appeared normal. In the rat, Krehbiel (1941) found that 6 sucklings delayed implantation for varying periods of time, whereas Brumley and De Feo (1964) reported that nidation in rats could be delayed for 2 to 13 days if the female nursed 8 young. When a large number of mice were allowed to suckle their entire litters, usually comprising 9 or more young, many of the pregnancies were not prolonged, but the young were frequently either stillborn or, if alive at parturition, too weak to survive (Bruce and East 1956).

The reason for the delay in the shedding of the zonae pellucidae in the experimental group has not been determined; it may be related to a combination of factors. Loss of zonae may be regarded either as the result of the normal maturation process of the blastocyst (Dickmann and Noyes 1961; Noyes et al. 1963; Mayer 1963) or as affected by environmental factors controlled by hormones (Yasukawa and Meyer 1966; Alloiteau and Psychoyos 1966; Dickmann 1968). When Orsini and McLaren (1967) retained mouse blastocysts within ligated oviducts, loss of zonae was delayed. These experiments emphasize that the environment plays an important role both in the rate of development of the blastocysts and loss of their zonae.

The metabolism of delayed blastocysts has been shown to alter during their stay within the cornual lumina. Using ^{35}S methionine as an index of protein synthesis, Weitlauf and Greenwald (1965) reported that mouse blastocysts whose implantation was delayed by the suckling stimulus did not incorporate the radioactive material until the 9th day, whereas normal blastocysts showed intense reactivity shortly before implantation on the 5th day.

Although blastocysts can be maintained within the cornual lumina with progesterone, it has been demonstrated repeatedly that small amounts of estrogens are required to initiate implantation (Bloch 1958; Psychoyos 1961; Humphrey 1967). It has been suggested that failure of delayed blastocysts to implant within normal intervals is the result of estrogen secretion in the milk, leaving insufficient hormone available for the initiation of nidation. The size of the nursing litter then appears to be an important factor; the larger the litter, the stronger the suckling stimulus; the stronger the suckling stimulus, the more estrogen is diverted to the young (Nalbandov 1964). Whitten (1955) has given concrete evidence that the suckling stimulus affects estrogen secretion by reporting a significant weight loss of the maternal ovaries as well as an alteration of their histologic appearance.

Implantation of delayed blastocysts which began on the 7th day *post coitum* and continued during the following 6 days suggested that there may be a gradient in the nidation process. It would be interesting to know whether the time of implantation of each embryo is related to its size, its maturity, its position within the cornua, or to some other factor. In discussing endocrine control of endometrial sensitivity in the mouse during induction of the decidual cell reaction, Finn (1966) has said that before blastocysts can initiate this reaction, a second wave of estrogen secretion, which he calls the nidatory surge, is necessary. In more explicit terms Greenwald (1958) described an estrogen surge which occurs at the 11th day of gestation in the lactating mouse, initiated by a neurohormonal stimulus. Between 10 and 15 days *post partum* the levels of both prolactin and progesterone drop; additional gonadotrophins are released, leading to an accelerated secretion of estrogen.

In a recent study of the fine structure of uterine tissues from mice delayed in implantation Enders (1967) reported that the luminal epithelium is almost devoid of glycogen. Acid phosphatase activity was demonstrable in the glandular epithelium and at the apical ends of the luminal epithelium. Succinic dehydrogenase activity was prominent in the glandular epithelium also. Enders stated that delay in implantation may be due to a temporary inadequacy of the uterine cellular environment or possibly to an elaboration of an inhibitor.

Thyroid and adrenal hormones appear to be involved also in influencing implantation. Holland et al. (1967) reported that certain levels of hyperthyroidism are beneficial in maintaining delayed blastocysts if progesterone is administered in minimal amounts to pregnant animals. When ACTH is injected into hypophysectomized-ovariectomized rats, the adrenals are stimulated to elaborate progesterone (Lyons et al. 1953). As Nutting and Meyer (1963) have suggested, the adrenals may be capable of maintaining unimplanted blastocysts during conditions of stress.

The data which have been presented here describe the natural history of the blastocysts whose implantation is delayed by the suckling stimulus. As the preimplanting embryo begins to expand within the perivitelline space and presses upon the zona, there is no biophysical evidence of zonal depolymerization. When the zonae begin to fracture, there is significant variation in size and shape of the blastocysts.

We conclude that the "hatching" of the mouse blastocyst is related more to its expansion than to any change in the environment which alters the physical characteristics of the zona pellucida. There is as yet much to be learned about the environmental factors which determine the state of development of the blastocysts *in utero*. Since these blastocysts may remain free within the cornual lumina for extended periods, consideration must be given to the fluid environment that bathes them, the state of the endometrial cells that determine the environment, and finally to the hormonal milieu which affects all components within this area. There is a great need for a more precise evaluation of the interrelationships between the embryo and endometrium if the process of implantation is to be understood.

IV. Summary

Nidation of blastocysts was delayed in mice in order to determine when the blastocysts under these conditions lose their zonae pellucidae and how long they remain within the cornual lumina before implanting. Data were obtained also on the time of parturition of these mice if gestation was not interrupted, together with the stage of development, body weight, and sex ratio of their young.

During the postpartum estrus, females were remated and allowed to suckle just 6 of their young. The mice were killed from the 3d through the 15th day after mating and their ova examined under the phase microscope. From the cornual washings of 128 females, a total of 1,359 ova were recovered and 943 implantation sites were counted in the cornua of 68 more mice.

Zonae were retained by the delayed blastocysts for 24 to 48 hr longer than by the control blastocysts. Decidual reactions were first observed on the 7th day *post coitum*. The increment of implanting blastocysts increased gradually over a period of 6 days, until by the 14th day no free blastocysts appeared in the cornual washings.

Eighty-four percent of the 63 females that delivered their young had prolonged gestation periods varying from 22 to 30 days. The average litter size, average body weight, and sex ratio of the newborn were not significantly different from the young of stock mice.

Acknowledgments

The authors wish to acknowledge the professional photographic assistance of Mr. Roy Hayashi.

This work was supported by United States Public Health Service Grant HD 03464 from the National Institutes of Health.

References

Alloiteau, J.-J., and Psychoyos, A. 1966. Y a-t-il pour l'oeuf de Ratte deux façons de perdre sa zone pellucide? *C R Acad Sci [D] (Paris)* 262:1561.

Bindon, B. M., and Lamond, D. R. 1969. Effect of hypophysectomy on implantation in the mouse. *J Reprod Fertil* 18:43.

Blandau, R. J. 1961. Biology of eggs and implantation. In *Sex and internal secretions,* 3d ed., ed. W. C. Young, chap. 2, p. 798. Baltimore: Williams and Wilkins Co.

Bloch, S. 1958. Experimentelle Untersuchungen über die hormonalen Grundlagen der Implantation des Sängerkeimes. *Experientia* 14:447.

Brambell, F. W. R. 1937. The influence of lactation on the implantation of the mammalian embryo. *Amer J Obstet Gynec* 33:942.

Bruce, H. M., and East, J. 1956. Number and viability of young from pregnancies concurrent with lactation in the mouse. *J Endocr* 14:19.

Brumley, L. E., and De Feo, V. J. 1964. Quantitative studies on deciduoma formation and implantation in the lactating rat. *Endocrinology* 75:883.

Dickmann, Z. 1968. Does shedding of the zona pellucida by the rat blastocyst depend on stimulation by the ovarian hormones? *J Endocr* 40:393.

Dickmann, Z., and Noyes, R. W. 1961. The zona pellucida at the time of implantation. *Fertil Steril* 12:310.

Enders, A. C. 1967. The uterus in delayed implantation. In *Cellular biology of the uterus,* ed. R. M. Wynn, p. 151. New York: Appleton-Century-Crofts.

Enzmann, E. V.; Saphir, N. R.; and Pincus, G. 1932. Delayed pregnancy in mice. *Anat Rec* 54:325.

Finn, C. A. 1966. Endocrine control of endometrial sensitivity during the induction of the decidual cell reaction in the mouse. *J Endocr* 36:239.

Finn, C. A., and McLaren, A. 1967. A study of the early stages of implantation in mice. *J Reprod Fertil* 13:259.

Greenwald, G. S. 1958. A histological study of the reproductive tract of the lactating mouse. *J Endocr* 17:17.

Holland, J. P.; Dorsey, J. M.; Harris, N. N.; and Johnson, F. L. 1967. Effect of thyroid activity upon delayed implantation of blastocysts in the rat. *J Reprod Fertil* 14:81.

Humphrey, K. W. 1967. The induction of implantation in the mouse after ovariectomy. *Steroids* 10:591.

Krehbiel, R. H. 1941. The effects of lactation on the implantation of ova of a concurrent pregnancy in the rat. *Anat Rec* 81:43.

Lyons, W. R.; Li, C. H.; Johnson, R. E.; and Cole, R. D. 1953. Evidence for progestogen secretion by ACTH-stimulated adrenals. *Proc Soc Exp Biol Med* 84:356.

McLaren, A. 1967. Delayed loss of zona pellucida from blastocysts of suckling mice. *J Reprod Fertil* 14:159.

Mayer, G. 1963. Delayed nidation in rats: A method of exploring the mechanisms of ova-implantation. In *Delayed implantation,* ed. A. C. Enders, p. 213. Chicago: University of Chicago Press.

Nalbandov, A. V. 1964. Role of hormones in pregnancy. In *Reproductive physiology,* 2d ed., p. 271. San Francisco: W. H. Freeman.
Noyes, R. W.; Dickmann, Z.; Doyle, L. L.; and Gates, G. H. 1963. Ovum transfers, synchronous and asynchronous, in the study of implantation. In *Delayed implantation,* ed. A. C. Enders, p. 197. Chicago: University of Chicago Press.
Nutting, E. F., and Meyer, R. K., 1963. Implantation delay, nidation, and embryonal survival in rats treated with ovarian hormones. In *Delayed implantation,* ed. A. C. Enders, p. 233. Chicago: University of Chicago Press.
Orsini, M. W., and McLaren, A. 1967. Loss of zona pellucida in mice, and the effect of tubal ligation and ovariectomy. *J Reprod Fertil* 13:485.
Psychoyos, A. 1961. Nouvelles recherches sur l'ovoimplantation. *C R Acad Sci* [*D*] (*Paris*) 252:2306.
———. 1966. Influence of oestrogen on the loss of the zona pellucida in the rat. *Nature* (*London*) 211:864.
Rumery, R. E., and Blandau, R. J. 1966. The loss of the zona pellucida in delayed implantation. *Anat Rec* 154:485 (abstract).
Weitlauf, H. M., and Greenwald, G. S. 1965. A comparison of ^{35}S methionine incorporation by the blastocysts of normal and delayed implanting mice. *J Reprod Fertil* 10:203.
Whitten, W. K. 1955. Endocrine studies on delayed implantation in lactating mice. *J. Endocr* 13:1.
Yasukawa, J. J., and Meyer, R. K. 1966. Effect of progesterone and oestrone on the pre-implantation and implantation stages of embryo development in the rat. *J Reprod Fertil* 11:245.
Yoshinaga, K., and Adams, C. E. 1966. Delayed implantation in the spayed progesterone treated adult mouse. *J Reprod Fertil* 12:593.

8

The Chemistry of Complex Carbohydrates Related to Blastocyst Composition

Roger W. Jeanloz

Department of Biological Chemistry
Harvard University Medical School
Laboratory for Carbohydrate Research
Massachusetts General Hospital, Boston

For many years components of carbohydrate nature have been detected in the outer layers of eggs, for example in the jelly coat of the sea urchin (Vasseur and Immers 1949) and in the zona pellucida of rabbit eggs (Harter 1948). A vast variety of sugar components, including glucose, galactose, mannose, fucose, xylose, fructose, glucosamine, and galactosamine has been identified in addition to sulfate groups and a linkage of the carbohydrate component to a protein core (see reviews in Austin 1961; Dickmann 1965; Monroy 1965; and Metz and Monroy 1967). The carbohydrate complexes have been characterized as glycoprotein, mucoprotein, and mucopolysaccharide. In view of the different interpretations given to these names in the literature (Jeanloz 1960; Gottschalk 1966) this classification is of little value to the reader not familiar with this biochemical field.

Recent developments in the chemistry of complex polysaccharides, such as the elucidation of the carbohydrate sequence in the vicinity of the linkage between the protein core and the acidic glycosaminoglycan chains (mucopolysaccharides) and progress in the structure elucidation of the carbohydrate moiety of glycoproteins and mucins (Jeanloz 1963, 1969; Brimacombe and Webber 1964; Gottschalk 1966) allow a better interpretation of the results published in the past. In the following lines the present nomenclature and our knowledge of the chemical structure of complex polysaccharides of mammalian origin will be very briefly reviewed, and the results of past work on the zona pellucida will be discussed.

I. Structure of Complex Polysaccharides of Higher Organisms

High-molecular, complex carbohydrate compounds are distributed throughout all living organisms with the exception of the simplest forms of viruses. The great number of different chemical structures found for complex carbohydrates isolated from plants, bacteria, and lower animal organisms renders any systematic classification a very arduous task. In compounds isolated from higher organisms, however, the number of structures is more limited, and enough information has been obtained to form a more coherent picture.

II. Proteoglycans

Under this recently introduced term are classified the complex molecules containing one protein core covalently linked to numerous long, straight polysaccharide chains. These carbohydrate moieties, such as chondroitin sulfate and heparin, when separated from the protein core, were formerly known as mucopolysaccharides or acidic

TABLE 1

Acid Glycosaminoglycans (Mucopolysaccharides)

Compound	Repeating Unit	Protein Carbohydrate Linkage Region
Hyaluronic acid	(1→3)-2-Acetamido-2-deoxy-β-D-glucopyranosyl-(1→4)-β-D-glucopyranuronyl	Unknown
Chondroitin 4-sulfate	(1→3)-2-Acetamido-2-deoxy-β-D-galactopyranosyl 4-sulfate-(1→4)-β-D-glucopyranuronyl	(1→3)-β-D-Galactopyranosyl-(1→3)-β-D-galactopyranosyl-(1→4)-β-D-xylopyranosyl-(1→*O*)-L-serine
Chondroitin 6-sulfate	(1→3)-2-Acetamido-2-deoxy-β-D-galactopyranosyl 6-sulfate-(1→4)-β-D-glucopyranuronyl	Same as in chondroitin 4-sulfate
Dermatan sulfate	(1→3)-2-Acetamido-2-deoxy-β-D-galactopyranosyl 4-sulfate-(1→4)-α-L-idopyranuronyl	Unknown
Heparin	(1→4)-2-Sulfoamino-2-deoxy-α-D-glucopyranosyl 6-sulfate-(1→4)-α-D-glucopyranuronyl	Same as in chondroitin 4-sulfate
N-Acetylheparan sulfate (Heparitin sulfate)	(1→4)-2-Acetamido-2-deoxy-α-D-glucopyranosyl 6-sulfate-(1→4)-α-D-glucopyranuronyl	Same as in chondroitin 4-sulfate

mucopolysaccharides. Because of the similarity of the chemical structure of these various carbohydrate chains, it is convenient to collect all these compounds into the same class. It should be noted that, to the present time, no strong evidence has been presented for a linkage of hyaluronic acid chains with a protein core and, if it is present, its nature is totally unknown; nor has the linkage of dermatan sulfate with protein been completely elucidated as yet. Table 1 summarizes our present knowledge of the carbohydrate components of connective tissues, heparins, and other derivatives of the heparin family, although these are chemically quite different and may play a completely different biological role.

The structure of these compounds is rendered more complex by microheterogeneity, both in the nature of the components of the same carbohydrate chains and in the types of the carbohydrate chain attached to the same protein core. For example, both L-iduronic acid and D-glucuronic acid units are present in the chains of dermatan sulfate and heparin, but L-iduronic acid units are very preponderant in the dermatan sulfate chain whereas D-glucuronic acid units are very preponderant in the heparin chain. On the other hand complex molecules containing both chondroitin sulfate and keratan sulfate (a complex carbohydrate chain which will be discussed later) chains attached to the same protein core have been isolated. Finally in the heparin series both *N*-sulfated and *N*-acetyl derivatives of D-glucosamine are present in the same polysaccharide chain.

The characteristics of all the proteoglycan compounds are the straight-chain carbohydrate components. They contain both acidic uronic acid residues and basic *N*-substituted hexosamine residues in an alternating sequence, and sulfate groups (with the exception of hyaluronic acid). These chains have molecular weights ranging from about 10,000 for members of the heparin class to about 50,000 for chondroitin sulfates and much higher values for hyaluronic acid. The molecular weight of the total assembly, protein core and carbohydrate chains, reaches 500,000 to 1,000,000 for the chondroitin sulfates. Very little is known about the protein core except that it contains a high proportion of lyophilic amino acids and that it is degraded by proteolytic enzymes. With the exception of papain, which degrades the protein core in the vicinity of the linkage to the carbohydrate chains, the other proteolytic enzymes attack the chains in the regions where sequences of 10 to 20 amino acids are lacking a carbohydrate linkage.

The carbohydrate chain is attached to the protein core through a D-xylosyl-L-serine linkage, and two additional D-galactosyl residues are linked between the xylose and the first uronic acid residue. This very typical linkage has so far been found in all proteoglycans investigated. No similar specificity has been established for the peptide structure surrounding the protein-carbohydrate point of attachment. The linkage with L-serine is alkali-labile. The involvement of serine itself in the protein-carbohydrate linkage is not specific to proteoglycans since glycoproteins, especially mucins, having much shorter carbohydrate chains of very different composition, have also been found to possess serine residues linked to carbohydrate chains.

Keratan sulfate is generally included in the same class as the other proteoglycans. Chains of keratan sulfate are usually linked to the same protein core as chains of chondroitin sulfate. In the keratan sulfate molecule, however, the uronic acid component is absent, having been replaced by a D-galactose component, the D-glucosamine component is linked at C-4, and there is no strong evidence that the carbohydrate component forms a straight chain. Our knowledge of the structure of sulfated polysaccharides attached to protein is still fragmentary; as new sulfated polysaccharides are found in invertebrates, similar findings may be made in vertebrates.

III. Glycoproteins

This term covers a group of substances which are also formed by a protein core covalently linked to carbohydrate components. In this case, however, the carbohydrate components are of smaller molecular weight than those found in proteoglycans, and their structure does not possess the uniformity of the carbohydrate chain of proteoglycans and may be branched. Glycoproteins differ also from proteoglycans by the absence of uronic acid components and by a greater number of sugar components, since they are generally composed of 2 to 5 different monosaccharide units. Table 2 shows the various sugars identified in glycoproteins which include three hexoses, two hexosamines, one 6-deoxysugar, and sialic acid—a very acidic component which has also been identified in numerous glycolipids.

The molecular weight of glycoproteins ranges from about 10 to 15,000, as in beef ribonuclease B where only one carbohydrate chain is present, to more than 1,000,000 for the submaxillary mucins which contain more than 1,000 carbohydrate chains. There is a possibility, however, that these very high molecular weights are the result of aggregation between glycoprotein molecules of a smaller size. The ratio of carbohydrate component to protein component can be as low as a few percent, such as in ovalbumin, where only a short carbohydrate chain is present, to a proportion of 70–80% for the glycoproteins possessing A, B, or O blood-group specificity isolated from secretions. Even when they are isolated from the same source, for example blood plasma, the glycoproteins show as great a variation in their content of carbohydrates, from less than 1% to nearly 50%.

The great variations in size, structure, and number of carbohydrate components attached to the protein core should not mask the properties common to all glyco-

TABLE 2

MONOSACCHARIDE COMPONENTS OF MAMMALIAN GLYCOPROTEINS

Acid Component	Hexosamines	6-Deoxysugar	Hexoses
Sialic acid (*N*-Acetyl- and *N*-glycolylneuraminic acid)	2-Acetamido-2-deoxy-D-galactose 2-Acetamido-2-deoxy-D-glucose	L-Fucose	D-Galactose D-Glucose D-Mannose

proteins. The linkage with the protein core is mediated through an alkali-stable *N*-acetyl-D-glucosaminylasparagine linkage or through a glycosidic linkage with L-serine or L-threonine involving *N*-acetyl-D-galactosamine. With the exception of some mucins, where the chains are composed solely of *N*-acetyl-D-galactosamine and *N*-acetylneuraminic acid (sialic acid), most carbohydrate moieties include an inner carbohydrate core attached to the protein chain. The inner core is composed of *N*-acetyl-D-glucosamine and D-mannose units whereas the outer branches are generally composed of D-galactose, *N*-acetyl-D-glucosamine, L-fucose, and *N*-acetylneuraminic units, the last two components being always located at the outside extremity of the carbohydrate chain. Glycoprotein molecules containing the inner carbohydrate core only, or complete assemblies only, or both types attached to the same protein core are known.

Microheterogeneity is also present in glycoproteins, both in the nature of components and in the fine structure. For example, *N*-acetylneuraminic acid units may be replaced by *N*-glycolylneuraminic acid units. These units are generally glycosidically linked to the D-galactosyl residues which in turn are glycosidically linked to the *N*-acetyl-D-glucosaminyl residues. Various linkages of these two residues have been established, even in the same glycoprotein (table 3). In the study of the inner carbohydrate core it has been established recently that the *N*-acetyl-D-glucosamine unit which is linked to asparagine is itself linked to another *N*-acetyl-D-glucosamine unit; this residue is, in turn, linked to a D-mannose residue. Complete elucidation of these complex structures is, however, still lacking.

Although the presence of D-glucose units has been reported in many glycoproteins, definite identification of a structure containing this monosaccharide has been made only in collagen and basal membrane components where it is part of the D-glucosyl-D-galactosyl disaccharide attached to hydroxylysine.

Glycoproteins are generally degraded by proteolytic enzymes, but the presence of sialic acid at the extremity of the carbohydrate chain or that of numerous carbohydrate chains has a hindering effect. Since the proteolytic degradation of glycoproteins containing sialic acid (*N*-substituted neuraminic acid) releases carbohydrate material having acidic properties and containing hexosamines, it is not easy to differentiate this material from that released by the degradation of proteoglycans except by a careful chemical analysis of the components. This problem, additional to that resulting from the confusing nomenclature used in the past, renders any evaluation of earlier experiments difficult.

TABLE 3

PARTIAL LIST OF LINKAGES DETERMINED IN GLYCOPROTEINS

Linkage	Glycoprotein
O-*N*-Acetylneuraminosyl-(2→3)-D-galactose	α_1-Acid glycoprotein; fetuin
O-*N*-Acetylneuraminosyl-(2→4)-D-galactose	α_1-Acid glycoprotein
O-*N*-Acetylneuraminosyl-(2→6)-D-galactose	α_1-Acid glycoprotein
O-*N*-Acetylneuraminosyl-(2→6)-2-acetamido-2-deoxy-D-galactose	Submaxillary gland glycoprotein
O-β-D-Galactopyranosyl-(1→4)-2-acetamido-2-deoxy-D-glucose	α_1-Acid glycoprotein; blood-group A, B, H, Le^a substances; fetuin; mucin; ovomucoid
O-2-Acetamido-2-deoxy-β-D-glucopyranosyl-(1→4)-2-acetamido-2-deoxy-D-glucose	Ribonuclease B

NOTE: Numerous linkages found only in the blood-group active glycoproteins are not enumerated.

One of the fascinating aspects of glycoprotein chemistry is that structures very similar to or identical with those established for the carbohydrate moieties of glycoproteins have been also found for the carbohydrate moieties of the sphingoglycolipids. It has become evident in the studies of mammalian cells, especially of cancer cells, that the outer plasma membrane is covered by glycoprotein material. The presence of glycolipids at the surface of the cells has been similarly established for erythrocytes and for some cancer cells. Any future study of the structure of the "glycoprotein" or "mucopolysaccharide" coat of the blastocyst should take into consideration that the material released might have been attached to a lipid core instead of a protein core.

IV. Chemical Structure of the Outer Layer of the Blastocyst (Zona Pellucida)

In the following discussion no attempt is made to review exhaustively the past work on this subject; rather, effort is directed toward interpretation of some of the work in this field in the light of our present knowledge of proteoglycan and glycoprotein chemistry.

In one of the first biochemical studies of mammalian eggs Braden (1952) relied mostly on histochemical reagents and enzymic degradation for the characterization of the zona pellucida. The very low metachromasia observed indicated the probable absence of proteoglycans containing uronic acid and sulfate groups whereas a positive periodate-Schiff reaction was good evidence for the presence of sialic acid residues. Finally, the positive action of various proteolytic enzymes confirmed the probable presence of a glycoprotein coat. The ability of hyaluronidase to dissolve the zona pellucida (Srivastava, Adams, and Hartree 1965) is not well established, since some proteolytic activity was also present. It should be noted that hyaluronidase is active not only on hyaluronic acid but also on chondroitin 4- and 6-sulfate. The detection of ester-bound sulfate in the zona pellucida (Seshachar and Bagga 1963) seems to have been ascertained only by histochemical means which could easily be misinterpreted.

The very small size of the mammalian egg and the difficulty of collecting a sufficient quantity of eggs are without doubt the main reasons for our very limited knowledge of the chemical structure of components of the zona pellucida. Although histochemical reagents can be invaluable tools for studying the distribution of known, well-defined compounds, they may lead easily to incorrect interpretations when the chemical structure of the visualized components is unknown. Progress of microdissection techniques may allow in the near future the isolation of pure biological material. It is hoped that our present methods of investigation in which gas-liquid chromatography and mass spectroscopy are employed may be scaled down through the use of radioactive reagents enough to be applicable to these materials.

Eggs of lower animals are easily collected and are generally available in large quantities. For this reason the most comprehensive study of the chemistry of the outer layer of eggs is that performed on the jelly coat of the sea urchin egg by Vasseur and his associates. A polyfucose sulfate was isolated as one of the main components of this echinoderm egg (Vasseur 1952). It is of interest that an identical polysaccharide has been found recently (Katzman and Jeanloz 1969) in the connective tissue of another echinoderm, the sea cucumber *Thyone briareus.* However this polysaccharide, present in some algae, has not been found in any other animal phylum. The very limited observation showing that the same polysaccharide is present both in connective tissue and at the surface of the egg suggests that the zona pellucida may not contain any specific substance. If this is true, the zona pellucida might not be qualitatively different from the surface of other mammalian cells and would be characterized by the excess production of certain components. Elucidation of this intriguing question is of great importance for our understanding of the biochemistry of the reproductive cycle, and it should attract in the future the interest of biochemists specializing in the study of glycoproteins and proteoglycans.

References

Austin, C. R. 1961. *The mammalian egg.* Oxford: Blackwell Scientific Publications.

Braden, A. W. H. 1952. Properties of the membranes of rat and rabbit eggs. *Aust J Sci Res Ser B* 5:460.

Brimacombe, J. S., and Webber, J. M. 1964. *Mucopolysaccharides.* Amsterdam: Elsevier Publishing Co.

Dickmann, Z. 1965. Sperm penetration into and through the zona pellucida of the mammalian egg. In *The preimplantation stages of pregnancy,* ed. G. E. W. Wolstenholme and M. O'Connor, p. 169. Boston: Little, Brown & Co.

Gottschalk, A. 1966. Definition of glycoproteins and their delineation from other carbohydrate-protein complexes. In *Glycoproteins,* ed. A. Gottschalk, p. 20. Amsterdam: Elsevier Publishing Co.

Harter, B. T. 1948. Glycogen and carbohydrate-protein complexes in the ovary of the white rat during the oestrous cycle. *Anat Rec* 102:349.

Jeanloz, R. W. 1960. The nomenclature of mucopolysaccharides. *Arthritis Rheum* 3:233.

———. 1963. Mucopolysaccharides (acidic glycosaminoglycans). In *Comprehensive biochemistry,* ed. M. Florkin and E. H. Stotz, sect. 2, 5:262. Amsterdam: Elsevier Publishing Co.

———. 1970. Mucopolysaccharides of higher animals. In *The carbohydrates,* ed. W. Pigman and D. Horton, 2B:589. New York: Academic Press.

Katzman, R. L., and Jeanloz, R. W. 1969. Acid polysaccharides from invertebrate connective tissue: Phylogenetic aspects. *Science* 166:785.

Metz, C. B., and Monroy, A., eds. 1967. *Fertilization: Comparative morphology, biochemistry, immunology,* vol. 1. New York: Academic Press.

Monroy, A. 1965. *Chemistry and physiology of fertilization.* New York: Holt, Rinehart and Winston.

Seshachar, R. B., and Bagga, S. 1963. Cytochemistry of the oocyte of *Loris tardigratus lydekkerianus* (Cabr.) and *Maccaca mulatta mulatta* (Zimmermann). *J. Morphol* 113:119.

Srivastava, R. N.; Adams, C. E.; and Hartree, E. F. 1965. Enzymatic action of lipoglycoprotein preparations from sperm-acrosomes on rabbit ova. *Nature (London)* 200:498.

Vasseur, E. 1952. Periodate oxidation of the jelly coat substance of *Echinocardium cordatum. Acta Chem Scand* 6:376.

Vasseur, E., and Immers, J. 1949. Hexosamine in the sea urchin egg jelly coat: A misinterpretation of the method of Elson and Morgan. *Arkiv Kemi* 1:253.

9

Some Maternal Factors Affecting Physicochemical Properties of Blastocysts

E. S. E. Hafez

Departments of Gynecology and Physiology
Wayne State University School of Medicine
Detroit, Michigan

To save other investigators time I have summarized in table 1 the various methods used to study mammalian blastocysts *in vivo:* injections of macromolecules, antibodies, viruses, transplantation, and radiography.

TABLE 1

TECHNIQUES USED TO STUDY THE PHYSIOCHEMICAL CHARACTERISTICS OF THE UNIMPLANTED BLASTOCYST OF RABBITS AND RODENTS

Technique	Description	Objectives	Reference
	In Vivo Techniques		
Pontamine sky blue injection	Inject large dye molecules intravenously into female (colored bands appear at implantation sites in the uterus)	To visualize implantation sites before they are macroscopically evident and after loss of the zona pellucida	Psychoyos 1961
Macromolecule injection	Inject antibodies, viruses, or proteins intravenously into female	To estimate the permeability of the blastocyst by its absorption of macromolecules	Brambell 1954 Mayersbach 1958 Schechtman and Abraham 1958 Flamm et al. 1963 Zimmermann et al. 1963
Time-lapse cinematography	Photograph blastocyst cultures	To observe differentiation, pulsation, hatching, and changes in the zona pellucida	Cole and Paul 1965 Cole 1967 Mulnard 1967

TABLE 1—*Continued*

Technique	Description	Objectives	Reference
		In Vivo Techniques—*Continued*	
Sexing of live blastocysts	Identify sex chromatin in a few trophoblastic cells removed for fluorescence microscopy	To estimate sex ratios *in utero*	Rowson and Moor 1966 Edwards and Gardner 1967
Over- or undercrowding of blastocysts *in utero*	(*a*) transfer excessive numbers of embryos to the the uterus (*b*) transfer one embryo	To determine the effects of maternal environment on implantation, placentation, and fetal survival	Hafez 1964*c*, 1968
	(*c*) initiate superovulation with exogenous gonadotropins (superpregnancy follows)		Hafez, Jainudeen, and Lindsay 1965
	(*d*) X-irradiate whole body		Hahn and Ward 1967 Ward and Hahn 1967
Transuterine migration of blastocyst	(*a*) unilaterally ovariectomize the animal; then cause superovulation with exogenous gonadotropins	To determine the spacing and transuterine migration of blastocysts	Hafez 1964*b*
	(*b*) transfer color-tagged embryos to one horn of the uterus		Dziuk, Polge, and Rowson 1964
Intraspecific transfer of blastocyst	Transfer "delayed" blastocysts from ovariectomized donors to hormone-treated recipients	To determine the hormones required by the blastocyst and during fetal development	Weitlauf and Greenwald 1968
Interspecific transfer of blastocyst	See table 8	To study the implantation process	See table 8
Transfer of blastocysts to extrauterine sites	(*a*) transfer blastocysts to the anterior chamber of the eye, the kidney capsule, spleen, or testis (*b*) introduce eggs (with or without the zona pellucida) into the peritoneal cavity by direct transfer or by allowing them to escape from the oviduct	To study the implantation process	Nicholas 1934 Fawcett, Wislocki, and Waldo 1947 Runner 1947 Kirby 1960, 1963*a*, *b* McLaren and Tarkowski 1963
Radioautography	Inject ^{35}S-methionine intravenously into female	To study incorporation of labeled amino acids into protein and hence protein synthesis	Weitlauf and Greenwald 1965 Prasad, Dass, and Mohla 1968
Isotopes and scintillation counting	Inject radioactive substances intravenously into female	To determine (*a*) the penetrability and selective uptake of ions (*b*) the intermediate compounds in biosynthetic pathways	Bennett, Boursnell, and Lutwak-Mann 1958 Fridhandler 1968

TABLE 1—*Continued*

Technique	Description	Objectives	Reference
In Vivo Techniques—*Continued*			
Culture and subsequent transfer	Culture the ovum (with or without the zona pellucida) in a microdrop of synthetic medium; later transfer it to a recipient animal	To study the growth rate and nutritive requirements of blastocysts; metabolic pathways	Edwards 1964 Staples 1967 Brinster 1969
O_2 uptake and subsequent transfer	Measure O_2 consumption of the blastocyst in a Gilson oxygraph before and after transferring it to a recipient animal	To study the growth rate and metabolic pathways of the embryo	Wintenberger-Torres, unpublished data Wintenberger-Torres and Hafez, unpublished data
Experimental production of mosaics and chimeras	Fuse two embryos into a single chimeric blastocyst	Mosaic embryos facilitate the study of differentiation	Tarkowski 1963, 1965 Mintz 1964, 1965
In Vitro Techniques			
Flat mount and sexing of blastocysts	Split the blastocyst open; mount it flat and stain	To study the embryo's cytology, cytochemistry, mitotic patterns, and the effects of culture media	Moog and Lutwak-Mann 1958 Sundell 1962 Vickers 1967
Implantation *in vitro*	Explant blastocysts on strips of endometrium in organ culture	To study the implantation process	Glenister 1961 Gwatkin 1966
Culture *in vitro*	Cultivate blastocysts in media containing high or low concentrations of substrates	To define metabolic pathways and nutritional requirements of the embryo	Daniel 1965, 1967*b*, *c*, 1968 Mounib and Chang 1965 Gwatkin and Meckley 1966 Huff and Eik-Nes 1966
Cleared uteri	Clear gravid uteri in benzyl-benzoate	(*a*) To visualize the positions of rabbit ova *in utero* at all stages of development (*b*) To locate degenerating blastocysts (*c*) To determine the position of the blastocyst in relation to previous implantation sites (*d*) To study implantation in pseudopregnant hamsters	Orsini 1962

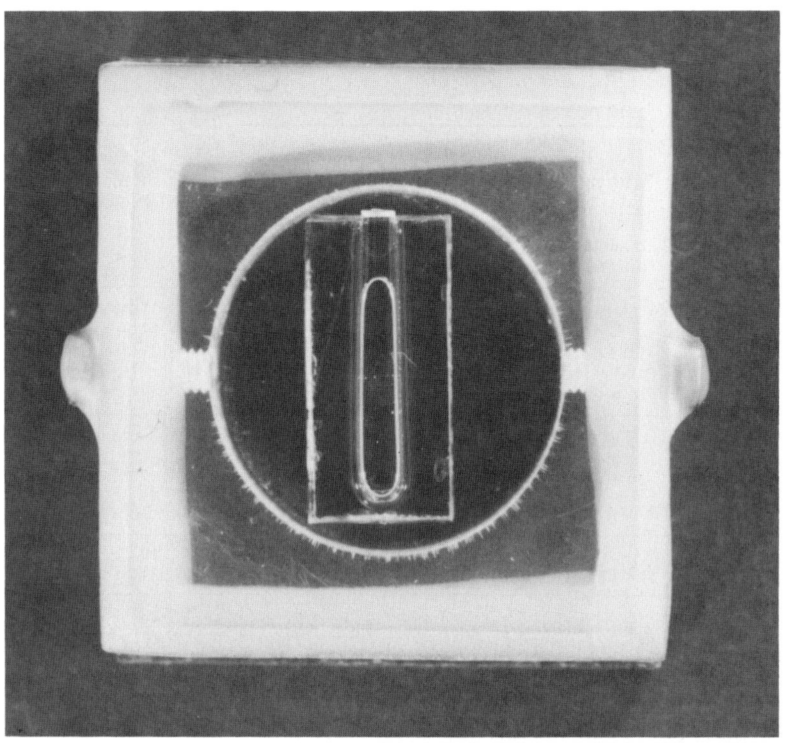

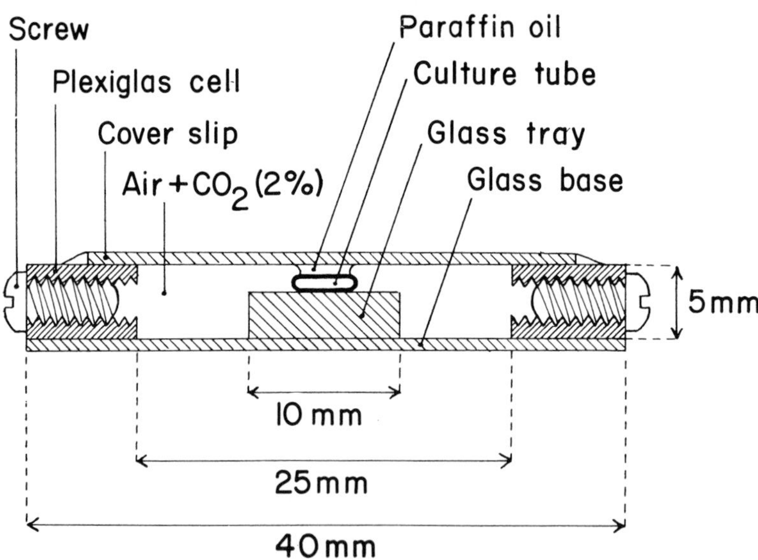

Fig. 1. Photograph (*top*) and diagram (*bottom*) of the cell used for cinematography of the blastocyst during blastulation. (By permission from J. G. Mulnard, *Arch Biol* [*Liège*] 78:107, 1967)

In vitro techniques include electron microscopy, flat-mount preparations, the benzyl-benzoate clearing of the pregnant uterus, and biochemical analysis of blastocoelic fluid. Time-lapse cinematography techniques are particularly useful for following the differentiation of blastocysts in culture (fig. 1). Blastocysts can be collected for *in vitro* studies by flushing the uterus with saline or by simply rupturing the uterine wall (fig. 2).

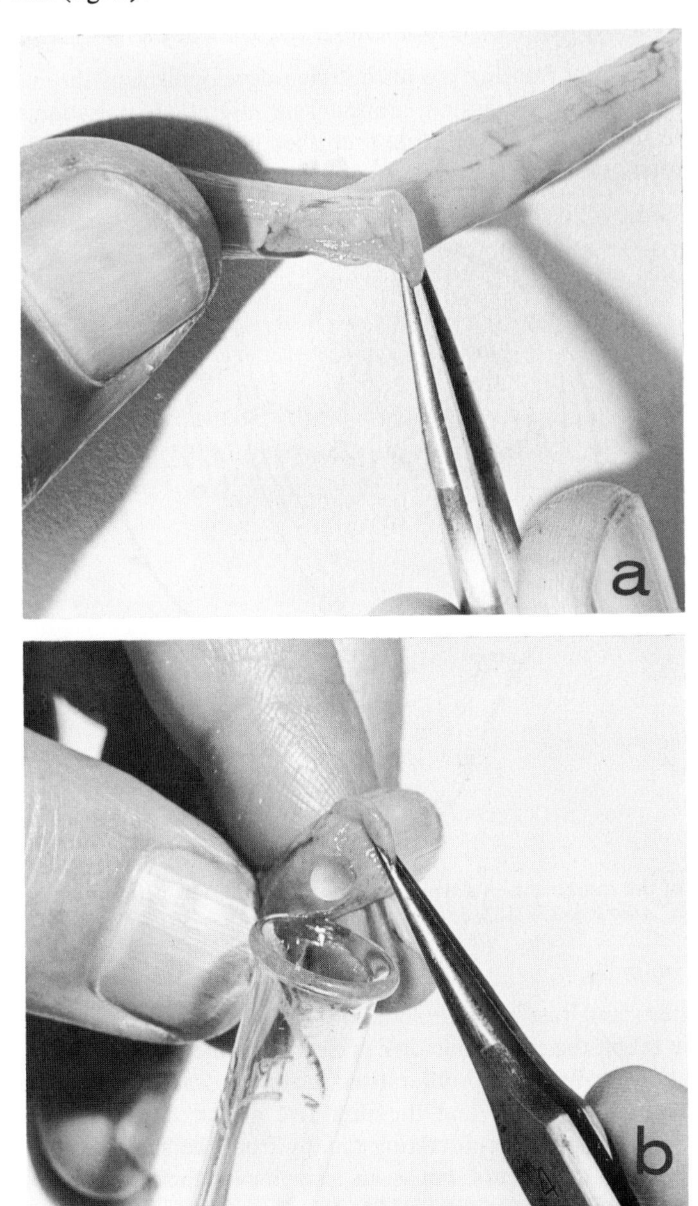

Fig. 2. Method of collecting 7-day-old rabbit blastocysts. The uterine horn is freed from the mesosalpinx and a strip of the uterus is peeled off (*a*) until the blastocyst is located. Once located, the blastocyst is transferred directly to a test tube (*b*).

The diversity of the techniques is impressive. Even with the large body of information garnered by utilizing them, many unanswered questions remain.

The present chapter deals with the patterns of growth and development of the blastocyst and with some of its physiological and biochemical properties in relation to the maternal environment.

I. Growth and Development of the Blastocyst

Dramatic changes occur during the intrauterine development of the blastocyst, for example, differentiation, expansion, transuterine migration, pulsation, orientation, and loss of the zona pellucida. The first four phenomena are discussed briefly in this section; the others are described in detail elsewhere.

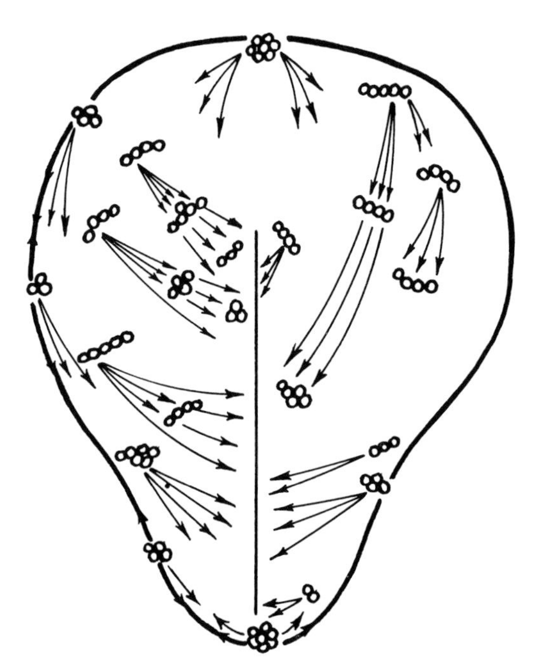

Fig. 3. A map of cell movements in a typical embryonic shield. (By permission from J. C. Daniel and J. D. Olson, *Anat Rec* 156:123, 1966)

A. *Differentiation*

The pattern and rate of differentiation of the blastocyst vary greatly among species. In the rabbit the embryonic disk is clearly visible 4 days *post coitum* (*p.c.*). Its differentiation involves the proliferation, movement, and death of cells. Formation of the embryonic shield and the primitive streak, for example, entails the migration of cells which are proliferating rapidly from the surface of the embryonic disk (fig. 3). Cell death occurs, but is an insignificant factor in this phase of prenatal development (Daniel and Olson 1966).

The blastocoelic cavity apparently develops from intercellular spaces which fuse into junctional complexes between adjacent cells. These spaces continue to coalesce and enlarge while the underlying cells from which they arose become the blastocoelic epithelium (Enders 1970). Junctional complexes appear in the rat at the 8-cell stage. However, small intercellular cavities occur also in the 16-cell stage of many species and may give rise to the blastocoel (Mulnard 1967).

The blastocyst is smaller in species with interstitial implantation than in species with superficial implantation. In the guinea pig, rat, mouse, and hamster the size of the implanting blastocyst and the oviductal egg is the same. The size of

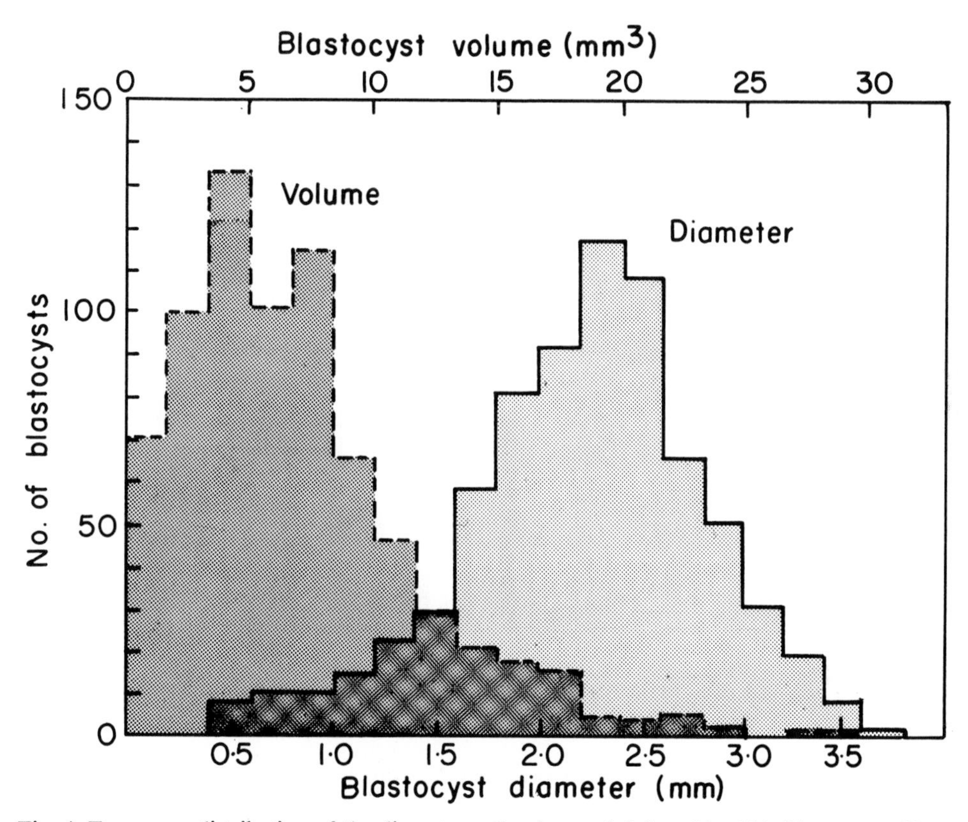

Fig. 4. Frequency distribution of the diameter and volume of 6-day-old rabbit blastocysts (N = 736). Histograms are plotted in the original mensuration units although the axes of the graph show actual mm and mm^3. (By permission from R. A. Beatty, *J Endocr* 17:248–60, 1958)

the blastocysts in a single uterine horn of the rabbit varies widely (fig. 4). It is not related to the weight of the dam or sire or to the age of the dam (Beatty 1958) but is apparently affected by the parity of the female, lactation (fig. 5), and the degree of uterine development.

The mucinous coat of rabbit oviductal embryos is thought to assist embryonic development, particularly through the blastocyst stage. Nonetheless blastocyst development and implantation still occur when only small amounts of mucin are present (Adams 1965) although the implantation rate declines (Greenwald 1962). The crystalloid polysaccharide-containing inclusions of rabbit blastocysts may be involved in the maintenance of the zona pellucida (Hadek and Swift 1960; Schlafke and Enders 1963; Enders and Schlafke 1965).

In most mammals the blastocyst remains free within the uterine lumen for only a few days. This free period is extended in certain species such as the bear, roe deer, armadillo, seal, many mustelids and marsupials, and lactating, pregnant rats and mice (Enders 1963). In these animals the blastocysts enter a dormant period during which they undergo little growth or cell division (fig. 6). They resume development when the maternal environment allows implantation (Baevsky 1963). The stage of differentiation in which the blastocyst diapauses varies with the species.

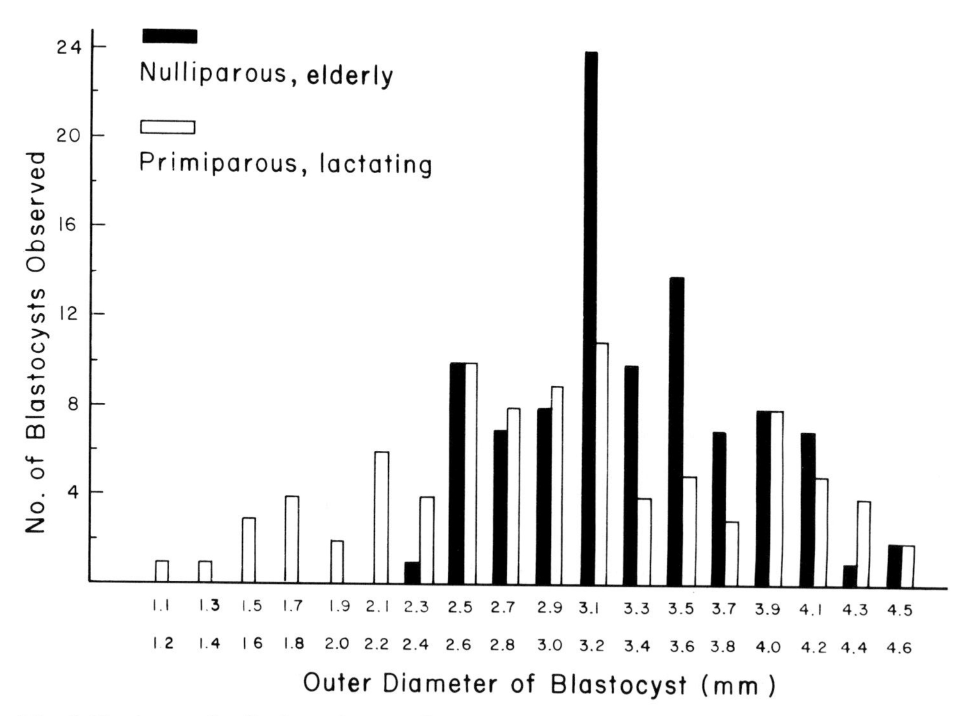

Fig. 5. Frequency distribution of outer diameters from blastocysts of nulliparous puberal and primiparous lactating does. Note the variation within each group and the reducing effect of lactation on the size of the blastocyst. (By permission from E. S. E. Hafez and I. Ishibashi, *Int J Fertil* 10:47, 1965)

B. *Expansion*

The pattern and rate of blastocyst expansion varies with the species (figs. 7, 8). In primates, rabbit, dog, cat, and ferret blastocoelic fluid accumulates rapidly, distends the blastocyst (50–100-fold), and thins out the zona pellucida (fig. 7). The rate of expansion is much slower in species with delayed implantation. Several biophysical factors may be involved in the accumulation of blastocoelic fluid, such as the membrane characteristics of the zona pellucida and trophoblastic cells and any intercellular material between them, as well as the osmotically active ingredients of the uterine fluid (Cole 1967). The process of expansion may involve passive water transport, electro-osmosis, doublet membrane effects, and pinocytosis.

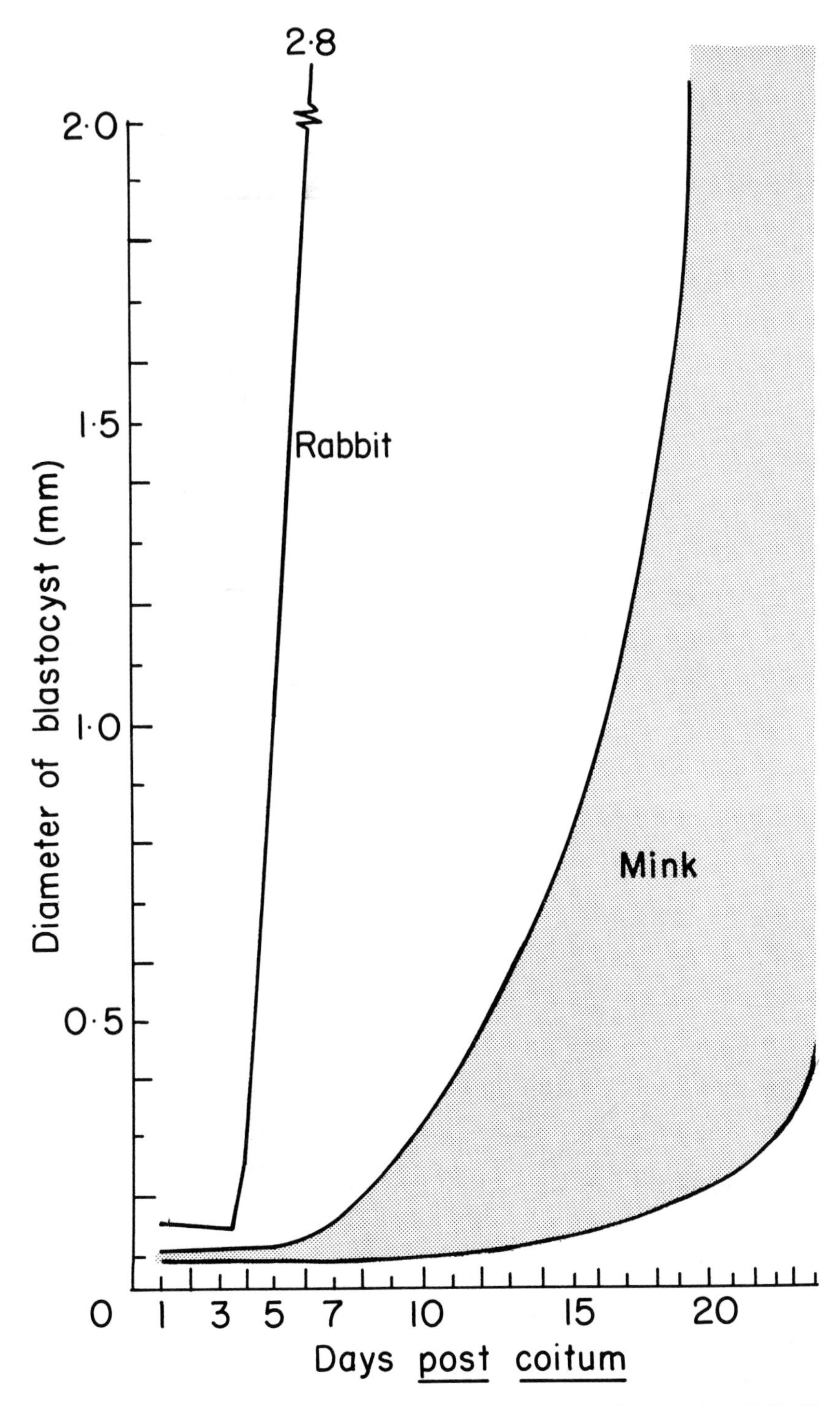

Fig. 6. Comparative growth of mink and rabbit blastocysts. (Data from Enders 1938; Hansson 1947; Baevsky 1963; and J. C. Daniel, *J Embryol Exp Morph* 17:293, 1967)

Blastulation seems to be governed by substances within the uterine fluid. Krishnan and Daniel (1967) isolated a protein fraction from this fluid, "blastokinin," which appears to regulate blastocyst development in the rabbit.

1. *Associated uterine changes.* The expansion of the blastocyst is associated with significant changes in uterine vascularity. The uterine tissues of the rabbit contain less than 2 ml of blood during estrus; by 5 days *p.c.*, its blood volume approaches 10 ml (Barcroft and Rothschild 1932). The capillaries of the subepithelial plexus increase in thickness and length during the period of blastocyst expansion (fig. 9).

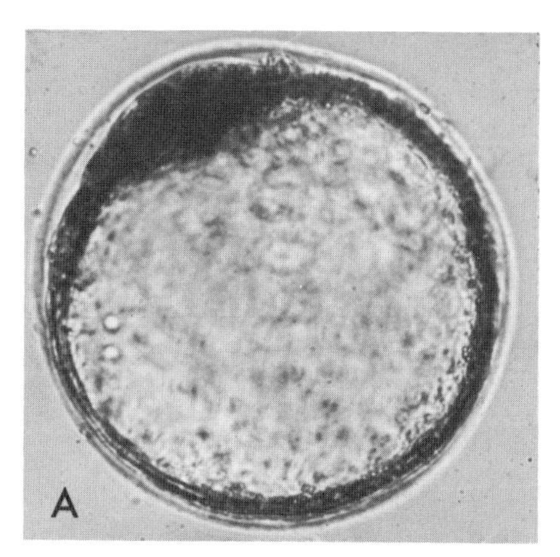

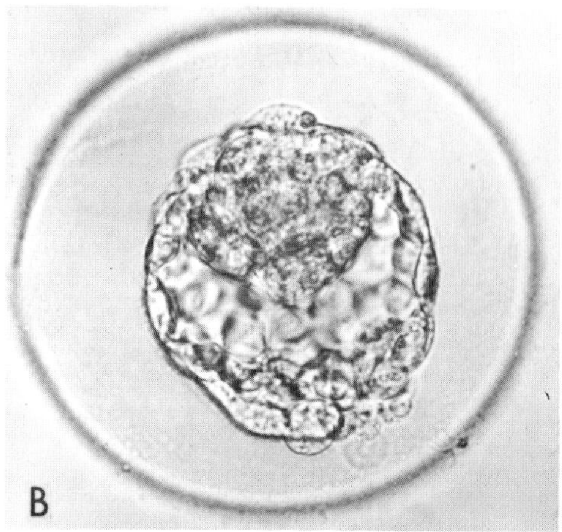

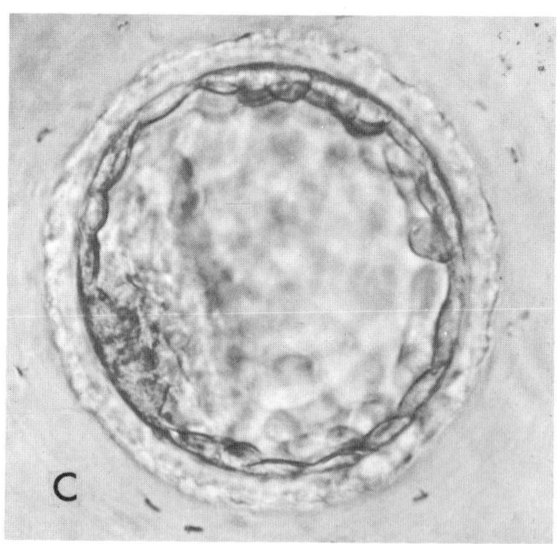

Fig. 7. Species differences in the morphology of the preimplanted blastocyst.
(*a*) 14-day-old bovine blastocyst; the embryonic disk is black.
(*b*) 5-day-old baboon (*Papio sp.*) blastocyst.
(*c*) 4-day-old rabbit blastocyst with a thick mucin coat. The thickness of the zona pellucida (20 μ) and the intrazonal diameter (125 μ) are relatively constant until 72 hr *p.c.* The expansion of the blastocyst which occurs 72–80 hr *p.c.* is associated with thinning of the zona pellucida to 5 μ.
(*a* from Greenstein, Murray, and Foley 1958; *b* from Moor and Rowson 1966; and *c* from Hendrickx and Kraemer 1968)

In the rat the number of patent uterine arterioles supplying the endometrium is primarily responsible for its vascularity and oxygen content. Ovarian steroid hormones influence the development of vascularity of the area and the oxygen-requiring processes of the uterine cells themselves (Mitchell and Yochim 1968). These steroids may modify the metabolic activity of the endometrium indirectly by regulating the amount of oxygen to which it is exposed (Yochim and Mitchell 1968).

In the rabbit the uterine epithelium develops into a syncytium, particularly in the conceptal area (Hafez and Tsutsumi 1966). Placental folds hypertrophy and obplacental folds atrophy at the implantation site (fig. 10).

C. *Transuterine Migration*

In 1845 Bischoff noted transuterine migration in bicornuate uteri and interpreted it as a means of adjusting differences in the number of ovulations from each

TABLE 2

MAMMALIAN SPECIES IN WHICH TRANSUTERINE MIGRATION OCCURS

Species in Which Implantation Occurs in Both Uterine Horns	Species in Which Implantation Occurs in Right Uterine Horn, But in Which Ovulation Is Bilateral or Sinistral
Cat (*Felis catus*)	Bat (*Miniopterus natalensis*)
Cattle (*Bos taurus*)	(*Myotis lucifugus*)
Dog (*Canis familiaris*)	(*Myotis myotis*)
Elk (*Alces alces*)	(*Pipistrellus pipistrellus*)
Ferret (*Mustela furo*)	(*Molossus m. crassicaudatus*)
Lemur (*Lemur rufipes*)	Duiker, common (*Sylvicapra grimmia*)
Mink (*Mustela vison*)	Impala (*Aepyceros melampus*)
Raccoon (*Procyon lotor*)	Lechwe (*Kobus lechee*)
Roe deer (*Capreolus capreolus*)	Uganda kob (*Adenota kob*)
Sheep (*Ovis aries*)	
Skunk, eastern (*Memphitis mephitis nigra*)	
Swine (*Sus domesticus*)	

SOURCES: Boyd and Hamilton 1952; Asdell 1964; Hafez 1964*b;* Chang 1968; van Tienhoven 1968.

ovary. Ungulates, carnivores, insectivores, and bats are subject to this phenomenon (Table 2). It occurs most commonly among those cattle and sheep whose ovaries ovulate at a variance of 1 or 2 eggs. Transuterine migration takes place in 80–100% of dogs and cats in whose cases one ovary ovulates 3 eggs more than the other. It is more frequent (13%) in unilaterally ovariectomized cows than in intact animals (6%) (Hafez 1964*b*).

When Dziuk, Polge, and Rowson (1964) transferred genetically marked pig eggs to a single oviduct or uterus they found that some implanted on the side opposite to that on which the eggs had been transposed, although definitely more of them implanted on the side of deposit (fig. 11). Migration, when it occurred, took place between day 8 or 9 and day 15 of pregnancy. The rate of migration and distribution of blastocysts was not affected by the number of embryos present, the number of corpora lutea, or the overall length of the uterus (Dhindsa, Dziuk, and Norton 1967).

Transuterine migration has been induced also in unilaterally ovariectomized

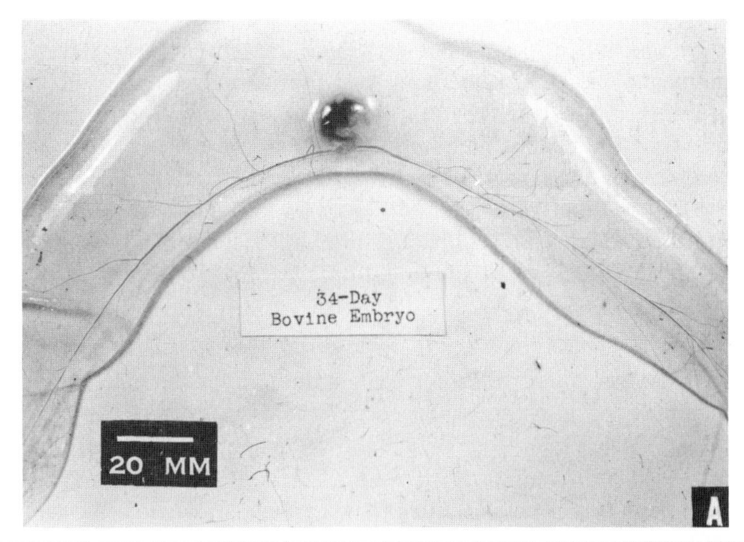

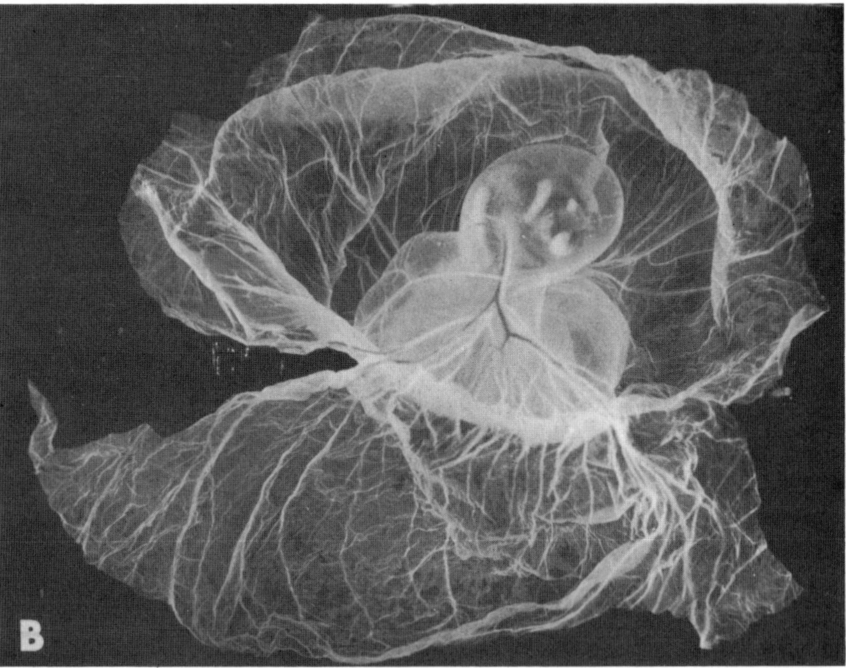

Fig. 8. Specific differences in the morphology of postimplantation blastocysts.

(*a*) 34-day-old bovine conceptus with differentiating chorioallantoic vascular system.

(*b*) 7-week-old equine conceptus. Note that the chorion is oval, rather than cylindrical as it is in the cow, and that it is more distended with fluid, which facilitates the early diagnosis of pregnancy. Also note the positions of the amnion, yolk sac, and chorioallantois.

(*c, d*) 13-day-old pig blastocyst (total length, 157 cm). The position of the embryonic disk is indicated by the arrow; *d* is an enlargement of the embryonic disk.

(*a,* by permission from L. C. Ulberg, in Reproduction in Farm Animals [Philadelphia: Lea & Febiger, 1969]; *b,* by permission from G. H. Arthur, *Wright Veterinary Obstetrics* [London: Baillière: Tindall and Cassell, 1964]; *c* and *d,* by permission from J. S. Perry and I. W. Rowlands, *J Reprod Fertil* 4:175, 1962)

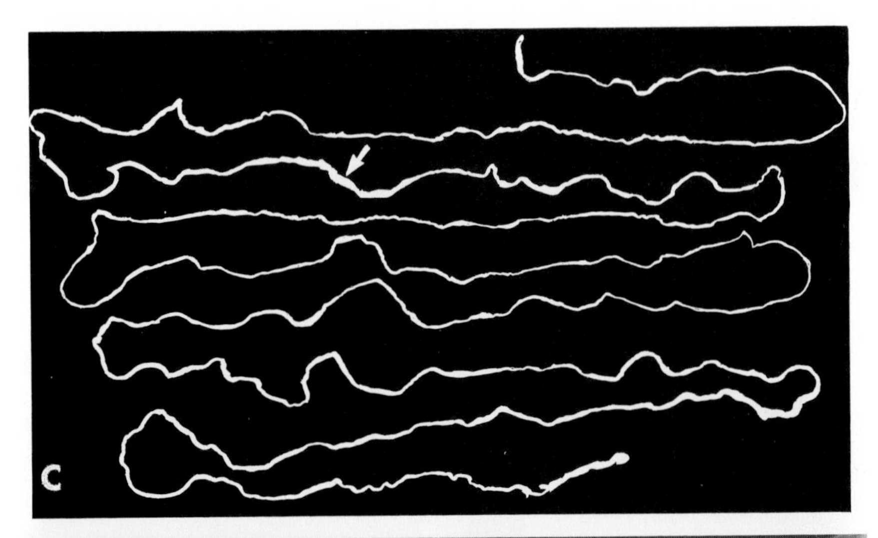
C
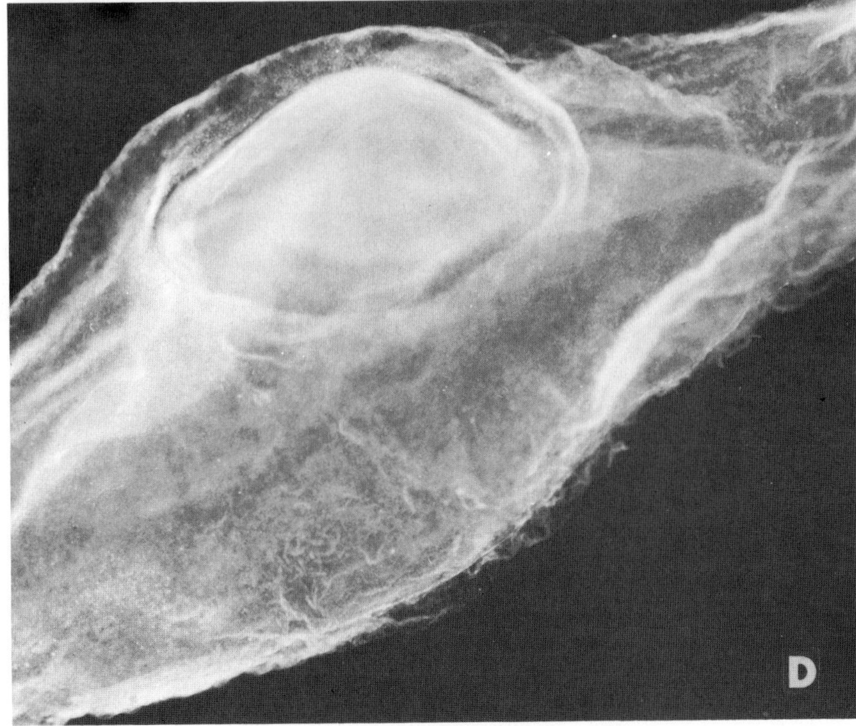
D

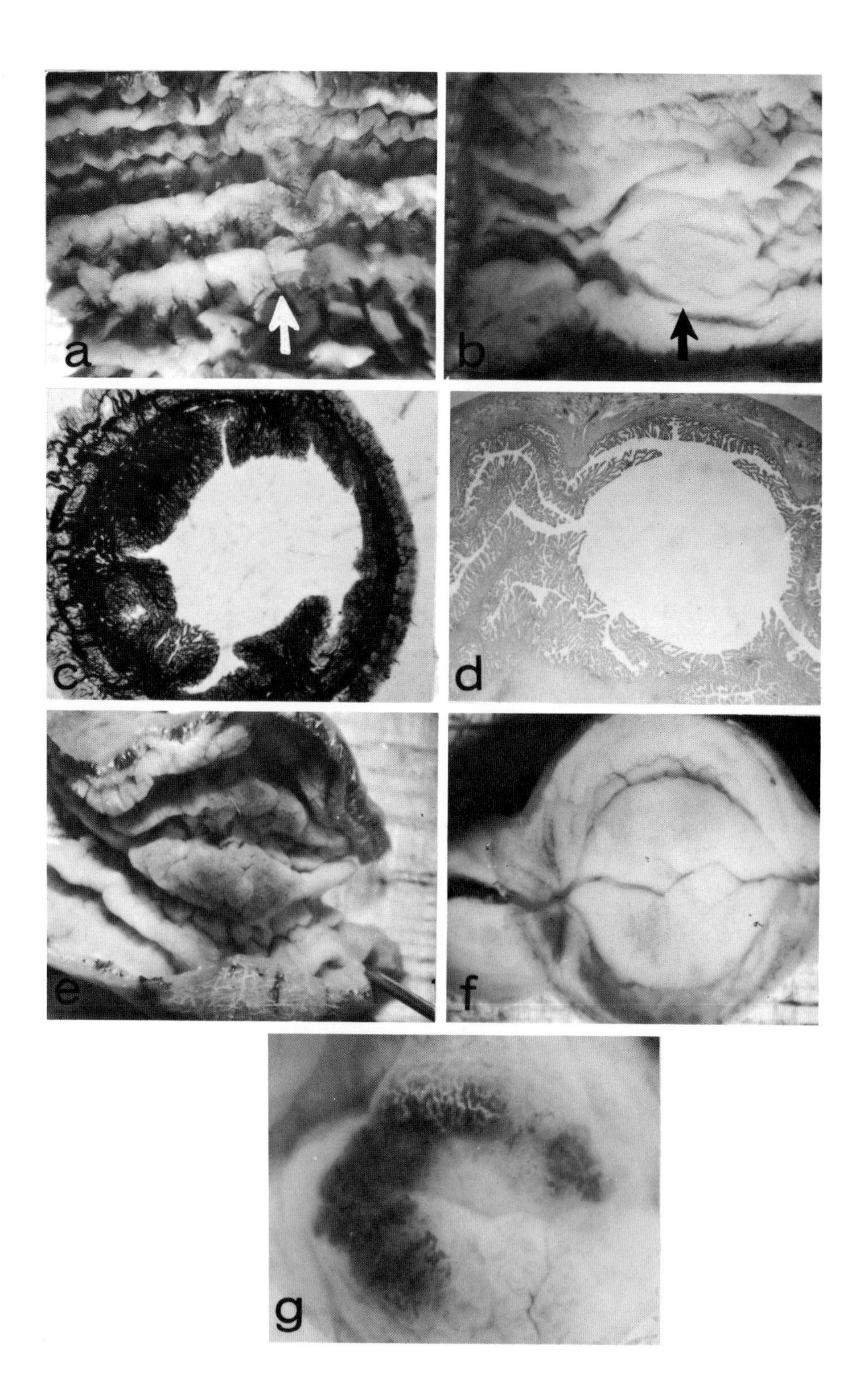
a
b
c
d
e
f
g

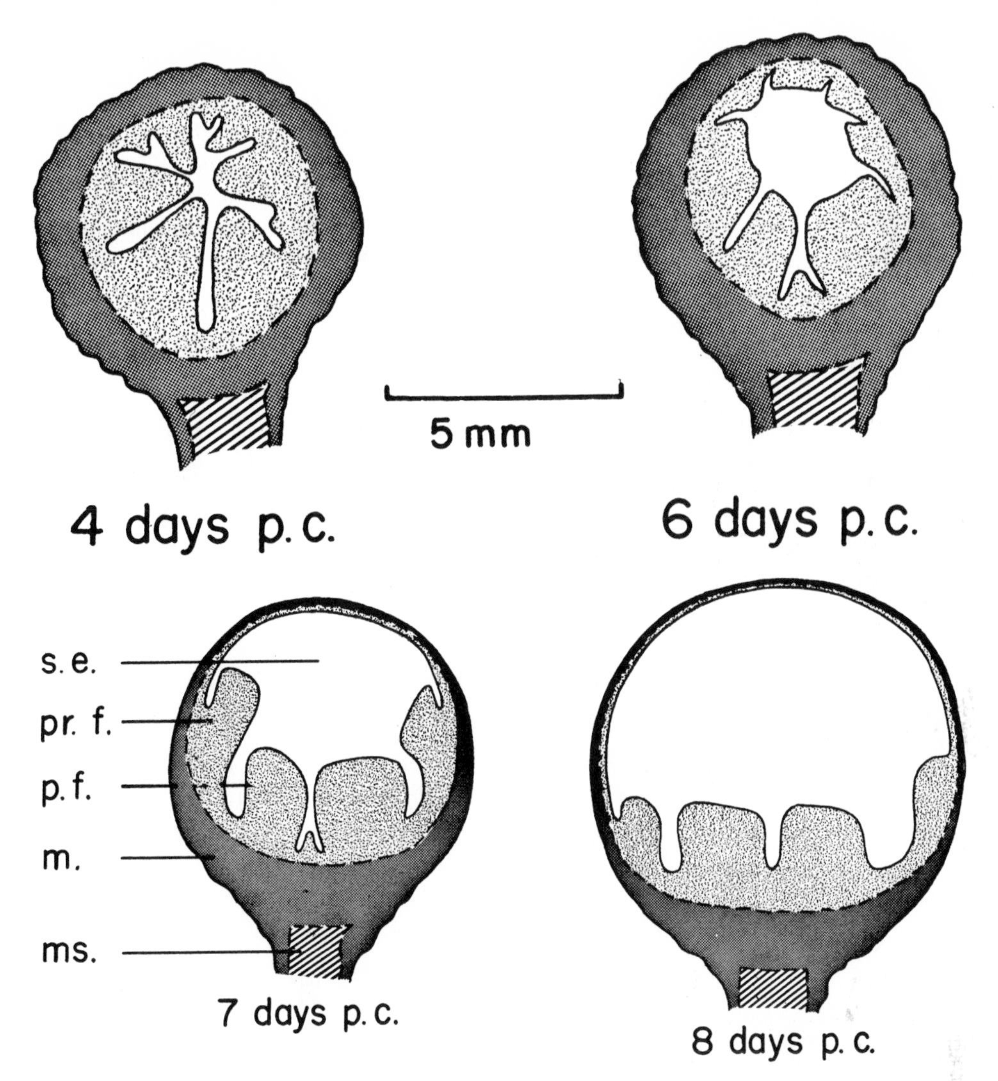

Fig. 10. The uterine changes which accompany the implantation of the rabbit blastocyst. The obplacental folds begin to disappear at 6 days *p.c.* The placental folds hypertrophy between 7 and 8 days *p.c.* Implantation occurs at 9 days *p.c.* *m,* musculature; *ms,* mesometrium; *pf,* placental fold; *prf,* preplacental fold; *se,* location of blastocyst. (By permission from E. S. E. Hafez and T. Tsutsumi, *Amer J Anat* 118:249, 1966)

Fig. 9. Changes in the endometrial vascular supply of the rabbit blastocyst during the early stages of implantation.

(*a*) The surface of the endometrium at 6 days *p.c.* showing the site of implantation (*arrow*) (latex injected specimen). ×3.6.

(*b*) Same as above. Note the obplacental folds (*above the arrow*) at the implantation site (*arrow*) (latex injected specimen). ×3.6.

(*c*) Cross section of the uterus at 6 days *p.c.* The folds were small, especially the obplacental ones (transparent section). ×9.

(*d*) Longitudinal section of the uterus showing conceptal and interconceptal sites at 6 days *p.c.* (histological section). ×9.

(*e*) The conceptal site at 7 days *p.c.* Note the marked development of the placental folds and the winding grooves (latex injected specimen). ×3.6.

(*f*) The placental folds at the conceptal site 8 days *p.c.* after removing the obplacental uterine wall. Note the distribution of capillaries on the surfaces of the folds (latex injected specimen). ×3.6.

(*g*) Well-developed "crescents" on the placental folds at 9 days *p.c.* Compare this with the previous picture. Blood vessels in the "crescents" were dilated (latex injected specimen). ×9.

(By permission from E. S. E. Hafez and T. Tsutsumi, *Amer J Anat* 118:249, 1966) rats whose uteri were anastomosed in various ways (Krehbiel 1946): (*a*) when one uterus was joined to the other just above the cervix, (*b*) when the cervical or the ovarian end of the sterile uterus was joined to the ovarian third of the fertile uterus, and (*c*) when the uteri were adjoined in tandem by either end. It did not occur when both uteri retained their cervical connections and were joined just above the cervix.

A. Eggs transferred to both sides

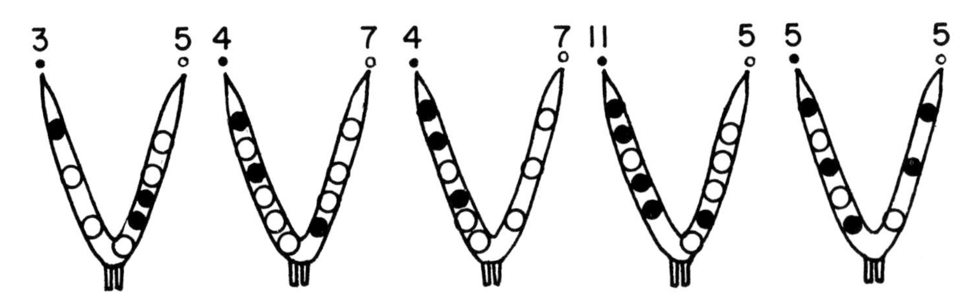

B. Eggs transferred to one side

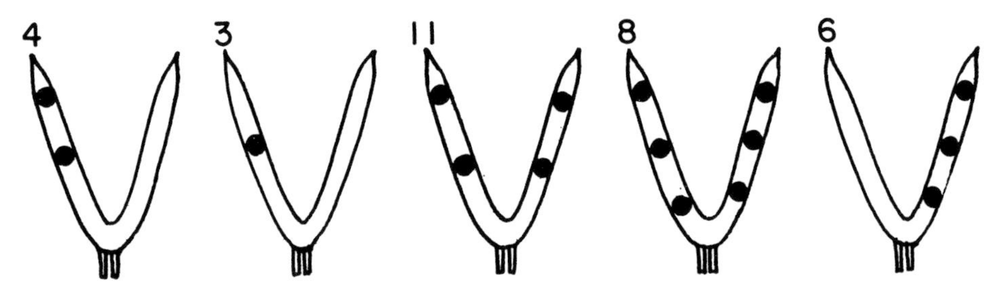

Fig. 11. Transuterine migration of genetically marked pig blastocysts after egg transfer.

(*a*) Eggs transferred to both uterine horns. ● = black eggs. ○ = white eggs. Numerals refer to the number of eggs transferred.

(*b*) Eggs transferred to one uterine horn. Numerals again refer to the number of eggs transferred to the one side.

When eggs were transferred to one or both uterine horns, part or all of them migrated to the opposite horn. (By permission from P. J. Dziuk, C. Polge, and L. E. Rowson, *J Anim Sci* 23:37, 1964)

D. *Pulsation*

The blastocyst of the mouse contracts and expands rhythmically before it sheds the zona pellucida (Kuhl and Friedrich-Freksa 1936; Borghese and Cassini 1963). Cycles of rapid contraction (4–5 min) followed by slower expansion (2–3 hr) occur before and after the trophoblast penetrates the zona. Vigorous contractions, in which the blastocoel almost disappears, cover 6–8 hr and are generally interspersed with 3 or 4 moderate contractions, each lasting 20–100 min (fig. 12). The vigorous contractions that occur immediately before the zona is sloughed lead one to believe that it may be necessary for the blastocyst to pulsate in order to shed the zona pellucida. This supposition is further supported by the degeneration of the quiescent mouse blastocysts without hatching (Cole and Paul 1965; Cole 1967).

The type and time sequence of pulsation may be different among other species. Collapsed mink blastocysts, for example, regain their original fluid-filled state *in vitro* (Daniel 1967*a*), whereas rabbit blastocysts initially contract *in vitro* and only later reexpand.

II. Biochemical Properties of the Blastocyst

Intricate biophysical and biochemical relationships exist between different parts of the blastocyst and endometrium (fig. 13). For example, not only does the pH vary in the blastocyst with each stage of development, it also alters within the parts

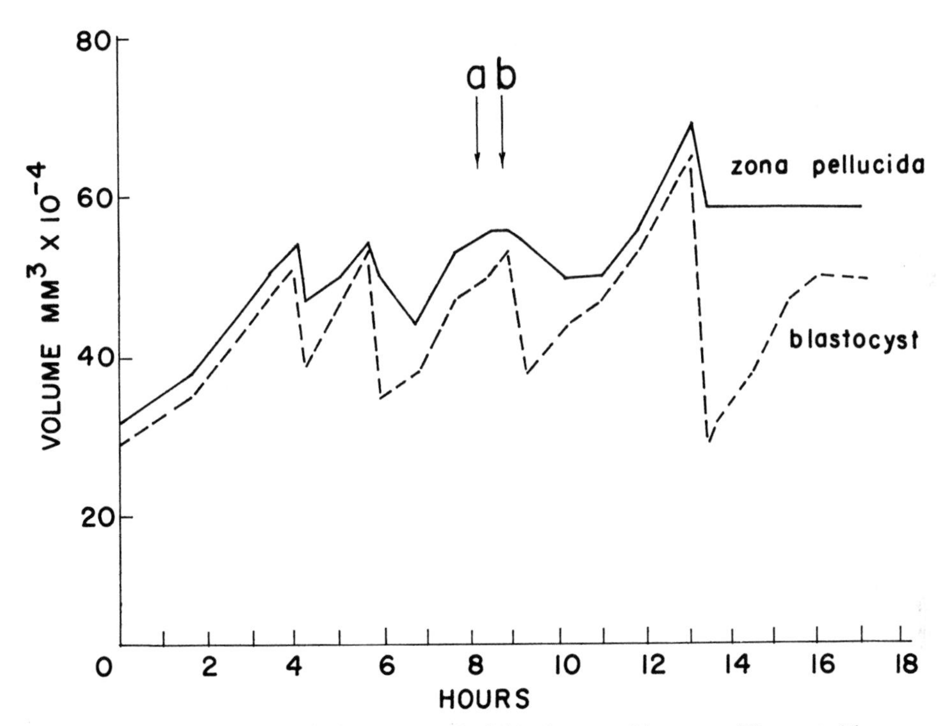

Fig. 12. Volume changes (pulsations) of an individual mouse blastocyst. The record began approximately 98 hr after ovulation. At (*a*) the blastocyst began to rotate and at (*b*) the trophoblast penetrated the zona pellucida. (By permission from R. J. Cole, *J Embryol Exp Morphol* 17:481, 1967)

for that stage of development (fig. 14). It is 7.8—8.0 in the 6- to 8-day-old rabbit blastocyst, but 7.3 thereafter (Zimmermann 1960). The blastocyst actively controls the rate of entry of substances from the uterine fluids (fig. 15). Some physiochemical characteristics of the rabbit blastocyst are summarized in table 3.

A. *Permeability*

The blastocyst exhibits a high degree of selective permeability (Brambell 1954; Mayersbach 1958). Preimplanted rabbit blastocysts, for example, are impermeable to trypan blue for 7–9 days *p.c.* (Ferm 1956), but molecules as large as

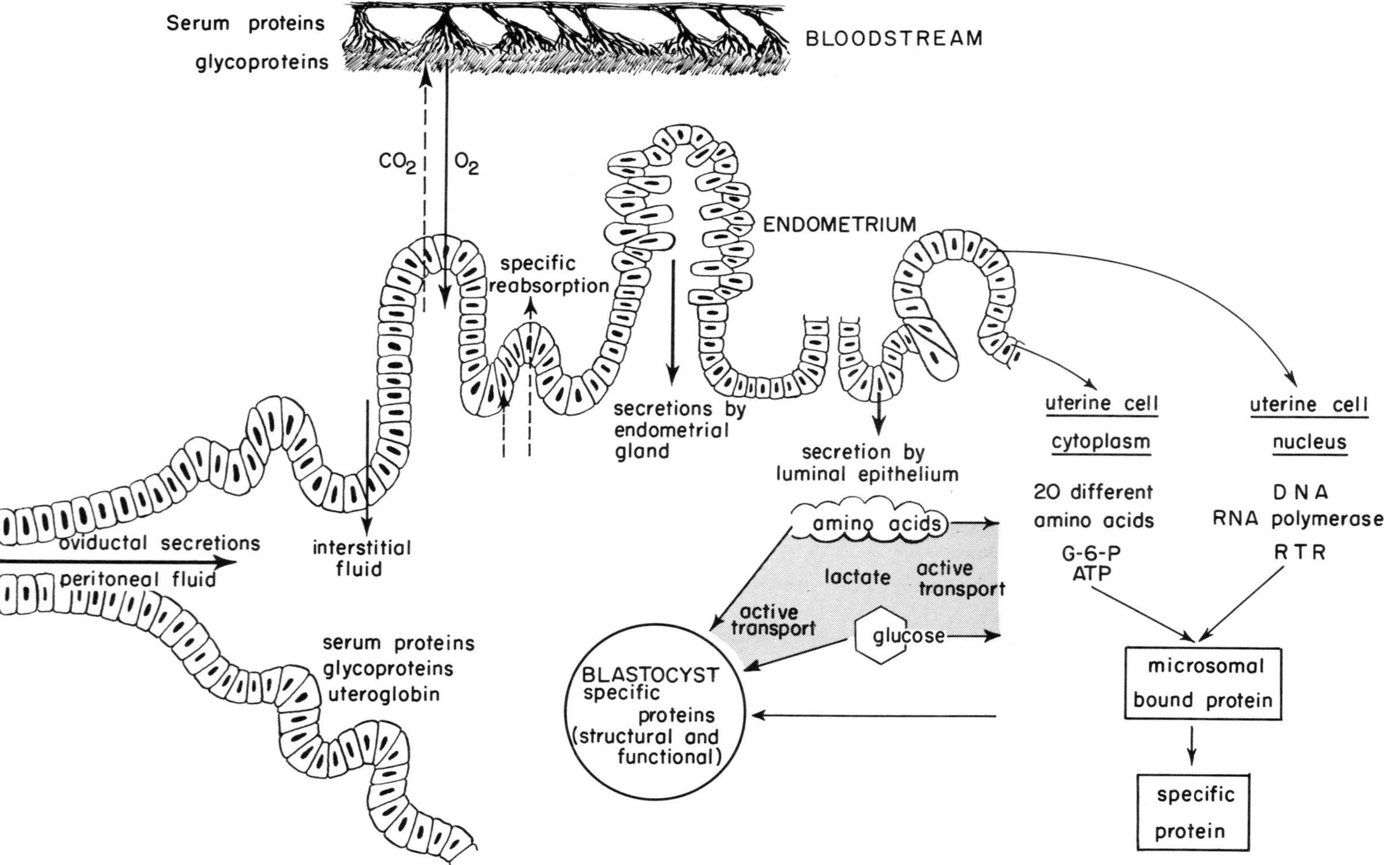

Fig. 13. Some of the physiological processes which maintain the intrauterine environment and provide the preimplantation blastocyst with the substrates necessary for growth.

antibodies can penetrate the rabbit blastocyst at 7 days *p.c.* (Zimmermann et al. 1963). The extraembryonic tissues of the blastocyst become more selective after the placenta forms.

B. *Protein Content and Synthesis*

A large supply of building material is essential to the rapid synthesis of cellular mass and blastocoelic fluid. The preimplanted blastocyst, unable to absorb all the proteins requisite to development directly, accumulates nitrogenous compounds of low molecular weight and synthesizes them into protein.

1. *Uterine Fluids.* The uterine fluid contains the secretions of endometrial glands, interstitial fluid expressed from capillaries, O_2, and CO_2. Its composition varies with species and by reproductive stage (Schwick 1965). Steroid hormones, for example, modify the protein content of endometrial secretions by stimulating the production of m-RNA (Karlson and Sekeris 1966) and inhibiting RNA-polymerase (reviewed by Gorski, Noteboom, and Nicolette 1965; Segal and Scher 1967;

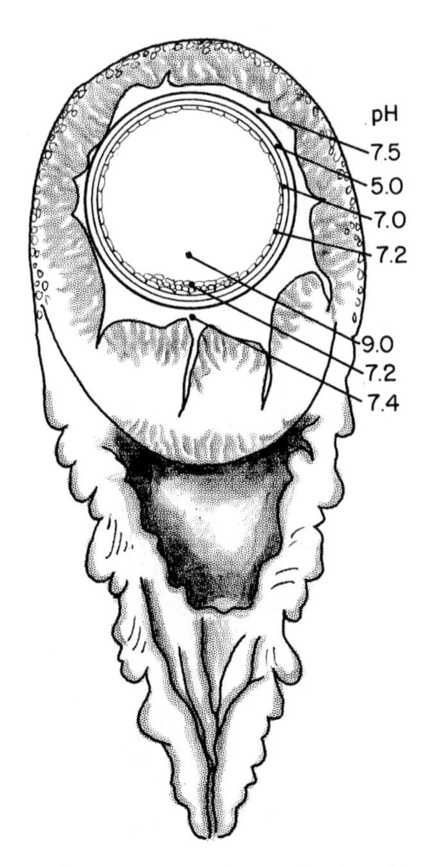

Fig. 14. The pH of the uterus and various areas of the preimplantation rabbit blastocyst. (Adapted by permission from H. M. Gottschewski, Naturwiss Rundschau 15:257, 1962)

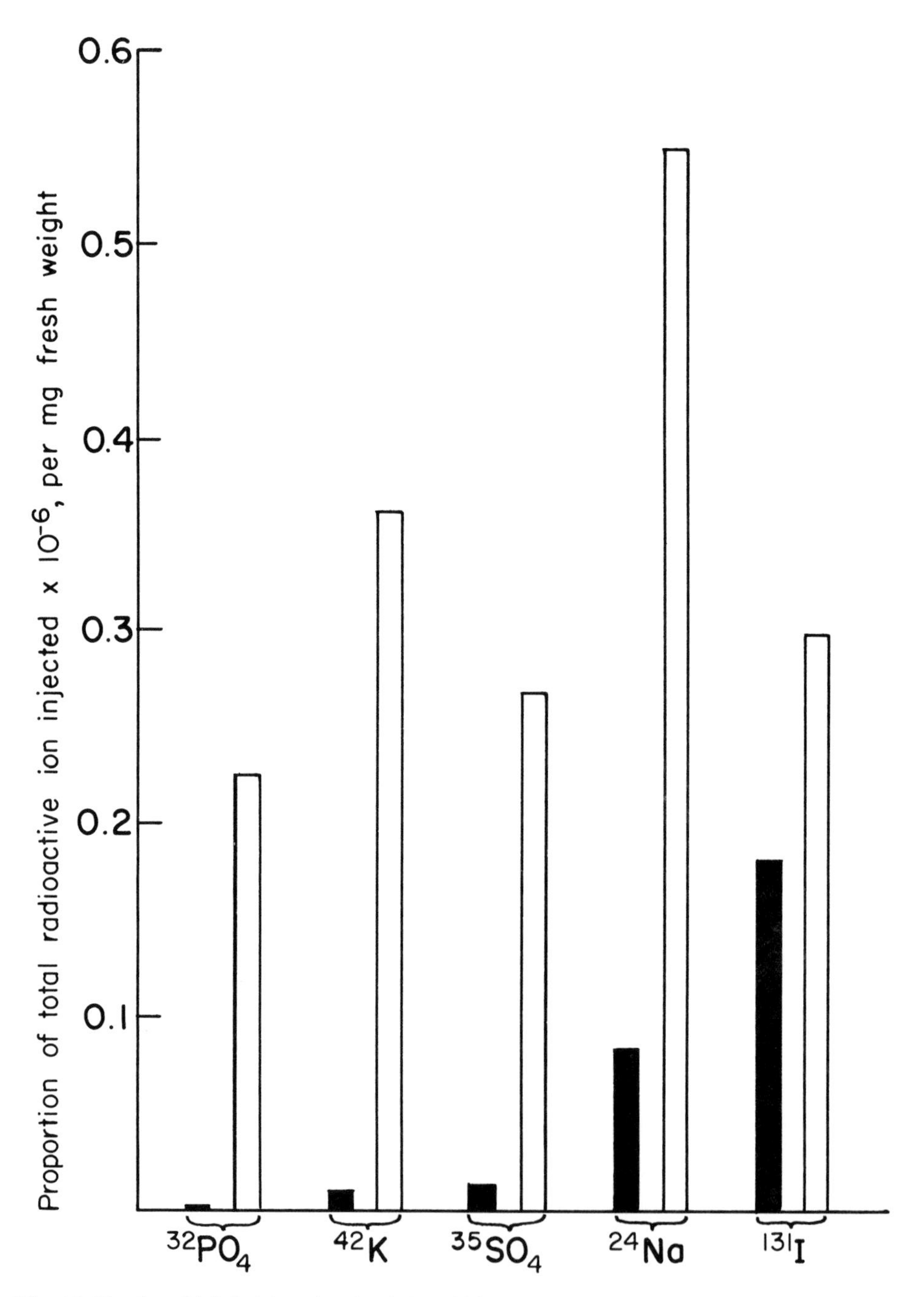

Fig. 15. Uptake of labeled ions by the 6-day-old blastocyst (■) and the endometrial secretions (□). Radioactivity was measured 45 min after injection. (By permission from C. Lutwak-Mann, J. C. Boursnell, and J. P. Bennett, *J Reprod Fertil* 1:169, 1960)

TABLE 3

SOME PHYSIOCHEMICAL CHARACTERISTICS OF 6-DAY-OLD RABBIT BLASTOCYSTS

Characteristic	Mean	Range	Reference
Diameter (mm)[a]	2.20	0.5– 3.5	Beatty 1958
Volume (mm^3)[a]	6.78	2.0–25.0	
Diameter (mm)	2.8	1.8– 3.9	Daniel 1965
Volume (mm^3)	11.5	3.1–31.0	
Surface area (mm^2)	24.8	10.2–47.6	
Mitotic index (%)[b]	4.45		
Time required to double in size (hr)	7.8		
Rate at which blastocoelic fluid accumulates (mm^3/hr)	0.8		
Weight of blastocyst (mg)	15.0	1.0–41.0	Hafez and Sugawara 1968
Weight of trophoblast (mg)	0.87	0.2– 2.0	
Volume of blastocoelic fluid (μl)	13.0	1.0–33.0	
Contents of blastocoelic fluid (mg/dl)			
Glucose	19.3	4.0–31.0	Sugawara and Takeuchi 1967
Lactic acid	67.0	32.0–92.0	
Pyruvic acid	5.2	2.3– 6.8	
α-keto glutaric acid	3.3	3.0– 3.5	
Total volatile fatty acids	5.2	2.3– 8.1	
Total amino acids	4.6	3.2– 9.0	
Free amino acids in blastocoelic fluid (μg/ml)			
Alanine	580	....	Lesinski et al. 1966
Glutamic acid+glutamine	420	....	
Glycine	296	....	
Histidine	108	....	
Leucine	44	....	
Phenylalanine	80	....	
Serine	216	....	
Threonine	108	....	
Valine	40	....	
Enzymatic activity (units/mg protein) of whole blastocyst homogenates			
Acid phosphatase ($\times 10^{-5}$)	12	....	Hafez and White 1967
Alkaline phosphatase ($\times 10^{-5}$)	393	....	
Amylase ($\times 10^{-5}$)	1616	....	
Glucose-6-phosphate dehydrogenase	Nil	....	
Glutamic-oxalacetic transaminase	36	....	
Lactic dehydrogenase	45	....	
Succinic dehydrogenase	Nil	....	
Developmental protein in blastocyst and uterine fluid, but not in maternal serum			
Blastokinin (uteroglobin)	Demonstrated by disk electrophoresis from 5.5 to 7 days *p.c.*		Hamana and Hafez unpublished data
Proteins in blastocyst, uterine fluid, and maternal serum			
Prealbumin	Immunological identity detected by PAAI, immunoreaction, immunoelectrophoresis, or Ouchterlony test; in cross reactions between all three fluids (blastocoelic fluid, uterine fluid, and maternal serum), and between the latter and specific antisera		Beier, 1968*a*, *b*
Albumin			
Uretoglobin			
β-globulin			
β_2-U-globulin			
γ-globulin			
Proteins in uterine fluid, not in blastocyst			
γ-transferrin macroglobulin			
γ_2-U-lipoprotein			
β-U-macroglobulin			
Proteins in maternal serum, not in blastocyst			
Postalbumin			
Posttransferrin			

[a] Crossbred rabbits.

[b] $\frac{\text{No. mitotic figures}}{\text{No. cells counted}} \times 100.$

Talwar et al. 1968). Most of the proteins in endometrial secretions have low molecular weights of 30,000–70,000, but a few of them (no more than 15–20%), including α-macroglobulin and β-U-macroglobulin, are much larger (160,000) (Beier 1968*a*).

The uterine fluid of the rabbit and rat contains characteristic proteins which do not occur in the blood. Beier (1968*a*, *b*) isolated large amounts of uteroglobin in the uterine fluids of pregnant and pseudopregnant rabbits which was not present in their blood serum. By means of gel electrophoresis (pH 8.7) uteroglobulin's R_F value was determined as 0.64 (R_F values for transferrin and prealbumin are 0.56 and 1.0, respectively); by means of agar gel immunoelectrophoresis it was demonstrated that its mobility resembles that of β_1-globulin (figs. 16, 17).

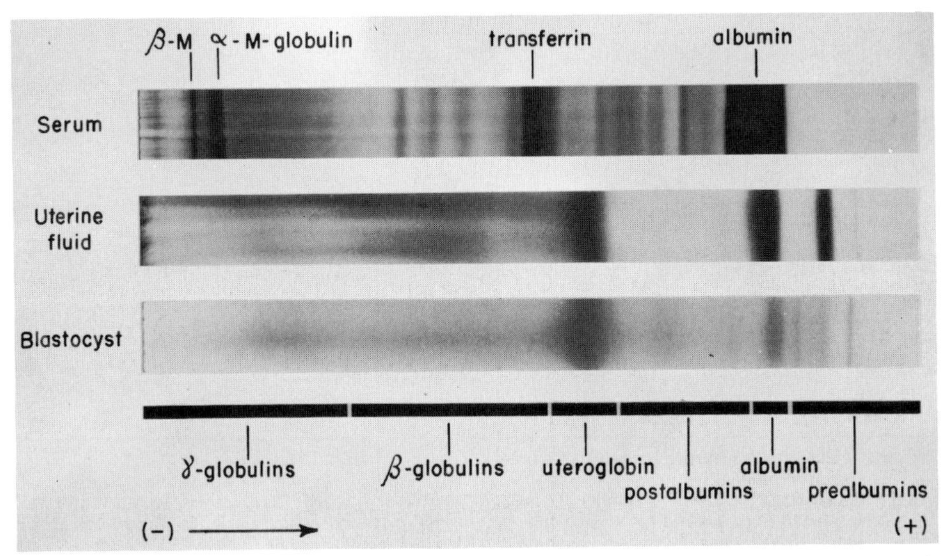

Fig. 16. Acrylamide gel electrophoresis of rabbit blood serum, uterine fluid, and blastocoelic fluid (*BL*) on the 6th day of pregnancy. (By permission from H. M. Beier, in *Ovo-implantation, human gonadotropins, and prolactin,* Brussels, 1968)

Blastokinin, a glycoprotein of low molecular weight (27,000) which induces blastulation and stimulates blastocyst development, is found in the uterine fluids of pregnant and pseudopregnant rabbits (Krishnan and Daniel 1967). By using acrylamide gel electrophoresis (pH 8) its R_F value is determined as 0.66–0.67. It does not occur in maternal blood, or in fetal serum, or in fetal amniotic fluid. After injections of progesterone and estradiol it appears in the uteri of immature rabbits (J. C. Daniel, Jr., personal communication).

2. *Blastocoelic Fluid.* The protein content of blastocoelic fluid varies with the species. The fluid of rabbits and other species with central implantation patterns contains blood protein as early as 7 days *p.c.* (i.e., before implantation). Scant amounts of such proteins occur in the blastocoelic fluid of the ferret, hedgehog, and other species with large vesicular blastocysts. The following protein fractions are common to both uterine and blastocoelic fluid: prealbumin, albumin, uteroglobin,

β_1-globulin, β_2-globulin, and γ-globulin. The albumin and γ-globulin fractions are similar to serum proteins (Beier 1968*a*, *c*). The fluid of the 6-day-old rabbit blastocyst contains 8–14 μg of protein whose electrophoretic patterns are quite different from those of maternal serum (figs. 16, 17) but nearly identical to those of uterine fluids.

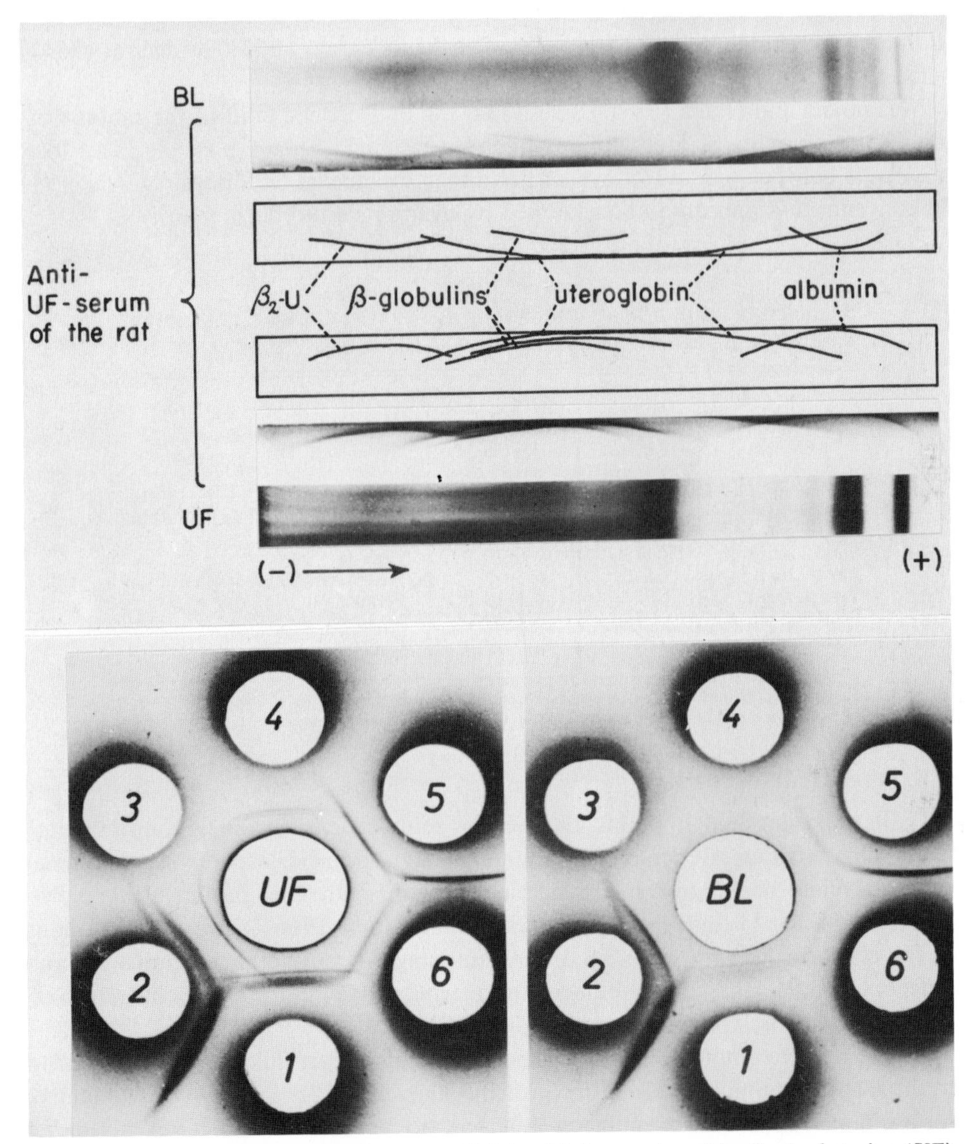

Fig. 17. *Top:* immunoelectrophoretic pattern (PAA-gel 1 Agar-gel combination) of uterine (*UF*) and blastocoelic (*BL*) fluid in the rabbit.

Bottom: immunodiffusion test (Ouchterlony method) of uterine (*UF*) and blastocoelic (*BL*) fluid in the rabbit. 1. Anti-rabbit serum (goat); 2. Anti-SE alb-serum (mouse); 3. Anti-SE α_1-serum (mouse); 4. Anti-SE α_2-serum (mouse); 5. Anti-SE β-serum (mouse); 6. Anti-SE γ-serum (goat). (By permission from H. M. Beier, in *Ovo-implantation, human gonadotropins, and prolactin,* Brussels, 1968)

The various protein fractions of the blastocoelic fluid appear at different stages of prenatal development. In rabbit blastocysts albumin appears first, followed by β-globulin, α-globulin, γ-globulin, and lastly fibrinogen (Sugawara and Hafez 1967*a*). The maximal concentrations of these protein fractions are reached at different embryonic ages (table 4).

Dramatic changes occur in the nitrogen metabolism of the rabbit blastocyst during the preimplantation period (Zimmermann et al. 1963; Sugawara 1964), some of which depend on hormonal factors (fig. 18).

Hamana and Hafez (1970) found that the blastocoelic fluid of the rabbit contains large amounts of uteroglobin which appears at 5.5 days *p.c.*, increases to a maximum between 6.5 and 7 days *p.c.*, and then disappears by 8 days *p.c.* (fig. 19). Its concentration appears to be governed by ovarian steroid hormones.

TABLE 4

THE EMBRYONIC AGE AT WHICH PROTEIN FRACTIONS APPEAR AND ARE MOST ABUNDANT IN THE BLASTOCOELIC FLUID OF THE RABBIT

Protein Fraction	Days *p.c.* When Protein First Detected	Days *p.c.* When Protein Reached Maximal Concentration	Persistence of Protein until 10 Days *p.c.*
Prealbumin	6	Weak	Disappears before implantation
Albumin	5	7.0–10.0	Persists
Postalbumin	7	9.0–10.0	Persists
Blastokinin	5.5	6.5– 7.0	Disappears before implantation
Transferrin	7	8.5–10.0	Persists
Posttransferrin	7	8.5–10.0	Persists
γ-globulin[a]	6	8.0–10.0	Persists

SOURCE: Unpublished disk electrophoresis data of Hamana and Hafez.
[a] A broad zone containing 3–6 bands according to the stage of development.

C. *Amino Acid Content*

Blastocoelic fluid contains the amino acids leucine, valine, phenylalanine, methionine, alanine, glutamic acid, glutamine, aspartic acid, asparagine, threonine, serine, glycine, histidine, lysine, ornithine, arginine, tyrosine, proline, hydoxyproline, and cystine (Lesinski et al. 1966; Lutwak-Mann 1966). The concentrations of the constituents vary with the species and stage of development (table 5) and reflect changes in the composition of the uterine fluids controlled by ovarian hormones (Gregoire, Gongsakdi, and Rakoff 1961).

The relative amounts of amino acids in the blastocyst fluid do not correspond to the relative amounts required for growth (Daniel and Krishnan 1967). Alanine, for example, occurs in abundance, but has no appreciable effect on blastocyst growth, at least *in vitro.* On the other hand, histidine, methionine. phenylalanine, threonine, and tyrosine occur only in trace amounts but are essential for *in vitro* growth (table 5).

Arginine and lysine are required in relatively high concentrations (10^{-2} and 10^{-3} M, respectively) for optimal growth (Daniel and Krishnan 1967). Alanine and glutamic acid benefit short-term growth but depress development when present in high concentrations for 24 hr or more (Daniel and Krishnan 1967). Complete omission of nonessential amino acids greatly reduces the rate of growth.

Specific differences in amino acid requirements are apparent. For example, valine is essential for the *in vitro* growth of rabbit blastocysts and isoleucine is not; the reverse is true for mouse blastocysts (Gwatkin 1966). Amino acid requirements for blastocyst growth generally correspond to those for postnatal growth.

The amino nitrogen content of the 7-day-old rabbit blastocyst is 5 times the concentration in the maternal blood serum (Lesinski, Jajszczak, and Choroszewska 1967). Implantation is associated with a progressive decrease in the amino nitrogen and a concurrent increase in the protein concentration of the blastocoelic fluid (fig. 20). The decline in amino nitrogen may be due to protein synthesis.

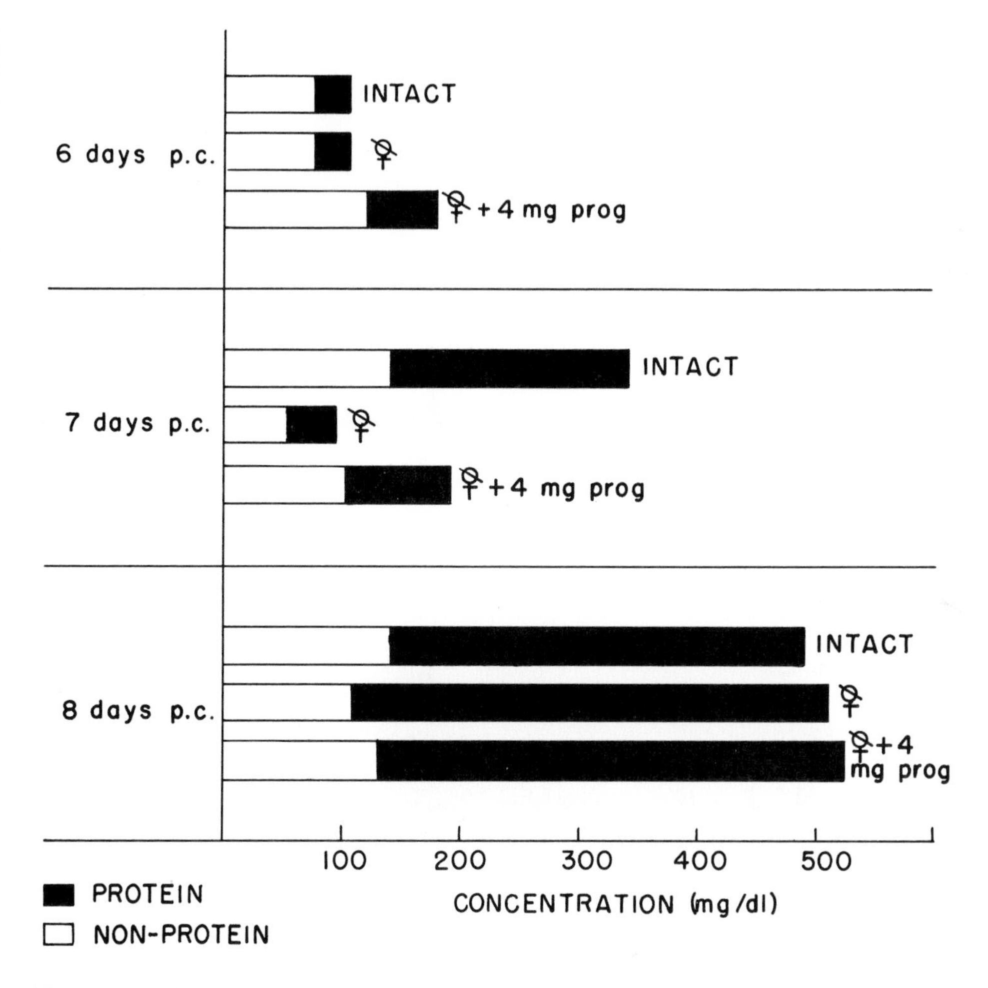

Fig. 18. The effect of progesterone administration on the concentration of protein and nonprotein nitrogen in the blastocoelic fluid of blastocysts ovariectomized on 5 (*top*), 6 (*middle*), and 7 (*bottom*) days *p.c.* Note that ovariectomy is only detrimental at 6 days *p.c.* (Data by permission from S. Sugawara and E. S. E. Hafez, *Anat Rec* 158:115, 1967)

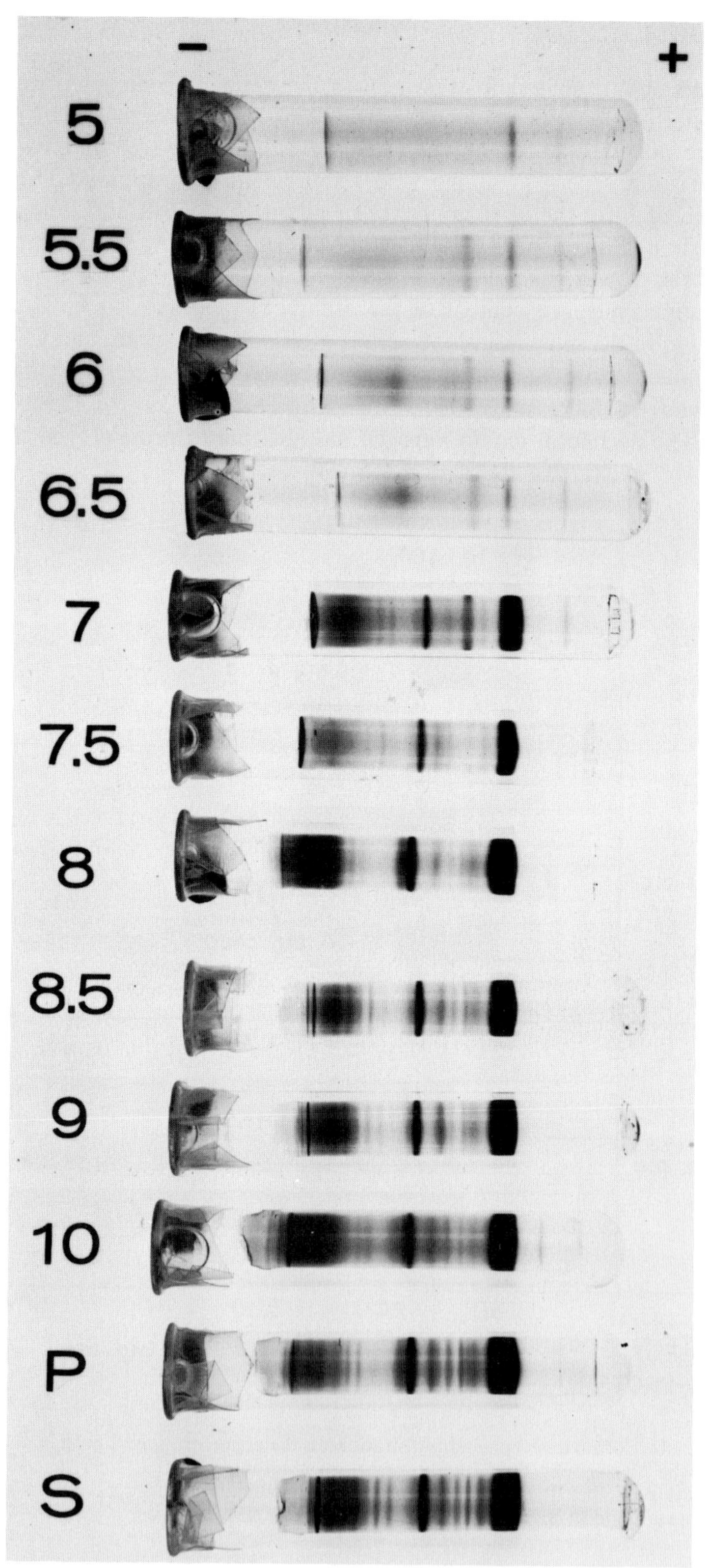

Fig. 19. Developmental changes in the protein fractions of blastocoelic fluid from the rabbit between 5 and 10 days *p.c.* Blastokinin appeared at 5.5 days *p.c.*, reached a maximum at 6.5–7 days *p.c.*, and then disappeared at 8 days *p.c.* (before implantation). (Hamana and Hafez, unpublished data)

D. *Carbohydrate Content and Energy Metabolism*

Characteristic and pronounced increases in the carbohydrate content of the blastocoelic fluid occur after implantation. Although they cannot penetrate the unimplanted blastocyst (Lutwak-Mann 1954, 1966), glucose, fructose, and sucrose pass freely from the rabbit maternal blood into the blastocoelic fluid of the implanted 8-day blastocyst. The glucose and lactate content of the fluid rises as the

TABLE 5

AMINO ACIDS IN BLASTOCOELIC FLUID AND REQUIRED FOR THE *in Vitro* GROWTH OF RABBIT BLASTOCYSTS

AMINO ACIDS IN BLASTOCOELIC FLUID		AMINO ACID REQUIREMENT *in Vitro*		
			Nonessential	
Amino Acid	Relative Quantity	Essential	Beneficial	Neutral (No Appreciable Effect)
Alanine Asparagine+aspartic acid Glutamine+glutamic acid Glycine Serine	Abundant (100–600 mg/ml)	Serine	Asparagine+aspartic acid Glutamine+glutamic acid Glycine	Alanine
Arginine Leucine Lysine+ornithine Valine	Moderate (100 mg/ml)	Arginine[a] Leucine[a] Lysine[b] Valine[c]		
Cystine Histidine Hydroxyproline Methionine Phenylalanine Proline Threonine Tyrosine	Detectable	Histidine[a] Methionine[b] Phenylalanine[b] Threonine[a] Tyrosine(?) Tryptophane[b]	Tyrosine[b] Cysteine[a] Isoleucine[c]	Hydroxyproline Proline

SOURCES: Five- and 6-day-old rabbit blastocysts (Lesinski et al. 1966).
Five-day-old rabbits (Daniel and Krishnan 1967).
Amino acid requirements of the mouse blastocyst *in vitro* (Gwatkin 1966).

[a] Amino acids essential for the growth of mouse blastocysts.

[b] Absence of these amino acids reduced but did not inhibit the growth of mouse blastocysts.

[c] Absence of these amino acids had no effect on the growth of mouse blastocysts.

embryo increases in age and weight (Lutwak-Mann 1962) and can be reduced experimentally by ovariectomizing the mother at 6 days *p.c.* (However ovariectomy has no effect at 5 or 7 days *p.c.*) Progesterone, alone or in combination with estradiol, partly restores the concentrations (Sugawara and Hafez 1967*c*).

The fact that mounts of glycogen are large during the cleavage of mouse eggs and diminish as the blastocyst expands (Thomson and Brinster 1966) suggests that glycogen is a source of energy during the preimplantation period.

Cleaving eggs oxidize glucose mainly via the hexose monophosphate pathway. However as the blastocyst develops, respiration suddenly rises and glucose oxidation is routed primarily through the citric acid cycle. In the rabbit the citric acid cycle is functional at the earliest stages of development, but the Embden-Meyerhof pathway does not appear for several days (Fridhandler, Hafez, and Pincus 1957; Fridhandler 1961; Daniel 1967*c*.)

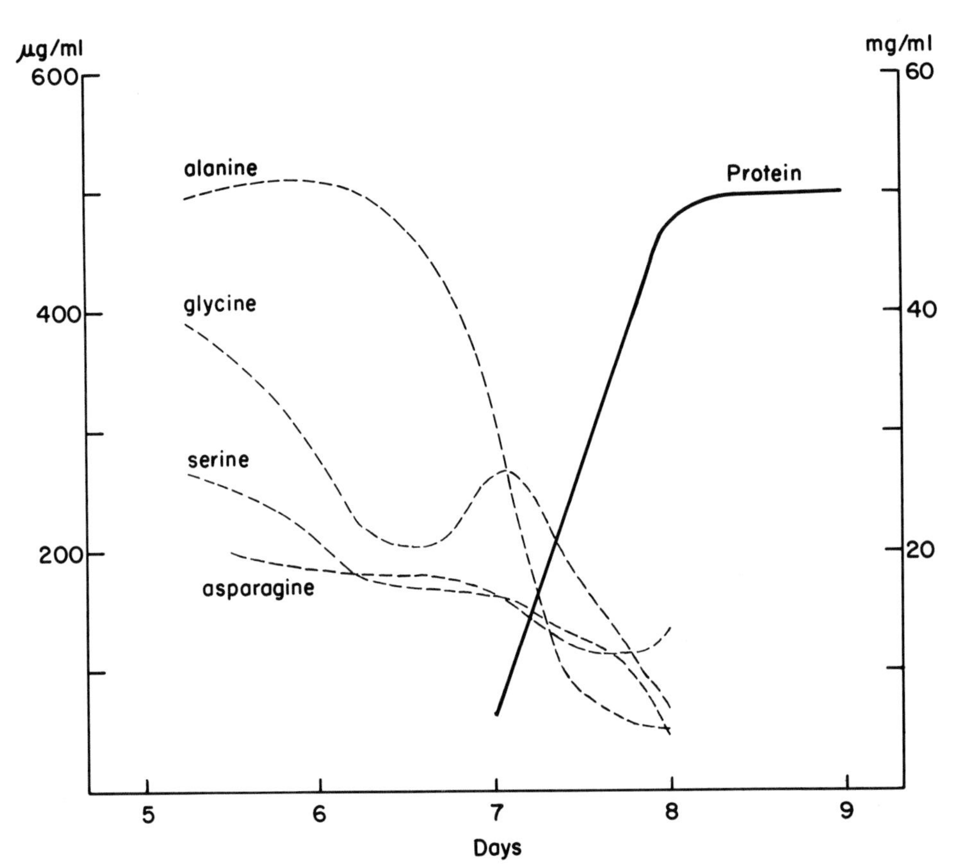

Fig. 20. Levels of free amino acids and protein in the blastocoelic fluid of the rabbit. Note that the decline in free amino acids coincides with a rise in protein at 7.5 days *p.c.* (Data from J. Lesiński, S. Jajszczak, K. Bentyn, and J. Janczarski, *Pol Med J* 6:1074, 1966; and J. Lesiński, S. Jajszczak, and A. Choroszewska, *Amer J Obstet Gynec* 98:48, 1967)

The respiratory rate of the blastocyst varies among species but does not differ appreciably between delayed and nondelayed blastocysts. Apparently the control of delayed implantation is not related to respiratory changes (Gulyas and Daniel 1967).

E. *Enzymes*

Enzymatic activity in the blastocyst varies with the species, stage of development, and regions of the organism. Phosphatases and dehydrogenases have been identified histochemically in the trophoblast (Dalcq 1957, 1959; Ishida and Chang 1965; Hafez 1968; Tondeur 1959). Phosphatase, transaminase, dehydogenase, and amylase activities have been measured spectrophotometrically in homogenates of rabbit blastocysts (Hafez and White 1967). The activities of most enzymes increase with development (table 8).

Many dehydrogenase systems (including those for glucose-1-phosphate, glucose-6-phosphate, glycerophosphate, glutamate, lactate, malate, isocitrate, and succinate) become dramatically active during formation of the rabbit blastocyst (4 days *p.c.*) (Sugawara and Hafez 1967*b*). A similar sharp rise occurs in the acid phosphatase content of ovine blastocoelic fluid 31–35 days after breeding (Hafez and White 1968). Except for amylase (2 to 5 times higher) the enzymatic activities in the rabbit's blastocoelic fluid are generally 10- to 100-fold lower than corresponding activities in the trophoblast (Hafez and White 1967).

The commonly encountered lactic dehydrogenase (LDH) isozymes of mammals are composed of two polypeptide subunits, A and B. Since the blastocyst of the mouse contains only B subunits before implantation but considerable amounts of A subunits thereafter, an abrupt change in genetic activity appears to accompany implantation (Rapola and Koskimies 1968).

F. *Vitamin, Mineral, and Hormone Content*

^{24}Na passes freely into blastocoelic fluid, but ^{32}P does not, at least until implantation, when both phosphorus and chloride increase noticeably in the fluid (Bennett, Boursnell, and Lutwak-Mann 1958). The potassium concentration of blastocoelic fluid is very high until implantation, when it drops to levels found in the maternal blood serum. Water-soluble vitamins (thiamin, riboflavin, vitamin B_{12}, nicotinic acid) are present also in the blastocoelic fluid before implantation (Brambell and Hemmings 1949; Jacobson and Lutwak-Mann 1956; Lutwak-Mann 1960, 1966). It is interesting that the rabbit blastocyst appears to grow in the absence of vitamin B_{12} (Daniel 1967*b*), while storing this vitamin in the blastocoelic fluid.

When grown *in vitro* 5-day-old rabbit blastocysts require hypoxanthine, inositol, pyridoxine, riboflavin, thiamidine, niacimide, and folic acid (Daniel 1967*b*). Exogenous choline chloride, thymidine, vitamin B_{12}, vitamin A, calcium pantothenate, biotin, ascorbic acid, ergosterol, and tocopherol are not essential at this stage of development, nor do they improve growth when added.

The rabbit blastocyst can synthesize cholesterol and pregnenolone from acetate. This may explain why LH and ACTH increase the rate of blastocyst differentiation *in vitro* (Huff and Eik-Nes 1966).

III. Maternal Factors and Experimental Manipulations Affecting the Blastocyst

Development of the blastocyst depends directly on the condition of the surrounding uterus. The factors which produce the maternal environment of the uterus—its glandular secretions, blood, lymph, and nerve supplies—not only are genotypically prescribed, but also are influenced by neighboring tissues and the mother's physiological status. The blastocyst induces local morphological changes in the nearby uterine wall before and after implantation; the ovary has local effects on adjacent uterine tissue; and the growing fetus undoubtedly alters the permeability of the endometrium to electrolytes and organic molecules. Parity, immunological reactions, lactation, maternal nutrition, disease, and complications of pregnancy also modify the physiological integrity of the uterus in relation to blastocyst development.

A. *Genetic Factors*

The genetic activity of developing embryos varies among breeds (Venge 1950), strains, individuals, and even among the progeny of the same female (Hafez 1962). C57BL mouse blastocysts take several hours longer than the blastocysts of Swiss Webster albino mice to reach a given stage of differentiation. Whitten and Dagg (1962) crossbred a fast-cleaving (129) with a slow-cleaving (BALB/C) strain of mice. At 50 and 77 hr the F_1 hybrid blastocysts were consistently further advanced in development than the BALB/C inbreds (table 6). On the other hand

TABLE 6

PERCENTAGE OF OVA WHICH HAD DEVELOPED INTO BLASTOCYSTS 77 HR AFTER OVULATION IN TWO STRAINS OF MICE AND THEIR RECIPROCAL CROSSES

Female Parent	Male Parent	Percentage of Blastocysts 77 Hr after Ovulation
129	129	26%
129	BALB/C	16%
BALB/C	BALB/C	11%
BALB/C	129	47%

SOURCE: Whitten and Dagg 1962.

no differences occur in the size of the 6-day-old blastocyst and the diameter of the embryonic disk among the nulliparous Chinchilla (live weight 4–5 kg), New Zealand White (3–4 kg), and Dutch-Belted (1.5–2 kg) breeds of rabbit (Hafez and Rajakoski 1964; Kodituwakku and Hafez 1969).

Delayed fertilization of the rabbit's ova causes polyploidy, mixoploidy, and trisomy in the blastocyst (Shaver and Carr 1967). Although triploid rabbit blastocysts have the same range of diameters as diploid blastocysts, they are mostly smaller (Bomsel-Helmreich 1965). The "hybrid blastocysts" which result from crossing two species of rabbits die immediately, whereas those of mink-ferret crosses continue to develop until the 22d day of pregnancy (Chang 1966).

B. *Endocrine Factors*

1. *Progesterone:* The quantitative effects of progesterone on the growth of the blastocyst are controversial and seem to vary with the species. Adams (1965) found no relationship between the number of corpora lutea in the rabbit ovary and blastocyst size. Nonetheless superovulation influences blastocyst development in the rabbit (fig. 21) and sheep (Wintenberger-Torres 1968). The administration

of progesterone to sheep increases the rate of blastocyst development between the 8th and 16th days of gestation (Wintenberger-Torres and Rombauts 1968). Superovulation in sheep increases the amount of progesterone secreted (from multiple corpora lutea) by the ovary (Stormshak et al. 1963; Hafez, Estergreen, and Foster 1965). Estrogen-progesterone imbalance also produces several types of blastocyst degeneration (fig. 22).

The process of implantation is influenced by the activity of the corpora lutea. Four corpora lutea are necessary to sensitize the rabbit uterus sufficiently to produce deciduomata, but only two are required to maintain pregnancy. Within limits a quantitative relationship exists between the number of blastocysts which implant

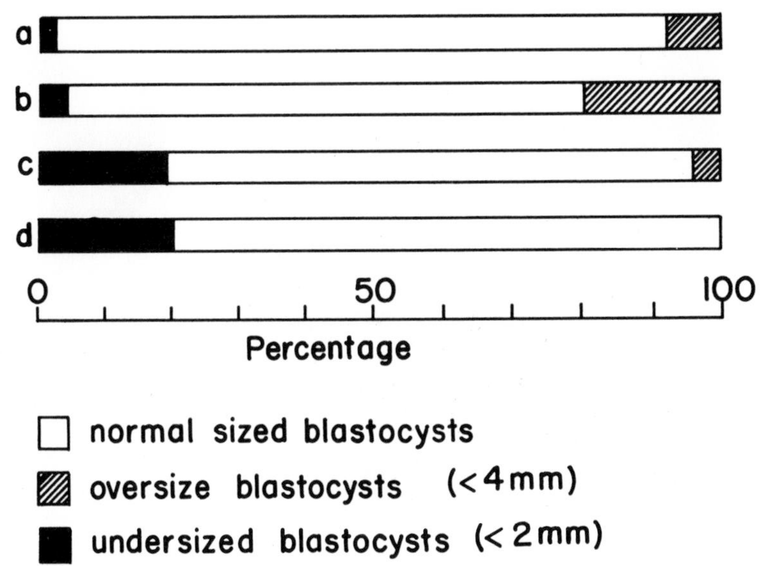

Fig. 21. The effect of superovulation on the size (outer diameter) of 6-day-old rabbit blastocysts: (*a*) normally ovulated doe; (*b*) superovulated doe (good response); (*c*) superovulated doe (poor response, but many corpora lutea); (*d*) superovulated doe (poor response, few corpora lutea). Note that a poor superovulatory response produces more undersized blastocysts and fewer oversized ones. A good superovulatory response with many corpora lutea yields large numbers of oversized blastocysts. (By permission from E. S. E. Hafez and E. Rajakoski, *J Reprod Fertil* 7:229, 1964)

and the dose of progesterone injected into ovariectomized rabbits. When 0.75 mg of progesterone is injected per day only one implantation occurs. The implantation rate then increases with graded doses up to 1.50 mg per day, above which it decreases (Kehl and Chambon 1948, 1949; Chambon 1960). A similar relationship exists between the amount of injected progesterone and the number of viable blastocysts in ovariectomized rabbits (Allen and Wu 1959; Hafez 1964*c*).

2. *Effects of Bilateral Ovariectomy.* The growth rate of the blastocyst is influenced by bilateral ovariectomy in such species as the rabbit (Hafez 1964*a*) and pig (Dhindsa, Dziuk, and Norton 1967). The timing of the operation seems to be the critical element in the effect. If a rabbit is ovariectomized at 6 days *p.c.*,

blastocyst growth and blastocoelic fluid volume are normal 24 hr later; ovariectomy at 5 or 7 days *p.c.* reduces blastocyst growth, the volume of blastocoelic fluid, and the latter's albumin content within 24 hr. Progesterone treatment corrects these changes (Sugawara and Hafez 1967*a*).

3. *Effects of Unilateral Ovariectomy and of the Uterus.* In some species (guinea pig, pig) the uterus contributes to the regular cyclical activity of the ovary by periodically shortening the functional life of the corpus luteum (du Mesnil du Buisson 1961; du Mesnil du Buisson and Dauzier 1959; du Mesnil du Buisson and

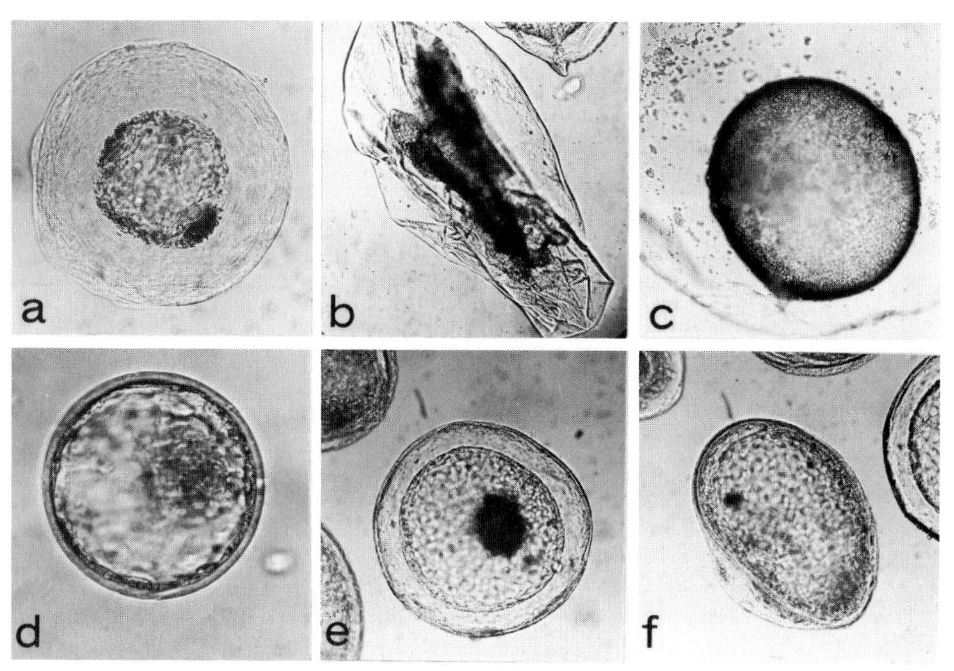

Fig. 22. Rabbit blastocysts undergoing different types of degeneration because of an estrogen-progesterone imbalance (*a, b, c, d*) or an unfavorable hormonal balance caused by superovulation and presence of luteinized follicles (*e, f*).

Rombauts 1963*a, b;* Fischer 1967; exhaustively reviewed by Barley, Butcher, and Inskeep 1966).

The ovary also has a local effect on the uterus of such species as the brush opossum (*Trichosurus vulpecula*) (Pilton and Sharman 1962) and the giant fruit bat (*Pteropus giganteus* Brünnich) (Marshall 1953). In the latter both ovaries and uterine horns are functional, but only one egg is released after copulation. A progestational reaction occurs in the uterine horn adjacent to the ovary which contains the ruptured follicle, whereas the opposite horn retains its estrous appearance (Marshall 1953). The site of this asymmetrical reaction (at which implantation subsequently takes place) is confined to the distal end of the horn where uterine and ovarian tissue lie in close proximity. When rabbits are unilaterally ovariectomized at 6 days *p.c.*, fewer blastocysts implant in the uterine horn on the ovariectomized side than on the opposite side (Lutwak-Mann, Hay, and Adams 1962). The same is true for rats which are unilaterally ovariectomized at 3.5 days *p.c.* (Lamming and Little 1968). However ovariectomy at 5.5 days *p.c.* has no effect on implantation, presumably because the remaining ovary can secrete enough hormone to cause endometrial proliferation in both uterine horns at that time. Embryonic mortality also increases in the uterine horn on the treated side of the animal after unilateral ovariectomy. Bilateral salpingemphraxis reduces embryonic survival (Little, Gray, and Lamming 1969). These facts suggest that the local effect of the ovary on the uterus may be mediated via the oviduct and may involve hormones secreted by the ovary before its corpora lutea are fully formed and functional.

TABLE 7

THE EFFECT OF LACTATION ON THE WEIGHT OF THE BLASTOCYST AND TROPHOBLAST AND ON THE VOLUME OF BLASTOCOELIC FLUID

DAY OF LACTATION WHEN BLASTOCYSTS RECOVERED	BLASTOCYST WT (mg)		TROPHOBLAST WT (mg)		BLASTOCOELIC FLUID VOLUME (μl)	
	Range	Mean±S.E.	Range	Mean±S.E.	Range	Mean±S.E.
17.5	54–152	97±30	19.3–3.2	2.6±0.7	50–146	90±29
22.5	106–198	154±29	3.6–4.4	4.0±0.1	92–186	142±29
27.5	141–294	204±51	3.2–6.4	4.8±1.9	136–286	198±28
Nonlactating (controls)	96–286	188±21	3.4–5.3	4.6±0.6	89–220	179±39

SOURCE: Hafez and Sugawara 1967.
NOTE: All the recovered blastocysts were 7.5 days old.

C. *Lactation*

Lactation seems to be one of the most important maternal factors influencing blastocyst growth (table 7; fig. 23). Blastocyst weight is subnormal in heavily lactating New Zealand rabbits (Hafez and Sugawara 1967). In some cases blastocysts undergo a degeneration associated with regression of the corpora lutea. Injections of progesterone partly alleviate the detrimental effects of lactation, even though the maintenance of corpora lutea in rabbits is estrogen-dependent (Keyes and Nalbandov 1967). In rabbits the effect of lactation on blastocyst mortality varies with the season of the year, the breed, strain, litter size, lactating ability of the mother, stage of lactation, and maternal caloric intake.

In most large populations of rats and mice blastocyst implantation is delayed 1-7 days, depending on the number of suckling young (table 8). In some populations it is delayed as long as 8-14 days (Mantalenakis and Ketchel 1966). Under normal conditions, however, no delay occurs unless the number of suckling young exceeds 4. It is worth noting that lactation also delays the loss of the zona pellucida from the blastocysts of mice (Rumery and Blandau; see chapter 7).

The blastocyst of the quokka (*Setonix brachyurus*) develops only to a unilaminar stage (0.3 mm in diameter and containing about 70 cells) as long as a young occupies the mother's pouch (Clark 1966). If the young is removed, the blastocyst resumes development and is born as a fetus 25 days later (Sharman, 1955; Tyndale-Biscoe 1963). This phenomenon of "lactation-controlled delayed implantation" also occurs in several other species of macropodid marsupials, including the potoroo (*Potorous tridactylus*), tammar wallaby (*Protemnodon eugenii*), euro (*Macropus robustus*), and red kangaroo (*Megalaio rufa*) (cf. Sharman 1963).

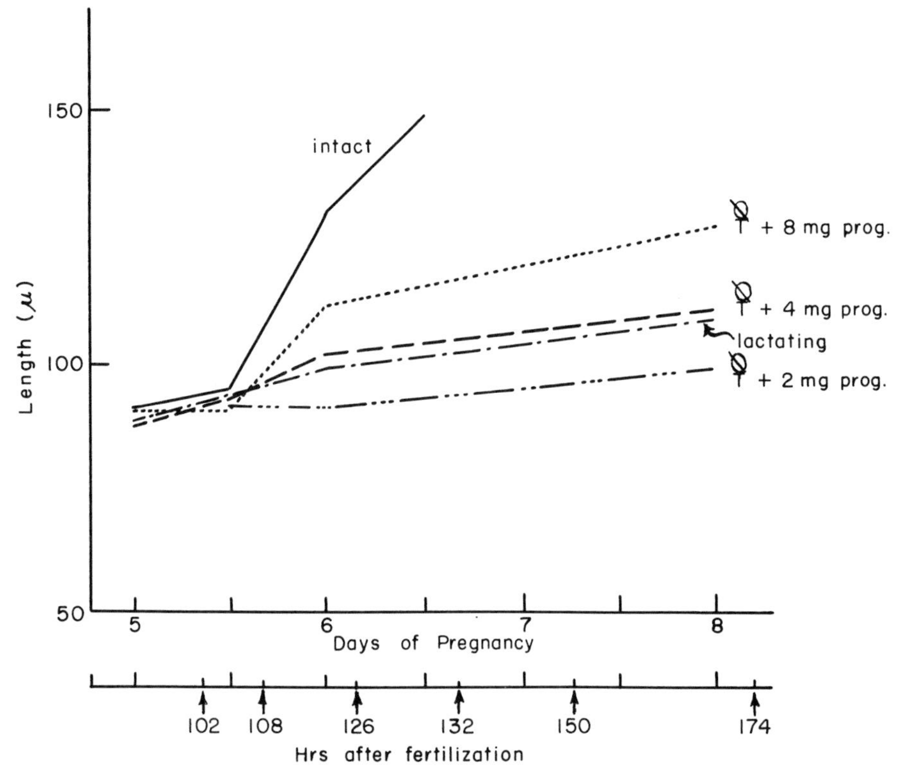

Fig. 23. The effects of lactation and progesterone on developmental changes in the major axis length of rat blastocysts. The adult females were ovariectomized prior to progesterone therapy. (By permission from S. Sugawara, S. Otake, and S. Takeuchi, *Tohoku J Agr Res* 18:107, 1967)

TABLE 8

RELATIONSHIP BETWEEN THE NUMBER OF SUCKLING YOUNG AND THE DELAY IN IMPLANTATION

Number of Suckling Young	Delay in Implantation (Days)	
	Rat	Mouse
4 or less	2.18	....
5–8	3.68	2.00
9–12	4.22	4.09
13 or more	4.76	4.39

SOURCE: Mantalenakis and Ketchel 1966.

The resumption of blastocyst development in the lactating red kangaroo is associated with the appearance of two specific proteins in the milk (Bailey and Lemon 1966).

Lactation also modifies the total nitrogen and protein content of blastocoelic fluid in the rabbit (Hafez and Sugawara 1968). The fluid contains no detectable α_1- or α_2-globulins if does are rebred 5–10 days *post partum* (Hafez and Sugawara 1967); however its α-globulin distributions approach normal values if rebreeding occurs more than 10 days *post partum*. The β-globulin content of blastocoelic fluid is elevated in blastocysts conceived at all stages of lactation, whereas the γ-globulin levels are higher in control blastocysts than in those of females which have lactated for 21 or 28 days. Lactation probably acts by way of the ovary whose hormones influence uterine enzymatic activity and protein synthesis. However little is known about uterine secretion during lactation or its role in the accumulation of proteins within blastocoelic fluid.

D. *Transfer of Embryos to an "Alien Environment"*

Blastocysts have been studied in three "alien environments": (*a*) the intrauterine environment of a different species or of another animal of the same species at a different stage of development; (*b*) the extrauterine environment of the same or a different animal of the same or a different species; and (*c*) tissue and organ cultures.

Since conditions in the uterus vary throughout the implantation period (cf. Schwick 1965), the blastocyst's stage of development and the luteal stage of the recipient must be synchronized for successful transfers (Chang 1950; Adams 1965). Cultured blastocysts develop normally in both uterine and ectopic sites (Billington, Graham, and McLaren 1968) and after being transferred to receptive uteri give rise to normal fetuses and living young.

Uterine factors apparently play a minimal role in the escape of murine blastocysts from the zona pellucida (Cole 1967), as shown by the shedding of the zonae pellucidae by blastocysts implanted in the anterior chamber of the eye, the peritoneal cavity (Bryson 1964), and under conditions of extreme hormone imbalance (hypophysectomy, ovariectomy, adrenalectomy).

1. *Interspecific Transfer.* Blastocysts from one species have been successfully transferred to the oviducts or uteri of other species and to the kidney capsule of the same species (table 9). Success depends on the species of the donor and recipient and the stage of embryonic development. Rat blastocysts (with or without the zona pellucida) survive for 2 days in the oviducts of the pseudopregnant rabbit (as demonstrated by their capacity to implant in rats), but few survive to term. Rabbit blastocysts begin to degenerate within 48 hr in the cornua of rats in a "delayed implantation" condition (Yoshinaga and Adams 1967). Most ferret eggs develop briefly in the uterus of the mink, but do not implant and later die (Chang 1968). In contrast, mink eggs will implant in the uterus of the ferret but fail to go through the normal dormant state and degenerate. Apparently, the mink egg's dormant state during delayed implantation, like the rat's, is controlled by the condition of the uterus (Dickmann and De Feo 1967).

Blastocysts are frequently unable to induce decidualization in a foreign uterus (cf. McLaren 1967): Rat blastocysts, for example, are incapable of inducing decidualization in hamster uteri (Blaha and De Feo, 1964; De Feo 1967). Hamster blastocysts, on the other hand, properly synchronized with the rat uterus, attach to the epithelium and induce normal decidualization (fig. 24).

2. *Transfer to Ectopic Sites.* Ova can develop *in vivo* to the blastocyst stage at

TABLE 9

SUCCESSFUL AND UNSUCCESSFUL INTRA- AND INTERSPECIFIC TRANSFER OF BLASTOCYSTS

Site of Blastocyst Transfer	Species	Reference
Oviduct or uterus	Sheep⇄goat	Warwick and Berry 1949
	Sheep→rabbit	Averill, Adams, and Rowson 1955
	Rat→mouse	Tarkowski 1962
	Rabbit⇄ferret	Chang 1966, 1968
	Rat⇄hamster	Blaha and De Feo 1964
	Rat→rabbit	Yoshinaga and Adams 1967
	Mink⇄ferret	Chang 1968
	Rabbit→rat (did not survive)	
	Mink→rabbit (did not survive)	Daniel 1967*a*
	Rat	
	Mouse→rabbit	Briones and Beatty 1954
	Guinea Pig (did not survive)	
	Rat⇄rat	
	(dormant blastocyst→active uterus) (active blastocyst→dormant uterus)	Dickman and De Feo 1967
Explanted oviduct	Rat⇄mouse	Whittingham 1968
Kidney capsule	Rat⇄mouse	Kirby 1962*a*, *b*; 1965
	Mouse→mouse kidney	Kirby 1967

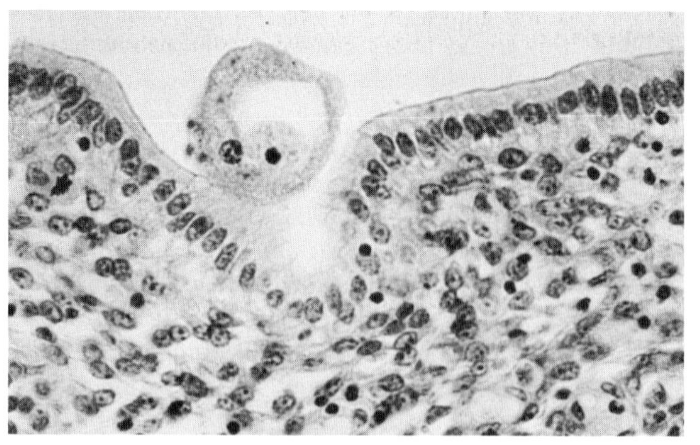

Fig. 24. Cross section of a rat uterus at noon on day 4 of pseudopregnancy. The blastocyst is that of a hamster and was introduced into the uterus as a morula on the previous day. Such blastocysts attach (as indicated), induce decidualization, and then die, often on the same day. However decidualization continues at the same rate as it does in a normal implantation until at least day 9. ×400. (Blaha and De Feo, unpublished)

ectopic sites (e.g., intact or explanted oviduct, kidney capsule) in the same or another species (table 9) (Whittingham 1968). There is evidence that those which develop outside the uterus have an impaired developmental capacity. Mouse eggs develop into blastocysts in the oviduct and appear to be histologically and morphologically normal (Kirby 1965); yet they fail to grow when transplanted to the kidney. If replaced in a hormonally receptive cornua they will implant and develop into normal embryos. If mouse blastocysts are transferred from the cornua to the kidney during delayed implantation, they "implant" and develop immediately, whereas those in the cornua remain in delay (Kirby 1967).

E. *Effects of Overcrowding in Utero*

Overcrowding *in utero* can be induced experimentally by superovulation, whole-body X-irradiation (Ward and Hahn 1967), or by transferring an excessive number of fertilized eggs into the oviduct of a receptive female.

Every species apparently has a maximum number of implantations that can be maintained adequately throughout pregnancy. The maximum is 4 in superovulated sheep and cattle (Hafez 1964*c*). Egg transfer experiments indicate that the implantation capacity (maximal number of implanted blastocysts) of the rabbit ranges between 1 and 50 (Hafez 1964*c*). Overcrowding *in utero* results in uneven spacing (fig. 25) and a high degree of postimplantation mortality (figs. 26, 27) but apparently has no effect on the development of the rabbit's blastocyst or embryonic disk (Hafez and Rajakoski 1964). The remarkably different implantation capacities of female rabbits are related to the physiological integrity of the uterus (maturity and development of the endometrium, uterine vascularity, age and parity of the female, and genetic factors).

F. *Effects of Undercrowding in Utero*

When only 1 or 2 eggs are transferred to the rabbit's oviduct, implantation is normal, but the incidence of postimplantation mortality (10–12 days *p. c.*) is unusually high (fig. 28). Recipients with a single implantation frequently resorb the fetus completely or give birth to giants (viable or dead); the dystocia associated with birth of the latter may cause fetal (perinatal) or maternal death (fig. 29). On the other hand 55% of the fetuses in recipients with 2 implants survive to full term; that is, a minimum of 2 implants is apparently required for fetal survival and normal parturition, at least in the New Zealand White breed.

The number of blastocysts can be reduced experimentally also in the pig by partial hysterectomy (du Mesnil du Buisson and Rombauts 1963*a*). Pregnancy continues in animals with only 1 embryo if all of the uterus is removed except that portion occupied by the embryo. If embryos are confined to a single uterine horn, the ovarian corpora lutea regress, thereby terminating pregnancy at an early stage unless the nongravid horn is removed before the 14th day of the estrous cycle (du Mesnil du Buisson 1961; Polge, Rowson, and Chang 1966). The presence of a small nongravid segment of uterus does not interrupt pregnancy, although it may cause ipsilateral regression of corpora lutea (Dhindsa and Dziuk 1968). These observations suggest that the presence of blastocysts in a large area of both uterine

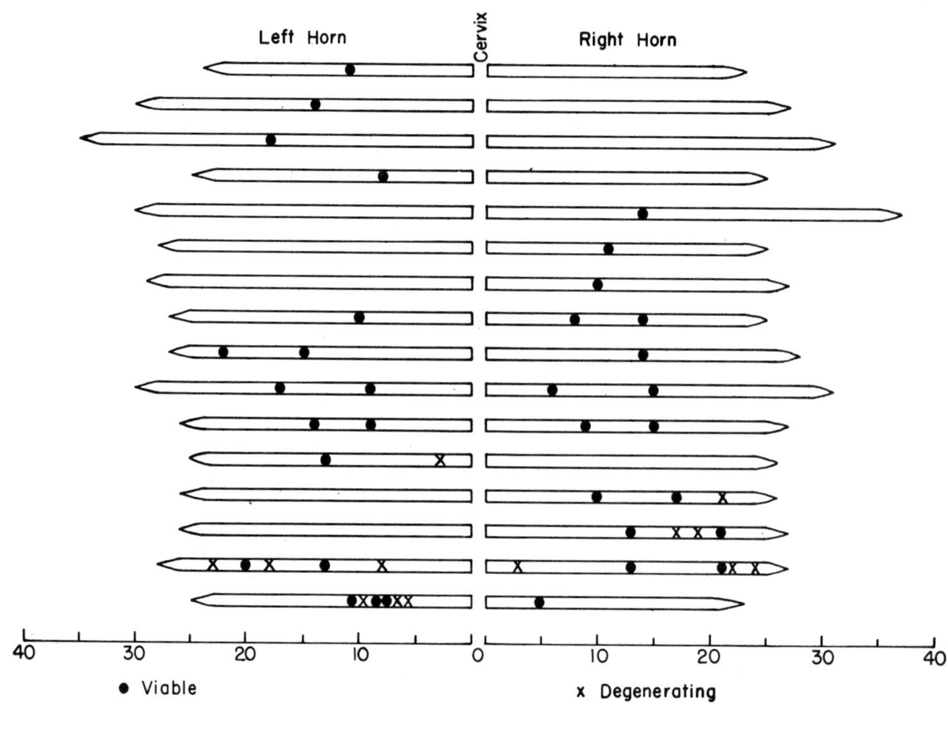

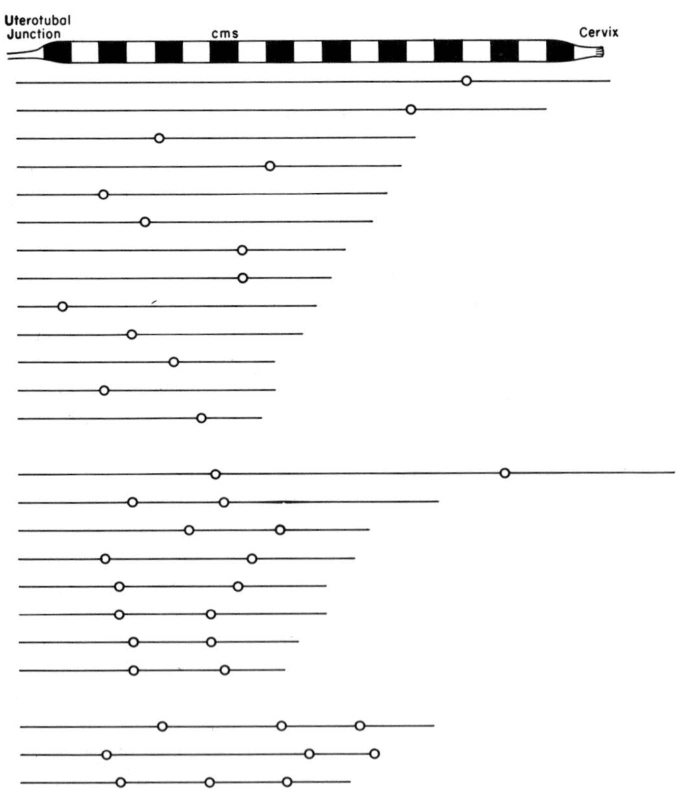

Fig. 25. *Top:* the spacing of blastocysts within the uterine horns during single and multiple pregnancy in cattle. In single pregnancies the embryo is midway between the uterotubal junction and the cervix. In multiple ones the embryos are unevenly distributed. (By permission from E. S. E. Hafez, *Anat Rec* 148:203, 1964)

Bottom: spacing of rabbit blastocysts *in utero;* 1, 2, or 3 eggs were transferred to one oviduct and the animals were autopsied at 9 days *p.c.*

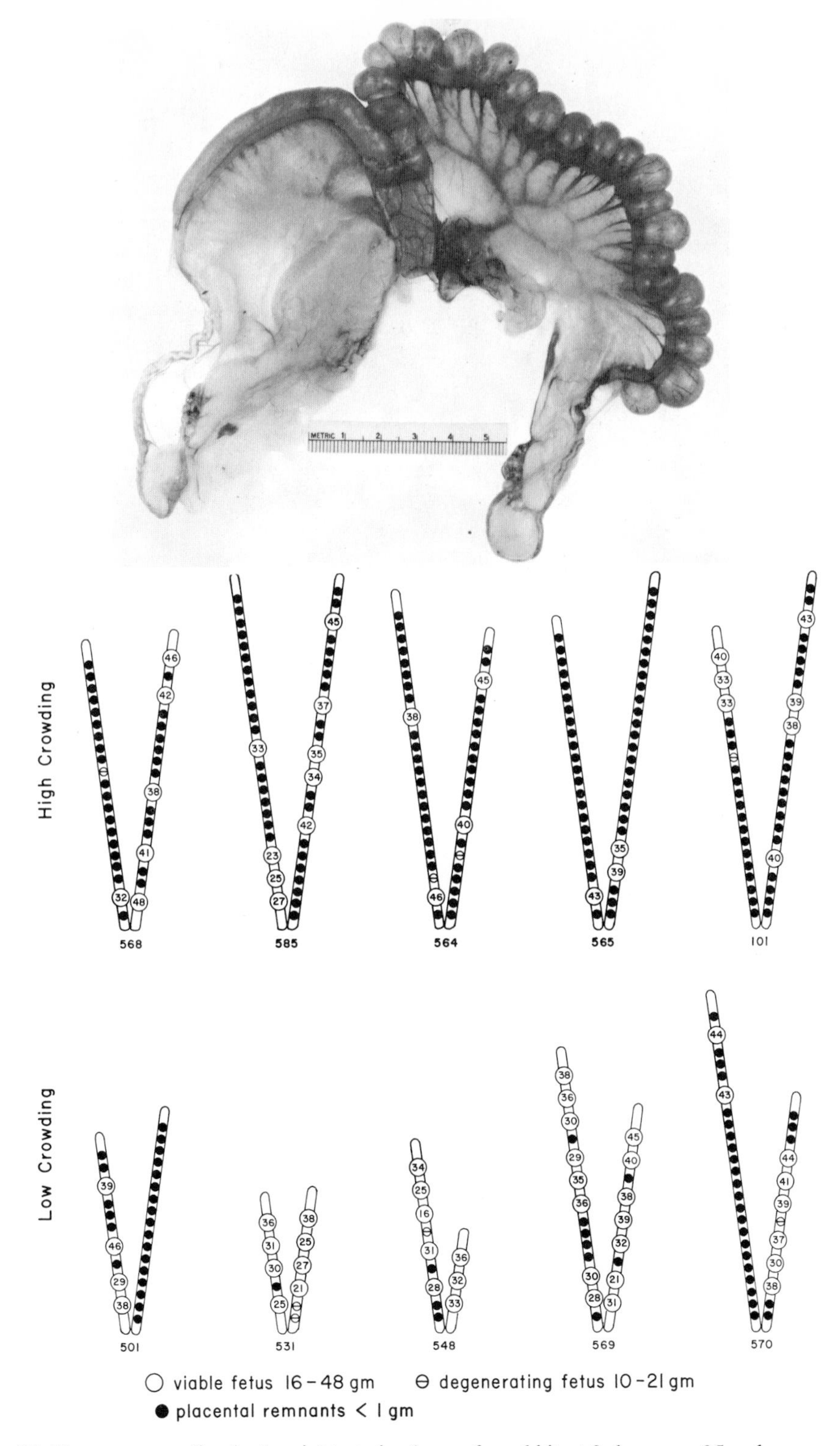

Fig. 26. *Top:* overcrowding in the right uterine horn of a rabbit at 9 days *p.c.*; 35 embryos were transferred to the recipient at 2 days of pseudopregnancy. Note the vascular supply of the implantation sites.

Bottom: the relationship between overcrowding *in utero* and fetal survival at 29 days *p.c.* Large numbers of embryos were transferred to both uterine horns at 1 day *p.c.* Note that fewer viable fetuses remain in cases of low crowding than in cases of high crowding. The location of a fetus in the uterine horn had no effect on its survival or weight at 29 days *p.c.* The figures within circles denote fetal weight at 29 days *p.c.* (By permission from E. S. E. Hafez, *J Exp Zool* 156:269, 1964)

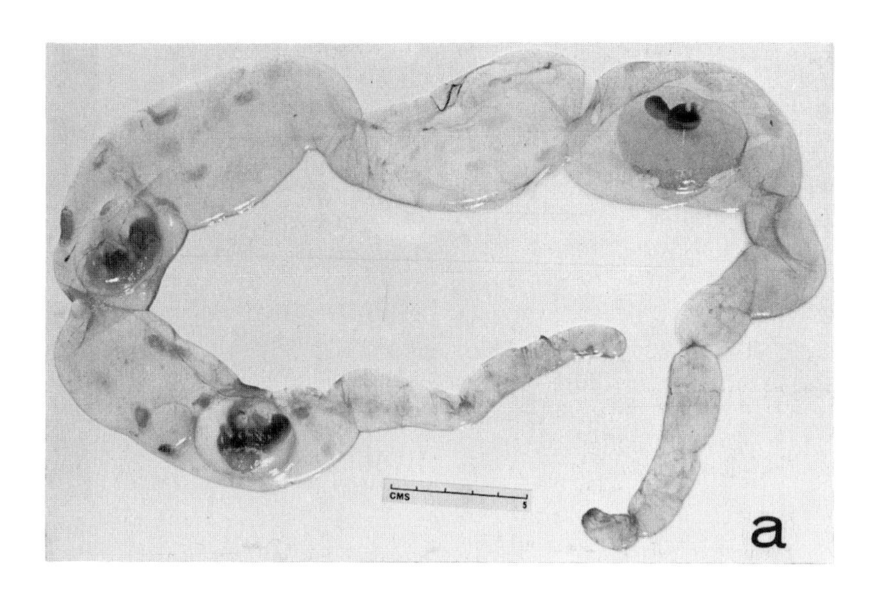
CMS
5
a

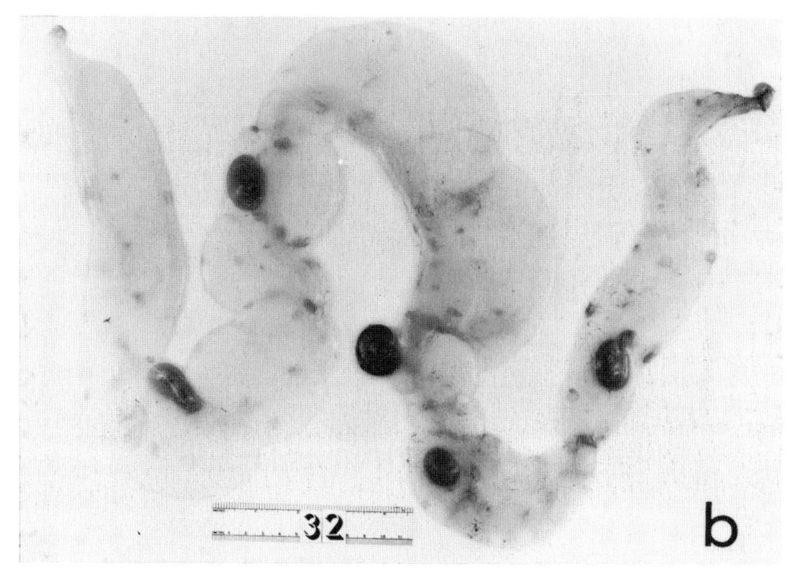
32
b

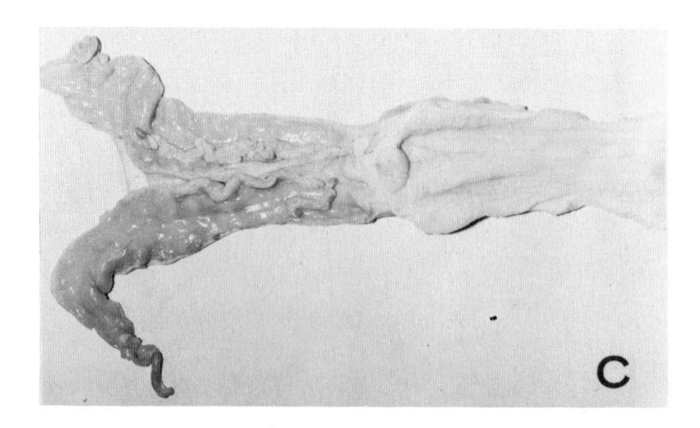
c

horns is necessary to establish and maintain pregnancy in certain polytocous species. The pregnancy-inhibiting action of the uterus is probably related to the presence of a thermolabile, nondialyzable "uterine luteolytic factor" (ULF) which has been extracted from the pseudopregnant endometrium of rodents.

IV. Concluding Remarks and Future Research

Repeated experiments with blastocysts frequently yield a variety of conflicting results. Many uncontrollable factors are probably responsible, including the variability

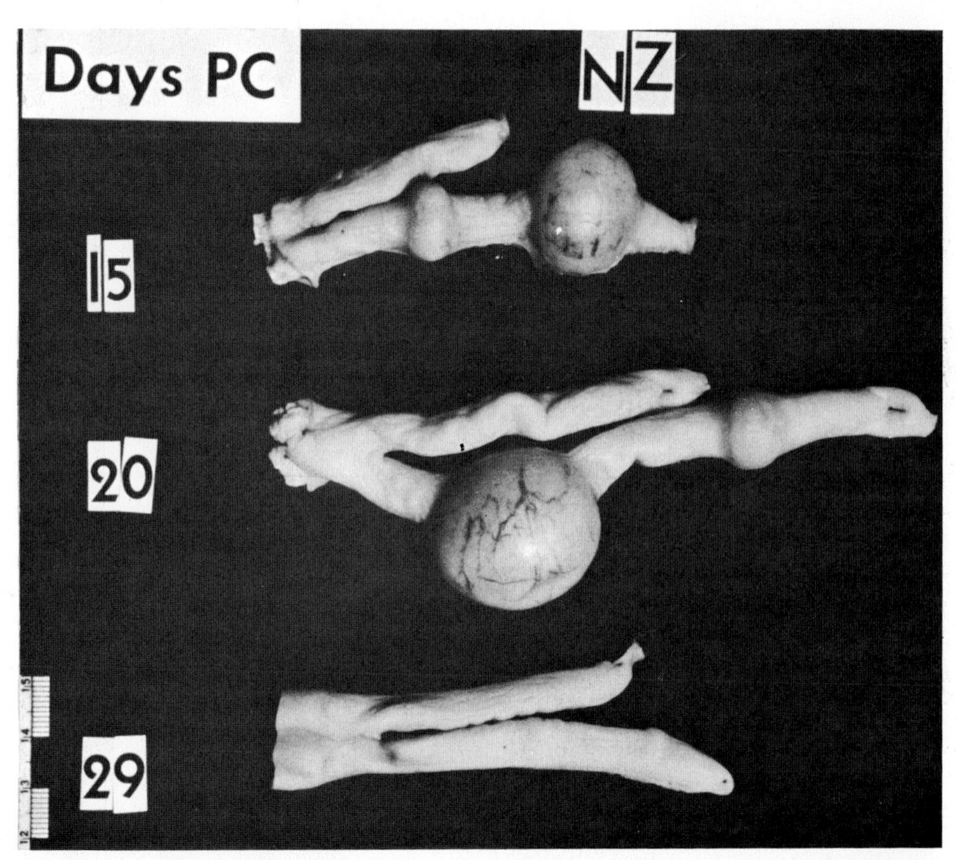

Fig. 28. The effect of undercrowding on postimplantation embryonic survival in the rabbit. In each case, 2 embryos were transferred to one oviduct of receptive does which were then autopsied at 15, 20, or 29 days *p.c.* Note that one of the implantations in the upper two specimens has degenerated.

Fig. 27. Postattachment embryonic degeneration in the overcrowded bovine uterus. Overcrowding was produced experimentally with injections of pregnant mare serum gonadotropin (1,500 i.u.) and human chorionic gonadotropin (2,000 i.u.). The animals were autopsied 5–6 weeks after breeding.

(*a*) Chorioallantoida containing 1 viable embryo (*left center*) and 2 embryos which have begun to degenerate.

(*b*) Chorioallantoida containing 5 degenerating embryos.

(*c*) Reproductive tract opened to show degenerating conceptuses in the final stages of resorption. Note the gray debris within the lumina of the uterine horns.

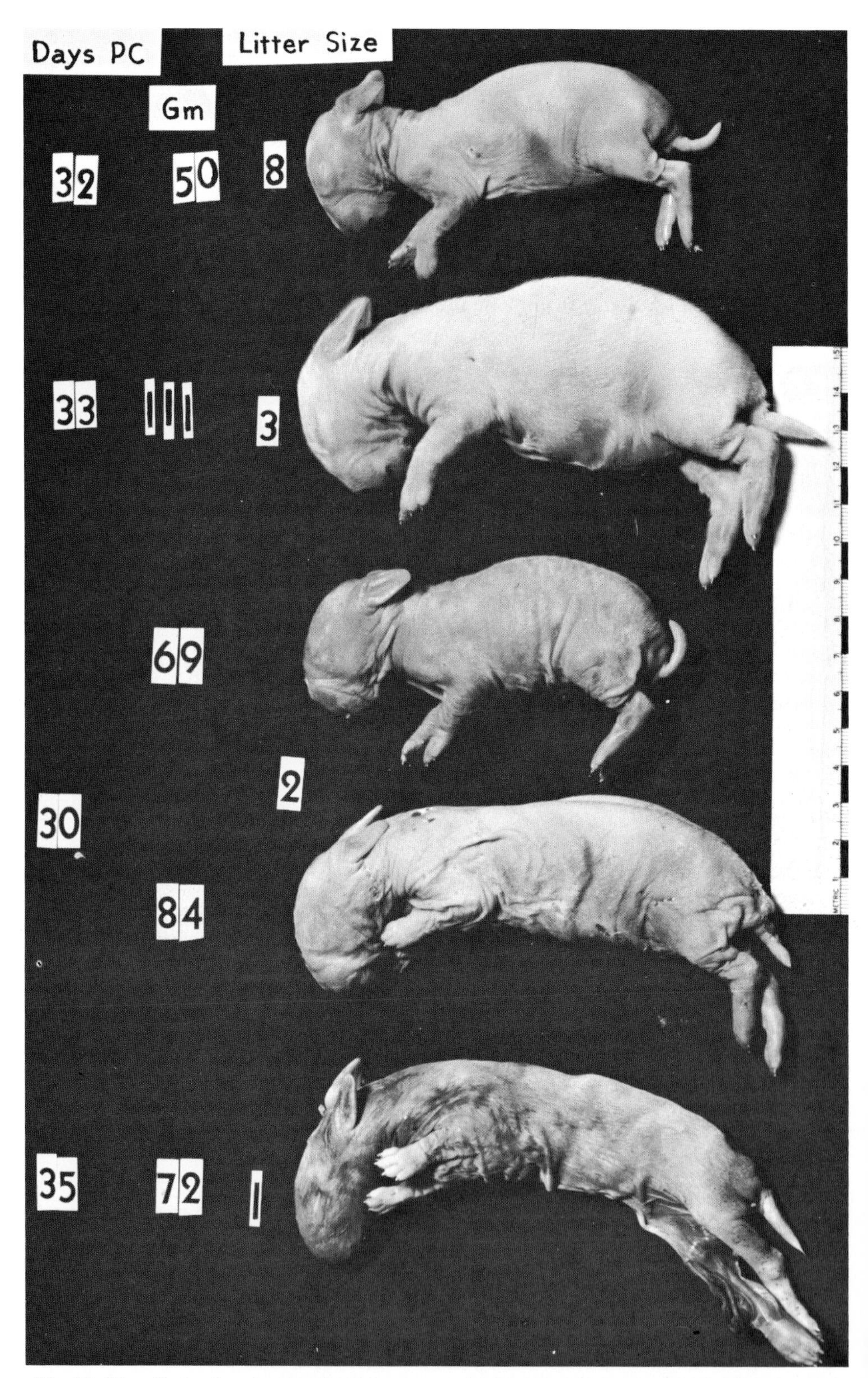

Fig. 29. The effects of undercrowding *in utero* on the fetal development of the rabbit. Neonates from small litters are overgrown compared to the control fetus (*top*) from a normal litter of 8. Fetal overgrowth produces dystocia and perinatal mortality, as shown by the fetus at the bottom which was not delivered until 35 days *p.c.*

in blastocyst size, the random genetic background of experimental animals, injury during transfer, the difficulty in washing blastocysts free of substances which are highly concentrated in uterine fluids, and the sensitivity of some compounds to changing pH, air, or light.

Rabbits have been employed in most physiochemical studies of the blastocyst because they are readily available and produce blastocysts with large volumes of blastocoelic fluid. Remarkable specific variations occur in the morphology and anatomy of blastocysts; consequently systematic comparative research is needed to determine how specific properties of the blastocyst relate to different types of implantation. The following inventory contains some of the more important unanswered questions about blastocyst biology.

1. What are the quantitative effects of exogenous steroids on development and differentiation of blastocysts?
2. What concentrations of amino acids and proteins occur in blastocoelic fluid during differentiation of the blastocyst? What is the physiological significance of those which occur in abundance but are not essential? What are the interrelationships of essential and nonessential amino acids in blastocoelic fluid?
3. Does pulsation occur in the blastocysts of species other than the mouse? How does this contraction occur in the absence of actinomycin? Is pulsation involved in shedding the zona pellucida?
4. Does loss of the zona pellucida increase the passage of metabolites from the uterine fluids into the blastocoelic fluid? What is the role of luminal epithelium in the reabsorption of metabolites?
5. Are cellular materials transferred from the trophoblast to the endometrium or vice versa?
6. How do the epithelial cells of the trophoblast process phagocytized material? Do these substances influence the development of the blastocyst?
7. What is the physiochemical basis for passage of metabolites to and from the uterine environment?
8. Why and what rate do metabolites pass to the uterus from the uterotubal junction and the cervix?
9. What is the physiological significance of the lymph supply to the uterus? Does lymph participate in blastocyst development? What does it contain?
10. Does the early uptake of protein by blastocoelic fluid lead to immunological tolerance?
11. What biochemical and cytogenetical factors cause "hybrid blastocysts" to degenerate at different stages of development?
12. What factors cause the blastocyst to degenerate in the uterus of some alien species but not others?
13. To what degree is the uterine epithelium a barrier to the diffusion of large molecules between uterine and blastocoelic compartments?
14. Are specific uterine factors responsible for blastulation, implantation, and fetal development? Or are they only involved in the development of the trophoblast and extraembryonic membranes?
15. What are the major biochemical pathways for the synthesis of blastokinin and uteroglobin? What is the role of these proteins in differentiation, implantation, and organogenesis?

16. What is the local effect of the ovary on the survival and subsequent development of the blastocyst? What, if any, hormones are involved? Are they transported by the blood vascular system, lymphatics, or both? Does a direct relationship exist between unusual anatomical arrangements (proximity of ovary, oviduct, and uterus to each other) and the degree of this local effect?

Acknowledgments

Previously unpublished results used in this paper were obtained with the support of United States Public Health Service research grant HD00585 from the National Institute of Child Health and Human Development and a Banta Research Fund grant.

References

Adams, C. E. 1965. The influence of maternal environment on preimplantation stages of pregnancy in the rabbit. In: *Preimplantation stages of pregnancy,* ed. G. E. W. Wolstenholme and M. O'Connor, p. 345. Boston: Little, Brown & Co.

Allen, W. M., and Wu, D. H. 1959. Effects of 17-alpha-ethinyl-19-nortestosterone on pregnancy in rabbits. *Fertil Steril* 10:424.

Arthur, G. H., ed. 1964. *Wright's veterinary obstetrics,* 3d ed. London: Baillière, Tindall & Cox.

Asdell, S. A. 1964. *Patterns of mammalian reproduction,* 2d ed. Ithaca, N.Y.: Cornell University Press.

Averill, R. L. W.; Adams, C. E.; and Rowson, L. E. A. 1955. Transfer of mammalian ova between species. *Nature (London)* 176:167.

Baevsky, U. B. 1963. The effect of embryonic diapause on the nucleic and mitotic activity of mink and rat blastocysts. In *Delayed implantation,* ed. A. C. Enders, p. 141. Chicago: University of Chicago Press.

Bailey, L. F., and Lemon, M. 1966. Specific milk proteins associated with resumption of development by the quiescent blastocyst of lactating red kangaroo. *J Reprod Fertil* 11:473.

Barcroft, J., and Rothschild, P. 1932. The volume of blood in the uterus during pregnancy. *J Physiol (London)* 76:447.

Barley, D. A.; Butcher, R. L.; and Inskeep, E. K. 1966. Local nature of utero-ovarian relationships in the pseudopregnant rat. *Endocrinology* 79:119.

Beatty, R. A. 1958. Variation in the number of corpora lutea and in the number and size of 6-day blastocysts in rabbits subjected to superovulation treatment. *J Endocr* 17:248.

Beier, H. M. 1968*a*. Biochemisch-entwicklungsphysiologische Untersuchungen am Proteinmilieu für die Blastozystenentwicklung des Kaninchens (*Oryctolagus cuniculus*). *Zool Jb Anat* 85:72.

———. 1968*b*. Unteroglobin: A hormone-sensitive endometrial protein involved in blastocyst development. *Biochim Biophys Acta* 160:289.

———. 1968*c*. Protein patterns of endometrial secretion in the rabbit. In *Ovo-implantation, human gonadotropins and prolactin.* Brussels.

Bennett, J. P.; Boursnell, J. C.; and Lutwak-Mann, C. 1958. Use of labelled ions in the study of blastocyst-uterine relationships in the rabbit. *Nature* (*London*) 181:1715.

Billington, W. D.; Graham, C. F.; and McLaren, A. 1968. Extra-uterine development of mouse blastocysts cultured *in vitro* from early cleavage stages. *J Embryol Exp Morphol* 20:391.

Bischoff, T. L. W. 1845. *Entwicklungsgeschichte des Hunde-Eies*. Brunswick.

Blaha, G. C., and De Feo, V. J. 1964. Interspecies ova transfer between hamsters and rats. *Anat Rec* 148:261.

Bomsel-Helmreich, O. 1965. Heteroploidy and embryonic death. In *Preimplantation stages of pregnancy,* ed. G. E. W. Wolstenholme and M. O'Connor, p. 246. Boston: Little, Brown & Co.

Borghese, E., and Cassini, A. 1963. Cleavage of mouse egg. In *Cinemicrography in cell biology,* ed. G. G. Rose, p. 274. New York: Academic Press.

Boyd, J. D., and Hamilton, W. J. 1952. Cleavage, early development and implantation of the egg. In *Marshall's physiology of reproduction,* ed. A. S. Parkes, 2:1. London: Longmans Green & Co.

Brambell, F. W. R. 1954. Transport of protein across the fetal membranes. *Sympos Quant Biol* 19:71.

Brambell, F. W. R., and Hemmings, W. A. 1949. The passage into the embryonic yolk-sac cavity of maternal plasma proteins in rabbits. *J Physiol* (*London*) 108: 177.

Brinster, R. L. 1969. Mammalian egg culture. In *The mammalian oviduct,* ed. E. S. E. Hafez and R. J. Blandau, p. 419. Chicago: University of Chicago Press.

Briones, H., and Beatty, R. A. 1954. Inter-specific transfers of rodent eggs. *J Exp Zool* 125:99.

Bryson, D. L. 1964. Development of mouse eggs in diffusion chambers. *Science* 144:1351.

Chambon, Y. 1960. Déficit alimentaire et ovoimplantation. *Bull Soc Roy Belg Gynec Obstet* 6:557.

Chang, M. C. 1950. Development and fate of transferred rabbit ova or blastocyst in relation to the ovulation time of recipients. *J Exp Zool* 114:197.

———. 1966. Reciprocal transplantation of eggs between rabbit and ferret. *J Exp Zool* 161:297

———. 1968. Reciprocal insemination and egg transfer between ferrets and mink. *J Exp Zool* 168:49.

Clark, M. J. 1966. The blastocyst of the red kangaroo, *Megaleia rufa* (Desm.) during diapause. *Aust J Zool* 14:19.

Cole, R. J. 1967. Cinemicrographic observations on the trophoblast and zona pellucida of the mouse blastocyst. *J Embryol Exp Morphol* 17:481.

Cole, R. J., and Paul, J. 1965. Properties of cultured preimplantation mouse and rabbit embryos, and cell strains derived from them. In *Preimplantation stages of pregnancy,* ed. G. E. W. Wolstenholme and M. O'Connor. Boston: Little, Brown & Co.

Dalcq, A. M., ed. 1957. *Introduction to chemical embryology*. New York: Oxford University Press.

Dalcq, A. M. 1959. La localisation cytochimique de l'adénosinetriphosphatase dans les oeufs des mammifères et sa relation avec leur organisation morphogénétique. *Bull Acad Roy Med Belg* 24:825.

Daniel, J. C., Jr. 1965. Studies on the growth of five-day-old rabbit blastocysts *in vitro*. *J Embryol Exp Morph* 13:83.

———. 1967*a*. Studies on growth of the mink blastocyst. *J Embryol Exp Morphol* 17:293.

———. 1967*b*. Vitamins and growth factors in the nutrition of rabbit blastocysts *in vitro*. *Growth* 31:71.

———. 1967*c*. The pattern of utilization of respiratory metabolic intermediates by preimplantation rabbit embryos *in vitro*. *Exp Cell Res* 47:619.

———. 1968. Oxygen concentrations for culture of rabbit blastocysts. *J Reprod Fertil* 17:187.

Daniel, J. C., Jr., and Krishnan, R. S. 1967. Amino acid requirements for growth of the rabbit blastocyst *in vitro*. *J Cell Comp Physiol* 70:155.

Daniel, J. C. Jr., and Olson, J. D. 1966. Cell movement, proliferation and death in the formation of the embryonic axis of the rabbit. *Anat Rec* 156:123.

De Feo, V. J. 1967. Decidualization. In *Cellular biology of the uterus,* ed. R. M. Wynn, p. 191. New York: Appleton-Century-Crofts.

Dhindsa, D. S., and Dziuk, P. J. 1968. Influence of varying the proportion of uterus occupied by embryos on maintenance of pregnancy in the pig. *J Anim Sci* 27:668.

Dhindsa, D. S.; Dziuk, P. J.; and Norton, H. W. 1967. Time of transuterine migration and distribution of embryos in the pig. *Anat Rec* 159:325.

Dickmann, Z., and DeFeo, V. J. 1967. The rat blastocyst during normal pregnancy and during delayed implantation, including an observation on the shedding of the zona pellucida. *J Reprod Fertil* 13:3.

Dziuk, P. J.; Polge, C.; and Rowson, L. E. 1964. Intra-uterine migration and mixing of embryos in swine following egg transfer. *J Anim Sci* 23:37.

Edwards, R. G. 1964. Cleavage of one- and two-celled rabbit eggs *in vitro* after removal of the zona pellucida. *J Reprod Fertil* 7:413.

Edwards, R. G., and Gardner, R. L. 1967. Sexing of live rabbit blastocysts. *Nature (London)* 214:576.

Enders, A. C., ed. 1963. *Delayed implantation.* Chicago: University of Chicago Press.

———. 1970. Fertilization, cleavage and implantation. In *Reproduction and breeding techniques for laboratory animals,* ed. E. S. E. Hafez, chap. 7. Philadelphia: Lea & Febiger.

Enders, A. C., and Schlafke, S. 1965. The fine structure of the blastocyst: Some comparative studies. In *Preimplantation stages of pregnancy,* ed. G. E. W. Wolstenholme and M. O'Connor, p. 29. Boston: Little, Brown & Co.

Enders, R. K. 1938. The ovum of the mink (*Mustela vison*). *Anat Rec* 72:469.

Fawcett, D. W.; Wislocki, G. B.; and Waldo, C. M. 1947. The development of mouse ova in the anterior chamber of the eye and in the abdominal cavity. *Amer J Anat* 81:413.

Ferm, V. H. 1956. Permeability of the rabbit blastocyst to trypan blue. *Anat Rec* 125:745.

Fischer, T. V. 1967. Local uterine regulation of the corpus luteum. *Amer J Anat* 121:425.

Flamm, H.; Kunz, C.; Zimmermann, W.; and Gottschewski, G. H. M. 1963. Experimentelle pränatale Infektion zur Zeit der Nidation. *Zbl Bakt (Orig)* 189:335.

Fridhandler, L. 1961. Pathways of glucose metabolism in fertilized rabbit ova at various pre-implantation stages. *Exp Cell Res* 22:303.

———. 1968. Intermediary metabolic pathways in preimplantation rabbit blastocysts. *Fertil Steril* 19:424.

Fridhandler, L., Hafez, E. S. E., and Pincus, G. 1957. Developmental changes in the respiratory activity of rabbit ova. *Exp Cell Res* 13:132.

Glenister, T. W. 1961. Organ culture as a new method for studying the implantation of mammalian blastocysts. *Proc Roy Soc (Biol)* 154:428.

Gorski, J.; Noteboom, W. D.; and Nicolette, J. A. 1965. Estrogen control of the synthesis of RNA and protein in the uterus. *J Cell Comp Physiol* 66:91.

Gottschewski, H. M. 1962. Probleme der Frühentwicklung beim Säugetier. *Naturwiss Rundschau* 15:257.

Greenstein, J. S.; Murray, R. W.; and Foley, R. C. 1958. Observations on the morphogenesis and histochemistry of the bovine pre-attachment placenta between 16 and 33 days of gestation. *Anat Rec* 132:321.

Greenwald, G. S. 1962. The role of the mucin layer in development of the rabbit blastocyst. *Anat Rec* 142:407.

Gregoire, A. T.; Gongsakdi, D.; and Rakoff, A. E. 1961. The free amino acid content of the female rabbit genital tract. *Fertil Steril* 12:322.

Gulyas, B. J., and Daniel, J. C., Jr. 1967. Oxygen consumption in diapausing blastocysts. *J Cell Physiol* 70:33.

Gwatkin, R. B. L. 1966. Amino acid requirement for attachment and outgrowth of the mouse blastocyst *in vitro*. *J Cell Physiol* 68:335.

Gwatkin, R. B. L., and Meckley, P. E. 1966. Chromosomes of the mouse blastocyst following its attachment and outgrowth *in vitro*. *Ann Med Exp Biol Fenn* 44:125.

Hadek, R., and Swift, H. 1960. A crystalloid inclusion in the rabbit blastocyst. *J Biophys Biochem Cytol* 8:836.

Hafez, E. S. E. 1962. Differential cleavage rate in 2-day litter mate rabbit embryos. *Proc Soc Exp Biol. Med NY* 110:142.

———. 1964*a*. Growth and survival of blastocysts in the domestic rabbit: II. Quantitative effects of exogenous progesterone following ovariectomy. *J Reprod Fertil* 7:241.

———. 1964*b*. Transuterine migration and spacing of bovine embryos during gonadotrophin-induced multiple pregnancy. *Anat Rec* 148:203.

———. 1964*c*. Effects of over-crowding *in utero* on implantation and fetal development in the rabbit. *J Exp Zool* 156:269.

———. 1968. Enzymology of the uterus and mammalian reproduction in *Perspectives of sexual behavior and reproduction,* ed. D. Milton. Bloomington: University of Indiana Press.

Hafez, E. S. E.; Estergreen, V. L.; and Foster, R. J. 1965. Progestin and nucleic acid content of corpora lutea during multiple pregnancy in beef cattle. *Acta Endocr (Kobenhavn)* 48:664.

Hafez, E. S. E., and Ishibashi, I. 1965. Effect of lactation and age at first breeding on size and survival of rabbit blastocysts. *Int J Fertil* 10:47.

Hafez, E. S. E.; Jainudeen, M. R.; and Lindsay, D. R. 1965. Gonadotropin-induced twinning and related phenomena in beef cattle. *Acta Endocr* (*Kobenhavn*) (Suppl. 102) 50:43.

Hafez, E. S. E., and Rajakoski, E. 1964. Growth and survival of blastocysts in the domestic rabbit. *J Reprod Fertil* 7:229.

Hafez, E. S. E., and Sugawara, S. 1967. Protein distribution in the blastocoelic fluid of lactating rabbits. *Fertil Steril* 18:566.

———. 1968. Maternal effects on some biochemical characteristics of the blastocyst in the domestic rabbit. *J Morph* 124:133.

Hafez, E. S. E., and Tsutsumi, T. 1966. Changes in endometrial vascularity during implantation and pregnancy in the rabbit. *Amer J Anat* 118:249.

Hafez, E. S. E., and White, I. G. 1967. Endometrial and embryonic enzymes in relation to implantation of the rabbit blastocyst. *Anat Rec* 159:273.

———. 1968. Endometrial and embryonic enzyme activities in relation to implantation in the ewe. *J Reprod Fertil* 16:59.

Hahn, E. W., and Ward, W. F. 1967. Increased litter size in the rat X-irradiated during the estrous cycle before mating. *Science* 157:956.

Hamana, K., and Hafez, E.S.E. 1970. Disc electrophoretic patterns of uteroglobin and serum proteins in rabbit blastocoelic fluid. *J Reprod Fertil* 21:555.

Hansson, A. 1947. The physiology of reproduction in mink (*Mustela vison,* Shreb.) with special reference to delayed implantation. *Acta Zool* 28:1.

Hendrickx, A. G., and Kraemer, D. C. 1968. Preimplantation stages of baboon embryos (*Papio sp.*). *Anat Rec* 162:111.

Huff, R. L., and Eik-Nes, K. B. 1966. Metabolism *in vitro* of acetate and certain steroids by six-day-old rabbit blastocysts. *J Reprod Fertil* 11:57.

Ishida, K., and Chang, M. C. 1965. Histochemical demonstration of succinic dehydrogenase in hamster and rabbit eggs. *J Histochem Cytochem* 13:470.

Jacobson, W., and Lutwak-Mann, C. 1956. The vitamin B_{12} content of the early rabbit embryo. *J Endocr* 14:19.

Karlson, P., and Sekeris, C. E. 1966. Biochemical mechanisms of hormone action. *Acta Endocr* (*Kobenhavn*) 53:505.

Kehl, R., and Chambon, Y. 1948. Besoins endocriniens comparés de l'ovoimplantation et du maintien de la gestation chez la lapine. *C R Soc Biol* (*Paris*) 142:676.

———. 1949. Synergie progestérofolliculinique d'ovoimplantation chez la lapine. *C R Soc Biol* (*Paris*) 143:1169.

Keyes, P. L., and Nalbandov, A. V. 1967. Maintenance and function of corpora lutea in rabbits depend on estrogen. *Endocrinology* 80:938.

Kirby, D. R. S. 1960. Development of mouse eggs beneath the kidney capsule. *Nature* (*London*) 187:707.

———. 1962*a*. Reciprocal transplantation of blastocysts between rats and mice. *Nature* (*London*) 194:785.

———. 1962*b*. The effect of the uterine environment on the development of mouse eggs. *J Embryol Exp Morph* 10:496.

———. 1963*a*. Development of the mouse blastocyst transplanted to the spleen. *J Reprod Fertil* 5:1.

———. 1963*b*. The development of mouse blastocysts transplanted to the scrotal and cryptorchid testis. *J Anat* 97:119.

———. 1965. The role of the uterus in the early stages of mouse development. In *Preimplantation stages of pregnancy,* ed. G. E. W. Wolstenholme and M. O'Connor, p. 325. Boston: Little, Brown & Co.

———. 1967. Ectopic autografts of blastocysts in mice maintained in delayed implantation. *J Reprod Fertil* 14:515.

Kodituwakku, G. E., and Hafez, E. S. E. 1969. Blastocyst size in Dutch-Belted and New Zealand rabbits. *J Reprod Fertil* 19:187.

Krehbiel, R. H. 1946. Distribution of ova in the combined uteri of unilaterally ovariectomized rats. *Anat Rec* 96:323.

Krishnan, R. S., and Daniel, J. C., Jr. 1967. "Blastokinin": Inducer and regulator of blastocyst development in the rabbit uterus. *Science* 158:490.

Kuhl, W., and Friedrich-Freksa, H. 1936. Richtungskörperbildung und Furchung des Eies sowie das Verhalten des Trophoblasten der weissen Maus (film). *Zool Anz* 9:187.

Lamming, G. E., and Little, L. S. 1968. Effects of post-coital unilateral ovariectomy on embryo survival in the rat. *J Physiol* (*London*) 196:14.

Lesinski, J.; Jajszczak, S.; Bentyn, K.; and Janczarski, J. 1966. Content of free amino acids in rabbit blastocyst fluid. *Pol Med J* 6:1074.

Lesinski, J.; Jajszczak, S.; and Choroszewska, A. 1967. The relationship between ·amino nitrogen and protein concentrations in rabbit blastocyst fluid. *Amer J Obstet Gynec* 98:48.

Little, S. L.; Gray, A. J.; and Lamming, G. E. 1969. Local effect of the ovary on the uterus of the rat. *J Endocr* 43:33.

Lutwak-Mann, C. 1954. Some properties of the rabbit blastocyst. *J Embryol Exp Morph* 2:1.

———. 1960. Some properties of the early embryonic fluids in the rabbit. *J Reprod Fertil* 1:316.

———. 1962. Glucose, lactic acid and bicarbonate in rabbit blastocyst fluid. *Nature* (*London*) 193:653.

———. 1966. Some physiological and biochemical properties of the mammalian blastocyst. *Bull Schweiz Akad Med Wiss* 22:101.

Lutwak-Mann, C.; Boursnell, J. C.; and Bennett, J. P. 1960. Blastocyst-uterine relationships: Uptake of radioactive ions by early rabbit embryo and its environment. *J Reprod Fertil* 1:169.

Lutwak-Mann, C.; Hay, M. F.; and Adams, C. E. 1962. The effect of ovariectomy on rabbit blastocyst. *J Endocr* 24:185.

McLaren, A. 1967. Delayed loss of the zona pellucida from blastocysts of suckling mice. *J Reprod Fertil* 14:159.

McLaren, A., and Tarkowski, A. K. 1963. Implantation of mouse eggs in the peritoneal cavity. *J Reprod Fertil* 6:385.

Mantalenakis, S. J., and Ketchel, M. M. 1966. Frequency and extent of delayed implantation in lactating rats and mice. *J Reprod Fertil* 12:391.

Marshall, A. J. 1953. The unilateral endometrial reaction in the giant fruit bat (*Pteropus giganteus* Brünnich). *J Endocr* 9:42.

Mayersbach, H. 1958. Zur Frage des Proteinübergangs von der Mutter zum Foeten: I. Fefunde an Ratten am Ende der Schwangerschaft. *Z Zellforsch* 48: 479.

Mesnil du Buisson, F. du. 1961. Possibilité d'un fonctionnement dissemblable des ovaires pendant la gestation chez la truie. *C R Acad Sci* (*Paris*) 253:727.

Mesnil du Buisson, F. du, and Dauzier, L. 1959. Contrôle mutuel de l'utérus et de l'ovaire chez la truie. *Ann Inst Nat Recherche Agron, Série D* (Suppl.) 8:147.

Mesnil du Buisson, F. du, and Rombauts, P. 1963*a*. Réduction expérimentale du nombre des foetus au cours de la gestation de la truie et maintien des corps jaunes. *Ann Biol Anim Biochim Biophys* 3:445.

———. 1963*b*. Effet d'autotransplants utérins sur le cycle oestrien de la truie. *C R Acad Sci* (*Paris*) 256:4984.

Mintz, B. 1964. Formation of genetically mosaic mouse embryos, and early development of "lethal (t^{12}/t^{12})-normal" mosaics. *J Exp Zool* 157:273.

———. 1965. Experimental genetic mosaicism in the mouse. In *Preimplantation stages of pregnancy,* ed. G. E. W. Wolstenholme and M. O'Connor. Boston: Little, Brown & Co.

Mitchell, J. A., and Yochim, J. M. 1968. Measurement of intrauterine oxygen tension in the rat and its regulation by ovarian steroid hormones. *Endocrinology* 83:691.

Moog, F., and Lutwak-Mann, C. 1958. Observations on rabbit blastocysts prepared as flat mounts. *J Embryol Exp Morph* 6:57.

Moor, R. M., and Rowson, L. E. A. 1966. Local maintenance of the corpus luteum in sheep with embryos transferred to various isolated portions of the uterus. *J Reprod Fertil* 12:539.

Mounib, M. S., and Chang, M. C. 1965. Metabolism of glucose, fructose and pyruvate in the 6-day rabbit blastocyst. *Exp Cell Res* 38:201.

Mulnard, J. G. 1967. Analyse microcinématographique de développement de l'oeuf de souris du stade II au blastocyste. *Arch Biol* (*Liège*) 78:107.

Nicholas, J. S. 1934. Experiments on developing rats: I. Limits of foetal regeneration; behavior of embryonic material in abnormal environments. *Anat Rec* 58: 387.

Orsini, M. 1962. Technique of preparation, study and photography of benzyl-benzoate cleared material for embryological studies. *J Reprod Fertil* 3:283.

Perry, J. S., and Rowlands, I. W. 1962. Early pregnancy in the pig. *J Reprod Fertil* 4:175.

Pilton, P. E., and Sharman, G. B. 1962. Reproduction in the marsupial *Trichosurus vulpecula. J Endocr* 25:119.

Polge, C.; Rowson, L. E. A.; and Chang, M. C. 1966. The effect of reducing the number of embryos during early stages of gestation on the maintenance of pregnancy in the pig. *J Reprod Fertil* 12:395.

Prasad, M. R. N.; Dass, C. M. S.; and Mohla, S. 1968. Action of oestrogen on the blastocyst and uterus in delayed implantation: An autoradiographic study. *J Reprod Fertil* 16:97.

Psychoyos, A. 1961. Perméabilité capillaire et décidualisation utérine. *C R Acad Sci* (*Paris*) 252:1515.

Rapola, J., and Koskimies, O. 1968. Embryonic enzyme patterns: Characterization of the single lactate dehydrogenase isozyme in preimplanted mouse ova. *Science* 157:1311.

Rowson, L. E. A., and Moor, R. M. 1966. Development of the sheep conceptus during the first 14 days. *J Anat* 100:777.

Runner, M. N. 1947. Development of mouse eggs in the anterior chamber of the eye. *Anat Rec* 98:1.

Schechtman, A. M., and Abraham, K. C. 1958. Passage of serum albumin from the mother to the foetus. *Nature (London)* 181:120.

Schlafke, S. J., and Enders, A. C. 1963. Observations on the fine structure of the rat blastocyst. *J Anat* 97:353.

Schwick, H. G. 1965. Chemisch-entwicklungsphysiologische Beziehungen von Uterus zu Blastozyste des Kaninchens (*Oryctolagus cuniculus*). *Roux' Arch* 156:283.

Segal, S. J., and Scher, W. 1967. In *Cellular biology of the uterus,* ed. R. M. Wynn, p. 114. New York: Appleton-Century-Crofts.

Sharman, G. B. 1955. Studies on marsupial reproduction: III. Normal and delayed pregnancy in *Setonix brachyurus. Aust J. Zool* 3:56.

———. 1963. Delayed implantation in marsupials. In *Delayed implantation,* ed. A. C. Enders, p. 3. Chicago: University of Chicago Press.

Shaver, E. L., and Carr, D. H. 1967. Chromosome abnormalities in rabbit blastocysts following delayed fertilization. *J Reprod Fertil* 14:415.

Staples, R. E. 1967. Development of 5-day rabbit blastocyst after culture at 37° C. *J Reprod Fertil* 13:369.

Stormshak, E. K.; Inskeep, J. E.; Lynn, J. E.; Pope, A. L.; and Casida, L. E. 1963. Progesterone levels in corpora lutea and ovarian effluent blood of the ewe. *J Anim Sci* 22:1021.

Sugawara, S. 1964. Appearance of metabolic substances in rabbit blastocyst fluid. *Jap J Zootech Sci* 36:2.

Sugawara, S., and Hafez, E. S. E. 1967*a*. Electrophoretic patterns of proteins in the blastocoelic fluid of the rabbit following ovariectomy. *Anat Rec* 158:115.

———. 1967*b*. Developmental changes in dehydrogenase activities in rabbit eggs. *Proc Soc Biol Med NY* 126:849.

———. 1967*c*. Effect of steroid hormones on biochemical characteristics of blastocysts in ovariectomized rabbits. *Anat Rec* 158:281.

Sugawara, S.; Otake, S.; and Takeuchi, S. 1967. Effect of ovarian hormone on the growth and differentiation of the rat blastocyst during delayed implantation: I. Progesterone. *Tohoku J Agr Res* 18:107.

Sugawara, S., and Takeuchi, S. 1967. Some metabolic intermediates and nitrogen compounds of the blastocoelic fluid in the rabbit blastocyst during pre-implantation and implantation. *Jap J Zootech Sci* 38:286.

Sundell, G. 1962. The sex ratio before uterine implantation in the golden hamster. *J Embryol Exp Morphol* 10:58.

Talwar, G. P.; Sopori, M. L.; Biswas, D. K.; and Segal, S. J. 1968. Nature and characteristics of the binding of oestradiol-17B to a uterine macromolecular fraction. *Biochem J* 107:765.

Tarkowski, A. K. 1962. Inter-specific transfers of eggs between rat and mouse. *J Embryol Exp Morphol* 10:476.

———. 1963. Studies on mouse chimeras developed from eggs fused *in vitro. Nat Cancer Inst Monogr* 11:51:

———. 1965. Embryonic and postnatal development of mouse chimeras. In *Pre-implantation stages of pregnancy,* ed. G. E. W. Wolstenholme and M. O'Connor. Boston: Little, Brown & Co.

Thomson, J. L., and Brinster, R. L. 1966. Glycogen content of preimplantation mouse embryos. *Anat Rec* 155:97.

Tondeur, M. 1959. Effets de l'imprégnation osmique in toto des oeufs vierges, fécondés ou segmentés du rat et de la souris. *Mededel Klin Wetenschap, Koninkl Belg Acad* 45:487.

Tyndale-Biscoe, C. H. 1963. The role of the corpus luteum in the delayed implantation of marsupials. In *Delayed implantation,* ed. A. C. Enders, p. 15. Chicago: University of Chicago Press.

Ulberg, L. C. 1969. Reproduction in cattle. In *Reproduction in farm animals,* ed. E. S. E. Hafez, p. 255. Philadelphia: Lea & Febiger.

Van Tienhoven, A. 1968. *Reproductive physiology of vertebrates.* Philadelphia: W. B. Sanders Co.

Venge, O. 1950. Studies of the maternal influence on the birth weight in rabbits. *Acta Zool* 31:1.

Vickers, A. D. 1967. A direct measurement of the sex-ratio in mouse blastocysts. *J Reprod Fertil* 13:375.

Ward, W. F. ,and Hahn, E. W. 1967. Latent effects of prefertilization X-irradiation on the reproductive performance of the female rat. *Radiat Res* 32:125.

Warwick, B. L., and Berry, R. O. 1949. Inter-generic and intra-specific embryo transfers in sheep and goats. *J Hered* 40:297.

Weitlauf, H. M., and Greenwald, G. S. 1965. A comparison of ^{35}S-methionine incorporation by the blastocysts of normal and delayed implanting mice. *J Reprod Fertil* 10:203.

———. 1968. Survival of blastocysts in the uteri of ovariectomized mice. *J Reprod Fertil* 17:515.

Whitten, W. K., and Dagg, C. P. 1962. Influence of spermatozoa on the cleavage rate of mouse eggs. *J Exp Zool* 148:173.

Whittingham, D. G. 1968. Intra- and inter-specific transfer of ova between explanted rat and mouse oviducts. *J Reprod Fertil* 17:575.

Wintenberger-Torres, S. 1968. Modification du milieu utérin chez les superovulées et développement des blastocystes. *Proc VI Int Cong Reprod A I Paris.*

Wintenberger-Torres, S., and Rombauts, P. 1968. Relation entre la mortalité embryonnaire et la quantité de progésterone sécrétée chez la brebis. *Proc VI Int Cong Reprod A I Paris.*

Yochim, J. M., and Mitchell, J. A. 1968. Intrauterine oxygen tension in the rat during progestation: Its possible relation to carbohydrate metabolism and the regulation of nidation. *Endocrinology* 83:706.

Yoshinaga, K., and Adams, C. E. 1967. Reciprocal transfer of blastocysts between the rat and rabbit. *J Reprod Fertil* 14:325.

Zimmermann, W. 1960. Investigations on the domestic rabbit: III. The hydrogen ion concentration in the female genital tract and in the blastocyst before, during and after implantation. *Verh Deutsch Zool Ges* 24:143.

Zimmermann, W.; Gottschewski, G. H. M.; Flamm, H.; and Kunz, C. 1963. Experimentelle Untersuchungen über die Aufnahme von Eiweiss, Viren und Bakterien während der Embryogenese des Kaninchens. *Develop Biol* 6:233.

10

The Surface Charge on the Five-Day Rat Blastocyst

C. A. B. Clemetson/ M. M. Moshfeghi/ V. R. Mallikarjuneswara

Research Laboratory
Department of Obstetrics and Gynecology
Methodist Hospital of Brooklyn
Brooklyn, New York

Implantation of the mammalian blastocyst cannot occur until attachment has been accomplished. One ponders whether contact between unexpanded mammalian blastocysts and the endometrium may not be conditioned by their electrical charges; delayed implantation is perhaps due to mutual repulsion resulting from like charges on the blastocyst and endometrium. If this is so, implantation would require the reduction, removal or reversal of one of the charges.

We therefore decided to find out whether the blastocyst possesses a surface charge similar to other cells and tissues of the body.

Abramson (1965) discussed the electronegativity of blood platelets, red cells, and leukocytes; Sawyer and Pate (1953) showed that there is normally a negative charge on the vascular endothelium, and they believe that these like charges normally prevent contact between the blood cells and the vessel walls by mutual repulsion. The blastocysts of the rat and the human are much larger than red or white blood cells but they are still small enough to move in an electrical field.

Tyler et al. (1956), using microelectrodes, studied the electrical potential of the eggs of the starfish, *Asterias forbesii,* and found a negative charge inside the egg relative to that of seawater. These investigators demonstrated that the potential decreases as the potassium content of the seawater increases; they concluded that the charge inside the egg was related closely to the potassium excess of the egg.

It should be noted that the charge inside the egg, measured by a microelectrode, is not necessarily the same as the surface charge which controls electrophoretic movement.

Maéno (1959) studied the electrical characteristics of the eggs of the toad, *Bufo vulgaris formosus,* and reviewed the work concerning the eggs of lower forms. Search of the literature failed to disclose studies on the charge or the electrophoretic mobility of mammalian ova. We decided to study the charge of the rat blastocyst on the 5th day, both before and after loss of the zona pellucida, by observing its movement in a charged field in an aqueous medium. Our choice of the rat was influenced by the fact that implantation may be delayed for 11 days in lactating rats or postponed artificially by bilateral oophorectomy followed by medroxy-progesterone acetate injections for even as long as 45 days (Barnes and Meyer 1964).

I. Methods

A. *General Description of Biological Techniques and of Electrophoretic Cell*

Daily vaginal cytology studies were carried out on 200 to 240 gm female, albino, Wistar rats. Each was put with a male on the evening of the day on which an estrous smear was obtained. If spermatozoa were found in the vagina on the following morning, successful mating was assumed. This day was counted as the first day of pregnancy.

On the fifth day (4½ to 5 days after mating) the rats were killed by intraperitoneal injection of Nembutal (Abbott); the uterine horns were removed and flushed with isotonic dextrose solution into a watch glass. Five percent dextrose adjusted to pH 7.4 by a few drops of 0.28 molar THAM (trishydroxymethyl aminomethane) was used as the recovery medium.

The blastocysts were found by phase contrast microscopy; each was placed in a glass capillary tube along with a droplet (0.01 ml) of the solution. For each experiment an ovum was blown out of a capillary tube into the center chamber of a specially prepared electrophoresis cell which was separated from two side chambers by two cellulose acetate (dialysis tubing) membranes sealed in with paraffin wax (fig. 1). A centimeter scale, divided into 10 numbered millimeters by vertical lines, was placed at the focus of the eyepiece of a phase contrast microscope: each millimeter interval corresponded to 100 μ of the $\times$ 10 objective. The images were inverted by the optical system; an apparent movement to the right was in fact a movement to the left. (This must be remembered in the determination of polarity.)

All three chambers were filled with 8.1% Dextran 75 (Abbott) in 5% glucose (adjusted to pH 7.4 with 0.28 M. THAM), an isotonic solution which has a specific gravity of 1.053 and floats the eggs (sp. gr. approximately 1.038) just beneath the surface. The two side chambers contained iron wire mesh electrodes close to the outer sides of each of the cellophane membranes. A 32-v d.c. potential could be applied across these electrodes. A reversing switch was incorporated in the circuit so that the electrical field could be changed from the on to the off position or reversed instantly. The center chamber was slightly overfilled with fluid to give a convex meniscus so that the ovum would not be moved to the sides of the chamber by surface tension but would remain stable.

In a typical experiment the blastocysts were studied for stability under the surface of the center chamber for 100 sec before turning on the electrical current, for 150 sec with the electrical current in one direction, and for 150 sec after reversing the current. Each experiment was immediately duplicated by a control study

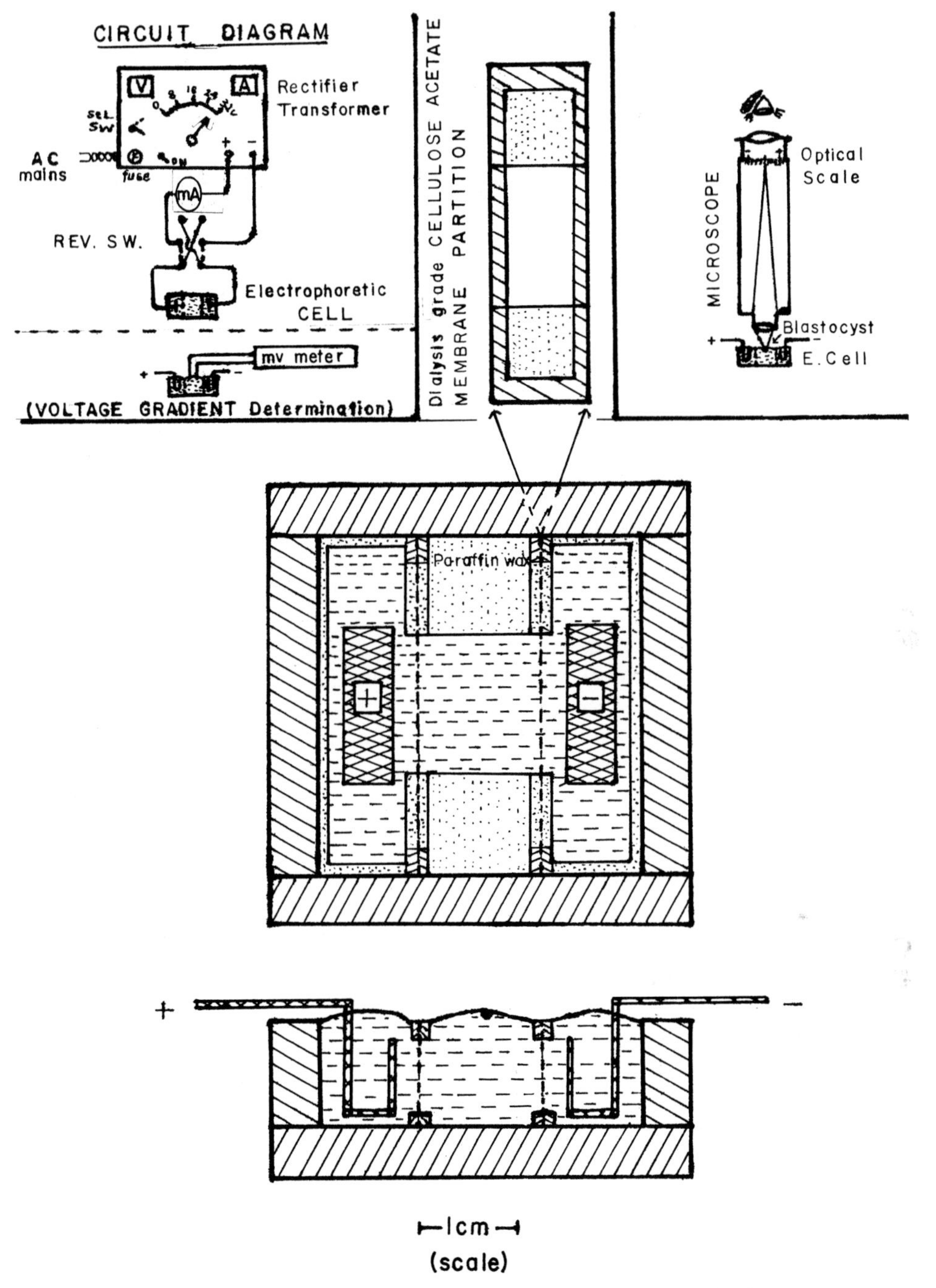

Fig. 1. Diagrammatic representation of the electrophoretic cell and its circuit diagram.

Top left insert: Overall circuit diagram and voltage gradient measurement.

Top center insert: Cellophane membrane partition.

Top right insert: Optical arrangement for observing and measuring the movement of the ovum in an electric field.

Bottom insert: Cross-sectional view of the electrophoretic cell.

in which paraffin particles of about the same diameter, prepared by heating, emulsifying, and cooling in a Dextran solution, were used in the same fluid under identical conditions. The paraffin particles generally were quite stationary for as long as 500 sec in a potential gradient of 3 to 8 v/cm in our pH adjusted medium. Only those experiments in which there was paraffin stability were considered valid. In every instance the velocity of uniform movement of the blastocysts was measured in both directions and the mean of the two directions calculated. The paraffin studies were necessary to make sure that adventitious movements of the fluid due to convection or electroendosmosis did not occur.

B. *Details of Instruments and Procedures*

Since to our knowledge this is the first attempt at measuring the electrical charge on a mammalian blastocyst, it may be worthwhile to mention the special equipment and methods used for the following purposes:

1. *Temperature.* The rate of temperature change in various solutions in the observation chamber during illumination and during passage of electrical current was measured with a Tele-Thermometer (Yellow Springs Instrument Company, Yellow Springs, Ohio), as this instrument has a conveniently small disk probe.

2. *Time.* Accurate timing was achieved with a "Time It" precision stopwatch (Precision Scientific Company) with digital read-out.

3. *Microscopy.* A Nikon model S-Ke phase contrast microscope was used for observation of the blastocysts and paraffin particles.

4. *pH of Solutions.* pH was measured with a Digicord Photovolt pH meter (Photovolt, New York).

5. *Electrical Power Source.* d.c. current was provided by a rectifier-transformer (Electro-Products Laboratories, Inc., Chicago, Illinois, model D-612T) attached to 115 v a.c. mains.

6. *Input Voltage.* The potential across the electrodes (32 v) was measured by a multimeter (Triplett Electrical Instrument Company, Bluffton, Ohio, model 310-VOM).

7. *Amperage.* The current was measured by use of the same multimeter.

8. *Potential Gradient.* The potential gradient just under the surface of the fluid in the center cell was measured by the use of two fine copper wires set 1 mm apart parallel to the plane of the electrical field; the ends of these were dipped just under the surface of the fluid after each ovum study and before its control paraffin study. A high-resistance volt meter was essential for an accurate assessment of the voltage gradient and was provided by the photovolt pH meter with the knob set to the millivolt position.

9. *Specific Gravity.* The specific gravity of the Dextran-dextrose solution was determined by weighing in a specific gravity bottle made to hold exactly 10 gm of water at 25° C.

10. *Conductivity.* The conductivity meter of a Bantam demineralizer (Barnstead Still & Sterilizer Company, Boston, Massachusetts) was used. Solutions of higher conductivity were diluted 10-fold or 100-fold (as appropriate) with low conductivity distilled water to give a measurable reading, and the conductivity was calculated by multiplying by the dilution factor.

11. *Viscosity.* Viscosity was measured by the conventional capillary flow method. The constant for the apparatus was determined by using distilled water at 25° C and cross-checked with ethyl alcohol.

The equation $V = \frac{3.142 \times r^4 \times g \times SG \times h \times t}{8v \times L}$ was used;

with the value of $\frac{3.142 \times r^4 \times g \times h}{8 \times v \times L}$ = the apparatus constant.

V = viscosity in Poises
r = radius of the capillary in cm
g = gravitational constant 981 cm/sec/sec
SG = specific gravity of the liquid
h = height of the liquid column in cm
v = volume of fluid flow in cc between graduations
t = time in sec taken for v cc of fluid to flow
L = length of the capillary in cm

12. *Dielectric Constant.* Attempts were made to measure the dielectric constant of the Dextran-dextrose solution by a high frequency alternating current technique in which a simple condenser of known dimensions was used; anomalous results were obtained owing to the well-known difficulties inherent in making such measurements in conducting solutions. We were therefore compelled to use the dielectric constant of 78.5 given for water at 25° C on the assumption that it does not differ too much from the value for an aqueous solution of Dextran and dextrose.

C. *Special Problems Encountered*

The problems encountered testify to the unfitness of conventional equipment and the usual physiological fluids for this work.

The Zeta Meter, used for measuring the charge on platelets and red blood cells, could not be employed because it depends on focus at exactly 14.7% of the tube diameter (believed to be the null-point for "electroendosmosis") and therefore requires a multitude of particles; blastocysts were available only in limited numbers. The rat blastocysts were heavier than isotonic glucose or saline solutions and sank to the bottom of the cell, where surface charge effects and adhesion interfered with charge measurements. Floating the ova in an albumin and electrolyte (NaCl or KCl) solution proved unsuitable; both paraffin particles and ova moved to the positive pole and reversed when the current was reversed, owing either to the albumin coating or to impulsion by charged protein molecules. When Dextran-dextrose solutions were used the paraffin particles remained stationary during passage of the current unless there was a mechanical leak in the cell. The evolution of gas at the electrodes caused bubbling in salt solutions which, in turn, brought about a disturbance of particles under study. Isolation of the central chamber by two cellophane membranes overcame this problem. All particles floating under the surface of the liquid were unstable until a convex meniscus was provided by overfilling the central (and side) chambers. Paraffin lining of the cell also helped to provide a uniform convex meniscus.

A potential gradient in excess of three v per cm or 30 mv/100 μ seemed to be needed to cause an appreciable movement of the blastocysts; when this voltage gradient was applied across a Dextran solution containing electrolytes, such as 0.15 M sodium or potassium chloride, the high conductivity of the medium caused a high current flow; heating and rapid convection currents in the cell resulted, as evidenced by the movement of paraffin wax particles which did not reverse when the current was reversed.

Oil droplets were found to be unsuitable for studies of the stability of the fluid medium; they floated on the surface instead of under it; they tended to spread and did not move when the fluid moved as in the electrophoresis of albumin electrolyte solutions.

The voltage gradient in the center cell changed greatly during electrophoresis in electrolyte solutions; less so in Dextran-dextrose solutions.

The pH of an aliquot of the Dextran-dextrose solution was adjusted from 5 to 6, 7, 8 or 9 with THAM, to 4 with acetic acid or to 3 with HCl before each experiment: if a quantity of a buffer sufficient to prevent change of pH during a series of experiments was added, it increased the conductivity and caused heating and convection currents. Electrophoretic studies were therefore carried out at different pH values while the solutions were fresh and before the pH had changed appreciably. A new solution was used for each experiment. Studies at pH 7.4 receive special mention because this was approximately the pH found by Hall (1936) for the uterine fluid of rats in early pregnancy.

II. Results

A. *Observations*

Studies of the physical constants of the Dextran-dextrose solution at room temperature (25° C) gave the following results:

Specific gravity —1.053
Viscosity —0.061 Poises
Specific conductance—240 μmhos per centimeter cube (at pH 7.4)

Blastocysts obtained on the 5th day with their zonae pellucidae still intact (fig. 2*a*) behaved as uncharged bodies in this solution and did not move in a charged field at pH 7.4. After loss of the zona pellucida (fig. 2*b*), they were negatively charged and moved consistently toward the positive pole at pH 7.4, reversing almost immediately when the current was reversed. Figure 3 shows that a 5-day blastocyst was relatively motionless until it escaped from its zonal coat.

The zona pellucida of the 5-day blastocyst may sometimes be retained for as long as 24 hr in isotonic saline at pH 7.4. Those blastocysts in this solution which lost their zonae appear to have done so by a rent or rupture in this coat; however in saline below pH 5 the zona pellucida swells and dissolves away in a few minutes.

In isotonic solutions of nonelectrolytes, such as glucose, dextran, glycerol, or

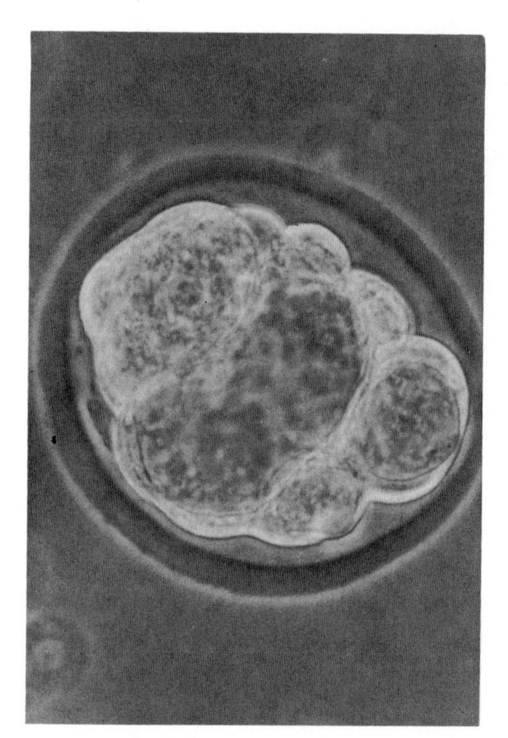

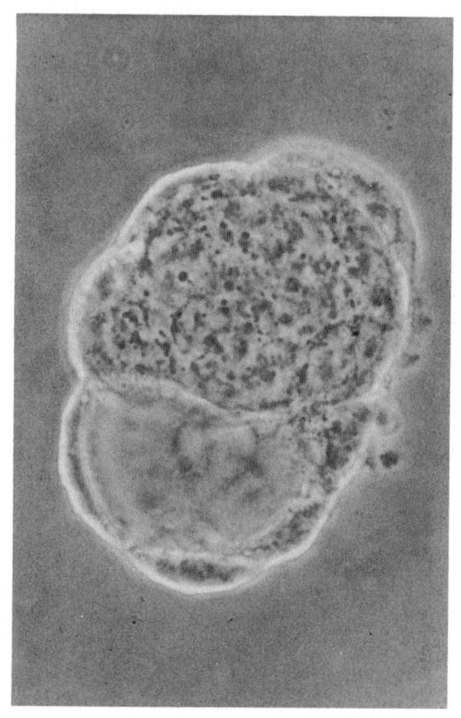

Fig. 2. 5-day rat blastocyst: (*a*) The blastocyst with zona pellucida; (*b*) the blastocyst after the loss of zona pellucida. Phase contrast microscopy of fresh, unstained specimens.

ethanol, we found that the zona pellucida swells and is lost in 5 to 15 min even when the pH has been adjusted to 7.4 with the requisite but small amount of THAM or other base.

Figure 3 represents the only occasion on which we have seen the zona pellucida lost by rupture in the Dextran-glucose solution.

The results of studies of the electrophoretic mobility of 5-day rat blastocysts in Dextran-dextrose solutions at different pH values are shown in table 1 and figure 4. Two of the actual experiments from which these values were derived are depicted in figure 5.

In table 1 and figure 4 we see that the mean electrophoretic mobility of naked

5-day rat blastocysts after attaining uniform velocity was 1.1 μ/sec per v/cm at pH 7.4.

The diameter of these blastocysts without their zonae pellucidae was approximately 75 μ.

B. *Calculations*

The quantity of charge on the blastocyst was calculated from the equation:

$$
\begin{aligned}
Q &= 6\pi r \cdot V \cdot (EM) \\
\text{where } Q &= \text{Charge in stat-coulombs} \\
r &= \text{Radius of blastocyst in cm} \\
V &= \text{Viscosity of suspending medium in Poises} \\
EM &= \text{Electrophoretic mobility in cm/sec per stat-volt/cm} \\
&\quad (1 \text{ stat-volt} = 300 \text{ practical volts}) \\
\text{therefore } Q &= 6 \times 3.1416 \times 37.5 \times 10^{-4} \times 0.061 \times 1.1 \times 10^{-4} \times 300 \\
\text{or } Q &= 142 \times 10^{-6} \text{ stat-coulombs} \\
\text{or } Q &= 47 \times 10^{-15} \text{ coulombs} \\
&\quad (1 \text{ coulomb} = 3 \times 10^{9} \text{ stat-coulombs}).
\end{aligned}
$$

Since the charge on one electron is stated to be 1.6×10^{-19} coulombs, the charge on the blastocyst is equivalent to an electron excess of

$$\frac{47 \times 10^{-15}}{1.6 \times 10^{-19}} = 29.4 \times 10^{4} \quad \text{or} \quad 294{,}000 \text{ electrons.}$$

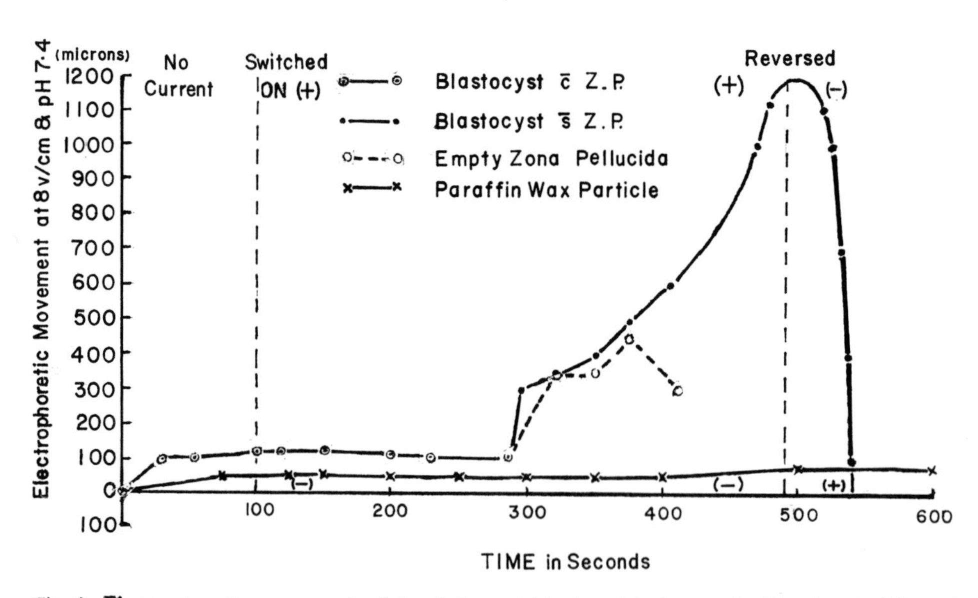

Fig. 3. Electrophoretic movement of the 5-day rat blastocyst before and after the shedding of its zona pellucida. Note the relative stability of the blastocyst with the zona on and its consistent movement toward the positive pole after losing the zona pellucida.

TABLE 1

ELECTROPHORETIC MOBILITY OF 5-DAY RAT BLASTOCYSTS WITHOUT ZONAE PELLUCIDAE IN 8.1% DEXTRAN–5% DEXTROSE SOLUTION

#	pH	EM	#	pH	EM	#	pH	EM	#	pH	EM	#	pH	EM
1	3.0	0.0	5	7.0	0.99	8	7.4	1.37	11	8.0	2.34	14	9.0	1.02
2	4.0	0.01	6	7.0	0.42	9	7.4	1.42	12	8.0	0.42	15	9.0	1.83
3	5.0	0.21	7	7.0	1.02	10	7.4	0.55	13	8.0	1.22			
4	6.0	0.85	Av:	7.0	0.81	Av:	7.4	1.11	Av:	8.0	1.33	Av:	9.0	1.43

NOTE: EM—Electrophoretic mobility of the blastocysts toward the positive pole in μ/sec per (practical) v/cm.
pH—Hydrogen ion concentration of the Dextran-dextrose solution (−log).
#—Serial number of the experiment.
Av:—Numerical average of the electrophoretic mobilities at the stated pH.

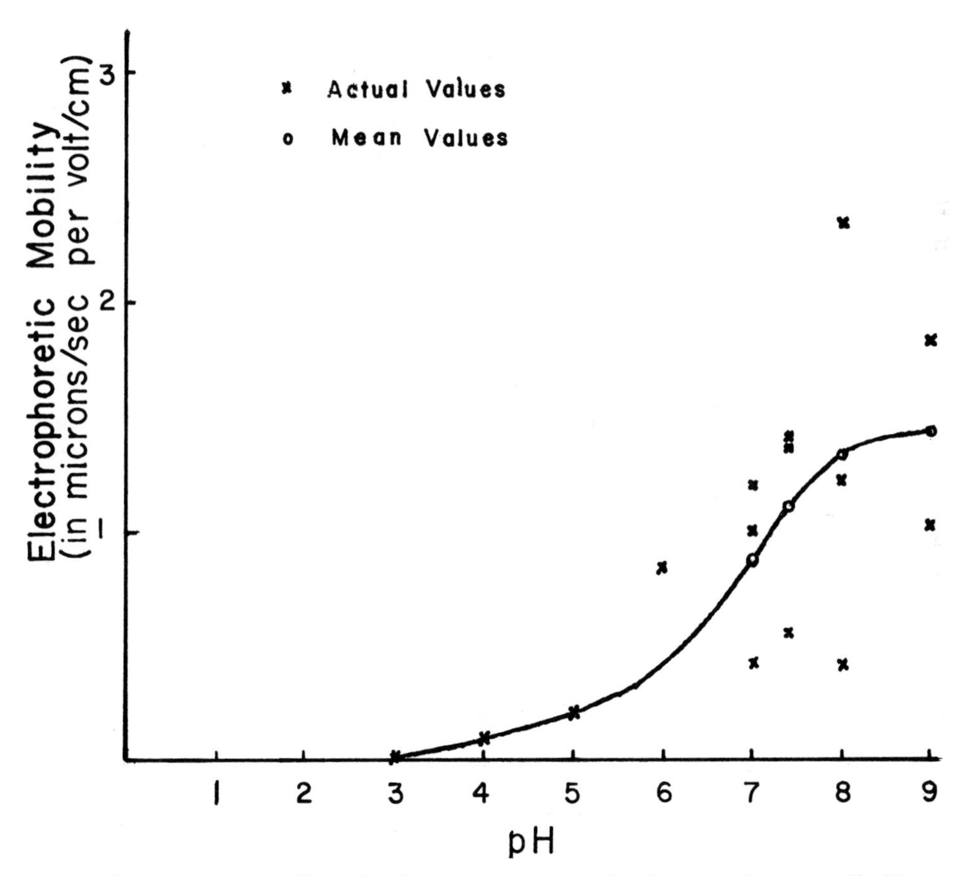

Fig. 4. Electrophoretic mobility of 5-day rat blastocysts after the loss of zonae pellucidae at various pH values. Note the pH dependence of the electrophoretic mobility and also its disappearance at pH 3.

The zeta potential of the blastocysts was calculated by the Helmholtz-Smoluchowski formula:

$$ZP = \frac{4\pi V \times EM}{D}$$

where ZP = (H.S.) Zeta-potential in stat-volts
V = Viscosity of suspending solution (Dextran-dextrose) in Poises at 25° C
EM = Electrophoretic mobility in cm/sec/stat-volt/cm
D = Dielectric constant of water at 25° C

$$ZP = \frac{4 \times 3.1416 \times 0.061 \times 1.1 \times 10^{-4} \times 300}{78.5}$$

= 97 mv (practical).

In round figures we find that the 5-day rat blastocyst after loss of the zona pellucida has an electron excess of about ¼ million electrons and a zeta potential of the order of 100 mv when studied in Dextran-dextrose solution at pH 7.4 and 25° C.

III. Discussion

There is no doubt that the 5-day rat blastocyst after loss of the zona pellucida behaves as a negatively charged body in Dextran-dextrose at pH 7.4, but its behavior in the natural medium of uterine fluid is unknown.

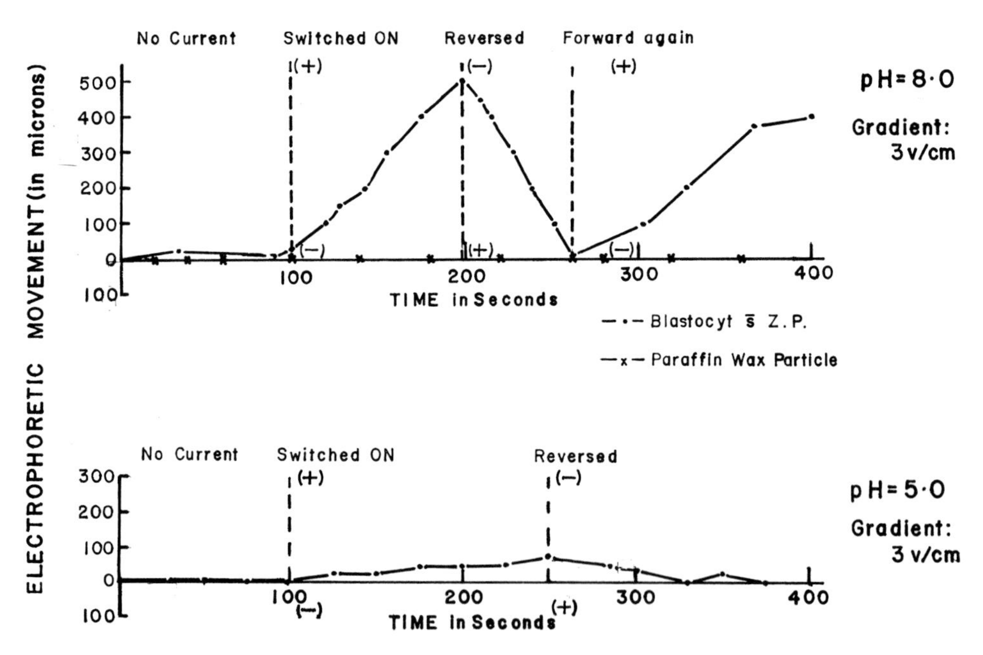

Fig. 5. Electrophoretic movement of 5-day rat blastocysts after the loss of zonae pellucidae at pH 5 and 8. Note the impressive rate of movement at pH 8 and the relatively negligible rate at pH 5. The stability of the blastocysts in the absence of current and that of paraffin wax particles with and without the current must also be noted.

Similar studies of red blood cells in plasma by Bernstein and Castenada (1965) showed that the addition of 10% Dextran caused a 25% increase in the relative negative charge on the cells; hence we appreciate that the true value of the charge on the blastocyst in 5% glucose might be something of the order of 80% of our observed value. Glucose itself, even up to 50% dilution, did not appear to affect the charge on red cells in plasma.

It is interesting to observe that 30 mv is required to prevent coalescence and precipitation of colloid particles. This happens to be of the same order as the potential difference across the diameter of the blastocyst necessary for electrophoretic movement, namely 30 mv/100 μ or 3 v per cm.

Schlafke and Enders (1963) have shown that the rat blastocyst loses its zona pellucida sometime during day 6 of delayed implantation but remains free and unattached in the uterine cavity despite this loss.

It is possible that the endometrium when under the influence of progesterone alone may produce a chemical substance capable of inhibiting attachment or implantation, but the fact that the blastocysts remain free and unattached during delayed implantation makes it seem more probable that it is embryo-endometrial "contact" that is delayed. One has to consider two aspects of cell contact: the first concerns chemical reactions which may be as precise as the specific interaction of complementary macromolecules, such as glycoproteins in the manner of antibody and antigen (Crandall and Brock 1968); the other is the physical aspect of cell contact including the influence of electrical forces. One must bear in mind that cell contact due to chemical bonding cannot occur when the two cell surfaces are held apart by the mutual repulsion of like electrical charges.

We have studied the charge on only the preimplanting blastocysts. It is clear that much work is required to investigate the potentials of the endometrium and uterine fluid and the changes in these potentials during various hormonal conditions.

It would be interesting to know whether the intrauterine device alters the charge on the endometrium and whether this might account for its contraceptive effect. The only common denominator in all the animal studies of intrauterine devices is a low-grade inflammation in the uterus; such a condition might disturb the charge relationships of the endometrium and the uterine fluid relative to the blastocyst.

If the charge on the blastocyst were due to the difference in the concentration of electrolytes, such as potassium or sodium on the inside and on the outside, one would not expect a disappearance of charge below pH 5, as is the case. We are forced to conclude that there are acidic groups on the surface of the blastocyst which are responsible for the observed negative charge, as this seems to be a pH-dependent property. If the charge were due to protein, we would expect reversal of the sign of the charge below the isoelectric point, normally between pH 4 and pH 5. Since there was no reversal even at pH 3, we conclude that the charge is due not to a protein but to acidic radicals such as the carboxyl or perhaps less likely to phenolic or sulfonic groups.

Dickmann and Noyes (1961) reported that the zona pellucida is shed from the rat blastocyst during the afternoon of the 5th day of pregnancy (also see chapter 7, Rumery); they concluded that removal of the zona pellucida precedes and is probably a prerequisite for attachment and implantation in the rat. The zona pellucida is retained on the guinea pig ovum until implantation has begun and the cells of the abembryonal pole have already penetrated the zona pellucida to form a naked attachment cone (Blandau 1949*a, b*). The situation in primates is not as clear. Böving (1959) reported that the zona pellucida of Macaque blastocyst shows degeneration at the embryonic and abembryonic poles at 8 days, just before implantation. There is still much debate about this point in the human, but no zona pellucida has been demonstrated to be still intact in the attached or implanted human ovum.

It seems that in most mammalian species the zona pellucida is either lost, locally digested, or penetrated by the trophoblastic cells before attachment. If embryonic attachment is favored by the attraction of dissimilar charges, then the charge on the exposed area of the blastocyst could account for the orientation of this area toward the endometrium in many species.

The tendency for trophoblastic adhesion in the rabbit to occur close to one of the endometrial capillaries rather than over an intervening space (Böving 1962) might well be explained by local differences in the tissue oxidation-reduction potential due to the state of oxygenation of the blood in these blood vessels. We have not studied the charge on the endometrium, but it is reasonable to suppose that an increased oxygen tension may create a local positivity in the electrical charge.

IV. Summary

Studies of the rat blastocyst in a Dextran-dextrose solution at pH 7.4 and 25° C have shown that it behaves as an uncharged body as long as the zona pellucida remains intact. After loss of the zona pellucida the 5-day rat blastocyst behaves as a negatively charged body and moves to the positive pole of an electrophoretic cell when the voltage gradient exceeds 30 mv/100 μ. The quantity of the negative charge on the blastocyst is approximately 47×10^{-15} coulombs and represents an electron excess of about ¼ million electrons: the zeta potential is of the order of 100 mv.

The possible significance of these findings with regard to contact for attachment of the ovum has been discussed. It is also pointed out that these findings may provide a clue as to the cause of delayed implantation in some species.

References

Abramson, H. A. 1965. Electrophoresis of blood cells. In *Biophysical mechanisms in vascular homeostasis and intravascular thrombosis,* ed. P. N. Sawyer, p. 3. New York: Appleton-Century-Crofts.

Barnes, L. E., and Meyer, R. K. 1964. Delayed implantation in intact rats treated with medroxyprogesterone acetate. *J Reprod Fertil* 7:139.

Bernstein, E. F., and Castaneda, A. R. 1965. Alterations in erythrocyte electrical charge associated with mannitol, polyvinylpyrrolidone (P.V.P.) dextrose and various dextran compounds. In *Biophysical mechanisms in vascular homeostasis and intravascular thrombosis,* ed. P. N. Sawyer, p. 105. New York: Appleton-Century-Crofts.

Blandau, R. J. 1949*a*. Observation on implantation of guinea pig ovum. *Anat Rec* 103:19.

———. 1949*b*. Embryo-endometrial interrelationship in the rat and guinea pig. *Anat Rec* 104:331.

Böving, B. G. 1959. Implantation. *Ann NY Acad Sci* 75:700.

———. 1962. Anatomical analysis of rabbit trophoblast invasion. *Contrib Embryol* 254:35.

Crandall, M. A., and Brock, T. D. 1968. Molecular aspects of specific cell contact. *Science* 161:473.

Dickmann, Z. and Noyes, R. W. 1961. The zona pellucida at the time of implantation. *Fertil Steril* 12:310.

Hall, B. V. 1936. Variation in acidity and oxidation-reduction potential of rodent uterine fluids. *Physiol Zool* 9:471.

Maéno, T. 1959. Electrical characteristics and activation potential of *Bufo* eggs. *J Gen Physiol* 43:139.

Sawyer, P. N., and Pate, J. W. 1953. Electrical potential differences across the normal aorta and aortic grafts of dogs. *Amer J Physiol* 175:113.

Schlafke, S., and Enders, A. C. 1963. Observations on the fine structure of the rat blastocyst. *J Anat Lond* 97:353.

Tyler, A.; Monroy, A.; Kao, C. Y.; and Grundfest, H. 1956. Membrane potential and resistance of the starfish egg before and after fertilization. *Biol Bull* 111:153.

11

Proteolytic Activity of the Rat and Guinea Pig Blastocyst *in Vitro*

Noel O. Owers/Richard J. Blandau

Department of Anatomy
Medical College of Virginia
Virginia Commonwealth University, Richmond

Department of Biological Structure
School of Medicine
University of Washington, Seattle

From lysis of tissues and the invasiveness associated with trophoblast, inferences have been drawn regarding its possible proteolytic activity (Mossman 1937; Wimsatt 1962; Wislocki and Padykula 1961). Some experimental evidence exists implicating trophoblast with proteolysis (Hagerman 1964; Huggett and Hammond 1952). Work done on the implanting blastocyst of the guinea pig and rat, in which unfixed gelatin, fibrin, and plasma films were used, indicated that a proteinase was present in the guinea pig trophoblast and absent in the rat trophoblast (Blandau 1949, 1961).

The purpose of this investigation was to determine (*a*) whether a proteolytic enzyme is present in the rat and guinea pig trophoblast; (*b*) what the nature of it is; and (*c*) whether the live trophoblast in tissue culture is capable of digesting a gelatin membrane.

The method of investigating these problems is new and is being reported for the first time. It consists of (1) treating trophoblast by hypo-osmotic shock at varying hydrogen ion concentrations and observing the reaction on a fixed gelatin membrane; and (2) growing trophoblast in tissue culture on a fixed gelatin membrane to determine whether living trophoblast is capable of depolymerizing the gelatin membrane.

Guinea pig trophoblast, on which the first method was used, treated at low pH shows high proteolytic activity, whereas rat trophoblast shows very little. Guinea pig trophoblast treated by the second method shows very weak proteolytic activity.

The presence of high proteolytic activity in both the rat and guinea pig yolk sac at low pH at certain stages of development was an unexpected finding.

I. Materials and Methods

A. *Forms Used*

Pregnant Holtzman rat embryos beginning at days 8, 9, and 10 were used most often in whole-mount preparations, although sometimes younger or older specimens were employed. (The time when spermatozoa were found in the vagina was counted as day 1.) Guinea pig embryos beginning at days 8, 9, 10, 13, and 14 were used for whole-mount preparations, guinea pig embryos beginning at day 8 for tissue culture preparations. Observations were made on the whole-mount preparations by using bright-field microscopy as well as time-lapse cinematography. Tissue culture preparations were examined by bright-field and phase-contrast microscopy.

B. *Preparation of Gelatin Membranes*

1. 408 mg of dry gelatin granules were placed in 6 ml of distilled water and allowed to stand for 15 min. The mixture was heated to 40° C in a water bath till the gelatin dissolved.
2. 0.33 ml of india ink was added to 6 ml distilled water. The mixture was warmed to 40° C in a water bath.
3. 6 ml of gelatin solution as given in 1 and 6 ml as given in 2 were mixed and warmed to 50° C.
4. 0.1 ml of the warmed mixture was immediately applied to clean, dry glass microscope slides or coverslips, covering an area 45 $\times$ 25 mm, and allowed to air dry for 18 hr.

C. *Fixation of Gelatin Membranes*

1. A solution of 0.05% glutaraldehyde was prepared in veronal acetate buffer pH 7.0 $\pm$ 0.05. The slides were placed with the dry gelatin membranes in the glutaraldehyde for 2 min, then transferred to each of 2 changes of buffer for 10 sec each. Excess buffer was shaken from the slides, and they were air dried for 1 hr or more. The slides were rinsed in 3 changes of distilled water for 10 sec each. Excess water was shaken off and the slides were allowed to air dry for 1 hr or more.
2. The above procedures were repeated with these changes: coverslips instead of slides were used, and the membranes were fixed in the glutaraldehyde solution at pH 9.0 $\pm$ 0.05 for 15 min instead of pH 7.0 $\pm$ 0.05 for 2 min.

D. *Tests for Membrane Stability*

Neither the 2 nor the 15 min fixed membrane dissolved in distilled water at 60° C if left therein for 30 min, or in 0.1 M acetate acetic buffers, pH 3.6 to 5.6, or in veronal acetate buffers, pH 2.6 to 9.0, if left at 37° C for 30 min or longer. These buffers were used also in control preparations where only the buffer was applied to the membrane.

E. *Test for Membrane Sensitivity*

Trypsin (Worthington) 1 mg/ml in 0.1 M tris buffer pH 8.1 was prepared. A drop about 10 μl in volume was placed on the gelatin membrane and covered with a coverslip. The coverslip was supported by double-stick tape which was applied to the two ends of the gelatin membrane prior to the application of the drop of trypsin. The appearance of a "hole" in the gelatin signified completion of digestion. A clear hole was produced in the 2 min fixed membrane in about 16 min after the solution was first applied and in the 15 min fixed membrane in approximately 32 min. (Calculation of the end point, the production of a clear hole, depends upon the discriminatory abilities of the observer.)

F. *Preparation of a Whole Mount of a Blastocyst in Which the 2 Min Fixed Gelatin Membrane Was Used*

Two strips of double-stick tape were applied about 15 mm apart on the gelatin surface. A drop of buffer of the required pH was placed between the two strips and a blastocyst put in the drop. The gelatin surface was then covered with a 22 mm square coverslip and gently pressed to make firm contact with the adhesive strips. Mineral oil was applied and allowed to run under the coverslip and surround the drop of buffer. The preparation was placed in an oven at 37° C for up to 4 hr and examined frequently by bright-field microscopy. Proteolysis of the gelatin membrane was indicated by (1) a lightening of the gelatin membrane from gray to white where erosion occurred, and (2) aggregation of the released carbon particles which formed gray to black spots.

G. *Preparation for* in Vitro *Culture in Which a Rose Chamber and a Fixed Membrane on a Coverslip Were Used*

The coverslip containing the membrane was sterilized for 24 hr in 95% alcohol. Usually an 8-day guinea pig embryo was placed on the gelatin surface of the coverslip and covered with a dialysis membrane. The Rose chamber was reassembled; a clear coverslip was used to enclose the opposite end of the chamber. Nutrient medium consisting of Locke-Lewis salt solution, guinea pig serum, and guinea pig or chick embryo extract, penicillin, and streptomycin was then introduced into the chamber, and the preparation was placed in an incubator. The culture was maintained for about 10 days, during which time daily observations were made. At the end of the period the Rose chamber was dismantled. The gelatin membrane coverslip was placed in alcohol-formalin-acetic mixture for 1 hr; it was then dehydrated, mounted in Permount, and examined with either bright-field or phase-contrast microscopy. In some instances the culture was treated with the appropriate buffer for 1 to 4 hr and then fixed, dehydrated, and mounted.

II. Observations

A. *Rat: Whole Mounts of Embryos*

Embryos were removed from rats on days 8, 9, and 10 of pregnancy. Each embryo was immediately placed on a fixed gelatin membrane with a drop of buffer added. A coverslip was placed over the drop, and the edges were sealed with paraffin oil. The preparation was incubated at 37° C and examined frequently under low power of the microscope for signs of proteolysis. Samples of maternal decidua and cornual epithelium were processed similarly and the slides were examined for evidence of proteolysis. The time when proteolysis was clearly discernible was noted. The amount of time required for proteolysis to begin was remarkably constant for a given tissue on a given day of pregnancy.

Of the four tissues examined for proteolytic activity the yolk sac was the most active and the decidua the least, with the ectoplacental cone or trophoblast and uterine epithelium in between.

The pH of the buffers used was 2.6, 4, 7, and 9. Activity was observed with buffers 2.6 and 4, none with buffers 7 and 9. The times of proteolysis mentioned below are given for the pH 4 experiments.

On day 8 the proteolytic activity of the yolk sac was observed at about 35 min after incubation (fig. 1). On day 9 the activity speeded up; it was detectable within 5 to 10 min (fig. 2). It again slowed down; on days 10 and 12 activity began at approximately 20 min (fig. 3).

The trophoblast showed no proteolytic activity during a 4 hr incubation period on day 8 (fig. 1) or day 9 (fig. 2), but on days 10 (fig. 3) and 12 weak reactions were obtained within 30 to 40 min of incubation time.

The uterine epithelium near the site of implantation was tested for proteolytic activity on days 8 and 9. Activity first appeared within 40 to 50 min.

The decidua was the least reactive of the tissues examined by the whole-mount method; activity first appeared within 50 to 60 min.

B. *Guinea Pig: Whole Mounts of Embryos*

Guinea pig embryos were dissected from their respective gestation chambers and placed on a fixed gelatin membrane, together with a drop of buffer, over which a coverslip was mounted. The preparation was placed in an incubator maintained at 37° C. The membrane was observed at varying intervals of time until proteolytic activity could be discerned under the dissection microscope. Proteolytic activity was identified by increase in blackness or increase of lightening of the uniformly gray gelatin membrane. The increased blackness represented the pooling of carbon black particles in a particular spot, and the increased lightness represented decreased number of carbon particles and a thinning of the gelatin membrane. The release of carbon from the gelatin membrane was a result of the depolymerization of the membrane.

In guinea pig tissues of high proteolytic activity it took as little as 5 min to detect proteolysis; in tissues of low proteolytic activity it took as much as 60 min or even longer.

The 8-day guinea pig embryo demonstrated high proteolytic activity in the trophoblast when incubation buffers pH 2.6 to 5.6 (figs. 4, 5) were used. The yolk sac and the embryonic cell mass were inactive at this time. No proteolytic activity was observed in preparations in which buffers having a pH of 7 or 8.6 were used.

The 9-day guinea pig embryo continued to demonstrate proteolytic activity in the trophoblast at pH 2.6 to 5.6. The activity associated with the yolk sac varied

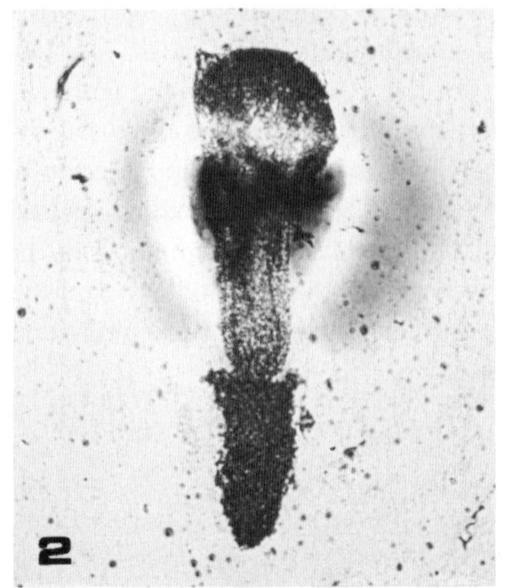

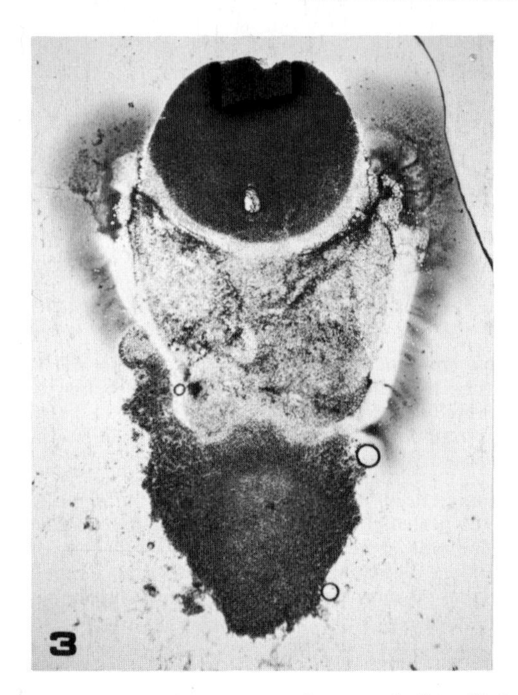

Fig. 1. Eight-day rat embryo mounted in a drop of acetate buffer pH 4 on a fixed gelatin membrane covered with a coverslip, and incubated for about 30 min. Note the absence of proteolytic activity associated with the ectoplacental cone, or Träger (*lower part*), and the formation of a halo of carbon particles around the yolk sac (*upper part*). ×28.

Fig. 2. Nine-day embryo treated like the embryo in fig. 1 and incubated about 35 min. Observe the absence of activity in the trophoblast and considerable activity indicated as halos of carbon black particles forming around the yolk sac. ×28.

Fig. 3. Ten-day embryo treated like the embryo in fig. 1 and incubated about 30 min. The large dark disk in the upper part of the figure is the embryonic mass. The yolk sac is outlined by the diffusion of carbon black released from the fixed gelatin membrane and the lower dark triangular mass is the ectoplacental cone, which is proteolytically almost inert. ×28.

from slight to considerable (figs. 6–10). At pH values of 7 and 8.6, the trophoblast, yolk sac, and embryonic mass were inactive.

In the 10-day guinea pig embryo the trophoblast had greatly reduced proteolytic activity, whereas the yolk sac, which was completely negative at day 8 and largely negative at day 9, appeared to be strongly positive (fig. 11). The trophoblast and yolk sac continued to show inactivity at pH 7 and 8.6.

This pattern of enzyme activity persisted through days 11, 12, 13. At day 14 the activity of the trophoblast persisted but took longer to appear than at day 8, showing that there was a progressive slowing of activity (figs. 12, 13). No proteo-

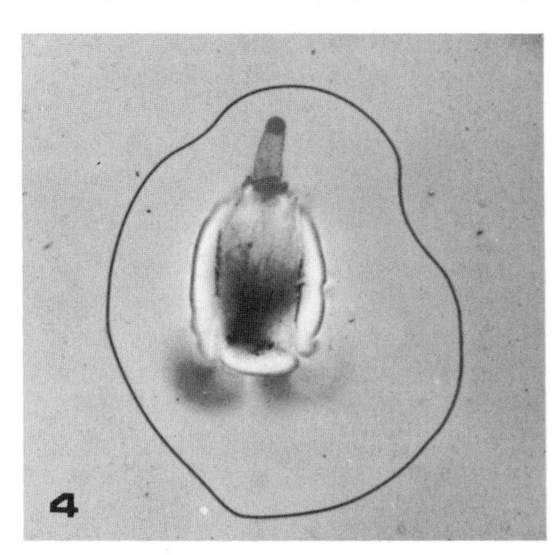

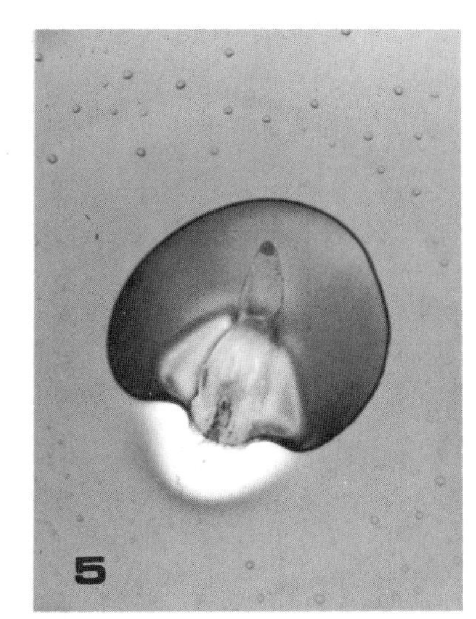

Fig. 4. Eight-day guinea pig embryo, incubated in a drop of buffer pH 4.5 at 37° C for about 60 min. The dark line encircling the embryo is the outline of the buffer drop. The white zone surrounding the trophoblast represents the area of dissolution of the gelatin membrane. The dark zone within the area of the trophoblast is formed by the accumulation of carbon particles released by proteolysis. The yolk sac is inactive. ×25.

Fig. 5. Eight-day guinea pig embryo, incubated in a drop of buffer pH 3.6 at 37° C for approximately 45 min. The zone of proteolysis of the gelatin membrane is seen as a clear area encompassing the trophoblast. The yolk sac is inactive. Carbon particles released by proteolysis have floated into the buffer solution, darkening it. In the photograph the buffer drop moved upward slightly away from its original position covering the area of proteolysis. Approximately ×25.

lytic activity was observed at pH 7 and 8.6. The embryonic mass continued to remain inactive at all pHs.

C. *Guinea Pig: Tissue Culture Preparations Subjected to Hypo-osmotic Shock at Low pH*

The newly formed syntrophoblast close to the original tissue implant had a moderate amount of proteolytic activity as compared with older syntrophoblast which had grown and migrated toward the periphery of the culture. This peripheral syntrophoblast exhibited the maximal amount of proteolytic activity (figs. 14, 15). There was considerable variation in syntrophoblast proteolysis in various areas of

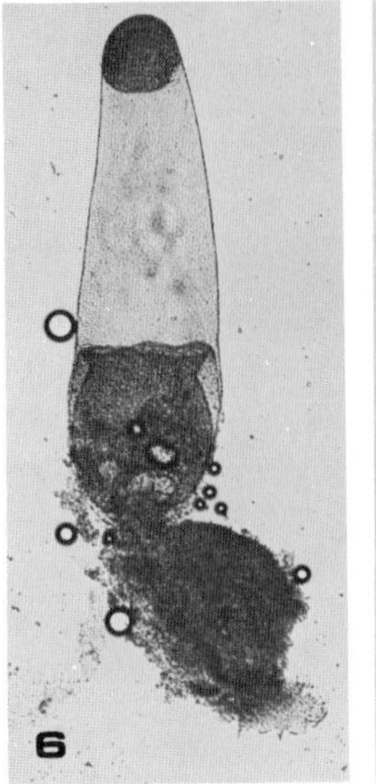

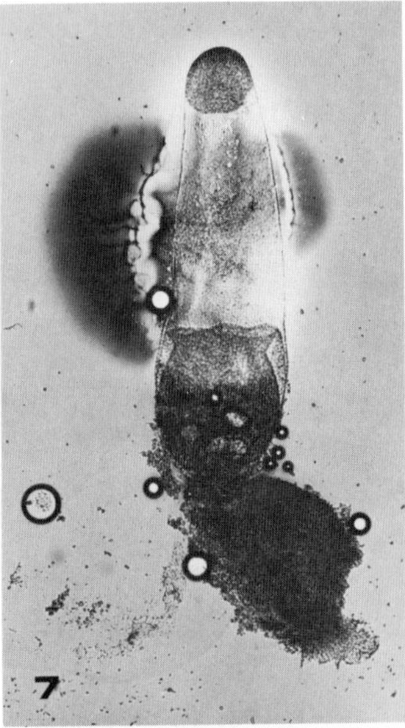

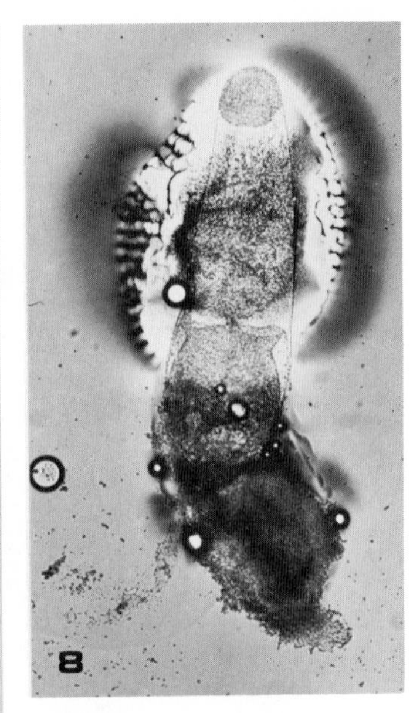

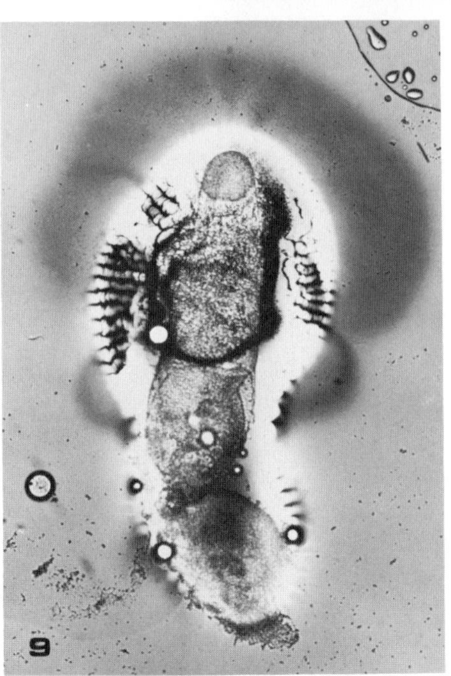

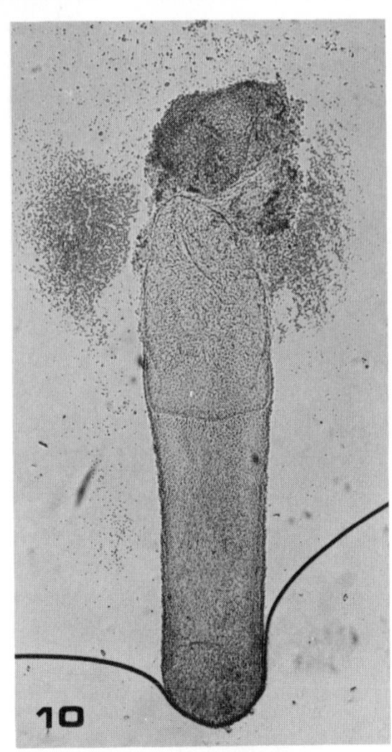

Fig. 6–10. Nine-day guinea pig embryo mounted on a fixed gelatin membrane, with a drop of buffer pH 4 and a coverslip, and incubated at 37° C. All magnifications ×18. Photographs are time-sequence exposures of the same embryo.

(6) No activity observed after 45 min.

(7) The yolk sac displays very much proteolytic activity in this specimen after 75 min; no activity in the trophoblast.

(8) The trophoblast becomes active after about 1½ hr.

(9) Proteolysis of the fixed gelatin membrane after about 3 hr of incubation.

(10) *Control specimen:* incubated at pH 8.96 in veronal acetate buffer for 90 min shows no activity. This specimen was inactive even after 24 hr incubation.

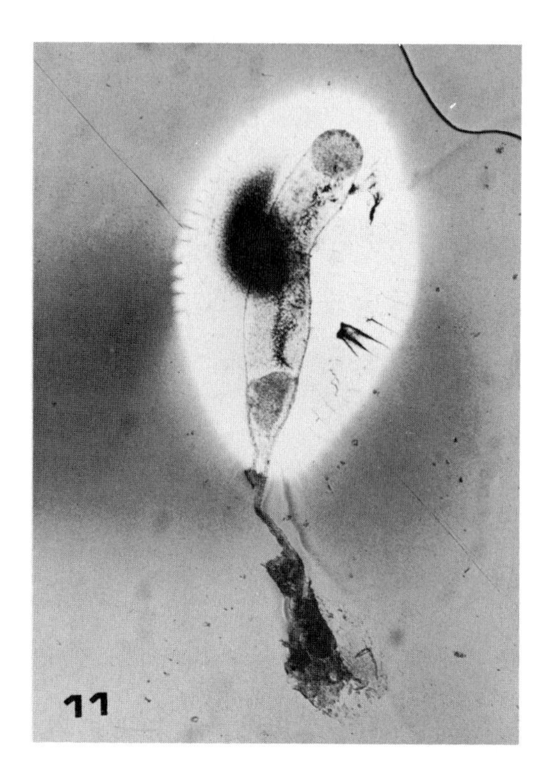

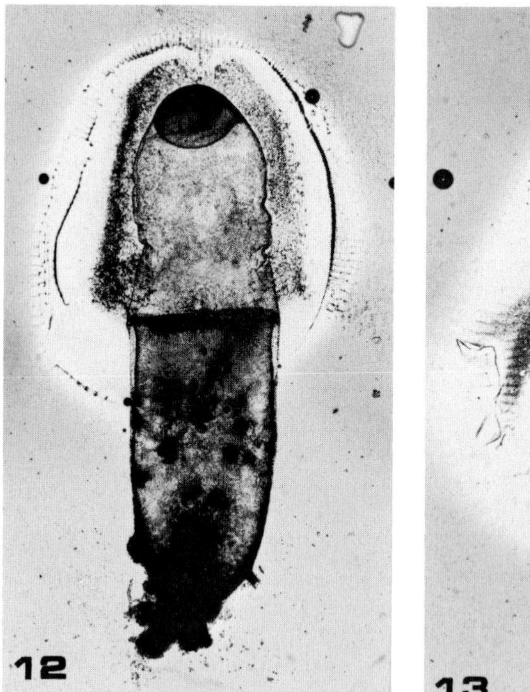

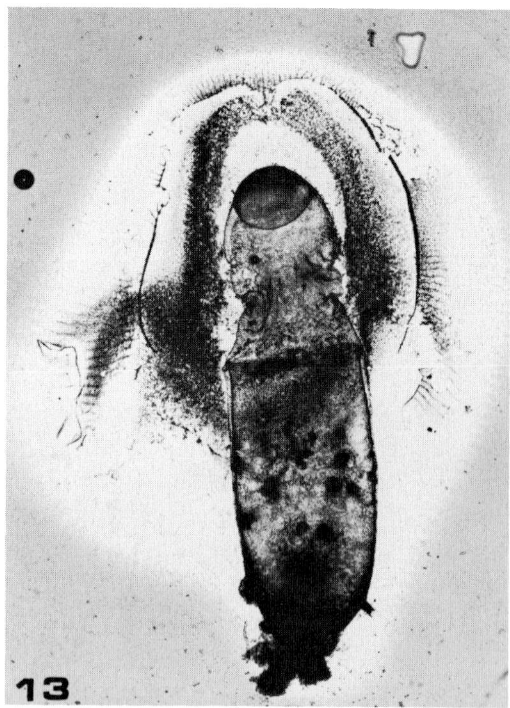

Fig. 11. A 10-day guinea pig embryo incubated in a drop of acetate-acetic buffer pH 3.6 on a fixed gelatin membrane. The Träger at this time shows little or no activity, whereas the yolk sac is strongly positive. Note the zone of proteolysis surrounding the yolk sac, and the aggregation of carbon particles at its middle. Photograph taken after 15 min incubation. ×24.

Figs. 12, 13. A 13-day embryo incubated in a drop of buffer pH 4 and photographed at 1 hr intervals. Note the proteolytic activity associated with the yolk sac and no observable activity in the trophoblast in fig. 12. In fig. 13, the trophoblast too appears to show activity but it is not known whether this occurs secondarily by diffusion of enzyme from the yolk sac or is residual activity in the trophoblast. ×14.

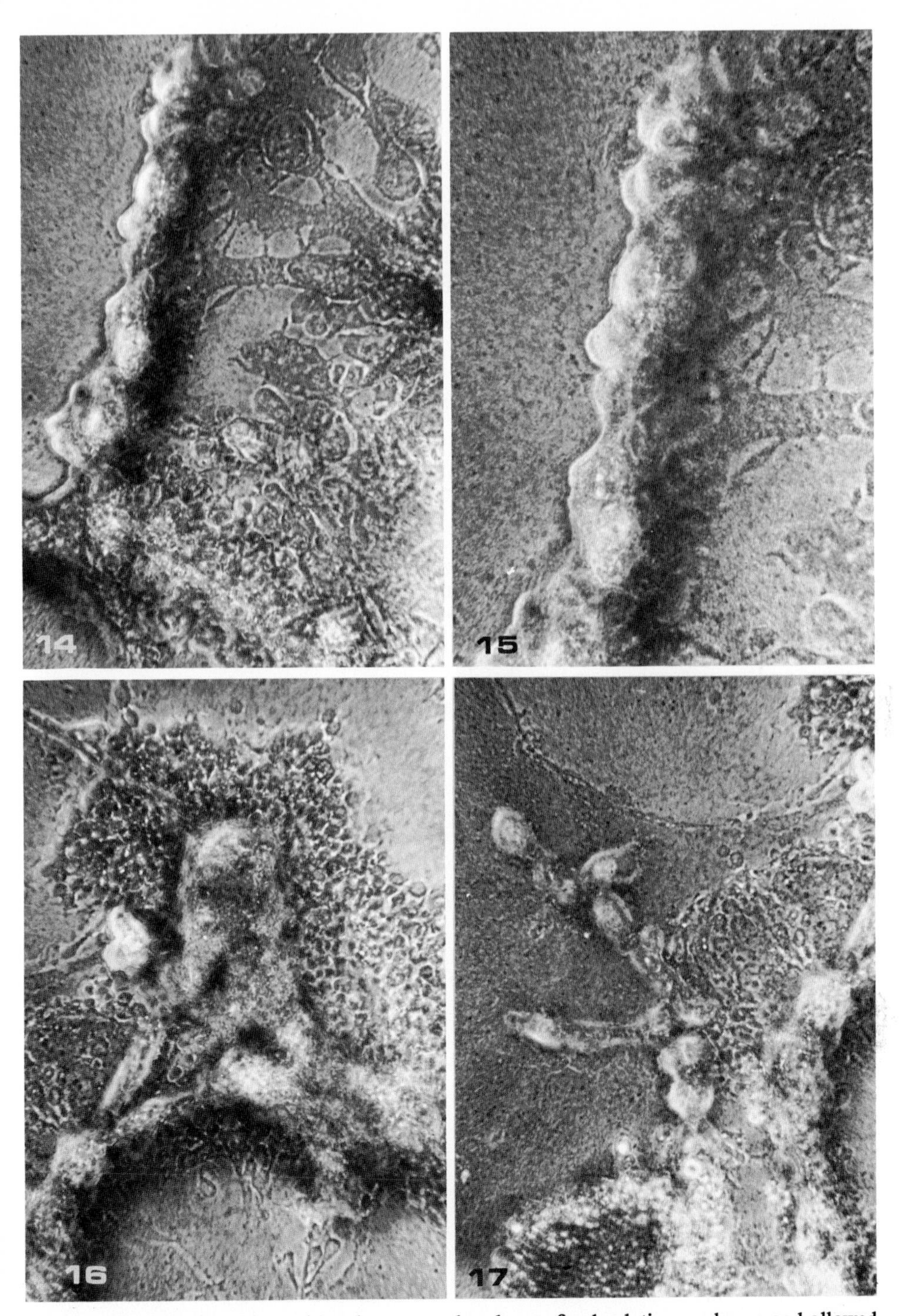

Figs. 14–17. An 8-day guinea pig embryo was placed on a fixed gelatin membrane and allowed to grow in a Rose chamber for 10 days. Locke-Lewis salt solution containing chick or guinea pig embryo extract and guinea pig serum was the nutritive solution used and was replenished 3 times in 10 days. The chamber was then dismantled and a drop of buffer pH 4.4 placed over the outgrowth and allowed to incubate for 4 hr at 37° C. The tissue was then flooded with alcohol formalin acetic acid mixture and dehydrated and mounted in Permount. The photographs in figs. 14, 15, 16, and 17 were then taken; phase-contrast objectives were used.

(14) The white zone lies along the leading edge of syncytium and indicates maximum proteolysis in this region. The dark band parallel to the white zone represents aggregation of carbon black. The fibroblastlike cells and the cytotrophoblast have not reacted with the gelatin membrane to any great extent. Approximately ×208.

(15) The previous figure taken at higher magnification showing that the leading edge contained much proteolytic activity. Approximately ×320.

(16) Proteolysis, as indicated by a light background, seems to be associated with a central zone surrounded by cytotrophoblast cells and fibroblasts which do not show much activity. Approximately ×208.

(17) The syntrophoblast coincides with areas of proteolysis and both are clearly seen in this photograph. The cellular components are relatively nonreactive. ×208.

the culture. Precisely what this implies is unclear. It may mean that there was insufficient contact between the syntrophoblast and the gelatin membrane, or that the enzyme was not present; or that, if present, it was inhibited in some way. The proximity of active areas and inactive areas of the syntrophoblast could be the result of varying times of differentiation in the two areas. Although there may be many explanations for inactivity, there is but one explanation for activity: the gelatin membrane dissolves because a proteinase is present. The evidence presented indicates that the enzyme diffuses out of the syntrophoblast and at low pH dissolves the fixed gelatin membrane.

The cytotrophoblast possessed little proteolytic activity; sometimes none was observed (figs. 16, 17).

D. *Guinea Pig: Tissue Culture Preparations Not Subjected to Hypo-osmotic Shock*

After 24 hr in culture the syntrophoblast had grown out to a remarkable degree from the junctional zone of the yolk sac and Träger. Proteolytic activity, if present, was difficult to detect because of the presence of phase contrast halos (figs. 18, 19). Occasionally slight erosion of the gelatin membrane occurred, making the area apposed to the syntrophoblast look brighter.

At the end of 5 days in culture the light-dark changes in the fixed gelatin

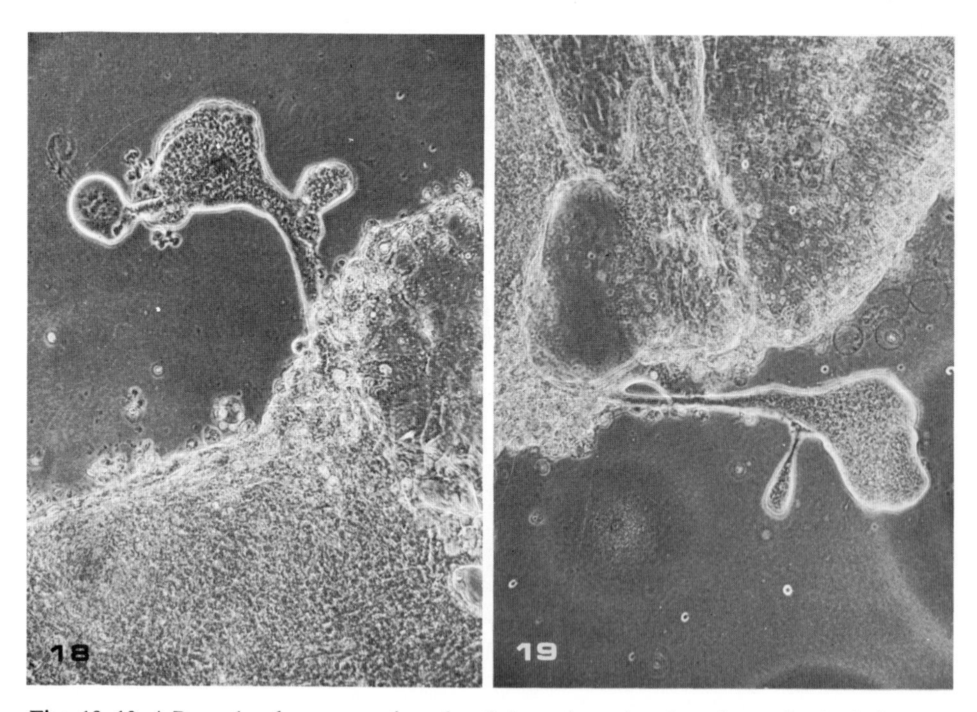

Figs. 18, 19. A Rose chamber preparation of an 8-day guinea pig cultured on a fixed gelatin membrane for 24 hr. The syntrophoblast is beginning to sprout branches. Proteolysis of the gelatin membrane associated with syntrophoblast is equivocal and cannot be distinguished with certainty since the phase-contrast halos of light overpower any increased brightness due to proteolysis of the membrane. ×185.

membrane were somewhat better observed, although they were difficult to photograph. The changes in light intensity were sometimes associated with syntrophoblast.

The light-dark changes due to proteolysis and migration of carbon particles were seen most clearly in 7-day cultures. There was a generalized darkening or lightening underlying the syntrophoblast, particularly below the leading edge.

The overall observations indicated that living trophoblast growing on a gelatin membrane in tissue culture did not digest the gelatin uniformly. After the tissues had been fixed and mounted, areas of proteolysis associated with the syntrophoblast were seen clearly (figs. 22, 23).

III. Discussion

A. *Membrane Substrates*

Adams and Tuqan (1961) used blackened panchromatic photographic plates to detect the proteolytic activity of some tissues fixed in formaldehyde with some success. Daoust (1957) developed a technique using a gelatin substrate for detecting a variety of enzymes in frozen sections of fresh tissue. The limitations of his technique were reviewed later (Daoust 1965). Cunningham (1967) prepared a pre-stained substrate film for use with cryostat sections of fresh tissue which gave reasonably good resolution when incubated for brief periods at 73° C.

Rat tail collagen was used by Ehrmann and Gey (1956) as a nutritive substrate for growing cells in tissue culture. Gross, Lapiere, and Tanzer (1962) have shown that a collagenase is elaborated and secreted by slices of tadpole tail cultured for 2–3 days on reconstituted collagen gel.

In the work reported here a new method of studying the proteolytic activity of trophoblast tissue in culture has been developed. The proteinase-sensitive gelatin membrane used was insolubilized by glutaraldehyde (Sabatini, Bensch, and Barnett 1963). To provide opacity carbon black was mixed with the gelatin prior to its fixation. The membrane so produced shares the advantage of adequate contrast with exposed photographic plates and is stable under tissue culture conditions.

B. *Comparison of Rat and Guinea Pig Blastocyst Proteolytic Activity*

The difference in proteolytic activity of the trophoblast of the 8-day guinea pig embryo and 8-day rat embryo was substantial. The guinea pig trophoblast was highly active whereas the rat trophoblast was highly inactive. Presence of proteolytic activity in the guinea pig may be related to the lysis in the adjacent maternal tissue as observed by Sansom and Hill (1931) by light microscopy. In the rat the absence of proteolytic activity may be related to the absence of extensive lysis in maternal tissues as observed by Enders and Schlafke (1967) in electron micrographs.

Changes observed in histological sections, from which inference is drawn concerning the proteolytic activities of trophoblast, correlate to some degree with experimental evidence presented, showing that a proteolytic enzyme is actually present in this tissue. This clear picture of cause and effect becomes clouded when other animals are considered. Owers (1960), for example, has shown by light microscopy

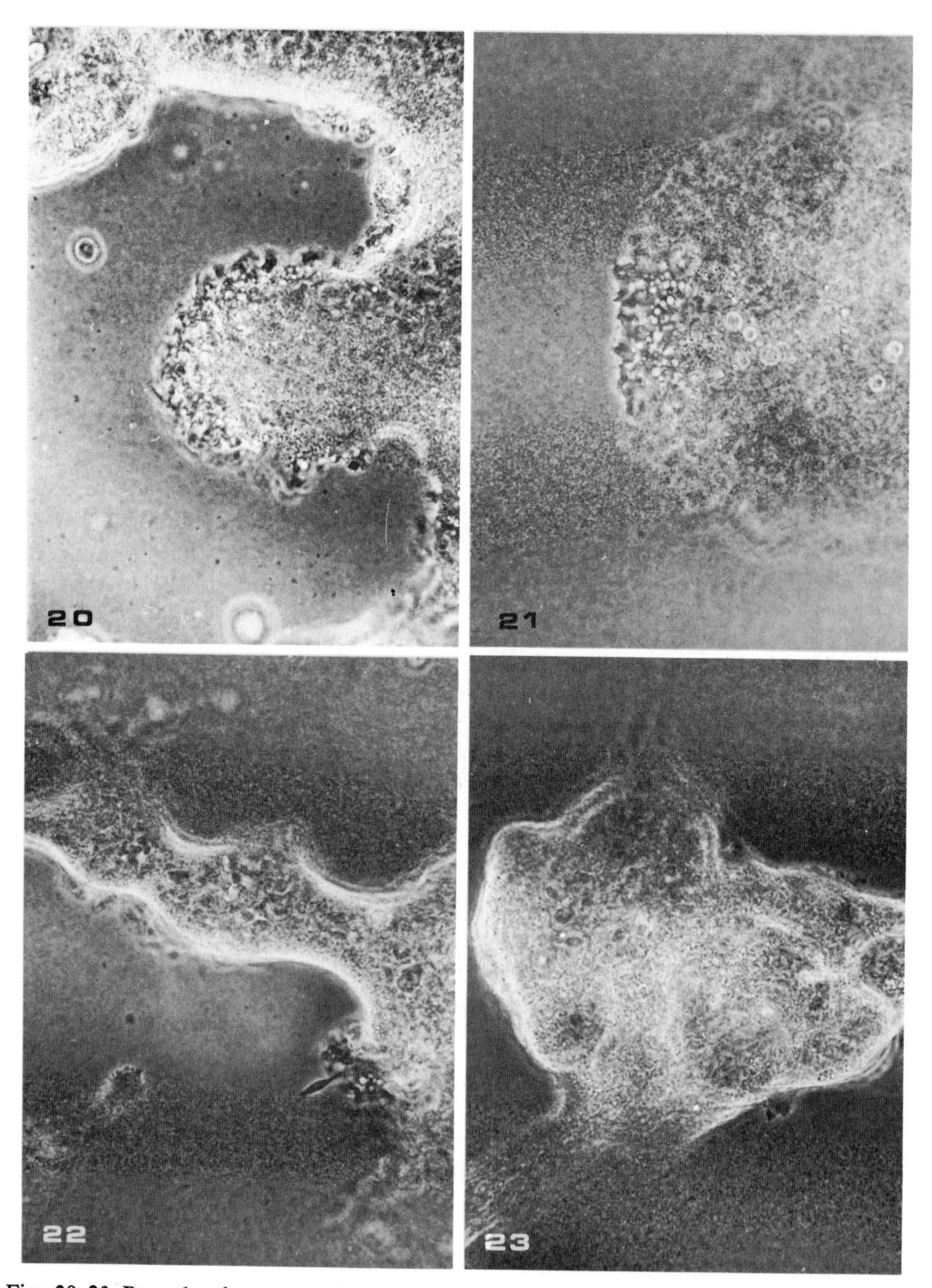

Figs. 20–23. Rose chamber preparations of 8-day guinea pig embryos grown in culture for 7 days. All magnifications ×180.

(20) The profusion of syntrophoblast shows active pinocytosis occurring at its margin; there may be some proteolysis but it is difficult to ascertain. In the more quiescent zone away from and below the protuberance the leading edge of syntrophoblast seems to be proteolytically active, causing a band of brightness which is the result of dissolution of the gelatin membrane adjacent to it. Live preparation.

(21) Pinocytotic vesicles are more clearly seen at the leading edge of syntrophoblast with a generalized lightening of the gelatin membrane. Live preparation.

(22) This protuberance of syntrophoblast is associated with aggregation of carbon particles forming dark bands; the area of erosion of the gelatin membrane appears bright. The phase-contrast halos of light no longer mask the brightness caused by erosion of the membrane. Fixed and mounted preparation.

(23) The increased brightness of the syntrophoblast is largely due to erosion of the gelatin membrane in contact with it. Fixed and mounted preparation.

that the trophoblast of the Indian musk shrew is lost at implantation and extensive cytolysis of maternal epithelium occurs when the trophoblast is no longer present. It is evident that drawing inferences regarding the proteolytic activity of trophoblast from the presence of lysis and necrosis observed in histological sections of maternal tissue should be done with caution.

Preliminary experiments indicate that a proteinase develops in rat trophoblast after day 10. The significance of the time lapse in the finding of proteinase in the trophoblast of the rat and the guinea pig is unknown. Comparative studies might be helpful in increasing our understanding of the role of proteinase activity of the trophoblast at this vital stage of development.

Equally unexplained differences in proteolytic activities took place in the rat and guinea pig yolk sac under hypo-osmotic shock at acid pH. The guinea pig yolk sac was inactive on day 8, in contrast to the trophoblast's activity: only slight activity, if any, was present on day 9; on day 10 strong proteolytic activity was observed. The reaction in the rat yolk sac was also the reverse of that of its trophoblast. Proteolytic activity was strongly positive on day 8, at which time the trophoblast was inactive, and persisted through days 9 and 10.

Nor is the contrast between the high proteolytic activity of guinea pig trophoblast toward a gelatin substrate under hypo-osmotic shock at acid pH and is relative inactivity in tissue culture whose pH is near neutrality clearly understood. Perhaps an acid proteinase, inactive at neutral pH, requires an acid pH for activation. The situation in tissue culture may not be a true reflection of the activity *in vivo,* for preliminary experiments showed that the uterine tissues at implantation are in fact quite acid. Or possibly the tissue culture lacks some stimulating factor derived from maternal tissues. That such an induction factor may exist is shown by the experiment of Houck and Sharma (1968), who were able to induce collagenolytic and proteolytic activities by the use of anti-inflammatory drugs in rat and human fibroblasts grown in tissue culture.

C. *Proteinases*

A brief review of proteinases obtained from other tissues is in order so that the experimental evidence presented here demonstrating the existence of proteinases in trophoblast and yolk sac may be better understood. Proteinases are found in both animals and plants. Consideration in this paper is limited to animal proteinases. They may be present as extracellular enzymes, where they are concerned with breakdown of protein products which can be absorbed by the cell, or as intracellular enzymes where their function may be related to homeostasis.

1. *Extracellular Proteinases.* The first crystalline proteolytic enzyme obtained was pepsin, from bovine gastric juice (Northrop 1930). Subsequently the other proteolytic enzymes were extracted from bovine pancreas, including trypsin (Northrop and Kunitz 1932), chymotrypsin (Kunitz and Northrop 1935), chymotrypsin B (Keith, Kazenko, and Laskowsky 1947), carboxypeptidase A (Anson 1935), carboxypeptidase B (Folk 1956), and aminopeptidases and dipeptidases.

The digestive enzymes are actually secreted in the form of enzymatically inactive precursors known by the general term of zymogen. Zymogens become active by a process of "unmasking" which requires the presence of an unmasking agent. Aminopeptidases and dipeptidases require the presence of traces of specific metals such as cobalt or manganese for activation. Rennin (Berridge 1943), thrombin (Waugh and Baughman 1960), and plasmin (Ablondi and Hagan 1960) are formed from inactive precursors. Thrombin, which is directly involved in the conversion of fibrinogen to fibrin, and plasmin, which liquefies blood clots by digesting fibrin, are both obtained from plasma. Plasminogen is deposited on fibrin during the process of clot formation and is activated easily to form plasmin, a nonspecific proteinase. For this reason plasma clots should not be used in tissue culture in which proteinase activity is to be evaluated.

2. *Intracellular Proteinases.* The major intracellular proteinases in animals are cathepsins. The term "cathepsin" was introduced by Willstaetter and Bamann in 1928 after they observed autolytic protein breakdown in hog intestines. Hedin and Rowland (1901) observed autolytic activity in beef, horse, and sheep spleen. Fruton (1960) devised the first classification of these enzymes into three groups, cathepsins A, B, and C, on the basis of their specificity toward synthetic substrates and the differences in their pH optima. Press, Porter, and Cebra (1960) discovered cathepsin D in beef spleen. Cathepsin D constitutes 2/3 of beef spleen catheptic activity. It has an optimum pH of 3.0 when hemoglobin is used as substrate. Fractionation studies suggest that this enzyme may exist in 10 different forms.

Cathepsins differ from the digestive proteinases in not having to be unmasked; they differ from the dipeptidases in not demanding the presence of cobalt or manganese. Some cathepsins require no activator at all; some can be activated by the addition of trace quantities of cyanide, hydrogen sulphide, cysteine, or glutathione; a few are activated by ascorbic acid.

3. *Consideration of 1 and 2 Above.* Tissue cultures of guinea pig embryos grown on fixed gelatin membranes were assembled with and without plasma clots. In both instances lysis of some cells was observed at the beginning of each culture experiment, followed by a localized lysis of the fixed gelatin membrane wherever these cells had become attached. After this initial period and once the remaining cells had spread over the surface of the fixed gelatin membrane, there was little evidence that the trophoblast actually secreted a proteolytic enzyme in any great quantity in its living state. The fixed gelatin membrane showed unequivocal traces of dissolution only after the tissues were fixed and mounted. Much more dissolution of the membrane occurred after the tissue culture preparation was subjected to hypo-osmotic shock at low pH, then fixed and mounted.

These observations indicate that additional enzymes were extracted from within the cell and that these enzymes were responsible for the enhanced proteolysis. The extracellular secretion of proteinase by guinea pig trophoblast is miniscule in tissue culture. This suggests another possible explanation for the absence of proteolytic activity in tissue culture: the near neutral pH culture medium negates the action of any acid proteinases that may be secreted.

Since it appears that the guinea pig trophoblast *in vitro* does not secrete an extracellular proteinase, the proteolysis observed after hypo-osmotic shock at low pH must be caused by an intracellular enzyme originating within the cells, possibly a cathepsin.

4. *Exopeptidases and Endopeptidases.* Bergman and Fruton (1937) divided the proteinases and peptidases into two groups:

(*a*) exopeptidases include carboxypeptidase, leucine aminopeptidase, dipeptidase and tripeptidases, and cathepsin C. These enzymes catalyze the hydrolysis of peptide bonds only when either one or both the amino acid residues taking part in the peptide bond have their other alpha-amino or carboxyl group free.

(*b*) endopeptidases include all other proteolytic enzymes as well as cathepsins A, B, and D. They catalyze the splitting of proteins into polypeptide fragments and act on synthetic substrates when the latter have their alpha-amino and carboxyl groups blocked.

Since the guinea pig trophoblast digests gelatin and, in a preliminary experiment, hemoglobin, the enzyme that causes the initial digestion could be classified as an endopeptidase. The presence of exopeptidases that act after the initial cleavage of the gelatin molecule has begun cannot be excluded.

5. *Serine, Thiol, Acid, and Metal Proteinases.* A more recent classification of proteinases, based on functional considerations, is given by Hartley (1960): (*a*) serine proteinases—enzymes which contain a serine residue which reacts uniquely with organophosphorus compounds and which include, among others, trypsin, chymotrypsin, thrombin, and subtilisin; (*b*) thiol proteinases—enzymes for which the presence of a sulfhydryl group is essential for activity and which include some cathepsins and some plant proteinases; (*c*) acid proteinases—enzymes which have their optimum activity at rather acidic pH and which include pepsin, rennin, and most cathepsins; (*d*) metal proteinases—enzymes which require a metal ion for activity and which include carboxypeptidases, aminopeptidases, some dipeptidases, and tripeptidases.

Guinea pig trophoblast has an enzyme that is active between pH 2.8 and 6.5. The enzyme is an acid proteinase according to Hartley's scheme (1960). It does not require a metal activator since it is active in the presence of ethylenediaminetetraacetic acid (EDTA).

IV. Summary and Conclusions

1. The trophoblast of the guinea pig, but not the trophoblast of the rat, shows high proteolytic activity on fixed gelatin substrate when treated with hypo-osmotic shock at acid pH on days 8, 9, and 10.
2. The yolk sac visceral layer of both the guinea pig and the rat displays high proteolytic activity when treated with hypo-osmotic shock at low pH on days 8, 9, 10, 13, and 14.
3. The trophoblast and yolk sac of the 8-day guinea pig embryo show little proteolytic activity, if any, when grown on a fixed gelatin membrane. No experiment was done on the 8-day rat trophoblast and yolk sac; no comparison could therefore be made.

4. The enzyme (singular for convenience) is intracellular in location, is an endopeptidase, acts at acidic pH, and is possibly a cathepsin.

Acknowledgments

This work was supported in part by United States Public Health Service grant HD 03464 from the National Institutes of Health.

References

Ablondi, F. B., and Hagan, J. J. 1960. In *The enzymes,* ed. P. D. Boyer, H. Lardy, and K. Myrback, 4:175. New York: Academic Press.

Adams, C. W. M., and Tuqan, N. A. 1961. The histochemical demonstration of protease by a gelatin-silver film substrate. *J Histochem Cytochem* 9:469.

Anson, M. L. 1935. Crystalline carboxypolypeptidase. *Science* 81:467.

Bergman, M., and Fruton, J. S. 1937. On proteolytic enzymes: XII. Regarding the specificity of aminopeptidase and carboxypeptidase; A new type of enzyme in the intestinal tract. *J Biol Chem* 117:189.

Berridge, N. J. 1943. Pure crystalline rennin. *Nature* 151:473.

Blandau, R. J. 1949. Embryo-endometrial interrelationships in the rat and guinea pig. *Anat Rec* 104:331.

———. 1961. In *Sex and internal secretions,* ed. W. C. Young, 2:863. Baltimore: Williams and Wilkins.

Cunningham, L. 1967. Histochemical observations of enzymatic hydrolysis of gelatin films. *J Histochem Cytochem* 15:292.

Daoust, R. 1957. Localization of deoxyribose nuclease in tissue sections: A new approach to the histochemistry of enzymes. *Exp Cell Res* 12:203.

———. 1965. Histochemical localization of enzyme activities by substrate film methods: Ribonucleases, deoxyribonucleases, proteases, amylase and hyaluronidase. *Int Rev Cytol* 18:191.

Ehrmann, R. L., and Gey, G. O. 1956. The growth of cells on transparent gel of reconstituted rat tail collagen. *J Natl Cancer Inst* 16:1375.

Enders, A. C., and Schlafke, S. 1967. A morphological analysis of the early implantation stages in the rat. *Am J Anat* 120:185.

Folk, J. E. 1956. A new pancreatic carboxypeptidase. *J Am Chem Soc* 78:3541.

Fruton, J. S. 1960. In *The enzymes,* ed. P. D. Boyer, H. Lardy, and K. Myrback, 4:233. New York: Academic Press.

Gross, J.; Lapiere, C. M.; and Tanzer, M. L. 1962. In *Cytodifferentiation and macromolecular synthesis,* ed. M. Locke, p. 175. New York: Academic Press.

Hagerman, D. D. 1964. Enzymatic capabilities of the placenta. *Fed Proc Part I.* 23:785.

Hartley, B. S. 1960. Proteolytic enzymes. *Ann Rev Biochem* 29:45.

Hedin, S. G., and Rowland, S. 1901. Ueber ein proteolytisches Enzym in der Milz. *Z Physiol Chem* 32:341.

Houck, J. C., and Sharma, V. K. 1968. Induction of collagenolytic and proteolytic activities in rat and human fibroblasts by anti-inflammatory drugs. *Science* 161: 1361.

Huggett, A. St. G., and Hammond, J. 1952. In *Marshall's physiology of reproduction,* ed. A. S. Parkes, 2:312. London: Longmans.

Keith, C. K.; Kazenko, A.; and Laskowsky, M. 1947. Studies on proteolytic activity of crystalline protein B prepared from beef pancreas. *J Biol Chem* 170:227.

Kunitz, M., and Northrop, J. H. 1935. Crystalline chymotrypsin and chymotrypsinogen: I. Isolation, crystallization and general properties of a new proteolytic enzyme and its precursor. *J Gen Physiol* 18:433.

Mossman, H. W. 1937. Comparative morphogenesis of the fetal membranes and accessory uterine structures. *Contr Embryol Carnegie Inst Washington* 26:129.

Northrop, J. H. 1930. Crystalline pepsin: I. Isolation and tests of purity; II. General properties and experimental methods. *J Gen Physiol* 13:739.

Northrop, J. H., and Kunitz, M. 1932. Crystalline trypsin: I. Isolation and tests of purity. *J Gen Physiol* 16:267.

Owers, N. O. 1960. The endothelio-endothelial placenta of the Indian musk shrew, *Suncus murinus:* A new interpretation. *Amer J Ant* 106:1.

Press, E. M.; Porter, R. R.; and Cebra, J. 1960. The isolation and properties of a proteolytic enzyme, cathepsin D, from bovine spleen. *Biochem J* 74:501.

Sabatini, D. D.; Bensch, K.; and Barnett, R. J. 1963. Cytochemistry and electron microscopy: The preservation of cellular ultrastructure and enzymatic activity by aldehyde fixation. *J Cell Biol* 17:19.

Sansom, G. S., and Hill, J. P. 1931. Observations on the structure and mode of implantation of the blastocyst of Cavia. *Trans Zool Soc Lond* 21:295.

Waugh, D. F., and Baughman, J. D. 1960. In *The enzymes,* ed. P. D. Boyer, H. Lardy, and K. Myrback, 4:215. New York: Academic Press.

Willstaetter, R., and Bamann, E. 1928. Uber Magenlipase. *Z Physiol Chem* 173:17.

Wimsatt, W. A. 1962. Some aspects of the comparative anatomy of the mammalian placenta. *Amer J Obstet Gynec* 84:1568.

Wislocki, G. B., and Padykula, H. A. 1961. In *Sex and internal secretions,* ed. W. C. Young, 2:905. Baltimore: Williams and Wilkins.

12

Ingestive Properties of Guinea Pig Trophoblast Grown in Tissue Culture: A Possible Lysosomal Mechanism

Noel O. Owers

Department of Anatomy
Medical College of Virginia
Virginia Commonwealth University, Richmond

From observations that the trophoblast in histological preparations of some placental tissues contains cell debris and maternal red blood cells, it has been inferred that the trophoblast possesses phagocytic properties (Fawcett, Wislocki, and Waldo 1947; Alden 1948; Amoroso 1952; Wimsatt 1962). Little experimental evidence, if any, exists showing that the trophoblast actively phagocytoses particulate material.

This study was an attempt to devise an experimental model that would be helpful in observing live trophoblast activity *in vitro* by light microscopy. A nontoxic protein-dye complex of colloidal dimensions was used to make the phagocytic vacuoles visible; carbon particles of about ½ μ in diameter were used to mark the uptake of the larger particles by the trophoblast. The possibility that lysosomes were present in the cyto- and syntrophoblast was investigated. Freshly fixed trophoblasts obtained from embryos, as well as trophoblasts grown in tissue culture, were subjected to histochemical procedures in which acid phosphatase substrates were used. The trophoblasts were obtained from 8-day-old guinea pig embryos.

The results of this investigation indicate that under tissue culture conditions guinea pig trophoblast engages actively in uptake of a protein-dye complex as well as carbon particles. Both remain in granules in the cells for up to 10 days, at which time the experiments were usually terminated. The acid phosphatase reaction product in discrete cytoplasmic particles points to the presence of lysosomes. Obviously the *in vitro* method is practicable to demonstrate not only uptake of particulate material but also other aspects of trophoblastic phagocytosis.

I. Materials and Methods

A. *Calculation of Embryo Age*

Mature female guinea pigs were mated overnight. The formation of a vaginal plug next morning (counted as day 0) was taken as an indication of pregnancy. Embryos were removed on the 8th day and were placed in Hank's balanced salt solution. They were then either placed in tissue culture or processed by histochemical methods.

B. *Preparation of Dye-Gelatin Membrane*

The uptake of particulate material by guinea pig trophoblast was demonstrated in the following manner: 0.1 ml of a 3.6% gelatin solution was placed on a coverslip and allowed to dry. The gelatin membrane so formed was stained either in freshly tetrazotized benzidine at 4° C for 15 min (Pearse 1961) or in 0.2% aqueous solution of Fast blue B salt in veronal-acetate buffer at pH 9.2 for 15 min at room temperature. It was washed in 3 changes of veronal acetate buffer at pH 9.2 for 2 min each; immersed in a saturated solution of H-acid in veronal acetate buffer at pH 9.2 for 15 min; washed in 2 changes of distilled water for 5 min each; and air dried overnight. Excess stain was removed by immersion in 70% alcohol for 24 hr. The membrane was left to air dry. Gelatin films so treated appeared orange with the tetrazotized benzidine and brown with the Fast blue B salt.

C. *Tissue Culture Preparation*

The embryos were placed on the gelatin surface of the coverslip. A dialysis membrane was placed over the embryo; the coverslip was then added to the other components of the Rose chamber.

The same procedure was followed as stated above except that a few drops of india ink were added to the gelatin sol prior to its application on a coverslip. After staining, a carbon-gelatin-dye complex formed from which both the carbon and the gelatin-dye were slowly released in the tissue culture medium. The tissue culture medium consisted of Locke-Lewis salt solution, chick embryo extract, guinea pig serum, penicillin, and streptomycin. The accumulation of the carbon and the gelatin-dye complex in trophoblast was observed for 10 days in the Rose chamber.

D. *Histochemical Preparation*

For histochemical studies embryos were processed in two ways.

1. Freshly removed embryos were fixed for 24 hr in 10% formalin containing 1% calcium chloride and a small quantity of calcium carbonate to remove excess acidity. A few specimens were fixed in 2% glutaraldehyde in 0.1 M cacodylate buffer pH 7.2 for 1 or 2 hr. After fixation the specimens were placed in 0.88 M sucrose for 1 hr. They were stained for acid phosphatase by the method of Gomori (1952) as modified by Novikoff (1963). Beta-glycerophosphate was used as substrate. Some specimens were stained for acid phosphatase by Burstone's method

(1958) with napthol ASBI phosphate as substrate and Fast red violet LB salt as a simultaneous coupling reagent. Other specimens were stained for acid phosphatase by the method of Barka and Anderson (1962) with napthol ASTR phosphate and hexazonium pararosanilin as the coupler. Comparison of azo dye methods with the lead salt Gomori method facilitated identification of spurious artifacts. In control experiments either the substrate was omitted from the incubation medium or 0.01 M sodium fluoride was added as inhibitor.

2. Each freshly removed embryo was placed on a coverslip (nongelatinized) and covered by a dialysis membrane. The coverslip and the other parts of the Rose chamber were assembled; nutrient medium was added. Rose chamber preparations were incubated for either 3, 5, 8, or 16 days. At the end of each period the coverslip to which the embryo had attached was fixed and stained for acid phosphatase as given above. In control experiments substrate was omitted in the histochemical procedure or 0.01 M NaF was added to the incubation medium.

II. Observations

A. *Particulate Material Ingestion* in Vitro

Guinea pig embryos grown in Rose chambers with the gelatin-dye complex as substrate showed active uptake of the gelatin-dye complex; the embryos grown on carbon-gelatin-dye complex showed active ingestion of that substrate (figs. 1, 2, 3). The syntrophoblast emerging from the embryo between 24 and 48 hr is capable of significant uptake of particulate material (fig. 1). Excessive uptake of carbon or gelatin-dye complex early in the experiment usually caused inhibition; consequently no further formation of syncytium took place (fig. 1). Once the process of ingestion began it proceeded fairly rapidly; overloading of cytoplasm with the particulate material occurred within a few days; the formation of new cytotrophoblast and syncytium slowed down; and the culture remained constant in terms of cell numbers and syntrophoblastic outgrowth. In the initial phases of ingestion the carbon or gelatin-dye complex appeared in cytoplasmic granules which resembled the acid phosphatase-positive granules described next.

B. *Histochemical Reactions for Acid Phosphatase*

1. *Activity in Whole Embryos Freshly Fixed and Stained* (figs. 4–6).

The reaction product of both the lead salt and azo dye methods was found in the trophoblast of the Träger or ectoplacental cone (figs. 4, 5, 6). The visceral yolk sac showed little or no acid phosphatase activity with the modified Gomori method (fig. 5). With the azo dye methods a diffuse stain which was not resolvable into granules appeared (fig. 4).

2. *Activity in Embryos Placed in Tissue Culture for 3, 5, 8, and 16 Days* (figs. 7–15).

a) Three and 5-day embryo cultures (figs. 7–10). An intense reaction was observed in the Träger and in the yolk sac with both the Gomori and azo dye methods (figs. 7, 9); a weak reaction in the syncytium. Newly formed syntrophoblast

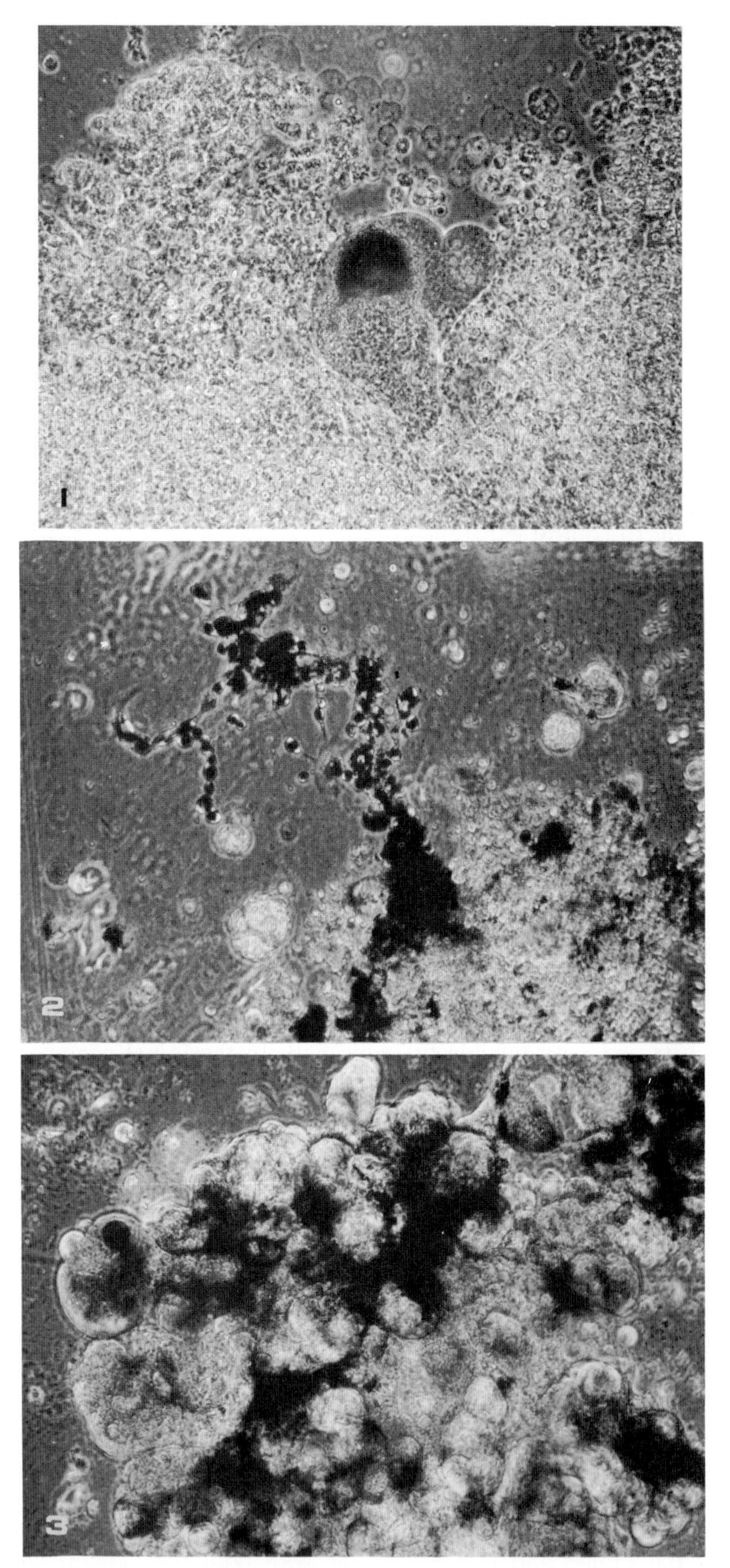

Fig. 1. An 8-day guinea pig embryo cultured for 3 days. The first sprout of syntrophoblast is filled with an opaque substance which could not be clearly identified as either carbon particles or the dye-gelatin. This particular specimen did not develop further and was considered to have died. The specimen was grown on a fixed gelatin membrane in a Rose chamber. ×177.

Fig. 2. An 8-day guinea pig embryo 7 days in culture. The syntrophoblast is filled with carbon particles. The carbon particles were released from the gelatin membrane by its gradual dissolution in the tissue culture medium. ×177.

Fig. 3. Same as fig. 2. ×274.

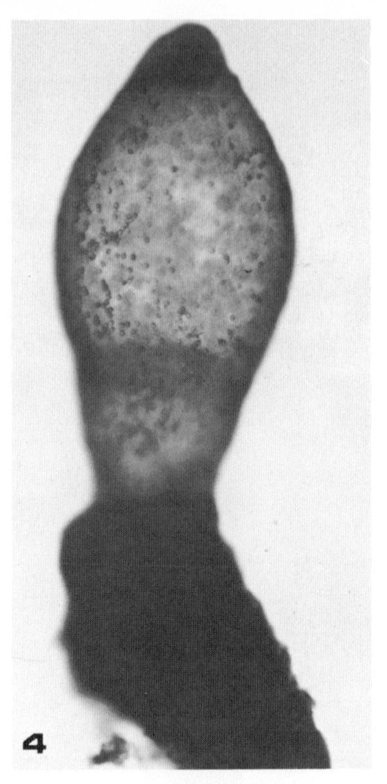

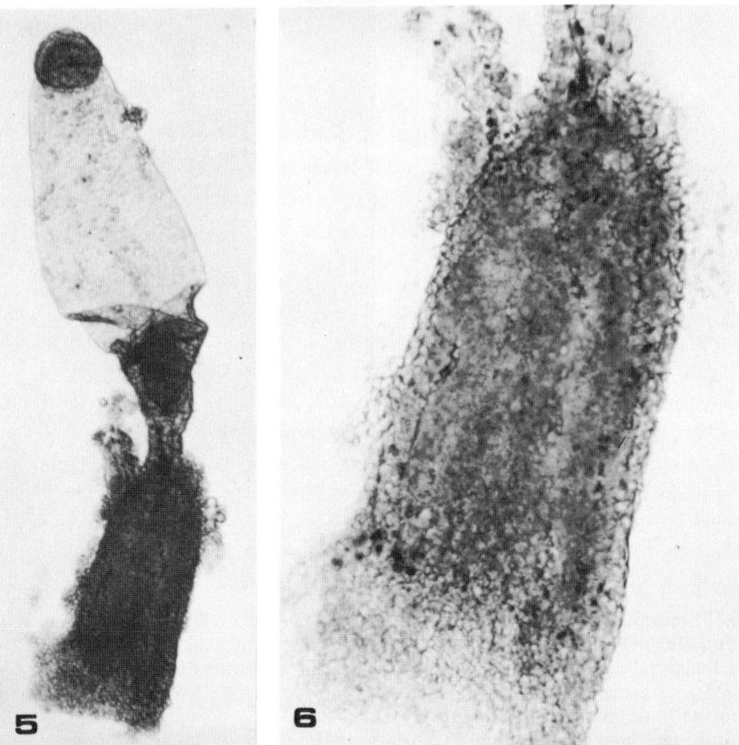

Figs. 4–6. Eight- and 9-day guinea pig embryos fixed and stained for acid phosphatase.

(4) Day 8 embryo fixed in glutaraldehyde and stained by Burstone's method for acid phosphatase. The reaction for acid phosphatase is most intense in the Träger (*lower half of figure*). The yolk sac also shows appreciable activity in the form of a diffuse stain. ×50.

(5) An embryo 9 days old fixed in formaldehyde and stained by the Gomori method showing a strong reaction product in the lower part or Träger. Little activity, if any, is present in the yolk sac. The round structure at the top end is the embryonic mass and shows up dark in the photograph because of its thickness rather than because of the presence of reaction product. ×50.

(6) An enlarged view of the Träger from fig. 1. The larger granules are probably artifacts, but the fine stippling represents the true location of acid phosphatase positive granules. ×207.

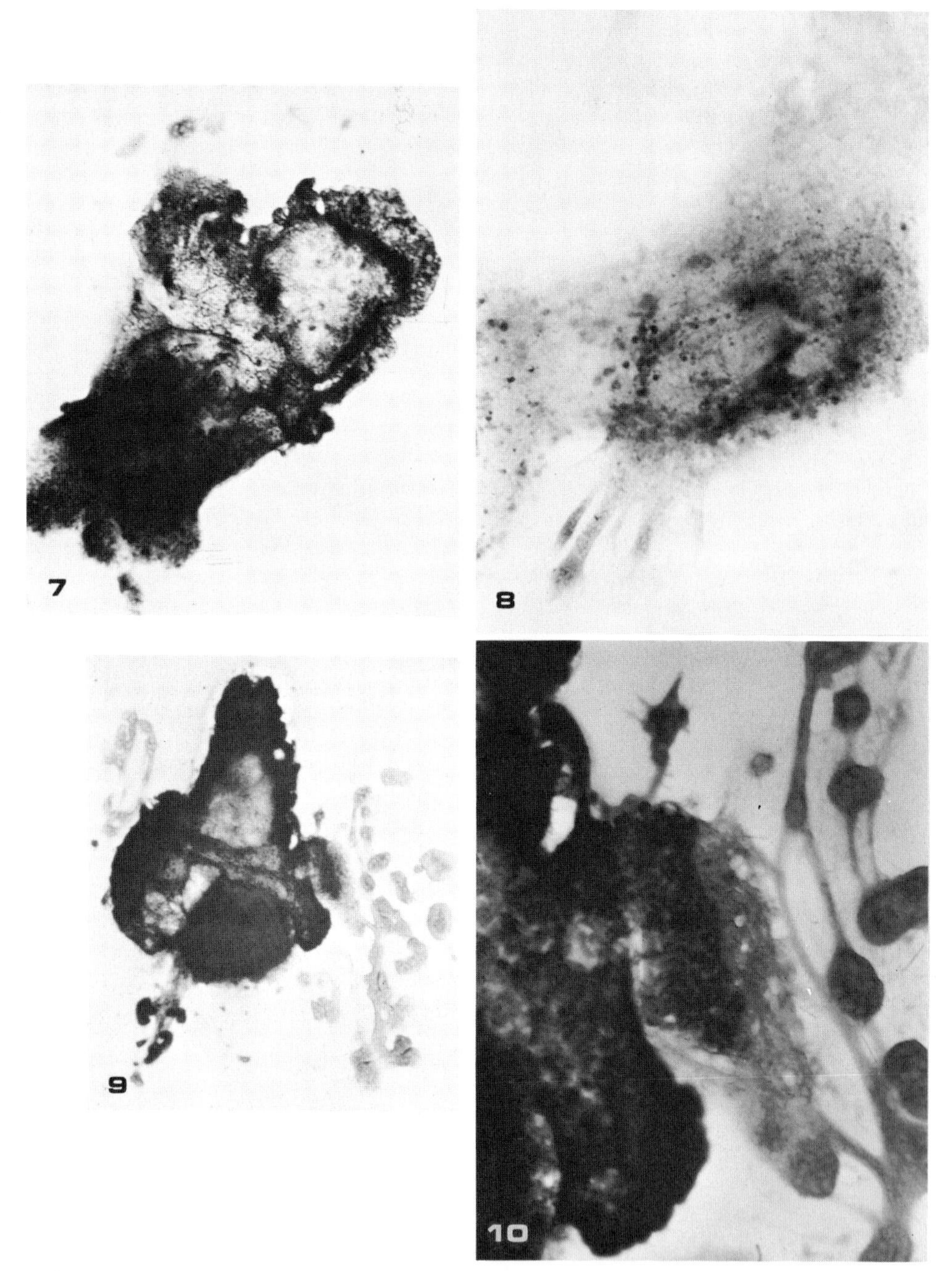

Figs. 7–10. Photomicrographs of 8-day guinea pigs placed in Rose chambers and allowed to incubate for 3 days. The tissue was then fixed and stained for acid phosphatase.

(7) The reaction product in the Träger represented by the large dark mass has increased over that observed in freshly fixed and stained 8-day embryos. An increase is also seen in the yolk sac cells. Formalin fixation. Gomori method. ×88.

(8) An enlarged view of syntrophoblast outgrowth which shows an abundance of acid phosphatase-positive granules. Glutaraldehyde fixation. Gomori method. ×292.

(9) The yolk sac (*upper half of specimen*) and Träger (*lower half of specimen*) both show strong positive reactions for acid phosphatase, but activity in the syntrophoblast outgrowths by comparison is very weak. ×88.

(10) An enlarged view from specimen in fig. 9 showing the junction between the original tissue placed in culture (*left half of photomicrograph*) and new outgrowths of syncytium (*right half*). Acid phosphatase activity is strong in the older tissue and weak in the new. ×234.

showed discrete particulate localization of acid phosphatase (fig. 8), possibly indicative of the presence of lysosomes. The granules ranged in size from 1 to 3 μ. No significant difference in distributional pattern or in intensity of reaction product between 3- and 5-day cultures was observed.

b) Eight-day embryo cultures (figs. 11, 12, 13). By the 8th day the syntrophoblast outgrowths had become widespread and profuse (fig. 11). The acid phosphatase reaction product was localized within discrete intracellular particles in both the modified Gomori and the azo dye methods (fig. 12). The number of such particles had increased somewhat over that in the 3- and 5-day cultures, owing to the augmented mass of cyto- and syntrophoblast and the intensified concentration of particles in a given unit of cytoplasm. By comparing figure 11 and figure 2 the similarity in distribution of acid phosphatase-containing particles and carbon particles in the syntrophoblast becomes evident.

c) Sixteen-day embryo cultures (figs. 14, 15). These were the oldest cultures maintained. They were examined to determine whether acid phosphatase reactivity increased with time. It did not; no increase in intensity was observed in the 16-day culture over the 8-day culture. Some of the cytotrophoblast cells and some parts of the syncytium were so filled with the reaction product that it was difficult to observe individual acid phosphatase-positive particles (fig. 15). A comparison of figure 15 and figure 3 shows the similarity in the distribution of acid phosphatase activity and carbon particles in syntrophoblast. In any given culture there were some areas of syntrophoblast that were only weakly reactive, and in these areas discrete particles containing reaction product were seen (figs. 13, 14).

III. Discussion

A. *The Lysosome Concept*

The name "lysosomes" or lytic bodies was proposed (de Duve et al., 1955) for a group of particles isolated from rat liver. These particles behave essentially like membrane-bounded sacs filled with soluble acid hydrolases which show "latency." Under normal conditions the membrane of the sac is assumed to prevent efflux of enzymes as well as the influx of external substrates. More than 20 enzymes capable of hydrolyzing proteins, mucopolysaccharides, and nucleic acids have been identified in lysosomal fractions (Allison 1968). Originally the identification of lysosomes as structural entities was problematical because of their polymorphism and contamination by other cell components. Recently the use of acid phosphatase tests for light microscopy (Gomori 1952; Burstone 1962) and for electron microscopy (Holt and Hicks 1961; Essner and Novikoff 1961) has facilitated the identification of lysosomes in tissues and cells. In the interpretation of such tests it is usually assumed that acid phosphatase is a reliable marker and that it is present only in lysosomes accompanied by other acid hydrolases. Although these assumptions are not entirely correct (de Duve 1963), acid phosphatase tests have verified the presence of lysosomes in a wide variety of cells, in single organisms and in many phyla of the animal kingdom (Novikoff 1961; de Duve and Wattiaux 1966). They have been associated frequently with phagocytosis (endocytosis) (Novikoff 1961, 1963). Through these and cell fractionation studies carried out on phagocytic cells (Cohn

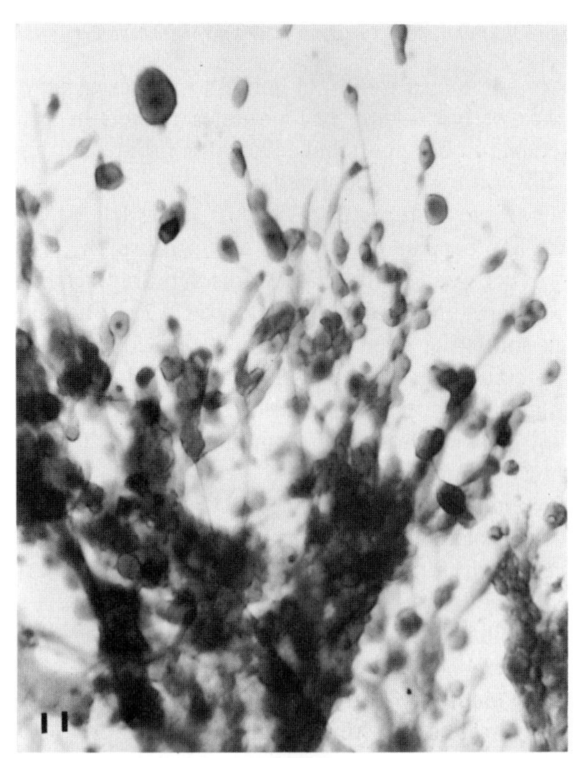
11

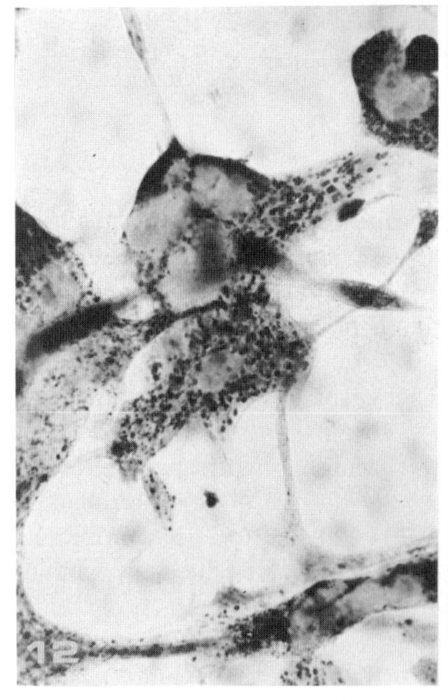
12

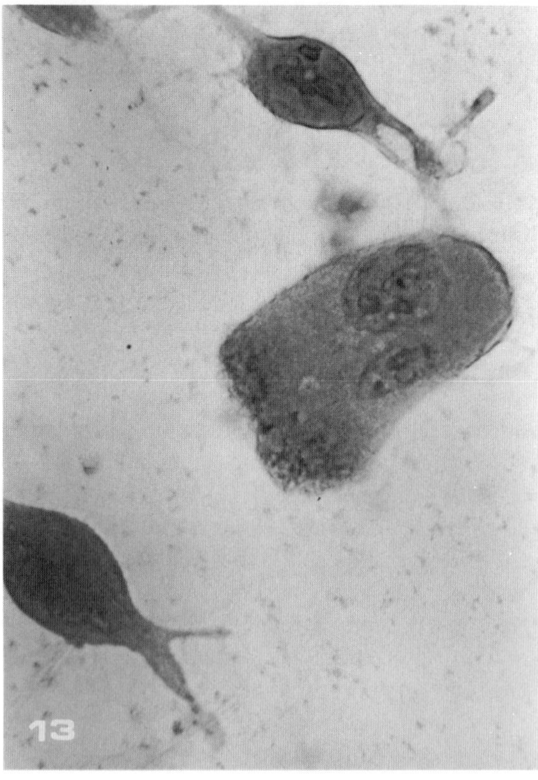
13

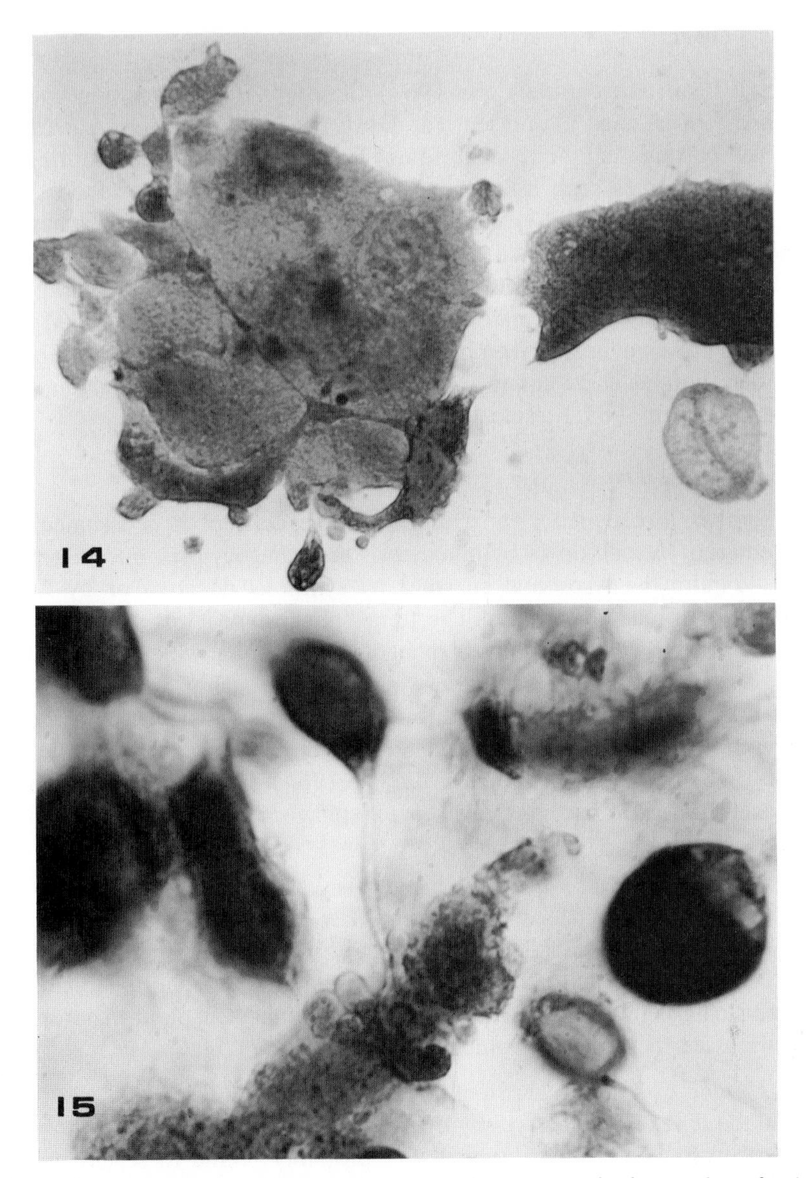

Figs. 14, 15. Photomicrographs of 8-day guinea pig embryos grown in tissue culture for 16 days, then fixed and stained for acid phosphatase by Burstone's method. The acid phosphatase activity varies in the different areas of syntrophoblast of the same culture from weak to strong.

(14) Area of weak activity. ×346.

(15) Area of strong activity. ×346. The cytoplasm is filled with granules so that individual particles cannot be distinguished.

Figs. 11–13. Photomicrographs of 8-day guinea pig embryos placed in Rose chambers and allowed to incubate for 8 days. The beta-glycerophosphatases are generally more active than the naphthol phosphatases at this time. The naphthol phosphatase activity is less in one culture and more in another.

(11) Syntrophoblast fixed in glutaraldehyde and stained by Burstone's method shows very little phosphatase activity whereas in another culture (fig. 13) the activity is greater but not as much as in fig. 12. ×173.

(12) Syntrophoblast fixed in glutaraldehyde and stained by the Gomori method shows an abundance of acid phosphatase-positive granules. Nuclei are seen as clear areas. ×346.

(13) Syntrophoblast from another culture, fixed and stained as in fig. 11 showing strong acid phosphatase reaction product in the form of discrete granules. ×346.

and Hirsch 1960; Hirsch and Cohn 1960) the original lysosome concept has been modified and broadened. At present (de Duve 1963) four distinct, but functionally related, cell particles can be recognized as lysosomes: (1) Enzyme-storing granules or primary lysosomes such as seen in polymorphonuclear neutrophilic leukocytes (Cohn and Hirsch 1960); (2) digestive vacuoles or secondary lysosomes in which exogenous materials are digested; (3) autolytic vacuoles in which endogenous parts of a cell are segregated and digested; and (4) residual bodies containing remnants of exogenous or endogenous materials. This systematization is supported by an impressive body of evidence derived from many cell types. A unified concept of intracellular digestion of endogenous and exogenous substances is emergent (Gordon, Miller, and Bensch 1965; Strauss 1967; Allison 1968).

B. *Fusion Phenomena*

Within the broadened view of the lysosome concept it is assumed that substances entering the cell do so by pinocytosis and phagocytosis, and that phagocytic vacuoles acquire the lysosomal enzymes by synthesis de novo or by fusion with preexisting or primary lysosomes. The term "phagosome" was suggested originally to indicate the segregating ability of phagocytic vacuoles which contain lysosomal enzymes (Strauss 1958, 1959). Later, when the relationship was better understood, the term "phagosome" was suggested to designate the phagocytic or pinocytic vesicles or vacuoles alone, before their fusion with lysosomes (de Duve 1963; Strauss 1964*a, b,* 1967). Novikoff (quoted in Essner 1960) suggested the term pinosomes for pinocytic vesicles and phagosomes for phagocytic vacuoles containing solid materials.

C. *Ingestion Study* (*Tissue Culture*)

1. *Pinocytosis.* Lewis (1931, 1937) first observed the ingestion of fluid droplets by cells in tissue culture. Rose (1957) made a cinephotographic record of the process using Hela cells. Holter (1959, 1961) and Chapman-Andresen (1957, 1963) discussed the process of pinocytosis.

2. *Uptake of Vital Dyes.* Stockinger (1964) and Allison and Young (1964) discussed the uptake of vital dyes by living cells. Acid dyes, such as trypan blue and lithium carmine, injected into animals are probably bound to serum proteins and taken up by pinocytosis as protein-dye complexes which accumulate in phagosomes and phagolysosomes (Kojima and Imai 1962; Schmidt 1962).

3. *Uptake of Azo Dye-Proteins.* A number of investigators have used proteins labeled with azo dyes in their studies on cellular uptake. Smetana (1947) labeled egg albumin, serum albumin, and serum globulin with diazotized 2-naphthol 3:6 sulphonic acid (R salt). The protein-R salt was injected intravenously in mice, rats, guinea pigs, rabbits, and dogs. It concentrated in the kidney tubules as well as in other tissues; the cells retained the dye-protein in unaltered form, in some cases for years. Sabin (1939) previously had used a similar dye bound to egg albumin and had observed the indigestibility of the dye-protein. Kruse and McMaster (1949) injected mice intravenously with a diazotized dye coupled to serum albumin or egg albumin and observed the uptake of dye-protein by cells of the reticuloendothelial

system. Latta, Gitlin, and Janeway (1951) injected rabbits intravenously with dye-azo-human serum albumins and observed the uptake of the dye-protein in many body tissues. When injected alone, the dye was not taken up. Intracellular localization of the dye-protein complex took the form of discrete granules.

In the present study diazotized compounds were coupled to a gelatin membrane for use in tissue culture. A survey of the literature indicates that the present work may be the first attempt to use a dye-protein complex prepared in the form of a membrane for tissue culture work.

Such a procedure has an advantage in showing uptake over the method of suspension of an azo dye-protein complex in the tissue culture medium, since a dialysis membrane is necessary for the initial support of the embryo within the chamber. Most azine dyes or dye-protein complexes do not penetrate the dialysis membrane; a suspension of these substances could not reach the embryo; consequently no uptake by the embryonic tissues could be observed. The method may suffer a disadvantage: growing tissues come into intimate contact with the membrane. What effect, if any, this has on tissue outgrowth is not known.

The dye-protein was slowly released in the tissue culture medium; it was taken up by both syn- and cytotrophoblast so rapidly that the cells were filled with the substance and complete cytoplasmic engorgement was observed occasionally. Subsequent growth of the tissues was notably reduced when compared with controls. Growth inhibition in the present experiment may be due to the inhibiting effects of the membrane or the excessive engorgement of the dye-protein by the trophoblast cells. An assessment of the inhibitive response must await further experimentation. Azo dye-protein complexes are not digested by naturally occurring enzymes (Sabin 1939; Cunningham 1967). The azo dye-gelatin used in this experiment was undigested by trophoblast; consequently the substance accumulated in discrete cytoplasmic droplets or vacuoles; carbon particles accumulated similarly. In some cells undigestible residues (e.g., lipofuscin) are stored in residual bodies (lysosomes) over long periods of time (Goldfischer, Villaverde, and Forschirm 1966) and continue to exhibit acid phosphatase activity. The azo dye-protein droplets in trophoblast were not tested for acid hydrolase activity; therefore no definitive statement can be made regarding their phagosomal or lysosomal nature. Further discussion of intracellular digestive processes is given by Essner (1960), Gordon, Miller, and Bensch (1965), Ehrenreich and Cohn (1968), and Coffey and de Duve (1968).

4. *Other Tissue Culture Studies*. Other substances used in uptake studies of cells in tissue culture are ferritin and ferritin-antibody complexes (Easton, Goldberg, and Green 1962), ^{131}I labeled albumin (Ryser 1963), bovine serum albumin, and rabbit antibovine serum albumin (Tahaba, Kinuwaki, and Hayashi 1964), fluorescent labeled proteins (Cormack and Ambrose 1962; Easty, Yarnell, and Andrews 1964), and horseradish peroxidase (Maeir 1961). A list of materials taken up by lysosomes is given by Allison (1968).

In this experiment no attempt was made to demonstrate the degradation of macromolecules by trophoblast; efforts were directed toward showing that trophoblast is capable of taking up macromolecules in the form of an azo dye-protein and colloidal carbon and storing them in intracytoplasmic granules. That trophoblast has the ability to digest macromolecules may have been demonstrated by Galassi

(1967). He showed that labeled maternal DNA is incorporated into trophoblastic DNA. Such a transfer implies that maternal DNA was broken down to smaller molecules, either within or without the trophoblast, and then reassembled in embryonic tissue nuclei.

D. *Cytochemical Staining for Lysosomes*

Caro and Palade (1964) established that secretory enzymes were formed in the endoplasmic reticulum of the pancreatic acinar cells and assembled in the Golgi region. The zymogen granules so formed were later discharged at the periphery of cells. Brandes (1965) showed that acid phosphatase-positive bodies were formed in the rough endoplasmic reticulum. Novikoff (1961) and Essner and Novikoff (1961), using electron microscopy, demonstrated the existence of acid phosphatase in Golgi cisternae and proposed the GERL (Golgi, Endoplasmic Reticulum, Lysosome) sequence of events for the production of acid phosphatase and other lysosomal hydrolases. The Gomori (1952) method (beta-glycerophosphate as the substrate for acid phosphatase) was used to demonstrate the existence of acid phosphatase in lysosomes at both the light- and electron-microscope level. Critical discussions of acid phosphatase histochemistry are given by Burstone (1962), Lison (1960), Barka and Anderson (1962), and Gahan (1967). Previously de Duve et al. (1955) had shown by biochemical means the presence of acid phosphatase and some other acid hydrolases in the lysosomal fractions of centrifuged preparations. The presence of acid phosphatase in the supernatant may reveal either a solubilized acid phosphatase from lysosomes or an acid phosphatase with properties different from the lysosomal enzyme. Neil and Horner (1964*a, b*) demonstrated by biochemical methods that formaldehyde inhibited acid phosphatase in supernatant (endoplasmic reticulum) more than in the particulate (lysosomal) fraction. This may explain why cytochemical staining for acid phosphatase after formaldehyde fixation reveals lysosomal acid phosphatase rather than cytoplasmic in certain tissues. Neil and Horner (1964*a, b*) pointed out, as had de Duve and his associates (1955), that beta-glycerophosphatase activity was localized mainly in the particulate (lysosomal) fraction. It should be mentioned that acid phosphatase may not always be a reliable marker for lysosomes in cytochemistry; a positive cytochemical reaction for acid phosphatase in a cell particle may point only to its possible lysosomal nature. In order to establish the lysosomal nature of a cell particle, it has to be proved that several acid hydrolytic enzymes which show "latency" are present in relatively high concentrations (de Duve 1963, de Duve and Wattiaux 1966). A quick estimation of the numbers and location of lysosomes in different cell types can be obtained in many cases by the use of formaldehyde fixed frozen sections with beta-glycerphosphate, as well as other naphthol phosphates, in the incubating medium for acid phosphatase.

1. *Other* in Vitro *Studies.* Cells grown in tissue culture containing acid phosphatase-positive granules are monocytes and macrophages (Weiss and Fawcett 1953; Ehrenreich and Cohn 1968; Hirsch, Fedorko, and Cohn 1968; Allison, Harrington, and Birbeck 1966), heterophil leukocytes (Baggiolini, Hirsch, and de Duve

1969), fibroblasts (Ogawa, Mizuno, and Okamoto 1961), Hela cells (Green and Verney 1956; Jervis and Labrec 1966), and lymphocytes (Brittinger et al. 1969). The acid phosphatase-positive granules in the photomicrographs in this chapter are similar in size and staining characteristics to published photomicrographs. Many of the acid phosphatase-positive granules in macrophages, lymphocytes, and leukocytes are lysosomes. It seems reasonable to assume, on the basis of this investigation, that the acid phosphatase-positive granules are lysosomes in guinea pig trophoblast.

2. *Guinea Pig Trophoblast in Tissue Culture.* In more extensive investigations acid phosphatase-positive granules were identified as lysosomes in certain cells by using formalin fixation and acid phosphatase as the marker enzyme (Fedorko, Hirsch, and Cohn 1968). By the same method lysosomes were identified in the trophoblast freshly removed from ectoplacental cone of the guinea pig and in the cyto- and syntrophoblast grown in tissue culture. The uptake of dye-gelatin complexes and carbon particles increased in tissue culture with time; in a few days the cytoplasm of the syn- and cytotrophoblast was completely filled with these materials. In parallel experiments, in which a nonstained gelatin membrane was used, a similar increase in acid phosphatase-positive granules was observed. It seems reasonable to deduce that one is observing phagolysosomes or secondary lysosomes or residual bodies. It is not known whether the increase in number and size of the acid phosphatase-positive bodies occurs only in trophoblastic tissue culture or whether a parallel increase occurs *in vivo*. Lieberman and Ove (1958) observed that induction of lysosomal enzymes takes place in a variety of cultured tissues leading to similar distributional patterns of acid phosphatase and other hydrolases. An increase in acid phosphatase in the trophoblast may therefore represent a response to the conditions in tissue culture, which may not parallel the *in vivo* situation. Work remains to be done to determine whether the trophoblast *in vivo* shows increased enzyme activity with time, and whether this activity is directly or indirectly related to the uptake of particulate material.

IV. Summary and Conclusions

1. When grown in tissue culture with an azo dye-gelatin complex, guinea pig trophoblast shows remarkable phagocytic activity, and colored intracytoplasmic deposits are extensive.
2. The azo dye-gelatin complex is not digested by natural enzymes, including those of the trophoblast, but remains bound intracellularly at least up to 10 days, the duration of the experiments.
3. Freshly removed, fixed and stained guinea pig trophoblast, 8 days old, possesses acid phosphatase-positive granules which are probably lysosomes.
4. Guinea pig trophoblast grown in tissue culture possesses acid phosphatase-positive granules.
5. In parallel experiments the distribution and size of the colored vacuoles in the dye-gelatin uptake study is similar to the acid phosphatase-positive particles obtained in trophoblast cultures on gelatin alone.
6. The acid phosphatase-positive granules in the trophoblast are similar in size,

shape, and density to published photomicrographs of comparable cytoplasmic granules which are lysosomes.

7. It seems likely, therefore, that the guinea pig trophoblast has a lysosomal component similar to that found in various other cells.

Acknowledgments

This work was supported in part by United States Public Health Service grant HD 03464 from the National Institutes of Health.

References

Alden, R. H. 1948. Implantation of the rat egg: III. Origin and development of primary trophoblast giant cells. *Amer J Anat* 83:143.

Allison, A. C. 1968. Lysosomes. In *The biological basis of medicine,* ed. E. E. Bittar and N. Bittar, 1:209. New York: Academic Press.

Allison, A. C.; Harrington, J. S.; and Birbeck, M. 1966. An examination of the cytotoxic effects of silica on macrophages. *J Exp Med* 124:141.

Allison, A. C., and Young, M. R. 1964. Uptake of dyes and drugs by living cells in culture. *Life Sci* 3:1407.

Amoroso, E. C. 1952. Placentation. In *Marshall's physiology of reproduction,* ed. A. S. Parkes. London: Longmans.

Baggiolini, M.; Hirsch, J. G.; and de Duve, C. 1969. Resolution of granules from rabbit heterophil leukocytes into distinct populations by zonal sedimentation. *J Cell Biol* 40:529.

Barka, T., and Anderson, P. J. 1962. Histochemical methods for acid phosphatase, using hexazonium pararosanilin as coupler. *J Histochem Cytochem* 10:741.

Brandes, D. 1965. Observations on the apparent mode of formation of "pure" lysosomes. *J Ultrastruct Res* 12:63.

Brittinger, G.; Hirschhorn, R.; Hirschhorn, K.; and Weissman, G. 1969. Effect of pokeweed mitogen (PWM) on lymphocyte lysosomes. *J Cell Biol* 40:843.

Burstone, M. S. 1958. Histochemical demonstration of acid phosphatases with naphthol AS phosphates. *J Nat Cancer Inst* 21:523.

———. 1962. *Enzyme histochemistry and its application in the study of neoplasms.* New York: Academic Press.

Caro, L. G., and Palade, G. E. 1964. Protein synthesis, storage and discharge in the pancreatic exocrine cell. *J Cell Biol* 20:473.

Chapman-Andresen, C. 1957. Some observations on pinocytosis in leukocytes. *Exp Cell Res* 12:397.

———. 1963. Pinocytosis in amebas. *CR Lab Carlsberg* 33:73.

Coffey, J. W., and de Duve, C. 1968. Digestive activity of lysosomes: I. The digestion of proteins by extracts of rat liver lysosomes. *J Biol Chem* 243:3255.

Cohn, Z. A., and Hirsch, J. G. 1960. The isolation and properties of the specific cytoplasmic granules of rabbit polymorphonuclear leukocytes. *J Exp Med* 112:983.

Cormack, D. H., and Ambrose, E. J. 1962. The cellular uptake of proteins labelled with fluorescent dyes. *J Roy Micr Soc* 81:11.

Cunningham, L. 1967. Histochemical observations of enzymatic hydrolysis of gelatin films. *J Histochem Cytochem* 15:292.

De Duve, C. 1963. The lysosome concept. In *Ciba Foundation symposium on lysosomes,* ed. A. V. S. de Reuck and M. P. Cameron, p. 1. Boston: Little, Brown & Co.

De Duve, C.; Pressman, B. C.; Gianetto, R.; Wattiaux, R.; and Appelmans, F. 1955. Tissue fractionation studies: 6. Intracellular distribution pattern of enzymes in rat-liver tissue. *Biochem J* 60:604.

De Duve, C., and Wattiaux, R. 1966. Function of lysosomes. *Ann Rev Physiol* 28: 435.

Easton, J. M.; Goldberg, B.; and Green, H. 1962. Demonstration of surface antigens and pinocytosis in mammalian cells with ferritin-antibody conjugates. *J Cell Biol* 12:437.

Easty, C. C.; Yarnell, M. M.; and Andrews, R. D. 1964. The uptake of protein by normal and tumour cells in vitro. *Brit J Cancer* 18:354.

Ehrenreich, B. A., and Cohn, Z. A. 1968. Fate of hemoglobin pinocytosed by macrophages in vitro. *J Cell Biol* 38:244.

Essner, E. 1960. An electronmicroscopic study of erythrophagocytosis. *J Biophys Biochem Cytol* 7:329.

Essner, E., and Novikoff, A. B. 1961. Localization of acid phosphatase activity in hepatic lysosomes by means of electronmicroscopy. *J Biophys Biochem Cytol* 9:773.

Fawcett, D. W.; Wislocki, G. B.; and Waldo, C. M. 1947. The development of mouse ova in the anterior chamber of the eye and in the abdominal cavity. *Amer J Anat* 81:413.

Fedorko, M. E.; Hirsch, J. G.; and Cohn, Z. A. 1968. Autophagic vacuoles produced in vivo: I. Studies on macrophages exposed to chloroquine. *J Cell Biol* 38:377.

Gahan, P. B. 1967. Histochemistry of lysosomes. *Int Rev Cytol* 21:1.

Galassi, L. 1967. Reutilization of maternal nuclear material by embryonic and trophoblastic cells in the rat for synthesis of deoxyribonucleic acid. *J Histochem Cytochem* 15:573.

Goldfischer, S.; Villaverde, H.; and Forschirm, R. 1966. The demonstration of acid hydrolase, thermostable reduced diphosphopyridine nucleotide tetrazolium reductase and peroxidase activities in human lipofuscin pigment granules. *J Histochem Cytochem* 14:641.

Gomori, G. 1952. *Microscopic histochemistry.* Chicago: University of Chicago Press.

Gordon, G.; Miller, L.; and Bensch, K. 1965. Studies on the intracellular digestive process in mammalian tissue culture cells. *J Cell Biol* 25:41.

Green, M. H., and Verney, E. L. 1956. A simple method for the study of intracellular acid phosphatase in Hela cells. *J Histochem Cytochem* 4:106.

Hirsch, J. G., and Cohn, Z. A. 1960. Degranulation of polymorphonuclear leukocytes following phagocytosis of microorganisms. *J Exp Med* 112:1005.

Hirsch, J. G.; Fedorko, M. E.; and Cohn, Z. A. 1968. Vesicle fusion and formation at the surface of pinocytic vacuoles in macrophages. *J Cell Biol* 38:629.

Holt, S. J., and Hicks, R. M. 1961. Studies on formalin fixation for electronmicroscopy and cytochemical staining purposes. *J Biophys Biochem Cytol* 11:31.

Holter, H. 1959. Pinocytosis. *Int Rev Cytol* 8:481.

———. 1961. *Pinocytosis Proc V Int Cong Biochem Moscow,* 2:248. New York: Pergamon Press.

Jervis, H. R., and Labrec, E. H. 1966. Correlation of acid phosphatase activity with degree of bacterial infection in Hela cells. *J Histochem Cytochem* 14:196.

Kojima, M., and Imai, Y. 1962. Mechanism of phagocytosis of reticulo-endothelial cells. *Tohoku J Exp Med* 76:161.

Kruse, H., and McMaster, P. D. 1949. The distribution and storage of blue antigenic azoproteins in the tissue of mice. *J Exp Med* 90:425.

Latta, H.; Gitlin, D.; and Janeway, C. A. 1951. Experimental hypersensitivity in the rabbit: The cellular localization of soluble azoproteins (dye-azo-human serum albumins) injected intravenously. *AMA Archs of Path* 51:260.

Lewis, W. H. 1931. Pinocytosis. *Johns Hopkins Hospital Bull* 49:17.

———. 1937. Pinocytosis by malignant cells. *Amer J Cancer* 29:666.

Lieberman, I., and Ove, P. 1958. Enzyme activity levels in mammalian cell cultures. *J Biol Chem* 233:634.

Lison, L. 1960. *Histochemie et cytochimie animale.* 3d edition. Paris: Gauthier-Villars.

Maeir, D. M. 1961. A technique for the study of protein uptake by cells in tissue culture. *Exp Cell Res* 23:200.

Neil, M. W., and Horner, M. W. 1964*a*. Studies on acid hydrolases in adult and foetal tissues. *Biochem J* 92:217.

———. 1964*b*. Studies on acid hydrolases in adult and foetal tissues. *Biochem J* 93:220.

Novikoff, A. B. 1961. Lysosomes and related particles. In *The cell,* ed. J. Brachet and A. E. Mirsky, 2:423. New York: Academic Press.

———. 1963. Lysosomes in the physiology and pathology of cells: Contributions of staining methods. In *Ciba Foundation symposium on lysosomes,* ed. A. V. S. de Reuck and M. P. Cameron, p. 36. Boston: Little, Brown & Co.

Ogawa, K.; Mizuno, N.; and Okamoto, M. 1961. Lysosomes in cultured cells. *J Histochem Cytochem* 9:202.

Pearse, A. G. E. 1961. *Histochemistry: Theoretical and applied.* 2d ed., p. 795. Boston: Little, Brown & Co.

Rose, G. G. 1957. Microkinetospheres and V.P. satellites of pinocytic cells observed in tissue cultures of Gey's strain Hela with phase contrast cinematographic techniques. *J Biophys Biochem Cytol* 3:697.

Ryser, H. 1963. The measurement of I^{131} serum albumin uptake by tumor cells in tissue culture. *Lab Invest* 12:1009.

Sabin, F. R. 1939. Cellular reactions to a dye-protein with a concept of the mechanism of antibody formation. *J Exp Med* 70:67.

Schmidt, W. 1962. Licht-und elektronenmikroskopische untersuchungen über die intrazellulare verabeitung von vitalfarbstoffen. *Z Zellforsch* 58:573.

Smetana, H. 1947. The permeability of the renal glomeruli of several mammalian species to labelled proteins. *Amer J Path* 23:255.

Stockinger, L. 1964. Vitalfarbung und vitalfluorochromierung tierischer Zellen. *Protoplasmatologia* 2, D1, 1.

Strauss, W. 1958. Colorimetric analysis with N, N-Dimethyl-p-phenylenediamine of the uptake of intravenously injected horseradish peroxidase by various tissues of the rat. *J Biophys Biochem Cytol* 4:541.

———. 1959. Rapid cytochemical identification in various tissues of the rat and their differentiation from mitochondria by the peroxidase method. *J Biophys Biochem Cytol* 5:193.

———. 1964*a*. Cytochemical observations on the relationship between lysosomes and phagosomes in kidney and liver by combined staining for acid phosphatase and intravenously injected horseradish peroxidase. *J Cell Biol* 20:497.

———. 1964*b*. Occurrence of phagosomes and phago-lysosomes in different segments of the nephron in relation to the reabsorption, transport, digestion and extrusion of intravenously injected horseradish peroxidase. *J Cell Biol* 21:295.

———. 1967. Lysosomes, phagosomes and related particles. In *Enzyme cytology,* ed. D. B. Roodyn, p. 239. New York: Academic Press.

Tahaba, Y.; Kinuwaki, Y.; and Hayashi, H. 1964. Cellular damage by soluble antigen-antibody complexes in tissue culture. *Proc Soc Exp Biol Med* 115:906.

Weiss, L. P., and Fawcett, D. W. 1953. Cytochemical observations on chicken monocytes, macrophages, and giant cells in tissue culture. *J Histochem Cytochem* 1:47.

Wimsatt, W. A. 1962. Some aspects of the comparative anatomy of the mammalian placenta. *Amer J Obstet Gynec* 84:1568.

13

The Rabbit Blastocyst and Its Environment: Physiological and Biochemical Aspects

C. Lutwak-Mann

Unit of Reproductive Physiology and Biochemistry
(Agricultural Research Council)
University of Cambridge, England

Mammalian embryology is nowadays a very active area of research endeavor, notably in the sectors concerned with physiology and biochemistry of the early embryo and various aspects of maternal-embryonic association. Progress in this field owes much to the following energetically pursued lines of investigative approach: establishment of requirements for sustained development of embryos *in vitro;* analysis of the chemical composition of young embryos; elucidation of maternal-embryonic relationships at consecutive stages of embryonic growth; comparison of the action of various agents on embryos *in vivo* and *in vitro;* characterization of secretory and other products in the endometrium in relation to ovarian hormonal control; and morphological investigations at light- and electron-microscopic levels of conditions arising at the stage of uterine attachment of blastocysts.

The scientific reasons for selecting the rabbit blastocyst as an object of study are manifold and will be duly set out below. But first let us record the fact that, whatever its advantage or disadvantage to the experimenter, the rabbit blastocyst has considerable aesthetic appeal; whether encountered *in situ* or *in vitro,* it never fails to arouse universal admiration. There is no doubt (though statistical proof is lacking) that the research worker's morale and experimental zeal are thereby significantly enhanced.

In the subsequent discussion of the rabbit blastocyst, embryonic age will be defined by referring to the day of mating as day 0. The term "blastocyst stage" will comprise embryos which lie free in the uterine horns, aged 5 and 6 days, as well as embryos which have undergone antimesometrial uterine attachment, aged 7, 8, and 9 days. Once the definitive placenta has taken over the function of the yolk sac, the conceptus will be referred to as *fetus.* Reference will be made hereafter to uterine *attachment* to indicate the termination of the free-lying blastocyst period (at about 168 hr *post coitum*), whereas *implantation* will signify the stage concomitant with the establishment of the *fetoplacental unit* (about day 10 of gestation).

I. Landmarks in Rabbit Blastocyst Development

In the rabbit embryo the blastocoelic cavity forms on day 4. Early on day 5 the blastocysts lie, usually in a cluster, at the ovarian uterine end; by day 5½ their average diameter is 1.0 mm, and fresh weight 1.0 mg. In the course of day 6 a series of consecutive changes culminates in the formation of primitive streak, outgrowth of endoderm, and formation of trophoblastic knobs. Shortly before their antimesometrial attachment the blastocysts are more or less evenly spaced along the uterine horns, and their fresh weight reaches 100 mg. This developmental phase is accompanied by a steep rise in cell number (5,000–10,000 on day 5; 60,000–70,000 on day 6). Insofar as dimensions are concerned the increase is greatest in the surface area of the entire blastocyst and is less pronounced in the area of the embryonic disk (Lutwak-Mann and Hay 1965*a*). There is much variability in these parameters in blastocysts recovered from normal rabbits of approximately the same weight.

At the postattachment stage reliable data on blastocyst size and weight are harder to come by. On the whole it is easiest with reference to 7-, 8-, and 9-day-old blastocysts to give (average) values for (*a*) the diameter, which is 0.4, 1.2, and 1.9 mm, and (*b*) the amount of blastocyst fluid that can be aspirated from the blastocoelic cavity, up to 0.3, 0.9, and 1.3 ml, respectively. Concerning the quantitative aspects of fluid in attached blastocysts, it is well to remember that it can fluctuate: for example, maternally administered hypertonic sugar solutions cause a marked rise, whereas maternal treatment with estrogen substantially reduces the volume of fluid in individual blastocysts (Lutwak-Mann 1962*a*).

II. The Rabbit Blastocyst as an Experimental Object: Advantages and Disadvantages

The blastocyst stage in the rabbit lends itself exceptionally well to physiological and biochemical studies for the various reasons commented upon below.

1. The age of the embryo can be defined with some degree of precision because, under normal conditions, ovulation takes place at 10 ± 2 hr from the time of mating. In view of the speed of development at the blastocyst stage and accompanying changes in differentiation and metabolism exact definition of embryonic (and endometrial) chronology is mandatory, and in experimental protocols ought to be narrowed down to hours, not merely days.

2. Numerical availability of rabbit blastocysts in fertile animals permits the subdivision of embryos within each litter into directly comparable experimental groups, particularly useful in experiments involving *in vitro* treatment of free-lying blastocysts. The number of blastocysts can be increased further by superovulation; but it is still a moot point whether blastocysts from superovulated rabbits are perfectly normal in every respect, particularly insofar as their developmental potential is concerned. That the endometrium can be overstimulated by superovulation is

shown, for example, by the fact that values for carbonic anhydrase activity tend to differ from those established under physiological conditions of mating (Lutwak-Mann and McIntosh 1969). Such an abnormal response may apply equally to other endometrial products, potentially altering the blastocysts' physiological properties.

3. The large size of rabbit blastocysts, even at the preattachment stage, is a self-evident boon to the experimenter, and offers exceptional opportunities to the morphologist, physiologist, and biochemist.

We have utilized the large size of the rabbit blastocyst at 5–7 days to work out the flat-mount procedure which we have used successfully in numerous studies dealing with the effect of foreign agents transmitted by the maternal route or acting directly. A full account of the method and its applicability to diverse research problems has been published elsewhere (Lutwak-Mann and Hay 1967); in outline the method consists of the following steps. Free-lying blastocysts are lifted from the exposed surface of the endometrium of nembutalized rabbits and fixed for several hours in methanol. The flat-mount is prepared so as to place the embryonic disk (embryo proper) centrally, in correct relationship to the trophoblast, and the preparation is stained; intact cells of the different regions can thus be viewed independently. Formation and development of trophoblastic knobs, the outgrowth of extra-embryonic endoderm, sex chromatin, and abnormal mitotic figures can also be observed. Mitotic counts can be made, especially in the abembryonic region, where the trophoblast consists of a single cell layer unlined with endoderm.

The free-lying rabbit blastocyst, in particular between day 6 and day 7, is sufficiently large to make microdissection of morphologically different regions an attractive proposition. Technically the microdissection of fresh blastocysts is awkward because it causes collapse of the blastocoelic cavity. However for purely chemical analysis dissection may be aided by fixation of blastocysts, either in methanol or in acetone, as we have done in determinations of RNA and DNA respectively.

4. The inclusion for experimental purposes of the postattachment stage, days 7–9, has much to commend it. Because the blastocysts before and after attachment differ not only in the degree of differentiation but also in chemical composition and permeability to exogenous factors, a great deal can be learned by comparative assessment of these developmental stages. Experiments of several days' duration can be undertaken. Experimental interference with uterine attachment, such as inhibition or temporary delay, permits comparisons to be made between blastocysts that (very punctually) attach at the normal time and those that have been prevented from doing so, for instance, by ovariectomy (Lutwak-Mann, Hay, and Adams 1962) or estrogen treatment (Lutwak-Mann and Hay 1964).

What of the disadvantages inherent in the use of the rabbit as provider of embryos? Most are obvious, but a brief survey may be helpful to potential users. Rabbits, if kept in a colony, should be housed singly to prevent an undue incidence of pseudopregnancy; this is costly in terms of space and labor. Before they can be successfully mated, rabbits should be 5–6 months old. Since pregnancy lasts 30–32 days, and first pregnancies are notoriously unreliable, the animals are nearly 8 months old before coming into experimental use. The rabbits' weight is a drawback in experiments involving administration of scarce and expensive materials on weight basis. As there is no reliable method by which to detect pregnancy in the rabbit before day 8, experimental wastage results, especially when the incidence of pregnancy in a colony is low (late spring and summer in England).

In a discussion of experimental disadvantages one other very real problem might as well be taken up here; though by no means confined to the rabbit it is probably more obtrusively evident in this species because of the large size and advanced degree of differentiation of the embryos at the blastocyst stage. It is the variability encountered within and between litters in respect to blastocyst dimensions, their developmental advance, and response to treatment. It will no doubt be argued that statistical treatment is the only way to resolve this difficulty. However randomization makes sense only if applied to meaningful parameters. To evaluate one's data statistically, say on the uptake of labeled compounds or concentration of various constituents, is reasonable; but in experiments involving exposure of blastocysts to agents that may either inhibit or enhance their growth, quantifiable parameters are sadly lacking. Criteria such as "onset of cavitation" or "increase in overall dimensions" tell us little about the blastocysts' condition, since it is well known that a purely trophoblastic vesicle devoid of even a vestigial embryonic disk can maintain blastocoelic inflation and continue to expand. Therefore neither of these two phenomena is a true index of blastocyst development; analyzing them statistically does not take us very far.

What we really need are rigorous metabolic parameters applicable to consecutive stages of embryonic development. For the present, dividing blastocyst litters into directly comparable groups (control and treated), coding them, and subjecting each individual embryo to detailed and unbiased microscopic scrutiny, as in our flat-mounts, provides a purposeful and at the same time informative experimental procedure.

III. Physiological and Biochemical Properties of the Rabbit Blastocyst and Its Environment

Specific aspects of the morphology and metabolism of the rabbit blastocyst and its environment will be touched upon elsewhere in this volume. Here comment will be made upon results obtained with rabbit blastocysts in experiments with which the author has had first-hand experience. An attempt will be made to draw up a list of characteristic blastocyst features which, though primarily noted in rabbit embryos, in all probability apply to the blastocyst stage of other mammalian species.

The interest of the author and her co-workers has centered over the years upon the following problems: (1) the chemical composition of the rabbit blastocyst, (2) the passage into and clearance from the blastocysts of extraneous substances introduced by the maternal route, (3) the action of various classes of agents on rabbit blastocysts under conditions *in vivo* and *in vitro,* and (4) the behavior of the endometrial carbonic anhydrase and uterine zinc in relation to the blastocyst stage of pregnancy.

A. *Some Chemical Constituents of the Rabbit Blastocyst*

Information on the chemical composition of the rabbit blastocyst is woefully incomplete, and it will be some time before the gaps are filled. Most of the analyses described here were done on entire free-lying blastocysts at 6–7 days of age; only a few relate to fluid withdrawn from the older, attached blastocysts. The embryonic values were compared as far as possible with analytical data pertaining to maternal whole blood, blood plasma, peritoneal fluid, and uterine secretion.

Unless otherwise stated the values for blastocyst constituents are expressed as average content per single embryo; the average fresh weights of blastocysts will be found in table 1. When chemical determinations were done on free-lying blastocysts, these were removed directly from the endometrial surface and used without rinsing; for enzyme assays, on the other hand, blastocysts were carefully rinsed with saline. The analyses at the postattachment stage were carried out on fluid aspirated from the blastocyst cavity. In the rabbit there are two quite different kinds of uterine secretion, depending upon the hormonal conditions: (*a*) the thin, clear, relatively abundant fluid which accumulates in the pre- and postovulatory stages (Lutwak-Mann 1962*b*), and (*b*) the thick, sticky, and scanty material produced by the progestational endometrium (and deposited on the zona pellucida of blastocysts). In the text below the former is referred to as uterine estrous fluid and the latter as progestational secretion. These "native" endometrial products must not be mistaken for or equated with artificial uterine wash-fluids or fluids accumulated in ligated uterine horns.

TABLE 1

PROTEIN CONTENT (TOTAL N×6.25) IN RABBIT BLASTOCYSTS

Age of Blastocysts (Days *p.c.*)	Fresh Weight of Blastocysts (mg)	Protein/Single Blastocyst (mg)
6 days 4 hr	25	0.20[a]
7 days	100	0.50[a]
7 days 5–11 hr	300	7.56
8 days 3–6 hr	500	40.20
9 days 2 hr	1200	66.15

NOTE: The value for amino N has been deducted from total N in 6- and 7-day blastocysts only. Values are averages based on 6–10 determinations.

[a] Blastocysts were analyzed with intact zona pellucida. All other analyses were done on fluid withdrawn from attached blastocysts.

1. *Proteins and Amino Acids.* Values available for total nitrogen in the free-lying and attached blastocysts make possible an evaluation of protein content as shown in table 1. The evaluation is only approximate, as in the calculations, except for deduction of amino nitrogen in the free-lying blastocysts, no allowance has been made for nitrogen constituents other than protein. So far data are available only for amino nitrogen and nucleic acid content in unattached blastocysts; we have little information in this respect for the later blastocyst stage.

In rabbit blastocysts the protein content rises abruptly from the time of uterine attachment; by day 9 it is roughly 4/5 of the protein content of rabbit blood plasma (7.2 %–7.8%). The values listed agree by and large with those reported by others (Lesiński, Jajszczak, and Choroszewska 1967). The most striking variability in protein content was encountered in the course of day 7: at this stage approximately 50% of blastocyst samples exhibited clotting of fluid, indicating thereby the presence of maternal fibrinogen which is invariably present in blastocyst fluid on days 8 and 9 (Brambell 1948). In connection with the protein content of blastocysts it may be of interest to mention observations made with blastocysts that had been exposed to toxic agents *in vivo*. When fixing such blastocysts with methanol to prepare flat-mounts, we noticed that the protein precipitate in the cavity was frequently much heavier than in blastocysts recovered from untreated animals, an indication presumably that the permeability of the embryos had been altered. It might be worthwhile to follow up this phenomenon quantitatively, perhaps in blastocysts kept in protein-containing media *in vitro,* in the presence of various agents.

Although up to day 7 the protein content of rabbit blastocysts is low, it appears strikingly high when compared with the value 20 ng recorded per single mouse blastocyst (Brinster 1967). However this disparity is not surprising in view of the much larger number of cells in the rabbit embryo enveloped in a thick, secretion-coated zona pellucida as compared to that of the tiny mouse blastocyst which (for technical reasons) was washed exhaustively before analysis and had often been divested of its zona.

Recently a protein fraction issuing from the rabbit endometrium and capable of entering free-lying blastocysts has acquired a certain notoriety, not least because of the christening fervor exhibited by the research workers concerned who have bestowed upon it individual names, uteroglobin (Beier 1968) and blastokinin (Krishnan and Daniel 1967), and have added to the confusion by claiming for it as yet uncertain embryotrophic properties.

This complicated problem is receiving the full treatment that it amply deserves (Gulyas, Krishnan, chapter 14). It suffices here to mention that uteroglobin, alias blastokinin, is a glycoprotein of relatively low molecular weight (27,000), which does not occur in rabbit blood. In connection with its occurrence in rabbit blastocysts, it may be worth recalling that we have detected in 6- and 7-day-old blastocysts very small amounts of carbonic anhydrase, an enzyme typical of the rabbit endometrium, of a molecular weight similar to the uterine glycoprotein. It seems therefore that molecules of a size up to approximately 30,000 can penetrate zona-enveloped rabbit blastocysts. One would like to know if these proteins, detectable in unattached blastocysts, are exclusively of endometrial origin, as one cannot ignore the possibility (until proved otherwise) that they may be produced by the embryos themselves. To solve this problem it would perhaps be best to culture rabbit morulae in a synthetic medium up to a late blastocyst stage, and then look for the presence of the above mentioned two proteins in blastocysts that have never been in contact with endometrium in which these proteins originate.

Amino nitrogen represents about ¼ of the total nitrogen in unattached blastocysts; 10 μg (8.1–12.4) has been found in 6½–7-day old embryos. Using the Technicon autoanalyzer procedure for chromatographic separation and detection of amino acids we found (Lutwak-Mann 1966) that at this stage the rabbit blastocyst has at its disposal practically every known amino acid. The finding of free amino acids in rabbit blastocysts is wholly in agreement with observations made by

Gregoire, Gongsakdi, and Rakoff (1961), and with Lesiński et al. (1967) except that they claim that the quantitatively predominant amino acid is alanine, whereas we (and Gregoire) have found that glycine occurs in the highest concentration. Lesiński et al (1967) have extended the analyses of amino acids to later stages and state that as protein content in the attached blastocysts rises the content of free amino acids decreases.

The abundance of free amino acids in blastocyst fluid at the preattachment stage inspires speculation about their role in embryonic development. We have made additional amino acid analyses in rabbit blood plasma, peritoneal fluid, and estrous uterine fluid. We have also determined the amino acid content of blastocysts recovered from rabbits treated with either 6-mercaptopurine or thalidomide in such a way as to induce distinct changes in the embryonic disk and (with mercaptopurine) in the trophoblast. Except for minor differences the amino acid pattern was the same in blastocyst fluid and the maternal body fluids examined; moreover, no major change occurred in the amino acid content of blastocysts taken from drug-treated animals. It seems therefore that the amino acid composition recorded in blastocysts is representative of the rabbit as an animal species rather than the blastocysts specifically. It still remains uncertain what significance to ascribe to the occurrence of amino acids in rabbit blastocysts. The absence of a marked gradient in their level in blastocyst fluid and uterine fluid suggests that the amino acids are conveyed to the blastocysts by diffusion rather than by a more complex form of transport characteristic of certain other constituents.

2. *Nucleic Acids.* Both DNA and RNA determinations were done in unattached blastocysts (Lutwak-Mann 1966). Analyses of DNA showed that between day 5 and day 7 of rabbit blastocyst development the DNA content rises 100-fold, from 0.031 μg to 3.160 μg per single embryo. In confirmation of our morphological observations the DNA content showed marked variability in chronologically identical blastocysts. When determinations of DNA were done on blastocysts recovered from mercaptopurine-treated rabbits, the values in these seriously damaged embryos were much lower than those recorded for coeval blastocysts from untreated rabbits. Thus the DNA content served to substantiate phenomena previously established in rabbit blastocysts by histological examination.

Analyses of RNA were done by the micromethod of Edström (1960), which permitted separate determinations on embryonic disks and extraembryonic material, chiefly the abembryonic pole area of the trophoblast. It was found that during the time span of 5–6½ days of pregnancy the total RNA content of the rabbit blastocyst increased approximately 3-fold, from 0.7 μg to 1.9 μg per single embryo. Of the two morphologically distinct embryonic regions the embryonic disk showed the larger relative increase; even so, the total RNA content of the disk was lower than that of the trophoblast layer. Microelectrophoretic RNA analyses did not reveal any significant differences between the disk and trophoblast. Ribonuclease-digested extracts obtained from these regions contained some contaminating material which interfered with the determination of the pyrimidine nucleotides; consequently reliable values were obtained only for the purines. Nevertheless it was possible to judge from some particularly good separations that the base composition was one typical for mammalian ribosomal RNA. No significant differences could be recorded in the adenine to guanine ratio between the embryonic disk and trophoblast, nor was there evidence of any significant change during days 5–6½ of development of the embryos.

3. *Glucose, Lactic Acid, and Bicarbonate.* It is convenient to tackle these three metabolites together for a number of reasons: their pathways meet and cross at certain points; they are an illustrative example of the speed with which metabolite concentrations rise and fall during a phase of fast embryonic growth; and in contrast

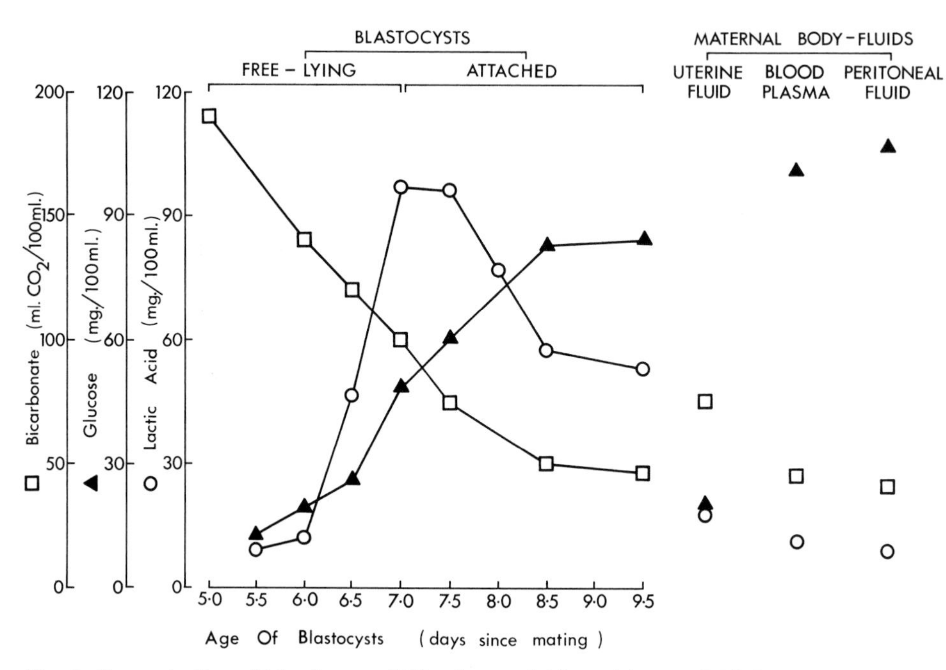

Fig. 1. Concentration of bicarbonate (□), glucose (▲), and lactate (○) in rabbit blastocysts, uterine fluid, blood plasma, and peritoneal fluid. Average values are given, based on 8–12 determinations. Blood plasma and peritoneal fluid were obtained from pregnant rabbits (days 5–9); uterine fluid was obtained from gonadotrophin-treated nonpregnant rabbits at the time of ovulation.

to the amino acids they occur in the embryos in concentrations strikingly different from those in maternal body fluids (Lutwak-Mann 1962*a*).

Figure 1 represents the concentration of glucose, lactate, and bicarbonate during the blastocyst stage with maternal values for blood plasma, peritoneal fluid, and estrous uterine fluid serving as background. It can be seen that the glucose concentration in 5–6-day-old blastocysts is low: on day 5½ it corresponds per single embryo to about 0.1–0.2 μg, at 6 days about 5–6 μg. A rise is evident late on day 6, resulting in 45–55 μg per single blastocyst shortly before uterine attachment; but even after attachment the glucose level in blastocysts remains distinctly below maternal blood glucose.

Lactate concentration at the onset of blastocyst development and up to day 6 is low; a steep increase sets in late on day 6, culminating in up to 100 mg/100 ml around the time of attachment; this means that about 100 μg is present in a single 7-day blastocyst. Thereafter the lactate concentration declines, but in the attached blastocyst its level still markedly exceeds peripheral maternal values.

Bicarbonate on the other hand shows on days 5 and 6 a remarkably elevated concentration exceeding by several times the maternal level; this amounts to a content of 1.8 μg on day 5, and 35 μg on day 6, per single blastocyst. From day 7 onward the concentration of bicarbonate decreases to near maternal level.

The causes for the striking changes in glucose and lactate concentration must presumably be sought in the blastocysts' metabolism, which switches from glucose-independent to glucose-dependent pathways coincidentally with the onset of the blastocyst stage (Fridhandler, Wastila, and Palmer 1967). Since the endometrial contribution of these metabolites or their precursors to the blastocyst makeup at each of the investigated points is still obscure, further speculation would not be very fruitful at this time. Although we are probably justified in assuming that the endometrial carbonic anhydrase is responsible for the presence of bicarbonate in rabbit blastocysts, we are in this instance, too, far from comprehending the extent of the endometrial involvement in the regulation of the blastocysts' internal milieu.

It is appropriate to mention here that carbohydrates other than glucose, namely fructose, inositol, glycogen, and sialic acid, have been looked for in the free-lying and attached rabbit blastocysts. Only glucose was found at this stage though some authors (Gregoire, Gongsakdi, and Rakoff 1962) have identified in blastocysts small amounts of inositol as well; this sugar alcohol is, of course, a well-known constituent of fetal fluids in the rabbit.

The presence of glycogen has been histochemically demonstrated in the mouse zygote (Thomson and Brinster 1966); however, the mouse blastocyst was found to contain only very small amounts, restricted to the area of the embryonic disk. In the past we have done chemical glycogen analyses, using gram quantities of 6½-day-old rabbit blastocysts, with negative results. Probably it would have been better to attempt a sensitive histochemical visualization of glycogen; if it could be confirmed that in blastocysts of species other than mouse, the embryonic disk, unlike the trophoblast, contains glycogen in however small a quantity, light might be thrown on the differences in the behavior of these two regions.

4. *Sodium, Potassium, and Chloride Ions.* Table 2 summarizes results, expressed as $m\ E_q/l$, in rabbit blastocysts, blood serum, and uterine estrous fluid, obtained earlier (Lewis and Lutwak-Mann 1954) and in later unpublished experiments. Points of interest are (*a*) the high content of K in free-lying blastocysts and in uterine fluid, (*b*) the small but distinct decrease in Na content at 7 days, and (*c*) the rather good agreement between the 6-day-old blastocysts and uterine fluid (also evident for glucose and bicarbonate; see Lutwak-Mann 1966, table 6).

5. *Coenzymes and Enzymes.* The occurrence in blastocysts of nicotinic acid, thiamine, riboflavin (Kodicek and Lutwak-Mann 1957), vitamin B_{12}, and folinic acid (Jacobson and Lutwak-Mann 1956) indicates the presence of essential coenzymes and attests to the embryos' advanced metabolic endowment. Several enzymes

have been investigated in rabbit blastocysts although the chief work has been with rabbit endometrium. ATPase, RNAse, and carbonic anhydrase were common to both endometrium and blastocysts, but not purine nucleosidase or 5-nucleotidase (Leone, Libonati, and Lutwak-Mann 1963). Other endometrial enzymes, studied mainly in relation to ovarian hormonal control, were certain dehydrogenases and transaminases (Murdoch and White 1969) and NADP-linked steroid reductases (Macartney and Thomas 1969).

6. *Hormones.* Although *in vivo* the mammalian embryo depends for its survival on a subtle hormonal balance, little is known about the occurrence of hormones in the preimplantation conceptus. A preliminary (unpublished) series of analyses (Eik-Nes, Mead, and Lutwak-Mann) showed per single 6½-day rabbit blastocyst 0.25 ng progesterone; the concentration of progesterone (ng/ml) in 7½-

TABLE 2

CONTENT OF SODIUM, POTASSIUM, AND CHLORIDE IONS IN RABBIT BLASTOCYSTS, BLOOD SERUM, AND UTERINE ESTROUS FLUID

Days after Mating		$m\ E_q/l$		
		Na	K	Cl
Blastocysts	6–6½	138.5	11.9	72.8
	7	125.8	13.2	79.3
	8–9	145.7	5.9	90.4
Blood serum	0	148.0	7.1	101.2
	7	150.0	6.7	105.6
Uterine estrous fluid	0	110	10.7	55.0

NOTE: Values are averages based on 4–6 determinations.

day blastocyst fluid was 0.0125—0.0133 as against 0.0046 in maternal blood plasma and 0.0019 in peritoneal fluid. The presence of progesterone in rabbit blastocysts is interesting, especially in the light of earlier experiments (Huff and Eik-Nes 1966) which demonstrated biosynthesis of pregnenolone from acetate and biotransformation of certain steroids by rabbit blastocysts *in vitro.*

B. *Passage into and Clearance from Rabbit Blastocysts of Substances Administered to the Mother*

Information on these fundamental aspects of maternal-blastocyst relationship derives from two types of experimental approach, biochemical and morphological. The former involves the administration to rabbits of sugars, labeled ions, vitamin A, and certain drugs with subsequent determination of their presence (concentration, uptake) simultaneously in maternal body fluids and tissues, and in the blastocysts and their environment. The latter is based on histological observations concerning the action of exogenous agents on free-lying blastocysts and is discussed separately.

1. *Simple Sugars.* In free-lying blastocysts there is no rise in sugar concentration following i.v. administration to rabbits of either glucose, fructose, or sucrose, coincidental with maternal hyperglycemia or fructosemia (Lutwak-Mann 1954). From day 7 onward, and concomitantly with uterine attachment, maternally introduced hexoses substantially raise reducing sugar concentration in attached blastocysts. Moreover, in experiments of longer duration the sugar levels in blastocysts remain elevated at a time when maternal blood levels have already returned to normal. This typical delay in clearance of glucose from the blastocyst system on day 8 is illustrated in figure 2.

We have also noticed that glucose levels in blastocysts are relatively independent of major changes in the maternal organism in short-term experiments with insulin; in spite of severe maternal hypoglycemia the glucose concentration in 8–9-

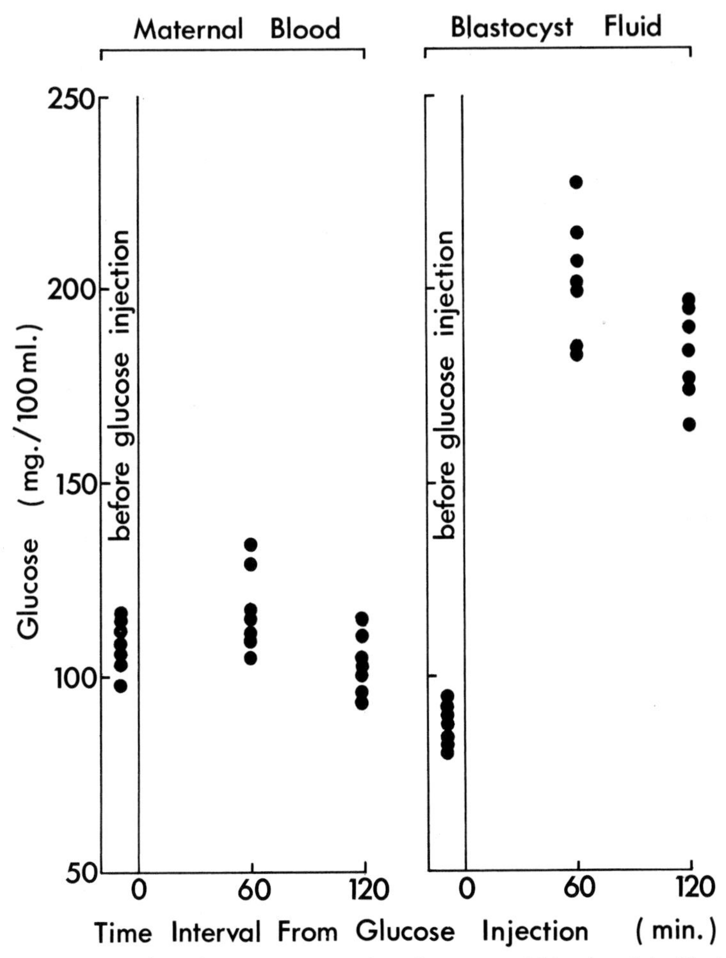

Fig. 2. Scattergram showing glucose concentrations in maternal blood and in blastocyst fluid, prior to, as well as at 60 and 120 min after intravenous glucose injection of 1 g/kg to rabbits on day 8 of pregnancy. Values at 30 min after glucose injection, not shown here, were (average from 7 determinations): maternal blood 397 mg/100 ml, blastocyst fluid 128 mg/100 ml.

day blastocysts was not markedly depressed. On the other hand alloxan diabetes (Lutwak-Mann 1954) or cortisone treatment of rabbits (Lutwak-Mann 1962*a*) caused distinct and lasting glucose elevation in blastocysts.

2. *Labeled Ions.* Following parenteral administration to rabbits of labeled ions such as ^{32}P, ^{42}K, ^{24}Na, ^{35}S, and ^{131}I, a series of informative patterns of maternal-blastocyst distribution has been established (Lutwak-Mann, Boursnell, and Bennett 1960). In experiments of 45 min duration all these ions entered the endometrium and progestational secretion in relatively high and roughly equal amounts for each ion investigated. In marked contrast to the uniform uptake in the blastocysts' environment the extent of incorporation in the blastocysts themselves was quite specific for each ion on each day of the blastocyst stage. In particular, early on day 6 there was practically no ^{32}P uptake in the blastocysts (paralleling the behavior of glucose); but with incipient uterine attachment, ^{32}P incorporation, especially into the nucleic acid fraction, became evident (Vittorelli, Harrison, and Lutwak-Mann 1967). In what form phosphate actually reaches blastocysts (inorganic, phosphorylated?) is as yet unknown.

3. *Vitamin A.* The concentration of vitamin A in blood plasma of untreated rabbits is about 2 to 3 times that of blastocyst fluid at 8–9 days (expressed as i.u. vit. A/100 ml, 250–400 was found in plasma, and 125–180 in blastocyst fluid). Twelve hours after single injections of 100,000–400,000 i.u. vit. A per kg weight on day 6½–7, maternal blood plasma levels more than doubled; they gradually reverted to normal by 48 hr. In contrast, throughout this long experimental period only minor increases at 12 and 24 hr from the time of injection, occurred in 8- and 9-day blastocyst fluid in spite of the great permeability at that stage of the yolk sac placenta (Baronos and Lutwak-Mann unpublished).

4. *Drugs.* Parenterally or orally administered drugs not only rapidly find their way into the endometrial progestational secretion and into free-lying and attached blastocysts but also tend to persist for considerably longer periods in the blastocysts than in maternal blood. This has been convincingly demonstrated in experiments based on chemical identification involving sulphonamides, salicylates (Lutwak-Mann 1954, 1962*a*), thalidomide (Fabro et al. 1964; Fabro, Smith, and Williams 1965), caffeine and nicotine (Fabro and Sieber 1969), and a host of other substances such as DDT, isoniazid, barbital, and thiopental (Fabro and Sieber 1968). It is a pity that no information exists, as it does for salicylates, sulphonamides, and thalidomide, as to whether these pharmacological agents cause damage to blastocysts.

C. *The Response of Rabbit Blastocysts to the Action of Different Classes of Agents* in Vivo *and* in Vitro

As an extension of earlier studies on embryotoxic agents (Adams, Hay, and Lutwak-Mann 1961; Lutwak-Mann and Hay 1962; Hay 1964; Lutwak-Mann and Hay 1965*b*), Lutwak-Mann, Hay, and New (1969) compared the effect of the following classes of agents on 6-day rabbit blastocysts *in vivo* and *in vitro:* metabolites and antimetabolites, enzyme inhibitors, antimitotic and cytostatic compounds, and hormones. Observations were made also on the effect of oxygen deprivation, chiefly in relation to temperature. In addition, experiments were carried out to test the feasibility of repairing *in vitro* damage sustained by blastocysts *in vivo,* and of promoting blastocyst growth *in vitro* by hormonal pretreatment of donor animals.

It was found that some of the agents examined caused distinctly recognizable effects in preattachment rabbit blastocysts both *in vivo* and *in vitro;* others acted either *in vivo* or *in vitro;* and a few were inert regardless of mode of treatment.

The "combined" *in vivo–in vitro* type of experiment indicated that the recovery *in vitro* of blastocysts damaged *in vivo* is indeed a distinct possibility depending on the nature of the maternally administered agent and the degree of damage sustained. Enhancement of blastocyst development *in vitro* by stimulating the embryos' maternal environment with a uterotrophic hormone also seems a practical proposition. Other observations made in the course of these and older (l.c.) experiments are included in the discussion of blastocyst characteristics.

IV. Carbonic Anhydrase and Zinc in the Rabbit Uterus: Endometrial-Blastocyst Links

The occurrence of bicarbonate in a high concentration in young blastocysts led to the discovery of carbonic anhydrase (CA) in the rabbit uterus (Lutwak-Mann and Laser 1954); further work (Lutwak-Mann 1955; Lutwak-Mann and Adams 1957*a, b*) established the dependence of this enzyme upon progesterone (and other luteoids); current experiments (McIntosh and Lutwak-Mann 1967) are concerned with isolation of CA isoenzymes in the endometrium of different species.

Attempts to correlate rabbit uterine CA with blastocyst development (l.c.) met with frustration, for it proved impossible to block CA activity effectively with maternally administered sulphonamides; in sublethal doses these drugs exerted no discernible effect on free-lying blastocysts.

Because CA is a Zn-linked enzyme (Keilin and Mann 1940) we turned our attention to the uterine Zn content (Lutwak-Mann and McIntosh 1969) and found that, as with CA activity, a marked rise in Zn concentration, both in rabbit endometrium and uterine tissue as a whole, accompanies the progestational phase. But even at the maximum of CA activity (postovulatory days 5–8) the enzyme's Zn content amounts to no more than 1% of the total uterine Zn. The blastocysts at 6 and 7 days contain minute amounts of Zn; the presence in blastocysts of traces of CA has been mentioned earlier.

We have tried to fit our observations on uterine CA and Zn into what is known of the hormonal and metabolic background of early pregnancy and pseudopregnancy in the rabbit. The ovary-regulated interplay between progesterone and 20αOH seems well established (Hilliard, Penardi, and Sawyer 1967; Hilliard, Spies, and Sawyer 1968). The most significant event in endometrial metabolism may be the marked rise in RNA content which sets in on day 2, extends at least to day 8, and is distinguished by the appearance of a stable 10 S RNA in addition to the usual 28 S and 18 S components (Vittorelli, Harrison, and Lutwak-Mann 1967). This early increase in RNA (conjecturally integrated with the preovulatory spurt of 20αOH), might be responsible subsequently for setting in motion the enzyme-synthesizing machinery in the rabbit endometrium; this, in turn, might lead to a progressive increase in CA activity from day 4 on, synchronous with the onset of the postovulatory release of the two rabbit progestins. How Zn, or indeed any other trace metal, is mobilized and shifted to the pregnant female reproductive tract is not known in mammals. Its accumulation during the endometrium-dependent, so-called histiotrophic phase of embryonic growth is certainly intriguing. The role envisaged recently for uterine Zn, apart from CA and other Zn-linked enzymes, namely binding of steroids to target tissue (Emanuel and Oakey 1969), is interesting but needs corroboration.

V. Characteristic Features of the Blastocyst Stage

Experience gained in our biochemical and morphological studies on rabbit blastocysts permits us to enumerate and briefly summarize certain outstanding features which mark this developmental stage.

A. *Blastocyst Variability*

Blastocysts of an identical gestational age differ in their degree of development; the phenomenon is encountered between and within litters, and is presumably also responsible for variability seen in blastocyst response to extraneous agents *in vivo* and *in vitro*.

B. *Blastocyst Wastage*

A relatively high incidence (10%) of abortive embryos in fertile animals is evident at this stage; their elimination involves little injury to the mother. The occurrence of these nonviable blastocysts may be due in part to aging of gametes and chromosomal abnormalities (Austin 1967; Shaver and Carr 1967).

C. *Accessibility and Vulnerability of Blastocysts*

The free-lying and, to an even greater extent, the attached blastocysts are accessible to maternally transmissible agents which can penetrate them very rapidly. The magnitude of damage depends in each case on maternal and embryonic genotype; it is difficult, therefore, to predict embryonic vulnerability. However maternally administered agents need not always be injurious to blastocysts because (1) some substances may, for various reasons, be unable to enter the embryos, (2) others may be efficiently detoxified by the maternal organism, and (3) the blastocysts may lack target systems for some of the potentially toxic substances. Certain agents, notably of hormonal nature, tend to act on embryos secondarily, rather than primarily, by inducing major changes in their environment (e.g., altering endometrial secretory products, affecting myometrial control of blastocyst positioning).

D. *Regulatory Maternal-Blastocyst Exchange Mechanisms*

A "barrier" is evident between the maternal and blastocyst compartments, presumably vested in specialized elements and operative jointly at the endometrial, yolk sac, and blastocyst level. It regulates quantitatively the influx into embryos of metabolites and foreign substances. Consequently, even if penetration of the blastocoelic space occurs, the embryo receives only a fraction of the load present in terms of concentration in the maternal bloodstream. There are some indications that maternally applied estrogen interferes with this control mechanism (Lutwak-Mann 1962*a;* Lutwak-Mann, Boursnell, and Bennett 1960).

E. *Delay in Clearance of Extraneous Agents*

The clearance of maternally introduced substances from the embryonic compartment lags significantly behind the rate of maternal elimination. This persistence in the blastocyst system carries with it some danger as, owing to its prolonged stay, a drug may exert damage not at the time of entry but at a later, and therefore much more susceptible, stage of embryonic development.

F. *Differential Sensitivity*

Morphologically distinct regions of the blastocyst react quite differently to the same agent; as a rule the embryo proper is very sensitive and therefore easily damaged whereas the trophoblast withstands injury much better. Within the trophoblast, however, a gradient is discernible; resistance is greatest at the abembryonic pole and declines toward the equatorial and near-embryonic disk zones. Differential sensitivity in the embryo proper as opposed to the trophoblast is evident equally *in vivo* and *in vitro*. Presumably it stems from a fundamentally different type and rate of metabolism in these two disparate blastocyst regions.

G. *Adaptational Capacity of Blastocysts*

Survival and continuation of quasi-normal development in extrauterine sites or synthetic nutrient media reflect the remarkable adaptational powers of young embryos. This adaptability enables us to experiment freely with mammalian embryos *in vitro;* at the same time the experimental results must be viewed with appropriate skepticism, as these malleable creatures can be made to perform acts *in vitro* that they might not dream of under physiological conditions.

H. *Recuperative Capacity of Young Embryos*

The ability of blastocysts to overcome *in vivo* or *in vitro* the effects of damage or developmental delay is an outstanding embryonic attribute. It needs stressing as a counterpart to the much-emphasized sensitivity. The resumption of growth and extent of recovery after exposure are bound to depend not only upon the degree of damage but also on the nature of the injurious factor. A hazard inherent in this embryonic feature is that the damage incurred may be incompletely repaired, resulting in gross or discrete malformation.

VI. Investigations in Prospect

Certain prospectively profitable areas of research, delineated by the experimental approach described above, may be listed in the following tentative order of priority:

1. The chemical composition of blastocysts and associated maternal tissues.

2. The metabolic properties specific for (*a*) the embryo proper and (*b*) the extraembryonic tissues, especially the trophoblast.

3. The nature of the regulatory mechanism which controls the entry into and exit from the blastocyst compartment of maternally introduced substances.

4. The mobilization from depots and shift into the female reproductive tract of trace elements at the preplacental stage of pregnancy.

5. The chemical and biological properties of endometrial products (proteins, nucleic acids) in relation to (*a*) ovarian hormonal control and (*b*) blastocyst development.

6. The recuperative capacity (*in vitro, in vivo*) of blastocysts after exposure to different classes of agents inherent in the embryo proper and trophoblast, and the enhancement of blastocyst development *in vitro* by appropriate (hormonal, nutritional) stimulation of the maternal environment in donor animals.

References

Adams, C. E.; Hay, M. F.; and Lutwak-Mann, C. 1961. The action of various agents upon the rabbit embryo. *J Embryol Exp Morph* 9:468.

Austin, C. R. 1967. Chromosome deterioration in ageing eggs of the rabbit. *Nature* (*London*) 213:1018.

Beier, H. M. 1968. Uteroglobin: A hormone-sensitive endometrial protein involved in blastocyst development. *Biochim Biophys Acta* 160:289.

Brambell, F. W. R. 1948. Studies on sterility and prenatal mortality in wild rabbits: 2. The occurrence of fibrin in the yolk sac contents of embryos during and immediately after implantation. *J Exp Biol* 23:332.

Brinster, R. L. 1967. Protein content of the mouse embryo during the first five days of development. *J Reprod Fertil* 13:413.

Edström, J. E. 1960. Extraction, hydrolysis and electrophoretic analysis of ribonucleic acid from microscopic tissue units (microphoresis). *J Biochem Biophys Cytol* 8:39.

Emanuel, M. B., and Oakey, R. E. 1969. Effect of Zn^{++} on the binding of oestradiol- 17 β to a uterine protein. *Nature* (*London*) 223:66.

Fabro, S.; Schumacher, H.; Smith, R. L.; and Williams, R. T. 1964. Identification of thalidomide in rabbit blastocysts. *Nature* (*London*) 201:1125.

Fabro, S., and Sieber, S. M. 1968. Excerpta Medica Int Cong Series no. 183. *The Foeto-placental unit* Sympos, Milan.

———. 1969. ^{14}C Caffeine and ^{3}H nicotine penetrate pre-implanting blastocysts. *Nature* (*London*) 223:410.

Fabro, S.; Smith, R. L.; and Williams, R. T. 1965. The persistence of maternally administered ^{14}C-thalidomide in the rabbit embryo. *Biochem J* 97:14.

Fridhandler, L.; Wastila, W. B.; and Palmer, W. M. 1967. The role of glucose in metabolism of the developing mammalian preimplantation conceptus. *Fertil Steril* 18:619.

Gregoire, A. T.; Gongsakdi, D.; and Rakoff, A. E. 1961. The free amino acid content of the female rabbit genital tract. *Fertil Steril* 12:322.

———. 1962. The presence of inositol in genital tract secretions of the female rabbit. *Fertil Steril* 13:432

Hay, M. F. 1964. Effects of thalidomide on pregnancy in the rabbit. *J Reprod Fertil* 8:59.

Hilliard, J.; Penardi, R.; and Sawyer, C. H. 1967. A functional role for 20α-hydroxypregn-4-en-3-one in the rabbit. *Endocrinology* 80:901.

Hilliard, J.; Spies, H. G.; and Sawyer, H. C. 1968. Cholesterol storage and progestin secretion during pregnancy and pseudopregnancy in the rabbit. *Endocrinology* 82:157.

Huff, R. L., and Eik-Nes, K. B. 1966. Metabolism *in vitro* of acetate and certain steroids by six-day old rabbit blastocysts. *J Reprod Fertil* 11:57.

Jacobson, W., and Lutwak-Mann, C. 1956. The vitamin B_{12} content of the early rabbit embryo. *J Endocr* 14:xix.

Keilin, D., and Mann, T. 1940. Carbonic anhydrase: Purification and properties. *Biochem J* 34:1163.

Kodicek, E., and Lutwak-Mann, C. 1957. The pattern of distribution of thiamine, riboflavin, and nicotinic acid in the early rabbit embryo. *J Endocr* 15:liii.

Krishnan, R. S., and Daniel, J. C. 1967. "Blastokinin": Inducer and regulator of blastocyst development in rabbit uterus. *Science* 158:490.

Leone, E.; Libonati, M.; and Lutwak-Mann, C. 1963. Enzymes in the uterine and cervical fluid and in certain related tissues and body fluids. *J Endocr* 25:551.

Lesiński, J.; Jajszczak, S.; Bentyn, K.; and Janczarski, I. 1967. Content of free amino acids in rabbit blastocyst fluid. *Pol Med J* 6:1074.

Lesiński, J.; Jajszczak, S.; and Choroszewska, A. 1967. The relationship between amino nitrogen and protein concentration in rabbit blastocyst fluid. *Amer J Obstet Gynec* 98:48.

Lewis, P. R., and Lutwak-Mann, C. 1954. The content of sodium, potassium and chloride in rabbit blastocysts. *Biochim Biophys Acta* 14:589.

Lutwak-Mann, C. 1954. Some properties of the rabbit blastocyst. *J Embryol Exp Morph* 2:1.

———. 1955. Carbonic anhydrase in the female reproductive tract: Occurrence, distribution and hormonal dependence. *J Endocr* 13:26.

———. 1962*a*. Glucose, lactic acid and bicarbonate in rabbit blastocyst fluid. *Nature (London)* 193:653.

———. 1962*b*. Some properties of uterine and cervical fluid in the rabbit. *Biochim Biophys Acta* 58:637.

———. 1966. Some physiological and biochemical properties of the mammalian blastocyst. *Bull Swiss Acad Med Sci* 22:101.

Lutwak-Mann, C., and Adams, C. E. 1957*a*. Carbonic anhydrase in the female reproductive tract: II. Endometrial carbonic anhydrase as indicator of luteoid potency; correlation with progestational proliferation. *J Endocr* 15:43.

———. 1957*b*. The effect of methyloestronolone on endometrial carbonic anhydrase and its ability to maintain pregnancy in the castrated rabbit. *Acta Endocrin* 25:405.

Lutwak-Mann, C.; Boursnell, J. C.; and Bennett, J. P. 1960. Blastocyst-uterine relationships: Uptake of radioactive ions by the early rabbit embryo and its environment. *J Reprod Fertil* 1:169.

Lutwak-Mann, C., and Hay, M. F. 1962. Effect on the early embryo of agents administered to the mother. *Brit Med J* 2:944.

Lutwak-Mann, C. and Hay, M. F. 1964. Effect of certain water-soluble oestrogens on rabbit blastocysts. *J Endocr* 30:ix.

———. 1965*a*. Maternally transmitted embryotropic agents. In *Biological Council Symposium on "Agents affecting fertility,"* ed. C. R. Austin and J. S. Perry, p. 261. London: J. & A. Churchill Ltd.

———. 1965*b*. Effect of 2-deoxyglucose on the rabbit blastocyst. *J Reprod Fertil* 10:133.

———. 1967. The blastocyst flat-mount technique in studies on embryotropic agents. In *Advances in teratology II,* ed. D. H. M. Wollam, p. 229. London: Academic Press.

Lutwak-Mann, C.; Hay, M. F.; and Adams, C. E. 1962. The effect of ovariectomy on rabbit blastocysts. *J Endocr* 24:185.

Lutwak-Mann, C.; Hay, M. F.; and New, D. A. T. 1969. Action of various agents on rabbit blastocysts *in vivo* and *in vitro. J Reprod Fertil* 18:235.

Lutwak-Mann, C., and Laser, H. 1954. Bicarbonate content of the blastocyst fluid and carbonic anhydrase in the pregnant rabbit uterus. *Nature* (*London*) 173:268.

Lutwak-Mann, C., and McIntosh, J. E. A. 1969. Zinc and carbonic anhydrase in the rabbit uterus. *Nature* (*London*) 221:1111.

Macartney, T. C., and Thomas, G. H. 1969. NADP-linked 17β- and 20α-steroid reductase activity in the rabbit uterus. *J Endocr* 43:247.

McIntosh, J. E. A. and Lutwak-Mann, C. 1967. Carbonic anhydrase isoenzymes in certain tissues of the male and female reproductive tract and red blood cells. *J Reprod Fertil* 14:344.

Murdoch, R. N., and White, I. G. 1969. The activity of enzymes in the rabbit uterus and effect of progesterone and oestradiol. *J Endocr* 43:167.

Shaver, E. L. and Carr, D. H. 1967. Chromosome abnormalities in rabbit blastocysts following delayed fertilization. *J Reprod Fertil* 14:415.

Thomson, J. L., and Brinster, R. L. 1966. Glycogen content of preimplantation mouse embryos. *Anat Rec* 155:97.

Vittorelli, M. L.; Harrison, R. A. P.; and Lutwak-Mann, C. 1967. Metabolism of ribonucleic acid in the endometrium of the rabbit during early pregnancy. *Nature* (*London*) 214:890.

14

Current Status of the Chemistry and Biology of "Blastokinin"

Bela J. Gulyas/R. Sivarama Krishnan

Department of Anatomy
School of Medicine
Georgetown University, Washington, D.C.

Department of Biochemistry
Colorado State University, Fort Collins

The fertilized mammalian eggs spend several days in the female reproductive tract while they undergo cleavage and are transported through the oviduct to the uterus, where they become stationary at their site of implantation. Although these free-living embryos contain some energy source in a stored form, they also rely on the oviductal and uterine secretions for nutrients. During the past decade an extensive effort has been made by reproductive biologists to mimic the maternal environment in culture techniques *in vitro*. These culture techniques have been used to determine the optimum pH, osmolarity, nitrogen, and energy requirements of some mammalian eggs at different stages of their development. It became evident very early in these pioneering studies that the maternal milieu is much more complicated than it was thought to be. Until recently sperm-penetrated mouse eggs could not be cultured *in vitro* before completion of first cleavage; rabbit eggs could be cultured up to 3 days *post coitum* and rabbit blastocysts could be cultured for 2 to 3 days, but blastulation could not be achieved. Cultured mammalian eggs fall behind their *in vivo* counterparts in development, as measured by the number of cell divisions they undergo. It is now known that the physical conditions employed and the media used to culture embryos were inadequate; these deficiencies led some investigators to examine the role of the reproductive tract secretions in the development of the preimplanting embryos.

In this chapter we shall discuss the role of a component, blastokinin, isolated from the uterine secretions of the rabbit, in the development of the rabbit blastocyst. Brief consideration will be given also to other as yet uncharacterized components, whether of maternal origin or present in the body fluids of other species, that may play a role in the regulation of blastocyst development analogous to that of the uterine component under discussion here.

I. Discovery of Blastokinin

A number of investigators have predicted a dependent relationship between the developing embryo and the maternal milieu, other than nutritional (Hammond 1928; Adams 1965; Kirby 1965). Adams (1965) reported *in vitro* culturing of rabbit eggs at practically all stages of development. He noted that it had not been possible to maintain the same embryos through blastulation until the time of implantation. In a number of studies only initial cavitation of the morulae could be accomplished even under the most favorable conditions of culture. Occasionally a limited expansion of these embryos took place during subsequent culturing, but they never resulted in blastocysts. It seemed highly probable that "the uterine environment in some way provides for the regulation" (Krishnan and Daniel 1967) of blastocyst development in the rabbit and that the regulator is probably in the uterine secretions. When the uterine fluid components from several mammals were compared by electrophoretic techniques (Daniel 1968) a particular protein band was found between the 4th and 9th days of pregnancy of the rabbit. This band was absent prior to the 4th day in the uterine secretions, in rabbit serum, and in animals with delayed implantation with the possible exception of the mink. These results prompted further investigation of the rabbit uterine secretions to learn more about the properties of this protein (Krishnan and Daniel 1967). Gel filtration studies have extended earlier findings (fig. 1). Since this protein stimulated blastocyst formation in the late rabbit morulae *in vitro,* it was named "blastokinin" pending further characterization.

The presence of a similar specific protein in endometrial secretions of pregnant rabbits was demonstrated independently by other workers (Schwick 1965; Beier 1968*a, b*). One of the many proteins was initially recognized by Schwick, who named it B_1-U-globulin because of its uterine specific occurrence and precipitation in the B_1-globulin area on agar immunoelectrophoresis. This was later renamed "uteroglobin" in 1966 (Beier personal communication; Beier 1968*b*).

II. Isolation and Purification

Isolation of blastokinin was accomplished by relatively simple means. Uteri from 5-day pregnant New Zealand white rabbits were extirpated and their lumina flushed with either saline solution or a suitable tissue culture medium. Pooled uterine flushings were chromatographed by gel fitration through Sephadex G-200 in 0.02 M phosphate-citrate buffer, pH 7.4, and the crude blastokinin fraction was isolated. A typical elution profile of the macromolecular components in uterine secretions from 5-day pregnant rabbits is shown in figure 2. Further purification of the crude blastokinin was then accomplished by rechromatography through Sephadex G-75 in the same phosphate-citrate buffer as above. Dialysis against distilled water followed by lyophilization yielded a product which was judged to be homogeneous and composed of a single polypeptide chain based on the following characteristics: (1) a predominant single band in acrylamide disk electrophoresis in 8 M urea-β-alanine-acetate buffer, pH 4.6, and in electrophoresis on cellulose acetate at various pH values, using barbitone-barbitone sodium buffer, I 0.05, for the pH range 7.4–9.0 and glycine-sodium glycinate buffer, I 0.05 for the pH range 8.8–10.6 (fig. 3); (2) a single peak on reductive cleavage with 2-mercaptoethanol in 8 M urea followed by alkylation with iodoacetate and rechromatography through Sephadex G-75 equilibrated with phosphate-citrate buffer (fig. 4); and (3) a single sharp precipitation line in microimmunodiffusion tests against antiserum for the protein component of rabbit uterine secretions (fig. 5).

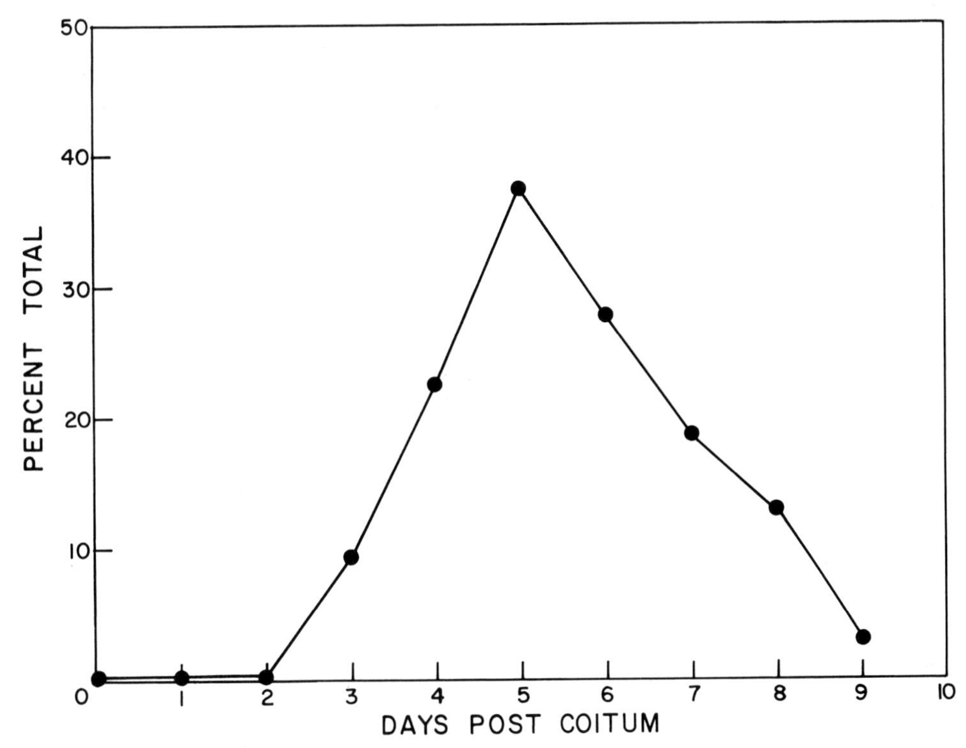

Fig. 1. Relative occurrence of blastokinin in the proteins of rabbit uterine secretions between 0 and 9 days of pregnancy.

III. Chemical Composition

Blastokinin is a glycoprotein with a molecular weight of approximately 27,000. Carbohydrates make up approximately 6% of the total weight of the molecule. Tables 1 and 2 describe its chemical composition in detail. It is rather interesting that it is devoid of any sialic acid and contains both glucosamine and galactosamine as well as the three neutral sugars, mannose, galactose, and glucose. Fucose (6-deoxygalactose) is the only 6-deoxyhexose present. Amino acids account for about 74% of the weight of the glycoprotein. As is the case in most glycoproteins, the content of aromatic amino acids in blastokinin is rather low; that of sulfur amino acids is relatively high.

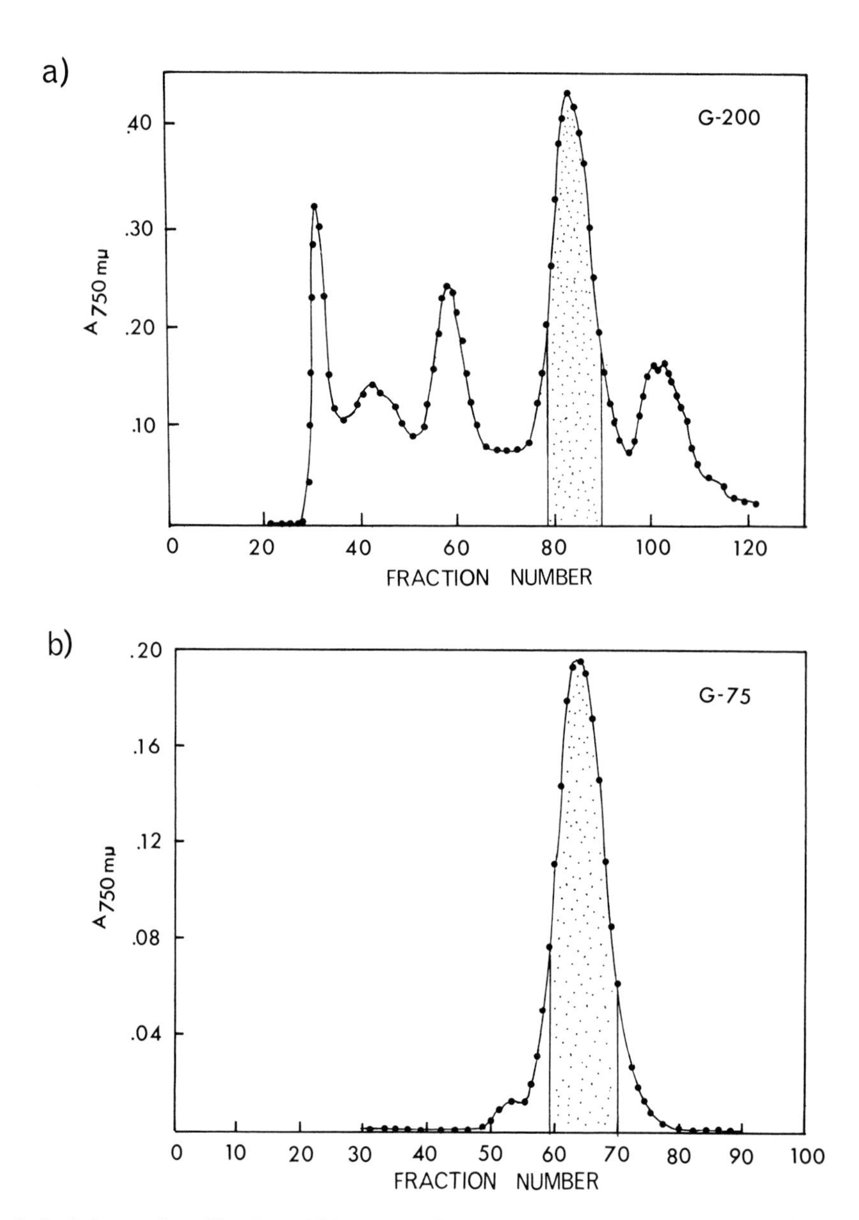

Fig. 2. Isolation and purification of blastokinin from 5-day pregnant rabbit uterine flushings by gel filtration. (*a*) Column (1.2 × 80 cms) was loaded with 5–10 mg of dialyzed, lyophilized uterine flushing concentrate, and eluted with 0.02 M phosphate-citrate buffer, pH 7.4. Fractions (1 ml) were collected at a flow rate of 4 ml/h. Aliquots of 0.2 ml were reached with Lowry reagent and protein determined at 750 mμ. Shaded area indicates the fractions pooled in the preliminary isolation. (*b*) Rechromatography and final purification of the blastokinin. Experimental conditions were similar to those described above.

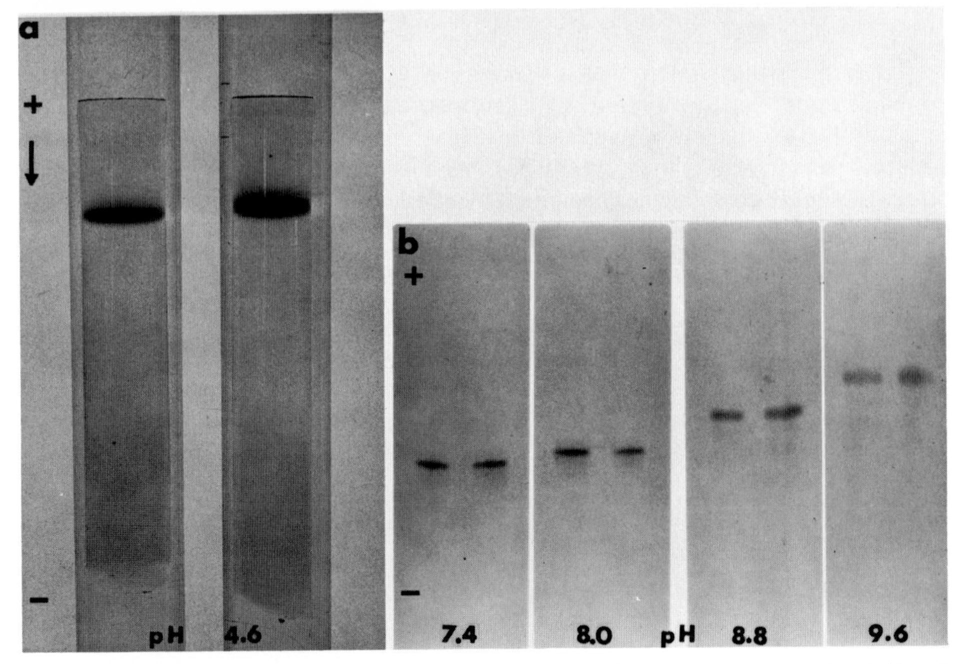

Fig. 3. Electrophoresis of purified blastokinin. (*a*) Acrylamide disk electrophoresis of 10 and 20 μg of blastokinin in 8 M urea-β-alanine-acetate buffer, pH 4.6, 120 V, 3 mA. (*b*) Electrophoresis on cellulose acetate strips (Millipore-Phoro Slide). Sample (2 μg) was applied at the middle of the strips. Barbitone-barbitone sodium buffer, 0.025 M, I 0.05, was used at pH values of 7.4 and 8.0; glycine-sodium glycinate buffer, I 0.05, was used at pH values of 8.8 and 9.6. Running time in each case was 20 min. (From Krishnan and Daniel 1968)

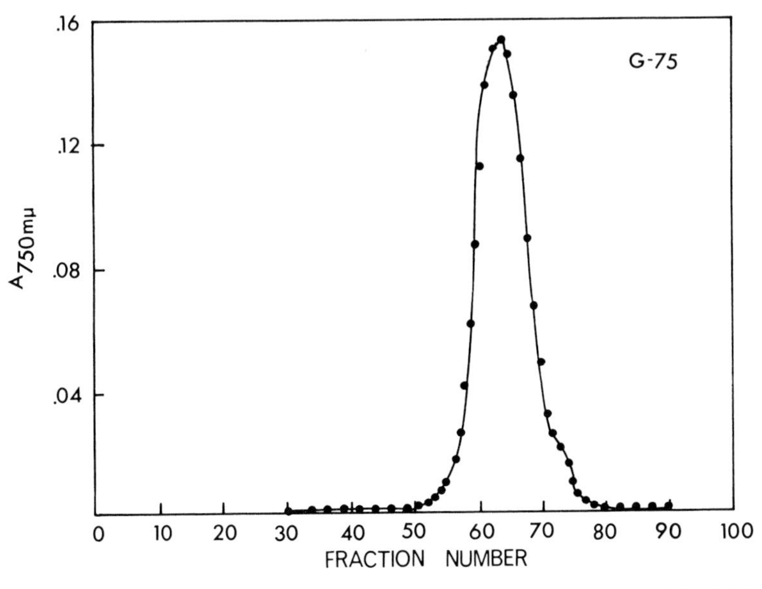

Fig. 4. Gel filtration of purified blastokinin treated with 2-mercaptoethanol in 8 M urea and alkylated with iodoacetate. Column (1.2 $\times$ 80 cms) was equilibrated with 0.02 M phosphate-citrate buffer, pH 7.4, the entire reaction mixture was applied to the column and eluted with the same buffer. Fractions (1 ml) were collected and protein determined as described earlier.

It is interesting to note that the molecular weight of uteroglobin has been reported as approximately 30,000. Its electrophoretic mobility appears to be similar to that of blastokinin. However, as distinct from blastokinin, it does not seem to be a glycoprotein (Beier 1968*b*). It is possible that blastokinin and uteroglobin are otherwise similar components. The ultimate relationship between the two must, however, await further chemical and biological characterization.

IV. Biological Effects

In order to determine the biological effects of the isolated rabbit blastokinin, Krishnan and Daniel (1967) cultured 3-day rabbit morulae under various conditions. Ham's F10 medium (Ham 1963) was supplemented with different concen-

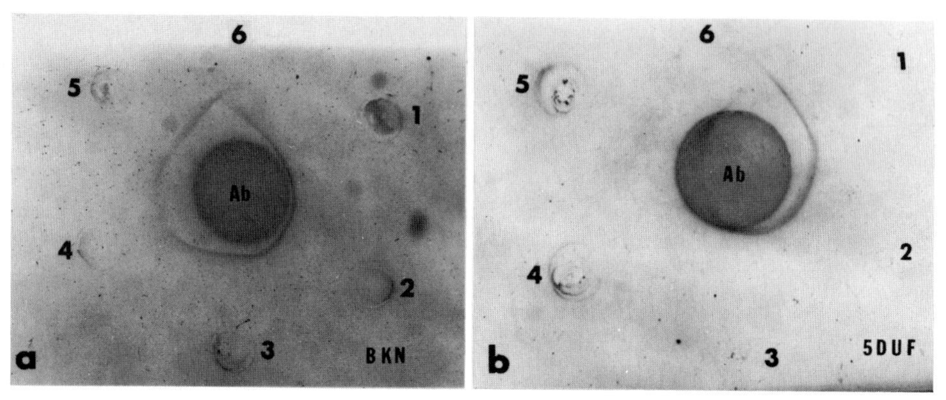

Fig. 5. Agar immunodiffusion of purified blastokinin (*BKN*) and of rabbit uterine secretions from 5-day uteri (*5 DUF*) against antiblastokinin preparation from chicken. Outer wells, 1–5, contained 1, 2, 4, 6, and 10 μg purified BKN respectively in *a;* and 1/50, 1/20, 1/10, 1/2, and 1/1 dilutions of whole 5-day uterine secretions respectively in *b*. Antibody preparation was placed in center well in both *a* and *b*.

TABLE 1

COMPOSITION OF PURIFIED RABBIT "BLASTOKININ"

Constituent	g/100 g	Molar Ratio	Residues/ Mole
Polypeptide	73.6		
Total nitrogen	11.8		
Carbohydrate	5.7		
Total hexoses	4.1		
Mannose	0.8	2.0	2
Galactose	1.6	3.8	4
Glucose	1.6	3.9	4
6-Deoxygalactose (Fucose)	0.4	1.0	1
Total amino sugar[a]	1.3		
2-Amino-2-deoxyglucose	0.4	1.0	1
2-Amino-2-deoxygalactose	0.9	2.1	2
Sialic acid	0.0	0.0	0

SOURCE: Krishnan and Daniel 1968.
[a] Expressed as the nonacetylated free bases.

trations of blastokinin, maternal serum, and complete uterine secretion complement of the 5-day pregnant rabbit. In the presence of blastokinin in concentrations up to 0.4 mg/ml nearly all the embryos underwent cavitation; at 0.2 mg/ml, 72% of the embryos expanded and began growth as blastocysts (table 3). The complete protein component of the uterine secretions at a concentration of 0.5 mg/ml caused 78% of the embryos to expand. Maternal serum proteins, on the contrary, did not promote comparable expansion of embryos to these stages at the concentrations used. The authors concluded that blastokinin from rabbit could induce and regulate blastocyst development. However, the blastocysts which developed from the 3-day

TABLE 2

AMINO ACID COMPOSITION OF "BLASTOKININ"

Constituent	g/100 g	Molar Ratio	Residues/ Mole
Alanine	2.1	3.4	7
Arginine	3.1	2.6	5
Aspartic acid	7.6	8.1	16
Half-cystine	1.6	1.9	4
Glutamic acid	10.2	10.4	21
Glycine	2.3	4.3	9
Histidine	1.6	1.4	3
Isoleucine	3.5	3.9	8
Leucine	8.9	9.7	19
Lysine	7.5	7.3	15
Methionine	4.2	4.0	8
Phenylalanine	3.0	2.6	5
Proline	4.7	6.0	12
Serine	4.1	5.6	11
Threonine	5.6	6.7	13
Tryptophan	1.4	1.0	2
Tyrosine	1.8	1.4	3
Valine	2.9	3.6	7
Ammonia	0.8	6.7	13
Total nitrogen	11.8		
Nitrogen recovered	10.7		
Weight recovered as amino acids	67.3		
Number of residues			181
Minimum molecular weight	26,390		

SOURCE: Krishnan and Daniel 1968.

NOTE: No corrections have been made for hydrolytic losses. Molar ratio: with respect to tryptophan taken as 1.0. Minimum molecular weight was calculated from the sum of the individual constituents from which the water of bond formation was subtracted.

morulae, when cultured in the presence of blastokinin, were not exactly comparable to normal 4-day blastocysts. There appeared to be extra layers or clusters of cells in the trophoblast, presumably resulting from the fact that the rigid zona pellucida did not allow the blastocyst to expand (fig. 6). The ultimate test of the normality of blastocysts cultured in the presence of blastokinin would be their transfer to suitably prepared pseudopregnant recipient does and further development.

V. Role of Blastokinin in Nucleic Acid and Protein Synthesis

If blastokinin is truly involved in the regulation of blastocyst development, what, then, is its role in the synthesis of nucleic acids and proteins? The answer to this question is far from complete. Certain interesting observations resulted from the

TABLE 3

EFFECT OF SUPPLEMENTATION OF HAM'S F10 WITH "BLASTOKININ" ON GROWTH AND DEVELOPMENT OF 3-DAY RABBIT MORULAE INTO BLASTOCYSTS, COMPARED TO THAT OF SUPPLEMENTATION WITH OTHER PROTEINS

Protein Concentration (mg/ml)	Embryos Used (No.)	Cavitated in 24 Hr (No.)	Expanded in 24 Hr (No.)
	Control (Ham's F10)		
	24	12	(1?)
	"Blastokinin"		
0.05	10	10	0
0.10	10	10 (2?)	2 (1?)
0.20	14	13	10
0.40	10	10	6
0.60	11	6	4 (1?)
1.00	10	5	3
2.00	6	0	0
10.00	6	0	0
	Complete Uterine-Fluid Protein, 5 Days after Coitus		
0.30	10	5	0
0.50	14	12	11
1.00	12	4	2
3.00	8	5	0
10.00	11	8	0
	Maternal Serum Proteins		
0.30	10	8	0
3.00	12	10	(2?)

SOURCE: Krishnan and Daniel 1967. Copyright 1967 by the American Association for the Advancement of Science.

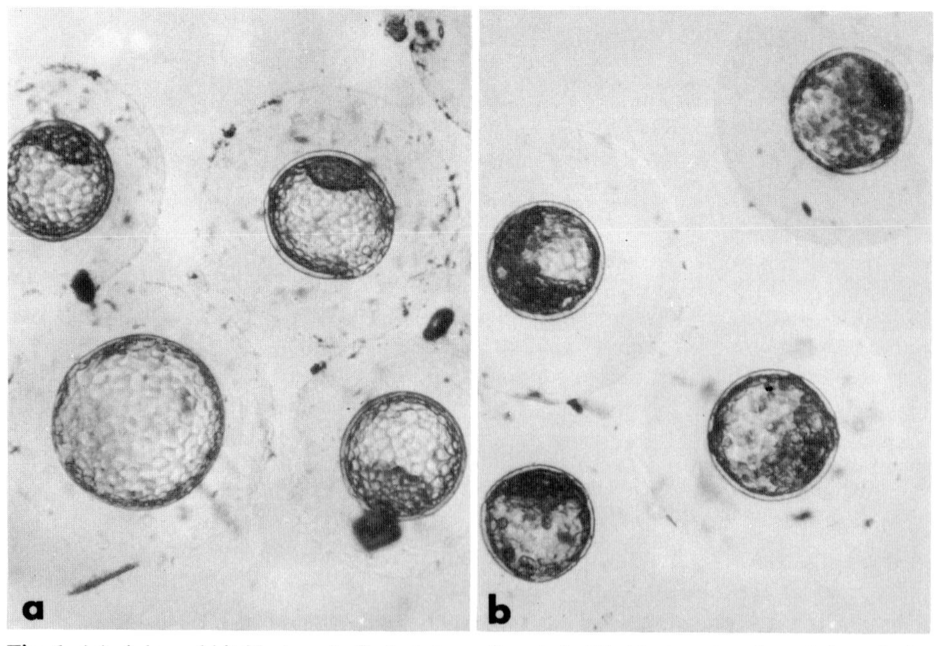

Fig. 6. (*a*) 4-day rabbit blastocysts flushed from the uteri; (*b*) blastocysts cultured from 3-day morulae in the presence of blastokinin *in vitro* for 24 hr. (From Krishnan and Daniel 1967. Copyright 1967 by the American Association for the Advancement of Science)

studies of Gulyas, Daniel, and Krishnan (1969) and Manes and Daniel (1969) in which radioactive precursors were incorporated into embryonic macromolecules *in vitro*. Gulyas, Daniel, and Krishnan (1969) precultured rabbit embryos for 2 hr in Ham's F10 medium (Ham 1963), which partially deprives them of their endogenous blastokinin by the absence of any macromolecular component, and then exposed them to a medium containing 0.2 mg/ml of blastokinin. Under these con-

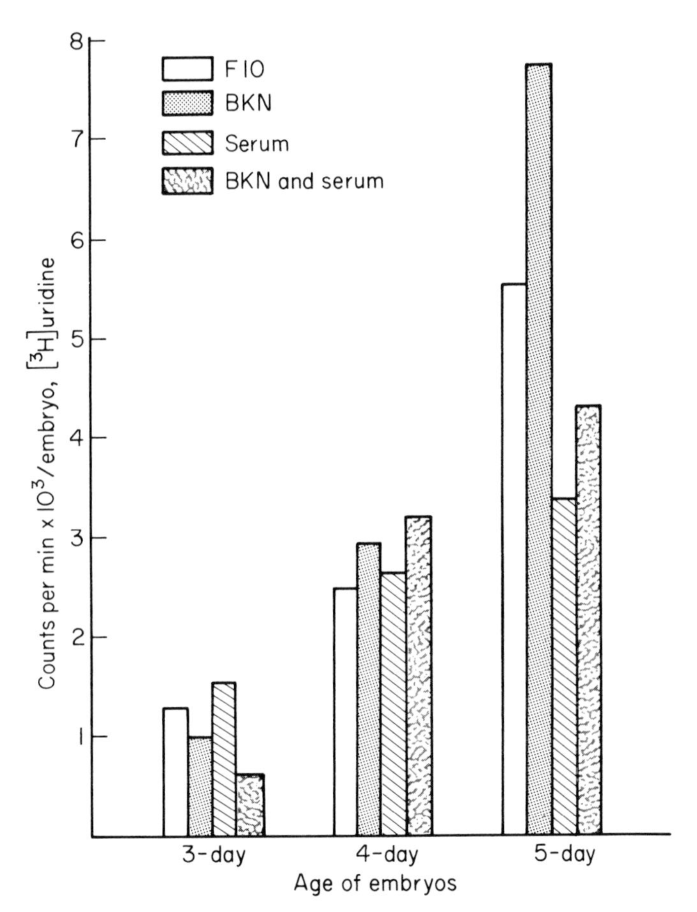

Fig. 7. Uridine-^{3}H incorporation into rabbit morulae and blastocysts in F10 medium alone and when supplemented with blastokinin or serum or both. Incorporation was measured by liquid scintillation counting. (From Gulyas, Daniel, and Krishnan 1969)

ditions uridine-^{3}H incorporation was significantly stimulated by blastokinin in 5-day rabbit blastocysts whereas rabbit serum inhibited the incorporation of uridine-^{3}H into the embryonic RNA (fig. 7). Thymidine-^{3}H incorporation, on the contrary, was not significantly enhanced by blastokinin; however, in serum containing media thymidine, uptake was somewhat greater in 4- and 5-day rabbit blastocysts than in the controls in the defined medium. The stimulatory effect of blastokinin on uridine uptake is directly reversed by actinomycin D, an inhibitor of DNA-dependent RNA synthesis, and also, at least partially, by puromycin. These observations seem to indicate that at least one (probably more) of the protein products (for example, RNA polymerase) normally synthesized by the embryos is essential for further RNA synthesis in the presence of blastokinin and that blastokinin possibly acts at the transcriptional level in the role of a derepressor.

In experiments using the same culture and technique Manes (personal communication) observed that these embryos when reexposed to blastokinin and ^{3}H-leucine for 2 hr incorporated the labeled amino acid into proteins at a significantly accelerated rate. Actinomycin D effectively blocked this effect, again suggesting that the action of blastokinin may be at the transcriptional level.

Manes and Daniel (1969) demonstrated that the transformation of rabbit morula into a blastocyst is associated with not only a 10-fold increase in the rate of protein synthesis per cell but also a qualitative change in protein synthesis, and proposed that the rate-limiting factor is genetic expression. Whether blastokinin is directly involved at this level is not established at this time.

VI. Relationship of Rabbit Blastokinin to Other Species

In a large number of mammals the blastocysts enter a diapause of variable duration, the phenomenon referred to as delayed implantation (for a review, see Enders 1963). The development of the blastocysts is slowed, and the blastocysts in some animals may remain in this state for as long as 9–10 months. These blastocysts have a mitotic index of 0 to 0.5, they do not incorporate ^{3}H-thymidine (Gulyas and Daniel 1969) or ^{35}S-methionine (Weitlauf and Greenwald 1965), they consume O_2 at rates similar to those blastocysts which do not delay (Gulyas and Daniel 1967), and they remain unattached to the uterus. Daniel (1968) suggested that the uterine secretions of animals with delayed implantation are either deficient in some essential components or contain an inhibitor which is the cause of the developmental arrest. Other hypotheses concerning delayed implantation will not be discussed here as they are beyond the scope of this chapter.

Acrylamide gel electrophoresis studies by Daniel (1968) revealed a protein with an average Rf value of 0.67 which was present only in the pregnant rabbit uterine secretions at days 4–7. A protein of the same mobility in acrylamide gel later proved to be identical to blastokinin, was observed in the 5- to 7-day rabbit blastocoelic fluid (Gulyas, unpublished observations). A macromolecule with similar Rf value was not detected in animals which exhibited delayed implantation, with the possible exception of armadillo and mink (Daniel 1968).

Sephadex G-200 gel filtration studies by Daniel and Krishnan (1969) confirmed and extended these earlier findings. "Blastokinin-like" protein was demonstrated in the 20-day mink and possibly in the 8-day dog uterine secretions. The term "blastokinin-like" protein refers to those uterine secretion components which are eluted at the same rate as rabbit blastokinin. The chemical identity of these blastokinin-like fractions and their biological role in blastocyst development have not yet been determined. Uterine secretion samples from armadillo, black bear, 13-day dog, fur seal, delaying and nondelaying rat failed to show a blastokinin-like fraction in Sephadex G-200 gel filtration studies (table 4).

Blastocysts from certain mammals with obligate diapause responded favorably when cultured in the presence of rabbit blastokinin (table 4). Armadillo, fur seal, and mink blastocysts showed varied expansion. As mentioned previously the limited expansion in some cases has been attributed to the rigid zona pellucida (Daniel and Krishnan 1969). The expansion of the blastocysts paralleled a severalfold increase of the mitotic index in the case of fur seal and mink and a smaller but significant increase in the armadillo blastocysts.

A somewhat different picture was presented by diapausing rat blastocysts taken from animals where a facultative delay in implantation had been induced by heavy lactation. Whereas these blastocysts showed no expansion when cultured in media containing rabbit blastokinin, the mitotic index of the blastocyst cells was significantly higher than in diapausing blastocysts examined immediately after ex-

TABLE 4

A TABLE SUMMARIZING THE DATA ON THE PRESENCE OF BLASTOKININ (BKN) OR "BKN-LIKE" PROTEINS IN DIFFERENT SPECIES OF MAMMALS AND THE EFFECT OF RABBIT BKN ON GROWTH AND MITOTIC INDEX OF BLASTOCYSTS FROM THESE ANIMALS DURING *in Vitro* CULTURE

ANIMAL		SAMPLING TIME; DAYS *Post Coitum*	PRESENCE OF BKN OR "BKN-LIKE" PROTEIN	STIMULATION OF BLASTOCYST GROWTH BY RABBIT BKN	AVERAGE MITOTIC INDEX	
					Without Rabbit BKN	With Rabbit BKN
Nondelaying	Rabbit	4	+	+	4.16	3.83
	Rat	5	−	+	1.22	1.98
Obligate delay	Armadillo	50–70	−	+	<1.0[a]	1.52
	Black bear	35–70	−			
	Fur seal	14–28	−	±	0.50	2.03
	Mink	20	±	±	0.32	2.15
Facultative delay	Rat (lactating)	9	−	−	0.20	1.15

SOURCE: Daniel and Krishnan 1969.
NOTE: + = Yes
− = No
± = Questionable
[a] From Enders 1962.

plantation. It is interesting to note, however, that diapausing rat blastocysts not only expanded in media containing either serum or the macromolecular components of uterine secretions taken from a 5-day pregnant rat, but also exhibited accelerated mitosis. Obviously rabbit blastokinin alone does not entirely satisfy the needs of diapausing rat blastocysts to expand and develop at the same time.

Blastocyst formation in the rabbit takes place about 3½ days *post coitum.* Blastulation and expansion of rabbit morulae is enhanced *in vitro* by blastokinin at the same stage of development, although the resulting blastocysts do not measure up to normal 4-day blastocysts. Maximum concentration of blastokinin in the rabbit uterine secretion occurs at 5 days *post coitum,* and it is at this stage of development that ^{3}H-uridine incorporation is the greatest in the presence of blastokinin. In those animals in which delayed implantation occurs blastocyst formation is initiated before the embryos enter diapause. If these animals have an essential protein which has a regulatory function in blastulation, it must appear before the onset of diapause. Although blastokinin has a function in early blastocyst development in the rabbit, it is possible also that a high concentration of it in the uterine secretions 5 days *post coitum* triggers a chain of reactions in the blastocysts which could eventually lead to their preparation for implantation.

VII. Other Factors with Blastokinin Activity

Recently Onuma, Maurer, and Foote (1968) reported that both bovine and rabbit serum facilitated blastulation and development *in vitro* of 2- and 4-cell rabbit eggs to the hatching blastocyst stage (table 5). Rabbit eggs were maintained under light paraffin oil (Brinster 1963) in whole rabbit or bovine serum supplemented with 0.05% glucose, 0.5 mg/ml streptomycin sulphate, and 100 units/ml of penicillin G. After 52 to 60 hr of culturing the early blastocysts were treated with a 1% pronase solution for 5 min at 22° C and reintroduced into fresh medium. A higher percentage of them hatched (table 5). Hatched or unhatched blastocysts after 6 days of

TABLE 5

DEVELOPMENT OF 2- AND 8-CELL RABBIT OVA TO BLASTOCYSTS *in Vitro*

Treatment	Total No. of Ova	Developed into Blastocysts	
		Number	%
Rabbit serum	40	29	72
Rabbit serum+glucose	38	28	74
Rabbit serum+pronase	35	17	49
Rabbit serum+glucose+pronase	35	21	60
Bovine serum	40	36	90
Bovine serum+glucose	45	36	80
Bovine serum+pronase	40	27	68
Bovine serum+glucose+pronase	35	30	86

SOURCE: By permission from H. Onuma, R. R. Maurer, and R. H. Foote, *J Reprod Fertil*, 16:491, 1968.

culture attained a size equivalent to normal 4-day blastocysts. Blastocysts cultured for 5 days in this manner failed to implant when transferred to synchronized pseudopregnant recipients.

It may be pointed out that rabbit blastocysts do not normally hatch prior to implantation, and that the failure of the blastocysts cultured *in vitro* to implant may possibly be due to their reduced rate of growth *in vitro*. If this is true and the blastocysts are otherwise normal, asynchronous transfer could give more promising results.

When one considers the complex nature of the media employed by Onuma, Maurer, and Foote (1968), their observations, though highly significant, provide little direct information as to the nature of the factor(s) involved in the process of blastulation and development. So far blastokinin has not been found in rabbit serum using gel filtration, acrylamide gel electrophoresis, and immunochemical techniques (Krishnan, unpublished observations). Nevertheless, the fact that blastulation and further development of the rabbit eggs *in vitro* is facilitated by whole rabbit serum and bovine serum seems to indicate that substances with blastokinin-like activity are present in these components.

Earlier in this chapter, we considered a uterine specific protein, uteroglobin, occurring in the endometrial secretions of pregnant rabbits. It has been shown to be present in the blastocoelic fluid from late blastocysts by electrophoretic and immunological methods (Beier 1968*a*). Uteroglobin has been postulated as involved in blastocyst development also.

VIII. Concluding Remarks

We concluded that blastokinin is possibly one of several factors through which the maternal environment regulates early mammalian development. That a relatively small uterine specific glycoprotein (molecular weight 27,000) facilitates blastulation and development of early rabbit embryos *in vitro* in a chemically defined medium is significant. The evidence presented in this discussion seems to indicate that blastokinin influences the embryos at the transcriptional or "posttranscriptional" level, possibly in the role of a derepressor. Even though there is some evidence to indicate that rabbit blastokinin may play a positive role in the development of diapausing blastocysts of other species of mammals, there seem to be qualitative and quantitative differences with respect to its effect. One might infer from these studies that blastokinin is at least partially species specific. It is not yet determined whether uterine secretions of other mammals contain blastokinin or blastokinin-like components.

The paucity of information on the biological activity of blastokinin itself and its specificity does not permit one to draw any far-reaching conclusions. Some of the areas where research is immediately needed to fill the gaps in our knowledge are:

1. Specific requirement for blastokinin for blastocyst formation and development *in vitro* in chemically defined media.
2. Permeability of both cleavage and blastocyst embryos to blastokinin.
3. A suitable bioassay for blastokinin in order to determine and compare the effectiveness of blastokinin and other factors with blastokinin activity in blastocyst formation and development in various mammalian species.
4. The role of blastokinin in differentiation of embryonic cells and blastocyst development.
5. The relationship, if any, between blastokinin or blastokinin-like components and implantation in general.

This is not to imply that other areas of blastokinin research are not important. The fact that pseudopregnancy induced by chorionic gonadotropin administration (Krishnan and Daniel 1967), or administration of estrogen-progesterone combinations to female rabbits during the preimplantation period (Beier 1968*b*), causes the occurrence of blastokinin or blastokinin-like components in uterine secretions suggests that the formation of these compounds is under ovarian hormonal control. Such observations will add to the understanding of the mechanism of delayed implantation, should it be established that "active" blastokinin or a blastokinin-like substance is the "limiting factor" in blastocyst diapause.

Little is known concerning the role of other components in uterine secretions on blastocyst development and implantation. Even though blastokinin may be a primary factor in blastocyst formation and development, the ultimate key to our knowledge of the mechanism of implantation may very well lie in our understanding of the interrelationships among the various components in uterine secretions during the preimplantation stages. Perhaps, for a start, one could examine in detail the relative proportions and absolute quantities of these various components and their turnover rates at various periods during this stage.

Acknowledgments

This work was supported in part by National Institutes of Health postdoctoral traineeship GM 00136 (B. J. G.) and faculty improvement grant 21-2128-774 from the Graduate School of Colorado State University.

References

Adams, C. E. 1965. The influence of maternal environment on preimplantation stages of pregnancy in the rabbit. In *Preimplantation stages of pregnancy,* CIBA Foundation Symposium, ed. G. E. W. Wolstenholme and M. O'Connor, p. 345. Boston: Little, Brown and Company.

Beier, H. M. 1968*a*. Biochemisch-entwicklungsphysiologische Untersuchungen am Proteinmilieu für die Blastozystenentwicklung des Kaninchens (*Oryctolagus cuniculus*). *Zool Jb Anat Bd* 85:72.

———. 1968*b*. Uteroglobin: A hormone-sensitive endometrial protein involved in blastocyst development. *Biochim Biophys Acta* 160:289.

Brinster, R. L. 1963. A method for *in vitro* cultivation of mouse ova from two-cell to blastocyst. *Exp Cell Res* 32:205.

Daniel, J. C., Jr. 1968. Comparison of electrophoretic patterns of uterine fluid from rabbits and mammals having delayed implantation. *Comp Biochem Physiol* 24: 297.

Daniel, J. C., Jr., and Krishnan, R. S. 1969. Studies on the relationship between uterine fluid components and the diapausing state of blastocysts from mammals having delayed implantation. *J Exp Zool* 172:267.

Enders, A. C. 1962. The structure of the armadillo blastocyst. *J Anat* 96:39.

———., ed. 1963. *Delayed implantation.* Chicago: University of Chicago Press.

Gulyas, B. J., and Daniel, J. C., Jr. 1967. Oxygen consumption in diapausing blastocysts. *J Cell Physiol* 70:33.

———. 1969. Incorporation of labeled nucleic acid and protein precursors by diapausing and nondiapausing blastocysts. *Biol Reprod* 1:11.

Gulyas, B. J.; Daniel, J. C., Jr.; and Krishnan, R. S. 1969. Incorporation of labelled nucleosides *in vitro* by rabbit and mink blastocysts in the presence of blastokinin or serum. *J Reprod Fertil* 20:255.

Ham, R. G. 1963. An improved nutrient solution for diploid Chinese hamster and human cell lines. *Exp Cell Res* 29:515.

Hammond, J. 1928. Die Kontrolle der Fruchtbarkeit bei Tieren. *Zuechtungskunde* 3:523.

Kirby, D. R. S. 1965. The role of the uterus in the early stages of mouse development. In *Preimplantation stages of pregnancy,* CIBA Foundation Symposium, ed.

G. E. W. Wolstenholme and M. O'Connor, p. 325. Boston: Little, Brown and Company.

Krishnan, R. S., and Daniel, J. C., Jr. 1967. "Blastokinin": Inducer and regulator of blastocyst development in the rabbit uterus. *Science* 158:490.

———. 1968. Composition of "blastokinin" from rabbit uterus. *Biochim Biophys Acta* 168:579.

Manes, C., and Daniel J. C., Jr. 1969. Quantitative and qualitative aspects of protein synthesis in the preimplantation rabbit embryo. *Exp Cell Res* 55:261.

Onuma, H.; Maurer, R. R.; and Foote, R. H. 1968. *In-vitro* culture of rabbit ova from early cleavage stages to the blastocyst stage. *J Reprod Fertil* 16:491.

Schwick, H. G. 1965. Chemisch-entwicklungsphysiologische Beziehungen von Uterus zu Blastozyste des Kaninchens (*Oryctolagus cuniculus*). *Arch Entw Mech Org* 156:283.

Weitlauf, H. M., and Greenwald, G. S. 1965. A comparison of ^{35}S methionine incorporation by the blastocysts of normal and delayed implanting mice. *J Reprod Fertil* 10:203.

15

Influence of Ovarian Hormones on the Incorporation of Amino Acids by Blastocysts *in Vivo*

Harry M. Weitlauf

Departments of Anatomy and Obstetrics and Gynecology
University of Kansas Medical Center, Kansas City

Labeled amino acids and nucleosides have been used extensively to study patterns of protein and nucleic acid synthesis in developing embryos. The justification for this approach to embryology lies in the generally accepted concept that differentiation is the expression of specific genetic information through the production of particular proteins. The incorporation of the labeled precursors into new proteins or nucleic acids offers a means by which chemical events in development can be observed and the effect of experimental manipulations determined.

A large body of information has been collected from experiments on developing marine invertebrates and amphibians (see reviews by Monroy 1965 and Deuchar 1965). However radioactive precursors have not been used widely to study differentiation in mammalian embryos; progress in this realm has been correspondingly less striking. The complexity of reproduction in the viviparous species presents several difficulties that discourage the study of their development: (1) the eggs are available in comparatively small numbers even with the use of superovulation techniques; (2) the embryos may be maintained *in vitro* under rigidly controlled conditions only in the preimplantation period and then in only a few species; (3) *in vitro* culture can be done only at the expense of placing the embryos in a milieu manifestly different from their natural environment; (4) there are comparatively few morphological changes in the preimplantation embryo which might be correlated with biochemical changes; and (5) the effect of maternal amino acid or nucleoside pools on the uptake of labeled compounds by the embryos is largely unknown.

Despite the difficulties, experiments with preimplantation mammalian embryos have been done. Tagged precursors have been given both *in vivo* and *in vitro* to preimplantation embryos, and the embryos have been evaluated for incorporation of the label by radioautographic or liquid scintillation-counting techniques. Two general approaches have been taken: (1) administration of labeled compounds during the course of normal development; and (2) administration of the tracers when development has been altered by experimental procedures.

I. Normal Pregnancy

The feasibility of using tagged amino acids to study protein synthesis by mammalian eggs *in vivo* was demonstrated by the early experiments with ^{35}S-methionine (Lin 1956) and ^{14}C-glycine (Edwards and Sirlin 1956). The tagged amino acids were injected intraperitoneally and ovulation was induced several hours later. The reproductive tracts were removed and prepared for radioautography. It was shown that labeled amino acids were taken up by the developing mammalian eggs. Lin (1956) labeled follicular eggs with ^{35}S-methionine and transferred them before fertilization to recently mated recipients; he found that the radioactive eggs could be fertilized and would develop into normal fetuses which went to term. Thus, the incorporation of detectable amounts of radioactive amino acid is not necessarily harmful to the eggs and may be compatible with normal development.

Edwards and Sirlin (1956) instilled ^{14}C-glycine directly into the uteri of mice at the time of mating. The animals were killed at progressively longer intervals thereafter and the embryos were examined for radioactivity. Blastocysts were more heavily labeled than pronuclear or 2-cell eggs and it was concluded that the amount of amino acid incorporated was related to the length of time between injection and autopsy (i.e., several days longer for the blastocysts). Greenwald and Everett (1959) reexamined this problem by giving single injections of ^{35}S-methionine to pregnant mice on various days of the preimplantation period. They, too, found that the blastocysts incorporated more labeled amino acid than did tubal eggs. Their experimental design ruled out the time factor since all animals were killed 6 hr after the injection of tracer. They suggested that the "cellular processes" in uterine blastocysts may be substantially different from those of tubal eggs.

The complex nature of the preimplantation period makes it particularly difficult to assign responsibility for observed differences in amino acid or nucleoside incorporation by the embryos to a specific cause. Any of several aspects of the preimplantation period could be responsible: (1) the eggs themselves may undergo maturational changes; (2) the eggs move from the oviduct to the uterus—observed differences may be simply the result of "environmental" changes; (3) the hormonal status of the mother changes during the preimplantation period from one of progesterone domination to one of estrogen and progesterone interaction, and this change may influence either or both of the above. If precursors are not incorporated at any stage of development, it is necessary to determine whether this is an artifact caused by the interference of endogenous pools of unlabeled compound. In designing experiments and interpreting experimental results these aspects of the preimplantation period must be taken into account.

The technique of egg transfer has been used to explore some of these aspects. Embryos have been placed in the uteri of recipients under various experimental conditions, and the effect on ^{35}S-methionine incorporation has been determined. It is my purpose in this chapter to review and discuss the experimental work pertinent to this aspect of development during the cleavage and blastocyst stages and during delayed implantation.

II. Incorporation of ^{35}S-Methionine by Blastocysts in the Oviduct

The observed difference in ^{35}S-methionine incorporation between tubal eggs and uterine blastocysts reported by Greenwald and Everett (1959) and Weitlauf and Greenwald (1965) reflects a metabolic difference in the embryos only if the labeled amino acid is secreted into the oviduct and the uterus at the appropriate time and is thus available to the eggs. If a large amino acid pool exists in the oviduct, the relatively small amount of labeled amino acid might be excluded from the vicinity of the egg and a bona fide synthetic reaction could be obscured. This question has

TABLE 1

COMPARISON OF ^{35}S-METHIONINE INCORPORATION BY BLASTOCYSTS AND 2-CELL EGGS

DOSE OF ^{35}S-METHIONINE AND ROUTE OF INJECTION	NO. OF TRANS-FERS	BLASTOCYSTS			2-CELL EGGS[b]	
		No. Trans-ferred	No. Re-covered	Reactivity[a] of Blastocysts (No.)	No. Re-covered	Reactivity[a] of Eggs (No.)
Group 1, 30 μc (subcutaneously)	12	77	48	+(48)	50	−(50)
Group 2, 30 μc (intraperitoneally)	8	62	40	+(40)	59	−(50) ±(9)
Group 3, 60 μc (intraperitoneally)	17	117	87	+(87)	199	−(181) ±(18)

SOURCE: By permission from H. M. Weitlauf and G. S. Greenwald, *Anat Rec* 159:249, 1967.
[a] See figures 1–3.
[b] 2-cell eggs from right and left oviducts were similar in reactivity and the data are therefore combined.

been raised by Mintz (1965) and suggested as a possible explanation for the observed differences between tubal and uterine eggs *in vivo*.

In order to determine whether ^{35}S-methionine is available within the oviducts Weitlauf and Greenwald (1967) recovered uterine blastocysts on day 5 of pregnancy (day 1 = the day of vaginal plug formation) and transferred them to the oviducts of mice on day 2 of pregnancy. The transferred blastocysts and the recipient's own 2-cell eggs were exposed to the same environment. The recipients were injected either subcutaneously or intraperitoneally with ^{35}S-methionine immediately after the transfer procedure and were killed 6 hr later. The blastocysts and 2-cell eggs were recovered and prepared for radioautography and scored as shown in figures 1–3. It was found that the blastocysts were labeled heavily whereas the 2-cell eggs incorporated little if any ^{35}S-methionine (table 1; figures 4, 5). It was concluded that failure of tubal eggs to incorporate ^{35}S-methionine whereas the transferred blastocysts became labeled demonstrated a metabolic difference between eggs at the two stages; that is, protein synthesis occurred at a relatively high level in the blastocysts as compared to the tubal eggs. This interpretation is supported by cytochemical data (Alfert 1950) and more recent biochemical data (Brinster

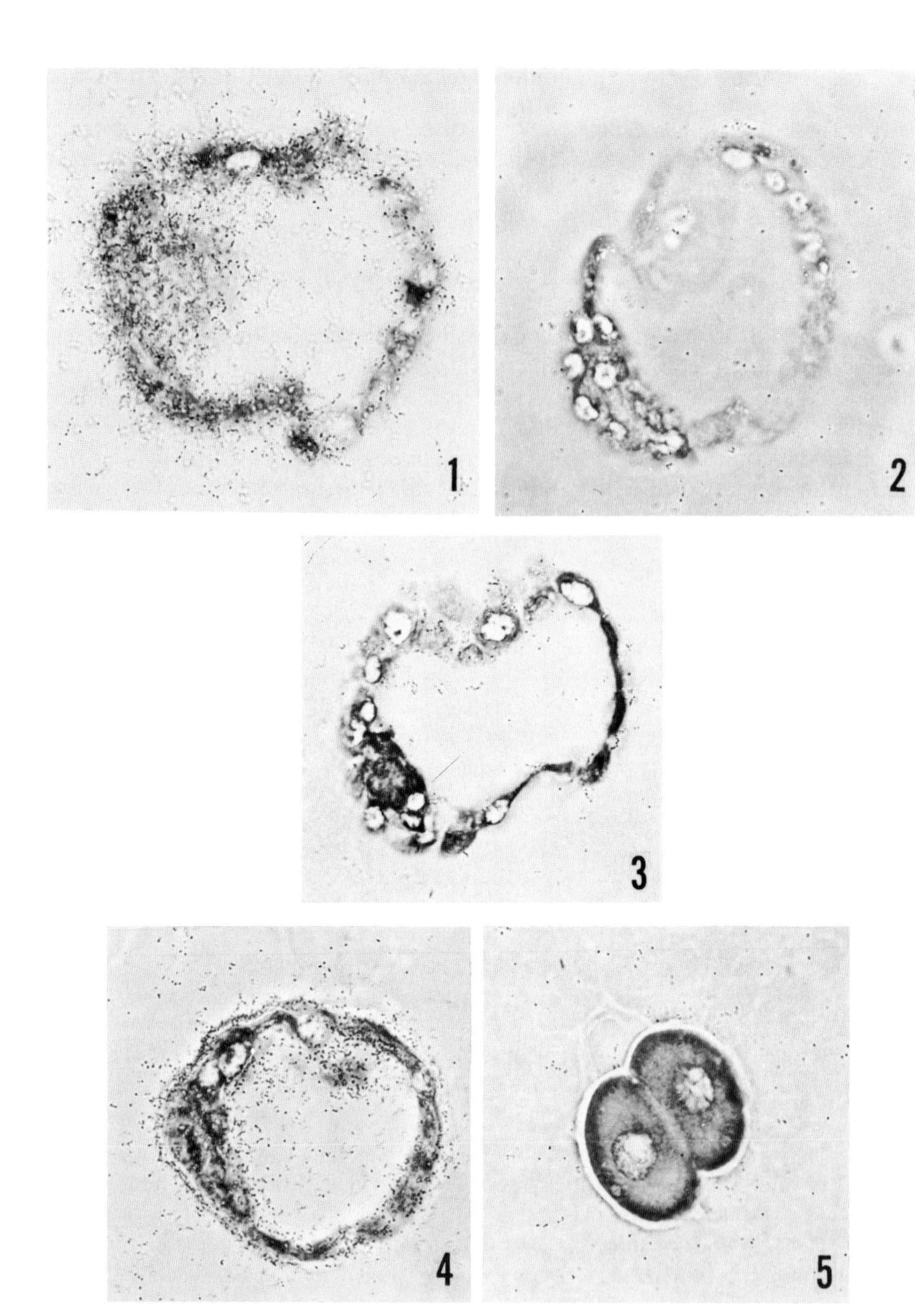

Fig. 1. Labeled blastocyst. The reduction over the inner cell mass is more than twice background. Designated + in tables 1–7. ×618.

Fig. 2. Unlabeled blastocyst. The reduction over the inner cell mass is equal to background. Designated — in tables 1–7. ×618.

Fig. 3. Lightly labeled blastocysts. The reduction over the inner cell mass is greater than background but not twice background. Designated ± in tables 1–7. ×618.

Fig. 4. Five-day-old blastocyst transferred into the same oviduct which contained the 2-cell egg in figure 5. The reduction over the blastocyst is +. ×618.

Fig. 5. Two-cell egg recovered from a female on the 2d day of pregnancy after 30μc ^{35}S-methionine was given intraperitoneally. The amount of reduction over the egg is —. ×618.

1967) showing that the protein content decreases (by 25%/embryo) between day 1 and day 4 of development and increases with formation of the blastocyst.

III. Delayed Implantation

The normal developmental sequence is arrested in mice and rats during delayed implantation; mitotic activity and blastocyst growth are decreased (Baevsky 1963; Yasukawa and Meyer 1966); and the blastocysts appear to be dormant. Incorporation of ^{35}S-methionine by blastocysts from normal and delayed implanting pregnancies was compared (Weitlauf and Greenwald 1965). Mice were injected with ^{35}S-methionine on day 5 of pregnancy and the blastocysts were recovered and prepared for radioautography. Mice mated at the postpartum estrus and suckling a standard litter of 10 young (i.e., conditions for delayed implantation) were injected with ^{35}S-methionine on day 5 or day 9 of pregnancy. The blastocysts were

TABLE 2

INCORPORATION OF ^{35}S-METHIONINE BY FERTILIZED MOUSE OVA

Condition of Mother	Day of Recovery of Eggs	No. Animals	Reactivity of Ova[a] (No.)
Control: normal pregnancy	Day 1	4	−(35)
	Day 2	4	−(20)
	Day 3	4	−(25)
	Day 4	4	−(21)
	Day 5	4	+(30)
Delayed implanting pregnancy	Day 5	6	−(42)
	Day 9	5	−(34)
	Young removed day 8, killed on day 9	7	+(30)

SOURCE: Modified by permission from H. M. Weitlauf and G. S. Greenwald, *J Reprod Fertil* 10:203, 1965.
[a] See figures 1–3.

recovered 6 hr after the injection of tracer and prepared for radioautography. The blastocysts from the normal pregnancies incorporated the ^{35}S-methionine; those from delayed implanting pregnancies did not (table 2). The delayed implanting eggs could be stimulated to incorporate ^{35}S-methionine on day 9 by removing the suckling young or injecting estrogen on day 8. Similar results have been reported in the rat (Prasad, Dass, and Mohla 1968). It was concluded that the synthesis of protein in the eggs is controlled by the same maternal factors that delay and eventually induce nidation. The results of those experiments demonstrated that neither formation of the blastocysts per se nor shedding of the zona pellucida is sufficient to initiate protein synthesis in mouse embryos.

IV. Influence of Ovarian Hormones

Delayed implanting blastocysts do not incorporate labeled amino acids unless the suckling young are removed or estrogen is injected. In either case the blastocysts are exposed to estrogen and progesterone before protein synthesis begins. In normal pregnancies the blastocysts first incorporate ^{35}S-methionine early on day 5 (i.e.,

0200–0600 hr; table 3). Thus the initiation of protein synthesis closely follows the release of estrogen as reported by Finn (1965).

Although estrogen is implicated as the hormone responsible for stimulating amino acid incorporation by the blastocysts, the possibility that the stimulus is actually provided by a combination of estrogen and progesterone is not ruled out; in each of the above mentioned experiments estrogen was provided in the presence of progesterone. Several experiments were carried out with ^{35}S-methionine in ovariectomized mice in an attempt to determine the effect of ovarian hormones on amino acid incorporation by blastocysts (Weitlauf and Greenwald 1968*b*).

Preliminary studies were undertaken to determine whether preimplantation mouse embryos would survive in the absence of ovarian hormones. This was necessary because the viability of blastocysts retained in ovariectomized and untreated females was disputed. Nutting and Meyer (1963) had suggested that blastocysts do not survive in the ovariectomized rat unless progesterone is provided. On the other

TABLE 3

TIME OF ^{35}S-METHIONINE INCORPORATION BY BLASTOCYSTS OF INTACT MICE

Group	Injection of ^{35}S-Methionine	(Hours *p.c.*)	Blastocysts Recovered (Hours *p.c.*)	Number of Animals	Reactivity of Blastocysts[a] (Number)
1	0900 hours day 4	76	78	3	−(21)
2	1400 hours day 4	81	83	5	−(47)±(21)
3	2200 hours day 4	89	91	7	−(65)±(17)
4	0200 hours day 5	93	95	5	−(21)±(49)+(9)
5	0600 hours day 5	97	99	5	+(40)

SOURCE: By permission from H. M. Weitlauf and G. S. Greenwald, *J Exp Zool* 169:463, 1968.
[a] See figures 1–3.

hand Smithberg and Runner (1960) had found that blastocysts do survive in immature mice in the absence of ovarian hormones. In those experiments, however, normal young were not obtained after hormones were injected to induce implantation and thus the quality of the blastocysts was open to question. Therefore, mature mice were mated and bilaterally ovariectomized on day 4 of pregnancy; hormone replacement was not given (Weitlauf and Greenwald 1968*a*). The blastocysts did not implant and could be recovered after various periods of delay and transferred to pseudopregnant recipients. These transferred blastocysts developed into normal young. Thus although mouse eggs do not implant, they do remain viable in the absence of ovarian hormones. This form of experimental delayed implantation offers a convenient model with which to test the effects of ovarian hormones on amino acid incorporation by blastocysts.

The first step was to determine whether the blastocysts would incorporate ^{35}S-methionine in the absence of ovarian hormones. Mice were ovariectomized on day 4 of pregnancy and given no further treatment until day 10, when 30 μc of ^{35}S-thionine was injected intraperitoneally. The blastocysts were recovered 2 hr later and prepared for radioautography; they did not incorporate the tracer (table 4, group 1). It appears then that in experimentally delayed implantation as in delayed implantation of lactation the blastocysts are dormant with respect to ^{35}S-methionine incorporation.

The second step was to test the effect of both estrogen and progesterone on ^{35}S-methionine incorporation by the eggs. Progesterone (2.0 mg/day) for 2 days followed by daily injections of progesterone (2.0 mg) plus estradiol (0.025 μg) induces implantation and maintains pregnancy in ovariectomized mice (Weitlauf unpublished results). Pregnant females were ovariectomized on day 4 and started on the hormone regimen. ^{35}S-methionine was injected on day 10 (i.e., 24 hr after

TABLE 4

EFFECT OF ESTRADIOL AND PROGESTERONE ON ^{35}S-METHIONINE INCORPORATION BY BLASTOCYSTS IN OVARIECTOMIZED MICE

GROUP	HORMONE TREATMENT		NUMBER OF ANIMALS	REACTIVITY OF BLASTOCYSTS[a] (NUMBER)
	Progesterone 2.0 mg/Day	Estradiol-17-β 0.025 μg/Day		
1	None	None	11	−(61)
2	Oil	Oil	5	−(50)
3	Days 7–10	Days 9–10	14	+(57)
4	Days 7–10	None	9	−(58)
5	None	Days 9–10	7	−(50)

SOURCE: By permission from H. M. Weitlauf and G. S. Greenwald, *J Exp Zool* 169:463, 1968.
[a] See figures 1–3.

TABLE 5

INCORPORATION OF ^{35}S-METHIONINE BY BLASTOCYSTS (DAY 5) TRANSFERRED FROM INTACT DONORS INTO THE UTERI OF OVARIECTOMIZED RECIPIENTS

GROUP	HORMONE TREATMENT OF RECIPIENTS		NUMBER OF RE-CIPIENTS	REACTIVITY OF BLASTOCYSTS[a]
	Progesterone 2.0 mg/Day	Estradiol-17-β 0.025 μg/Day		
1	None	None	9	+(37) ± (11)
2	Days 7–10	None	5	+(21) ± (2)
3	None	Days 9–10	7	+(20) ± (4)

SOURCE: By permission from H. M. Weitlauf and G. S. Greenwald, *J Exp Zool* 169:463, 1968.
[a] See figures 1–3.

the first injection of estrogen), and blastocysts were recovered 2 hr later and prepared for radioautography. The blastocysts did incorporate the tracer (table 4, group 3). Thus following the injection of both ovarian hormones the blastocysts were stimulated to begin protein synthesis. In a timed study it was shown that activation occurs approximately 12–14 hr after the injection of estrogen (table 6).

The next step was to test estrogen and progesterone individually. Following only the progesterone portion of the hormone treatment the blastocysts did not incorporate ^{35}S-methionine (table 4, group 4). This was expected since progesterone is present during the delayed implantation of lactation and the eggs are inactive. To our surprise, the blastocysts also failed to incorporate ^{35}S-methionine after receiving only the estrogen portion of the treatment (table 4, group 5). It was concluded that both estrogen and progesterone are necessary to stimulate protein synthesis by eggs *in utero.*

Because of the extensive effects of ovarian hormones on various aspects of uterine metabolism, including amino acid transport (Roskoski and Steiner 1967), it seemed possible that under certain conditions (i.e., no hormones; progesterone alone; or estrogen alone) the uterus might selectively restrict amino acids. This has been suggested as a possible explanation for the mechanism of delayed implantation (Gwatkin 1966). If true, then the failure of the delayed implanting blastocysts to incorporate amino acids would be simply a passive effect of uterine metabolism. To test this possibility normal 5-day blastocysts were transferred into the uteri of pseudopregnant, ovariectomized females that were either (1) untreated; (2)

TABLE 6

TEMPORAL RELATIONSHIP BETWEEN ESTRADIOL AND THE INCORPORATION OF ^{35}S-METHIONINE BY BLASTOCYSTS IN OVARIECTOMIZED, PROGESTERONE-TREATED MICE

Group	Number of Hours Post Estradiol Injection[a]		Number of Animals	Reactivity[b] of Blastocysts (Number)
	Methionine Injected	Blastocysts Recovered		
1	0	2	3	−(22) ±(1)
2	6	8	4	−(8) ±(10)
3	12	14	3	+(32)
4	18	20	4	+(34)
5	24	26	14	+(50)

SOURCE: By permission from H. M. Weitlauf and G. S. Greenwald, *J Exp Zool* 169:463, 1968.
[a] Injection of 2 mg progesterone + 0.025 μg estradiol on day 9.
[b] See figures 1–3.

given progesterone only on days 7–10; or (3) given estrogen only on days 9–10. The blastocysts were transferred on day 10 and ^{35}S-methionine was injected immediately thereafter. The results are shown in table 5; blastocysts incorporated ^{35}S-methionine in each type of uterus. Hence the failure of delayed implanting blastocysts to incorporate the amino acid is not due to its restriction by the uterus. It appears that changes within the eggs themselves are involved in regulating the level of amino acid incorporation.

In view of those observations it was of interest to determine whether the embryos need only an activating stimulus or require continued hormonal support to sustain protein synthesis. Do the hormones trigger a series of reactions in the eggs which then proceed autonomously to complete the embryos' preparations for implantation (i.e., similar to the events at activation of the egg after sperm penetration); or are the hormones required to sustain the protein synthesis and the embryos' preparations for implantation (i.e., as in the maintenance of a typical hormonal target organ)?

To answer these questions blastocysts were removed from donors on day 5 of normal pregnancy and transferred to the uteri of ovariectomized females on day 5 of pseudopregnancy (Weitlauf 1969). Thus normally active eggs were placed in uteri that were under hormonal conditions of delayed implantation. Progesterone injections (2.0 mg) were given on days 6 and 7, and as might be expected the blastocysts did not implant, but remained free in the uterine lumina (see fig. 6). After a delay of 4 days the eggs were induced to implant by injecting estrogen and progesterone and carried to term (same regimen described earlier); they developed into normal fetuses. Therefore the transfer procedures and subsequent period of delay were not harmful to the eggs.

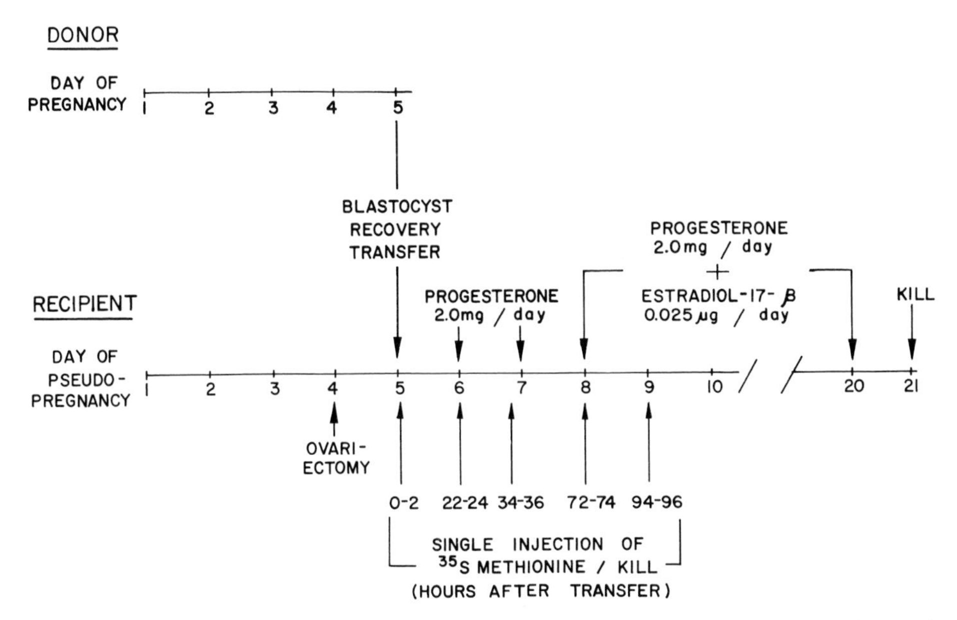

Fig. 6. Schematic representation of procedures used to obtain active blastocysts and their transfer to ovariectomized recipients. The recipients were either injected with estrogen and progesterone and allowed to go to term, or given a single injection of ^{35}S-methionine or ^{35}S-methionine in addition to hormones and killed as indicated at various intervals after the transfer procedure.

To determine the effect of such treatment on protein synthesis in the eggs the experiment was repeated and the recipients were divided into several groups and given hormones according to the regimen shown in figure 6. ^{35}S-methionine was injected at progressively longer intervals after the transfer procedure (see schedule in table 7). The eggs were recovered 2 hr after injection of the tracer in each case. Blastocysts recovered from recipients injected immediately after the transfer (0–2 hr) incorporated tracer (group 2). On the other hand the response by eggs recovered 24 hr after the transfer procedure was mixed; a few eggs incorporated the ^{35}S-methionine but the remainder did not (group 3). Blastocysts were no longer incorporating the amino acid after remaining 34–36 hr in the ovariectomized recipients (group 4); similar results were obtained at 72 hr (group 5).

That this changing level of methionine incorporation represented a change in the blastocysts and not simply one of uterine secretion of the tracer was shown by transferring active blastocysts to ovariectomized recipients on day 8 (i.e., into uteri that were comparable to the 72-hr group). ^{35}S-methionine was injected immediately thereafter; the eggs incorporated the tracer (group 6). Therefore ^{35}S-methionine reaches the uterine lumen and again the failure of blastocysts to incorporate it represents an embryonic change.

That the eggs were still viable had been evidenced earlier by the pregnancy's going to term; that protein synthesis was reactivated before implantation was shown by giving injections of estrogen and progesterone as for the induction of implantation. Twenty-four hr later ^{35}S-methionine was given and the eggs were recovered; they were incorporating the tracer (table 7, group 7).

These results suggest that protein synthesis in the blastocysts, initially at a high

TABLE 7

INCORPORATION OF ^{35}S-METHIONINE BY BLASTOCYSTS TRANSFERRED TO THE UTERI OF OVARIECTOMIZED MICE

Group	Number of Hours after Blastocyst Transfer		Number of Animals	Reactivity[a] of Blastocysts (Number)
	^{35}S-Methionine Injected	Blastocysts Recovered		
1	(Nontransferred Controls)		7	+(63) ±(1) −(4)
2	0	2[b]	10	+(53) ±(1) −(1)
3	22	24	9	+(13) ±(12) −(35)
4	34	36	13	+(0) ±(8) −(65)
5	72	74	17	+(0) ±(1) −(55)
6[c]	0	2	8	+(28) ±(0) −(4)
7[d]	94	96	12	+(43) ±(2) −(0)

[a] See figures 1–3.
[b] See text figure 6.
[c] Active eggs transferred to ovariectomized recipients on day 8 and immediately tested for ^{35}S-methionine incorporation.
[d] Group 7 injected with estrogen and progesterone at 72 and 94 hr (days 8 and 9; text figure 6).

level, either dropped to low levels or was "switched off" within 24–36 hr after the eggs were placed in hormonally unconditioned uteri.

V. Discussion

Fawcett, Wislocki, and Waldo (1947) suggested that neither the embryo nor the uterus is "chiefly" responsible for the events at implantation but that both act in concert. This view was restated a decade later as a guiding principle for research in reproduction in the opening paper of the 1957 Conference on Implantation (Eckstein, Schelesnyak, and Amoroso 1959): "There is a mutual relationship between the ovum and the endometrium, neither being solely responsible nor entirely active or passive." In the decade since that conference a large body of literature has appeared attempting to define the respective roles of the uterus and the blastocyst in accomplishing implantation. The uterus has received the greatest emphasis to date. A possible reason for this predilection is the response of the endometrium to the blastocyst, the so-called decidual reaction. This response can be elicited by experimental means and appears remarkably similar to that induced by the embryo during actual implantation (reviewed by De Feo 1967). Such a model system makes it unnecessary for the embryo to be present and allows great freedom in designing experiments to test the control of endometrial "receptivity." There are arguments for and against the validity of using an artificially induced decidual response as a substitute for that evoked by the implanting embryo, but overall it must be recognized as having provided an extremely fruitful approach to the study of the physiology and biochemistry of the endometrial response to the implanting embryo.

Although the physiological studies have provided much information about the preimplanting embryos, they have been confined to testing the effect of experimental procedures on blastocysts by either inducing implantation *in situ* with exogenous hormones or transferring the eggs to timed uteri which are at the proper stage for implantation (see reviews by Mayer 1963; Psychoyos 1966; Dickmann 1969). This limitation has hampered study of regulatory mechanisms in development and maturation of the blastocyst. The experiments of Noyes and Dickmann (1960) and Dickmann and Noyes (1960) are notable exceptions. Chemical determinations of preimplanted embryos and their environmental fluids have also provided important information about the embryo (Lowenstein and Cohen 1964; Brinster 1967, 1968; Stern and Biggers 1968). However such studies are limited by technical difficulties to relatively few substances.

If the burst of protein synthesis seen in the preimplanting embryo represents a requirement for implantation, their utilization of amino acids and nucleosides opens the way to study the role of the embryo in implantation without the necessity of placing the embryos in suitable hosts for final determination of the effects of test treatments.

The results of experiments on incorporation of amino acids by mouse and rat blastocysts *in utero* suggest that estrogen and progesterone are necessary to stimulate the eggs to incorporate amino acids. They demonstrate also that the difference in amino acid incorporation by delayed implanting blastocysts when compared with those from normal pregnancies is controlled by a hormonally induced, reversible change within the eggs themselves and not by the level of amino acid secreted by the uterus. The experiments do not differentiate between a direct hormonal effect on the eggs as suggested by Smith (1968) and an indirect effect induced by the uterus.

In vitro studies of the incorporation of amino acid or nucleoside have been done with eggs recovered from intact females which were cultured immediately in a medium containing one or more radioactive compounds; in some cases this was followed by a period of culture in nonradioactive medium. Under such *in vitro* conditions the blastocysts are not subjected to direct hormonal influences; nevertheless amino acids are incorporated (Mintz 1965; Monesi and Salfi 1968), as are nucleosides (Mintz 1965; Izquierdo and Roblero 1965; Fridhandler and Palmer 1968; Wilson and Smith 1968). The results of *in vivo* experiments, discussed above, demonstrate that once the blastocysts have been stimulated by estrogen and progesterone they remain active for several hours. In the *in vitro* experiments blastocysts were tested within a few hours of their recovery from intact females. The results obtained do not rule out the possibility that a direct effect of ovarian hormones on the eggs had already occurred before their recovery.

Blastocysts "implant" and undergo some "differentiation" in extrauterine sites in the absence of gonadal or pituitary hormones (Runner 1947; Fawcett, Wislocki, and Waldo 1947). That eggs are capable of developing beyond the blastocyst stage in the absence of hormones in extrauterine sites but are repressed when left in the uterus implies that the uterus has a repressive effect on the eggs in the absence of estrogen and progesterone.

The critical experiment to test this hypothesis appears to have been done by Psychoyos and Bitton-Casimiri (1969). In their experiments rat blastocysts from normal pregnancies were found to incorporate ^{3}H-uridine *in vitro.* In contrast, blastocysts recovered from delayed pregnancies did not incorporate the nucleoside until they had been *in vitro* for several hours. It seems clear that delayed blastocysts began incorporating ^{3}H-uridine without hormonal stimulation. They refer to a supernatant fraction from uteri of ovariectomized, progesterone-treated rats (i.e., conditions for delayed implantation) that will inhibit the incorporation of ^{3}H-uridine by blastocysts from normal pregnancy. If this fraction has not killed the blastocysts and if it disappears when estrogen and progesterone are given, it may be responsible for repressing blastocysts *in utero* during delayed implantation.

It seems possible, therefore, that regulation of preimplanting mouse and rat embryos may be dependent upon some kind of repression and derepression. The blastocysts may simply tend to differentiate autonomously and to complete their preparations for implantation unless they are inhibited from doing so by a uterus that is deprived of hormonal stimulation.

Further study with radioisotopes at the level of genetic transcription and translation should provide a clearer insight into the development of the preimplantation embryo and the regulation of its maturation.

Acknowledgments

This work was supported by grants from the Atomic Energy Commission (report C00-1801-5), the Ford Foundation, and the National Institutes of Health.

References

Alfert, M. 1950. A cytochemical study of oogenesis and cleavage in the mouse. *J Cell Comp Physiol* 36:381.

Baevsky, U. B. 1963. The effect of embryonic diapause on the nucleic and mitotic activity of mink and rat blastocysts. In *Delayed implantation,* ed. A. C. Enders, p. 141. Chicago: University of Chicago Press.

Brinster, R. L. 1967. Protein content of the mouse embryo during the first five days of development. *J Reprod Fertil* 13:413.

———. 1968. Lactate dehydrogenase activity in the oocytes of mammals. *J Reprod Fertil* 17:139.

De Feo, V. J. 1967. Decidualization. In *Cellular biology of the uterus,* ed. R. M. Wynn, p. 191. New York: Appleton-Century-Crofts.

Deuchar, R. M. 1965. Biochemical patterns in early developmental stages of vertebrates. In *The biochemistry of animal development,* ed. Rudolf Weber. New York: Academic Press.

Dickmann, Z. 1970. Egg transfer. In *Methods in mammalian embryology,* ed. J. C. Daniel, Jr. San Francisco: W. H. Freeman & Co.

Dickmann, Z., and Noyes, R. W. 1960. The fate of ova transferred into the uterus of the rat. *J Reprod Fertil* 1:97.

Eckstein, P.; Schelesnyak, M. D.; and Amoroso E. C. 1959. A survey of the physiology of ovum implantation in mammals. In *Implantation of ova,* ed. P. Eckstein. Cambridge: at the University Press.

Edwards, R. G., and Sirlin, J. L. 1956. Studies in gametogenesis, fertilization and early development in the mouse, using radioactive tracers. *Proc II World Cong Fertil Steril, Naples.*

Fawcett, D. W.; Wislocki, G. B.; and Waldo, C. M. 1947. The development of mouse ova in the anterior chamber of the eye and in the abdominal cavity. *Amer J Anat* 81:413.

Finn, C. A. 1965. Oestrogen and the decidual cell reaction of implantation in mice. *J Endocr* 32:223.

Fridhandler, L., and Palmer, W. M. 1968. Biosynthesis of RNA and DNA in rabbit preimplantation blastocysts: Effects of halogenated deoxyuridines. *Fertil Steril* 19:707.

Greenwald, G. S., and Everett, N. B. 1959. The incorporation of S^{35} methionine by the uterus and the ova of the mouse. *Anat Rec* 143:171.

Gwatkin, R. B. L. 1966. Amino acid requirements for attachment and outgrowth of the mouse blastocyst *in vitro. J Cell Physiol* 68:335.

Izquierdo, L., and Roblero, L. 1965. The incorporation of labelled nucleosides by mouse morulae. *Experientia* 21:532.

Lin, T. P. 1956. DL-methionine (sulphur-35) for labelling unfertilized mouse eggs in transplantation. *Nature* (*London*) 178:1175.

Lowenstein, J. E., and Cohen, A. I. 1964. Dry mass, lipid content and protein content of the intact and zona-free mouse ovum. *J Embryol Exp Morph* 12:113.

Mayer, G. 1963. Delayed nidation in rats: A method of exploring the mechanisms of ovo-implantation. In *Delayed implantation,* ed. A. C. Enders, p. 213. Chicago: University of Chicago Press.

Mintz, B. 1965. Nucleic acid and protein synthesis in the developing mouse embryo. In *Preimplantation stages of pregnancy,* ed. G. E. W. Wolstenholme and M. O'Connor. Boston: Little, Brown and Company.

Monesi, V., and Salfi, V. 1967. Macromolecular synthesis during early development in the mouse embryo. *Exp Cell Res* 46:632.

Monroy, A. 1965. *Chemistry and physiology of fertilization.* New York: Holt, Rinehart and Winston.

Noyes, R. W., and Dickmann, Z. 1960. Relationship of ovular age to endometrial development. *J Reprod Fertil* 1:186.

Nutting, E. H., and Meyer, R. K. 1963. Implantation delay, nidation and embryonic survival in rats treated with ovarian hormones. In *Delayed implantation,* ed. A. C. Enders, p. 233. Chicago: University of Chicago Press.

Prasad, M. R. N.; Dass, C. M. S.; and Mohla, S. 1968. Action of oestrogen on the blastocyst and uterus in delayed implantation: An autoradiographic study. *J Reprod Fertil* 16:97.

Psychoyos, A. 1966. Recent researches on egg implantation. In *Egg implantation,* ed. G. E .W. Wolstenholme and M. O'Connor. Boston: Little, Brown and Co.

Psychoyos, A., and Bitton-Casimiri, V. 1969. Captation *in vitro* d'un précurseur d'acide ribonucléique (ARN) (uridine-5-^{3}H) par le blastocyste du Rat; Différences entre blastocystes normaux et blastocystes en diapause. *C R Acad Sci (Paris)* 268:188.

Roskoski, R., Jr., and Steiner, D. F. 1967. The effect of estrogen on amino acid transport in rat uterus. *Biochim Biophys Acta* 135:727.

Runner, M. N. 1947. Development of mouse eggs in the anterior chamber of the eye. *Anat Rec* 98:1.

Smith, D. M. 1968. The effect on implantation of treating cultured mouse blastocysts with oestrogen *in vitro* and the uptake of (^{3}H) oestradiol by blastocysts. *J Endocrin* 41:17.

Smithberg, M., and Runner, M. N. 1960. Retention of blastocysts in non-progestational uteri of mice. *J Exp Zool* 143:21.

Stern, S., and Biggers, J. D. 1968. Enzymatic estimation of glycogen in the cleaving mouse embryo. *J Exp Zool* 169:61.

Weitlauf, H. M. 1969. Temporal changes in protein synthesis by mouse blastocysts transferred to ovariectomized recipients. *J Exp Zool* 171:481.

Weitlauf, H. M., and Greenwald, G. S. 1965. A comparison of ^{35}S methionine incorporated by blastocysts of normal and delayed implanting mice. *J Reprod Fertil* 10:203.

———. 1967. A comparison of the *in vivo* incorporation of S^{35} methionine by two-celled mouse eggs and blastocysts. *Anat Rec* 159:249.

———. 1968*a*. Survival of blastocysts in the uteri of ovariectomized mice. *J Reprod Fertil* 17:515.

———. 1968*b*. Influence of estrogen and progesterone on the incorporation of ^{35}S methionine by blastocysts in ovariectomized mice. *J Exp Zool* 169:305.

Wilson, I. B., and Smith, M. S. R. 1968. Isotopic labelling of the mouse blastocyst. *J Reprod Fertil* 16:305.

Yasukawa, J. J., and Meyer, R. K. 1966. Effect of progesterone and oestrone on the preimplantation and implantation stages of embryo development in the rat. *J Reprod Fertil* 11:245.

16

Permeability of the Mammalian Blastocyst to Teratogens

Vergil H. Ferm
Department of Anatomy and Cytology
Dartmouth Medical School
Hanover, New Hampshire

The mammalian blastocyst is an important stage in the development of the embryo. It represents the last claim to self-sufficiency, the transition to critical extraembryonic relationships, and finally, the complete dependence of the early embryo to increasingly complex fetal-maternal relationships. For the experimental mammalian teratologist these changing relationships of the blastocyst should represent a challenging area for critical research upon a critical phase in the life history of the embryonic system. However surprisingly little attention has been given to this particular stage of development. The reason for this is simply that the mammalian blastocyst is, in most forms, an extremely small mass, difficult to isolate and difficult to use as an experimental model. Nevertheless such a handicap should not preclude our attention to this stage or our examination of a variety of mammalian blastocyst forms as possible models for teratogenic investigation. The predominant morphologic characteristic of a blastocyst is the presence of a blastocyst cavity or blastocoele which comes to exist in the inner or embryonic side of the trophoblastic membrane. The blastocyst cavity, filled with fluid, is intimately apposed to the embryonic disk proper; such a relationship indicates that the constitution of the fluid must have important consequences for the developing embryo.

The blastocyst of the rabbit has received the most attention from experimental teratologists because it is unusually large and can be directly approached experimentally. The rabbit blastocyst then merits some special attention.

I. The Rabbit Blastocyst

A. *Development and Structure*

Some 70–80 hr after coitus the fertilized rabbit egg passes from the oviduct into the uterus (Gregory 1930). At this point development is in the morula stage and the cells are firmly compressed within the zona pellucida. The total diameter of the morula at this stage is about 0.3–0.5 mm. Although the initial development of a blastocoele usually takes place at the time the morula enters the uterus, this is not always the case in the rabbit. Gregory has pointed out that the length of the oviduct varies in different strains of rabbits and that in those strains with longer oviducts the blastocyst stage may begin to form within the oviduct, an indication that time rather than location is probably the important factor in this phase of development.

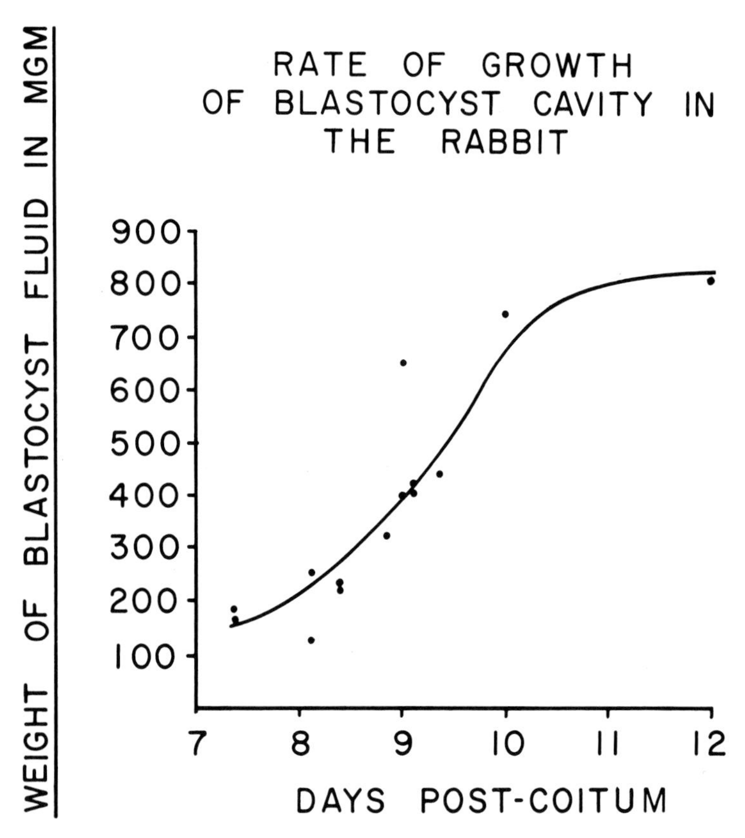

Fig. 1. Increase in the amount of blastocyst fluid in rabbit blastocysts from days 7 through 12 *post coitum.*

From this point on growth of the blastocyst is extremely rapid. The enlargement of the blastocyst cavity accounts for most of the growth. Assheton (1894) concluded that fluids actually collected in this cavity and caused the expansion. As the blastocyst expands the zona pellucida thins out. A definite inner cell mass appears and the trophoblast expands rapidly. Fixed specimens of 4-day-old blastocyst measure approximately 0.5 mm.

Adams, Hay, and Lutwak-Mann (1961) have made careful measurements of the size of fresh and fixed rabbit blastocyts. On day 5 the mean diameter of the blastocyst ranges from 1.3 to 2.2 mm. On day 6 the diameter ranges from 2.2 to 4.8 mm. Figure 1 shows the increase in the amount of blastocyst fluid in rabbits from day 7 through 12 (Ferm 1955).

Embryonic development begins at the time of inner cell mass formation. Gregory (1930) has suggested that even as early as the 18-cell stage, beginning at the 3d day *post coitum,* certain cells can be identified as embryoblastic. He refers to these as primitive mass cells. Certainly the most regularly and easily identifiable embryonic cells appear at the time of the formation of the blastocoele. Again, measurements of the diameter of the embryonic disk within the blastocysts by Adams, Hay, and Lutwak-Mann (1961) and Lutwak-Mann, Hay, and Adams (1962) indicate a range in size of from 0.52 to 0.90 mm on day 5 to 1.3 mm on day 7. The first evidence of primitive streak formation does not begin until the overall blastocyst diameter reaches 4 mm, approximately midway in the 6th day of development. Further evidence of embryonic development occurs during the 7th day of gestation when the presomite and early somite stages first appear (Waterman 1943; Edwards 1968).

These very early somite stages of the developing rabbit embryo correspond to the beginning implantation of the blastocyst into the uterine epithelium. Most investigations have indicated that implantation occurs late in the 7th day of gestation, although implantations can occur earlier on that day (Lutwak-Mann 1954).

Beginning on the 8th day of gestation the rabbit embryo is in the 2-somite stage, and more somites develop rapidly. The neural tail folds, and otic and optic vesicles appear. Late on the 8th day the neural folds have closed (except for the neuropores), the brain has divided into its three primary divisions, and the primitive heart has begun to beat. During the 9th and 10th days the major stages in organ differentiation are completed (Waterman 1943). Another peculiarity of development, in addition to the unusually large size of the blastocyst in the rabbit, is the superficiality of implantation. Late in the 7th day of development the blastocyst attaches to the uterine epithelium. Mossman (1926) has discussed the morphological details of implantation and placentation in the rabbit, and Böving (1967) has described the physical and chemical events in its early placentation. In the rabbit only a decidua basalis forms, with a more or less temporary decidua parietalis.

The important stages in the development of the extraembryonic membranes in the rabbit as they relate to the problem of the permeability of the blastocyst to teratogens are the development of the bilaminar omphalopleure, the inverted yolk sac placenta, and the interaction of the trophoblast and uterine epithelium in the obplacental region. For our purposes it is sufficient to say that the bilaminar omphalopleure, or outer wall of the blastocyst cavity, is very temporary in the rabbit and begins to degenerate on the 10th day of gestation. The inverted yolk sac placenta develops rather early, and, upon the disintegration of the bilaminar omphalopleure it comes to lie in direct contact with the uterine endometrium.

It is necessary to note that the trophoblastic component of the bilaminar omphalopleure opposite the ectoplacenta undergoes thickening and specialization while the adjacent uterine epithelium swells, becomes vacuolated, and forms a symplasma. The trophoblastic cells invade the symplasma, thus forming a temporary bilaminar yolk sac and choriovitelline placenta (Amoroso 1958). It is through this surface that most of the fluid material that causes the blastocyst to expand so rapidly in the early stages of development must pass.

With these developmental details in mind let us examine a variety of experimental approaches to the study of the rabbit blastocyst as they relate to possible teratogenic influences.

B. *The Organ Culture of Rabbit Blastocysts*

New (1966) and Mintz (1967) have reviewed the various methodologies for obtaining mammalian blastocysts and the techniques utilized for their culture *in vitro.* The implications for several important research problems are obvious, but to date little attention has been paid to combining blastocyst culture with teratogenic studies. Glenister (1961*a*) has developed special culture techniques for studying implantation mechanisms and embryonic development of rabbit blastocysts placed upon sensitized strips of rabbit endometrium. By utilizing these techniques it has been possible to explant 7-day-old rabbit blastocysts in primitive streak and early somite stages and to keep these alive to the 10th day of development while observing neural tube, somite, gut, and heart formation (Glenister 1961*b*). Such techniques offer exciting possibilities for the experimental teratologist and should be exploited.

C. *The Use of Rabbit Blastocysts Prepared as Flat Mounts*

Moog and Lutwak-Mann (1958) developed techniques for observations of the flat-mounted rabbit blastocyst which maintained the intact cells in their correct topographical relationships. Their initial studies showed that the cellular mitotic rates were apparently different in the embryonic and abembryonic trophoderm and that the rabbit blastocysts maintained in a variety of culture media for short intervals were extremely sensitive to the media as reflected by changes in the mitotic rates of these areas.

Studies utilizing their techniques have been expanded to include the parenteral treatment of pregnant rabbits with a variety of chemical and teratogenic agents (Adams, Hay, and Lutwak-Mann 1961). They have included observations on the rate of permeability of the blastocyst to the various agents even prior to implantation, the relative susceptibility of the pre- and postimplantation blastocysts, and the variability of response within single and different litters. The authors have suggested that this technique might be utilized as a rapid screening device for cytostatic and even potentially teratogenic agents.

It has been postulated that certain chemical substances may stimulate the growth of embryonic areas disproportionately to the rest of the blastocyst (Lutwak-Mann and Hay 1962). Thus vitamin A, a potent teratogen, caused an overall stimulation of growth in blastocysts which were advanced for their age, and other substances caused a significant enlargement of the embryonic disk in comparison to the overall size of the blastocyst. Estradiol benzoate produced developmentally advanced embryonic disks in otherwise normal-sized blastocysts. Lutwak-Mann and Hay suggest that relative stimulation of growth may have as much significance for teratogenesis as do agents which cause cell growth inhibition or cell death.

Hay (1964) has utilized the flat-mount technique to study the effects of thalidomide on the rabbit blastocyst. Maternal treatment with thalidomide resulted in the retardation of cell migration and normal formative movements in the preimplantation blastocyst, particularly in the embryonic disk area.

An important extension of this approach has been the recent report of Lutwak-Mann, Hay, and New (1969) concerning the effect of a wide variety of embryotoxic agents including antimetabolites, inhibitors, antimitotic agents, cytostatic agents, metabolites, and hormones on the rabbit blastocyst. Experiments were conducted both *in vivo* and *in vitro* utilizing organ culture techniques. The results suggest three possibilities: some agents are toxic to rabbit blastocysts when acting directly but are harmless *in vivo* where they are apparently metabolized by the maternal system; some compounds which are harmless to the implanting blastocyst may become deleterious in the postimplantation period; some blastocysts which are damaged in *utero* by an embryotoxic agent apparently recuperate when cultured in a nutrient medium for a limited length of time. As the authors suggest, their findings have important potential implications for accidental teratogenic exposure. Further intensive investigation is needed.

D. *The Effect of Various Agents upon the Rabbit Embryo and Blastocyst*

The important work of Brambell, Hemmings, and Henderson (1951) on the methods for the study of the transfer of maternal antibodies to rabbit blastocysts was the first direct approach to experimental studies on this fundamental phase of embryonic development. Essentially their method consisted of rapidly freezing the uterus of rabbits containing blastocysts and collecting the blastocyst fluid as a spherule of ice. Measurements of antibody titers following the administration of various amounts of homologous and heterologous antibodies to the mother revealed the degree of permeability of the blastocyst wall to these substances.

Observations by several investigators then followed on the permeability of the rabbit blastocyst to a variety of chemical agents and their subsequent embryopathic effects. Lutwak-Mann (1954) studied the chemical composition of blastocyst fluid by measurements upon the whole unimplanted blastocyst and upon blastocyst fluid obtained from implanted blastocysts by needle aspiration directly through the uterine wall. It is important to note that in these studies the permeability of the blastocyst to glucose, fructose, and sucrose was negligible when the blastocyst was unimplanted. Following implantation, however, the permeability increased considerably.

In the first use of this technique for studying the permeability of the blastocyst to known teratogens Ferm (1956) injected pregnant rabbits with trypan blue and then recovered blastocysts on days 7, 8, 9, and 10 by rapidly freezing the uterus in ethyl alcohol supercooled with dry ice. Spectrophotometric measurements of the amount of trypan blue in the blastocyst cavity revealed that the blastocyst wall was very permeable to the dye at the time of implantation. Permeability decreased slightly later in the 8th day of gestation; by the middle of the 9th day the blastocyst was not permeable to trypan blue. This was the first direct evidence of the placental transport of trypan blue; the phenomenon could only have been studied in the rabbit. The teratogenic action of trypan blue on the rabbit embryo was described. Ferm has suggested that the change in permeability of the blastocyst to the dye represents a change from the embryotrophic type of nutrition to that of a histotrophic stage characterized by the development of an embryonic vascular system in the extraembryonic membranes. Adams, Hay, and Lutwak-Mann (1961) have confirmed the findings of the permeability of the rabbit blastocyst to trypan blue as well as its teratogenic effect on the rabbit embryo. They also noted the impermeability of the unimplanted blastocyst to this dye, a finding comparable to that for the sugars previously described.

Brinsmade and Rubsaamen (1957) have noted the teratogenic and embryopathic effect of insulin on rabbit embryos when insulin is injected into the pregnant rabbit on the 6th to the 13th days of gestation. Curry and Ferm (1962) studied the equilibration of glucose in the rabbit maternal serum and blastocyst fluid by the above techniques. Their experiments showed that alterations (either hyper- or hypoglycemia) in the maternal glucose levels resulted in similar changes in the glucose content of the blastocyst fluid and that severe maternal hypoglycemia induced by insulin caused a marked drop in glucose levels within the rabbit blastocyst fluid. This might well explain the teratogenic action of insulin.

Interesting observations have been made on the permeability of the rabbit blastocyst to ^{14}C-thalidomide and its metabolites. Fabro, Smith, and Williams (1965) have found that thalidomide, administered as a single dose to pregnant rabbits on the 8th day of gestation, was able to penetrate the blastocyst and persist there for over 58 hr, mainly in the form of metabolites of thalidomide. Keberle et al. (1965) reported that ^{14}C-thalidomide was able to penetrate into the blastocyst of the preimplantation, implanting, and postimplantation blastocyst, but that two of the ^{14}C metabolites of thalidomide were not able to penetrate into the blastocyst cavity until implantation had occurred. The extensive studies done by these authors on the permeability of the blastocyst to thalidomide and its metabolites offer an important example of the sophisticated techniques than can be utilized in the study of blastocyst permeability to teratogens.

Beaudoin and Ferm (1961) have examined the alteration of proteins in rabbit maternal serum and blastocyst fluid following administration of teratogenic and nonteratogenic azo dyes during early pregnancy. In both the maternal serum and blastocyst fluid there was a significant increase in the beta globulin fraction accompanied by a decrease in albumin. The changes in protein constitution were directly correlated with the teratogenic capabilities of the azo dyes.

The rabbit blastocyst has been used also to study the early transplacental passage of viruses from mother to embryo. Flamm (1966) injected Coxsackie virus A9 intravenously into pregnant rabbits on the 7th day of pregnancy and was able to detect it within 1 min in the blastocyst fluid of the embryos. Oxford and Sutton (1968) injected rubella virus into pregnant rabbits and found that it had no detectable histological effect upon the preimplantation rabbit blastocyst. The intravenous injection of herpes virus hominis, however, did increase the fetal resorption rate. Although none of these viruses was teratogenic in the rabbit, the observations on them nevertheless suggest numerous other possibilities for the investigation of viral infections and their potential teratogenic influence in early gestation.

II. Teratogenesis and Other Mammalian Blastocysts

Because of the small size of most mammalian blastocysts much of the work done on the effects of teratogenic agents on this particular stage of development has been restricted to histological studies. There are, however, a few experimental techniques which have shown some promise of leading to important observations on the effect of teratogens on the mammalian blastocyst.

A. *Culture and Transplantation of Mammalian Eggs and Blastocysts*

New (1966) and Mintz (1967) have reviewed in detail many of the *in vitro* culture techniques for mammalian embryos. Turbow (1966) has removed 5–14 somite rat embryos and maintained them in tissue culture with an intact amnion and yolk sac. By injecting trypan blue through the yolk sac he has been able to induce malformations. Utilizing essentially the same technique Turbow and Chamberlain (1968) studied the teratogenic effect of the antimetabolite 6-aminonicotinamide as well as the protective effect of nicotinamide. Skalko and Morse (1969) cultured mouse blastocysts in media containing varying concentrations of another teratogen, actinomycin D, and found that this substance affected several developmental processes, including blastocyst formation as well as blastocyst survival and differentiation.

Runner (1965) summarized the techniques available for the transplantation of mammalian blastocysts. Such studies might be used for the exposure of mammalian blastocysts *in vitro* to a variety of teratogenic agents and their subsequent analysis following transplantation into recipient mothers. McLaren and Biggers (1958) removed fertilized mouse eggs 2½ days after mating and maintained them in culture media for 2 days until they developed into blastocysts. They were then transplanted into recipient mothers and 20% survived and developed until the 17th day of pregnancy, a few even until term. Chang and Hunt (1960) have irradiated eggs and blastocysts of rabbits ranging from 1 to 6 days of age with a radiocobalt source and then transplanted them at these stages into recipient animals. They were then examined later in gestation. No abnormal fetuses were found and there was no significant increase in degeneration after resorption. These authors suggest that the fertilized ovum and blastocyst are less sensitive to irradiation than unfertilized ova. On the other hand Inman and Markivee (1963) treated rabbit blastocysts with 150–200 r at 3.5 days of gestation and then examined these, along with controls, at 9.5 days. Stunting, retardation in the development of the allantois, and damage to the hematopoietic tissue were all increased by X-irradiation. A few malformations were noted in the irradiated groups. Although no observations were made on the permeability of these blastocysts to any compounds in these studies it would be desirable to know if irradiation does alter blastocyst permeability. Although Runner (1965) cautions against undue optimism concerning the ultimate value of this transplantation technique, such studies might lead to the partial separation of maternal, placental, and embryonic factors in teratogenesis and give additional valuable information concerning blastocyst permeability to teratogenic agents.

Another potentially valuable technique for the study of blastocyst permeability to teratogens *in vitro* is that described by Bryson (1964), who has cultivated preimplantation mouse eggs in intraperitoneal diffusion (Millipore) chambers. Under these conditions of growth on a two-dimensional surface no progressive embryonic differentiation or organogenesis is noted. However the possibilities are intriguing; further efforts should be made to evaluate this method of culture and its potential for teratogenic testing.

The production of chimeric blastocysts by fusion of two or more cleaving mouse eggs (Tarkowski 1965; Mintz 1965) affords one other interesting possibility for study of teratogenic effects upon blastocysts. Blastocysts formed following such fusion are larger than normal and twinning and hermaphroditism have been reported from such experiments. Cells of the inner cell mass of mouse blastocysts have been removed by microsurgical techniques (Lin 1969) and then transferred to uteri of recipient females. No abnormal young were found in survivors, but the technique offers a direct approach to the study of injected teratogens into the blastocoele of experimental animals.

B. *Histological and Radioisotopic Studies of Teratogenic Effects upon Mammalian Blastocysts*

Although some teratogenic studies have included histological observations on the effects of various agents upon the implanting blastocysts, not much direct information concerning the actual permeability of the blastocyst to teratogenic agents can be inferred unless radioisotopic tracers are employed. However some interesting observations have been made on the direct effect of known teratogens on mammalian blastocysts. Lucey and Behrman (1963) have reported indirect evidence for the effect of thalidomide on the monkey blastocyst. They treated monkeys daily with thalidomide by mouth immediately after mating and observed a significant decrease in the pregnancy rate as compared to controls. They suggested that thalidomide affected the blastocyst, killing it prior to implantation. In support of this possibility Marin-Padilla and Benirschke (1963) have described the effect of thalidomide on the unimplanted blastocyst of the armadillo (*Dasypus novemcinctus*). In this species there appears to be a differential effect of thalidomide. Although there was clear damage to the cells of the inner cell mass as revealed by clumping of nuclear chromatin, nuclear pyknosis, and cytoplasmic eosinophilia, the cells forming the blastocyst wall were not damaged.

The findings above correlate well with the rabbit blastocyst permeability studies on thalidomide previously mentioned (Fabro, Smith, and Williams 1965) in which the unimplanted blastocyst was quite permeable to thalidomide. Further direct evidence of the permeability to thalidomide is suggested by the experiments of Fabro, Hague, and Smith (1967), who administered ^{14}C-thalidomide orally to golden hamsters late on the 8th day of gestation. Analysis of these embryos some 4 and 12 hr later revealed that they contained relatively high amounts of both thalidomide and its metabolites.

The teratogenic effects of specific heavy metals, lead (Ferm and Carpenter 1967), cadmium (Ferm and Carpenter 1968*a*), and arsenic (Ferm and Carpenter 1968*b*) afford an opportunity to study directly the permeability of the implanting blastocysts to molecular teratogens. Cadmium causes a rather specific teratogenic effect on hamster embryos, inducing cleft lips and palates. The simultaneous administration of zinc inhibits the teratogenic effect of cadmium (Ferm and Carpenter 1968*a*). Ferm, Hanlon, and Urban (1969) have studied the permeability of the implanting hamster blastocyst to ^{109}Cd and have found that it is indeed permeable to this teratogen and that the simultaneous administration of zinc does not prevent the transfer of this teratogen to the embryo.

An interesting teratogenic effect, the production of the congenital adrenogenital syndrome in rats, has been produced by the single injection of an inhibitor of 3β

hydroxysteroid dehydrogenase during the preimplantation stage of the rat blastocyst or even before ovulation (Goldman 1969). This suggests the possibility that a specific teratogen may be bound to maternal tissues and released to become effective when the end-organ, in this case adrenal and Leydig cells, undergoes development.

Weitlauf and Greenwald (1965) have found that ^{35}S-methionine was incorporated only very slightly in the early stages of cleaving mouse ova but that the incorporation of this compound increased substantially just prior to implantation. Delay of blastocyst implantation induced by a suckling stimulus caused a significant reduction in ^{35}S incorporation in the free blastocyst. This strongly suggests an important metabolic gradient in blastocyst metabolism which may help to explain a differential teratogenic sensitivity to a variety of agents. Jollie (1968) has reported that intratubal rat morulae do not incorporate tritiated thymidine after intraperitoneal injection of this labeled compound into pregnant rats. However the nuclei of the implanting blastocyst, including both embryonic and extraembryonic cells, do become labeled, indicating the development of an important maternoembryonic transport mechanism during early stages of implantation.

III. Summary and Conclusions

It is apparent that a variety of techniques and approaches are available for the study of blastocyst permeability to teratogenic agents. Most of the basic research in this area has utilized the blastocyst of the rabbit because it is unusually large, it implants superficially into the endometrium, and a relatively large quantity of blastocyst fluid is available for analysis. Through the use of this form as an experimental model important and valuable information concerning normal development and biochemical composition has been learned. Valuable studies have been made on the effects of a variety of teratogenic agents including among others azo dyes, viruses, antimitotic agents, growth inhibitors, and irradiation.

Only a few such studies have embodied detailed analysis of blastocyst permeability, however.

The use of other mammalian blastocysts for the study of permeability to teratogens is limited mainly by the small size of these blastocysts. Therefore most studies on early stages of teratogenesis have been confined to histological observations. However the use of radioisotopically labeled teratogens, heavy metal teratogens, and *in vitro* culture techniques of mammalian blastocysts has opened up new possibilities for the study of blastocyst permeability to experimental teratogens in mammalian blastocysts other than the rabbit.

There is little doubt that further intensive study of the permeability of the mammalian blastocyst to teratogenic agents will yield exciting and important results. Such studies will add an extremely important dimension to our knowledge and understanding of the complex problems of congenital malformations.

Acknowledgments

This work was supported in part by United States Public Health Service grants GM 10210 and HD 02616.

References

Adams, C. E.; Hay, M. F.; and Lutwak-Mann, C. 1961. The action of various agents upon the rabbit embryo. *J Embryol Exp Morph* 9:468.

Amoroso, E. C. 1958. Placentation. In *Marshall's physiology of reproduction,* ed. A. S. Parkes, p. 127. London: Longmans, Green and Co.

Assheton, R. 1894. A re-investigation into the early stages of the development of the rabbit. *Quart J Micro Sci* 37:113.

Beaudoin, A. R., and Ferm, V. H. 1961. The effect of disazo dyes on protein metabolism in the pregnant rabbit. *J Exp Zool* 147:219.

Böving, B. G. 1967. Chemo-mechanics of implantation. In *Comparative aspects of reproductive failure,* ed. K. Benirschke, p. 142. New York: Springer-Verlag.

Brambell, F. W. R.; Hemmings, W. A.; and Henderson, M. 1951. *Antibodies and embryos.* London: Athlone Press.

Brinsmade, A. B., and Rubsaamen, H. 1957. Zur teratogenetischen Wirkung von unspezifischem Fieber auf den sich entwickelnden Kaninchenembryo. *Ziegler's Beitr zur Path Anat and zur Allgem Path* 117:154.

Bryson, D. L. 1964. Development of mouse eggs in diffusion chambers. *Science* 144:1351.

Chang, M. C., and Hunt, D. M. 1960. Effects of *in vitro* radiocobalt irradiation of rabbit ova on subsequent development *in vivo* with special reference to the irradiation of maternal organism. *Anat Rec* 137:511.

Curry, H. F., Jr., and Ferm, V. H. 1962. Blastocyst sugar concentration following maternal glucose changes. *Anat Rec* 142:21.

Edwards, J. A. 1968. The external development of the rabbit and rat embryo. In *Advances in teratology,* ed. D. H. M. Woollam, 3:239. New York: Academic Press.

Fabro, S.; Hague, D.; and Smith, R. L. 1967. The fate of [14C] thalidomide in the pregnant hamster. *Biochem J* 103:26 (abstr.).

Fabro, S.; Smith, R. L.; and Williams, R. T. 1965. The persistence of maternally administered ^{14}C-thalidomide in the rabbit embryo. *Biochem J* 97:14 (abstr.).

Ferm, V. H. 1955. Studies on the teratogenic activity of trypan blue. Ph.D. thesis, University of Wisconsin.

———. 1956. Permeability of the rabbit blastocyst to trypan blue. *Anat Rec* 125: 745.

Ferm, V. H., and Carpenter, S. J. 1967. Developmental malformations resulting from the administration of lead salts. *J Exp Mol Path* 7:208.

———. 1968*a*. The relationship of cadmium and zinc in experimental mammalian teratogenesis. *Lab Invest* 18:429.

———. 1968*b*. Malformations induced by sodium arsenate. *J Reprod Fertil* 17: 199.

Ferm, V. H.; Hanlon, D. W.; and Urban, J. 1969. The permeability of the hamster placenta to radioactive cadmium. *J Embryol Exp Morph* 22:107.

Flamm, H. 1966. Some considerations concerning the pathogenesis of prenatal infections. In *The prevention of mental retardation through control of infectious diseases,* ed. H. F. Eichenwald, p. 79. Public Health Service Publication no. 1692.

Glenister, T. W. 1961*a*. Organ culture as a new method for studying the implantation of mammalian blastocysts. *Proc Roy Soc* (*Biol*) 154:428.

———. 1961*b*. Observations on the behaviour in organ culture of rabbit trophoblast from implanting blastocysts and early placentae. *J Anat* 95:474.

Goldman, A. S. 1969. Congenital effectiveness of an inhibitor of 3β hydroxysteroid dehydrogenase administered before implantation of the rat blastula. *Endocrinology* 84:1206.

Gregory, P. W. 1930. The early embryology of the rabbit. *Contr to Embryol Carnegie Inst* 21 (125):141.

Hay, M. F. 1964. Effects of thalidomide on pregnancy in the rabbit. *J Reprod Fertil* 8:59.

Inman, O. R., and Markivee, C. R. 1963. Gross effects on rabbit embryos and membranes of X-irradiation in the blastocyst stage. *Anat Rec* 147:139.

Jollie, W. P. 1968. Radioautographic evidence of materno-embryonic transport of thymidine into implanting rat embryos. *Acta Anat* 70:434.

Keberle, H.; Faigle, J. W.; Fritz, H.; Knusel, F.; Loustalot, P.; and Schmid, K. 1965. Theories on the mechanism of action of thalidomide. In *Embryopathic activity of drugs,* ed. J. M. Robson, F. M. Sullivan, and R. L. Smith, p. 210. Boston: Little, Brown and Co.

Lin, T. P. 1969. Microsurgery of inner cell mass of mouse blastocysts. *Nature* (*London*) 222:480.

Lucey, J. F., and Behrman, R. E. 1963. Thalidomide: Effect upon pregnancy in the rhesus monkey. *Science* 139:1295.

Lutwak-Mann, C. 1954. Some properties of the rabbit blastocyst. *J Embryol Exp Morph* 2:1.

Lutwak-Mann, C., and Hay, M. F. 1962. Effect on the early embryo of agents administered to the mother. *Brit Med J* 2:944.

Lutwak-Mann, C.; Hay, M. F.; and Adams, C. E. 1962. The effect of ovariectomy on rabbit blastocysts. *J Endocr* 24:185.

Lutwak-Mann, C.; Hay, M. F.; and New, D. A. T. 1969. Action of various agents on rabbit blastocysts *in vivo* and *in vitro*. *J Reprod Fertil* 18:235.

McLaren, A., and Biggers, J. D. 1958. Successful development and birth of mice cultivated *in vitro* as early embryos. *Nature* (*London*) 182:877.

Marin-Padilla, M., and Benirschke, K. 1963. Thalidomide induced alterations in the blastocyst and placenta of the armadillo, *Dasypus novemcinctus mexicanus,* including a choriocarcinoma. *Amer J Path* 43:999.

Mintz, B. 1965. Experimental genetic mosaicism in the mouse. In *Preimplantation stages of pregnancy,* ed. G. E. W. Wolstenholme and M. O'Connor, p. 194. Boston: Little, Brown and Co.

———. 1967. Mammalian embryo culture. In *Methods in developmental biology,* ed. F. H. Wilt and N. K. Wessells, p. 379. New York: Thomas Y. Crowell Co.

Moog, F., and Lutwak-Mann, C. 1958. Observations on rabbit blastocysts prepared as flat mounts. *J Embryol Exp Morph* 6:57.

Mossman, H. W. 1926. The rabbit placenta and the problem of placental transmission. *Amer J Anat* 37:433.

New, D. A. T. 1966. *The culture of vertebrate embryos.* London: Logos Press Ltd.

Oxford, J. S., and Sutton, R. N. P. 1968. The effect of rubella and herpes virus hominis on the pre- and post-implantation stages of pregnancy in laboratory animals. *J Embryol Exp Morph* 20:285.

Runner, M. 1965. Transplantation of mammalian ova: Blastocyst and earlier stages. In *Teratology: Principles and techniques,* ed. J. G. Wilson and J. Warkany, p. 104. Chicago: University of Chicago Press.

Skalko, R. G., and Morse, J. M. D. 1969. The differential response of the early mouse embryo to actinomycin D treatment *in vitro. Teratology* 2:47.

Tarkowski, A. K. 1965. Embryonic and postnatal development of mouse chimeras. In *Preimplantation stages of pregnancy,* ed. G. E. W. Wolstenholme and M. O'Connor, p. 183. Boston: Little, Brown and Co.

Turbow, M. M. 1966. Trypan blue induced teratogenesis of rat embryos cultivated *in vitro. J Embryol Exp Morph* 15:387.

Turbow, M. M., and Chamberlain, J. G. 1968. Direct effects of 6-aminonicotinamide on the developing rat embryo *in vitro* and *in vivo. Teratology* 1:103.

Waterman, A. J. 1943. Studies of normal development of the New Zealand white strain of rabbit. *Amer J Anat* 72:473.

Weitlauf, H. M., and Greenwald, G. S. 1965. A comparison of ^{35}S methionine incorporation by the blastocysts of normal and delayed implanting mice. *J Reprod Fertil* 10:203.

17

Mammalian Embryo Metabolism

R. L. Brinster

Laboratory of Reproductive Physiology
School of Veterinary Medicine
University of Pennsylvania, Philadelphia

During the last ten years a considerable amount of information has been accumulated about mammalian embryo metabolism (see Austin 1961 for a review of earlier work). Despite this increasing body of knowledge there is still a paucity of facts about the intricacies of metabolism during the early stages of mammalian development. At present we know most about energy metabolism and next most about protein metabolism; outside these two areas very little is known. Therefore the following discussion will deal primarily with energy and protein metabolism.

For most studies on mammalian embryos laboratory animals are used, and consequently we must rely on information from these species for what we know about mammalian embryo metabolism in general. It seems likely that the mouse and the rabbit will continue to supply us with most of the information. Fortunately it seems quite likely that much of this can be extrapolated and applied to the embryos of domestic animals and primates, including the human. In fact recent evidence in several areas suggests that in certain metabolic parameters there are indeed similarities among the species studied. A particularly striking example of this is the preference of the embryo for oxidation of pyruvate to glucose. There is evidence the pyruvate may be the energy source of choice for all mammalian embryos during the first few days of development.

I. Energy Metabolism

A. *Oxygen Consumption of the Embryo*

One of the first questions to ask relative to energy metabolism is, What is the overall capability of the embryo to produce energy for use in metabolic processes?

It is partly answered by determining the oxygen consumption of the embryo, already defined for both the rabbit (Fridhandler 1961) and the mouse (Mills and Brinster 1967). The results of these experiments are summarized in table 1. For the preblastocyst stage the oxygen uptake of the rabbit is about 3.5 times that of the mouse, a reasonable ratio since the volume of the rabbit embryo is 3.5 times that of the mouse. However later embryonic stages cannot be compared easily, since the rabbit blastocyst expands much more than the mouse blastocyst.

In order to relate the embryos to other body tissues, one needs to know the amount of protein or tissue present in the embryo. For the mouse embryo the protein content has been determined, and we can therefore calculate the Q_{O_2} of the embryo. We find that the early stages have a low Q_{O_2} comparable to bone or skin, whereas, the blastocyst has a high Q_{O_2} comparable to whole brain. Thus the embryo begins as a relatively inactive tissue and implants as a rapidly metabolizing tissue.

TABLE 1

RESPIRATORY ACTIVITY OF PREIMPLANTATION EMBRYOS

STAGE OF DEVELOPMENT	MOUSE EMBRYO		RABBIT EMBRYO
	Uptake mμl/Hr/Embryo	Q_{O_2} μl/mg Dry Wt.	Uptake mμl/Hr/Embryo
Unfertilized	0.156	3.73	....
Fertilized	0.157	3.77	0.64
Day 2	0.150	3.84	0.60
Day 3	0.191	5.44	0.84
Day 4	0.405	12.13	5.45
Day 5	0.535	16.32	53.00
Day 6	....	....	214.20

SOURCE: Mouse oxygen uptake from Mills and Brinster (1967). Rabbit oxygen uptake from Fridhandler (1961).

NOTE: Dry weight of the stages was calculated from values for the protein content of each stage assuming protein = 0.66 × dry weight (Brinster 1967*d*).

B. *Culture of the Embryo*

One excellent method used in the past to acquire information about metabolism in early embryos is to remove the embryos from the reproductive tract and determine what nutritional substances and what type of environment the embryos need to maintain normal development. Most of these studies have been done on embryos from the mouse and the rabbit. Whitten (1956) cultured 8-cell mouse embryos and was able to show that they develop into blastocysts in a modified Krebs-Ringer bicarbonate solution containing bovine serum albumin and glucose. The ionic composition of Krebs-Ringer bicarbonate approximates very closely the ionic composition of blood serum. Whitten was unable to obtain the development of 2-cell embryos in this solution; however when he included lactate in the medium, some 2-cell embryos did develop into blastocysts (Whitten 1957). Brinster (1963, 1965*b*, 1969*a*) extended these studies and was able to identify those compounds which allow *in vitro* development of 2-cell mouse embryos into blastocysts and those compounds which were ineffective. A summary of these results is shown in table 2. Pyruvate appears to be an essential nutrient for the 2-cell mouse embryo. Biggers, Whittingham, and Donahue (1967) have also shown that pyruvate seems to be the key compound in the support of the development of the mouse embryo from the 1-cell to the 2-cell stage and is essential for *in vitro* maturation of mouse oocytes. Glucose, the compound that most cells use as an energy source, is not able to support the development of the mouse embryo until the 4- to 8-cell stage is reached (Brinster 1965*d*). After the 8-cell stage is reached, in sharp contrast to its previous requirements for both an amino nitrogen and an energy source, the mouse embryo is able to develop into a blastocyst in very simple culture media containing only

TABLE 2

POSSIBLE ENERGY SOURCES FOR THE MOUSE EMBRYO

I. Compounds which will not support development of the 2-cell mouse embryo	
Malate[a]	Glucose[a]
Fumarate	Fructose[a]
Succinate	Ribose
Iso-citrate	D-glyceraldehyde
Citrate[a]	Glucose-6-phosphate
Acetate[a]	Fructose-1,6-diphosphate
Cis-aconitate	Bovine serum albumin[a]
α-ketoglutarate	
(All compounds tested at 10^{-2}, 10^{-3}, 10^{-4}, 10^{-5} molar except glucose and fructose, which were tested at 2.78×10^{-2}, 5.56×10^{-3}, and 2.78×10^{-3} molar; ovine serum albumin tested at 1 mg/ml).	

II. Compounds which will support development of the 1-cell mouse embryo and the optimum concentration of the compound	
Pyruvate	$(5 \times 10^{-4}M)$[b]
Lactate DL	$(5 \times 10^{-2}M)$
Oxaloacetate	$(5 \times 10^{-4}M)$[b]
Phosphoenolpyruvate	$(1 \times 10^{-2}M)$

SOURCE: Brinster 1965*c*.

[a] Compounds which will support development of the 8-cell but not the 2-cell embryo. Fumarate, Iso-citrate, and Cis-asconitate were not tried with 8-cell embryos (Brinster and Thomson 1966).

[b] Compounds which will support development of the 1-cell mouse ovum and oocyte (Biggers, Whittingham, and Donahue 1967).

certain single amino acids or carbohydrate sources (Whitten 1957; Brinster and Thomson 1966). Studies on preimplanting rabbit embryos have likewise shown that pyruvate and lactate have a beneficial effect during the first 2 to 3 days of development (Brinster 1970). Thus there is definite evidence in the mouse embryo and suggestive evidence in the rabbit embryo that environmental requirements of the embryo change with developmental age. The change in requirements is probably indicative of gene activity in the early cleavage stages of the mammalian embryo.

Until recently it has not been possible to cultivate the embryos of the mouse and rabbit continuously from the 1-cell through the blastocyst stage. Whitten and Biggers (1968) were able to do so with the embryos of certain F_1 hybrid mice in a medium similar to that shown in table 3 but with the sodium chloride concentration reduced to give the medium an osmolarity of approximately 0.256, a value similar to that found optimum for the *in vitro* development of 2-cell embryos (Brinster

1965*a*). In the experiments with the hybrids the albumin concentration of the medium was raised to 4 mg per ml. More recent studies suggest that it is the albumin rather than the osmolarity which exerts the beneficial effect in these experiments (see Biggers, chapter 18 of this volume). Successful cultivation of the rabbit embryo from the early cleavage stages through the expanding blastocyst has recently been achieved (Onuma, Maurer, and Foote 1968) by employing heat-treated bovine or rabbit serum.

C. *Substrate Oxidation by the Embryo*

Culture studies suggest that three substrates, pyruvate, lactate, and glucose are important sources of energy during the preimplantation development of mammalian embryos. Studies have been made to determine the rate of oxidation of these

TABLE 3

BRINSTER'S MEDIUM FOR OVUM CULTURE (BMOC-2)

	mM	Gm/l	ml of 0.154 M Stock in 13 ml
NaCl	94.88	5.546	5.90
NaLactate	25.00	2.253[a]	2.10
NaPyruvate	0.25	0.028	2.10[b]
KCl	4.78	0.356	0.40
$CaCl_2$	1.71	0.189	0.20[c]
KH_2PO_4	1.19	0.162	0.10
$MgSO_4.7H_2O$	1.19	0.294	0.10
$NaHCO_3$	25.00	2.106	2.10
Pen strep	100 U/ml of Pen, 50 μg/ml of strep		
Bovine serum albumin	1 mg/ml = 1 gm/l		

SOURCE: Brinster 1965*d*.

NOTE: It is desirable to add 1 mg/ml glucose to this medium (Brinster 1969*c*).

[a] NaLac added as liquid prepared as follows:
Add 1.82 ml of concentrated lactic acid (85–90%) to 200 ml of double distilled H_2O. Neutralize to pH 7.4 (about 15–20 ml of 1N NaOH). This is enough lactic acid to make one liter of medium.

[b] NaPyr stock is 0.00154 M pyruvate in 0.154 NaCl (17 mg/100 ml).

[c] $CaCl_2$ stock is 0.11 M.

substrates in both the mouse and rabbit embryo. A summary of the results is shown in table 4. Carbon dioxide formation from pyruvate during the first 2 days of development in the mouse is very high and accounts for 100% of the oxygen consumption in unfertilized ova (Brinster 1967*b*); in contrast, carbon dioxide formation from glucose at the same stage is very low, representing only about 2% of that formed from pyruvate. About the time of implantation CO_2 formed from pyruvate accounts for about 65% of oxygen consumption. However glucose oxidation rises dramatically during the preimplantation period; at the time of implantation it is as readily oxidized as pyruvate. Meaningful comparisons of CO_2 production from labeled substrates cannot be made between species during the blastocyst stages owing to their different rate of expansion, or tissue increase, as already mentioned. However all the stages of the embryo can be compared by relating CO_2 formed to oxygen consumed. This comparison shows that the oxidation of the substrates by the rabbit blastocyst is similar to the mouse blastocyst (Brinster 1969*b*). The requirement for pyruvate during the first 2 to 3 days of development seems to be a general characteristic of the preimplantation stages of mammalian embryos. In later stages either pyruvate or glucose can be used.

The mouse and rabbit embryos handle what little glucose oxidation there is in the preblastocyst stages very differently. In the rabbit embryo the CO_2 formed from carbon 1 of glucose is much greater than that formed from carbon 6. The ratio of C_1 to C_6 is 9.3, which indicates a very active pentose shunt oxidation (Fridhandler 1961; Brinster 1967*a*). After blastocyst formation the C_1 to C_6 ratio reduces to 1. In the mouse embryo the C_1 to C_6 ratio is approximately 1.6 throughout the preimplantation period. A ratio close to 1 of C_1 to C_6 suggests that the oxidation is predominantly by the Krebs cycle. Throughout the preimplantation period in the mouse, glucose oxidation seems to be through the Krebs cycle, whereas in the rabbit the early cleavage stages oxidize a substantial part of their glucose through the pentose shunt (Fridhandler 1961; Brinster 1967*a*, 1968*a*).

Recently it has been possible to determine the amount of lactate formed from

TABLE 4

COMPARISON OF CARBON DIOXIDE PRODUCTION FROM GLUCOSE, PYRUVATE, AND LACTATE BY THE PREIMPLANTATION MOUSE AND RABBIT EMBRYO

STAGE OF DEVELOPMENT	GLUCOSE		PYRUVATE		LACTATE	
	Mouse	Rabbit	Mouse	Rabbit	Mouse	Rabbit
Unfertilized	0.13	0.61	7.24	16.68	3.09	14.71
Fertilized	0.68	0.81	6.95	14.47	3.31	11.28
Day 2	1.19	5.98	6.03	13.42	2.77	22.11
Day 3	2.16	13.71	7.25	22.31	4.54	23.06
Day 4	8.84	50.33	11.96	201.19	11.20	178.2
Day 5	14.69	1335.15	15.73	2032.91	15.06	655.2
Day 6	....	4904.98	....	9882.36	....	3844.2

SOURCE: The glucose values for the rabbit are from Brinster 1968*a*, and the values for the mouse are from Brinster 1967*a*, *b*.
NOTE: All values are $\mu\mu$moles of CO_2 produced/embryo/hr. The substrate concentration was for glucose, 1 mg/ml; for pyruvate, 5×10^{-4}M; and for lactate, 5×10^{-2}M.

glucose (Wales 1969). In general it was found that lactate formation changes in a manner similar to glucose oxidation (tables 4, 5). Lactate formation from glucose rose from 0.06 $\mu\mu$moles per 1-cell embryo per hr to 11.00 $\mu\mu$moles per blastocyst per hr. Major increases in lactate formation as well as in CO_2 formation appear to occur at approximately the same time in development, that is, at fertilization and about the time of blastocyst formation (Brinster 1967*a;* Wales 1969). It has been suggested that the varying rate at which the embryo utilizes substrates in general and oxidizes glucose and forms lactate specifically is a result of changes in enzyme activities along metabolic pathways involved (Brinster 1967*a*, 1969*a*). Changes in cell membrane permeability during the development of the embryo may very well play a role in substrate utilization also, as is discussed later.

D. *Enzyme Activity in the Embryo*

The activity of a number of important enzymes has been measured during the preimplantation period of the mouse embryo. Lactate dehydrogenase (LDH) was the first enzyme to be measured: it was found that the mouse embryo contained a considerable amount (Brinster 1965*e*). The activity in the newly ovulated ovum and in the oocyte is 4.6×10^{-8} moles of substrate oxidized per hr per ovum, about 10 times the activity contained in mouse skeletal muscle, which contains the highest activity of all adult mouse tissue. By using turnover numbers that have been determined for other tissues for the enzyme LDH it can be calculated that LDH represents approximately 5% of the protein in the mouse ovum. It is difficult to justify such a large amount of LDH (about 10,000 times that required at optimum activity to convert sufficient pyruvate from lactate for the 1-cell embryo [Brinster 1965*c*, 1967*b*]) on the basis of the need to convert lactate to pyruvate. Perhaps LDH plays an important role during the growth and development of the oocyte before ovulation.

Although the rabbit embryo has a considerable ability to oxidize lactate, the LDH activity in the rabbit embryo is only about 1/100 that found in the mouse embryo (Brinster 1967*c*). This suggests that all the lactic dehydrogenase present in

TABLE 5

GLUCOSE METABOLISM BY THE PREIMPLANTATION MOUSE EMBRYO

Stage of Development	Hexokinase Activity $\mu\mu$moles of NADP Reduced/Hr/Embryo	CO_2 Produced[a]	Lactate Produced[a]	Carbon Incorporated[a]	Total[a]
Unfertilized	1.23 (7.38)	0.13	0.15	....	....
Fertilized	1.76 (10.56)	0.68	0.72	....	....
2-cell	1.70 (10.20)	1.19	2.10	1.65[b]	4.94
8-cell	2.24 (13.44)	2.16	3.75	1.72	7.63
Morula	5.63 (33.78)	6.73	27.96	....	....
Blastocyst	7.94 (47.64)	10.94	33.30	7.01	41.70
Late blastocyst	9.40 (56.40)	14.69	42.30	20.01	77.00

NOTE: Parentheses indicate maximum intracellular glucose carbon available to the embryo.
[a] Values are $\mu\mu$moles of carbon from glucose per hr per embryo.
[b] Values are average of previous 24 hr.

the mouse embryo is not needed for the conversion of lactate to pyruvate by the embryo after ovulation. It has been found that there is considerable variation among the oocytes of various species in their LDH activity (Brinster 1968*c*). Oocytes of the mouse have the highest activity; the oocytes of the rabbit and human the lowest. The reasons for the differences in LDH activity among the oocytes of various species is not known but may very well be related to preovulatory needs.

Hexokinase is known to be an important regulatory enzyme of glucose metabolism in many cells. In the mouse embryo at ovulation the hexokinase activity is only 1.2 $\mu\mu$moles of substrate converted per hr per embryo (Brinster 1968*b*), about 1/10,000 the activity of LDH. During the preimplantation period there is a 7-fold increase in activity of hexokinase in the mouse embryo (table 5). In the early stages of development there is sufficient hexokinase activity to account for the glucose metabolism, but at the blastocyst stage the total glucose used in various manners actually exceeds by about 50% the glucose that should be available on the basis of hexokinase activity (table 5). Perhaps the hexokinase operates at a faster rate under the more favorable conditions which are likely to be found within the cell. At any rate it seems clear that hexokinase activity is probably an important regulatory enzyme in the mouse embryo, since its activity is close to the rate of utilization of glucose throughout the preimplantation period. It will be interesting to see if the activity of phosphofructokinase, also shown to be a regulatory enzyme in a number of cell systems, accounts for any regulation of glucose utilization in the mouse embryo.

In the preimplantation mouse embryo both glucose-6-phosphate dehydrogenase (G6PDH) and malic dehydrogenase activity have been measured (Brinster 1966*a*, *b*). The activity of glucose-6-phosphate dehydrogenase is only about 3% of the LDH activity and shows a significant decrease during the preimplantation period. Despite its low activity compared to LDH, the G6PDH is still very active relative to levels found in other tissues. One wonders if this enzyme may also be important during oocyte development, considering the slight activity of the pentose shunt in the mouse embryo. Malic dehydrogenase activity is about ½ the glucose-6-phosphate dehydrogenase activity and is approximately the level found in other cell types within the mouse body. Malic dehydrogenase shows a slight but significant increase during the preimplantation period at about the time of blastocyst formation, when Krebs cycle activity is also increasing, whereas the other two enzymes show a decrease in activity.

E. *Permeability of the Embryo*

Changes in enzyme activity appear to be important in determining changes in the utilization of glucose. However, permeability of the ovum to substrates has been examined also as a possible explanation for the changes in utilization during development. Glucose will support the development of the 8-cell but not the 2-cell embryo (Whitten 1957; Brinster 1965*d*); yet there was no significant difference in permeability to glucose between them (Wales and Brinster 1968). Uptake of the 1-cell stage was considerably less than that of the 2-cell stage, suggesting that glucose permeability might play a role in differential utilization between the 1- and 2-cell stages. Unfortunately it is quite difficult to separate the contribution of permeability changes and enzyme activity changes for at least two reasons. First, the period of incubation is long in the uptake studies (30 min), and it is not possible to determine what part of the uptake is in the form of glucose and what part in the form of other intermediates. Total uptake of labeled substrate could be greatly affected by enzyme activities. Second, hexokinase activity is very closely tied to glucose permeability in many cells.

Permeability of the embryo to malate as studied by uptake studies shows a 7-fold increase between 2-cell and 8-cell stages of development (Wales and Biggers 1968). This seems to be a possible reason why malate will support the development of the 8-cell embryo but not the 2-cell embryo. However here again the long incubation period and inability to determine intracellular forms of the radioactive substrate suggest that differences in enzyme activity may be important. The evidence that permeability changes affect utilization in the case of malate seems much stronger than in the case of glucose since malate dehydrogenase is unchanged during the period when uptake was studied.

The uptake of pyruvate and lactate has been determined for 1- and 2-cell mouse embryos (Wales and Whittingham 1967). The fact that no difference exists between 1- and 2-cell embryos in their ability to accumulate intracellular products of pyruvate or lactate suggests that the difficulty experienced in culturing 1-cell embryos completely through to blastocysts may not be related to pyruvate or energy metabolism. The ability of the 1-cell embryo to develop in a medium containing pyruvate but not in a medium containing lactate may be due to the greater oxidation of pyruvate compared to lactate by the 1- and 2-cell embryo (Brinster 1967*b;* Wales and Whittingham 1967). The 1-cell stage may require a more readily available energy source than required by the 2-cell stage.

F. *Incorporation of Substrates by the Embryo*

Incorporation of carbon from glucose and pyruvate into the mouse embryo has been determined for the preimplantation mouse embryo over relatively long periods of time (20 to 24 hr). These studies show that the incorporation of glucose carbon is greater than the incorporation of pyruvate carbon for all the preimplantation developmental stages (Brinster 1969*c*). However oxidation of pyruvate carbon is much greater than glucose carbon up to the blastocyst stage, after which both substrates seem to be oxidized equally well. Pyruvate is oxidized more than it is incorporated and glucose is incorporated more than it is oxidized during the preimplantation period.

Previous studies have shown that the mouse embryo contains a considerable amount of glycogen during the preimplantation stage (Thomson and Brinster 1966; Stern and Biggers 1968). Some of this glycogen may arise from incorporated pyruvate or glucose carbon. The exact amount of glycogen present in the embryo is subject to debate. Histochemical studies (Thomson and Brinster 1966) suggest that a considerable amount of glycogen is present in the oocyte at ovulation and that additional glycogen is present at the 2- and the 8-cell stage. During blastocyst formation the stored glycogen is utilized. Biochemical studies (Stern and Biggers 1968) show very little glycogen present in the 1-cell embryo and a rapid accumulation of glycogen between the 1- and 8-cell stages (2.1 $\mu\mu$grams in 48 hr), and suggest that only about 20% of the glycogen is used during blastocyst formation. The studies on incorporation of pyruvate and glucose carbon (Brinster 1969*c*) show very little incorporation of carbon into the glycogen fraction of the mouse embryo before the blastocyst stage, but a rapid turnover of glycogen in the blastocyst. Only about 0.023 $\mu\mu$atoms per hr of carbon from glucose and pyruvate are incorporated into the glycogen of the embryo between the 1-cell and 8-cell stages. However 1.46 $\mu\mu$atoms of carbon per hr are required to accumulate the 2.1 mμgrams of glycogen found present after 48 hr. Obviously then pyruvate and glucose do not account for the required accumulation into glycogen, in fact for only 1.6% of it. The rest must come from other sources (perhaps endogenous protein) or from other precursors which might be detected histochemically but not biochemically. What these precursors are is unknown. One must remember that there is always the possibility that the embryos grown *in vitro* do not synthesize glycogen in as large quantities as the embryos grown *in vivo*.

II. Protein and Nucleic Acid Metabolism

The amount of work which has been done on protein and nucleic acid metabolism is somewhat less than that on energy metabolism, not because the interest in protein and nucleic acid metabolism is less but rather because the techniques are somewhat more tedious and the information somewhat more difficult to obtain. The mouse embryo has provided us with most of our information about macromolecule metabolism in the early mammalian embryo.

A. *Culture of the Embryo*

It has been shown that amino acids and proteins or both are important constituents of the culture medium for maintaining early mammalian embryos *in vitro*. Mouse embryos younger than the 8-cell stage will die in 24 hr if the medium does not contain bovine serum albumin or a complement of amino acids (Brinster 1965*c*). Despite the need for a fixed nitrogen source it is not possible to show a requirement for a specific amino acid. Any of the essential amino acids can be omitted from the culture medium, and the embryos will develop from the 2-cell stage to the blastocyst. Only the absence of cystine significantly reduces development (Brinster 1965*c*). It can be demonstrated that the mouse embryo will develop from the 8-cell stage to the blastocyst on single amino acids (Whitten 1957; Brinster and Thomson 1966) and from the 2-cell stage on glutathione as the only amino nitrogen source (Brinster 1968*d*). Thus culture studies suggest that the mouse embryo does not have an essential amino acid requirement. It seems likely however that endogenous stores in the mouse embryo are able to substitute the amino acids missing from the culture medium, making it impossible to demonstrate the essential amino acid requirement. After the embryo reaches the morula or blastocyst stage, the inclusion of serum or amino acids or both is beneficial to blastocyst expansion. It can be said that the culture requirements for the cleavage stages (preblastocyst stages) of the mouse embryo are rather simple but specific. The embryos seem to grow best with an energy source such as pyruvate and an amino nitrogen source such as bovine serum albumin.

The rabbit embryo requires an amino nitrogen source in the culture medium, but the nature of this amino nitrogen source varies, depending upon the experimental conditions. Daniel and Olson (1968) found that cystine, tryptophan, phenylalanine, lysine, arginine, and valine were essential for the cleavage of the early rabbit embryo. Mauer et al. (1968) could show no decrease in development when glycine, serine, glutamine, glutamic acid, tryptophan, or methionine were omitted from culture medium for 2-cell rabbit embryos. Brinster (1968*d*) found that the early rabbit embryo showed normal cleavage for 2 days when the only amino nitrogen source in the medium was oxidized glutathione or some single amino acids, such as alanine or glutamine. Presumably then there is no amino acid essential for the preblastocyst rabbit embryo, a condition similar to that found in the mouse embryo. During blastocyst formation and expansion there is no doubt that protein or amino acids or both are beneficial.

Since both mouse and rabbit embryos have been cultivated outside the body in media which contained no nucleic acid precursors, it seems likely that an exogenous source of these is not required for development. Since we know that DNA synthesis takes place throughout the cleavage stages and that RNA synthesis occurs also, in both late cleavage and blastocyst stages (Mintz 1964), it seems likely that the embryo is able to synthesize the required nucleic acids from simple exogenous carbon and nitrogen sources or endogenous precursors. TenBroeck (1968) found that the inclusion of the bases adenine, cytosine, thymine, and uracil had no effect on development of 2-cell mouse embryos into blastocysts. Likewise the nucleosides, adenosine, cytidine, guanosine, thymidine, and uridine had no effect. There was no evidence of a beneficial effect regardless of whether the compounds were employed singly or in combination. Since we know from autoradiographic studies (Mintz 1964) that these compounds can enter the embryo, the lack of effect cannot be due to impermeability of the embryos.

B. *Protein and Nucleic Acid Content*

During the preblastocyst cleavage period the embryo increases very little in size. Volume studies (Lewis and Wright 1935) and actual protein determinations (Brinster 1967*d*) have shown that the mouse embryo actually loses approximately 25% of its volume and 25% of its protein during the period from fertilization to blastocyst expansion. Reamer (1963), using biochemical techniques, was able to

TABLE 6

DNA AND RNA CONTENT OF THE MOUSE EMBRYO

Stage of Development	DNA	RNA
Unfertilized	45.8	1750
Fertilized	49.0	1680
2-cell	24.0	905
4-cell	21.0	500
8-cell	20.4	200
16-cell	19.3	50

SOURCE: Reamer 1963.
NOTE: All values are $\mu\mu$grams per cell.

determine the quantity of DNA and RNA in the mouse embryo from the time of ovulation up to the 16-cell stage. The results of these experiments are shown in table 6.

The DNA content in a fertilized mouse ovum is 49 $\mu\mu$grams, considerably more than the 2.5 $\mu\mu$grams one would expect in a haploid mouse cell. The possibility that the exact stage of DNA synthesis of the 1-cell ovum was not determined in these assays (Mintz 1964) may account for the high DNA content. If the polar bodies were both formed, the total DNA could be 4N and possibly slightly higher. Even so the large amount of DNA found by Reamer suggests very strongly that there is some extra chromosomal DNA present in the early mouse embryo, but the exact amount in the embryo proper is still not definite. Whether some of this extra chromosomal DNA is mitochondrial DNA is not known; it seems probable that the amount of mitochondrial DNA, if present at all, must be rather small since the number of mitochondria during the early cleavage stages is very low. The total DNA per embryo increases during the early cleavage stages as the total number of cells increases but the DNA per cell decreases toward the normal diploid quantity of 5 $\mu\mu$grams.

The RNA content of the newly ovulated mouse ovum is quite high compared to the average content of postimplantation embryonic cells (20 $\mu\mu$grams per cell). The total RNA per embryo decreases considerably from the 1-cell to the 16-cell stage so that at the 16-cell stage each cell contains only about 50 $\mu\mu$grams of RNA. The large quantity of RNA found in the newly ovulated mouse ovum in the early cleavage stages is surprising considering the small number of ribosomes generally present in these developmental stages.

C. *Incorporation of Precursors by the Embryo*

The uptake and incorporation of radioactive precursors into RNA and into protein of the mouse embryo grown *in vitro* has been studied quantitatively by several workers. Monesi and Salfi (1967) demonstrated that the uptake of uridine, lysine, and leucine into the mouse embryo was relatively constant until the 8-cell stage, following which the radioactivity per embryo increased rapidly. Claque and

TABLE 7

UPTAKE AND INCORPORATION OF ^{14}C-LEUCINE AND ^{3}H-URIDINE INTO PROTEIN AND RNA OF MOUSE EMBRYOS

STAGE OF DEVELOPMENT	LEUCINE INTO PROTEIN		URIDINE INTO RNA	
	Uptake	Incorporation	Uptake	Incorporation
2-cell	2508	1039	251	19
4-cell	2631	1376	658	114
8-cell	2727	2048	2081	443
Morula	3636	3029	9155	2774

SOURCE: Tasca and Hillman 1970.
NOTE: Values are CPM/100 embryos.

Glass (1968) have shown that the uptake of arginine by the mouse embryo is 20 times as great as to uptake of leucine. Tasca and Hillman (1970) have measured separately the uptake and incorporation of ^{3}H-uridine and ^{14}C-leucine into the mouse embryos grown *in vitro* both in the presence and absence of metabolic inhibitors. Their findings show a gradual increase in incorporation of both uridine and leucine with increasing age of the embryo (table 7). At the 8-cell and morula stages of development there is a sharp increase in uptake for both compounds.

The effect of inhibitors of nucleic acid and protein metabolism on the *in vitro* development of the preimplantation mouse embryo has been examined by several workers, and they have found that these compounds at various doses will inhibit development and the uptake of precursors, depending on the methods employed and the length of time of incubation. It is interesting to note that the dose of actinomycin (which inhibits DNA-dependent RNA synthesis(required to prevent cleavage of the mouse embryo is about one-millionth that required to prevent cleavage of the sea urchin embryo, suggesting that the mouse embryo relies much more than the sea urchin on DNA-dependent RNA synthesis during early cleavage stages.

Tasca and Hillman (1970) measured quantitatively the effect of actinomycin and cycloheximide (which inhibits protein synthesis) on the uptake and incorporation of ^{3}H-uridine and ^{14}C-leucine into the mouse embryo. Actinomycin inhibited 20 to 55% of the uridine incorporation but, surprisingly, did not affect leucine incorporation during the 3-hr incubation period. They found that the early embryos which lack nucleoli are more resistant to actinomycin than those in later stages which have fully developed nucleoli. The extranucleolar nuclear localization of the RNA synthesis at the 2- and 4-cell stages and the predominately nucleolar localization of synthesis at later stages strongly suggest that qualitative changes in RNA synthesis as well as quantitative changes in RNA synthesis occur during the preimplantation period in the mouse. Since leucine incorporation continued for 3 hr in the absence of new RNA synthesis, protein synthesis at the translational level must continue in the absence of new RNA synthesis. This suggests the embryo depends during the early cleavage stages to some extent upon previously synthesized RNA.

In the sea urchin protein synthesis during the cleavage stages up to gastrulation depends almost entirely on preovulatory synthesized RNA (Gross and Cousineau 1963*a, b,* 1964). Protein synthesis in the mouse embryo is much more dependent on new RNA than is protein synthesis in the sea urchin embryo as shown in long-term experiments (24 to 72 hr) with mouse embryos when ^{14}C-leucine incorporation and cleavage were inhibited with much lower levels of actinomycin than is required to inhibit protein synthesis and cleavage in the sea urchin embryo (Thomson and Biggers 1966; Mintz 1964; Gross and Cousineau 1964). Tasca and Hillman's studies (1970) show that cycloheximide inhibits 45 to 70% of leucine incorporation and 40 to 65% of the uridine incorporation into early cleavage stages of the mouse embryo. The inhibition of uridine incorporation by cycloheximide suggests that the inhibitor is blocking the formation of protein necessary for RNA synthesis and uridine incorporation.

Recently Ellem and Gwatkin (1968) measured the rate of nucleic acid synthesis in the preimplantation mouse embryo. They incubated the embryo for 5 hr in labeled precursors and then separated DNA, soluble RNA, ribosomal RNA, and DNA-like RNA (messenger RNA). By their system they were not able to detect nucleic acid synthesis in the 2-cell embryo; they did find incorporation into all nucleic acid fractions at the 8-cell and subsequent developmental stages. The increase in DNA, soluble RNA, and messenger RNA was gradual during the period from the 8-cell stage to the blastocyst. The gradual increase in messenger RNA suggests a gradual increase in the number of genes being transcribed in the genome as the demands for synthetic diversity accompanying differentiation increase. The synthesis of ribosomal RNA was greatly increased from the 8-cell stage to the morula; the rate at the 8-cell stage was only 12% of that at the morula stage. The increases in ribosomal RNA and in soluble RNA are severely inhibited by low concentrations (10^{-7}M) of actinomycin; acceleration in DNA and messenger RNA are unaffected.

Woodland and Graham (1969) have shown that synthesis of high molecular weight and low molecular weight RNA in the mouse embryo can be detected first during the 2-cell stage. Using sucrose density gradient centrifugation, they identified the high molecular weight RNA in the 28S and 18S regions of the gradient at the 8-cell stage. The low molecular weight RNA separated primarily with the 4S fraction on Sephadex columns at the 8-cell stage.

The sharp increase in RNA synthesis between the 8-cell stage and the morula found in the above studies and its identification as ribosomal RNA confirms and extends the investigations of Mintz (1964), Monesi and Salfi (1967), and Tasca and Hillman (1970). Reamer's analyses have shown that by the time the embryo has become a morula the RNA content of the blastomeres is approaching that of other tissue cells. It is reasonable to expect to see in the embryo at this stage of development the initiation of RNA synthesis at a rate which would avoid further depletion of RNA content below a level adequate for protein synthesis.

III. Summary

There appears to be a morphological, biochemical, and developmental similarity among the preimplantation stages of the eutherian mammals. This is particularly true for the newly ovulated ovum and for the embryo in the first 2 to 3 days of development. Unlike most other cells grown *in vitro* the early mouse embryo will not survive on glucose as the only exogenous energy source but requires pyruvate and perhaps lactate in its environment. Exogenous protein or amino acids are necessary for development, but a requirement for the essential amino acids has not been definitely demonstrated. There is some indication that internal stores of glycogen are first synthesized by the early embryo and then utilized by the blastocyst. There is a progressive development or change in the importance of major energy pathways. The Krebs cycle appears to be a major energy-supplying pathway throughout the preimplantation period, whereas the Embden-Meyerhof pathway has a very low capability until about the time of blastocyst formation. The pentose shunt is very active in the early rabbit embryo but not in the mouse embryo. It seems probable that changes in enzyme activity play an important role in the regulation of metabolic pathway activity.

Incorporation of radioactive precursors into protein and RNA is very low before the 4-cell stage. True nucleoli appear between the 4- and 8-cell stages, after which protein and RNA synthesis is accelerated. This acceleration, which is much accentuated at the morula and blastocyst stages, can be inhibited by actinomycin D in low concentration; it has now been shown that the largest increase in RNA synthesis is in the ribosomal-RNA fraction. The large increase in RNA synthesis occurs in the morula about the time when the store of RNA present in the newly ovulated ovum has fallen to a low level on an individual cell basis.

There is considerable evidence that metabolism of the mammalian embryo is different from that of amphibian and invertebrate embryos such as the sea urchin.

Acknowledgments

Financial support for the author's research is currently from the National Institutes of Health grant HD 03071. In the past financial assistance has been received from the National Institutes of Health, National Science Foundation, and the Population Council.

References

Austin, C. R. 1961. *The mammalian egg*. Springfield, Ill.: Charles C. Thomas.

Biggers, J. D.; Whittingham, D. G.; and Donahue, R. P. 1967. The pattern of energy metabolism in the mouse oocyte and zygote. *Proc Nat Acad Sci* 58:560.

Brinster, R. L. 1963. A method for *in vitro* cultivation of mouse ova from two-cell to blastocyst. *Exp Cell Res* 32:205.

———. 1965*a*. Studies on the development of mouse embryos *in vitro:* I. The effect of osmolarity and hydrogen ion concentration. *J Exp Zool* 158:49.

———. 1965*b*. Studies on the development of mouse embryos *in vitro:* II. The effect of energy source. *J Exp Zool* 158:59.

———. 1965*c*. Studies on the development of mouse embryos *in vitro:* III. The effect of fixed-nitrogen source. *J Exp Zool* 158:69.

———. 1965*d*. Studies on the development of mouse embryos *in vitro:* IV. Interaction of energy sources. *J Reprod Fertil* 10:227.

———. 1965*e*. Lactic dehydrogenase activity in the preimplanted mouse embryo. *Biochim Biophys Acta* 110:439.

———. 1966*a*. Malic dehydrogenase activity in the preimplanted mouse embryo. *Exp Cell Res* 43:131.

———. 1966*b*. Glucose 6-phosphate-dehydrogenase activity in the preimplantation mouse embryo. *Biochem J* 101:161.

———. 1967*a*. Carbon dioxide production from glucose by the preimplantation mouse embryo. *Exp Cell Res* 47:271.

——— 1967*b*. Carbon dioxide production from lactate and pyruvate by the preimplantation mouse embryo. *Exp Cell Res* 47:634.

———. 1967*c*. Lactate dehydrogenase activity in the preimplantation rabbit embryo. *Biochim Biophys Acta* 148:298.

———. 1967*d*. Protein content of the mouse embryo during the first five days of development. *J Reprod Fertil* 13:413.

———. 1968*a*. Carbon dioxide production from glucose by the preimplantation rabbit embryo. *Exp Cell Res* 51:330.

———. 1968*b*. Hexokinase activity in the preimplantation mouse embryo. *Enzymologia* 34:304.

———. 1968*c*. Lactate dehydrogenase activity in the oocytes of mammals. *J Reprod Fertil* 17:139.

———. 1968*d*. Effect of glutathione on the development of two-cell mouse embryos in vitro. *J Reprod Fertil* 17:521.

———. 1969*a*. Mammalian embryo culture. In *The mammalian oviduct,* ed. E. S. E. Hafez and R. Blandau, p. 419. Chicago: University of Chicago Press.

———. 1969*b*. Radioactive carbon dioxide production from pyruvate and lactate by the preimplantation rabbit embryo. *Exp Cell Res* 54:205.

———. 1969*c*. The incorporation of carbon from glucose and pyruvate into protein and glycogen of the preimplantation mouse embryo. *Exp Cell Res* 58:153.

———. 1970. Culture of two-cell rabbit embryos to morulae. *J Reprod Fertil* 21:17.

Brinster, R. L., and Thomson, J. L. 1966. Development of eight-cell mouse embryos *in vitro. Exp Cell Res* 42:308.

Claque, B. H., and Glass, L. E. 1968. The quantitation and localization of arginine and leucine uptake in the preimplantation mouse embryo. Presented at the Third International Congress of Cytology, Rio de Janeiro, Brazil, 19–22 May 1968.

Daniel, J. C., Jr. and Olson, J. D., 1968. Amino acid requirements for cleavage of the rabbit ovum. *J Reprod Fertil* 15:453.

Ellem, K. A. O., and Gwatkin, R. B. L. 1968. Patterns of nucleic acid syntheses in the early mouse embryo. *Devel Biol* 18:311.

Fridhandler, L. 1961. Pathways of glucose metabolism in fertilized rabbit ova at various pre-implantation stages. *Exp Cell Res* 22:303.

Gross, P. R., and Cousineau, G. H. 1963*a*. Synthesis of spindle-associated proteins in early cleavage. *J Cell Biol* 19:260.

———. 1963*b*. Effects of actinomycin D on macromolecule synthesis and early development in sea urchin eggs. *Biochem Biophys Res Commun* 10:321.

———. 1964. Macromolecule synthesis and the influence of actinomycins on early development. *Exp Cell Res* 33:369.

Lewis, W. H., and Wright, E. S. 1935. On the development of the mouse. *Contr Embryol Carneg Inst* 25:113.

Mauer, R. E.; Hafez, E. S. E.; Ehlers, M. H.; and King, J. R. 1968. Culture of two-cell rabbit eggs in chemically defined media. *Exp Cell Res* 52:293.

Mills, R. M., Jr., and Brinster, R. L. 1967. Oxygen consumption of preimplanted mouse embryos. *Exp Cell Res* 47:337.

Mintz, B. 1964. Synthetic processes and early development in the mammalian egg. *J Exp Zool* 157:85.

Monesi, V., and Salfi, V. 1967. Macromolecular syntheses during early development in the mouse embryo. *Exp Cell Res* 46:632.

Onuma, H.; Maurer, R. R.; and Foote, R. H. 1968. *In vitro* culture of rabbit ova from early cleavage stages to the blastocyst stage. *J Reprod Fertil* 16:491.

Reamer, G. R. 1963. The quantity and distribution of nucleic acids in the early cleavage stages of the mouse embryo. Ph.D. thesis, Boston University.

Stern, S., and Biggers, J. D. 1968. Enzymatic estimation of glycogen in the cleaving mouse embryo. *J Exp Zool* 168:61.

Tasca, R. J., and Hillman, N. 1970. Effects of actinomycin D and cycloheximide on RNA and protein synthesis in cleavage stage mouse embryos. *Nature (Lond)* 225:1022.

TenBroeck, J. T. 1968. Effect of nucleosides and nucleoside bases on the development of preimplantation mouse embryos *in vitro*. *J Reprod Fertil* 17:571.

Thomson, J. L., and Biggers, J. D. 1966. The effect of inhibitors of protein synthesis on the development of mouse embryos *in vitro*. *Exp Cell Res* 41:411.

Thomson, J. L., and Brinster, R. L. 1966. Glycogen content of preimplantation mouse embryos. *Anat Rec* 155:97.

Wales, R. G. 1969. Production of carboxylic acids by the preimplantation mouse embryo. *Aust J Biol Sci* 22:701.

Wales, R. G., and Biggers, J. D. 1968. The permeability of two- and eight-cell mouse embryos to L-malic acid. *J Reprod Fertil* 15: 103.

Wales, R. G., and Brinster, R. L. 1968. The uptake of hexoses by mouse embryos. *J Reprod Fertil* 15: 415.

Wales, R. G., and Whittingham, D. G. 1967. A comparison of the uptake and utilization of lactate and pyruvate by one- and two-cell mouse embryos. *Biochim Biophys Acta* 148:703.

Whitten, W. K. 1956. Culture of tubal mouse ova. *Nature* (*Lond*) 176:96.

———. 1957. Culture of tubal ova. *Nature* (*Lond*) 179:1081.

Whitten, W. K., and Biggers, J. D. 1968. Complete development *in vitro* of the preimplantation stages of the mouse in a simple chemically defined medium. *J Reprod Fertil* 17:399.

Woodland, H. R., and Graham, C. F. 1969. RNA synthesis during early development of the mouse. *Nature* (*Lond*) 221:327.

18

New Observations on the Nutrition of the Mammalian Oocyte and the Preimplantation Embryo

J. D. Biggers
Division of Population Dynamics
School of Hygiene and Public Health
The Johns Hopkins University
Baltimore, Maryland

Although rabbit embryos have been cultured *in vitro* for many years (Brachet 1912; Pincus 1930), the new era of mammalian ovum culture began with a paper published by Whitten in 1956. He showed that a very simple chemically defined medium, Krebs-Ringer bicarbonate supplemented by glucose and bovine plasma albumin, would support the development of 8-cell mouse embryos to the blastocyst stage. Subsequently McLaren and Biggers (1958) showed that blastocysts produced in this way are capable of developing into normal mice after surgical transfer into uterine foster-mothers. Whitten also reported that mouse embryos, prior to the 8-cell stage, could not develop in his medium. One year later he (Whitten 1957) discovered the singular fact that when lactate was added to his initial medium, late 2-cell mouse embryos could develop into blastocysts. In the same year Bishop (1957) found that high concentrations of lactate occur in the secretions of the Fallopian tubes of rabbits. Thus, more than a decade ago, evidence had already been accumulated to indicate that the metabolism of the early mammalian embryo might be unusual.

In 1962 Biggers, Gwatkin, and Brinster demonstrated that newly fertilized mouse ova could develop at a normal rate in organ cultures of Fallopian tubes maintained in medium BGJ, a chemically defined medium originally described by Biggers, Gwatkin, and Heyner (1961). These results, reviewed in the light of those of Whitten (1957), suggested that the Fallopian tube secretions contribute something essential for the development of the mouse embryo prior to the late 2-cell stage. Recently Whittingham (1968) showed that fertilized mouse ova can develop only in organ cultures of the ampullary region of the Fallopian tube, and that organ cultures of the isthmal region will not support development. Thus it seems likely that the ampullary region of the Fallopian tube provides a specialized environment for the early development of the mouse. The purpose of this chapter is to review the factors involved in the light of these results.

I. Theoretical Aspects

Providing that mating has been successful, a newly ovulated ovum and a sperm will fuse in the ampullary region of the oviduct to form a new individual. Two basic events occur at this time: first, the association in one cell of two haploid sets of chromosomes whose interaction will create the genome of the new individual; second, activation which leads to the development of a series of mitotic divisions, known as cleavage divisions, in the ovum. In oviparous forms, such as *Arbacia,* these events appear to be entirely supported by the newly fertilized ovum since cleavage will occur in artificial seawater consisting only of inorganic salts (Shapiro 1941).

From a functional point of view the following types of substance in the Fallopian tube secretions which may regulate the development of embryos can be postulated:

1. Specific maternal regulators of development
   a) Maternal hormones
   b) Specific substances of maternal origin whose exclusive function is to regulate embryonic development
2. Compounds of maternal origin which provide exogenous substances of nutritive value to the embryo.

A static classification of this kind falls short of conveying the complexities which may occur as pregnancy develops. Instead, kinetic concepts must be adopted in which we visualize changes in the concentrations of the various substances in the Fallopian tube with time, and simultaneous changes in the ability of the developing embryos to respond to or utilize the various substances.

How can such a complex system be studied? One approach is to analyze the biochemical processes occurring in the embryos during the early stages of development and the composition of the Fallopian tube secretions which form the immediate environment of the developing embryo. The progress in both these areas has been reviewed elsewhere—mammalian embryo metabolism (Brinster, chapter 17, this volume; Biggers and Stern 1970; Whittingham 1969*a*); Fallopian tube secretions (Restall 1966). Another approach is to utilize the methods of embryo culture in which attempts are made to develop simple chemically defined media which support development. A detailed account of these methods has been given by Biggers, Whitten, and Whittingham (1970). In this chapter I wish to describe very recent work on the mouse in this area and reemphasize some of the problems in the interpretation of these experiments. At present our knowledge is largely confined to the mouse, and it is with this species that I will be concerned.

II. Study of the Nutritional Requirements of Mouse Embryos *in Vitro*

A. *Late 2-Cell Embryo*

One application of chemically defined media is for the determination of a set of essential substances we assume to be the minimum necessary for development. Although this approach is simple in concept, it has problems of interpretation because of interactions which may occur between the effects of different components in the medium, and because alternative components may exist (see Biggers, Rinaldini, and Webb 1957).

Some of these phenomena are well illustrated in the quantitative studies done on the minimal energy-source requirements of the late 2-cell mouse embryo (Biggers and Brinster 1965; Brinster 1963, 1965*a, b*). These studies, starting with the medium of Whitten (1957), showed that only four compounds could supply energy

TABLE 1

EFFECT OF DIFFERENT CONCENTRATIONS OF LACTATE AND PYRUVATE ON THE DEVELOPMENT OF LATE 2-CELL MOUSE EMBRYO TO BLASTOCYSTS

		Pyruvate (X10⁻⁴M)		
		0	2.5	5.0
Lactate (X10⁻²M)	0	0	46.9	46.9
	2.5	29.2	66.7	63.5
	5.0	50.0	66.7	68.8

SOURCE: After Brinster 1965*b*.
NOTE: The result is expressed as the percentage number of blastocysts. There are 96 embryos per group.

which would support the development of late 2-cell mouse embryos to blastocysts: lactate, pyruvate, oxaloacetate, and phosphoenolpyruvate. Surprisingly, glucose would not support development. If no energy source was included in the medium, the 2-cell embryos disintegrated within 24 hr. These experiments clearly showed that alternative compounds may have similar functions in a minimal chemically defined medium. The problem of studying the minimal requirements is even more complex as shown by the investigation of the joint action of lactate and pyruvate in the medium (Brinster 1965*b*). The percentage of 2-cell mouse embryos which develop into blastocysts in various combinations of lactate and pyruvate concentrations is given in table 1. The results demonstrate that the maximum incidence of development occurs when both lactate and pyruvate are present. Both compounds apparently have a synergistic effect. Thus studies of the minimal requirements in culture may reveal not only a set of alternative compounds but also combinations which are more effective than any one alone. The interaction at a biochemical level of lactate and pyruvate in 2-cell mouse embryos has been confirmed recently by Whittingham and Wales (1970).

B. *1-Cell Embryo*

Let us now turn to the problem of the culture requirements of the newly fertilized mouse ovum. The organ culture experiments of Biggers, Gwatkin, and Brinster (1962) and Whittingham (1968), in comparison with the totally *in vitro* studies of Whitten (1957), could suggest that the Fallopian tube provides some critical conditions for the initial stages of development by supplying protection or some essential substance.

A major breakthrough occurred as the result of the recognition that often 1-cell mouse embryos cleave to the 2-cell stage and remain arrested and yet do not degenerate. This was first made explicit in informal discussions at the Ciba symposium on the preimplantation stages of pregnancy (Wolstenholme and O'Connor 1965). Soon after, Whittingham and Biggers (1967) showed that in a medium containing pyruvate and lactate this first cleavage division was normal since the arrested 2-cell stage embryos when placed in organ cultures of the ampullary region of the Fallopian tube developed into blastocysts. When transferred to pseudopregnant uterine foster-mothers by the methods described by Biggers, Moore, and Whittingham (1965), some of these blastocysts developed into fetuses. These experiments demonstrated two important facts: (1) that the 1st cleavage division can occur in simple defined media and does not depend on a specific oviductal substance acting during formation and fusion of the pronuclei and the events leading to the 1st division; and (2) that young 2-cell mouse embryos may experience a block to development in the intermitotic period between the 1st and 2d cleavage division. The latter fact could be interpreted as an indication that there is a specific maternal regulator at this stage of development of the mouse that is not present in the chemically defined media. As will be shown, there is now little evidence for such a factor.

III. Requirements of the 1-Cell Embryo

The minimum energy source requirements of the 1st cleavage division were studied by Biggers, Whittingham, and Donahue (1967). The results showed that only pyruvate and oxaloacetate could support the 1st cleavage division. Lactate, phosphoenolpyruvate, and glucose were ineffective, and, according to recent work of Whittingham (1969*b*), in the presence of these substances abnormal pronuclei develop which do not fuse.

IV. Requirements of the Oocyte

When mouse oocytes which are in the arrested dictyate stage are removed from the ovary and placed in culture, they spontaneously undergo meiotic maturation (Edwards 1965). The first polar body is extruded and the ovum eventually becomes arrested at the metaphase II stage. This spontaneous maturation happens also in the simple chemically defined media which allow the 1st cleavage to occur (Biggers, Whittingham, and Donahue 1967; Donahue 1968). This fact was exploited to determine the minimum energy requirements for oocyte maturation in the mouse. The results showed that only pyruvate and oxaloacetate were effective and that lactate, phosphoenolpyruvate, and glucose were not. Thus the energy requirements of the oocyte appear to be the same as those of the 1-cell embryo.

V. Summary of the Minimum Energy-Source Requirements

Table 2 summarizes the work which has been done on the minimum energy requirements of various stages of mouse development. Two main tentative conclusions emerge: (1) the metabolism of the early stages is restricted, and, as development proceeds, metabolic pathways emerge so that a wider spectrum of substances can be utilized; and (2) the metabolic characteristics of the early stages are determined by the ovum and are, therefore, produced during the differentiation of the oocyte

TABLE 2

COMPARISON OF COMPOUNDS WHICH SUPPORT THE DEVELOPMENT *in Vitro* OF MOUSE OOCYTES, FERTILIZED OVA, 2-CELL AND 8-CELL EMBRYOS

Substrate	Oocyte[a]	1-Cell[a]	2-Cell[b]	8-Cell[c]
Lactate	−	−	+	+
Pyruvate	+	+	+	+
Oxaloacetate	+	+	+	+
Phosphoenolpyruvate	−	−	+	+
Malate	?	?	−	+
Citrate	?	?	−	+
α-ketoglutarate	?	?	−	+
Acetate	?	?	−	±
d-glyceraldehyde	?	?	−	−
Glucose-6-phosphate	?	?	−	−
Fructose-1, 6-diphosphate	?	?	−	−
Glucose	−	−	−	+
Fructose	?	?	−	±

[a] Biggers, Whittingham, and Donahue 1967.
[b] Brinster 1965*a*.
[c] Brinster and Thomson 1966.

within the mother. The little knowledge available concerning the mechanisms which control these changes in metabolism has been discussed by Biggers, Whittingham, and Donahue (1967).

VI. Problem of the Intermitotic 2-Cell Block

New insight has recently been obtained concerning the nature of the block to development at the 2-cell stage in the routine culture of mouse embryos in chemically defined media (Whitten and Biggers 1968). A modified medium was developed which allowed the complete normal development of a fertilized ovum from an F_1 hybrid between two inbred lines (C57 $\times$ SJL). The same medium, however, would not support development of ova from the parental inbred strains or from random-bred mice. These initial experiments showed that fertilized ova from the F_1 hybrid develop completely from the 1-cell stage, prior to pronuclear fusion, to the blastocyst stage.

Recently some of the components of the medium were studied systematically in factorially designed experiments (unpublished). The importance of bovine plasma albumin and sodium chloride emerged from the experiments. The work led to the formulation of a new medium rich in crystalline bovine serum albumin with a higher concentration of sodium chloride than that described by Whitten and Biggers (1968) (table 3).

One-cell ova were obtained from several genetic strains of mice and placed in microdroplets of the medium shown in table 3. The percentage of these 1-cell ova which developed to the 2-cell stage was recorded at 24 hr and the percentage which developed to blastocysts was recorded at 72 hr. The results are summarized in table 4. It is clear that the medium supports the 1st cleavage division of a variety ova derived from inbred and F_1 hybrid mothers. However the subsequent development of these 2-cell embryos to blastocysts shows considerable variation among the genotypes of the ova which were fertilized. The contrast can be seen by comparing the response of the 1-cell ova from B6AF_1 mice in which 90.3% and 58.3% develop to the 2-cell stage and the blastocyst respectively with the response from DBA mice in which 93.0% and 0% develop to the 2-cell stage and blastocyst respectively. Thus a simple chemically defined medium has now been found which allows the

TABLE 3

MEDIUM USED FOR THE CULTURE OF PRE-IMPLANTATION MOUSE EMBRYOS FROM THE 1-CELL STAGE

Compound	mM
Sodium chloride	94.6
Potassium chloride	4.78
Potassium dihydrogen phosphate	1.19
Magnesium chloride	1.19
Calcium lactate	1.71 (DL)
Sodium lactate (DL)	21.58
Sodium pyruvate	0.33
Sodium bicarbonate	25.07
Glucose	5.55
Crystalline bovine albumin	4 mg/ml
Penicillin	100 U/ml
Streptomycin	50 μg/ml

TABLE 4

RESPONSE OF DIFFERENT STRAINS OF MICE CULTURED FROM THE 1-CELL STAGE

Strain	% 2-Cell	% Blastocyst
B6AF_1	90.3	58.3
C57×SJL	91.5	50.0
C3H	77.8	36.1
B6 D2 F_1	97.0	29.2
C3H×BDA	66.5	8.3
C57[a]	70.8	4.2
Swiss (Random bred)	91.5	2.8
DBA	93.0	0

NOTE: There are 72 ova per group in 2 replicates.
[a] 48 ova only.

complete development *in vitro* of 1-cell embryos of several strains of mice to the blastocyst stage. This refutes the idea that a specific maternal regulator of embryonic development at the 2-cell stage is required for all strains of mice; the possibility still remains, however, that the block in development which occurs in some strains *in vitro* may be due to the lack of a specific maternal regulator. Preliminary data on these strains suggest that arrested development occurs not only at the 2-cell stage but also at the 8-cell and morula stages. If this work is confirmed, there will be little reason to believe that a unique control of development occurs at any stage of the preimplantation period of the mouse; the blocks to development which have been observed previously will be considered as artifacts of the various media used at different times.

VII. Conclusions

The combined results of several investigators who have studied the conditions necessary for the culture of mouse embryos *in vitro* show that special environmental conditions are required, and that these conditions may change with the stage of development. The primary question is, Are these results strong evidence for what occurs when the embryos develop under natural conditions *in vivo?* At best the results should only be regarded as clues to events *in vivo,* and every attempt should be made to obtain independent corroboration. The danger of relying solely on the results of culture methods is illustrated by two problems of interpretation described in this chapter: (1) when interactions occur between the effects of different components of the medium, and (2) when apparent blocks to certain stages of development occur. Unfortunately at present we have no techniques available for the direct study of mammalian embryos *in situ.* Instead we must rely on indirect evidence supplied by biochemical studies on the embryos and Fallopian tube secretions. Provided the results obtained from the various sources are mutually consistent, we may have some confidence in the overall findings.

Acknowledgments

The work described in this paper has been supported in part by grants from the Population Council, the Ford Foundation, and the National Institute for Child Health and Human Development.

References

Biggers, J. D., and Brinster, R. L. 1965. Biometrical problems in the study of early mammalian embryos *in vitro. J Exp Zool* 158:39.

Biggers, J. D.; Gwatkin, R. B. L.; and Brinster, R. L. 1962. Development of mouse embryos in organ cultures of Fallopian tubes on a chemically defined medium. *Nature* 194:747.

Biggers, J. D.; Gwatkin, R. B. L.; and Heyner, S. 1961. Growth of embryonic avian and mammalian tibiae on a relatively simple chemically defined medium. *Exp Cell Res* 25:41.

Biggers, J. D.; Moore, B. D.; and Whittingham, D. G. 1965. Development of mouse embryos *in vivo* after cultivation from two-cell ova to blastocysts *in vitro. Nature* 206:734.

Biggers, J. D.; Rinaldini, L. R.; and Webb, M. 1957. The study of growth factors in tissue culture. *Sympos Soc Exp Biol* 11:264.

Biggers, J. D., and Stern, S. 1970. Metabolic aspects of early mammalian development. *Advances Reprod Physiol.* In press.

Biggers, J. D.; Whitten, W. K.; and Whittingham, D. G. 1970. The culture of mouse embryos *in vitro.* In *Methods of mammalian embryology,* ed. J. C. Daniel. San Francisco: Freeman. In press.

Biggers, J. D.; Whittingham, D. G.; and Donahue, R. P. 1967. The pattern of energy metabolism in the mouse oocyte and zygote. *Proc Nat Acad Sci USA* 58:560.

Bishop, D. W. 1957. Metabolic conditions within the rabbit oviduct. *Int J Fertil* 2:11.

Brachet, A. 1912. Developpement *in vitro* de blastomeres et jeunes embryons de mammifères. *C R Acad Sci [D] (Paris)* 155:1191.

Brinster, R. L. 1963. A method for *in vitro* cultivation of mouse ova from two-cell to blastocyst. *Exp Cell Res* 32:205.

———. 1965*a*. Studies on the development of mouse embryos *in vitro:* II. The effect of energy source. *J Exp Zool* 158:59.

———. 1965*b*. Studies on the development of mouse embryos *in vitro:* IV. Interaction of energy sources. *J Reprod Fertil* 10:227.

Brinster, R. L., and Thomson, J. L. 1966. *In vitro* culture requirements of the 8-cell mouse ovum. *Exp Cell Res* 42:308.

Donahue, R. P. 1968. Maturation of the mouse oocyte *in vitro:* I. Sequence and timing of nuclear progression. *J Exp Zool* 169:237.

Edwards, R. G. 1965. Maturation *in vitro* of mouse, sheep, cow, pig, rhesus monkey and human ovarian oocytes. *Nature* 208:349.

McLaren, A., and Biggers, J. D. 1958. Successful development and birth of mice cultivated *in vitro* as early embryos. *Nature* 182:877.

Pincus, G. 1930. Observations on the living eggs of the rabbit. *Proc Roy Soc [Biol], Ser B* 107:132.

Restall, B. J. 1966. The biochemical and physiological relationships between the gametes and the female reproductive tract. *Advances Reprod Physiol* 2:181.

Shapiro, H. 1941. Centrifugal elongation of cells, and some conditions governing the return to sphericity, and cleavage time. *J Cell Comp Physiol* 18:61.

Whitten, W. K. 1956. Culture of tubal mouse ova. *Nature* 177:96.

———. 1957. Culture of tubal ova. *Nature* 179:1081.

Whitten, W. K., and Biggers, J. D. 1968. Complete development *in vitro* of the preimplantation stages of the mouse in a simple chemically defined medium. *J Reprod Fertil* 17:399.

Whittingham, D. G. 1968. Development of zygotes in cultured mouse oviducts: I. The effect of varying oviductal conditions. *J Exp Zool* 169:391.

———. 1969*a*. Biochemical aspects of early gestation. *Proc III Int Cong Congenital Abnormalities, The Hague.*

———. 1969*b*. The failure of lactate and phosphoenolpyruvate to support development of the mouse zygote *in vitro. Biol Reprod* 1:381.

Whittingham, D. G., and Biggers, J. D. 1967. Fallopian tube and early cleavage in the mouse. *Nature* 213:942.

Whittingham, D. G., and Wales, R. G. 1970. The accumulation and utilization of substrate carbon by 2-cell mouse embryos in media containing various combinations of lactate and pyruvate. *Biochim Biophys Acta.* In press.

Wolstenholme, G. E. W., and O'Connor, M., eds. 1965. *Preimplantation stages of pregnancy*. London: Churchill.

19

Recent Progress in Investigations of Fertilization *in Vitro*

Benjamin G. Brackett
Division of Reproductive Biology
Department of Obstetrics and Gynecology
School of Medicine
Department of Animal Biology
School of Veterinary Medicine
University of Pennsylvania

Fertilization is the union of the gametes. It begins with sperm penetration through the zona pellucida and includes the development and syngamy of male and female pronuclei with alignment of their respective chromosome groups on the first cleavage spindle. Fertilization is completed with the metaphase of the 1st cleavage mitosis or cleavage to the 2-cell stage of the ovum. Efforts have been made to observe mammalian fertilization *in vitro* (literally, in glass) for almost a century. This has been the subject of many reviews (Austin 1951, 1961; Austin and Bishop 1957; Brackett 1969*b;* Chang 1957, 1968; Chang and Pincus 1951; Smith 1951; Thibault and Dauzier 1961). The most commonly studied mammalian species has been the rabbit, although success of *in vitro* fertilization has been claimed also in the guinea pig, rat, hamster, mouse, and man. Discovery of sperm capacitation in 1951 (Austin 1951; Chang 1951) enhanced the development of this field greatly. Capacitation is some kind of physiological change that the spermatozoon must undergo before it is capable of penetrating the ovum (Austin 1952). There is sufficient evidence to support the authenticity of fertilization *in vitro* for only three mammalian species: the hamster, the mouse, and the rabbit. The purpose of this chapter is to review the literature on *in vitro* fertilization of mammalian ova with special emphasis on recent investigations.

I. Early Investigations and Criteria for *in Vitro* Fertilization

Before the discovery of sperm capacitation it was presumed that fertilization would take place if mammalian ova and spermatozoa were brought together. Many of the early reports therefore consist of descriptions of observations made on rabbit, rat, and guinea pig ova which were exposed to ejaculated or epididymal spermatozoa (Frommolt 1934; Krassovskja 1934, 1935; Long 1912; Onanoff 1893; Schenk 1878; Yamane 1935). The ova were recovered from the oviduct, uterus, or, more commonly, the ovarian follicles. Extensive experiments on *in vitro* fertilization were carried out by Pincus (1930, 1939) and Pincus and Enzmann (1934, 1935). After the addition of ejaculated or epididymal spermatozoa to ovarian or tubal rabbit ova they noted shrinkage of the vitellus, extrusion of the second polar body, presence of spermatozoa in the perivitelline space and vitellus as seen in histological sections, formation of pronuclei, segmentation of ova, and birth of young. However Pincus and Enzmann (1936), Pincus (1936, 1939), and Pincus and Shapiro (1940) have recorded extrusion of the second polar body, formation of pronuclei, segmentation of ova and birth of young when rabbit ova, in the absence of spermatozoa, were subjected to supra- or subnormal temperatures, hyper- or hypotonic solution, butyric acid, or culturing in a moist chamber. Thibault (1947, 1948, 1949) has studied artificial activation by cold and noted the formation of a single, diploid nucleus or two nuclei closely resembling pronuclei. Following cold treatment the first polar body may divide so that the ovum appears to have two normal polar bodies. The action of the microtome in histological sectioning of ova can lead to translocation of sperm heads; as a result the ovum may appear to have been penetrated by a sperm cell. In possible refutation of the claim that live young resulted from the transfer of their *in vitro* fertilized ova, Chang and Pincus (1951) point out that the development of transferred ova may have been due to adherent spermatozoa effecting fertilization within the oviducts. This might also explain how Venge (1953) succeeded in obtaining young of both sexes following transfer of rabbit ova which were washed after a 3½ hr incubation with semen. That the vasectomized males used to induce ovulation in the recipient were not sterile is another possible explanation for reports of successful development of experimentally treated ova to the birth of live young.

Pincus and Saunders (1939) reported that about 30% of human ovarian ova cultured in serum for intervals ranging between 8½ and 24 hr showed polar body formation and theoretically at least became susceptible to fertilization. Rock and Menkin (1944) and Menkin and Rock (1948) cultured human oocytes *in vitro* for 24 hr to induce maturation and then cultured them for an additional 45 hr in the presence of human spermatozoa. Four of 138 ova cleaved into 2 to 3 cells. It is doubtful that these ova were actually fertilized since the oocytes were allowed to remain at room temperature for several hours before the beginning of the maturation incubation and spermatozoa were added at room temperature for an hour before the second incubation period. Two of the 3 cleaved ova which were described in detail were recovered from cultures that were grossly contaminated by bacteria (Menkin and Rock 1948).

Interesting experiments on rabbit ova were carried out in two different laboratories at about the time that sperm capacitation was discovered. Moricard and Bossu (1949) and Moricard (1949, 1950) pointed out that if ova contained in pieces of oviduct were treated with spermatozoa under a layer of mineral oil and incubated for 7 hr sperm heads could be identified in the zona pellucida, perivitelline space, and vitellus. No penetration was observed under aerobic conditions in which methylene blue remained oxidized or in the absence of oviduct. They concluded that a reducing potential existed in the normal oviduct as an important requirement for sperm penetration. Smith (1951, 1953) also found the presence of pieces of oviduct to be beneficial to sperm penetration through the zona pellucida.

The possibility remains that at least some spermatozoa might have been capacitated *in vitro* in some of these experiments. Perhaps human spermatozoa do not have to undergo a lengthy capacitation process prior to fertilization. The early investigations were essential to the development of acceptable criteria for subsequent claims of *in vitro* fertilization.

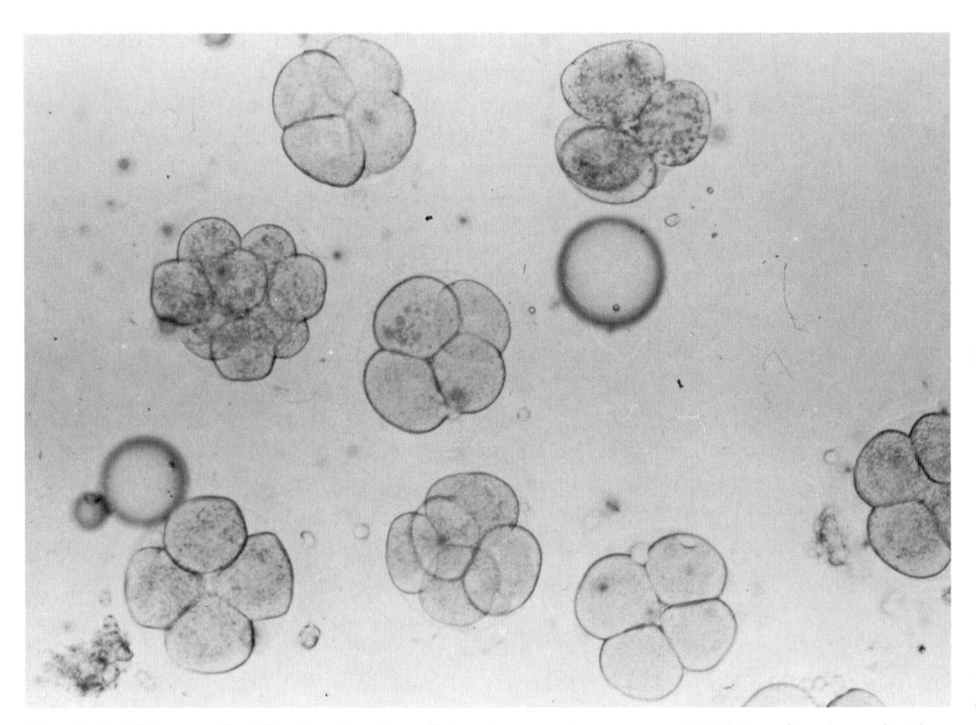

Fig. 1. Rabbit ova fertilized and cultured *in vitro* as they appear 23½ hr after insemination. ×77.

Since the discovery of sperm capacitation investigators have become more conscious of the difficulty of selecting suitable criteria for *in vitro* fertilization. Cleavage of ova has been used widely as a criterion for the achievement of fertilization *in vitro* (fig. 1). However, unfertilized mammalian ova, both *in vivo* and *in vitro,* may undergo degenerative fragmentation, superficially resembling the cleavage of fertilized ova (Austin 1950; Bacsich and Wyburn 1945; Chang 1950; Kingery 1914; Thibault 1949). Chromatin, which can be stained, is present in each blastomere of cleaved ova that are fertilized, whereas fragments of cytoplasm that may appear to be blastomeres are usually devoid of chromatin.

A simple control can eliminate the possibility of confusing fertilization with parthenogenetic activation. Ova are subjected to exactly the same experimental treatment as those to be fertilized but with no exposure to sperm cells. This type of control is especially important when an experimental procedure is under development.

An excellent way to prove that *in vitro* fertilization has occurred is to recover genetically marked young of both sexes after transfer of the fertilized ova into a host. The possibility of parthenogenetic development (of the males at least) is thereby excluded. Ova should be in at least the 2-cell stage before transfer to eliminate the possibility of fertilization by spermatozoa in the oviduct of the host animal. Chang (1959) was the first to provide proof of *in vitro* fertilization; development of young followed the transfer to a host doe of *in vitro* fertilized rabbit ova. Other investigators have subsequently confirmed this achievement using both rabbit (Bedford and Chang 1962; Brackett 1969*a;* Thibault and Dauzier 1961; Seitz, Brackett, and Mastroianni 1970) and mouse ova (Whittingham 1968).

Although it is desirable to prove that fertilization has occurred *in vitro* by development of ova to live young, it is not always practicable, especially in the case of human experiments. Direct observation of the process, documented by time-lapse microcinematography, would provide indisputable evidence.

II. Progress since the Discovery of Sperm Capacitation

Sperm capacitation appears to be a general phenomenon in all mammals studied (Braden and Austin 1954; Chang and Sheaffer 1957; Chang and Yanagimachi 1963; Mattner 1963; Noyes 1953; Yanagimachi 1966). However the time as well as the environmental requirements for the accomplishment of sperm capacitation vary greatly among species. The events of the fertilization process, sperm penetration, pronuclear development, syngamy, and cleavage, occur in different time sequences in different species. Generalizations regarding conditions required for *in vitro* fertilization of mammalian ova will become more apparent as progress is made in investigations on different species.

A. *Rabbit*

Thibault, Dauzier, and Wintenberger (1954) recorded that 3.8% of the ova incubated with fresh ejaculated spermatozoa were activated compared with 86.9% when uterine spermatozoa were used. The same workers (Dauzier, Thibault, and Wintenberger 1954), operating at 38° to 39° C in order to avoid parthenogenetic activation, placed ova and spermatozoa in Locke's solution and cultured some ova in serum to observe cleavage. They confirmed the need for sperm capacitation as a prerequisite of fertilization *in vitro*. Dauzier and Thibault (1956) reported that 50 of 197 ova (25.4%) were fertilized by uterine spermatozoa recovered 8 hr after mating; and no fertilization resulted from the use of uterine spermatozoa recovered 6 hr after mating.

Results from the author's laboratory have indicated that rapidly respiring 6 hr *in utero* incubated rabbit spermatozoa were unable to fertilize ova *in vitro* (Brackett 1968). However extraordinarily large numbers of washed sperm cells were introduced directly into ligated uterine horns in those experiments. Recently a few tubal rabbit ova have been fertilized by spermatozoa recovered from the uterus of mated rabbits 4 hr after mating (Brackett, Rocha, and Seitz, unpublished data). Freshly ovulated ova recovered from the ovarian surface have failed to undergo fertilization when incubated with 6 hr uterine spermatozoa but have been fertilized when incubated with 8 hr uterine spermatozoa (Seitz, Rocha, Brackett, and Mastroianni 1970).

Dauzier and Thibault (1956) disclosed that capacitation could be obtained in any part of the female reproductive tract including the vagina. Bedford found that the cervix must be patent for capacitation to occur in the rabbit vagina (Bedford 1967). In recent experiments in which he tested sperm capacitation by ability to fertilize ova which were put into the uterus, he concluded that the time interval required for complete sperm capacitation in the uterus was increased from 11 hr to at least 15 hr by separating the uterus from the oviduct at the uterotubal junction (Bedford 1969). Since capacitation is normally achieved within 6 hr in spermatozoa exposed both to the uterus and to the oviduct, Bedford concluded that a second phase of capacitation was normally completed in the oviduct. The conditions found in the oviduct have been at least partially duplicated *in vitro* in our laboratory; 21 out of 54 recently ovulated ova recovered from the ovarian surface of rabbits whose oviducts had been surgically removed 1 month previously were fertilized *in vitro* by spermatozoa recovered from the uterine horns of does mated 8 hr before (Seitz, Rocha, Brackett, and Mastroianni 1970). The authenticity of the accomplishment of *in vitro* capacitation, in the opinion of the author, must be documented by successful fertilization *in vitro* in the absence of the complications posed by *in vivo* assays.

Suzuki and Mastroianni (1965) noted that 64% of 79 rabbit ova incubated in tubal fluid with capacitated spermatozoa were fertilized *in vitro* when the gametes were incubated under paraffin oil equilibrated with 5% CO_2 in air. Recently these investigators reported successful *in vitro* fertilization in tubal fluid of 29% of 74 ova recovered from ovarian follicles 12 hr after injection of the rabbit with HCG (Suzuki and Mastroianni 1968). *In vitro* fertilized ova have been cultured to the morula stage (Suzuki 1968; Brackett 1969*c*). With our present knowledge it appears that a follicular rabbit oocyte could be taken from the ovary, fertilized, cultured to the blastocyst stage *in vitro,* then surgically transplanted into the uterus of a hormonally prepared recipient doe where development could be continued to birth.

Dauzier and Thibault (1959) showed that best results were obtained when spermatozoa were added to the ova within 5 to 15 min after recovery. They (Thibault and Dauzier 1960; Dauzier and Thibault 1961) later postulated a sperm-repulsive substance on ova analogous to the "fertilizin" of certain invertebrate ova but different from the "fertilizin" of mammalian ova described by Bishop and Tyler (1956). By washing the ova in Locke's solution for 25 to 157 min before adding capacitated spermatozoa, they succeeded in fertilizing 353 (66%) of 532 ova. Some fertilized ova were obtained in 95% of such experiments (Thibault and Dauzier 1961). Without the washing period there was a 2–3 hr latent period before the spermatozoon could penetrate the ovum. They proposed that "antifertilizin" in the female reproductive tract neutralized the "fertilizin" of the ova, enabling sperm penetration to occur immediately after ovulation *in vivo*. Bedford and Chang (1962) were unable to increase the proportion of ova fertilized by washing the ova. In the author's experience washing both ova and spermatozoa caused a statistically significant decrease in the proportion of ova that could be fertilized *in vitro,* an effect which could be explained by the washing of the sperm cells alone. No statistically significant increase in the proportion of ova fertilized was afforded by washing only the ova (Brackett 1969*a*). These findings are not necessarily in conflict since the experimental conditions of each laboratory were different.

Chang (1959) reported the fertilization of 21% of the ova incubated with capacitated spermatozoa in Krebs-Ringer bicarbonate solution with glucose added. By using acidic saline containing heated serum and glucose for the medium 42 of 74 ova (57%) were fertilized *in vitro.* The author achieved consistent fertilization of rabbit ova by the method of Chang (1959), and Bedford and Chang (1962) when the gametes were handled initially and incubated at 37° or 39° C under an atmosphere of 96%–97% relative humidity (Williams et al. 1964). Better results were obtained when the gametes were covered with paraffin oil. When 0.6–2.0 ml of fresh uterine fluid was recovered with capacitated spermatozoa and incubated in the medium, 53 of 73 ova (73%) were fertilized *in vitro* (Brackett and Williams 1965). The fresh uterine fluid accounted for 20–40% of the fertilization medium in these experiments.

Further experiments were directed toward eliminating the requirement of high relative humidity and the presence of biologic fluids of unknown composition in the culture medium (Brackett and Williams 1968). Crystalline bovine serum albumin was substituted for serum in the medium. Results of previous experiments indicated that a pH of 7.8 was optimal for *in vitro* fertilization. Sodium bicarbonate was added to maintain this pH under an atmosphere of 5% CO_2 in air at 37° C. Equilibration of the medium and paraffin oil in 5% CO_2 in N_2 seemed to enhance fertilization and early development *in vitro.* These innovations eliminated the need for high relative humidity as well as the need for serum and fresh uterine fluid in the medium. In experiments in which 18 to 36% of the volume of the medium was contributed by fresh uterine fluid, the only component of the fertilization medium of unknown composition, 67% of 45 ova were fertilized. In comparison 65% of 54 ova were fertilized *in vitro* in experiments in which no measurable volume of uterine fluid was recovered with the capacitated spermatozoa (Brackett and Williams 1968).

The medium for *in vitro* fertilization of rabbit ova in current use in our laboratory (table 1) consists of a simple salt solution (a modification of the salt solution originally used by Hammond [1949] and referred to as acidic saline by Chang [1959]) plus 50 units of penicillin-G per ml, crystalline bovine albumin, glucose, and sodium bicarbonate added just before use (Brackett 1969*a*). The pH of this isotonic synthetic medium at 37° C under a 5% CO_2 in air atmosphere is 7.8. The fertilization process can be initiated in this defined medium after removal, by washing, of soluble substances carried from the female reproductive tract by the ova and spermatozoa. Ova are incubated with capacitated spermatozoa for 5 hr in this medium. The ova are then removed from the sperm suspension and cultured through early cleavage stages in acidic saline to which has been added 50 units penicillin per ml, heated rabbit serum (10% by volume of the final solution), and 0.25% glucose. The sodium chloride content of acidic saline required for isotonicity is 8.8 gm/liter. The pH of this culture medium is 7.2. Serum can be eliminated by using Brinster's medium for rabbit ova, which contains lactate and pyruvate (Brinster 1970).

The execution of controlled *in vitro* fertilization experiments has been made possible only by the increase in knowledge which has enabled us to obtain and handle gametes, maintaining their viability in an appropriate milieu and developing a simple defined medium in which sperm penetration of the ovum can take place (fig. 2). Ova from one oviduct of each donor are placed in one tissue culture dish, and ova from the other oviduct in a second. The sperm suspension recovered in 4.0 ml of medium is divided; approximately ½ of the sperm cells are incubated with each group of ova. Distribution of gametes from the same animals allows comparison of results from two different treatments in each experiment. Experiments using this technique have been carried out in the author's laboratory to assess the effect of

TABLE 1

COMPOSITION OF THE MEDIUM FOR *in Vitro* FERTILIZATION OF RABBIT OVA

Component	g/l	mM
NaCl	6.550	112.00
KCl	0.300	4.02
$CaCl_2 \cdot 2H_2O$	0.330	2.25
$NaH_2PO_4 \cdot H_2O$	0.113	0.83
$MgCl_2 \cdot 6H_2O$	0.106	0.52
$NaHCO_3$	3.104	37.00
Glucose	2.500	13.90
Crystalline bovine albumin	3.000	
Penicillin, sodium salt	0.031	
Distilled water to 1,000 ml		

SOURCE: Brackett 1969*b*.

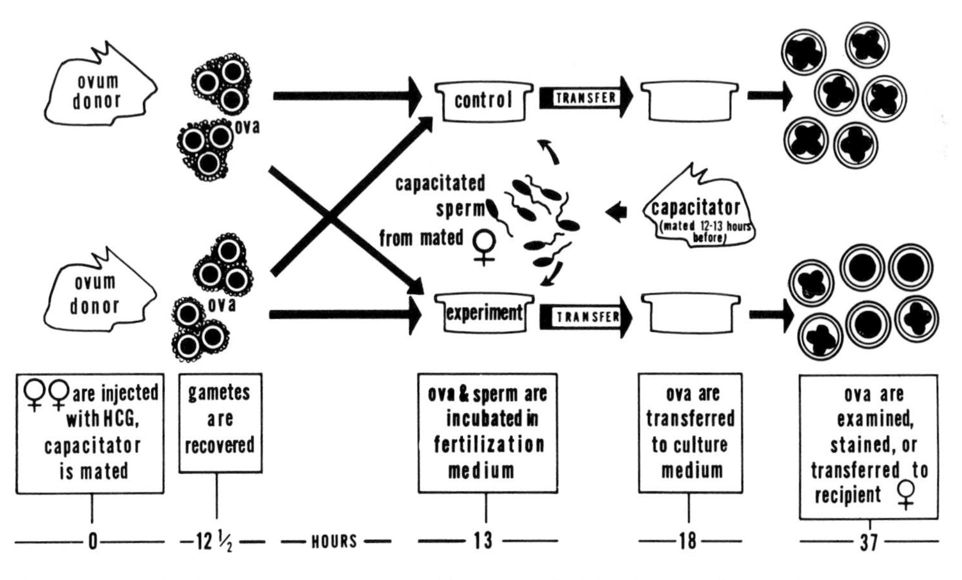

Fig. 2. Schematic diagram of controlled rabbit *in vitro* fertilization experiment. (From Brackett 1969*a*)

washing the secretions of the female reproductive tract from rabbit ova and spermatozoa prior to their incubation together (Brackett 1969*a*), to study the sequence of events in the fertilization process (Brackett 1969*c*), to test the ability of trypsin inhibitors to inhibit fertilization (Stambaugh, Brackett, and Mastroianni 1969), and to compare the fertilizing ability of uterine spermatozoa of different ages (Brackett, Rocha, and Seitz unpublished data).

A study of the time sequence of events in the *in vitro* fertilization process (Brackett 1969*c*) revealed that if ova are to be penetrated this process occurs during the 3 hr interval following insemination. On one occasion sperm penetration of the vitellus was observed and documented by phase contrast cinematography. The

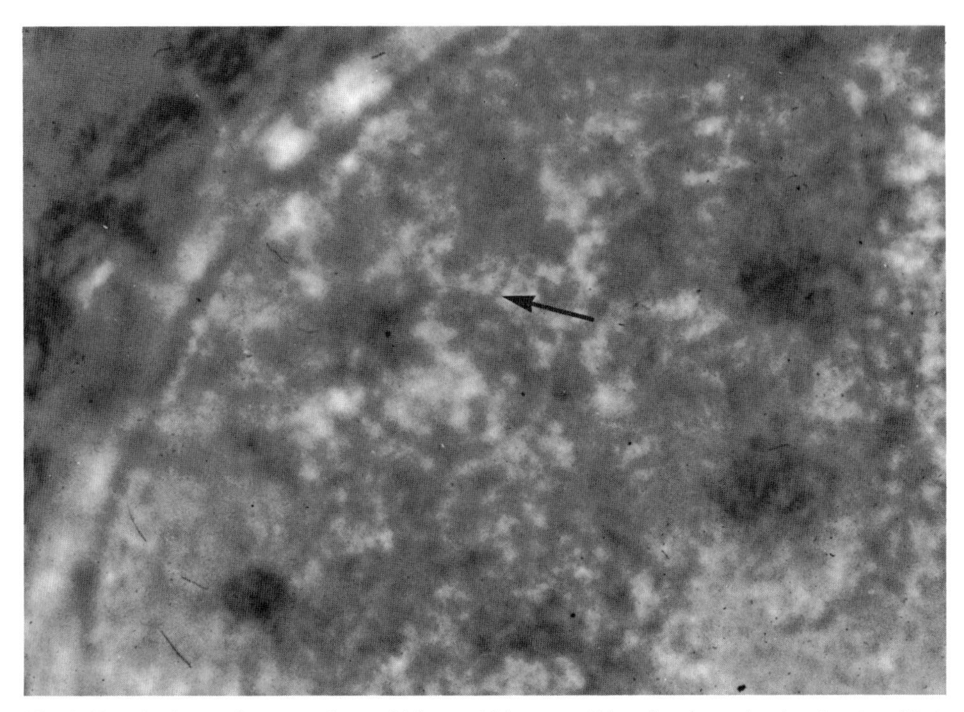

Fig. 3. Developing male pronucleus within a rabbit ovum 7 hr after insemination *in vitro*. Note the sperm tail still attached (*arrow*). ×750.

ovum in figure 3 was removed from the CO_2 incubator 7 hr after insemination, fixed in neutral formalin for 15 hr, held in 95% ethanol for 6 hr, and then stained with lacmoid. The developing male pronucleus can be seen with the sperm flagellum still attached. Male and female pronuclei have developed completely by 9 hr after insemination *in vitro*. Pronuclear ova, 9 hr after insemination, appear in figure 4. Figure 5 is an ovum which was cleaving when examined 15 hr and 50 min after insemination. Cleavage to the 2-cell stage, figure 6, was found to occur as early as 15½ hr after insemination. The duration of each stage of the fertilization process and of early cleavage are comparable for ova fertilized *in vivo* and *in vitro*.

Enzymes which dissolved the zona pellucida were isolated from the rabbit sperm head by Stambaugh and Buckley (1968). Experiments were conducted to test

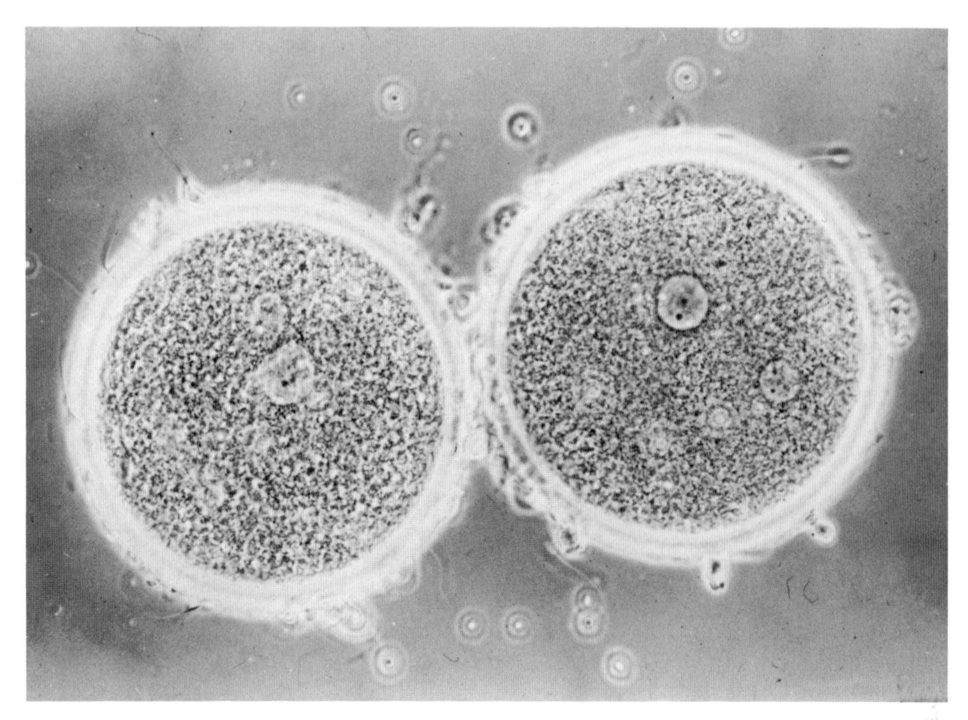

Fig. 4. Pronuclear stage rabbit ova 9 hr after insemination *in vitro*. ×191

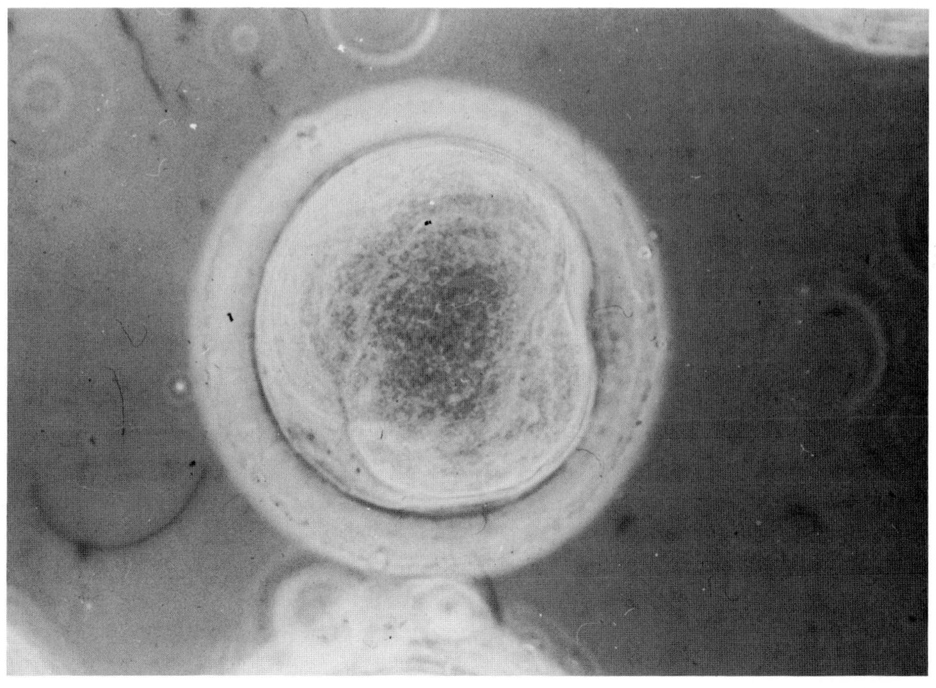

Fig. 5. *In vitro* fertilized rabbit ovum undergoing cleavage at 15 hr 50 min after insemination. ×246.

the ability of crystalline ovomucoid and soybean trypsin inhibitors to inhibit *in vitro* fertilization (Stambaugh, Brackett, and Mastroianni 1969). A direct relationship was found between the percent inhibitions by the inhibitor of the trypsin-like enzyme of the rabbit sperm acrosome and fertilization. Only 4 of 105 ova were fertilized when soybean trypsin inhibitor was added to the medium (1.0 mg of inhibitor per ml of medium) while 39 of 96 ova (41%) were fertilized in the control dish lacking the inhibitor. These results support the contention that the trypsin-like component of the acrosome is essential for sperm penetration of the zona pellucida.

Recent experiments have been carried out to compare the fertilizing ability

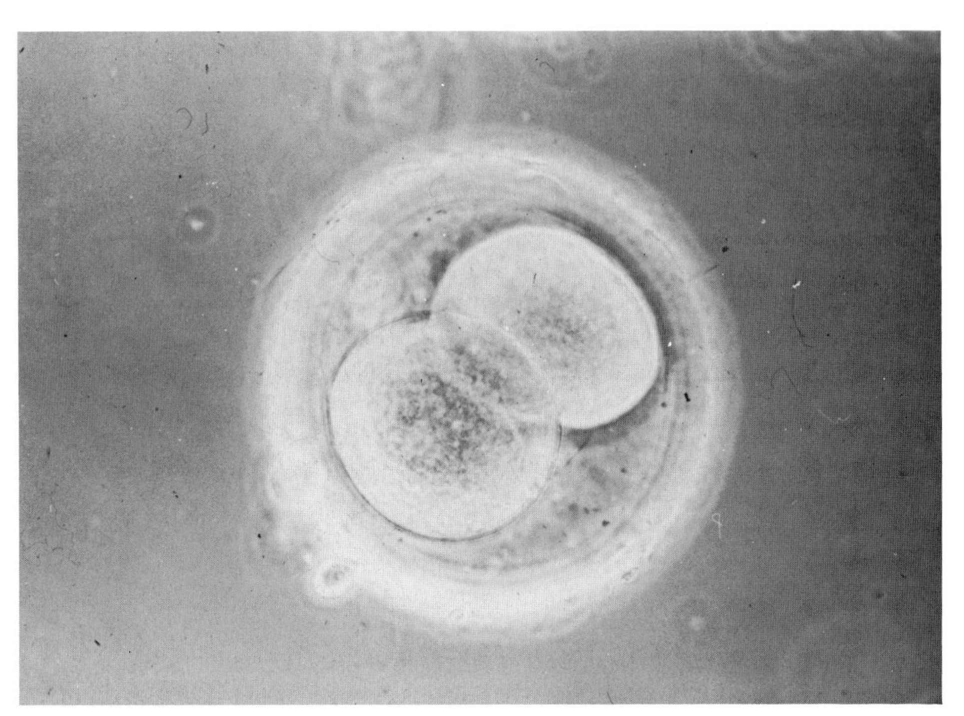

Fig. 6. Two-cell stage rabbit ovum 15 hr 50 min after *in vitro* insemination. ×246

of spermatozoa recovered from the uterus of mated does at varying intervals after coitus with that of spermatozoa recovered 13 hr after coitus (Brackett, Rocha, and Seitz unpublished data). Eighteen-hr uterine spermatozoa are capable of fertilizing a greater proportion of ova recovered from the oviducts than 13-hr uterine spermatozoa.

A similar trend has been observed when ovulated ova recovered from the ovarian surface were used. Heated serum was added to the defined medium (table 1) to the extent of 20% by volume. Twenty-nine of 32 ova (91%) were fertilized *in vitro* by spermatozoa recovered from uteri of does mated 16 to 18 hr before (Seitz, Rocha, Brackett, and Mastroianni 1970), a proportion comparable to that of ova fertilized *in vivo*. Fertilization was observed when 8- to 36-hr uterine spermatozoa were used; similar results were obtained when the spermatozoa were recovered from does in which the oviducts had been removed. Previous experiments using the same procedure with 12-hr uterine spermatozoa yielded the fertilization of 117 of 143 ova (81%) and demonstrated that any conditioning of rabbit ova prior to fertilization by direct physical contact with the oviduct is unnecessary under the experimental conditions employed. (Seitz, Brackett, and Mastroianni 1970).

Bomsel-Helmreich and Thibault (1962) have induced triploidy in rabbit ova with colchicine. Colchicine, if introduced before sperm penetration and removed 1 hr later, suppresses the expulsion of the second polar body during fertilization. Rabbit ova treated in this manner and surgically transferred to host does resulted in heteroploid embryos which died between days 12 and 16 of gestation (Bomsel-Helmreich 1965).

Investigations of *in vivo* and *in vitro* fertilization revealed that 14–50% of the fertilizations were abnormal when rabbit ova were fertilized 9 hr after ovulation (Thibault 1967). Most of the abnormal fertilizations were digynic. Digyny was observed when freshly ovulated ova were fertilized by aged spermatozoa; it was concluded that the aging of male gametes appears to have the same consequences on fertilization as the aging of ova.

B. *Hamster*

Yanagimachi and Chang (1963, 1964) were the first to fertilize rodent ova *in vitro.* Enlarged sperm heads or male pronuclei in the vitellus along with the flagella of the spermatozoa were taken as evidence of sperm penetration. Best results were obtained when Tyrode's solution under paraffin oil was used as the fertilization medium. Since fertilization was achieved with epididymal spermatozoa, and since hamster spermatozoa must be capacitated before sperm penetration (Chang and Sheaffer 1957), these investigators may be credited with achieving both sperm capacitation and fertilization in a single *in vitro* system. They reported *in vitro* fertilization of 134 of 207 ova (65%) when spermatozoa recovered from the uterus of females mated 4–5 hr previously were used. When spermatozoa were recovered from females mated 0.5 hr previously, 52 of 100 ova were fertilized. When epididymal spermatozoa were used, 80 of 180 ova (44%) were fertilized. Judging from the development of male pronuclei they concluded that epididymal spermatozoa penetrated ova less well and 0.5-hr uterine spermatozoa were slower in penetrating ova than were 4–5-hr uterine spermatozoa. The incidence of polyspermy and of supplementary spermatozoa was much higher than that normally found *in vivo* (Austin 1956; Austin and Bishop 1958).

Yanagimachi (1966) observed the process of sperm penetration through the zona pellucida on one occasion. He noted that the acrosome or at least the outer acrosome membrane of the fertilizing spermatozoon was absent before the sperm cell penetrated the zona pellucida. The spermatozoon passed through the zona at an angle to the surface in 3–4 min. Only 1–2 sec was required to traverse the perivitelline space. The head of the spermatozoon lay flat on the vitelline suface and sank into the vitellus without active movement of the sperm flagellum. In the rabbit the fertilizing spermatozoon may be motile within the perivitelline space for many minutes before penetrating the vitellus; the flagellum remains motile during the penetration process (Brackett 1969*c*).

Barros and Austin (1967*a, c*) found that only 5 of 109 ova (5%) washed in cumulus to remove tubal fluids before insemination were undergoing fertilization as opposed to 54 of 110 unwashed ova (49%) from the same donors. Epididymal spermatozoa were used in these experiments. Ninety-four of 141 ova (67%) recovered from large ovulatory follicles were undergoing fertilization when examined after incubation at 37° C for 6–10 hr. The incidence of polyspermy in these follicular ova was higher than that found in tubal ova.

The acrosomal reaction went to completion *in vitro* by using either tubal fluid with or without follicular contributions or follicular fluid without tubal fluid; 95% of randomly chosen vigorously motile spermatozoa completed the acrosomal reaction in follicular fluid alone (Barros and Austin 1967*b*). Follicular contents recovered 9–10 hr after HCG injection induced the acrosome reaction in 97% of epididymal spermatozoa; those recovered after 4 hr induced it in only 43% (Barros and Austin 1967*b*). This suggests that the agents responsible for the acrosomal reaction are produced or accumulate in the follicle in the late preovulatory phase. Barros and Austin (1967*c*) reported that secretions in the female tract were not necessary for fertilization. When Barros (1968) incubated half the spermatozoa with tubal fluid containing ova in saline for 4 hr, sperm penetration occurred 2.5–3 hr earlier than when they were incubated in saline or control media without tubal fluid. Yanagimachi (1969) has recently found that hamster epididymal spermatozoa could be capacitated fully by incubation for 3 hr in Tyrode's solution containing follicular fluid. After this treatment sperm penetration occurred most frequently between 30 and 50 min after insemination. Hamster follicular fluid was the best medium for effecting capacitation; mouse follicular fluid, better than rat; and rabbit follicular fluid, totally ineffective. Interesting observations on the fine structural aspects of the acrosome reaction and of the gamete membranes during sperm entry into the ovum have been possible by using the *in vitro* fertilization technique (Barros et al. 1967; Barros and Franklin 1968).

Bavister (1969) used Tyrode's solution, pH 7.6 (3 mg/ml sodium bicarbonate, 2.5 mg/ml bovine serum albumin and 10^{-4} M sodium pyruvate under 5% CO_2 in air), and found when 2 to 4 volumes of this medium were used to each volume of cumulus a very high fertilization rate with a low incidence of polyspermy could be attained.

C. *Mouse*

Brinster and Biggers (1965) obtained fertilization and development of mouse ova to blastocysts in explanted oviducts cultured in Brinster's medium (3.16×10^{-4} M pyruvate and 1 mg/ml glucose) (Brinster 1965). One of two procedures was used: the ampulla containing ova was incubated at 32° to 37° C in a suspension of epididymal spermatozoa for 15 to 30 min; or 2 to 5 μl of sperm suspension was introduced into the fimbrial end of the oviduct before the ampulla was cut away. The ampulla was cultured for 4 days before the ova were examined. Of 1,748 ova, 175 (10%) developed into morulae and blastocysts.

Pavlok (1968), using a modification of this method, found that a decrease of glucose from 500 mg% to 66 mg% in his medium produced a negative effect on fertilization and initial development of the ova. A lowering of pH from near 8.0 to approximately 7.0 caused a decrease in the fertilization of ova 16 to 17 hr after HCG injection; no such decrease was shown in ova fertilized 13 to 15 hr after HCG injection. Fertilizing capacity of spermatozoa recovered from the uterus 1 to 1½ hr *post coitum* was higher [99 (66%) of 143 ova] than that of spermatozoa recovered 6¼ to 7 hr *post coitum*. Ova were examined after either 18 to 22 hr or 48 hr in culture. Ova with 2 or more pronuclei in syngamy or those with 2 or more blastomeres were classified as fertilized.

Whittingham (1968) succeeded in fertilizing mouse ova *in vitro* in a simple defined medium. Spermatozoa were recovered from the cornua approximately 1 to 2 hr *post coitum*. Proof of fertilization was obtained by subsequent development of some ova to 17-day-old fetuses following their transfer to host females. The defined medium used was a slight modification of the Whitten and Biggers (1968) medium used to culture mouse zygotes to the blastocyst stage. The highest yield of 2-cell ova observed at the end of 24 hr of culture was obtained when oocytes from F_1 hybrids (357B1 × Balb/c) were inseminated with spermatozoa from Swiss males; 65 (41%) of 159 ova were fertilized *in vitro*. Whittingham concluded that with the fertilization of mouse ova *in vitro* and the development of mouse zygotes to the blastocyst stage (Whitten and Biggers 1968) it should be possible to obtain all stages of preimplantational development in chemically defined media.

D. *Rat*

Toyoda and Chang (1968) emphasize that authentic reports of successful *in vitro* fertilization of rat ova are still lacking. All attempts to fertilize rat ova *in vitro* by the procedures of Yanagimachi and Chang (1963) were unsuccessful. If the zona pellucida was dissolved by treatment with proteolytic enzymes, epididymal spermatozoa as well as spermatozoa recovered from the cornua could penetrate the vitelline membrane, activate the ovum, and lead to the formation of pronuclei. Following removal of the zona pellucida by treatment with 0.02% chymotrypsin, 34 (15%) of 235 zona-free ova of immature superovulated rats were penetrated by epididymal spermatozoa and 30 of 100 were penetrated by uterine spermatozoa; 17 (53%) of 32 zona-free spontaneously ovulated ova were penetrated by epididymal spermatozoa and 18 (47%) of 38 were penetrated by uterine spermatozoa. Such zona-free ova were unable to undergo development beyond the early pronuclear stages. Either capacitation of spermatozoa can be achieved in a short time in the presence of chymotrypsin or the capacitation of spermatozoa is necessary only for the penetration of the zona pellucida; this experiment demonstrated that the passing of spermatozoa through the zona pellucida is not a prerequisite for the fusion of gametes and the formation of pronuclei (Toyoda and Chang 1968).

E. *Human*

A need for sperm capacitation in the human has not been demonstrated. Shettles (1953, 1955) incubated human follicular and tubal ova in follicular fluid or in human serum with semen and pieces of tubal mucosa added. He reported sperm penetration through the zona pellucida and development of one ovum to the morula stage. However the published pictures are not convincing evidence of such an accomplishment.

Hayashi (1963) classified follicular oocytes into four groups: growing, retrogressing, parthenogenetic, and degenerative. Growing oocytes had no perivitelline space and accounted for 85% of 160 oocytes examined. Retrogressing oocytes had a perivitelline space as a result of shrinkage of ooplasm. Polyspermy was seen commonly after incubation of retrogressing follicular oocytes with semen and monospermy was noted when growing oocytes were used. In a medium at pH 7.5 which included hormones 20 ova cleaved after exposure to semen. Two of these developed to the morula stage, 1 to the 8-cell, 1 to the 3-cell, and 15 to the 2-cell stage. One cleavage was atypical. Parthenogenetic cleavage was observed in some oocytes that were not exposed to spermatozoa; these ova could not be distinguished from the cleaved ova resulting from incubation with spermatozoa. Perhaps all the cleavages were parthenogenetic. It is unfortunate that experimental details were not reported and that these interesting observations were not pursued further.

Edwards (1965) successfully effected maturation of human oocytes *in vitro* from the postdictyate stages to the extrusion of the first polar body by culturing them 36 to 43 hr. When human oocytes were matured and then exposed to experimentally treated spermatozoa, Edwards et al. (1966) reported that 4 of 56 oocytes were possibly fertilized by washed spermatozoa, 1 of 14 by spermatozoa after incubation in the rabbit uterus, and 2 of 20 by spermatozoa in the presence of endosalpinx. They considered the presence of pronuclei, sperm flagellum, or the second polar body as evidence of fertilization. These criteria are insufficient proof that the early part of the fertilization process was underway, and their published pictures of pronuclei without identification of the sperm flagellum in the vitellus as well as polar bodies are not convincing evidence of fertilization.

Edwards, Bavister, and Steptoe (1969) fashioned techniques to fertilize human ova *in vitro* after successful *in vitro* fertilization experiments in the hamster (Barros and Austin 1967*c;* Bavister 1969; Yanagimachi and Chang 1964). The medium employed was similar to that used by Bavister in the hamster (Bavister 1969). Follicular oocytes were cultured in pure follicular fluid or in follicular fluid diluted with various culture media. Spermatozoa were washed and sometimes treated with follicular fluid. Of 56 ova inseminated 20 remained in dictyotene, having failed to mature *in vitro;* spermatozoa were present inside the perivitelline space of 5 of the 34 mature ova; in 4 of these the spermatozoa were motile. Seven ova had pronuclei but only 2 of these could be considered as undergoing normal fertilization; that is, 2 pronuclei and 2 polar bodies present in each. The other 5 ova had at least 3 pronuclei. Structures of a size similar to midpieces were seen in 3 ova after fixation. This suggests that human follicular oocytes can be matured to metaphase 2 of meiosis and that human spermatozoa can be capacitated *in vitro* by exposure to follicular fluid. From this investigation and that of Shettles (1953) it seems that the mechanism of human sperm capacitation may be similar to that of the hamster. Although present evidence is suggestive (Dukelow and Chernoff 1968; Marston and Kelly 1968), capacitation of primate spermatozoa as a prerequisite for sperm penetration remains to be established. The evidence put forth by Edwards, Bavister, and Steptoe (1969) supporting the achievement of early stages of the fertilization process, that is, sperm penetration and formation of pronuclei, in at least a few human ova is the most convincing to date. However the achievement *in vitro* of complete fertilization, including syngamy and subsequent cleavage of human ova, remains to be documented in a scientifically acceptable manner.

III. Summary

Sufficient evidence has accumulated to support the authenticity of *in vitro* fertilization in three mammalian species—hamster, mouse, and rabbit. Mouse and rabbit *in vitro* fertilized ova have developed into live young following surgical transfer to host animals. Following the discovery of sperm capacitation in the rabbit in 1951 major contributions have been made in this field primarily through use of this animal.

The need for sperm capacitation in the rabbit was confirmed through the use of *in vitro* fertilization which provides a means by which sperm capacitation can be assayed in the absence of the complications of widely used *in vivo* assays. The favorable environment of the oviduct has been duplicated at least partially. Advances have been made through the development of simple chemically defined media and defined physical conditions which permit a consistently large proportion of ova to be fertilized. Meaningful controlled experiments including those designed to study effects of washing the gametes, inhibition of fertilization, time sequence of events, and optimal age of uterine spermatozoa have been concluded. A large proportion of recently ovulated rabbit ova recovered from the ovarian surface can be fertilized *in vitro* by spermatozoa recovered from mated does in which the oviducts have been removed.

In the hamster sperm capacitation and fertilization can occur in a single *in vitro* system. Sperm penetration through the zona pellucida and into the vitellus has been observed in this species. Investigations using the hamster, the mouse, and the rat may be expected to provide valuable information concerning mammalian sperm penetration and fertilization phenomena. The biochemical and physical conditions which are adequate for successful *in vitro* fertilization of the small laboratory animals should prove to be fruitful in experimentation with gametes of domestic animals and primates, including man.

References

Austin, C. R. 1950. Fertilization of the rabbit egg. *Nature (London)* 166:407.

———. 1951. Observations of the penetration of the sperm into the mammalian egg. *Aust J Sci Res* B4:581.

———. 1952. The capacitation of the mammalian sperm. *Nature (London)* 170:326.

———. 1956. Ovulation, fertilization and early cleavage in the hamster (Mesocricetus auratus). *J Roy Micr Soc* 75:1401.

———. 1961. Fertilization of mammalian eggs *in vitro*. *Int Rev Cytol* 12:337.

Austin, C. R., and Bishop, M. W. H. 1957. Fertilization in mammals. *Biol Rev* 32:296.

———. 1958. Capacitation of mammalian sperm. *Nature (London)* 181:851.

Bacsich, P., and Wyburn, G. M. 1945. Parthenogenesis of atretic ova in the rodent ovary. *J Anat* 79:177.

Barros, C. 1968. *In vitro* capacitation in golden hamster spermatozoa. *Anat Rec* 160:310.

Barros, C., and Austin, C. R. 1967*a*. *In vitro* fertilization of golden hamster ova. *Anat Rec* 157:209.

———. 1967*b*. *In vitro* acrosomal reaction of golden hamster spermatozoa. *Anat* Rec 157:348.

———. 1967*c*. *In vitro* fertilization and the sperm acrosome reaction in the hamster. *J Exp Zool* 166:317.

Barros, C.; Bedford, J. M.; Franklin, L. E.; and Austin, C. R. 1967. Membrane vesiculation as a feature of the mammalian acrosome reaction. *J Cell Biol* 34:C1.

Barros, C., and Franklin, L. E. 1968. Behavior of the gamete membrane during sperm entry into the mammalian egg. *J Cell Biol* 37:C13.

Bavister, B. D. 1969. Environmental factors important for *in vitro* fertilization in the hamster. *J Reprod Fertil* 18:544.

Bedford, J. M. 1967. Experimental requirement for capacitation and observations on ultrastructural changes in rabbit spermatozoa during fertilization. *J Reprod Fertil* Suppl. 2:35.

———. 1969. Limitations of the uterus in the development of the fertilizing ability (capacitation) of spermatozoa. *J Reprod Fertil* 8:19.

Bedford, J. M., and Chang, M. C. 1962. Fertilization of rabbit ova *in vitro. Nature* (*London*) 193:898.

Bishop, D. W., and Tyler, A. 1956. Fertilizins of mammalian eggs. *J Exp Zool* 132:575.

Bomsel-Helmreich, O. 1965. Heteroploidy and embryonic death. In *Preimplantation stages of pregnancy,* ed. G. E. W. Wolstenholme and M. O'Connor, p. 246. London: J. and A. Churchill.

Bomsel-Helmreich, O., and Thibault, C. 1962. Fécondation *in vitro* en présence de colchicine et polyploidie expérimentale chez le lapin. *Ann Biol Anim Biochem Biophys* 2:13.

Brackett, B. G. 1968. Respiration of spermatozoa after *in utero* incubation in estrus and pseudopregnant rabbits: VI *Int Congr Anim Reprod A I* (*Paris*) 1:43.

———. 1969*a*. Effects of washing the gametes on fertilization *in vitro. Fertil Steril* 20:127.

———. 1969*b*. *In vitro* fertilization of mammalian ova. Lecture given at the Schering Symposium, "Mechanisms in Conception," Berlin 1969. The text has been published in Raspé, G., ed., *Advances in the biosciences 4,* Schering Symposium on Mechanisms in Conception, Berlin 1969; Pergamon Press and Vieweg, Oxford, New York and Braunschweig.

———. 1969*c*. *In vitro* fertilization of rabbit ova: time sequence of events. *Fertil Steril* 21:169.

Brackett, B. G., and Williams, W. L. 1965. *In vitro* fertilization of rabbit ova. *J Exp Zool* 160: 271.

———. 1968. Fertilization of rabbit ova in a defined medium. *Fertil Steril* 19:144.

Braden, A. W. H., and Austin, C. R. 1954. Fertilization of the mouse egg and the effect of delayed coitus and of hot shock treatment. *Aust J Biol Sci* 7:552.

Brinster, R. L. 1965. Studies on the development of mouse embryos in vitro: III. The effect of energy source. *J Exp Zool* 158:59.

———. 1970. Culture of two-cell rabbit embryos to morulae. *J Reprod Fertil* 21:17.

Brinster, R. L., and Biggers, J. D. 1965. *In vitro* fertilization of mouse ova within the explanted Fallopian tube. *J Reprod Fertil* 10:277.

Chang, M. C. 1950. Cleavage of unfertilized ova in mature ferrets. *Anat Rec* 108:31.

———. 1951. Fertilizing capacity of spermatozoa deposited in the Fallopian tubes. *Nature (London)* 168:697.

———. 1957. Some aspects of mammalian fertilization. In *The beginning of embryonic development,* ed. A. Tyler, C. B. Metz, and R. C. Borstel, p. 109. Washington, D.C.: AAAS.

———. 1959. Fertilization of rabbit ova *in vitro. Nature (London)* 184:466.

———. 1968. *In vitro* fertilization of mammalian eggs. *J Anim Sci* 27, Suppl. 1:15.

Chang, M. C., and Pincus, G. 1951. Physiology of fertilization in mammals. *Physiol Rev* 31:1.

Chang, M. C., and Sheaffer, D. 1957. Number of spermatozoa ejaculated at copulation, transported into the female tract and present in the male tract of the golden hamster. *J Hered* 48:107.

Chang, M. C., and Yanagimachi, R. 1963. Fertilization of ferret ova by deposition of epididymal sperm into the ovarian capsule with special reference to the fertilizable life of ova and the capacitation of sperm. *J Exp Zool* 145:175.

Dauzier, L., and Thibault, C. 1956. Recherches expérimentale sur la maturation des gamètes mâle chez les mammifères, par l'étude de la fécondation *"in vitro"* de l'oeuf de lapine. *Proc III Int Congr Anim Reprod (Cambridge)* 1:58.

———. 1959. Données nouvelles sur la fécondation *in vitro* de l'oeuf de la lapine et de la brebis. *C R Acad Sci (Paris)* 248:2655.

———. 1961. La fécondation *in vitro* de l'oeuf de lapine. *Proc IV Int Cong Anim Reprod (The Hague).*

Dauzier, L.; Thibault, C.; and Wintenberger, S. 1954. La fécondation *in vitro* de l'oeuf de la lapine. *C R Acad Sci (Paris)* 238:844.

Dukelow, W. R., and Chernoff, H. N. 1968. Primate sperm capacitation. *Fed Proc* 27:567.

Edwards, R. G. 1965. Maturation *in vitro* of human ovarian oocytes. *Lancet* 2:926.

Edwards, R. G.; Bavister, B. D.; and Steptoe, T. C. 1969. Early stages of fertilization *in vitro* of human oocytes matured *in vitro. Nature (London)* 221:632.

Edwards, R. G.; Donahue, R. P.; Baramki, T. A.; and Jones, H. 1966. Preliminary attempts to fertilize human oocytes matured *in vitro. Amer J Obstet Gynec* 96: 192.

Frommolt, G. 1934. De Befruchtung und Furchung des Kanincheneies in Film. *Zbl Gynaek* 58:7.

Hammond, J. 1949. Recovery and culture of tubal mouse ova. *Nature (London)* 163:28.

Hayashi, M. 1963. Fertilization *in vitro* using human ova. In *Proc Conf Int Planned Parenthood Fed Singapore,* p. 505.

Kingery, H. M. 1914. So called parthenogenesis in the white mouse. *Biol Bull* 27:240.

Krassovskaja, O. V. 1934. Fertilization of the rabbit egg outside the organism. *Arkh Anat* 13:415. As cited by Smith (1951).

Krassovskaja, O. V. 1935. Cytological studies of the heterogenous fertilization of the egg of rabbit outside the organism. *Acta Zool* 16:449. As cited by Smith (1951).

Long, J. A. 1912. Studies on early stages of development in rats and mice. *Univ Cal Publ Zool* 9:105.

Mattner, P. E. 1963. Capacitation of ram sperm and penetration of the ovine egg. *Nature (London)* 199:772.

Marston, J. H., and Kelly, W. A. 1968. Time relationship of spermatozoon penetration into the egg of the rhesus monkey. *Nature (London)* 217:1073.

Menkin, M. F., and Rock, J. 1948. *In vitro* fertilization and cleavage of human ovarian eggs. *Amer J Obstet Gynec* 55:440.

Moricard, R. 1949. Pénétration *in vitro* du spermatozoïde dans l'ovule des mammifères et niveau du potentiel d'oxydoreduction tubaire. *C R Soc Franc Gynec* 19:226.

———. 1950. Penetration of spermatozoon into the mammalian ovum oxydopotential level. *Nature (London)* 165:763.

———. 1954. Observation of *in vitro* fertilization in the rabbit. *Nature (London)* 173:1140.

Moricard, R., and Bossu, J. 1949. Premières études du passage du spermatozoïde au travers de la membrane pellucida d'ovocytes de lapine fécondés *in vitro. Bull Acad Nat Med (Paris)* 133:659.

Noyes, R. W. 1953. Fertilizing capacity of spermatozoa. *West J Surg Obstet Gynec* 61:342.

Onanoff, J. 1893. Recherches sur la fécondation et la gestation des mammifères. *C R Soc Biol (Paris)* 45:719.

Pavlok, A. 1968. Fertilization of mouse ova *in vitro:* I. Effect of some factors on fertilization. *J Reprod Fertil* 16:401.

Pincus, G. 1930. Observations on the living eggs of the rabbit. *Proc Roy Soc (Biol)* 107:132.

———. 1936. The eggs of mammals. New York: Macmillan Co.

———. 1939. The comparative behavior of mammalian eggs *in vivo* and *in vitro:* IV. The development of fertilized and artificially activated rabbit eggs. *J Exp Zool* 82:85.

Pincus, G., and Enzmann, E. V. 1934. Can mammalian eggs undergo normal development *in vitro? Proc Nat Acad Sci USA* 20:121.

———. 1935. The comparative behavior of mammalian eggs *in vivo* and *in vitro:* I. The activation of ovarian eggs. *J Exp Med* 62:665.

———. 1936. The comparative behavior of mammalian eggs *in vivo* and *in vitro:* II. The activation of the tubal eggs of the rabbit. *J Exp Zool* 73:195.

Pincus, G., and Saunders, B. 1939. The comparative behavior of mammalian eggs *in vivo* and *in vitro:* VI. The maturation of human ovarian ova. *Anat Rec* 75:537.

Pincus, G., and Shapiro, H. 1940. The comparative behavior of mammalian eggs *in vivo* and *in vitro:* VII. Further studies on the activation of rabbit eggs. *Proc Amer Phil Soc* 83:631.

Rock, J., and Menkin, M. F. 1944. *In vitro* fertilization and cleavage of human ovarian eggs. *Science* 100:105.

Schenk, S. L. 1878. Das Säugethierei künstlich befruchtet ausserhalb des Mutterthieres. *Mitt Embryol Inst (Vienna)* 1:107.

Seitz, H. M., Jr.; Brackett, B. G.; and Mastroianni, L., Jr. 1970. *In vitro* fertilization of ovulated rabbit ova recovered from the ovary. *Biol Reprod* 2:262.

Seitz, H. M., Jr.; Rocha, G.; Brackett, B. G.; and Mastroianni, L., Jr. 1970. Influence of the oviduct on sperm capacitation in the rabbit. *Fertil Steril* 21:325.

Shettles, L. B. 1953. Observations on human follicular and tubal ova. *Amer J Obstet Gynec* 66:235.

———. 1955. A morula stage of human ova developed *in vitro. Fertil Steril* 6:287.

Smith, A. U. 1951. Fertilization *in vitro* of the mammalian egg. *Biochem Soc Sympos* 7:3.

———. 1953. In discussion after paper by Venge. In *Mammalian germ cells,* ed. G. E. W. Wolstenholme, M. P. Cameron, and J. S. Freeman, p. 243. London: J. & A. Churchill.

Stambaugh, R.; Brackett, B. G.; and Mastroianni, L. Jr. 1969. Inhibition of the *in vitro* fertilization of rabbit ova by trypsin inhibitors. *Biol Reprod* 1:223.

Stambaugh, R., and Buckley, J. 1968. Zona pellucida dissolution enzymes of the rabbit sperm head. *Science* 161:585.

Suzuki, S. 1968. *In vitro* cultivation of rabbit ova following *in vitro* fertilization in tubal fluid. *Cytologia (Tokyo)* 31:416.

Suzuki, S., and Mastroianni, L., Jr. 1965. *In vitro* fertilization of rabbit ova in tubal fluid. *Amer J Obstet Gynec* 93:465.

———. 1968. *In vitro* fertilization of rabbit follicular oocytes in tubal fluid. *Fertil Steril* 19:716.

Thibault, C. 1947. La parthénogénèse expérimentale chez le lapin. *C R Acad Sci (Paris)* 224:297.

———. 1948. L'activation et la régulation de l'ovocyte parthénogénétique de lapine. *C R Soc Biol (Paris)* 142:495.

———. 1949. L'oeuf des mammifères: Son développement parthénogénétique. *Ann Sci Nat Zool, 11th Ser* 11:136.

———. 1967. Analyse comparee de la fécondation et de ses anomalies chez la brebis, la vache et la lapine. *Ann Biol Anim Bioch Biophys* 7:5.

Thibault, C., and Dauzier, L. 1960. "Fertilisines" et fécondation *in vitro* de l'oeuf de lapine. *C R Acad Sci (Paris)* 250:1358.

———. 1961. Analyse des conditions de la fécondation *in vitro* de l'oeuf de la lapine. *Ann Biol Anim Bioch Biophys* 1:277.

Thibault, C.; Dauzier, L.; and Wintenberger, S. 1954. Etude cytologique de la fécondation *in vitro* de l'oeuf de la lapine. *C R Soc Biol (Paris)* 148:789.

Toyoda, Y., and Chang, M. C. 1968. Sperm penetration of rat eggs *in vitro* after dissolution of zona pellucida with chymotrypsin. *Nature (London)* 220:589.

Venge, O. 1953. Experiments on fertilization of rabbit ova *in vitro* with subsequent transfer to alien does. In *Mammalian germ cells,* ed. G. E. W. Wolstenholme, M. P. Cameron, and J. S. Freeman, p. 243. London: J. & A. Churchill.

Whitten, W. K., and Biggers, J. D. 1968. Complete development *in vitro* of the preimplantation stages of the mouse ova in a simple chemically defined medium. *J Reprod Fertil* 17:399.

Whittingham, D. G. 1968. Fertilization of mouse eggs *in vitro. Nature (London)* 220:592.

Williams, W. L.; Hamner, C. E.; Weitman, D. E.; and Brackett, B. G. 1964. Capacitation of rabbit spermatozoa and initial experiments on *in vitro* fertilization. *Proc V Int Cong Reprod* (*Trento*) 7:288.

Yamane, J. 1935. Kausal-analytische Studien über die Befruchtung des Kaninche-neies: I. Die Dispersion der Follikelzellen und die Ablösung der Zellen der Corona radiata des Eies durch Spermatozoon. *Cytologia* (*Tokyo*) 6:233.

Yanagimachi, R. 1966. Time and process of sperm penetration into hamster ova *in vivo* and *in vitro*. *J Reprod Fertil* 11:359.

———. 1969. *In vitro* capacitation of hamster spermatozoa by follicular fluid. *J Reprod Fertil* 18:275.

Yanagimachi, R., and Chang, M. C. 1963. Fertilization of hamster eggs *in vitro*. *Nature* (*London*) 200:281.

———. 1964. *In vitro* fertilization of hamster ova. *J Exp Zool* 156:361.

20

Chromosome Abnormalities in the Preimplanting Ovum

David H. Carr
Department of Anatomy
McMaster University
Hamilton, Canada

More data are available on the chromosome disorders of man than of any other mammal. However for the preimplantation zygote we have no cytogenetic data for man and variable information for other mammals. This chapter is a review of the published reports of chromosome analysis of the preimplantation stages of various mammals. Isolated examples of abnormalities will be avoided and the discussion restricted to populations of embryos. Evidence based only on nuclear detail, rather than chromosome counts, will be reviewed as supplementary information only. This is because the presence of nuclear disorders in the fertilized ovum does not invariably produce heteroploid embryos (Edwards 1963) and also because the only definite evidence of heteroploidy is obtained by counting the chromosomes (Bomsel-Helmreich 1965).

The material falls naturally into two categories: the spontaneous incidence of chromosome anomalies in the preimplantation conceptus and the results of attempts to induce chromosome disorders. The material may be further subdivided into two methods by which data have been collected: first, evidence of haploidy and polyploidy from simple squash preparations of the zygote, and second, the use of techniques, developed in the last decade, which provide reliable chromosome counts for aneuploid as well as polyploid conceptuses.

I. Euploid Anomalies in Squash Preparations

There is considerable information available on the incidence, in varying circumstances, of polyploidy and haploidy in early zygotes. Most of this work was carried out 15 to 20 years ago. The simple squash preparations then available enabled the investigator to answer the question, How many haploid sets of chromosomes are present? (Beatty 1957). As the whole blastocyst was normally used in this technique it is natural to wonder whether the cells of the embryo and trophoblast have a similar chromosome constitution. Indications are that they do, at least in the rabbit (Venge 1956).

A. *Spontaneous Haploids and Polyploids*

The spontaneous occurrence of haploidy in a dividing zygote appears to be a rare event. Chromosome counts in the haploid range (20) were found in 6 3½-day mouse embryos from crosses between "silver" females and "nonsilver" males (Beatty 1957). These occurred among about 3,700 conceptuses from various matings. The number of cells present indicated that the specimens had developed to the blastocyst stage.

Venge (1954) used various techniques in an attempt to induce polyploidy in the rabbit. They all failed. He then mated the apparently normal offspring from these experiments and studied the chromosomes in their 5-day progeny. Among 162 blastocysts from these parents 1 was found to be haploid with a count of 22 chromosomes.

Beatty (1957) noted that spontaneous triploidy in the mouse was a rare event except in crosses between "silver" females and "nonsilver" males. From that mating about 6% of blastocysts were triploid by chromosome count. Most of the specimens were examined at 3½ days, but the incidence was no lower at 9½ days, which is a postimplantation stage and about midterm in the mouse (Fishberg and Beatty 1951).

Tetraploidy appears to be even rarer than triploidy. Only 5 definite tetraploid mouse ova were found among 3,671 chromosomally analyzable specimens (Beatty 1957). The few tetraploid specimens found were not concentrated in the mating which produced the haploid and most of the triploid blastocysts ("silver" female : "nonsilver" male). In the same series hexaploidy was found in only 2 3½-day mouse embryos, and there were no examples of pentaploidy.

Squash preparations of blastocysts for spontaneous haploidy and polyploidy in animals other than the mouse are lacking. The only information in this respect is based on indirect evidence relating to polar bodies in rat and rabbit ova and will be discussed later in brief.

B. *Induced Haploids and Polyploids*

A number of chemical and physical agents have been used in an attempt to induce chromosome anomalies in mammals. Most of the attempts were based on experiments with amphibian eggs but it was soon found that these techniques are not necessarily applicable to the two classes of animals.

The various agents used and the results for different species are summarized in table 1. Several examples of haploid embryos were found among populations of mouse blastocysts fertilized by spermatozoa that had been treated with nitrogen mustard, or ultraviolet or X rays (Edwards 1954 and personal communication to Beatty 1957). Cleavage occurred with development to as many as 14 cells but these haploid embryos were more abnormal than spontaneously occurring specimens or those induced by colchicine.

The most successful techniques for inducing triploidy in the mouse were heating the uterine tubes at the estimated time of the second meiotic division and the use of colchicine. Cooling the egg, an effective method for production of amphibian polyploids, appeared to be relatively ineffective in mammals.

Heat treatment resulted in triploidy in 19 of 173 (11%) analyzable 3½-day mouse embryos (Beatty 1957). By a similar technique 7 out of 60 (12%) of 3½-day embryos were found to be tetraploid. In this case the eggs were warmed at the estimated time of the first cleavage division. All the tetraploid embryos showed cell division, and the farthest advanced had reached the blastocyst stage.

Colchicine appears to be an effective method of inducing polyploidy in the mouse. Triploid or tetraploid embryos may be produced depending on the time

TABLE 1

INDUCTION OF HAPLOIDY AND POLYPLOIDY IN MAMMALS

Author	Species	Inducing Agent	Haploid	Triploid	Tetraploid
Edwards 1954	Mouse	U.v. light	++	−	−
		X ray	+	−	−
		N-mustard	+	?	?
		Basic dyes	?+	?+	−
Beatty 1957	Mouse	Chilling	?	−	?
Beatty 1957	Mouse	Heat	?	++	++
Venge 1954	Rabbit	Colchicine, heat	−	−	−
Edwards 1958*a* and *b*	Mouse	Colchicine	?	++	++
Piko and Bomsel-Helmreich 1960	Rat	Colchicine	?	+	?
Bomsel-Helmreich 1965	Rabbit	Colcemid	?	++	?
Shaver and Carr 1969	Rabbit	Aging	−	++	−

NOTE: − Negative; + Abnormality found; ++ Incidence of abnormality over 10%; ? No information available; ?+ Doubtful if related to treatment

at which the colchicine is acting. The most successful method for producing triploid embryos was injection of colchicine solution into the uterine cervix. As many as 12% were triploid with the optimal concentration of colchicine (Edwards 1958*a*).

The use of colchicine at first cleavage to induce tetraploidy was less successful in producing classifiable embryos. From the small number of suitable 3½-day conceptuses, as many as 10% were tetraploid (Edwards 1958*b*).

It has been shown that both hyperthermia and colchicine treatment are capable of inducing polyspermy or suppression of the second polar body in rats (Piko and Bomsel-Helmreich 1960). By inference these same treatments should produce triploid offspring; the authors were able to present evidence of this by demonstrating triploidy, by chromosome count, on implanted rat embryos. Unfortunately no information is available on the percentage of polyploid embryos in the preimplantation stages in this species.

Venge (1954) attempted to induce polyploidy in rabbit blastocysts by treating the germ cells with colchicine and by hyperthermia. There were no examples of polyploidy among 203 blastocysts from these various experiments. These included 60 ova treated *in vitro* with colchicine before subsequent transfer to normal does.

In spite of the failure of the last mentioned technique a similar method was applied successfully in the rabbit by Bomsel-Helmreich (1965). She obtained 65% fertilized eggs; pronuclear findings showed 97% of these were potentially triploid. The chromosome counts were checked in various animals using hypotonic pretreatment squash technique. From this progressive study Bomsel-Helmreich (1965) was able to draw several conclusions about triploid blastocysts in the rabbit: the mean diameter of triploid blastocysts is smaller than that of diploid ones; triploid blastocysts implant later than normal ones; and they do not progress beyond midterm (15 days).

II. Heteroploidy in Air-dried Preparations

The main difference between the early work just described and more recent studies is purely technical. The introduction of hypotonic pretreatment of cells into mammalian cytogenetics and the use of air-dried, instead of squash, preparations has greatly improved the accuracy of chromosome counts. These two maneuvers have been incorporated into a variety of methods for chromosome preparations of high quality (Tarkowski 1966; McFeely 1966; Shaver and Carr 1967).

In view of the rapid advances in mammalian cytogenetics and the detailed study of man and other mammals it is surprising how few data are available for the preimplantation stages. It is of course impossible to imagine the procurement of such data for man but they should be readily available for a variety of laboratory and domestic animals.

A. *Control series*

McFeely (1967) studied 88 blastocysts collected from 7 gilts and sired by unrelated boars. The chromosomes of the somatic cells of the pigs were studied in leukocyte cultures and found to be normal. The blastocysts were collected on day 10, and the recovery rate, based on corpus luteum count, was 98%. Nine of the blastocysts (10%) showed a chromosome anomaly (these are listed in table 2). Two other blastocysts were degenerating and unsuitable for chromosome analysis.

In the rabbit only 1 of 58 6-day blastocysts was found to have a chromosome abnormality (Shaver and Carr 1969). This was a 44/45 mosaic, the extra element in the 45-cell line being one of the acrocentric chromosomes. In addition to this small study of blastocysts from untreated rabbits another type of control mating was studied. Seventy-three blastocysts were recovered from rabbits mated immediately after injection of 25 i.u. of human chorionic gonadotrophin (Shaver and Carr 1969). Five of these 73 blastocysts had a chromosome anomaly. This was greater than the incidence in the untreated rabbits, but the difference was not statistically significant. The 5 anomalies consisted of 2 mixoploids (2n/4n and 2n/4n/8n), 1 pentaploid, a trisomy involving chromosome no. 1, and a probable deletion in a diploid complement.

The pentaploid blastocyst was the largest 6-day blastocyst to be recovered. It is probably the only example of pentaploidy in a mammal, and the karyotype is shown in figure 1. The usual route for the origin of a pentaploid conceptus is by the suppression of both polar bodies during oocyte maturation, followed by fertilization with a single sperm. However such an origin is excluded in this case by the XXXYY sex chromosome complex. There are various means by which such an anomaly could arise. These include: fertilization of a haploid oocyte by 4 sperms, 2 of which are Y-bearing; fertilization of a diploid oocyte by 3 sperms; suppression of 1st cleavage in a normal XY zygote followed by incorporation of the second polar body or another sperm; and finally fusion of an XXY triploid ovum with a normal XY ovum (Shaver 1968).

The anomalies just described may have been related to the ovulation-inducing dose of HCG. This is uncertain as the increase in the abnormalities was not

TABLE 2

CHROMOSOME ANOMALIES IN BLASTOCYSTS OF PIG, RABBIT, AND MOUSE

Authors	Animal	Ovulation	Number of Blastocysts	Chromosomally Abnormal		Abnormalities
				Number	Percentage	
McFeely 1967	Pig	Spontaneous	88	9	10	4 triploid (3n) 3 tetraploid (4n) 1 mixoploid (2n/3n) 1 structural anomaly
Shaver and Carr 1969	Rabbit	Spontaneous	58	1	2	1 trisomy/normal mosaic (45/44)
Shaver and Carr 1969	Rabbit	Induced with 25 i.u. of HCG	73	5	7	1 pentaploid (5n) 2 mixoploids (2n/4n & 2n/4n/8n) 1 trisomy (45) 1 structural anomaly
Vickers 1967	Mouse (Albino, PDE strain)	Induced with 3 i.u. of HCG	115	4	3	1 haploid (n) 1 trisomy (41) 2 structural anomalies

significantly different from that found in controls. In order to test whether anomalies were increased by raising the dose of HCG, a small experiment was carried out. Twenty-eight blastocysts were recovered from 3 rabbits receiving 50–75 international units of HCG. Two of them were chromosomally abnormal. In each instance the specimen was a 43/44 mosaic. Of 24 blastocysts recovered from 4 rabbits after 100–300 i.u. of HCG none had a chromosome anomaly. Thus from this small study there was no indication of an increasing incidence of chromosome anomalies with an increasing dose of HCG (Shaver 1968).

Vickers (1967) studied the chromosomes in 3-day blastocysts of an albino strain of mice. As in the rabbits mentioned above ovulation was induced with HCG. Of 115 specimens in which chromosomes could be analyzed 4 were abnormal; 1 was haploid, 1 was trisomic, and 2 had structural abnormalities (1 with an isochromosome and 1 with a chromosomal fragment).

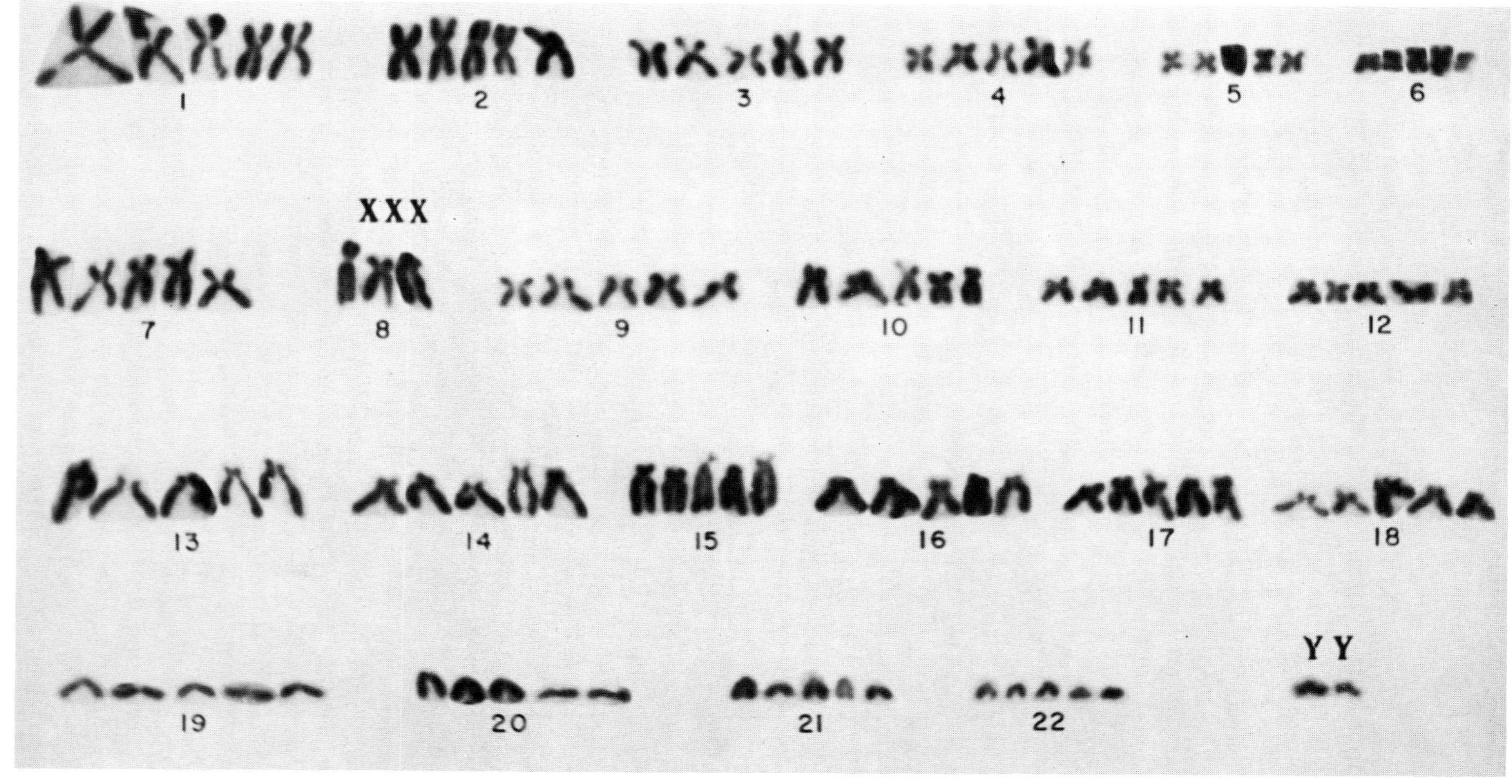

Fig. 1. Pentaploid karyotype with XXXYY gonosomes from a 6-day rabbit blastocyst. By permission from E. L. Shaver and D. H. Carr, "Chromosome abnormalities in rabbit blastocysts following delayed fertilization," *J Reprod Fertil* 14:415, 1967.

B. *Delayed Fertilization*

For over 30 years investigators have been interested in the relationship between aging of the mammalian ovum and abnormalities of the embryo. This interest was aroused by the classical experiments of Blandau and Young (1939) in guinea pigs and in the rat (Blandau and Jordan 1941). There has been indirect evidence that some of these structural anomalies might be due to chromosome disorders, especially polyploidy. Austin and Braden (1953) found a great increase in polyspermy associated with delayed mating in the rat. The possibility that these polyspermic eggs may give rise to triploid embryos was reinforced by the finding of cleavage up to 8 cells. The incidence of polyspermy with delayed mating shows strain differences in the rat (Odor and Blandau 1956). In the mouse delayed fertilization does not cause polyspermy (Braden and Austin 1954).

TABLE 3

CHROMOSOME ABNORMALITIES AND SEX RATIO OF RABBIT BLASTOCYSTS RECOVERED AFTER DELAYED MATING

Delay (Hrs)	Number of Blastocysts Karyotyped	Sex		Abnormalities
		♂	♀	
2	33	21	12	1 mixoploid 2n/4n 1 monosomy/normal mosaic (43/44)
4	24	13	11	1 monosomy/normal mosaic (43/44)
6	30	15	15	3 triploid, XXY 1 triploid, XXX
8	29	18	11	3 triploid, XXY 1 triploid, XXX
9	21	10	11	1 triploid, XXY 1 triploid, XXX
10	3	1	2	....
11	1	....	1	....
12	3	2	1	....
13	9	4	5	1 aneuploid/normal mosaic (44/47)
14	1	....	1	....

Witschi (1952) had demonstrated repeatedly that the fertilization of aged ova produced anomalies in amphibians. He later showed that the same treatment produced chromosome anomalies in *Rana pipiens* and *Xenopus laevis* (Witschi and Laguens 1963). Chang (1952) reported that, as in guinea pigs and rats, aging of ova increased the incidence of anomalies in early embryos of rabbits. Again indirect evidence suggested that some of these may be due to chromosome anomalies. Austin and Braden (1953) found 3 pronuclei in 16.4% of aged ova compared to 1.4% from normal control matings in rabbits.

Shaver and Carr (1967, 1969) studied the chromosomes of 6-day blastocysts, mating having been delayed by 2–14 hr after an injection of human chorionic gonadotrophin. The results are summarized in table 3. There were 3 chromosomally abnormal blastocysts when mating was delayed for 2–4 hr. Two were 43/44 mosaics and 1 was 2n/4n mixoploid. Among 11 chromosomally abnormal blastocysts from

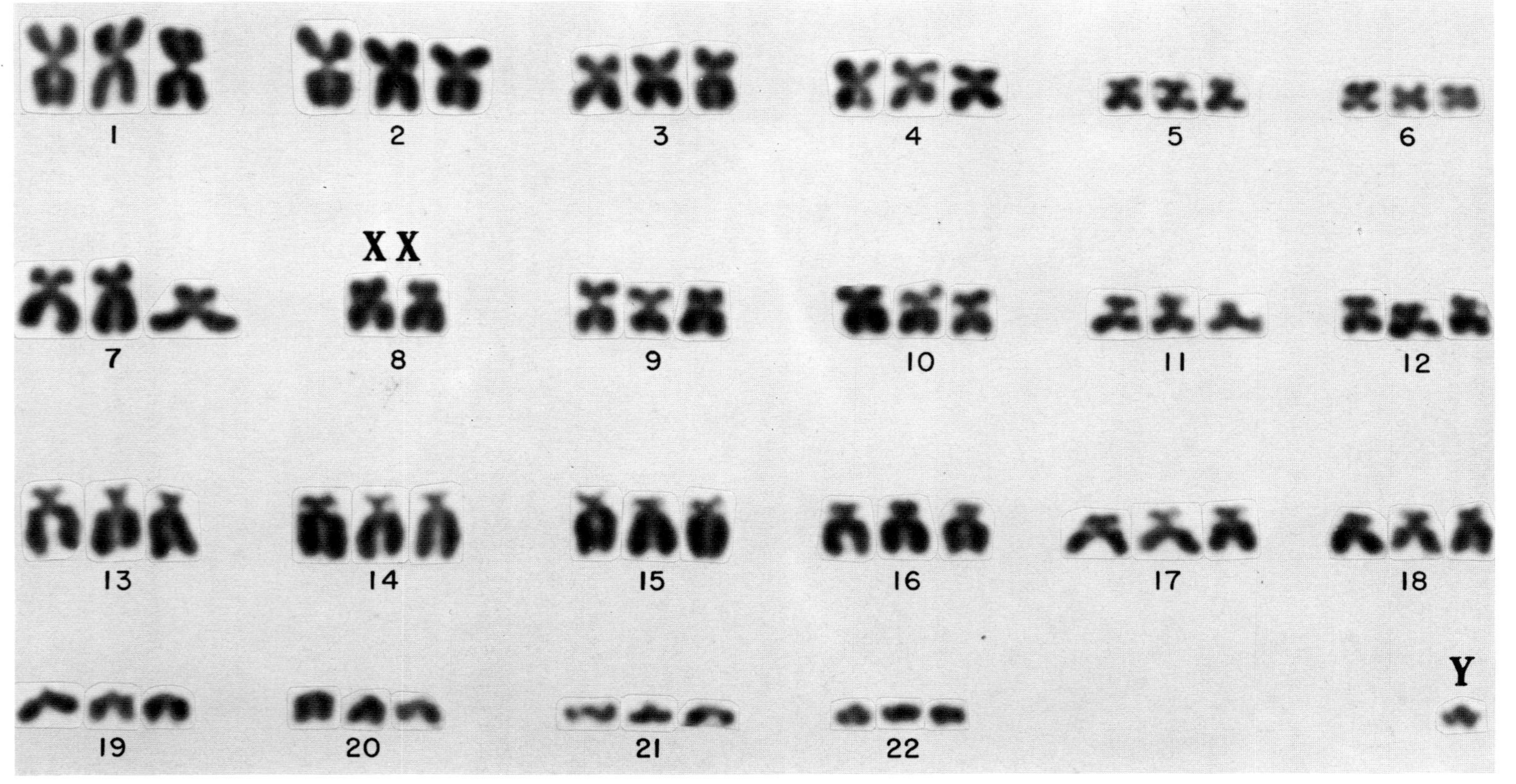

Fig. 2. Triploid karyotype from 6-day rabbit blastocyst the product of a delayed mating. By permission from E. L. Shaver and D. H. Carr, "Chromosome abnormalities in rabbit blastocysts following delayed fertilization," *J Reprod Fertil* 14:415, 1967.

matings delayed 6–14 hr all but 1 were triploid (fig. 2). This anomaly, present in 12½% of blastocysts from matings with a 6–9 hr delay, was not found in any other category. The incidence of triploidy in this category was significantly raised above that in controls.

III. Discussion

It is difficult to draw many conclusions about chromosome abnormalities in the blastocyst stage because of the very limited data available. A great deal of work was done before really accurate chromosome counts were possible. This gave valuable information regarding the occurrence of haploidy and polyploidy but no information on the incidence of aneuploidy. From this early work it appears that there is considerable strain difference in the occurrence of triploidy in the mouse (Beatty 1957). The tendency for strain differences in abnormalities of fertilization in the mouse was confirmed by Braden (1957).

Triploidy seems to be much more common and less lethal than tetraploidy. This is particularly interesting because the sex chromosome imbalance is presumably greater in the former than the latter. It suggests that lethality is related to some factor other than the genetic imbalance suggested by Melander (1963). The haploid conceptus, like the tetraploid, seems doomed to early death, although without doubt haploid, cleaving zygotes have been known to reach the blastocyst stage.

There are several possible modes of origin for the triploid conceptus. They have been discussed in detail by Austin (1960). The three possible mechanisms for syngamy involve 1 female and 2 male pronuclei (dispermy), 1 male and 2 female pronuclei (digyny) or 1 haploid and 1 diploid pronucleus, either being the male and the other the female (aneugamy). It seems likely that different mechanisms may be important in different animal species. For example polyspermy is increased by delayed mating in the rat, and this is compatible with cleavage (Austin and Braden 1953). The same treatment produces 3 pronuclei in a high percentage of rabbit ova, though it is not clear whether this is polyandry or polygyny (Austin and Braden 1953). There is evidence that the extra pronucleus in aged rabbit eggs is female (Thibault 1967; Chang and Hunt 1968). Perhaps further support for this view is drawn from the fact that none of the 10 triploid blastocysts resulting from delayed mating in the rabbit had an XYY sex chromosome constitution (Shaver and Carr 1969).

In contrast to the situation in the rabbit the mouse does not give evidence of extra male or female pronuclei with delayed fertilization (Braden and Austin 1954), nor does the incidence of triploidy increase in this animal (Gates and Beatty 1954).

There is now considerable evidence that, as far as polyspermy and polygyny are concerned, the effect of delayed mating is variable from species to species (Thibault 1967). Extra pronuclei were noted in rabbit ova subject to delayed mating but not in sheep or cattle. There is ample evidence of increased polyspermy in pigs when ova are aged. The literature on this subject has recently been summarized by Day and Polge (1968). These authors showed an increase in polyspermy in pigs treated with progesterone before ovulation, a process which also produced an increase in the rate of egg transport in the treated animals.

The survival of triploid conceptuses is of considerable interest. They are known to implant and to survive to midterm in mouse and rabbit (Beatty 1957; Bomsel-Helmreich 1965). However Ekins and Shaver (1969) recently investigated the rate of implantation of triploid rabbits induced by 8 hr delayed mating. Only 1 of 44 embryos implanted at day 10 was found to be triploid. All 41 embryos examined on days 12, 14, and 16 were normal diploid. There are various possible reasons for the differences in these two studies. The incidence of induced anomalies was potentially 97% in one series and only 12½% in the other. It is conceivable that after implantation triploid conceptuses fare less well in competition with normal diploids than in competition with other triploids. There is also the possibility of strain differences. Finally Edwards (1954) found that haploid blastocysts produced by different methods varied in their development when compared with one another. These differences are of great importance when considering the fate of heteroploid embryos.

It is interesting to speculate whether intrafollicular and extrafollicular aging of the ovum produce different types of chromosome disorder. Only triploidy was found among blastocysts from rabbit ova aged 6–9 hr. In this case the aging was postovulation (Shaver and Carr 1969). Fugo and Butcher (1966) used Nembutal to suppress ovum release in the rat. They studied fertilized ova and found a 3-fold increase in polyspermy following delayed ovulation. However when chromosomes were studied in 11-day embryos after the same treatment, there was a great increase in aneuploidy compared with controls but no increase in triploidy (Butcher and Fugo 1967). It would be most interesting to compare the types of chromosome anomaly in pre- and postovulation aging in the same animal species.

IV. Summary

1. Early squash techniques were only adequate to detect haploid and polyploid blastocysts. Most of this work was performed in the mouse.
2. It was found that triploid embryos could occur spontaneously, especially in certain strains, and that this anomaly could be induced by a variety of agents.
3. Haploid and tetraploid embryos were much rarer than triploid conceptuses but were capable of early cleavage.
4. The introduction of hypotonic pretreatment and air-dried techniques have made it possible to detect aneuploidy as well as euploidy.
5. There is relatively little basic information on the incidence of chromosome anomalies in the preimplantation stages in mammals.
6. Delayed mating of 6–9 hr causes an increase in triploidy in the rabbit blastocyst.

Glossary

Autosome. Any chromosome of the complement that is not a sex chromosome

Acrocentric. A chromosome whose centromere is situated near one end

Metacentric. A chromosome whose centromere is situated near the center

Karyotype. An array that is designed to reveal clearly certain features of a chromosome complement, including morphology

Haploid. The basic chromosome number typical of the gametes of an organism (n)
Diploid. The basic chromosome number typical of the somatic cells of an organism (2n)
Heteroploid. All chromosome numbers *not* haploid or diploid
Euploid. Haploid and exact multiples thereof
Polyploid. Exact multiples of the basic haploid number, other than diploid; e.g., triploid (3n), tetraploid (4n), pentaploid (5n), hexaploid (6n), octaploid (8n), decaploid (10n), duodecaploid (12n)
Mixoploid. The condition in which an organism contains cells with 2 or more euploid chromosome complements
Aneuploid. Chromosome numbers, *not* exact multiples of the haploid number (e.g., trisomy and monosomy)
Trisomy. Having 1, or more, chromosomes present 3 times in the somatic cells, the other chromosomes being present only twice
Monosomy. Having 1, or more, chromosomes present only once in the somatic cells, the other chromosomes being present twice
Mosaicism. The condition in which an organism contains 2 or more cell lines, each having a different chromosome constitution

Acknowledgments

Dr. Evelyn Shaver was responsible for the rabbit experiments reviewed in this paper. Her assistance and that of Mr. John Ekins is gratefully acknowledged. The work was supported by the Medical Research Council of Canada.

References

Austin, C. R. 1960. Anomalies of fertilization leading to triploidy. *J Cell Comp Physiol* 56: (suppl. 1) 1.

Austin, C. R., and Braden, A. W. H. 1953. An investigation of polyspermy in the rat and rabbit. *Aust J Biol Sci* 6:674.

Beatty, R. A. 1957. *Parthenogenesis and polyploidy in mammalian development.* London: Cambridge University Press.

Blandau, R. J., and Jordan, E. S. 1941. The effect of delayed fertilization on the development of the rat ovum. *Amer J Anat* 68:275.

Blandau, R. J., and Young, W. C. 1939. The effects of delayed fertilization on the development of the guinea pig ovum. *Amer J Anat* 64:303.

Bomsel-Helmreich, O. 1965. Heteroploidy and embryonic death. In *Preimplantation stages of pregnancy,* p. 246. London: J. & A. Churchill.

Braden, A. W. H. 1957. Variation between strains in the incidence of various abnormalities of egg maturation and fertilization in the mouse. *J Genet* 55:476.

Braden, A. W. H., and Austin, C. R. 1954. Fertilization of the mouse egg and the effect of delayed coitus and of hot-shock treatment. *Aust J Biol Sci* 7:552.

Butcher, R. L., and Fugo, N. W. 1967. Overripeness and the mammalian ova: II. Delayed ovulation and chromosome anomalies. *Fertil Steril* 18:297.

Chang, M. C. 1952. Effects of delayed fertilization on segmenting ova, blastocysts and fetuses in rabbits. *Fed Proc* 11:24.

Chang, M. C., and Hunt, D. M. 1968. Attempts to induce polyspermy in the rabbit by delayed insemination and treatment with progesterone. *J Exp Zool* 167:419.

Day, B. N., and Polge, C. 1968. Effects of progesterone on fertilization and egg transport in the pig. *J Reprod Fertil* 17: 227.

Edwards, R. G. 1954. The experimental induction of pseudogamy in early mouse embryos. *Experientia* 10: 499.

———. 1958*a*. Colchicine-induced heteroploidy in the mouse: I. The induction of triploidy by treatment of the gametes. *J Exp Zool* 137:317.

———. 1958*b*. Colchicine-induced heteroploidy in the mouse: II. The induction of tetraploidy and other types of heteroploidy. *J Exp Zool* 137:349.

———. 1963. In discussion, Session 1. *2d Int Conf Congenital Malformations.* p. 70. New York: Int Med Cong.

Ekins, J. G., and Shaver, E. L. 1969. Triploidy in post-implantation rabbit embryos following delayed fertilization. *Proc Canadian Fed Biol Soc Edmonton, Alta.*

Fishberg, M., and Beatty, R. A. 1951. Spontaneous heteroploidy in mouse embryos up to mid-term. *J Exp Zool* 118:321.

Fugo, N. W., and Butcher, R. L. 1966. Overripeness and the mammalian ova: I. Overripeness and early embryonic development. *Fertil Steril* 17:804.

Gates, A. H., and Beatty, R. A. 1954. Independence of delayed fertilization and spontaneous triploidy in mouse embryos. *Nature* (*London*) 174:356.

McFeely, R. A. 1966. A direct method for the display of chromosomes from early pig embryos. *J Reprod Fertil* 11:161.

———. 1967. Chromosome abnormalities in early embryos of the pig. *J Reprod Fertil* 13:579.

Melander, Y. 1963. A presumed obstacle to mammalian polyploidization. *Nature* (*London*) 197:152.

Odor, D. L., and Blandau, R. J. 1956. Incidence of polyspermy in normal and delayed matings in rats of the Wistar strain. *Fertil Steril* 7:456.

Piko, L., and Bomsel-Helmreich, O. 1960. Triploid rat embryos and other chromosomal deviants after colchicine treatment and polysperm. *Nature* (*London*) 186: 737.

Shaver, E. L. 1968. Chromosome abnormalities in blastocysts of the rabbit following delayed fertilization. Ph.D. thesis, University of Western Ontario.

Shaver, E. L., and Carr, D. H. 1967. Chromosome abnormalities in rabbit blastocysts following delayed fertilization. *J Reprod Fertil* 14:415.

———. 1969. The chromosome complement of rabbit blastocysts in relation to the time of mating and ovulation. *Can J Genet Cytol* 11:287.

Tarkowski, A. K. 1966. An air-drying method for chromosome preparations from mouse eggs. *Cytogenetics* 5:394.

Thibault, C. 1967. Analyse comparée de la fécondation et de ses anomalies chez la brebis, la vache et la lapine. *Ann Biol Anim Biochim Biophys* 7:5.

Venge, O. 1954. Experiments on polyploidy in the rabbit. *K Lantbr Högsk Annlr* 21:417.

———. 1956. Chromosome number in rabbit blastocysts. *Nature* (*London*) 177: 384.

Vickers, A. D. 1967. A direct measurement of the sex-ratio in mouse blastocysts. *J Reprod Fertil* 13:375.

Witschi, E. 1952. Overripeness of the egg as a cause of twinning and teratogenesis: A review. *Cancer Res* 12:763.

Witschi E., and Laguens, R. 1968. Chromosomal aberrations in embryos from overripe eggs. *Devl Biol* 7:605.

21

Intrauterine Oxygen Tension and Metabolism of the Endometrium during the Preimplantation Period

J. M. Yochim

Department of Physiology and Cell Biology
University of Kansas, Lawrence

During the preimplantation period of gestation the conceptus is dependent for survival upon substrates in its external environment. Some of these nutrients for metabolism have been defined and have been identified as products of oviductal and uterine metabolic activity (Bishop 1957; Brinster 1965*a, b;* Fridhandler 1961; Krishnan and Daniel 1967; Mastroianni and Wallach 1961; Saldarini and Yochim 1967, 1968). Nevertheless, the oviduct and the uterus are unable to support development or permit nidation unless very strict spatial-temporal conditions between the conceptus and the reproductive tract are maintained (Dickman and Noyes 1960; Noyes and Dickman 1960, 1961). Thus, it is apparent that a very close relationship exists between the conceptus and its mother, a relationship limited temporally by the microenvironment of the uterine lumen. Exactly how this intrauterine environment is established is unknown. It is to this problem that we have directed our efforts.

To understand how provision is made for survival of the conceptus *in utero* and for the initiation of its nidation, it is necessary to examine the metabolism of the uterus and its regulation by the changing hormonal milieu of early pregnancy. In the cornua of the rat on day 1 of progestation (within 36 hr after ovulation) the utilization of an endogenous substrate for the production of lactate is favored by the endometrium; uterine oxygen consumption is low (Saldarini and Yochim 1967, 1968). By day 4 of progestation when the blastocysts are in the cornua (within 108 hr after ovulation), the endometrium is dependent upon exogenous substrate, and the pool of lactate has declined; oxygen consumption is high (fig. 1). Similar changes occur in the myometrium (Yochim and Saldarini 1969). This shift in metabolic activity of the uterus appears to be a reflection of the change in hormonal stimulation accompanying progestation, a change from estrogen dominance at the time of ovulation to progestogen dominance with permissive estrogenic action at the time of implantation, when the blastocysts are established *in utero*.

From this general assessment of uterine metabolism during the preimplantation period several questions arise: What relation do these changes bear to decidualization? How might such metabolic changes be related to implantation? It is possible to develop a tentative hypothesis which relates progestational metabolism to progestational physiology, a hypothesis based upon evidence derived from morphologic, physiologic, and metabolic observations.

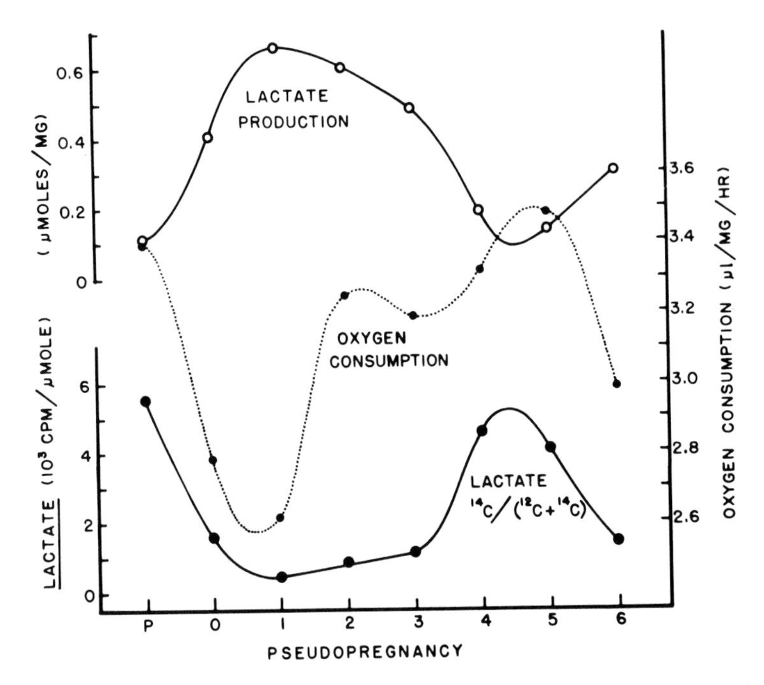

Fig. 1. Oxygen consumption (*dotted line*) and lactate production by the rat uterus during early pseudopregnancy. *Top line:* total lactate produced during incubation of endometrium with labeled glucose. *Bottom line:* extent of exogenous substrate contribution (^{14}C-glucose) to the total lactate pool. From Saldarini and Yochim 1967, 1968.

I. Metabolism of the Uterus during Progestation: Postulation of Its Relationship to Decidualization and Implantation

The phenomena of decidualization and implantation may be considered as physiologic events limited temporally by substrate availability. It might be hypothesized that in the intact rat both an extracellular and an intracellular source of energy are required for decidualization and implantation. The availability of these substrates may be regulated somewhat independently.

A. *Decidualization*

Oxidative activity in the cornua of the rat increases to a peak during the preimplantation period (Saldarini and Yochim 1967). In intact animals and in hormone experiments which permit maximum decidualization a depletion of the lactate pool (to feed the oxidative machinery?) is observed within a 12 hr period prior to the development of maximal decidual sensitivity. The phenomenon occurs in the same time sequence whether sensitivity, induced by endocrine manipulation, occurs on day 3 midnight, day 4 noon, at both times, or on day 4 midnight (Saldarini and Yochim 1968). The major contribution to the lactate pool during these times is from exogenous glucose, whose uptake and utilization appears to be regulated by progesterone (fig. 2). The data suggest that this exogenous substrate, undergoing complete oxidation, provides the energy necessary for the initiation of decidualization. Since the uptake of glucose may be progesterone regulated, it is

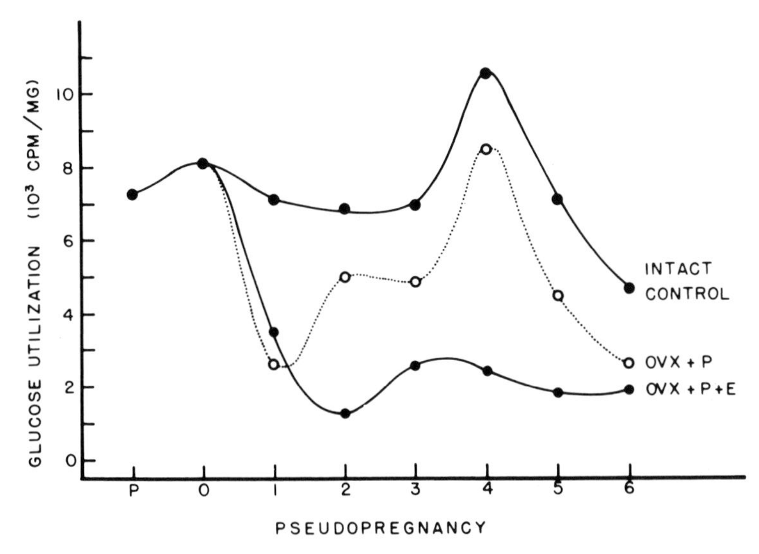

Fig. 2. Glucose utilization by the endometrium in intact rats and in ovariectomized rats treated daily with progesterone (*P*) ± estrone (*E*). In both experimental groups the hormone treatments were capable of eliciting maximal decidual sensitivity. From Saldarini and Yochim 1968.

consistent that the endometrium can respond to a traumatic stimulus at almost any time in ovariectomized rats treated with progesterone only. No temporal restrictions are imposed.

How then is such prolonged sensitivity to decidualization prevented from occurring in the intact rat? Studies of its hormonal control revealed that estrogen blocks the extended duration of sensitivity observed in ovariectomized, progesterone-treated rats (Yochim and De Feo 1963). Treatment of these rats with estrogen also blocks the progesterone-maintained glucose uptake in both the endometrium and the myometrium (Saldarini and Yochim 1968; Yochim and Saldarini 1969) (fig. 2). The evidence is consistent with the concept that the presence of a low level of estrogenic stimulation in the intact animal during the period preceding implantation impairs glucose uptake or utilization or both by the endometrium and in this fashion denies to the tissue much of the immediate potential energy for decidual sensitivity. A temporal restriction on decidualization is thus imposed.

B. *Implantation*

The accumulation in the uterus of an endogenous substrate such as glycogen is an estrogen-regulated phenomenon which occurs during the preovulatory period of the estrous cycle in the rat (Kostyo 1957; Rubulis, Jacobs, and Hughes 1965; Walaas 1952). After ovulation the content of glycogen in the uterus declines (Boettiger 1946; Connolly et al. 1962; Kostyo 1957; Walaas 1952), the result of a postovulatory deprivation of estrogen and perhaps an increased secretion of progestogen, which is known to be glycogenolytic (Boettiger 1946; Brody and Westman 1958). (The decline in glycogen content at this time is correlated with the increased utilization of endogenous substrate and the production of lactate by the uterus [Saldarini and Yochim 1968; Yochim and Saldarini 1969].) Throughout the preimplantation period the content of glycogen in the cornua of the rat is very low (Boettiger 1946; Connolly et al. 1962). However it is apparent that the rate of turnover of this substrate can be increased without changing its content by administration of appropriate dosages of estrogen and progestogen (Rosenbaum and Goolsby 1957). Moulton and Leonard (1966) have shown that during early pseudopregnancy in the rat the activity of the enzyme phosphorylase-a gradually increases to a peak on day 4. This enzyme is required for glycogenolysis; its increased activity is, in part, an indication that the metabolic machinery governing the rate of turnover of glycogen is operating. Such an increase in glycogen turnover, if it occurs, is a reflection of estrogen action on glycogenesis (Rubulis, Jacobs, and Hughes 1965; Walaas 1952) and of estrogen-progestogen interaction on glycogenolysis (Brody and Westman 1958; Moulton and Leonard 1966; Rosenbaum and Goolsby 1957). During the period preceding implantation in the rat, when both ovarian hormones are known to be present, this phenomenon would be expected. The increased glycogen turnover should coincide therefore with the period of high oxidative activity in the uterus.

Because the supply of a highly mobilizable pool of glycogen is regulated by estrogen, it is possible that it may be reserved in part for the nidatory process. When this endogenous store is impaired, as occurs in the absence of estrogen, nidation does not take place in the rat, although artificial decidualization is still possible (De Feo 1967). Conversely, when estrogen is present, the turnover of glycogen is increased within 12 hr (Bitman et al. 1965).

Thus in the rat the action of estrogen on a progestogen-dominated endometrium may (*a*) stimulate a rapid turnover in glycogen to provide a *potential* nutrient source for the invading blastocyst, and (*b*) limit the use of exogenous substrate to decrease the time during which the endometrium can respond to a decidualizing stimulus. The level of estrogen is critical: if it is too low, the highly mobile glycogen pool is not maintained for the implantation process, and the uterus will continue to respond to decidualizing stimuli; if too high, the glycogen pool may be available, but decidual sensitivity is blocked by too rapid a restriction of glucose uptake or utilization or both. In either case implantation is precluded.

By the use of a sensitive histochemical technique Christie (1966) determined the timing and distribution of glycogen accumulation in the endometrium of the rat during early pregnancy. Implantation began about 120 hr after ovulation, early on day 5 (day 0 = sperm-positive vaginal smear). Prior to this time no glycogen was found in the epithelium, decidua, or stroma. Immediately before implantation a few granules were detectable antimesometrially in some cells of the epithelium and primary decidua. Within the next 12 hr (by noon of day 5) glycogen accumulated at this antimesometrial site as implantation occurred. By midnight of day 5 (about 144 hr after ovulation) and afterward, although glycogen continued to accumulate laterally and mesometrially, a decrease in the intensity of stain was noted in the antimesometrial area.

Correlated with these changes was the histochemical localization of glucose-6-phosphatase (G-6-P-ase), an enzyme required for the conversion of G-6-P to glucose. Only trace activity was present in the epithelium lining the uterine lumen and glands by noon of day 4. Within 24 hr (day 5 noon) some staining for this enzyme was visible in the primary decidua and, with the appearance of the secondary decidua, it became more intense. Christie suggests that increased G-6-P-ase activity facilitates the pathway for conversion of glycogen → G-6-P → glucose; from the decrease in stain intensity of glycogen in the antimesometrial epithelium and subepithelial decidua after day 5 he infers possible increased glycogenolytic activity and a release of glucose into the extracellular space.

From the postulation above (I A, B), three basic questions arise: How is the impending site of decidualization and implantation localized in the uterus? How is the process of implantation triggered? How is the endometrium induced to relinquish its nutrient stores during invasion? Only the first and last questions will be considered.

Recent studies have shown that changes in oxygen tension can alter the metabolic activity of the cornua of the rat. Since the availability of oxygen within the lumen of the cornua is influenced by ovarian steroids, these hormones may be able to regulate metabolism by means different from their direct and ubiquitous actions on all uterine cells. They may modify indirectly the metabolic activity of *localized* areas of the endometrium by regulating the oxygen supply to which these areas are exposed. Such effects of oxygen, described below, have been demonstrated *in vitro* in the estrogen-primed cornua of the rat and rabbit.

II. Effects of Oxygen Tension on Carbohydrate Metabolism *in Vitro*

Oxygen tension has been shown to affect at least two key aspects of carbohydrate metabolism: the production of lactate and the metabolism of glycogen.

A. *Glycolysis and Lactate Production: Lactic Dehydrogenase (LDH) Activity*

In the laboratory of Kaplan (Cahn et al. 1962; Dawson, Goodfriend, and Kaplan 1964; Goodfriend and Kaplan 1964) it has been shown that under estrogen domination or stimulation the cornua of the rat and rabbit produce LDH isozymes which can maintain activity in the presence of high concentrations of pyruvate. When explants of cornua were maintained in tissue culture, the production of M-LDH was favored over that of H-LDH in a fashion similar to that observed after estrogen stimulation *in vivo*. Estradiol added to the medium did not affect these changes further; however elevation of the oxygen tension suppressed the overproduction of M-subunits by uterine explants.

On the basis of these and other experiments Kaplan has proposed a physiologic role for the two types of LDH subunits. The concept is based on the demonstration of a significant difference in the degree to which the activity of M- and H-LDH is inhibited by pyruvate. M-LDH, the muscle form, is able to maintain activity at relatively high pyruvate concentrations; thus under conditions of rapid glycolysis, as might occur in an oxygen-deprived environment, the conversion of pyruvate to lactate is assured. In contrast the H-subunit is maximally active at low concentrations of pyruvate and is strongly inhibited by an excess of this substrate; therefore the inhibition of H-LDH by pyruvate in tissues which contain this subunit favors the oxidative pathways for the metabolism of glucose.

The change in synthesis of subunits in the uterine explants which occurred with changes in oxygen tension are consistent with the proposed roles of M- and H-LDH (Cahn et al. 1962; Dawson, Goodfriend, and Kaplan 1964). The low oxygen tension favored M-LDH, the form best suited for anaerobic metabolism, whereas high pO_2 blocked the production of M-LDH, perhaps decreasing the extent to which lactate production could occur and shifting metabolism of pyruvate into oxidative pathways.

B. *Glycogenolysis*

It is well known that the rate of glycogenolysis in the cornua of the estrogen-primed rat is affected greatly by the availability of an exogenous energy source and oxygen in the medium (Shane and Leonard 1965; West and Cervoni 1955). Shane and Leonard (1965) have shown that under aerobic conditions two phases in glycogenolysis appear (fig. 3): an initial, rapid phase (half-life = 31 min) which lasts for about 15 min, and a slow phase beginning after 15 min (half-life = 153 min).

In the presence of low concentrations of glucose the initial rate of mobilization of glycogen was not affected, but the second phase was blocked (half-life = 47.5 hr). With higher concentrations of glucose in the medium the initial rate of glycogenolysis was slowed (half-life = 90 min) and persisted for about 40 min. Following this period, during the second phase, no glycogenolysis occurred.

In contrast, under anaerobic conditions the initial rate of depletion of glycogen was substantially increased. Within 10 min half of the glycogen was lost and the rate of loss was independent of the presence or absence of glucose in the medium. Thereafter (the second phase of glycogen breakdown) the presence of glucose in the medium retarded, but did not completely block the rate of glycogenolysis (fig. 3). These workers concluded that "if maintenance of a glycogen concentration approaching that reported for control horns is of importance to the normally functioning estrogen-primed uterus, then the availability of both a utilizable substrate and oxygen is essential."

The *in vitro* experiments by Shane and Leonard establish the concept of a rapidly mobilizable glycogen store which is induced by action of estrogen, which can be drawn upon within minutes, and whose rate of mobilization is regulated by the availability of extracellular glucose and oxygen.

These studies (II A, B) demonstrate effectively that the oxygen tension to which uterine tissue is exposed can modify to a significant extent the metabolism of glucose. However, to understand what metabolic changes are wrought within the uterine cavity and the lining of the endometrium, it is necessary to examine the level of oxygen to which the tissue is exposed *in vivo* and to study also those factors which may regulate PO_2 *in vivo*. Accordingly the following experiments were performed.

III. *In Vivo* Measurement of Intrauterine Oxygen Tension in the Rat

A. *Effects of Vasodynamic Activity and Hormone Stimulation*

Experiments designed to measure intrauterine oxygen tension revealed that: (*a*) pO_2 fluctuated in a rhythmic fashion similar to that observed for contractions

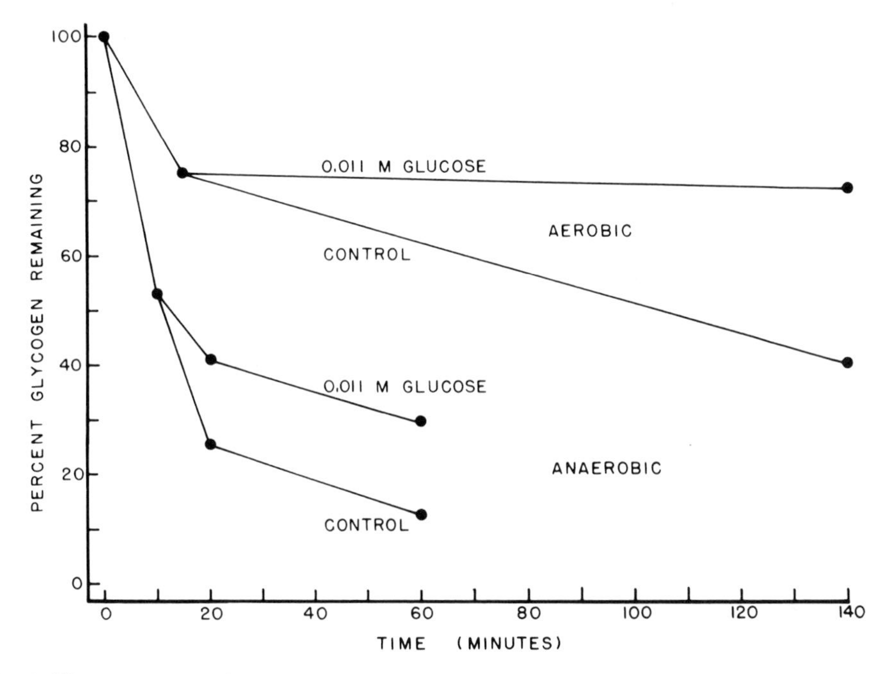

Fig. 3. The percentage of glycogen remaining in the uterus as a function of the time incubated *in vitro*. From Shane and Leonard 1965.

of the myometrium *in vitro* (fig. 4); (*b*) following ovariectomy the mean oxygen tension increased greatly over a 5-day period; (*c*) the increased pO_2 measured after ovariectomy could be prevented by treatment with estrogen, less effectively with progesterone, and (*d*) treatment with both steroids (fig. 5) had effects intermediate to those measured with either hormone alone (Mitchell and Yochim 1968*a*).

In an attempt to explain the origin of the rhythmic fluctuations in pO_2 direct observations of the uterine vasculature *in situ* were made while oxygen tension was recorded. It was noted that the periodicity was largely determined by fluctuations in the patency of the arterioles beneath the epithelial lining of the endometrium. With enhancement of blood flow through the capillary bed, a result of arteriolar dilation, oxygen tension increased; after vasoconstriction pO_2 decreased. No clear correlation could be made between locally observed myometrial activity and oxygen tension.

On occasion, however, uterine contractions shifted slightly the position of the organ with respect to the elecrode, which was held rigidly in its holder. This resulted in changes in pO_2 dependent upon the location of the nearest capillary bed and its arteriolar supply. Thus the rhythmic fluctuations in intrauterine oxygen tension were found to be related more to the contractions of the intrauterine vascular bed than to myometrial contractile activity.

Changes in mean pO_2 during the estrous cycle or after estrogen treatment could be related to a differential effect of estrogen on oxidative metabolism and on tissue and vascular growth. Injection of ovariectomized rats with estrogen prevented the postovariectomy increase in mean intraluminal pO_2 (fig. 5). The degree of re-

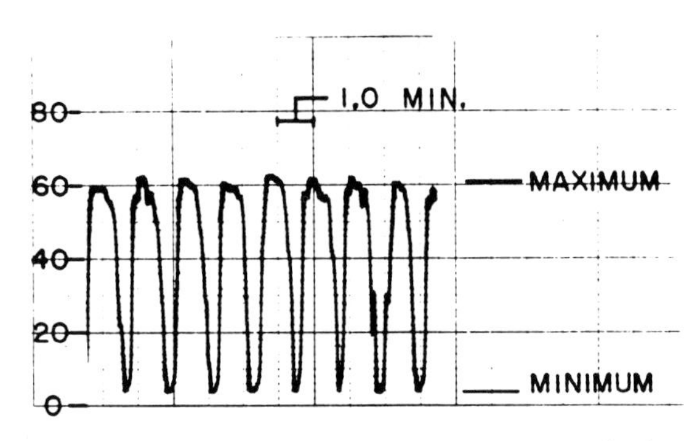

Fig. 4. Fluctuations in intrauterine oxygen tension measured *in vivo* in the rat. *Ordinate:* oxygen tension, mm Hg; *abscissa:* time, in min.

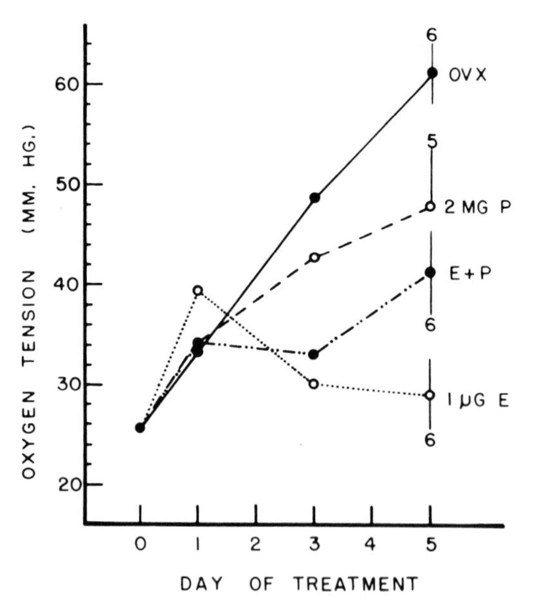

Fig. 5. Intrauterine oxygen tension in the rat following ovariectomy during estrus (day 0) and daily replacement therapy with 2.0 mg progesterone (*P*), 1.0 μg estrone (*E*) or the combination of hormones ($E + P$). Vertical bars represent ± 1 standard error of mean. From Mitchell and Yochim 1968*a*.

sponse was approximately dose-dependent. Although estrogen is known to stimulate increased blood flow and induce hyperemia (factors which would *favor* a pO_2 rise), the lack of response is accounted for in part by the reduced vascular density resulting from edema, hypertrophy, and hyperplasia, and by the increased respiration of the tissue (Mitchell and Yochim 1968*a, b*).

Progesterone, in contrast to estrogen, only partly prevented the postovariectomy increase in mean intrauterine pO_2 and in general mimicked the patterns of hormonally deprived animals (fig. 5). These effects were consistent with the morphological and physiological actions of progesterone. Tissue regression occurs to some degree, but the uterine vasculature shows no major alterations. Thus a relative increase in vascular density occurs, essentially like that following ovariectomy.

It was evident from this series of experiments that the partial pressure of oxygen in the lumen of the cornua of the rat is probably the result of a balance between those morphologic and metabolic factors which favor and those factors which militate against an increase in oxygen diffusion. The level of oxygen within the cornual cavity is relatively low in comparison with arterial tension, a situation which usually predominates in the intact animal. This relationship is maintained through steroid hormone stimulation since following hormone deprivation pO_2 rises.

B. *Intrauterine Oxygen Tension during Progestation*

Measurements of intraluminal oxygen during the preimplantation period revealed a sequence of changes fairly consistent with what is known of the endocrinology of this stage (Yochim and Mitchell 1968). In both pregnant and pseudopregnant rats average pO_2 increased from a minimal value after ovulation to levels somewhat higher than venous tension (fig. 6). A transient depression in pO_2 during day 4 might have been considered the result of a surge in estrogen secretion the previous day, but experimental evidence suggested that the entire pattern during the 5-day period was a reflection of a *relative* deficiency of estrogen, similar to that measured after ovariectomy with or without progestogen replacement.

Thus, from an endocrine standpoint, after initiation of progestation the level and pattern of oxygen tension in the uterus of the rat reflected an increased progestogenic stimulation, coupled with a decreased estrogenic stimulation. The result of such hormone interaction was the production of a relatively hypoxic environment in the uterine lumen (in relation to arterial tension), though the oxygen partial pressure was elevated above venous pO_2.

C. *Uterine Factors Which Affect Oxygen Tension during Progestation*

Intraluminal oxygen during the luteal phase appeared to vary inversely with respiration of the cornua (Saldarini and Yochim 1967; Yochim and Mitchell 1968). During the preimplantation period, when O_2 consumption was high, intrauterine pO_2 was relatively low. During the postimplantation period, as cornual respiration decreased, oxygen tension increased. A plot of oxygen utilization was made, based upon the pO_2 measurements, for it was thought that the pattern of activity might be similar to that for direct measurements of whole organ respiration. However, when the comparison was made, the rapid decline in the calculated respiratory rate during days 4–5 of pseudopregnancy did not parallel the constant respiratory rate actually measured during the period. This "discrepancy" indicated that diffusion of oxygen into the lumen was increased despite an elevated tissue respiration. Either the rate of blood flow to the endometrial area exceeded the rate of growth, differentiation, and oxygen utilization of the tissue, or permeability to oxygen was significantly increased. That the latter may be in part the case is suggested by the rapid decline in cornual weight which occurs during the day 4 to 5 interval (Saldarini and Yochim 1967). In addition Marley and Robson (1967) have reported that in the rat during the evening of day 4 of progestation a significant rise in the rate of transfer of ^{22}Na

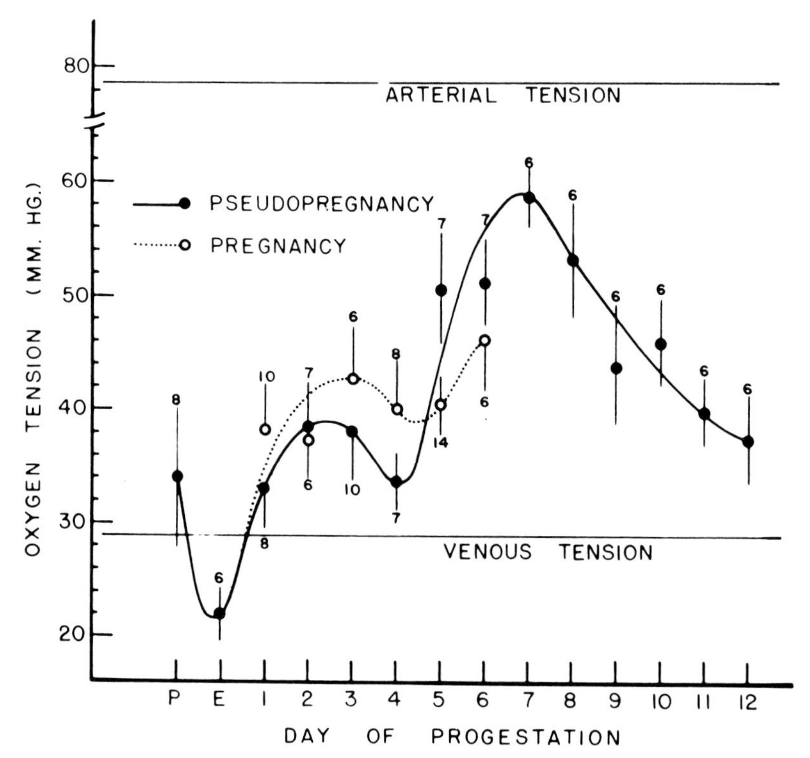

Fig. 6. Intrauterine oxygen tension in the rat during pseudopregnancy and the first 6 days of pregnancy. Vertical bars represent $\pm$ 1 standard error of mean. From Yochim and Mitchell 1968.

from the uterine extracellular space into the uterine lumen occurs; the permeability of the tissue to sodium increased about 3-fold in pseudopregnant rats (table 1).

Thus during progestation the level of oxygen presented to the conceptus is affected by the rate of oxygen consumption of the uterus, the relative vascular density in the endometrium, and the degree of vasodynamic activity. A rapid increase in intrauterine pO_2 between days 4 and 5 appeared to be the result of an increased permeability to oxygen.

Do modifications occur in endometrial glycogen metabolism and LDH activity in the rat as a result of changes in intrauterine oxygen? To our knowledge kinetic studies of glycogen turnover during progestation have not been done. Thus no correlation between glycogenolysis and intrauterine pO_2 is possible. However lactic dehydrogenase has been measured in the endometrium and myometrium of the cornua and a correlation between the activity of this enzyme and intrauterine pO_2 during progestation can be made.

IV. Relation of Intrauterine pO_2 to Lactic Dehydrogenase Activity during Progestation

By fluorometric measurement of the reduction of NAD with appropriate substrates, and by electrophoretic separation of the 5 isozymes on cellulose acetate strips, it was possible to determine changes in M- and H-LDH subunit activity in the endometrium and myometrium of the cornua of the rat. These measurements were made during the estrous cycle and the first 6 days of pseudopregnancy (Clark and Yochim 1969).

TABLE 1

RATE OF RECOVERY OF ^{22}Na FROM UTERINE LUMEN DURING EARLY PROGESTATION

DAY	PSEUDOPREGNANCY			PREGNANCY		
	N	$\bar{x}$[a]	SE	N	$\bar{x}$[a]	SE
3	6	4.91	0.90	6	3.50	0.36
4	8	3.75	0.34	6	4.89	0.53
4[b]	6	10.63	0.77	8	9.39	1.13
5	9	11.14	1.09	8	4.89	0.64

SOURCE: Marley and Robson 1967.
[a] Expressed as a percentage of the ^{22}Na in 1 ml of plasma.
[b] Perfused between 9 P.M. and 11 P.M.; all other experiments performed between 10 A.M. and 4 P.M. Copulation plug present on day 0 in pregnant animals.

A. *Measurement of LDH Activity during Progestation*

The specific activity of LDH was 2–5 times greater in endometrium than in myometrium, depending on the stage of the cycle (fig. 7). Marked changes in endometrial LDH were measured during the estrous cycle and pseudopregnancy, in contrast to the low, relatively stable pattern in the myometrium. Enzyme activity increased to a peak during estrus (day 0 of pseudopregnancy) and declined during days 1–6 of progestation. During estrus M-LDH subunits accounted for about 90% of the total LDH activity measured. By day 6 M-LDH declined to 80% of the total activity. In the myometrium the proportion of M-LDH decreased from 73% to 62% during the same period.

Within 5 days after ovariectomy the activity of M- and H-LDH subunits declined in the endometrium (table 2). Treatment of ovariectomized rats with estrogen prevented the decline in a dose-dependent fashion. Injection of progesterone did not prevent the postcastration drop in M-subunit activity, but low doses of the hormone did maintain H-LDH at the level measured on day 0 (table 3). When both hormones were injected together, progesterone blocked the action of estrogen (table 4). The pattern of activity of M-LDH during treatment with both steroids was similar to that measured in intact rats during early progestation (fig. 8).

In order to localize the areas in the endometrium which exhibited LDH activ-

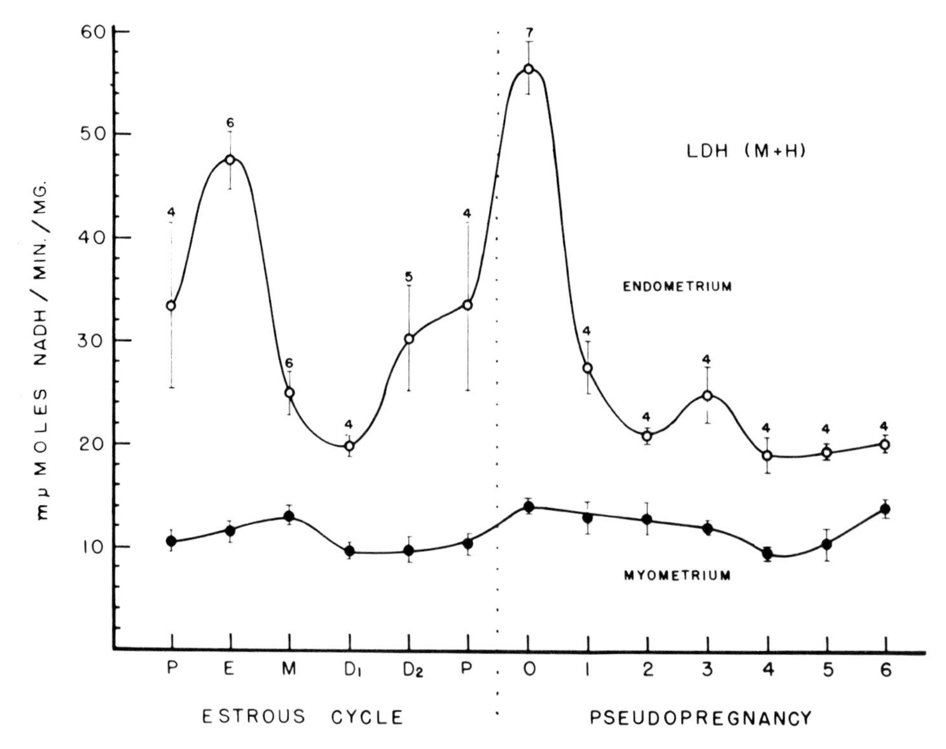

Fig. 7. Lactic dehydrogenase activity in the endometrium and myometrium during the estrous cycle and early pseudopregnancy in the rat. Vertical bars represent ± 1 standard error of mean.

TABLE 2

EFFECTS OF OVARIECTOMY AND REPLACEMENT WITH DAILY INJECTIONS OF ESTRONE ON M- AND H-LDH ACTIVITY IN THE ENDOMETRIUM OF THE RAT

Treatment	N[a]	M-LDH[b]	H-LDH[b]
Control (estrus, day 0)	5	51.8±3.7	6.1±0.8
0 μg estrone, 5 days	4	12.0±1.4	3.1±0.5
0.5 μg estrone, 5 days	4	19.4±1.3	2.5±0.4
1.0 μg estrone, 5 days	4	27.1±2.3	3.8±0.7
5.0 μg estrone, 5 days	4	49.7±3.2	5.1±0.6

[a] N = number of animals.

[b] LDH activity measured as mμmoles NADH formed/min/mg tissue ± std. error of mean. Subunit activity was calculated from isozyme electrophoretic patterns.

TABLE 3

EFFECTS OF OVARIECTOMY AND REPLACEMENT WITH DAILY INJECTIONS OF PROGESTERONE ON M- AND H-LDH ACTIVITY IN THE ENDOMETRIUM OF THE RAT

TREATMENT	N[a]	mμmoles LDH[b]				% OF DAY 0			
		M	SE[c]	H	SE	M	SE	H	SE
Control (estrus, day 0)	5	51.8	3.7	6.1	0.8	100	7	100	13
Prog. 5 days, 0 mg	4	12.0	1.4	3.1	0.5	23	3	51	8
2 mg	4	19.6	0.3	5.3	0.4	38	1	87	7
4 mg	4	17.0	0.8	3.6	0.7	33	2	59	16

[a] N = number of animals.

[b] LDH activity measured as mμmoles NADH formed/min/mg tissue. Subunit activity was calculated from isozyme electrophoretic patterns.

[c] SE = Standard error of mean.

ity, histochemical determinations were made. The most intense staining for the enzyme was localized in the epithelium in all sections. Granulation, less intense and variable, was observed also in the subepithelial stroma. In sections taken from animals during early progestation the density of granulation apeared to decline with time in both the epithelium and the subepithelial stroma.

TABLE 4

EFFECT OF PROGESTERONE ON ESTRONE-MAINTAINED M-LDH ACTIVITY IN ENDOMETRIUM OF THE UTERUS OF THE RAT

TREATMENT, 4 DAYS		M-LDH ACTIVITY[a]		% INHIBITION
Estrone	Progesterone	mμmoles	SE	
....	....	12.3	1.7	....
1 μg	....	24.0	2.6	....
1 μg	2 mg	16.7	0.3	30.4
5 μg	....	71.5	4.6	....
5 μg	2 mg	24.0	3.0	66.4

NOTE: N = 4 rats/expt.

[a] LDH activity measured as mμmoles NADH formed/min/mg tissue. M-subunit activity was calculated from isozyme electrophoretic patterns.

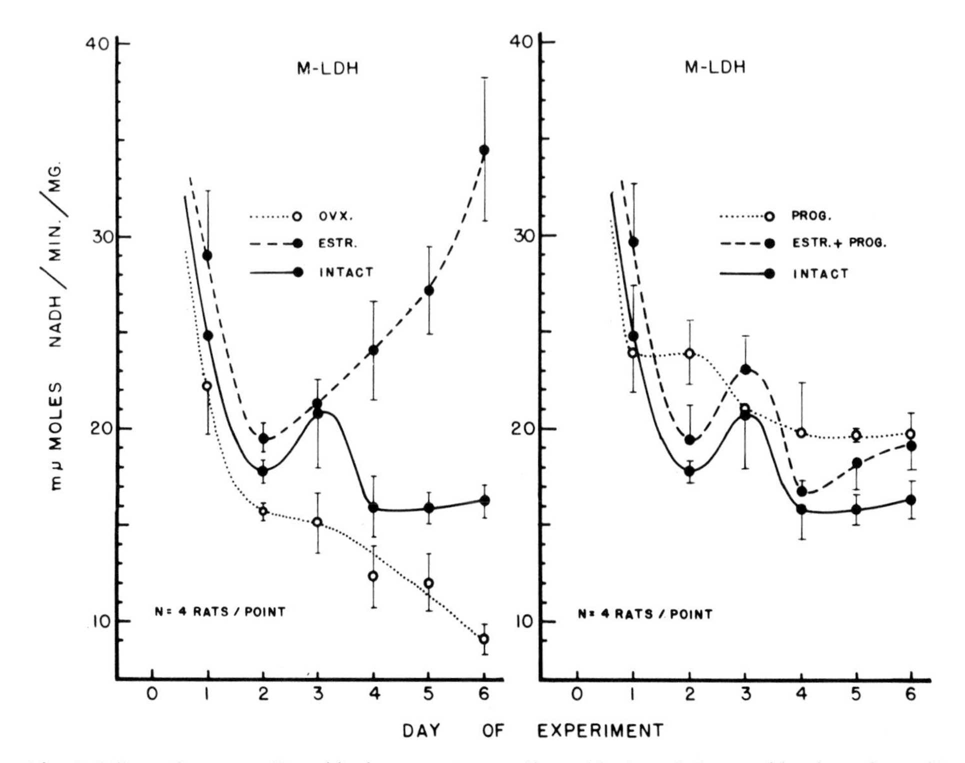

Fig. 8. Effect of estrone (1 μg/day), progesterone (2 mg/day) and the combination of steroids on M-LDH activity in the endometrium of the rat following ovariectomy during estrus (day 0). Data from intact pseudopregnant rats are shown for comparison. Vertical bars represent ± 1 standard error of mean.

B. *Relation of LDH Activity to Changes in pO_2*

The above studies reveal the distribution and hormonal regulation of M- and H-subunit activity in the endometrium. When endometrial M-LDH, plotted as specific activity (mμmoles NAD reduced/mg tissue), was compared with intrauterine oxygen tension, an almost perfect mirror image of patterns was observed (fig. 9). Fluctuations in intrauterine pO_2 were mirrored by changes in the opposite direction in M-subunit activity with one exception: no increase in specific activity of the enzyme occurred on day 4 to match the transient depression in pO_2 recorded on this day. However when endometrial M-LDH was plotted as total activity (mμmoles NAD reduced/endometrium), the expected rise in M-subunit activity on day 4 was

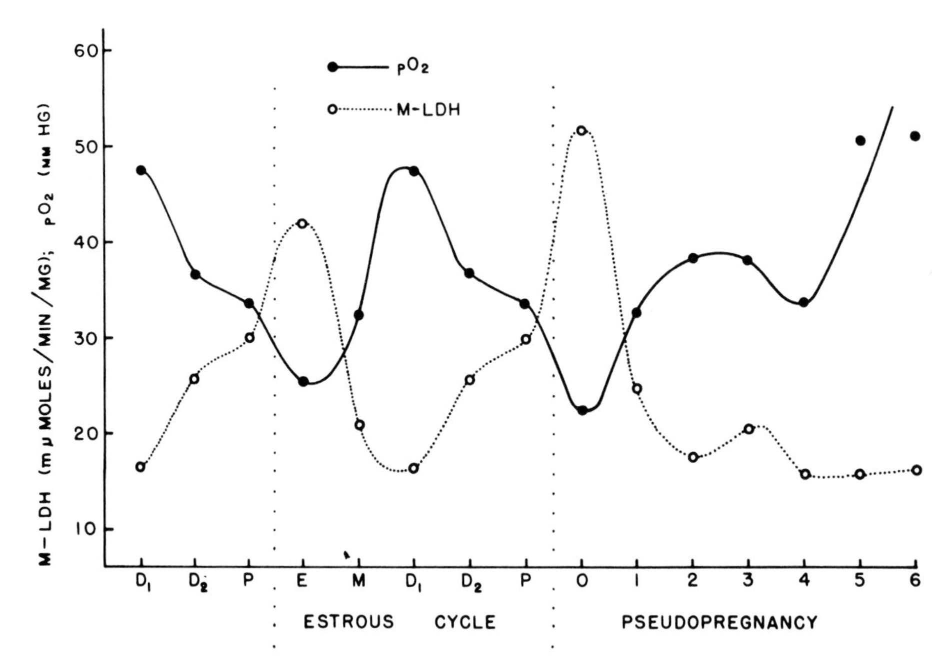

Fig. 9. Inverse relation between M-LDH activity in endometrium and intrauterine oxygen tension in the rat during the estrous cycle and early pseudopregnancy.

noted. It appears that the increase in M-subunit activity on day 4 of progestation was masked by a parallel increase in weight and water content of the endometrium at this time.

The inverse relationship between M-LDH and intrauterine oxygen tension was evident only in the endometrium. M-subunit activity in the myometrium did not reflect changes in luminal oxygen. It is possible therefore that intrauterine pO_2 may modify the activity of LDH in the epithelium and subepithelial stroma of the endometrium of the rat during early progestation.

The pO_2 and LDH experiments indicate one way in which sensitivity to decidualization may be established—by alteration of LDH activity with subsequent shifts in oxidative metabolism. How this phenomenon is localized to provide a clue to the site of implantation, is described below.

V. Relation of Intrauterine pO_2 to the Site of Implantation

Only a slight variation in pO_2 was shown during days 1–4 of the luteal phase in contrast to the very rapid increase which occurred between day 4 and day 5 of progestation (Yochim and Mitchell 1968). The free-floating blastocyst during days 3–4 of pregnancy must satisfy its oxygen requirement in an intrauterine environment which contains an average partial pressure of oxygen of only 40–42 mm Hg (5.2–5.5%). A comparison of the data between pregnant and pseudopregnant rats during days 4–5 was somewhat revealing. In pseudopregnant rats, despite a continued high oxy-

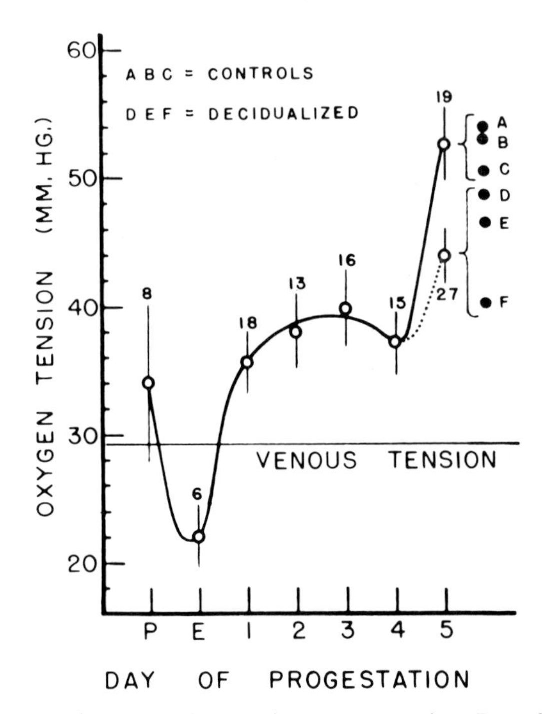

Fig. 10. Effect of decidualization on intrauterine oxygen tension. Data from pseudopregnant and pregnant rats (days 1–4) were pooled. *A* = control cornua from unilaterally decidualized rats; *B* = pregnant rats, unilaterally ligated oviduct; *C* = intact pseudopregnant rats; *D* = pregnant rats, patent oviduct; *E* = traumatized cornua from decidualized rats; *F* = intact, pregnant rats. Number of animals is listed above or below each standard error of mean. From Yochim and Mitchell 1968.

gen consumption by the uterus, intraluminal oxygen tension increased significantly from 37 mm Hg to 53 mm Hg during the day 4 to 5 interval, whereas *no significant change occurred in pregnant rats.* On day 5 oxygen tension was in general higher in nondecidualizing uteri than in uteri with deciduomata or with implanted blastocysts (fig. 10). The data indicated that during this period the uterus provides for a rapidly increased diffusion of oxygen into the lumen, *a process which is impaired should decidualization begin locally in the antimesometrial stroma.*

Examination of the data of Marley and Robson (1967) reveals that a rapidly increased permeability to sodium occurs in both pseudopregnant and pregnant rats late on day 4. In pregnant rats the phenomenon is reversed by day 5, an indication that the rate of transfer of sodium into the uterine lumen is impaired transiently by the decidualizing sites (table 1). Since the pO_2 *in utero* is low during the preimplantation period, local changes in vasculature and in permeability might provide areas along the antimesometrial surface of the endometrium with a relatively high oxygen concentration. Evidence that the endometrium in this region contains the greatest capillary density (Williams 1948; Young 1951), that oxygen utilization of the unimplanted blastocyst of several mammals is very high during this time (Boell and Nicholas 1948; Brinster 1968; Fridhandler 1961; Fridhandler, Hafez, and Pincus 1956*a, b,* 1957; Mills and Brinster 1967; Popp 1958), and that an elevated oxygen tension may facilitate aspects of the nidatory process *in vitro* (Glenister 1963) lends some support to a hypothesis that increased oxygen may be involved in localizing the sites of nidation in the rat.

The rise in pO_2 during days 4–5 of pseudopregnancy was not restricted to intact control animals; gonadectomized rats and animals receiving progesterone also responded in this fashion. Therefore if a change in oxygen tension was the *stimulus* for the nidatory process, implantation should occur after day 4 in all of these animals, since the experimental treatments produced similar patterns. Under these endocrine conditions, however, implantation is precluded (De Feo 1967; Mayer 1963; Nutting and Meyer 1963; Psychoyos 1965). It must be presumed, therefore, that the nidatory process is regulated by some means in addition to rapid changes in pO_2. A shift in tension may aid only in the localization of the site of attachment. That it requires 4–5 days for the pO_2 shift to occur (in the intact rats as well as in ovariectomized animals with or without progesterone replacement) may be a function of the level and timing of exposure to estrogen during ovulation rather than the hormonal milieu during the preimplantation period.

From all the evidence presented it is possible to describe some aspects of decidual sensitivity and implantation in terms of the hormonal regulation of carbohydrate metabolism and its modification by intrauterine oxygen tension. It should be stressed that such a description is, at best, a crude outline open to modification as more evidence (and thus a better interpretation) becomes available. Such a description is presented below as a summary.

VI. The Regulation of Decidual Sensitivity and Implantation in the Rat

1. Uterine metabolism during progestation shifts from glycolysis, with increased utilization of endogenous substrate, to aerobiosis, with increased dependence upon exogenous substrate. This phenomenon may be progestogen-regulated.
2. During transition into progestation (days 0–4) evidence suggests that the metabolic machinery for glycogenolysis is active and the content of glycogen is low and constant. A gradual increase in intrauterine oxygen tension (days 0–4) may prevent too rapid a glycogenolysis, establishing a turnover whose rate is dependent upon the presence or absence of estrogen.
3. The gradual increase in intrauterine oxygen tension (days 0–4) may limit the production of M-LDH subunits, a change which may permit increased oxidative

metabolism. The luminal linings of the cornua are now exposed to two environmental stimuli (oxygen and progesterone) which enhance the capacity for oxidative metabolism and restrict that for lactate production. Oxygen consumption rises (days 2–5) and lactate is gradually depleted (days 3–4).

4. The presence (or appearance) of permissive levels of estrogen during day 3 may initiate within 12–24 hr an increased rate of glycogen synthesis and thus accelerate the turnover of glycogen, even though the actual content of this intracellular product is low. The elevated glycogen turnover rate may provide a potential source of substrate for the invading blastocyst.

5. Glucose uptake (progesterone-regulated), maintained throughout the preimplantation period, may be a major source of substrate for sensitivity to decidualization. The uptake or utilization of this substrate can be blocked by estrogen, as can the sensitivity to decidualization. The estrogen secreted during day 3 will become effective late on day 4 and restrict the duration of maximal sensitivity to a 12 hr period between day 3 midnight and day 4 noon.

6. A rapid increase in permeability late on day 4, before implantation, flushes oxygen into the uterine lumen via the antimesometrial capillary beds. The rise in pO_2 may aid in the localization of potential sites of implantation, as described below. However the mechanisms involved in the process of attachment and invasion include factors in addition to changes in intrauterine oxygen tension.

7. The elevation of intraluminal oxygen tension may block the rate of glycogenolysis locally. This, in conjunction with a progesterone-stimulated increase in glucose uptake on day 4 (and an estrogen-stimulated increase in glycogenesis?), gradually shifts the turnover of glycogen in favor of glycogen synthesis. Glycogen now accumulates sporadically in the antimesometrial endometrium by early day 5.

8. The change in intrauterine oxygen tension late on day 4 also may cause an additional local restriction of M-LDH isozyme production by day 5. As a result substrate is shifted increasingly into oxidative pathways in areas where decidualization already is underway. Such activity could provide energy for growth and differentiation of the decidua.

9. The attachment of the blastocysts early on day 5 and the edema, hypertrophy, and hyperplasia in the decidualizing sites increase the tissue density in the antimesometrial endometrium. As a result oxygen tension is reduced in the uterine lumen on day 5 of gestation. Consequently the local rate of turnover of glycogen may be reversed gradually, favoring glycogenolysis with formation of glucose-6-phosphate. This product is potentially usable by the invading blastocysts or by the endometrial cells undergoing decidualization.

10. Locally increased G-6-P-ase activity at this time in the decidua provides the enzymatic machinery for converting G-6-P to glucose. Glucose may diffuse into the extracellular space in the decidua to provide a nutrient source for the invading blastocysts.

The question whether such processes do in fact operate as part of the mechanism of nidation will require much further study. In any event the experiments described, as well as those in the rabbit, rat, and mouse (Glenister 1963; Kirby 1965), suggest that the metabolic requirements for nidation are not unique to the uterus, since aspects of the nidatory process may be observed in tissues other than endometrium. Furthermore, these nidation-related metabolic processes may not require endocrine regulation unless the cornua are *in situ,* as in the intact animal. It may be precisely because of the unique structure of the uterus that endocrine-induced changes in the luminal environment (and thus in endometrial metabolism) can impose the time limits on decidualization and implantation in the rat, and perhaps in other mammals.

References

Bishop, D. W. 1957. Metabolic conditions within the oviduct of the rabbit. *Int J Fertil* 2:11.

Bitman, J.; Cecil, H. C.; Mench, M. L.; and Wrenn, T. R. 1965. Kinetics of *in vivo* glycogen synthesis in the estrogen-stimulated rat uterus. *Endocrinology* 76:63.

Boell, E. J., and Nicholas, J. S. 1948. Respiratory metabolism of the mammalian egg. *J Exp Zool* 109:281.

Boettiger, E. G. 1946. Changes in the glycogen and water content of the rat uterus. *J Cell Comp Physiol* 27:9.

Brinster, R. L. 1965*a*. Studies on the development of mouse embryos *in vitro:* II. The effect of energy source. *J Exp Zool* 158:59.

———. 1965*b*. Studies on the development of mouse embryos *in vitro:* IV. Interaction of energy sources. *J Reprod Fertil* 10:227.

———. 1968. *In vitro* culture of mammalian embryos. In Eighth Biennial Symposium on Animal Reproduction, ed. A. V. Nalbandov and D. E. Becker. *J Anim Sci* 27, Suppl. 1:1.

Brody S., and Westman, A. 1958. Effects of oestradiol and progesterone on the glycogen content of the rabbit uterus. *Acta Endocr* 28:39.

Cahn, R. D.; Kaplan, N. O.; Levine, L.; and Zwilling, E. 1962. Nature and development of lactic dehydrogenases. *Science* 136:962.

Christie, G. A. 1966 Implantation of the rat embryo: Glycogen and alkaline phosphatases. *J Reprod Fertil* 12:279.

Clark, S. W., and Yochim, J. M. 1969. Lactic dehydrogenase activity, its isozyme distribution in the uterus of the rat and its relation to intrauterine oxygen tension. *Fed Proc* 28:637 (abstr).

Connolly, M. C.; Bitman, J.; Cecil, H. C.; and Wrenn, T. R. 1962. Water, electrolyte, glycogen and histamine content of rat uterus during pregnancy. *Amer J Physiol* 203:717.

Dawson, D. M.; Goodfriend, T. L.; and Kaplan, N. O. 1964. Lactic dehydrogenases: Functions of the two types. *Science* 143:929.

De Feo, V. J. 1967. Decidualization. In *Cellular biology of the uterus,* ed. R. M. Wynn, p. 191. New York: Appleton-Century-Crofts.

Dickman, A., and Noyes, R. W. 1960. The fate of ova transferred into the uterus of the rat. *J Reprod Fertil* 1:197.

Fridhandler, L. 1961. Pathways of glucose metabolism in fertilized rabbit ova at various pre-implantation stages. *Exp Cell Res* 22:303.

Fridhandler, L.; Hafez, E.; and Pincus, G. 1956*a*. O_2 uptake of rabbit ova. *Third Int Cong Anim Reprod,* sec. 1 p. 48.

———. 1956*b*. Respiratory metabolism of mammalian eggs. *Proc Soc Exp Biol Med* 92:127.

———. 1957. Development changes in respiratory activity of rabbit ova. *Exp Cell Res* 13:132.

Glenister, T. W. 1963. Observations on mammalian blastocysts implanting in organ culture. In *Delayed implantation*, ed. A. C. Enders, p. 171. Chicago: University of Chicago Press.

Goodfriend, T. L., and Kaplan, N. O. 1964. Effects of hormone administration on lactic dehydrogenase. *J Biol Chem* 239:130.

Kirby, D. R. S. 1965. The role of the uterus in the early stages of mouse development. In *Preimplantation stages of pregnancy,* ed. G. E. W. Wolstenholme and M. O'Connor, p. 325. Boston: Little, Brown and Co.

Krishnan, R. S., and Daniel, J. C., Jr. 1967. "Blastokinin": Inducer and regulator of blastocyst development in the rabbit uterus. *Science* 158:490.

Kostyo, J. L. 1957. A study of the glycogen levels of the rat uterus and certain skeletal muscles during pregnancy. *Endocrinology* 60:33.

Marley, P. B., and Robson, J. M. 1967. The passage of ^{22}Na into the uterine lumen of the rat. *J Physiol* 189:67.

Mastroianni, L., and Wallach, R. 1961. Effect of ovulation and early gestation on oviduct secretions in the rabbit. *Amer J Physiol* 200:815.

Mayer, G. 1963. Delayed nidation in rats: A method of exploring the mechanisms of ova-implantation. In *Delayed implantation,* ed. A. C. Enders, p. 213. Chicago: University of Chicago Press.

Mills, R. M., Jr., and Brinster, R. L. 1967. Oxygen consumption of pre-implantation mouse embryos. *Exp Cell Res* 47:337.

Mitchell, J. A., and Yochim, J. M. 1968*a*. Measurement of intrauterine oxygen tension in the rat and its regulation by ovarian steroid hormones. *Endocrinology* 83:691.

———. 1968*b*. Intrauterine oxygen tension during the estrous cycle in the rat: Its relation to uterine respiration and vascular activity. *Endocrinology* 83:701.

Moulton, B. C., and Leonard, S. L. 1966. Phosphorylase activity in the pseudopregnant rat uterus. *Endocrinology* 78:383.

Noyes, R. W., and Dickman, Z. 1960. Relationship of ovular age to endometrial development. *J Reprod Fertil* 1:186.

———. 1961. Survival of ova transferred into the oviduct of the rat. *Fertil Steril* 12:67.

Nutting, E. F., and Meyer, R. K. 1963. Implantation delay, nidation, and embryonal survival in rats treated with ovarian hormones. In *Delayed implantation,* ed. A. C. Enders, p. 233. Chicago: University of Chicago Press.

Popp, R. A. 1958. Comparative metabolism of blastocysts, extra-embryonic membranes and uterine endometrium of the mouse. *J Exp Zool* 138:1.

Psychoyos, A. 1965. Neurohumoral aspects of implantation. *Proc Second International Congress Endocrinology London* 83:508.

Rosenbaum, R. M., and Goolsby, C. M. 1957. The histochemical demonstration of hormonally controlled, intracellular glycogen in the endometrium of the rat. *J Histochem Cytochem* 5:33.

Rubulis, A.; Jacobs, R. D.; and Hughes, E. C. 1965. Glycogen synthetase in mammalian uterus. *Biochem Biophys Acta* 99:584.

Saldarini, R. J., and Yochim, J. M. 1967. Metabolism of the uterus of the rat during early pseudopregnancy and its regulation by estrogen and progestogen. *Endocrinology* 80:453.

———. 1968. Glucose utilization by endometrium of the uterus of the rat during early pseudopregnancy and its regulation by estrogen and progestogen. *Endocrinology* 82:511.

Shane, H. P., and Leonard, S. L. 1965. Rate of glycogenolysis in the estrogen-primed rat uterus *in vitro*. *Endocrinology* 76:686.

Walaas, O. 1952. Effect of oestrogens on glycogen content of the rat uterus. *Acta Endocr* 10:175.

West, T. C., and Cervoni, P. 1955. Influence of ovarian hormones on uterine glycogen in the rat: Glycogen requirements for contractility under varying environmental conditions *in vitro*. *Amer J Physiol* 182:287.

Williams, M. F. 1948. The vascular architecture of the rat uterus as influenced by estrogen and progesterone. *Amer J Anat* 83:274.

Yochim, J. M., and De Feo, V. J. 1963. Hormonal control of the onset, magnitude and duration of uterine sensitivity in the rat by steroid hormones of the ovary. *Endocrinology* 72:317.

Yochim, J. M., and Mitchell, J. A. 1968. Intrauterine oxygen tension in the rat during progestation: Its possible relation to carbohydrate metabolism and the regulation of nidation. *Endocrinology* 83:706.

Yochim, J. M., and Saldarini, R. J. 1969. Glucose utilization by the myometrium during early pseudopregnancy in the rat. *J Reprod Fertil* 20:481.

Young, A. 1951. Vascular architecture of the rat uterus. *Proc Royal Soc (Edinburgh) Ser B* 64:292.

22

Endocrine Control of Implantation

A. V. Nalbandov

Department of Animal Science
University of Illinois, Urbana

The detailed role of hormones in implantation has been reviewed recently in great depth and a new review of the details appears redundant (see, for instance, De Feo 1967). Accordingly this contribution will limit itself to a short recapitulation and perhaps a reinterpretation of the present state of knowledge of the field together with presentation of newer evidence, some of which is yet unpublished.

I. Implantation in the Rat

A. *The Role of Progesterone*

The endocrine events preceding implantation are best known and understood in the rat. It is clear that in both the rat and the mouse decidualization of the cornual endometrium and implantation require the cooperative action of progesterone and estrogen. Progesterone alone is able to induce implantation only in a few ovariectomized rats and even here the suspicion arises that remnants of ovarian tissue may have secreted sufficient estrogen to make implantation possible. In the great majority of animals, however, it has been demonstrated repeatedly that progesterone alone is unable to cause implantation. The following generalizations are in order. If progesterone is injected into intact females 1 or 2 days after mating, implantation occurs at the normal time (Sammelwitz, Dziuk, and Nalbandov 1956). If, however, rats are castrated 3 days after mating and progesterone injections are begun, implantation may be delayed as long as 45 days (Cochrane and Meyer 1957). Similar results were obtained by Mayer (1963) except that he finds that day 4 is the critical period of castration after which implantation following progesterone treatment is delayed. (This difference of 1 day in the critical period between the experiments of Cochrane and Meyer on the one hand and of Mayer on the other may be explained by strain differences in the rats used in the two laboratories.) Mayer and his co-workers also find that castration on day 5 followed by progesterone treatment does result in implantation without delay. (The corresponding figure in Cochrane's data is 4 days.) Thus we see that the hormonal requirements for implantation differ: up to 3 or 4 days after mating progesterone alone fails to cause implantation; progesterone alone leads to implantation in rats castrated 5 days after mating.

B. *The Role of Estrogen*

It was found earlier by Weichert (1942) and others (Yoshinaga 1961) that systemic injection of as little as 1 μg of estrogen significantly shortens the duration of delayed implantation, leading to the inevitable conclusion that implantation was delayed in ovariectomized, progesterone-treated rats as a result of the absence of estrogen. These findings strongly suggested that progesterone treatment of rats castrated on day 3 or 4 led to delayed implantation because castration occurred prior to the time when the ovaries release the amount of estrogen considered essential for normal implantation; and that progesterone treatment of rats castrated on day 5 led to immediate implantation because the prerequisite estrogen has been secreted by that time. The release of estrogen by the ovaries, called the "estrogen surge" by Shelesnyak and Kraicer (1963), seemed to be the sine qua non for implantation.

Since then the experimental techniques of studying the role of estrogen in implantation have been refined to the point where estradiol benzoate injected locally can cause the implantation of individual, progesterone-maintained blastocysts in castrated or in lactating rats. Canivenc and Mayer (1955) demonstrated that the injection of estradiol into the wall of the uterus could cause implantation of a single blastocyst in the injected portion of the cornu without concomitant implantation of other blastocysts in the uninjected contralateral horn. Yoshinaga (1961) felt that the trauma inflicted on the horn by the insertion of the hypodermic needle was undesirable; consequently he injected estradiol into the adipose tissue next to the tufts of blood vessels near the wall of the uterus. He showed that as little as 0.05 to 0.1 μg of estradiol injected in this manner caused 40 to 80% of the treated rats to respond by local implantation and that 0.2 μg injected locally into progesterone-maintained rats abolished delay in implantation in all cases. To be sure, occasionally cornual areas adjacent to the site of estrogen injection were sensitized, and implantations occurred, apparently as a result of diffusion of estrogen from the injection site. However if the estrogen dose is small enough the effect produced is strictly local. Thus there can be no longer any doubt that estrogen plays a vital role in causing implantation in castrated rats in which the blastocysts are maintained with progesterone and probably in normal intact rats in which the estrogen surge occurs in the morning of day 4 of pregnancy.

The exact source of estrogen for the surge remains unknown although it could be from the unovulated smaller ovarian follicles or the corpora lutea which are able to secrete enough estrogen to cause implantation or vaginal mucification in the presence of progesterone when stimulated by LH (MacDonald, Armstrong, and Greep 1967). The hypophyseal-ovarian interrelation, which would control such an estrogen surge regardless of whether it is of follicular or luteal origin, is obscure.

II. Implantation in Other Species

As far as the other species are concerned, much less study has been devoted to the endocrine requirements of implantation. Nevertheless enough is known to allow the unequivocal statement that the rat and the mouse are the only two species in which there is a clear-cut and undisputed requirement for estrogen action as a prerequisite for implantation.

A. *Is Estrogen Obligatory or Facilitatory?*

In the hamster progesterone alone causes implantation according to Prasad, Orsini, and Meyer (1960) and Orsini and Psychoyos (1965). In rabbits, too, progesterone alone is able to cause implantation. A nice dose-response relationship has been demonstrated between the amount of progesterone and the number of implanting blastocysts in rabbits: 0.75 mg of progesterone causes 2 implantations, 1 mg 4 to 5, 1.25 mg 6 to 8, and 5 mg/day the maximal number of 8 to 9 (Chambon 1949*a, b*). Dosage can be substantially reduced if the animals are primed with estrogen. Larger doses are always, or at least in this and many other instances, found necessary when progesterone is called upon to do the task alone. It appears probable that estrogen normally synergizes with progesterone to lower the progesterone requirements. Kehl and Chambon (1949) postulate that this "implantation and postimplantation" estrogen may come from the trophoblast, an assumption for which there is no evidence. Accordingly I would like to propose an alternative possibility. Recently Keyes and Nalbandov (1967) have shown that in the rabbit estrogen is essential for the morphologic maintenance of corpora lutea of both pregnancy and pseudopregnancy and for their ability to synthesize progesterone. Their studies have indicated that if the ovaries of pregnant rabbits are X-irradiated, thus destroying the follicles but not the corpora lutea, the corpora lutea degenerate and the rabbits abort. (This can be prevented by the injection of 2–4 μg of the estradiol benzoate daily.) These data suggest that the unovulated smaller follicles in the ovaries of rabbits secrete enough estrogen to maintain the corpora lutea and effect steroidogenesis and may be the source of whatever estrogen is needed to facilitate implantation.

In the guinea pig progesterone alone is able to cause implantation. When the animals are pretreated with estrogen, deciduomata formation is totally or partially blocked (Weichert 1928; Loeb and Kountz 1928).

In a group of animals in which delayed implantation occurs, such as the mustelids, the endocrine requirements for implantation are not clear. According to Wright (1963) there is some suggestion that estrogen may be beneficial for implantation in some mustelids, but the number of observations is far too small to permit meaningful generalization.

No proper experiments have been performed in primates to permit judgment concerning the role of estrogen; all that can be said is that progesterone is essential for implantation. It appears that we can add at least one domestic animal, the pig, to the list of those in which no estrogen is required for implantation. In work nearing completion in my laboratory, both ovaries of pregnant pigs were irradiated 8–10 days after mating. X-irradiation is known to destroy primary and secondary follicles, leaving a very few primary follicles of questionable viability. Thus it can be surmised that by day 10 to 12 of pregnancy the amount of ovarian estrogen is reduced to insignificantly low levels. Even though one cannot speak of a definite time of implantation in the pig, at least not in the same sense in which this term is used in rats, it appears that females with destroyed follicles and thus no obvious source of estrogen are perfectly able to carry blastocysts, embryos, and fetuses through all stages of gestation.

On the basis of the available evidence it seems fair to conclude that estrogen plays an obligatory role in implantation only in rats and mice. In all other animals studied estrogen may act as a progesterone-sparing agent. We wish to emphasize that as far as implantation itself is concerned it can be accomplished by progesterone alone.

III. The Relationship between Estrogen and Histamine

Next I would like to turn to the problem of the role played by estrogen in the implantation of blastocysts in rats. To a great extent the available studies deal with the phenomenon of decidualization rather than implantation as such, but, as I will illustrate later, there is good reason to believe that the two are under essentially identical endocrine control mechanisms. The information obtained on one probably can be extrapolated to the other.

Shelesnyak (1952, 1957) and Shelesnyak and Kraicer (1963) were the first to suggest that histamine was an inducer of decidualization and, by implication, of implantation. This concept was hotly disputed by several workers, most strongly by De Feo (1961, 1962, 1963), who pointed out that the solvents for histamine and several other substances are able to act as inducers of decidualization. He suggested that histamine as such is not the specific inducer of decidualization (and, by implication, of implantation), and that these phenomena could be triggered by totally nonspecific substances.

In the meantime the evidence for some kind of involvement of histamine in the process of decidualization mounted steadily. It was found that cornual histamine content increased in deciduomata (Cecil, Wrenn, and Bitman 1962), and that a decrease in histamine concentration was closely correlated with depletion of endometrial mast cells (Shelesnyak 1960). These findings were promptly related by Shelesnyak to histamine release by the mast cell, the histamine being directly responsible for the induction of decidualization. This effect, he found (Shelesnyak 1959), could be inhibited by antihistamine. Most significantly, Spaziani and Szego (1958, 1959) showed that estrogen causes histamine release.

The Shelesnyak school proposed a chain of events in which the estrogen surge caused histamine release which sensitized certain areas of the cornua (or perhaps the whole uterine lumen), to respond to the physical presence of the blastocyst by becoming physically and biochemically receptive to the blastocyst seeking a hospitable site for implantation. As good as this argument appeared it could not suppress the contention of the opposition that a host of seemingly nonspecific substances, such as Hanks's solution or blood plasma, could induce decidualization. None of the opponents however presented any critical evidence as to whether these nonspecific intraluminal inducers are able to release histamine.

IV. The Role of Histamine in Implantation

Here the matter rested until a somewhat different experimental approach tilted the scales in favor of one of the two groups of disputants. Our attention was drawn to a paper by Zipper and his colleagues in which it was shown that the lumina of rat cornua could be exposed to freezing temperatures for about 5 min without obviously damaging the histological appearance of the cornual endometrium (Zipper et al. 1965). The statement that the rats so treated did not conceive even after prolonged recovery periods especially interested us. We decided to pursue this problem further, and the results of these studies were eventually published (Ferrando and Nalbandov 1968). The techniques used have been described in detail in the article by Zipper et al. (1965); only a brief summary will be presented here. Virginal, sexually mature Holtzman female rats with well-established normal estrous cycles as verified by vaginal smears were used in the study. Without regard to the stage of the estrous cycle, their cornua were exposed and a hypodermic needle was inserted through the lumen of one horn in such a way that it passed through the ovarian half, through the cervical half, or through the entire length of the lumen. The openings of the needle were connected to a vessel containing alcohol cooled to —25° C by dry ice. The temperature of the fluid after passing through the needle inserted into the cornual lumen was about —15° C. By means of a hand bulb the cooled alcohol was circulated through the needle for about 3 min. During the circulation of the fluid the cornu became grayish white in color and very hard. After a recovery period of 15 days, during which normal sequences of vaginal smears were found to persist in most rats, the endometrium from previously frozen horns of most animals was indistinguishable histologically from that of control rats or from that of unfrozen horns in the same rats, as seen in figures 1 and 2. (If the freezing was extended to 15 min, the endometrium sloughed off and the cornual lumen became lined with an abnormal layer of cells not resembling a normal endometrium.)

After the 15-day rest period the rats were exposed to a series of experiments designed to test the ability of treated horns or portions of horns to respond to decidualization or to implanting blastocysts. The techniques used were all routine and are described in the paper mentioned previously. If histamine was used it was injected intraluminally as histamine dichloride (1 mg in 0.1 ml of saline) and saline was used for control injections. Estrogen (estradiol benzoate) was injected systemically (0.2 mg/rat) or locally (Yoshinaga technique) at the rate of 0.03 μg in 0.05 ml of oil. The histamine content of the cornua was determined by the modified method of Marcus, Shelesnyak, and Kraicer (1964). The mast-cell population was estimated by counting the number of cells appearing in 5 different sections of different cornual segments cut at 5 μ.

The results of this study are summarized in table 1. First we confirmed (table 1A) that implantation does not occur in the previously frozen horns; it does occur in the untreated contralateral horns of the same animals. If either the upper or the lower half of one horn was frozen, normal implants were found in the unfrozen half (regardless of whether it was the lower or upper segment), whereas the frozen half remained barren (the limits of the portion of the frozen half could be deter-

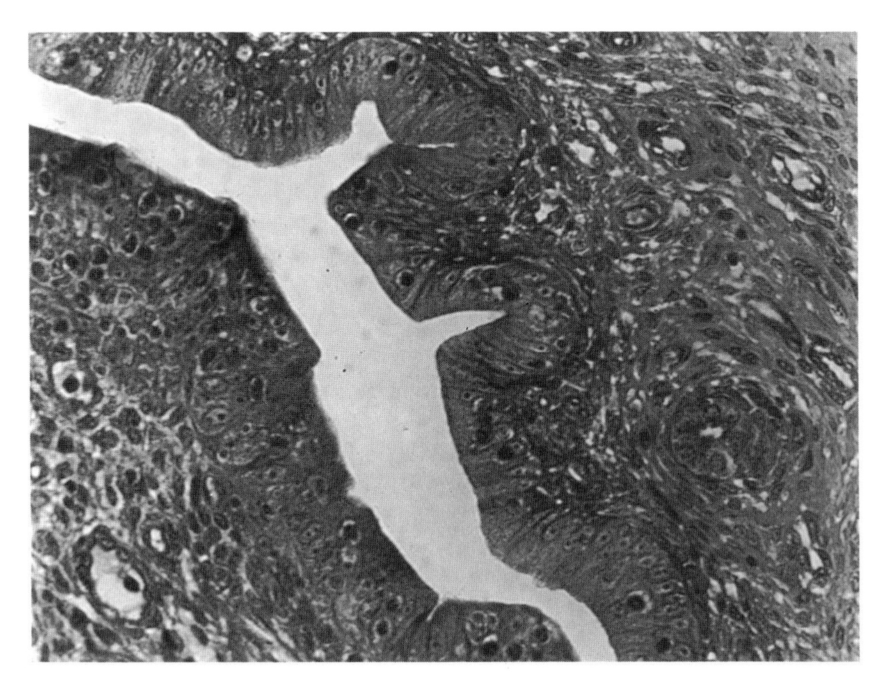

Fig. 1. Section through rat uterus which has been previously frozen. ×375

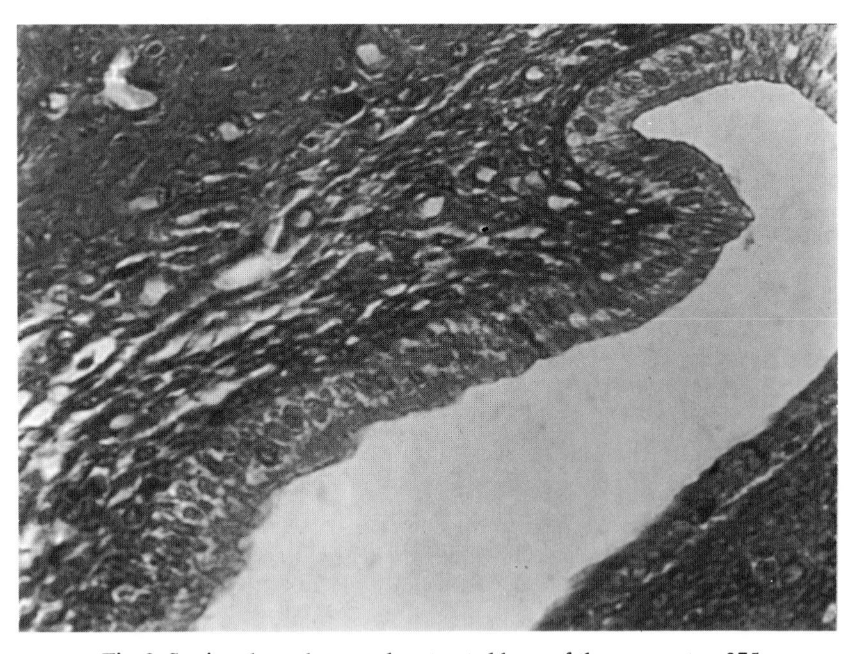

Fig. 2. Section through normal, untreated horn of the same rat. ×375

mined easily by finding the scar left by the hypodermic needle). Next we tested the effect of previous freezing on the ability of uterine horns to respond to decidualization. When we scratched the frozen portions of the horns no deciduomata were produced (table 1B); when we scratched the unfrozen portion, deciduomata were formed which extended into the previously frozen half, again regardless of whether the treated half was near the cervical or the oviductal end (table 1C). By using the Yoshinaga technique the effect of estrogen applied locally on implantation was tested. As can be seen (table D), local estrogen did not induce implantations in previously frozen horns although it was able to do so in the control portion of the same horn and in the contralateral untreated horn.

The fact that estrogen was unable to induce implantation in previously frozen segments of the horn regardless of whether it was given locally (table 1D) or systemically was puzzling, especially if we remember that deciduomata can and do spread to previously frozen portions provided the trauma is applied to the normal half. Keeping in mind the fact that estrogen was shown to release uterine histamine (Spaziani and Szego 1958), we asked whether freezing had abolished the ability of the endometrium to respond by histamine release when estrogen was injected. Histamine was infused intraluminally into whole frozen horns and into control horns. As shown in table 1F, the frozen horns responded to histamine infusion both by deciduomata and, what is more important, by implantation (table 1G). Next we asked whether freezing prevented cornua from responding to estrogen injection with increase in weight. This appeared to be an important question: an affirmative answer would meet the criticism that previous freezing may have modified cellular competence to respond to estrogen by enhanced protein synthesis, an apparent prerequisite for uterine response to estrogen; a negative response would assign the primary reason for failure of implantation to failure of protein synthesis. As can be seen (table 1H), both the frozen horn and the normal control horn responded

TABLE 1

EFFECTS OF VARIOUS TREATMENTS ON THE PHYSIOLOGICAL RESPONSES OF FROZEN AND UNFROZEN CORNUAL PORTIONS AND CONTROL HORNS OF THE RAT

	No. Rats	Frozen Portion	No. Rats	Unfrozen Portion	No. Rats	Control Horn
A Implantation	8/8	– – –[a]	6/8	+++[b]	8/8	+++
B Deciduoma (scratch frozen portion)	5/5	– – –	5/5	– – –		
C Deciduoma (scratch all along the horn)	5/5	+++	5/5	+++		
D Implantation and local estrogen	8/8	– – –	8/8	+++	8/8	+++
E Number of mast cells[c]		0.5		4.0		3.0
F Deciduoma and intraluminal histamine[d]	3/5	+++				
G Implantation and intraluminal histamine[d]	5/6	+++			6/6	+++
H Cornual growth and estrogen (mg)	6	339			6	333

SOURCE: Ferrando and Nalbandov 1968.

[a] – – – The expected phenomena do not occur in 8 out of 8 rats.

[b] +++ The expected phenomena do occur.

[c] Mean average of number of mast cells in 2 areas in each section, 5 sections for each portion.

[d] In sections F and G, control horns received intraluminal injection of saline.

to a given dose of estrogen by the same weight increases, demonstrating that the ability of the frozen horn to synthesize protein was not impaired. On the basis of these findings we can separate the action of estrogen on the uterus into two functions: one concerned with the overall growth-stimulating effect due to protein synthesis (which is not impaired by previous freezing); the other concerned with histamine release by its direct action on the mast cells.

Finally a comparison of the number of mast cells was made between treated and untreated horns. There were almost no mast cells left in the frozen portions (average of 0.5 cells/section), whereas the unfrozen portion of the same horn had 4 cells and the control horn 5. These differences in cell counts were reflected in the histamine content of frozen versus unfrozen horns. The frozen horn contained 0.84 ± 0.18 μg/g of histamine whereas the control horn had 1.06 ± 0.29 ($p < 0.05$).

V. Concluding Remarks

A survey of the literature shows clearly that the laboratory rat is in a class by itself in that it requires estrogen for implantation in an endometrium previously prepared for implantation by the action of progesterone. It is reasonably clear also that Shelesnyak's concept of an estrogen surge in the rat is correct and that it occurs on day 4 after mating. The source of this estrogen is not known. In a great many other species studied only progesterone is needed for implantation and no estrogen surge occurs after heat. In view of the practical certainty that some estrogen continues to be secreted after ovulation by the smaller, unruptured follicles, it is plausible to postulate that estrogen does participate synergistically with progesterone in implantation. Thus estrogen may not be obligatory for implantation as it is in the rat, but may be facilitating. The role of estrogen as a progesterone-sparing hormone seems to be strongly suggested by the available data on rabbits. However it must be emphasized again that under experimental conditions which may not accurately mimic the true physiological state of affairs, progesterone alone is perfectly able to cause implantation in the rabbit as well as in a great many animals other than the rat.

In the matter of the role of estrogen surge in the rat the newer data of Ferrando seem to lend support to the Shelesnyak school of thought that histamine is the actual trigger for implantation and the role of estrogen lies only in causing release of histamine. In the experiments cited the mast-cell population was drastically reduced and the amount of histamine correspondingly so. It was found that in histamine-poor cornua neither decidualization nor implantation was possible. Histamine-poor cornua were able to respond to estrogen by normal growth, but neither local nor systemic estrogen was able to repair their inability to respond by deciduomata or by permitting implantation of blastocysts. In striking contrast, both decidual formation and implantation became possible when such defective cornua were infused intraluminally with histamine. It should be noted that the repair of the cornual defect was accomplished with histamine *alone* without an estrogen surge (the rats were lactating) and without the administration of exogenous estrogen.

Although the argument on the respective roles of estrogen and of histamine in implantation may be considered settled, the question of the molecular role of histamine in implantation remains totally unresolved.

Acknowledgments

The preparation of this paper was in part supported by National Institutes of Health grant AM 3043, and part of the experimental data cited was collected while Dr. G. Ferrando held a Population Council fellowship at the University of Illinois.

References

Canivenc, R., and Mayer, G. 1955. Contribution à l'étude expérimentale de la superfétation chez la rate: Recherches basées sur la nidation retardée. *Ann Endocr (Paris)* 16:1.

Cecil, H. C.; Wrenn, T. R.; and Bitman, J. 1962. Uterine histamine in rat deciduomata. *Endocrinology* 71:960.

Chambon, Y. 1949*a*. Besoins endocrininens qualitatifs et quantitatifs de l'ovo-implantation chez la lapine. *C R Soc Biol (Paris)* 143:1172.

———. 1949*b*. Essais de réalisation du déciduome en phase progestative par l'administration locale de folliculine chez la lapine. *C R Soc Biol (Paris)* 143:1528.

Cochrane, R. L., and Meyer, R. K. 1957. Delayed nidation in the rat induced by progesterone. *Proc Soc Exp Biol Med* 96:155.

De Feo, V. J. 1961. Intraluminal histamine and massive diciduoma formation. *Anat Rec* 139:298.

———. 1962. Comparative effectiveness of several methods for the production of deciduomata in the rat. *Anat Rec* 142:226.

———. 1963. Determination of the sensitive period for the induction of deciduomata in the rat by different inducing procedures. *Endocrinology* 73:488.

———. 1967. Decidualization. In *Cellular biology of the uterus,* ed. R. M. Wynn, p. 191. New York: Appleton-Century-Crofts.

Ferrando, G., and Nalbandov, A. V. 1968. Relative importance of histamine and estrogen on implantation in rats. *Endocrinology* 83:933.

Kehl, R., and Chambon, Y. 1949. Synergie progestérofolliculinique d'ovoimplantation chez la lapine. *C R Soc Biol (Paris)* 143:1169.

Keyes, P. L., and Nalbandov, A. V. 1967. Maintenance and function of corpora lutea in rabbits depend on estrogen. *Endocrinology* 80:938.

Loeb, L., and Kountz, W. B. 1928. The effect of injection of follicular extract on the sex organs in the guinea pig and the interaction between the follicular substances and substances given off by the corpus luteum. *Amer J. Physiol* 84:283.

MacDonald, G. J.; Armstrong, D. T.; and Greep, R. O. 1967. Initiation of the blastocyst implantation by luteinizing hormone. *Endocrinology* 80:172.

Marcus, G. J.; Shelesnyak, M. C.; and Kraicer, P. F. 1964. Studies on the mechanism of nidation: X. The estrogen surge, histamine release and decidual induction in the rat. *Acta Endocr (Kbh)* 47:255.

Mayer, G. 1963. Delayed nidation in rats: A method of exploring the mechanism of ova-implantation. In *Delayed implantation,* ed. A. C. Enders, p. 213. Chicago: University of Chicago Press.

Orsini, M. W., and Psychoyos, A. 1965. Implantation of blastocysts transferred into progesterone-treated virgin hamsters previously ovariectomized. *J Reprod Fertil* 10:300.

Prasad, M. R. N.; Orsini, M. W.; and Meyer, R. K. 1960. Nidation in progesterone-treated, estrogen-deficient hamsters, *Mesocricetus auratus* (Waterhouse). *Proc Soc Exp Biol Med* 104:48.

Sammelwitz, P. H.; Dziuk, P. J.; and Nalbandov, A. V. 1956. Effects of progesterone on embryonal mortality in rats and swine. *J Animal Sci* 15:1211.

Shelesnyak, M. 1952. Inhibition of decidual cell formation in the pseudopregnant rat by histamine antagonists. *Amer J Physiol* 170:522.

———. 1957. Aspects of reproduction: Some experimental studies on the mechanism of ova implantation in the rat. *Recent Progr Hormone Res* 13:269.

———. 1959. Fall in uterine histamine associated with ovum implantation in pregnant rat. *Proc Soc Exp Biol Med* 100:380.

———. 1960. Nidation of the fertilized ovum. *Endeavour* 19:81.

Shelesnyak, M., and Kraicer, P. F. 1963. The role of estrogen in nidation. In *Delayed implantation,* ed. A. C. Enders, p. 265. Chicago: University of Chicago Press.

Spaziani, E., and Szego, C. M. 1958. The influence of estradiol and cortisol on uterine histamine of the ovariectomized rat. *Endocrinology* 63:669.

———. 1959. Further evidence of mediation of histamine or estrogenic stimulation of the rat uterus. *Endocrinology* 64:713.

Weichert, C. K. 1928. Production of placentomata in normal and ovariectomized guinea pigs and albino rats. *Proc Soc Exp Biol Med* 25:490.

———. 1942. The experimental control of prolonged pregnancy in the lactating rat by means of estrogen. *Anat Rec* 83:1.

Wright, P. 1963. Variations in reproductive cycles of North American mustelids. In *Delayed implantation,* ed. A. C. Enders, p. 77. Chicago: University of Chicago Press.

Yoshinaga, K. 1961. Effect of local application of ovarian hormones on the delay of implantation in lactating rats. *J Reprod Fertil* 2:35.

Zipper, J. G.; Ferrando, G.; Guiloff, E.; Saez, G.; and Tchernitchin, A. 1965. The response to *in vivo* freezing of uterus and ovaries in rats. *Amer J Obstet Gynec* 93:510.

23

Blastocyst-Uterine Relationship before and during Implantation

D. R. S. Kirby
Department of Zoology
University of Oxford, England

It is generally assumed that normally the mouse blastocyst has a short free-living existence in the uterus before it is implanted. In laboratory conditions this is so, but it is unlikely to be true in the natural environment. The female mouse living in a wild colony would probably be mated during each postpartum estrus. Since the ensuing lactation would induce delayed implantation, the free-living existence of the blastocyst would, of course, be greatly extended. Thus delayed implantation would be normal and immediate implantation unusual.

This paper presents the results of experiments performed to examine three related problems associated with delayed implantation and implantation: (*a*) the influence of progesterone on blastocyst viability; (*b*) the influence of progesterone on blastocyst/uterus relationship; and (*c*) the control of orientation of the implanting blastocyst.

I. The Influence of Progesterone on Blastocyst Viability

In both the rat and the mouse bilateral ovariectomy performed early in pregnancy prevents implantation. A review of the literature shows that it is general practice to administer progesterone during the ensuing period of delayed implantation. This procedure seems reasonable since in naturally occurring (lactation-induced) delayed implantation the uterus remains under progesterone control. However the studies of Smithberg and Runner (1956) show clearly that progesterone is not essen-

On his way to the conference Dr. Kirby's car was involved in an accident in which he suffered spinal injury. He was flown back to England, where he died on 11 November 1969. He had great gifts as a research worker; science has lost an enthusiastic and productive investigator.

tial for the survival of mouse blastocysts. They found in superovulated prepuberal mice (in which the corpora lutea do not develop) that blastocysts could be recovered from the uterus 22 days *post coitum*. In a subsequent paper (Smithberg and Runner 1960) they noted that mouse blastocysts could induce a decidual reaction after 31 days of hormone deprivation. There have been two recent reports dealing with blastocyst viability in the absence of progesterone. Dickmann (1968) observed that blastocysts can withstand 3 days in the uterus of a long ovariectomized rat and develop normally when transferred to a foster-mother. Weitlauf and Greenwald (1968), using the ovariectomized pregnant mouse, determined the recovery rate of blastocysts and their developmental capacity when transferred to a pseudopregnant host. Over a period of 40 days' delayed implantation they found that there was a progressive decline in both the number of blastocysts recovered and the proportion that developed normally after transfer. They did not seek to determine whether progesterone administration affected the fate of the blastocyst during delayed implantation.

To determine the effect of progesterone on the viability of blastocysts during delayed implantation, mice with copulation plugs were assigned to one of three groups: (1) delayed implantation with progesterone treatment "delay progesterone" (24 mice); (2) delayed implantation without progesterone treatment "delay no progesterone" (25 mice); (3) controls, sham ovariectomy (51 mice). Details of the treatment are shown in figure 1.

Samples of decidual swellings from each group were histologically examined to determine whether they contained embryos.

The numbers of implantation sites (decidual swellings) found in each group of mice are shown in figure 2. The decidual swellings at the time of autopsy are equivalent in size to a normal 7-day pregnancy. All 10 decidua from each group that were examined histologically contained embryos.

The effect on blastocyst viability of administering or withholding progesterone during delayed implantation can be assessed by the number surviving to implant once estrogen is given. The assumption is made that failure to implant is due to impairment of the blastocyst rather than impairment of uterine sensitivity. This seems reasonable since the hormonal therapy used (3 days of progesterone priming before estrogen administration) appears to restore the sensitivity of the cornua of the ovariectomized rat and mouse to various deciduogenic stimuli including the blastocyst (Psychoyos 1967; Kirby 1969).

The proportion of mice with no implantation sites in the three groups is statistically the same ($X_2^2 = 2.56$) and represents the natural failure rate in this colony. The proportion of infertile mice in the ZO colony, although it seems high, compares with some colonies investigated by Boshier (1968). The variance in numbers of mice showing implantation sites invalidates direct comparison between the two experimental groups. In the "delay progesterone" group one mouse had 1 implantation site, and in the "delay no progesterone" group one mouse showed 16 implantation sites. This latter number is rare and it is difficult to see how it could have resulted from the experimental treatment. Moreover if Dixon's (1953) criteria for the determination of statistical outliers is applied we find that both 1 and 16 implantation sites are outside the expected distribution ($p < 0.02$, $p = 0.05$ respectively) and may be excluded from further computations. If they are excluded then overall the data are heterogeneous $F = 9.78$, $P < 0.01$. By the "t" test ovariectomized mice not receiving progesterone show significantly fewer implantations (mean = 6.0) than either the progesterone maintained (mean = 8.60, $p < 0.005$) or the control mice (mean = 8.64, $p < 0.001$). There is then clearly no difference in the number of implantation sites between the ovariectomized mice receiving progesterone and the controls.

It seems then that if mice are ovariectomized early in pregnancy and given no further treatment the viability of the blastocysts falls off; fewer survive to implant. On the other hand mice given progesterone after ovariectomy show no obvious decline in viability during the period of delayed implantation.

Naturally the question arises of the role of progesterone during this period of pregnancy and its influence on the blastocyst. This was investigated by determining the influence of progesterone on the physical relationship between blastocyst and the uterine wall, and is considered in the next section.

DELAY-NO PROGESTERONE

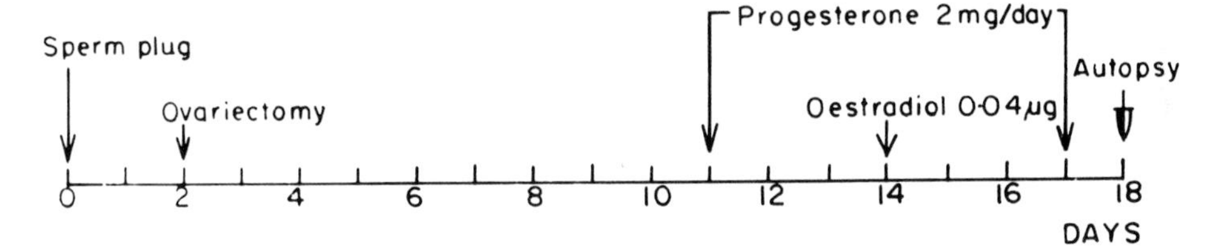

DELAY-PROGESTERONE

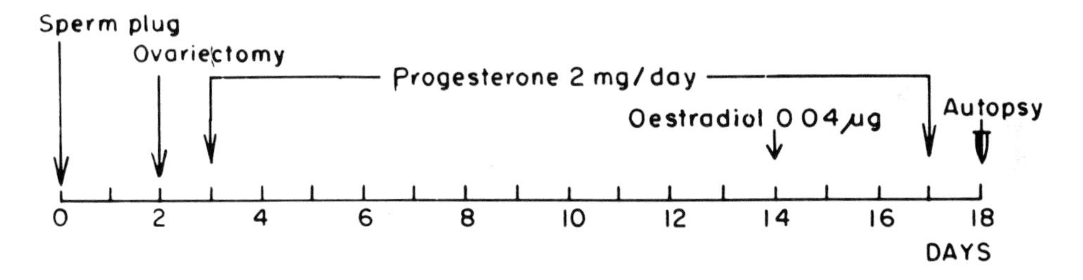

CONTROLS

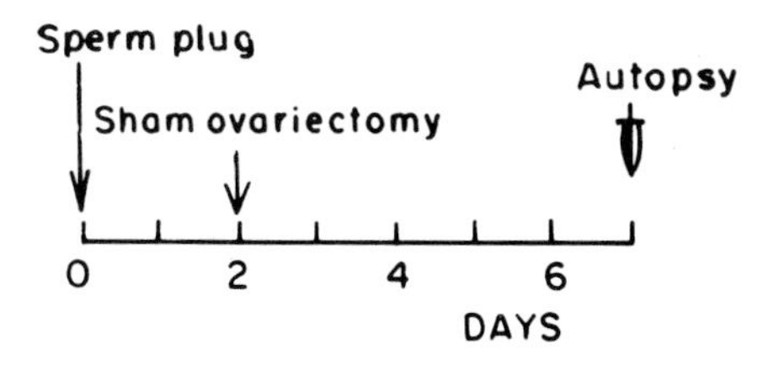

Fig. 1. Schedule of treatments

II. The Influence of Progesterone on Blastocyst-Uterine Relationship

For this histological investigation mice were bilaterally ovariectomized on day 2 of pregnancy (sperm plug = day 0) and beginning on day 3 given either 2 mg progesterone in oil daily or given no further treatment. On day 8 or 9 the mice were killed by cervical dislocation and the cornua removed, pinned out to its approximate *in vivo* length, and fixed in either AFA or Bouin's fluid. The cornua in entirety were cut into serial longitudinal sections (8 μ) in the plane of the mesometrium, and stained in hematoxylin and eosin. Each blastocyst was carefully examined in the serial sections to determine its true relationship to the cornual epithelium.

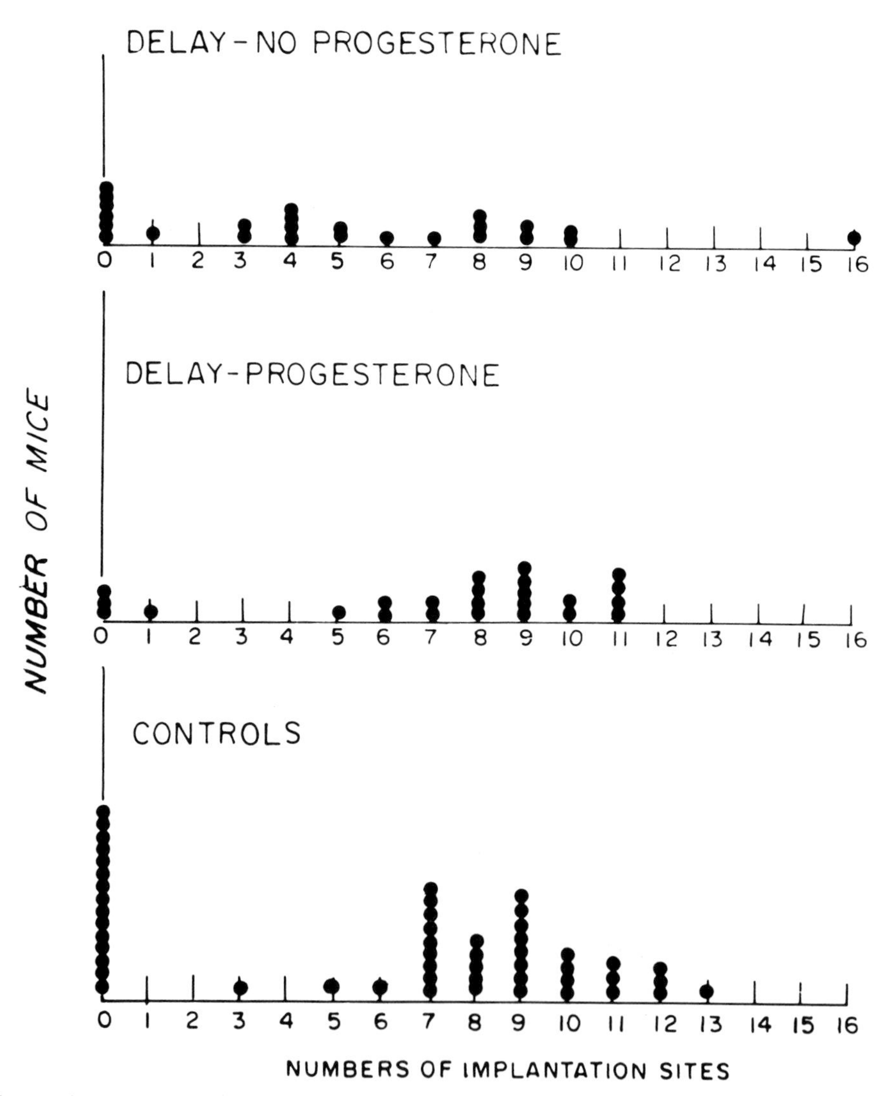

Fig. 2. Distribution of the number of implantation sites per mouse. Each dot represents 1 mouse

Mice Receiving No Progesterone. Twenty-nine blastocysts were located and examined. They were devoid of zonae pellucidae, elongated in shape and lying free in the lumina of the cornua with their long axes parallel to the long axes of the cornua (fig. 3). No blastocyst was attached to the cornual epithelium. There was a slight enlargement of the abembryonic trophoblast in about 2/3 of the blastocysts (fig. 3).

The cornual tissues showed regressive changes but the luminal epithelium was not conspicuously atrophic.

In one mouse two blastocysts were tightly attached to one another by their abembryonic poles (fig. 4), but apart from this the blastocysts appeared to be spaced out along the length of the cornua.

Progesterone-Treated Mice. All 34 blastocysts observed were lying more or less parallel to the long axes of the cornua and were attached by the abembryonic or

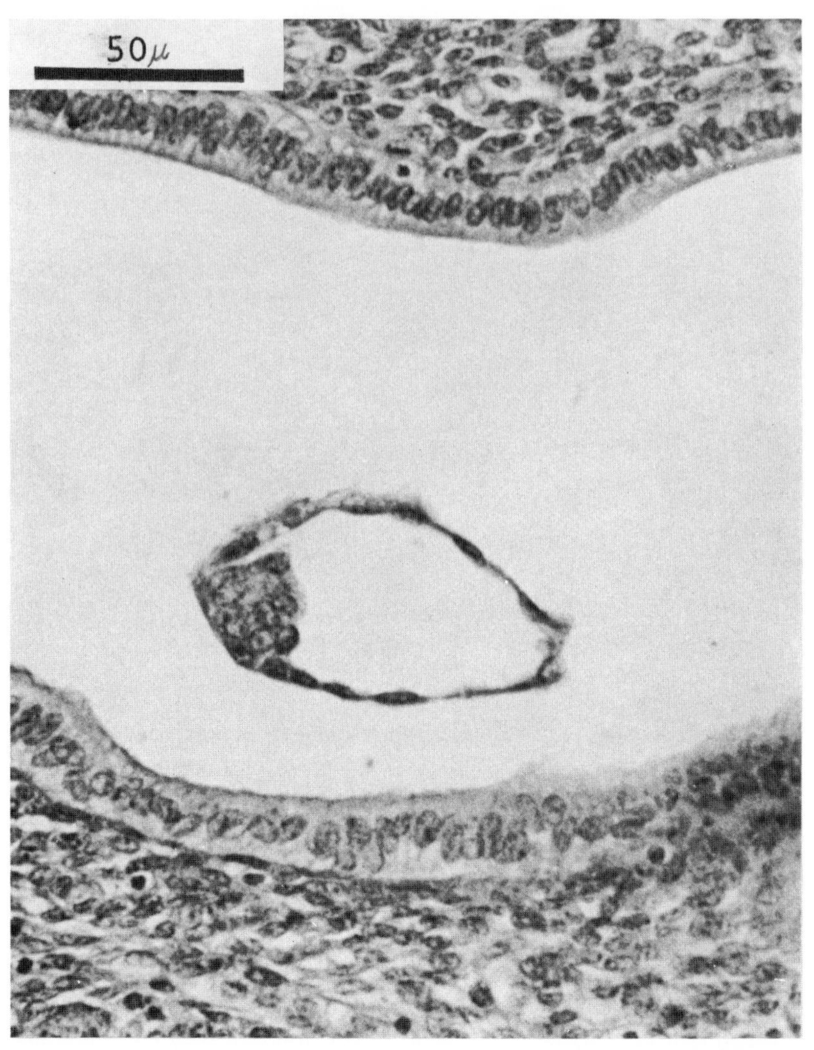

Fig. 3. Delay—no progesterone. Blastocyst lying free in the uterine lumen parallel to the long axis of the atrophic uterus. Note the enlargement of some of the lateral and abembryonic trophoblast cells.

lateral trophoblast to the antimesometrial uterine epithelium (figs. 5, 6). The union was so close that the maximum resolving power of the light microscope could not distinguish a space between trophoblast and the uterine epithelial cells. In many places wedge-shaped projections of the trophoblast had become inserted between adjacent epithelial cells (fig. 7).

The uterine epithelial cells underlying the blastocyst showed characteristic changes. They were slightly swollen. The cytoplasm appeared less dense and in the

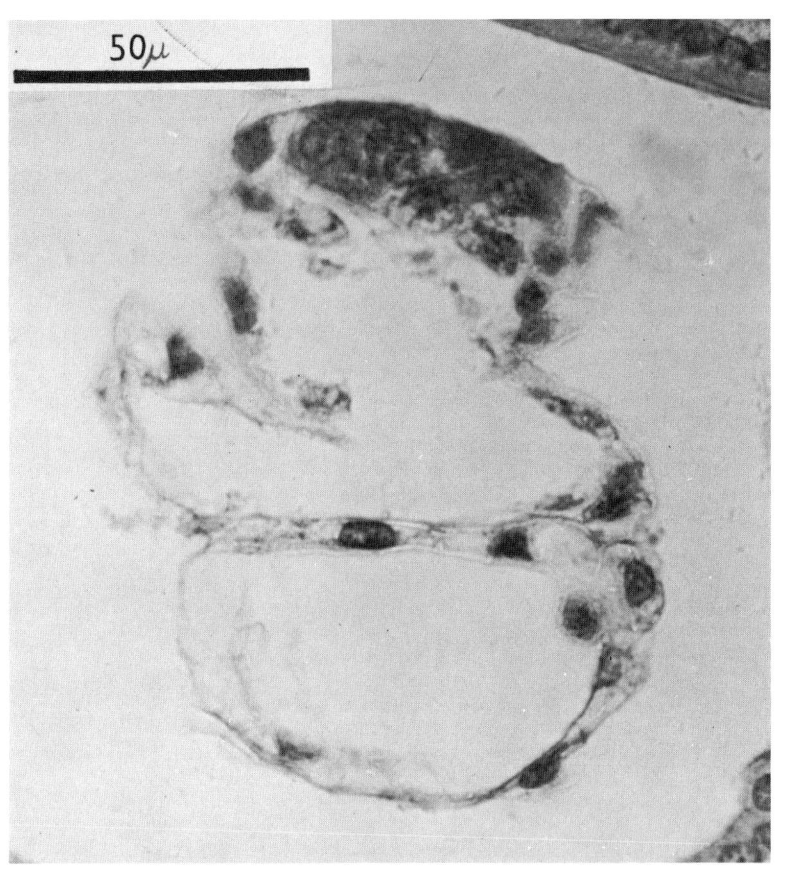

Fig. 4. Delay—no progesterone. Two blastocysts fused by the abembryonic trophoblast. Only 1 inner cell mass is showing.

subnuclear region was distinctly vacuolated. These changes made the demarcation between the epithelial and subepithelial cells more distinct than in most of the other regions of the cornua (fig. 5).

These results show that in ovariectomized pregnant mice receiving no further treatment the blastocysts lie free in the uterine lumen. If the ovariectomized mice are given progesterone the blastocysts attach by the abembryonic pole to the antimesometrial uterine wall. Using the electron microscope Potts and Psychoyos (1967) found that in progesterone-treated ovariectomized pregnant mice microvilli of the uterine epithelial cells were closely interdigitated with trophoblastic microvilli over the entire surface of the blastocysts. They did not investigate the situation in the absence of progesterone. McLaren (1967) noted that in lactation-induced delayed implantation the blastocyst of the mouse is in close contact with the uterine epithelium, except where fixation shrinkage had torn the walls of the lumen apart. In the present experiments the fixation process had also caused some shrinkage of the tissues, and since this would disturb the association between the blastocyst and

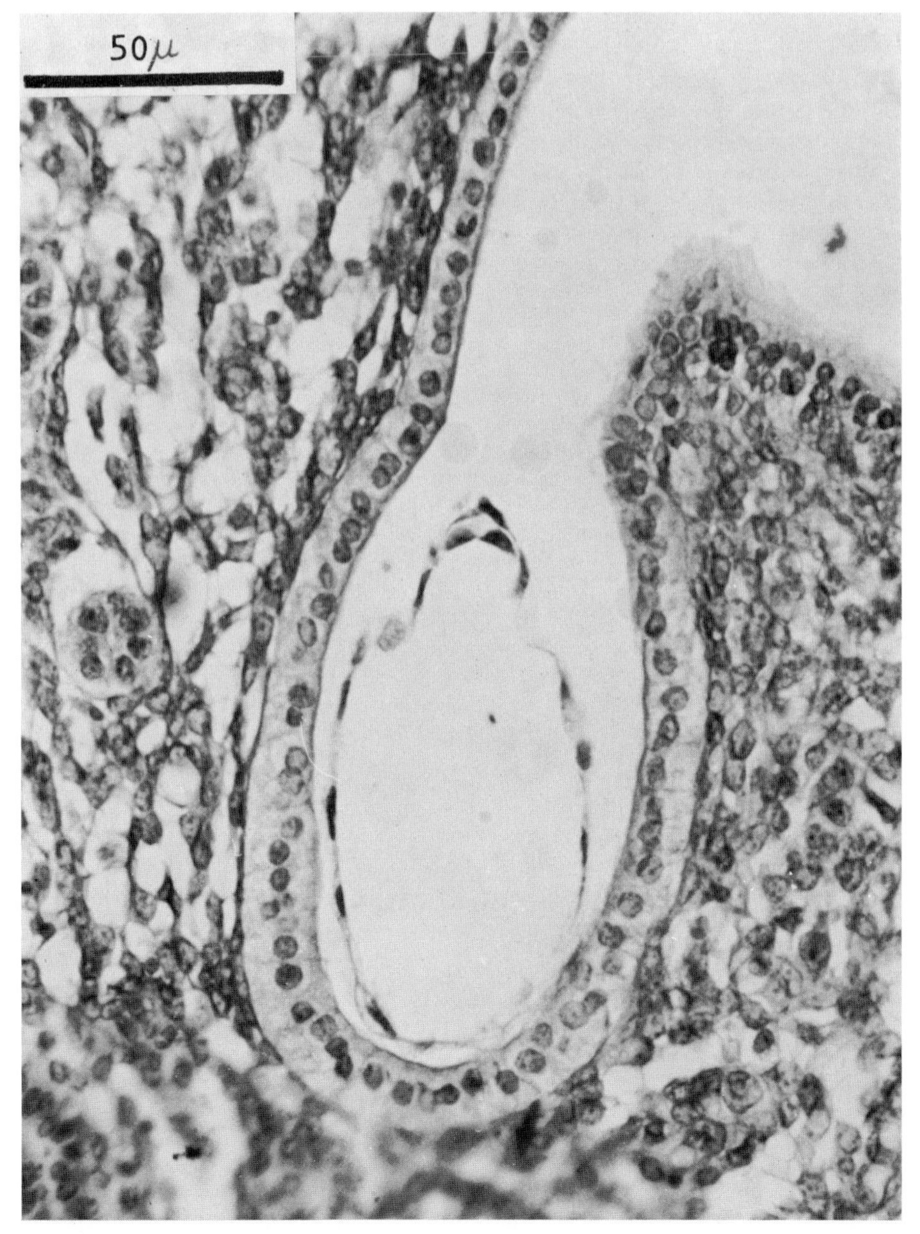

Fig. 5. Delay—progesterone. Blastocysts tightly attached to antimesometrial uterine epithelium

cornu the findings must be viewed accordingly. None of the 29 blastocysts observed in the absence of progesterone was attached to the cornua. This suggests that any connection which may have existed between blastocyst and cornu was very weak, or that the blastocysts were truly free-floating in the lumen.

Observations made with the electron microscope of the overall connection between blastocyst and uterine epithelium in the progesterone-treated animals (Potts and Psychoyos 1967) seemed to indicate that, owing to shrinkage of the tissues, it had been broken at every point except the abembryonic pole. This suggests that attachment at that pole is particularly strong. Psychoyos (1966) has studied blastocyst attachment in the ovariectomized progesterone-treated rat. He found in a cornu showing signs of shrinkage that the blastocyst was attached to the antimesometrial epithelium by the *embryonic* pole. Now in both the rat and mouse the definitive orientation of the implanting blastocyst is with the *embryonic* pole toward the uterine lumen. It follows then that at the termination of delayed implan-

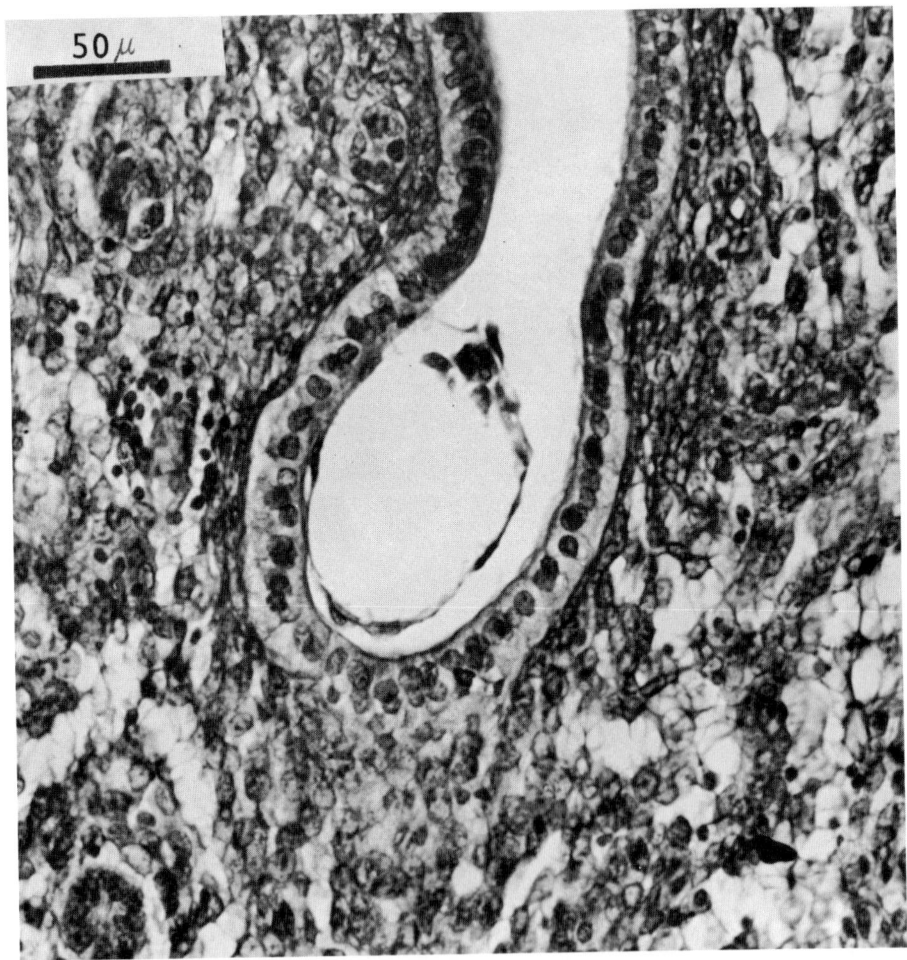

Fig. 6. Delay—progesterone. Blastocyst attached to uterine wall. Note the increase in size and light staining characteristics of the epithelial cells underlying the blastocyst.

tation in the mouse the blastocyst can (but not necessarily does) remain in the position it adopted during delay. The rat blastocyst on the other hand must reverse its orientation before implanting. The way in which this final orientation is achieved is considered in the next section (III).

The advantages derived from attachment are not known: it may be that transfer of nutrients from the mother to the blastocyst is expedited, or simply that the blastocysts, thus anchored to the uterine wall, are not lost through the cervix. Weitlauf and Greenwald (1968) found that ligation of the cornua near the cervical

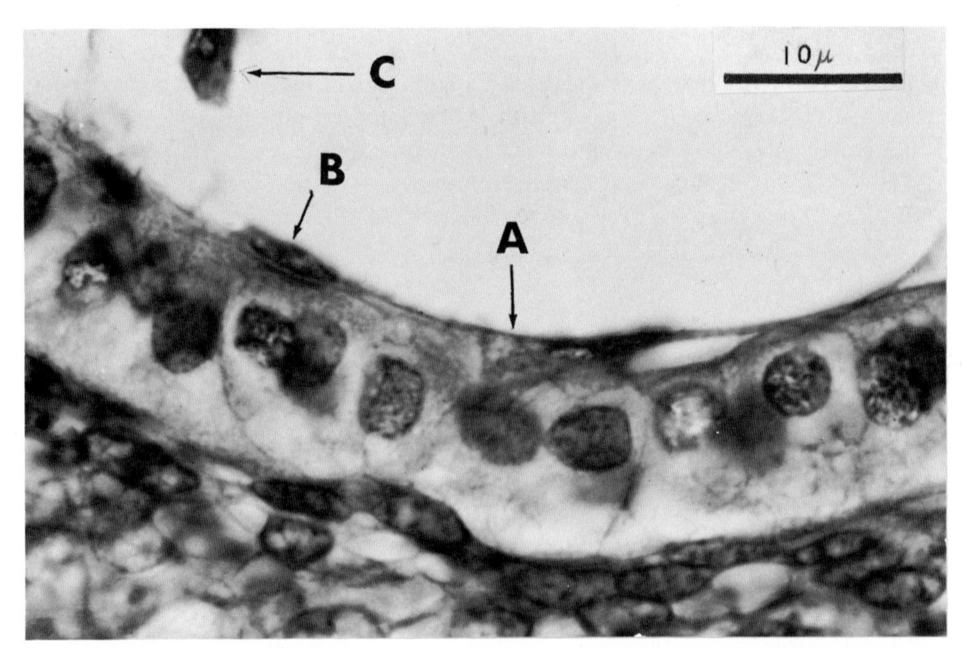

Fig. 7. Delay—progesterone. High-power photograph of attachment between lateral trophoblast and uterine epithelium. *Arrow A* points to a tongue of trophoblast cytoplasm infiltrating between epithelial cells. *Arrow B:* Flattened trophoblast nucleus. *Arrow C:* Inner cell mass.

Figures 6, 7, 8, and 9 are orientated with the mesometrium to the top of the plate. The tissue in figs. 6 and 7 was fixed in Clarke's fluid and strained with Heidenhain's iron hematoxylin; the space around the blastocysts is an artifact. The tissue in fig. 8 was fixed by freeze-substitution and stained with toluidine blue, that in fig. 9 by osmium tetroxide.

end prevented the progressive diminution in the number of blastocysts that could be recovered from the progesterone-deprived, ovariectomized pregnant mouse. Many of the blastocysts recovered in this way, however, were in a degenerate condition.

III. The Control of Orientation of the Implanting Mouse Blastocyst

The control of orientation of the implanting blastocyst has been reviewed by Blandau (1961). He concludes by saying that "the role of the blastocyst in determining the pole of attachment is unknown." We (Kirby, Potts, and Wilson 1967) recently considered the problem and decided there were six ways a priori in which the orientation of the blastocyst could be achieved (fig. 8). Each hypothesis except the first suggests that there is a mechanism for moving the blastocyst in response to a directional stimulus.

1. The blastocyst develops at its prospective implantation site with the position of the inner cell mass determined in response to the underlying maternal tissue.
2. While the blastocyst is being maneuvered by the myometrial contractions toward the antimesometrial side of the uterus, its correct orientation is established as a result of differential compressibility of the abembryonic pole and the inner cell mass regions.
3. Uterine epithelial cells have a capacity for movement independent of the underlying stromal tissue. The movement of the uterine epithelial cells rotates the blastocyst lying upon them into the correct orientation.
4. The initial attachment between the blastocyst and the uterine epithelium is random. Thereafter the trophoblast cells which differentiate from the wall of

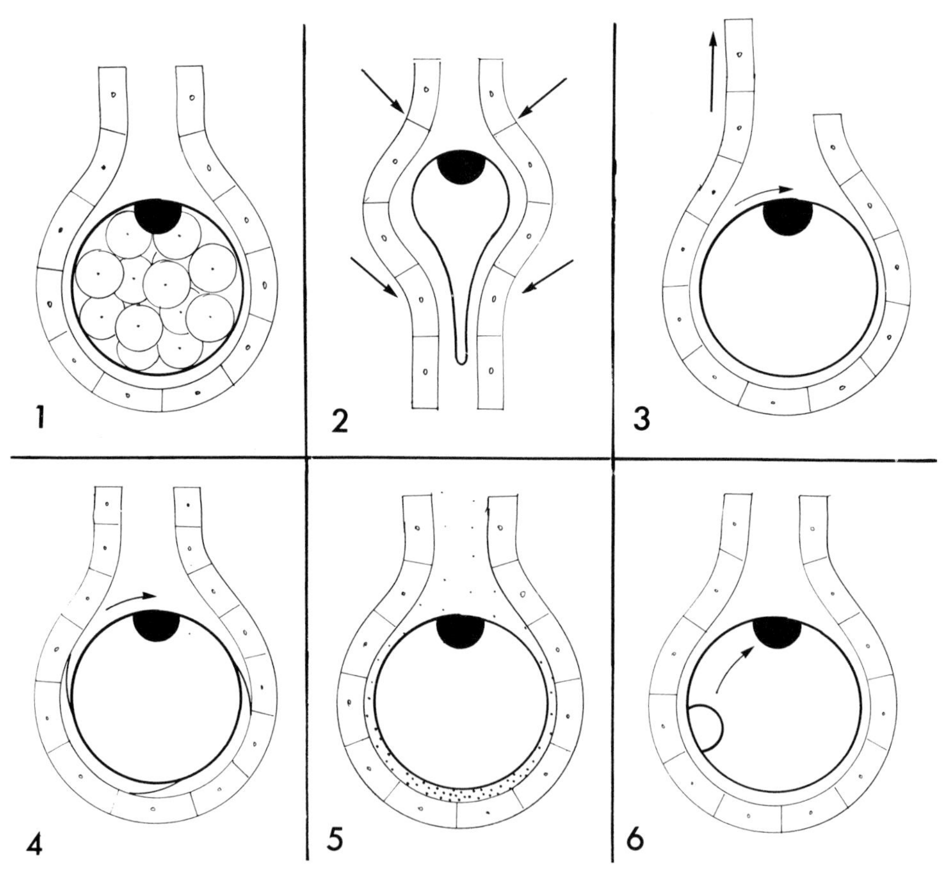

Fig. 8. Diagrams representing the six mechanisms by which the blastocyst could be correctly orientated.

the blastocyst anchor it to the underlying epithelium by cytoplasmic processes. As a result of selective shortening of these processes the blastocyst is rotated into its correct orientation.

5. Before implantation the blastocyst is subject to rotary movements by myometrial contractions. There is a morphogenetic field across the vertical axis of the uterus. When the inner cell mass, as a result of random movements, arrives in the correct orientation, attachment is initiated in all, or in a previously determined part, of the trophoblast wall.
6. The surface of the blastocyst has a uniform potential for attaching to the uterine epithelium, and attachment occurs at random immediately after the loss of the zona pellucida. The inner cell mass is free to move around the inside of the trophoblast shell. The final position of the inner cell mass is determined by a morphogenetic gradient across the vertical axis of the uterus, by changes in the trophoblast associated with its attachment to the underlying tissues, or by both.

To determine which of these hypotheses is correct the following observations of both normal and experimental material must be considered.

(*a*) Shortly before 86 hr *post coitum,* in a randomly bred colony of mice the blastocysts are distributed along the length of the cornua randomly positioned in the vertical axis of the uterine lumen. Rather abruptly, at about 86 hr, the uterine lumen closes up and the blastocysts appear to be forced down to the antimesometrial limit of the lumen (which generally lies along the central axis of the uterus). At approximately 90 hr the blastocysts may be found occupying the presumptive implantation sites. The zona pellucida is still intact but the inner cell mass has no constant orientation (fig. 9). The zona pellucida then disappears and there is a short interval during which specimens can be found with the trophoblast in intimate contact with the epithelium and the inner cell mass haphazardly positioned (fig. 10). By 92 hr the blastocysts are usually correctly orientated (fig. 11). Approximately 6 hr later trophoblast giant cells first appear at the abembryonic pole.

(*b*) Electron-microscopic studies reveal that immediately following the loss of the zona, when the inner cell mass is not always correctly orientated, the uterine epithelium embraces the blastocyst intimately (fig. 12), and microvillous processes from the uterine epithelium and trophoblast interlock.

(*c*) The junction between the inner cell mass and the trophoblast shell is loose (fig. 12). There is considerable extracellular space within the inner cell mass (fig. 11), and there are no desmosomes uniting the cells.

(*d*) Blastocysts in which the inner cell mass had already formed were transplanted to the anterior chamber of the eye. In this site the blastocysts can rotate freely before implantation. Sections were made of the anterior chamber when the blastocysts had implanted. In all 10 blastocysts recovered from the 16 transferred, the inner cell mass was opposite the point of attachment (fig. 13). Judging from the variation in the development of the implanted blastocysts it seems that attachment does not occur synchronously as it does in the cornua, but that this process is spread over a considerable period. This precludes the possibility of constructing a time schedule for implantation processes in the anterior chamber of the eye.

IV. Evaluation

It is clear that hypothesis 1 is excluded, by observations *a, b,* and *d*. The inner cell mass bears no constant relationship to the cornua until after the zona pellucida is lost, about 20 hr after the inner cell mass becomes delineated. Experiments on the transfer of blastocysts to the eye (observation *d*) eliminate hypotheses 2, 3, and 5. Electron microscope observations *b* on the trophoblast-maternal junction and the time sequence of normal development *a* exclude hypotheses 4 and 5. Only hypothesis 6 is compatible with all the observations.

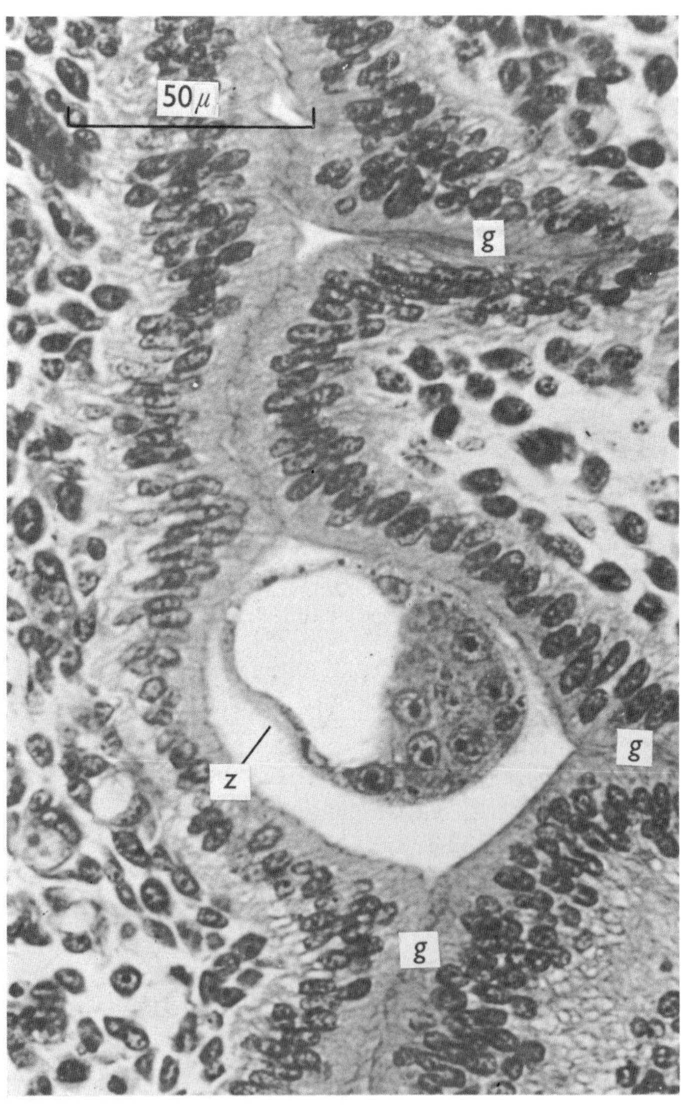

Fig. 9. The zona pellucida (*z*) is still present around this blastocyst which is almost "upside down" in the implantation chamber. Gland openings (*g*) are also shown. (Approximately 90 hr *post coitum.*)

Additional arguments are available for hypothesis 6—that the inner cell mass moves around the inside of the trophoblast shell of the blastocyst. Electron-microscopic observations of the stage under discussion show a discontinuous extracellular deposit on the inside of the trophoblast shell (which for most of its circumference is the precursor of Reichert's membrane). This deposit is always found between the inner cell mass and the trophoblast shell. Small segments of this material are found in other places on the trophectoderm, suggesting that these regions were

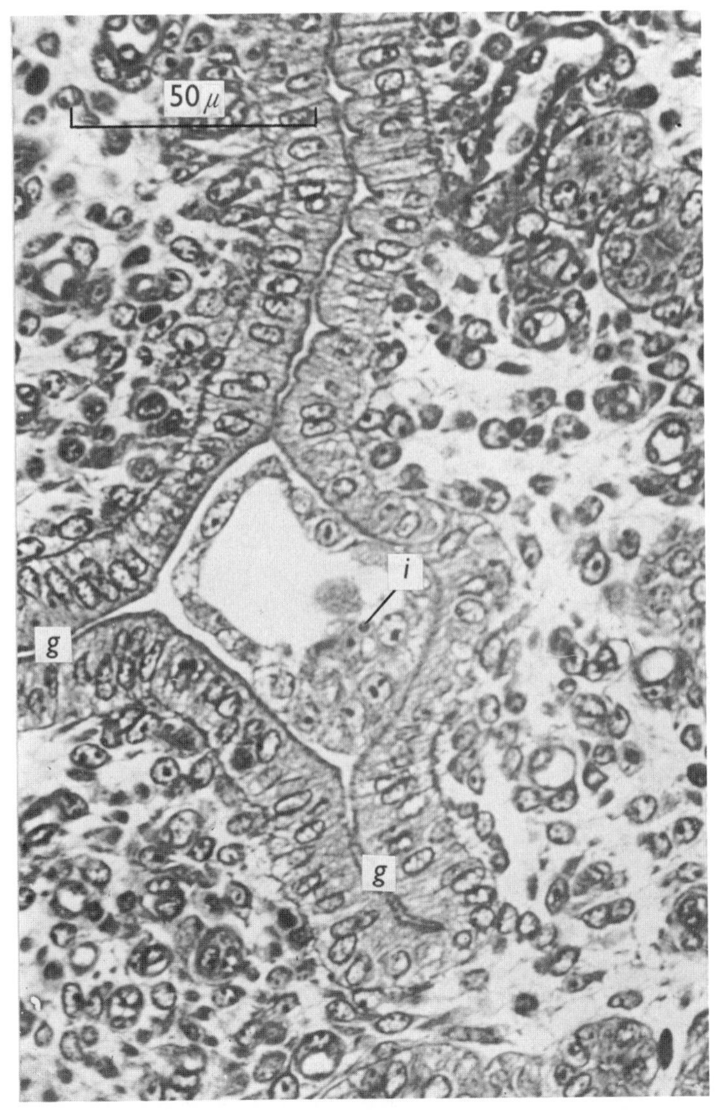

Fig. 10. The zona pellucida has gone but the blastocyst is still "upside down" though clearly it is firmly attached to the uterine epithelium, particularly to the side of the inner cell mass. There is an inclusion (*i*) of a primary invasive cell at the free border of the inner cell mass. (Approximately 90–92 hr *post coitum.*)

formerly occupied by underlying inner cell-mass cells (fig. 14). In experiments involving interspecific transfer of rat and mouse eggs Tarkowski (1962) found that the inner cell-mass cells of rat eggs seem to lose adhesion with one another and distribute themselves as a monolayer around the trophoblast shell. Clearly the cellular arrangement of the blastocyst is labile, as is further indicated by the work of Mintz on the fusion of mouse eggs (Mintz 1964*a, b*). The degree to which the trophoblast and inner cell-mass cells are morphogenetically independent of one another is also shown by the experiments of Mulnard (1965) in which isolated mouse blastomeres, cultured *in vitro,* could give rise to empty trophoblast vesicles.

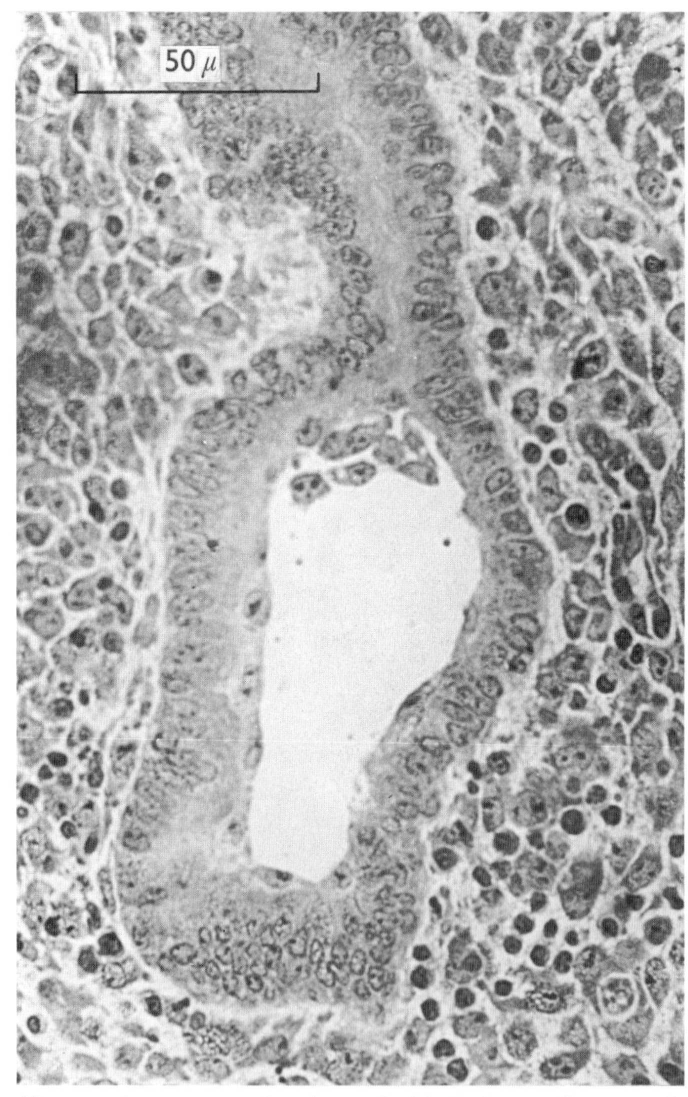

Fig. 11. The blastocyst is now correctly orientated with the inner cell mass at its mesometrial pole. (Approximately 98 hr *post coitum,* though this stage is reached usually by about 92 hr *post coitum.*)

The microsurgical manipulation of mouse blastocysts performed by Gardner (1968) has yielded interesting evidence in support of the hypothesis proposed here. The introduction of cells from the blastocyst of an albino strain into the blastocoele of a blastocyst from an agouti strain resulted in a chimeric individual of mottled coloration. Clearly the resident and introduced cells must have coalesced to form an inner cell mass. In subsequent *in vitro* studies Gardner found (personal communication) that cells of the inner cell mass were capable of independent movement in the culture vessel. Separate masses of cells quickly coalesced to form a single "inner cell mass."

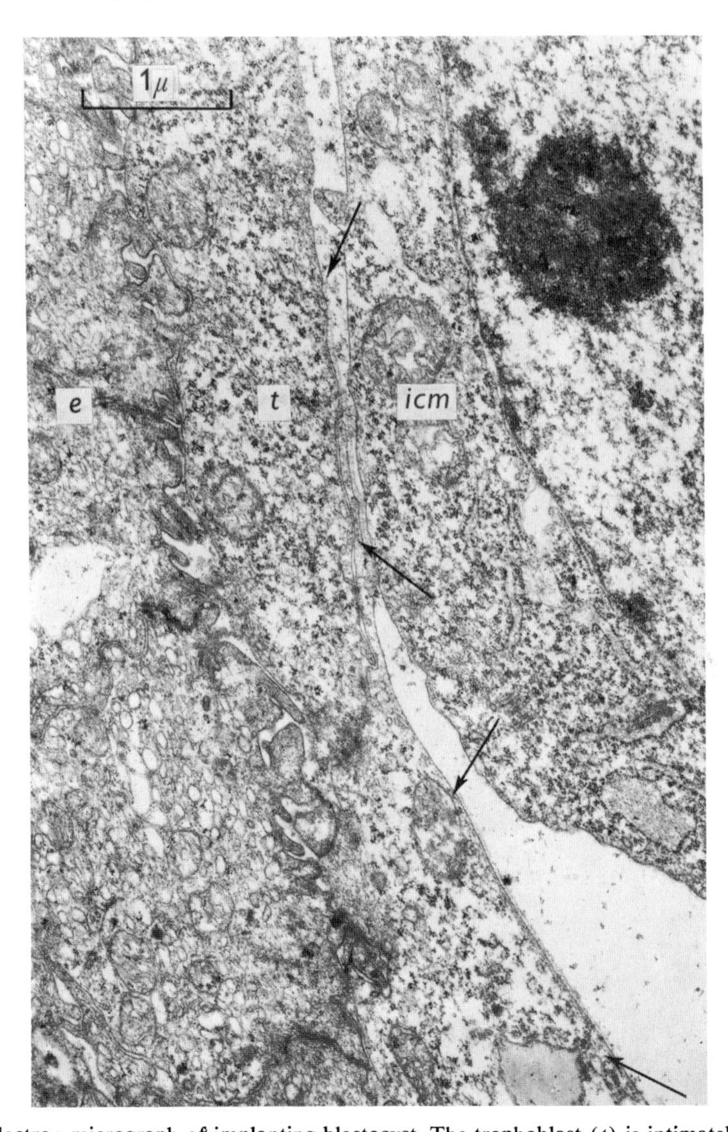

Fig. 12. Electron micrograph of implanting blastocyst. The trophoblast (*t*) is intimately attached to the underlying uterine epithelium (*e*). The inner cell mass (*icm*) is loosely related to the trophoblast shell. A discontinuous extracellular deposit (*arrows*) is present between the inner cell mass and trophoblast.

The mouse blastocyst appears to retain its ability to *reorientate* itself some time *after* implantation. Gardner (personal communication) noted that if the blastocyst is severed from the uterine wall it quickly heals and will successfully implant when transferred to a fresh pseudopregnant host.

The nature of the stimulus responsible for the final orientation of the inner cell mass is unknown. Change in oxygen tension might set up a gradient. The

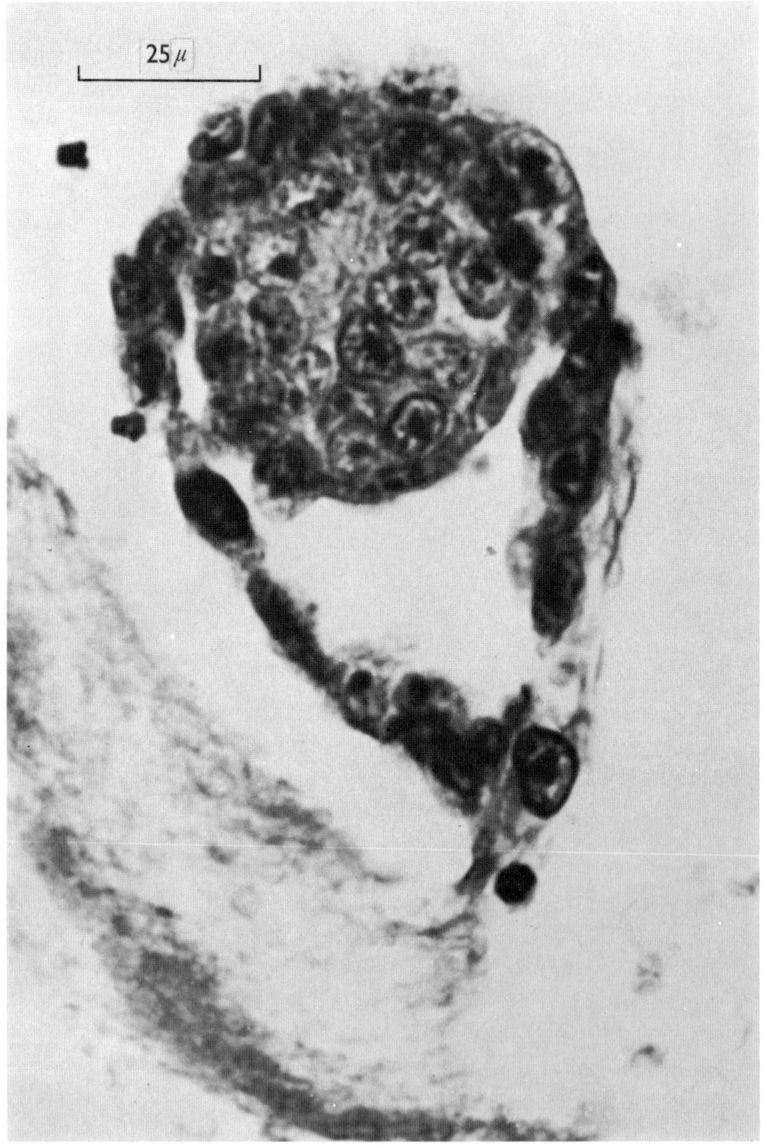

Fig. 13. Blastocyst firmly anchored to the cornea, which has become fragmented during histological preparation. The blastocyst was transplanted into the anterior chamber of the eye 48 hr before autopsy. Note that the inner cell mass is directed away from the point of attachment to the host tissue.

uterine epithelium underlying the abembryonic pole of the blastocyst has a rich capillary blood supply (fig. 14), and the inner cell mass might move away from the area of highest oxygen tension. Such a mechanism could also explain observation *d*. Cowell (see Kirby 1966) has shown that a mouse blastocyst will implant in an area mechanically denuded of epithelium even though this area is located on the mesometrial side of the uterus. In this case the inner cell mass migrates toward the antimesometrial aspect of the blastocyst, and orientation is upside down with reference to the vertical axis of the uterus. However these studies were carried out on ovariectomized mice, in which the vasculature of the cornua is unlikely to be the same as that of pregnant mice. Indeed the oxygen tension may be highest near the hyperemic area beneath the damaged epithelium.

At the time when the orientation of the inner cell mass is taking place the circumference of the trophoblast shell is uniform in structure. When, a few hours later, the giant cells appear they always have a constant relationship to the inner

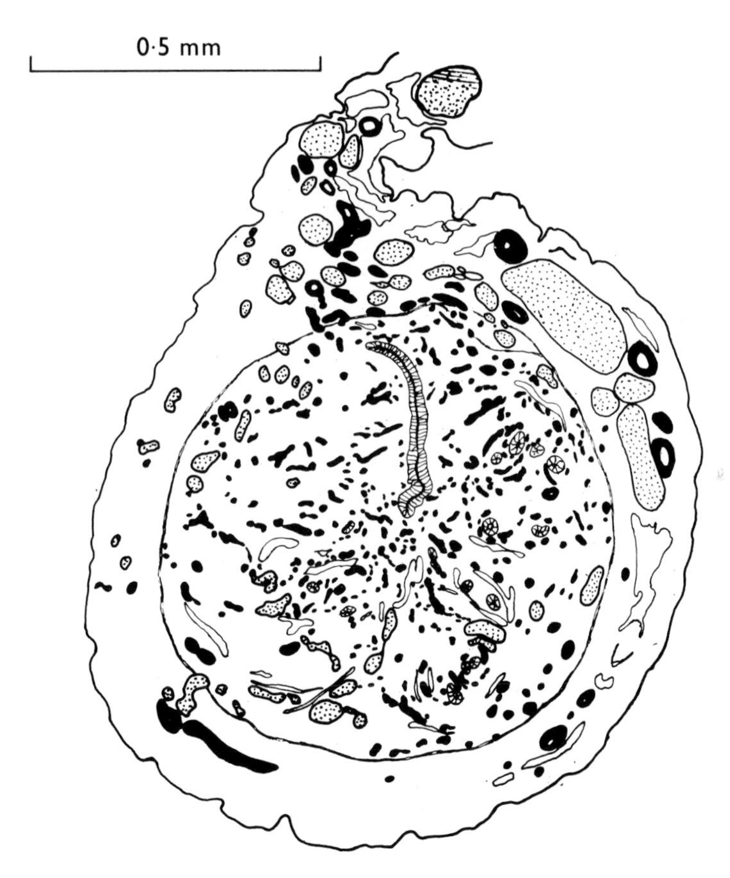

Fig. 14. A projection drawing of a 20-μ thick section of mouse uterus approximately 100 hr *post coitum*. The blood supply is shown in detail only for the endometrium. Main arteries and all smaller vessels are solid black, main veins are stippled; lymph vessels, as far as they could be distinguished, are unshaded. Note that the antimesometrial limit of the lumen is more or less central to the body of the uterus and that the capillary supply and the lymph vessels are concentrated in this region.

cell mass. It is not known how this relationship is mediated except that it is independent of any uterine influence. It is also found when blastocysts are transferred into the mesenchyme of organs other than the cornua, for example, the spleen (Kirby 1963).

The mechanism controlling the orientation of the inner cell mass may well apply to species other than the mouse.

V. Conclusions

If mice are ovariectomized early in pregnancy and given no further treatment there is a falloff in the viability of the blastocysts in that fewer survive to implant. However if progesterone is administered following ovariectomy, there is no obvious decline in the viability of the blastocyst during the period of delayed implantation. Histological examination showed that in the former situation (no progesterone) the blastocysts lie free in the cornual lumen, whereas progesterone causes the blastocyst to attach firmly to uterine epithelium. The benefits bestowed upon blastocysts through their attachment to the uterine wall are unknown. Passage of nutrients from the uterine epithelial cells may be facilitated, or it could perhaps be that blastocysts anchored to the cornua in this way are not lost through the uterine cervix.

When blastocysts attach to the uterine wall either preceding or at the time of implantation they exhibit a constant orientation with respect to underlying maternal tissue. It is suggested that orientation is achieved by the inner cell mass traveling around the inside of the trophoblast shell once the latter has become fixed to the uterine epithelium.

Acknowledgments

I am very grateful to Dr. Richard Gardner, Physiological Laboratories, Cambridge University, for allowing me to quote his unpublished results. This work was supported by the Medical Research Council.

References

Blandau, R. J. 1961. Biology of eggs and implantation. In *Sex and internal secretions,* ed. W. C. Young, 3d ed., p. 797. London: Baillière, Tindall and Cox.

Boshier, D. P. 1968. The relationship between genotype and reproductive performance before parturition in mice. *J Reprod Fertil* 15:427.

Dickmann, Z. 1968. Can the rat blastocyst survive in the absence of stimulation by ovarian hormones? *J Endocr* 42:605.

Dixon, W. J. 1953. Processing data for outliers. *Biometrics* 9:74.

Gardner, R. L. 1968. Mouse chimaeras obtained by the injection of cells into the blastocyst. *Nature* (*London*) 220:596.

Kirby, D. R. S. 1963. The development of mouse blastocysts transplanted to the spleen. *J Reprod Fertil* 5:1.

———. 1966. In discussion following paper by A. Psychoyos. In *Egg implantation,* ed. G. E. W. Wolstenholme and M. O'Connor, p. 19. London: J. & A. Churchill.

———. 1970. Immunological aspects of egg implantation. *Proc II Int Seminar on Reproductive Physiology and Sexual Endocrinology*. In press.

Kirby, D. R. S.; Potts, D. M.; and Wilson, I. B. 1967. On the orientation of the implanting blastocyst. *J Embryol Exp Morph* 17:527.

McLaren, A. 1967. Delayed loss of the zona pellucida from blastocysts of suckling mice. *J Reprod Fertil* 14:159.

Mintz, B. 1964*a*. Synthetic processes and early development in the mammalian egg. *J Exp Zool* 157:85.

———. 1964*b*. Formation of genetically mosaic mouse embryos, and early development of "Lethal(t^{12}/t^{12})-Normal" mosaics. *J Exp Zool* 157:273.

Mulnard, J. G. 1965. In *Preimplantation stages of pregnancy,* ed. G. E. W. Wolstenholme and M. O'Connor. London: J. & A. Churchill.

Potts, M., and Psychoyos, A. 1967. L'ultrastructure des relations ovo-endométriales au cours du retard expérimental de nidation chez la souris. *C R Acad Sci* (*Paris*) 264:956.

Psychoyos, A. 1966. Etude des relations de l'oeuf et de l'endométre au cours du retard de la nidation ou des premières phases du processus de nidation chez la rate. *C R Acad Sci* (*Paris*) 263:1755.

———. 1967. *Advances in reproductive physiology*. London: Logos Press.

Smithberg, M., and Runner, M. N. 1956. The induction and maintenance of pregnancy in prepuberal mice. *J Exp Zool* 133:441.

———. 1960. Retention of blastocysts in nonprogestational uteri of mice. *J Exp Zool* 143:21.

Tarkowski, A. K. 1962. Inter-specific transfers of eggs between rat and mouse. *J Embryol Exp Morph* 10:476.

Weitlauf, H. M., and Greenwald, G. S. 1968. Survival of blastocysts in the uteri of ovariectomized mice. *J Reprod Fertil* 17:515.

Wilson, I. B. 1963. A tumor tissue analogue of the implanting mouse embryo. *Proc Zool Soc Lond* 141:137.

24

Biological Aspects of Gestational Neoplasms Derived from Trophoblast

Roy Hertz
The Population Council
Rockefeller University
New York

This report describes certain features of the biological behavior of tumors derived from trophoblast either during or after pregnancy in women. These tumors are termed "gestational trophoblastic neoplasms" to distinguish them from such trophoblast-containing tumors as embryonal carcinomas arising in the testes, ovary, or in other sites, referred to collectively as "nongestational trophoblastic neoplasms" (Holland and Hreshchyshyn 1967).

Our conclusions regarding the nature and behavior of gestational trophoblastic neoplasms stem from direct clinical and laboratory observation of over three hundred instances of this type of tumor over the past two decades. Two hundred of these cases represented "metastatic trophoblastic neoplasms," a term used to indicate that the disease had spread beyond the uterus. The remainder were instances of "nonmetastatic trophoblastic neoplasms" with the disease limited to the uterine fundus (Hertz 1968).

Classical morphologists have regarded gestational neoplasms as including three distinct entities which can be separated readily on histopathological grounds (Novak and Seah 1954): (*a*) hydatidiform mole, (*b*) chorioadenoma destruens (invasive mole), and (*c*) choriocarcinoma, the degree of retention of recognizable villous structures in the neoplastic chorionic tissue being the distinguishing features. When the entire tumor mass is confined to the uterine cavity and is constituted of hydatidiform villi with limited trophoblastic overgrowth, the lesion is termed a "hydatidiform mole." Whenever the tumor has invaded the uterine wall or extrauterine tissues but retains residual villous structures it is termed an "invasive mole" or "chorioadenoma destruens." In the absence of any residual villous structures a gestational trophoblastic tumor further characterized by invasiveness, hemorrhage, and necrosis is termed a "choriocarcinoma." Despite these clearly defined morphological criteria a vast amount of confusion arises in daily clinical practice concerning the diagnostic and particularly the prognostic significance of these purely structural terms. Over 50% of patients with what ultimately becomes a "choriocarcinoma" have a clear history of having previously had a "hydatidiform mole," and numerous patients with a prior histological diagnosis of "invasive mole" subsequently present tissue clearly characteristic of "choriocarcinoma." We have therefore inferred that this group of tumors constitutes a biological continuum, each merging imperceptibly into the other.

The situation is further complicated by the fact that the maternal host possesses an inherent capacity to reject the invading fetal tissue, thus leading to rare but well-documented spontaneous regression of the tumor process associated with complete absence of residual disease in some cases (Brewer 1961). Any prognostic inferences or any therapeutic claims in relation to gestational trophoblastic tumors must involve a large number of cases carefully evaluated on both clinical and morphological grounds and must be made with a full appreciation of the highly varied and complex dynamic interrelationship between these several forms of tumor.

TABLE 1

INFLUENCE OF DURATION OF DISEASE ON OUTCOME

DURATION OF DISEASE (IN MONTHS)	COMPLETE REMISSIONS (TOTAL PATIENTS)	
	Series A	Series B
4 or less	18/25 (72%)	23/27 (85%)
4 or more	12/38 (31%)	14/23 (61%)
Totals	30/63 (48%)	37/50 (74%)

It is inappropriate in this setting to detail our clinical experience with the chemotherapy of choriocarcinoma and related trophoblastic tumors in women. The reader may avail himself of these purely clinical observations elsewhere (Hertz 1968) (Ross et al. 1965). Certain aspects of our chemotherapeutic experience have aided us in the biological characterization of these tumor processes. Table 1 summarizes our therapeutic results in 113 cases of metastatic trophoblastic disease in relation to the duration of disease prior to treatment. It is clear that a much higher remission rate is obtained in patients with a brief disease history than in those with a prolonged illness. Increasing resistance to the elimination of all trophoblastic tissue suggests that a progressive adaptation between tumor and host tissue occurs with time. Mutual adaptation between host and tumor tissue of differing genetic constitution may be interpreted as indicating either a lack of immunogenic potential on the part of the invading trophoblast or special instances of failure of such an immune mechanism.

We have now observed 70 subsequent pregnancies in women previously freed by chemotherapy of extensive metastatic trophoblastic disease. The massive infiltration of the lungs and other peripheral tissues with highly proliferative malignant trophoblast fails to impart any resistance of the host to the intrauterine growth of normal trophoblastic elements. In one instance such a normal pregnancy occurred within 3 months of the subsidence of the trophoblastic disease process. It seems that if any immune response had occurred, it would have had to be short-lived.

One is obliged to consider the intrauterine environment as a special locus for such tolerance. Oddly enough, malignant trophoblast has extreme difficulty in surviving for any length of time in the uterine cavity. That it does is evidenced by the almost uniform absence of tumor within the uterus at autopsy in untreated women dying with extensive metastatic disease. In 30 autopsies on previously treated women we have found no instance of intrauterine disease. Frequent subsidence of the primary tumor focus despite metastatic tumor dissemination is a unique phenomenon among all malignancies. It is especially noteworthy that intramural masses of tumor within the uterine wall are, on the contrary, very resistant to chemotherapy (Hertz 1968).

The functional comparison of malignant and normal trophoblastic tissue reveals many similarities and a few differences. Both malignant and normal trophoblast are highly proliferative and extremely vascular. The growth rate and mitotic index of both are very high. Each proliferates a vast endothelial bed permitting an enormous blood flow. This phenomenon manifests itself clinically in a marked predisposition to massive and intractable hemorrhage in patients with metastatic trophoblastic neoplasms.

The most prominent hormonal feature of both normal and malignant trophoblast is the production of chorionic gonadotrophin. Because of the astronomically high titers of chorionic gonadotrophin seen in some patients with relatively limited extent of disease, the question of the identity of tumor-hormone with normal pregnancy-hormone has been raised. Accordingly we undertook to compare the fractional distribution of chorionic gonadotrophin in the plasma of 12 normally pregnant women and 12 women with metastatic choriocarcinoma. We discovered that although hormonal activity is present in the gamma and beta globulin fractions in normal pregnancy, it is found mainly in the alpha globulin fraction in women with choriocarcinoma. We interpreted this as representing a difference in hormone transport rather than any significant difference in chemical composition of the chorionic hormone in the normal and neoplastic state (Reisfeld, Bergenstal, and Hertz 1959). Finding that the dose-response curve for the ovarian weight response in the rat is identical for hormone prepared from the urine of tumor-bearing and of normally pregnant women strengthened our interpretation. Despite extensive attempts to differentiate immunologically between tumor-hormone and pregnancy hormone, we have been able to detect no significant difference as manifested by comparative behavior in Ouchterlony plates, dose-response curves of respective antisera in bioassay systems, and red-cell agglutination inhibition tests of antisera to each type of hormone. It is therefore our present interpretation that tumor-hormone and pregnancy-hormone are very similar and probably identical.

Nevertheless the unusually high titers seen in some cases suggest that an especially potent form of the hormone may be arising from malignant trophoblast. Conversely, a relatively uniform rate of residual urinary gonadotrophin titer is observed in most patients who exhibit resistance to chemotherapy after an initial significant but incomplete regression of disease. It usually ranges between 250 and

500 i.u. per 24 hr excretion in the urine. The constancy of this finding in 20 out of 28 such cases suggests either that the chemotherapeutically resistant cells produce an atypical form of chorionic hormone or that they have a limited capacity to produce biologically active substance.

An equally puzzling phenomenon observed in the ovaries of tumor-bearing women merits discussion. In about 1/3 of women with recently developed trophoblastic neoplasms one observes a massive enlargement of the ovaries, consisting of huge follicles, corpora hemorrhagica, and corpora lutea. Their enlargement is interpreted as the result of a synergistic gonadotrophic effect of endogenous pituitary and chorionic gonadotrophin. As the disease process advances and despite great increments in urinary and plasma levels of chorionic gonadotrophin, the ovaries involute rapidly so that by 4 months the ovaries are of normal size as observed clinically and at autopsy. Such ovaries then present an entirely normal histological appearance with no residual evidence of their previous stimulation. This refractoriness to further gonadotrophic stimulation may reflect the failure of continued production of endogenous pituitary hormone required for synergistic action with the ever increasing amounts of tumor chorionic hormone.

An additional phenomenon of endocrine interest is the reestablishment of normal, cyclic vaginal bleeding despite the persistence of the low residual chorionic hormone titers described above in women in incomplete remission; the ovary appears to be responding to cyclic endogenous pituitary function without being affected by the reduced chorionic gonadotrophin.

The occurrence of hyperthyroidism in women with trophoblastic disease has been noted for some time. Since both these diseases occur somewhat selectively in young women, it has been difficult to determine whether this phenomenon may be attributed to coincidence. In our experience with 93 patients with metastatic trophoblastic disease, in whom the customary parameters of thyroid function were studied, we encountered 7 women with laboratory evidence of increased thyroid function. Assay of tumor tissue from 2 of these patients by Bates revealed the presence of TSH in amounts greatly exceeding that found in blood. In 2 other patients elevated plasma levels of TSH were demonstrable. Suppression of the disease in these patients was accomplished by administration of the folic acid antagonist Methotrexate; a prompt return of their thyroid function tests to normal followed without specific antithyroid therapy. Hence it appears that certain trophoblastic tumors produce thyrotrophin as well as gonadotrophin (Odell et al. 1963).

The development of an accurate and specific radio-immunoassay for placental lactogen (HPL) has permitted the determination of the comparative plasma levels of this hormone in normal pregnant women and in women with trophoblastic tumors. Whereas there is a progressive increment in the plasma level of HPL during the course of normal pregnancy, the initially low plasma level of HPL in tumor-bearing women fails to rise with time. The discrepancy is thought by some to provide a basis for a good differential diagnostic test between the two conditions (Saxena 1967) (Josimovich and MacLaren 1962).

Patients with hydatidiform mole frequently exhibit a striking toxemialike syndrome. It consists of hypertension, edema, albuminuria, and hyperemesis. All of these conditions subside promptly when the mole is removed. The humoral basis for syndrome is yet to be elucidated, but a derangement in steroid metabolism is suggested by the clinical findings.

We now direct our attention away from women and to certain animal studies on trophoblastic tumors.

It was Greene (1952) who initially described the survival and growth of human malignant tissue in the anterior chamber of the guinea pig's eye. His pioneer studies provided the basic experimental background for the observations reported here. Subsequently Toolan (1953) and Sommers, Chute, and Warren (1952) introduced the use of the hamster cheek-pouch as a site for such heterologous tumor transfer. They found that the frequency of tumor takes could be very much increased by prior suppression of the immune response of the host animal through X-irradiation or cortisone administration or through a combination of both these procedures. Pierce,

TABLE 2

SEVEN STRAINS OF HUMAN CHORIOCARCINOMA IN HAMSTER CHEEK-POUCH

Strain	Date Started	Source	Patient's Clinical Status
BO	7/22/57	Metastasis to breast (S)	Patient had incomplete remission on MTX; cerebral hemorrhage while off therapy; strain requires cortisone
WO	10/24/58	Brain metastasis (A)	Resistant to MTX after initial response
MA	11/12/58	Lung metastasis (A)	Resistant to MTX after initial response
JO	11/11/59	Lung metastasis	Resistant to MTX after initial response; then no response to VLB or Cytoxan
RE	1/21/60	Metastasis to cervix (S)	No chemotherapy before tissue obtained; subsequently had complete remission on MTX followed by actinomycin D although still potentially responsive to MTX
GR	11/ 2/60	Brain metastasis (A)	Limited response to MTX and VLB followed by resistance to both drugs
CA	9/26/61	Lung metastasis (A)	No initial response to MTX or actinomycin D

NOTE: (A) = Autopsy; (S) = Surgical specimen.
All hamsters treated daily subcutaneously for 4 to 6 days beginning on day 7 after transplantation; effective doses were: (MTX) Methotrexate—50.0 mgm/kg; (VLB) Vinblastine—0.75 mgm/kg; Actinomycin D—50 gamma/kg.

Berney, and Dixon (1957) showed that certain human testicular tumors could be maintained as heterologous grafts by serial passage through similarly conditioned hamsters. Such tumors continued to produce the chorionic type of gonadotrophin even after prolonged maintenance in the foreign host.

In our studies tumor tissue was initially obtained under aseptic conditions within 1 to 7 hr after surgical excision or necropsy (table 2). Female golden hamsters of the NIH strain varying from 1 to 3 months in age were used. Purina Checkers were fed as a basal diet and a daily ration of kale, apples, and carrots was provided. Under Nembutal anesthesia each hamster received by direct inoculation into the cheek-pouch a piece of freshly excised tissue about 0.05 cc in volume. Recipient hamsters were either left totally untreated or they received 3 mgm of cortisone acetate in aqueous suspension at the time of inoculation and every 3d day for the ensuing 2 weeks. The course of growth of the tumor inocula was recorded by freehand sketches of the size and shape of the growing tumor mass. Histological studies were performed on certain specimens by fixation in Bouin's solution, sectioning and staining with hematoxylin and eosin. Hamster plasma was prepared from heparinized blood drawn from the abdominal aorta under Nembutal anesthesia just prior to autopsy. Detailed observations were recorded at autopsy concerning the qualitative and quantitative effects of the tumor transplant on various endocrine organs.

All chemotherapeutic agents were administered subcutaneously daily in the form and dosage indicated in the footnote to table 2. The drugs were started on the 6th day after transplantation when initial growth of the transplant could be observed clearly. Repeated sketches of the size and form of the tumors of treated and untreated control animals provided an estimate of the extent of inhibitory effects obtained.

Upon discontinuing serial passage of each strain portions of tumor were preserved in 50% saline and 50% glycerol under liquid nitrogen refrigeration. The frozen tissue from each strain was subsequently reinoculated successfully into untreated hamsters as needed for further study.

The 7 successfully established strains of choriocarcinoma resulted from a total of 30 attempts. Five of these strains "took" initially in totally untreated hamsters, and all but one strain were maintained in such untreated animals. Cortisone conditioning had been continuously required for the transfer of one strain even after extended passage over a period of 12 years. However this strain differs in no other observable feature from those maintained without cortisone.

The growth behavior is essentially the same for all 7 strains. The initial host response consists of the appearance within 2 to 3 days of a turbid exudate about the tumor inoculum. In the following 3 days it is rapidly replaced by highly vascularized, bluish purple tissue which then extends and rapidly fills the cheek-pouch. Sixty to 80% takes are already apparent the first 6 days following inoculation; 90% within 10 to 14 days. The findings are the same for previously frozen tissue as well. In about 10 to 12 days the tumor reaches its maximum size of 1 to 1.5 cc. During the ensuing 5 to 10 days the tumor loses its bluish color and becomes successively pink, gray, and finally greenish yellow. At this point the necrotizing mass may slough and drain. Frequently by 30 days following inoculation only a small residual scar may be found in the cheek-pouch. Less frequently the liquefied tumor mass is retained indefinitely.

Histological study of all strains reveals a close similarity in morphology of the transplant to that seen in the original human material. The rich vascularity of the tumor is very striking, the vascular elements and extravasated blood constituting about half of the tumor mass. Both cytotrophoblast and syncytiotrophoblast are clearly distinguished. In some preparations the syncytiotrophoblast is scant.

It is noteworthy that although this heterologously maintained tissue derives from tumor which is highly invasive in its natural host, no evidence of invasion of surrounding host tissue is observed in the hamster cheek-pouch. When the tumor is implanted subcutaneously, intraperitoneally, or intracerebrally, it grows at the site of implantation but has never been observed to invade or metastasize in any of the thousands of animals observed. The application of massive cortisone dosage has not altered its behavior in this regard.

Galton, Goldman, and Holt (1963) have described the chromosomal makeup of one of our tumor strains. The tumor "reveal[s] an unstable and individually variable aneuploid karyotype. The modal chromosome number at the 56th passage was 80 and after an interval of nine months, at the 80th generation, it had risen to 88–92. However, there was no concomitant change in tumor histology."

Wynn and Davies (1964) have described the fine structure of the same tumor strain. Electron microscopy revealed cells which resembled normal syncytiotrophoblast in that the cytoplasm contained abundant endocytoplasmic reticulum, ribosomes, and distinctive Golgi bodies. Distinctly cytotrophoblastic cells were identified also. A type of transitional cell was described with structural resemblance in some features to both cytotrophoblast and syncytiotrophoblast.

Of major interest is the sustained hormonal activity of all of our tumor strains. From what is known of hormonal production by normal trophoblastic tissue the primary hormone would be chorionic gonadotrophin. Thyrotrophin might also be expected in view of the demonstrated association of hyperthyroidism with malignant trophoblastic disease and its suppression by oncolytic chemotherapy, as well as the presence of excessive quantities of thyrotrophin in patient's plasma and tumor as demonstrated by bioassay. The possible presence of placental lactogen as described above in normal human placenta should be considered in view of the frequency of persistent lactation in women with trophoblastic disease.

The intact female hamster bearing any of our tumor strains exhibits an extreme gonadotrophic response consisting initially if extensive follicular stimulation which subsequently gives way to extensive luteinization. Although hemorrhagic follicles are seen frequently, ovulation is not observed. The uterus shows great enlargement and hyperemia, reflecting the ovarian hormonal response. These ovarian and uterine effects occur in essentially the same degree in the hypophysectomized tumor-bearing hamster. No uterine effects are evident in the ovariectomized hamster, indicating a lack of steroid-type hormone production.

The mammary glands of the intact tumor-bearing hamster are not stimulated grossly. Detailed microscopic examination of such mammary tissue has not been carried out. Immunofluorescence study of the tumor tissue is indicated also since Sciarra, Kaplan, and Grumbach (1963) have demonstrated the presence of a "growth hormone-prolactin" in normal syncytiotrophoblast through the application of specific fluoresceinized antiserum.

In the hypophysectomized hamster there is neither gross nor histological evidence of thyrotrophic effect or adrenocorticotrophic effect from the tumor. These negative observations apply equally to all strains despite the fact that our CA strain was derived at autopsy from a patient who had been freed of prior evidence of hyperthyroidism by oncolytic therapy (Odell et al. 1963).

The distribution of the tumor-produced gonadotrophin in the host's tissues has been studied repeatedly for several strains by either rat ovarian or mouse uterine weight assay of body fluids or tissue homogenates. Fresh tumor homogenates contain about 50 i.u. per gm. Urine and blood contain about 20 to 30 i.u. per ml. It is interesting that homogenates of liver, kidney, muscle, and spleen contain no detectable biological activity. Apparently, although the hormone can be cleared by the kidney, it does not readily enter the cells of peripheral tissues.

The ultimate rejection of the tumor by the host indicates a delayed immune response to the presence of heterologous tissue. We have found that a hamster which has previously rejected a tumor after several weeks' growth is completely resistant to reinoculation for at least several months, and the daily injection of 2 ml of plasma from such a tumor-resistant animal will completely inhibit tumor growth in a previously untreated hamster. Hence both active and passive resistance to tumor inoculation may be demonstrated. Prior exposure of hamsters to human serum globulin will also impart resistance to tumor inoculation, indicating that resistance relates to the formation of species specific antibodies against antigens of human origin and not to tumor-specific antigens. This phase of induced resistance to tumor inoculation merits extensive study.

Another immunological aspect of these studies relates to the specifically antigenic effect of chorionic gonadotrophin in numerous species. Antibodies to this hormonal antigen can be demonstrated readily by classical immunological methods as well as by neutralization of hormonal effects in bioassay. Accordingly we have tested the sera of animals having previously borne hormone-producing choriocarcinoma by bioassay for evidence of such antihormonal activity. In no instance was it demonstrable. Prolonged administration of massive doses of purified human chorionic gonadotrophin to normal hamsters failed to yield antihormonal sera. It appears that, whereas the hamster shows good immune response to growing human choriocarcinoma as well as to human serum globulin, chorionic gonadotrophin is not an effective antigen in this species.

The effect of a wide variety of drugs upon the growth of well-established tumors during the 2d week following inoculation has been studied according to the procedures outlined above. We have done most extensive work with Methotrexate. We were impressed at the outset by the remarkably high tolerance of the hamster for Methotrexate. (Such unusual resistance to the effect of colchicine had been described previously [Orsini and Paustry 1952].) We found that the dose of Methotrexate required for complete tumor inhibition was about 50 mgm per kg when administered subcutaneously daily on each of 4 consecutive days (table 2), about 100 times that found to induce clinical toxicity and tumor response in women. In the hamster this very high dosage produced no apparent systemic toxicity as manifested by loss of body weight, change in coat quality, diarrhea, or ulceration of oral mucous membranes.

Profound tumor inhibition was observed uniformly in all 7 strains of tumor in response to this dose of Methotrexate. It should be emphasized that 5 strains originated from patients who had exhibited unequivocal evidence of unresponsiveness to previously effective doses of Methotrexate (table 2). The discrepancy may come about through the experimental paradigm: we were testing the growth and survival of freshly transplanted tumor cells from a heterologous source, whereas in the patient these same tumor cells were isologous and had been resident in the host for prolonged periods by the time drug resistance was encountered. That the quantitative difference in tolerated dosage between human and hamster is probably not a factor in the discrepancy is shown by the fact that the effective dose of the other drugs is comparable in man and hamster despite a similar discrepancy in responsiveness between the two species. This, then, is a good demonstration of the fallibility in considering animal experimental data as directly applicable to man despite certain broad similarities.

We have previously described how the Vinca alkaloids suppressed the growth of these tumor strains (Hertz 1960). In the case of vinblastine the effective and well-tolerated dose in the hamster closely approximated the effective but highly toxic dose in women (Hertz, Lipsett, and Moy 1960) and was effective in two strains of tumor which had been derived from patients in whom a prior therapeutic trial had induced no apparent tumor response.

Actinomycin D has similarly proved to be uniformly effective in inhibiting all 7 tumor strains, although 1 of these was derived from a patient who had been totally unresponsive to it (table 2).

In addition to the drugs mentioned we have screened a wide variety of steroidal, alkaloidal, and antibiotic preparations for possible inhibitory effect. The only significant inhibition not related to general systemic toxicity has occurred through such alkylating agents as cyclophosphamide and several preparations derived from podophyllotoxin. The activity observed in this latter group of agents warrants further investigation with a view to possible clinical trial.

In closing it may be emphasized that observations in women with hormone-producing tumors of trophoblastic origin provide a highly instructive body of endocrinological and immunological data of both practical and theoretical interest.

References

Brewer, J. I. 1961. *Textbook of gynecology,* 3d ed. Baltimore: Williams and Wilkins.

Galton, M.; Goldman, P. B; and Holt, S. 1963. Karyotypic and morphologic characterization of a serially transplanted human choriocarcinoma. *J Nat Cancer Inst* 31:1019.

Greene, H. S. N. 1952. The significance of the heterologous transplant ability of human cancer. *Cancer* 5:24.

Hertz, R. 1960. Suppression of human choriocarcinoma maintained in the hamster cheek-pouch by extracts and alkaloids of Vinca Rosea. *Proc Soc Exp Biol Med* 105:281.

———. 1968. Eigenschaften und Behandlung des Chorionkarzinoms und verwandter Trophoblast-Tumoren bei Frauen. *Geburtsh Frauenheilk* 28:810.

Hertz, R.; Lipsett, M. B.; and Moy, R. H. 1960. Effect of Vincaleukoblastine on metastatic choriocarcinoma and related trophoblastic tumors in women. *Cancer Res* 20:1050.

Holland, J. F., and Hreshchyshyn, M. M. 1967. *Choriocarcinoma.* U.L.C.C. monograph series, vol 3; Appendix I. New York: Springer-Verlag.

Josimovich, J. B., and MacLaren, J. A. 1962. Presence in the human placenta and term serum of a highly lactogenic substance immunologically related to pituitary growth hormone. *Endocrinology* 71:209.

Novak, E., and Seah, C. S. 1954. Choriocarcinoma of uterus: Study of 74 cases from Matthieu Registry. *Amer J Obstet Gynec* 67:933.

Odell, W. D.; Bates, R. W.; Rivlin, R. S.; Lipsett, M. B.; and Hertz, R. 1963. Increased thyroid function without clinical hyperthyroidism in patients with choriocarcinoma. *J Clin Endocr* 23:658.

Orsini, M. W., and Paustry, B. 1952. The natural resistance of the golden hamster to colchicine. *Science* 115:88.

Pierce, B.; Berney, E. L.; and Dixon, F. J. 1957. The biology of testicular cancer: 1. Behavior after transplantation. *Cancer Res* 17:134.

Reisfeld, R. A.; Bergenstal, D. M.; and Hertz, R. 1959. Distribution of gonadotrophic hormone activity in the serum proteins of normal pregnant women and patients with trophoblastic tumors. *Arch Biochem* 81:456.

Ross, G. T.; Goldstein, D. P.; Hertz, R.; Lipsett, M. B.; and Odell, W. D. 1965. Sequential use of Methotrexate and actinomycin D in the treatment of metastatic choriocarcinoma and related trophoblastic disease. *Amer J Obstet Gynec* 93:223.

Saxena, B. 1967. Serum placental lactogen levels in trophoblastic diseases. In *Transcript of the IV Rochester Trophoblast Conf,* p. 423. Rochester University Publications.

Sciarra, J. J.; Kaplan, S. J.; and Grumbach, M. M. 1963. Localization of anti-human growth hormone serum within the human placenta: Evidence for a human chorionic growth hormone prolactin. *Nature* 199:1005.

Sommers, S. C.; Chute, R. N.; and Warren, S. 1952. Heterotransplantation of human cancer: 1. Irradiated rats. *Cancer Res* 12:909.

Toolan, H. W. 1953. Growth of human tumors in cortisone-treated laboratory animals: The possibility of obtaining permanently transplantable human tumors. *Cancer Res* 13:389.

Wynn, R. M., and Davies, J. 1964. Ultrastructure of transplanted choriocarcinoma and its endocrine implications. *Amer Obstet Gynec* 88:618.

25

Biomechanics of Implantation

Bent G. Böving
Departments of Gynecology-Obstetrics and Anatomy
Wayne State University
Detroit

In unguarded moments I, and perhaps you too, have thought of implantation as a series of histological stages beginning with an unattached human blastocyst and ending with another of the Hertig-Rock specimens (fig. 1). For the missing stages we can be brutally honest and leave blanks, or we can fill in with *Macaca mulatta* information and cringe quietly at thoughts of the 3-day difference in implantation schedule and the array of homological discrepancies such as stromal decidual reaction versus epithelial plaque, conceptus lodgment in stroma versus uterine lumen, and formation of one placenta versus two. In spite of missing stages, we can be confident that a consistent sequence of anatomical changes does occur; and that implies that a consistent sequence of causes exists.

The anatomical changes of implantation and presumably their immediate mechanical causes exhibit a progression from the general to the minute if one may judge from rabbits and a few other mammals. First the whole conceptus is transported to a location definable in terms of the whole uterus. There follows a species-specific localization of the conceptus at the antimesometrial or other aspect of the uterus and an orientation of the conceptus with the embryonic pole toward or away from that aspect of the uterus. Subsequently the conceptus and mother proceed through mutual integrative development at the histological, cytological, and chemical levels.

The method of investigation may be chosen to accord with the order of magnitude of the anatomical change. The mechanical transport of a conceptus along the uterus is unlikely to be explained by investigations of capillaries. Conversely, trophoblastic penetration of uterine epithelium exclusively where there is a capillary is unlikely to be explained by chemical examinations of samples that represent the uterine lumen as a whole, for they would reflect conditions where the penetration does not occur and not necessarily the conditions where it does occur. The tendency for biochemical methods to leave the events of implantation unexplained may be attributed partly to difficulty in looking just where the action is as the order of magnitude changes—certainly not to unimportance of chemical conditions. Far from it!

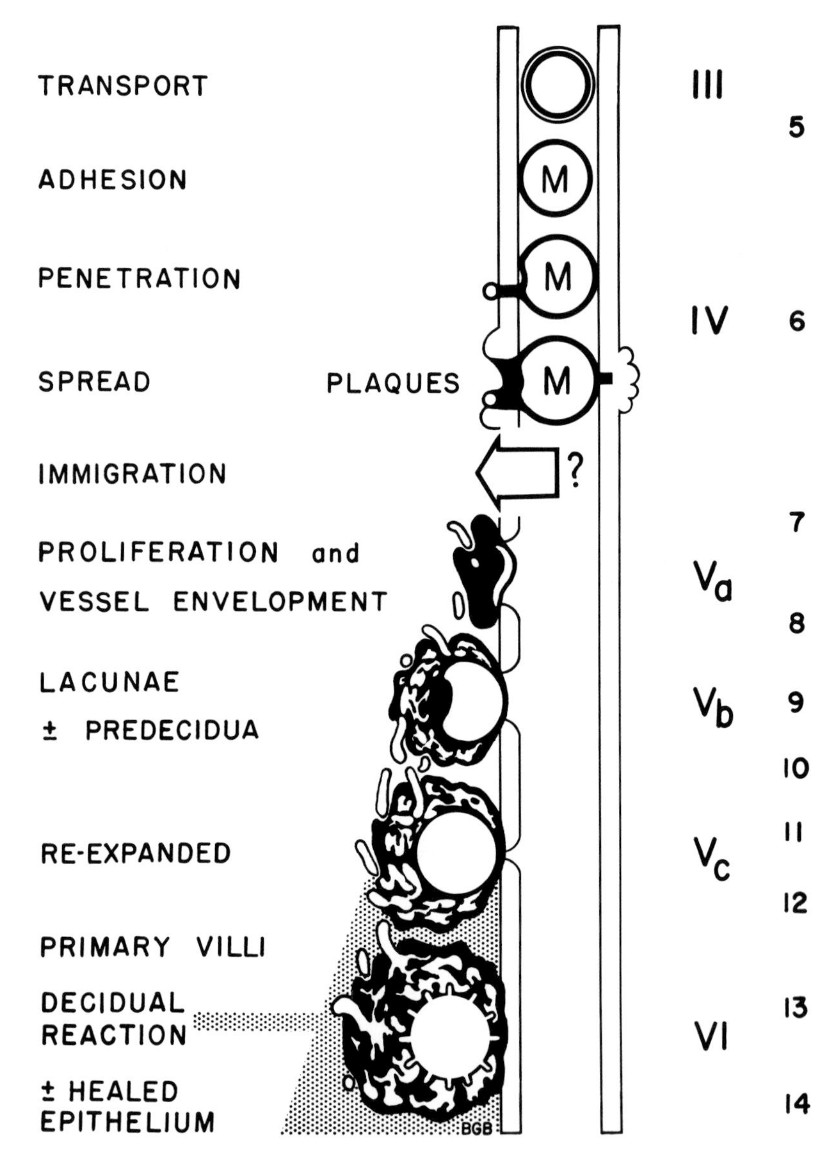

Fig. 1. Diagram of anatomical stages in implantation, human except where *Macaca mulatta* (marked *M*) has been used to fill gaps. The terms describing the stages suggest the underlying mechanics. The drawings show the relations of parts but are not drawn to scale. Roman numerals indicate the modified Streeter Horizon or stage (Böving 1965*b*). The arabic numbers give the estimated age in days since ovulation for *Homo;* for *Macaca* add about 3 days. (By permission from B. G. Böving, Anatomy of reproduction. In *Obstetrics,* ed J. P. Greenhill. Philadelphia: Saunders, 1965)

Implantation is in essence a progressive integration of the chemical and mechanical aspects of conceptus and mother. The point is that we need to bridge the gaps in understanding of cause and effect between (*a*) methods that preserve and examine structures but not their actions and (*b*) methods that preserve and examine some chemical actions but ignore or destroy structures and their mechanical actions. Physiology, histochemistry, and *in vitro* methods may fill some of the gaps, but there is obvious need for new methods and insights.

Besides the opportunity for technical and intellectual originality, there are practical reasons for study. The IUD is graced by the special merits of exerting long term yet reversible contraceptive action by presumably nonsystemic means. Yet its drawbacks are significant. Its empirical origin provided no rational basis for improvement, but simple design engineering of its materials, size, shape, and packaging has now been achieved, along with recognition of some of its biological consequences (Advisory Committee F.D.A. 1968). Ultimately one should know all the possible consequences. But to understand consequences, the normal situation should first be understood in all details. For the moment, we must content ourselves with consideration of just some of the more mechanical aspects of implantation.

I. Blastocyst Transport and Spacing

Among mammals there is a little variation in the age and stage at which conceptuses pass from oviduct to uterus. In man and rabbit, which will be our chief concern, it is a late morula or early blastocyst that enters the uterus, in the case of man about 3 or 4 days after estimated ovulation time (Hertig, Rock, and Adams 1956), in the case of rabbit 72–73 hr after observed coitus (Böving 1956). The progress of the blastocyst through the uterus is not known for man. The ignorance stems partly from the severe scarcity of human specimens in preimplantation stages and partly from the custom of collecting them by flushing. The day-by-day location of blastocysts in terms of percentage of uterine length is known accurately enough for statistical treatment only for the rabbit (Böving 1956). Such formal quantitation is unnecessary to justify the simple observation that the net progess of conceptuses through the uterus is slow. That conclusion could have been reached by noting that it was not until 5 days after mating that the fastest moving blastocyst had traveled about as far down the uterus as it was likely ever to go before implanting (fig. 2). But the simple observation speaks to a family of problems that do require quantitative analysis. The 2-day normal transit time casts doubt on the inference drawn by Markee (1944) that rabbit blastocysts are transported by the "peristalsis and recoil" that carry fixed sea urchin eggs suspended in 0.1 ml of salt solution throughout the full length of the uterus within 2 hr of the time they were injected. Since the experiments were considered to explain blastocyst spacing as well as blastocyst transport, the two matters require reexamination.

By the time rabbit blastocysts implant they have become separated and spaced out along the uterus at what appear to be nearly equal distances from each other and from the ends of the horn. To determine whether the spacing is significantly more regular than random, to measure the degree to which it is or is not equidistant, and to ascertain the time over which any change in regularity develops, a convenient

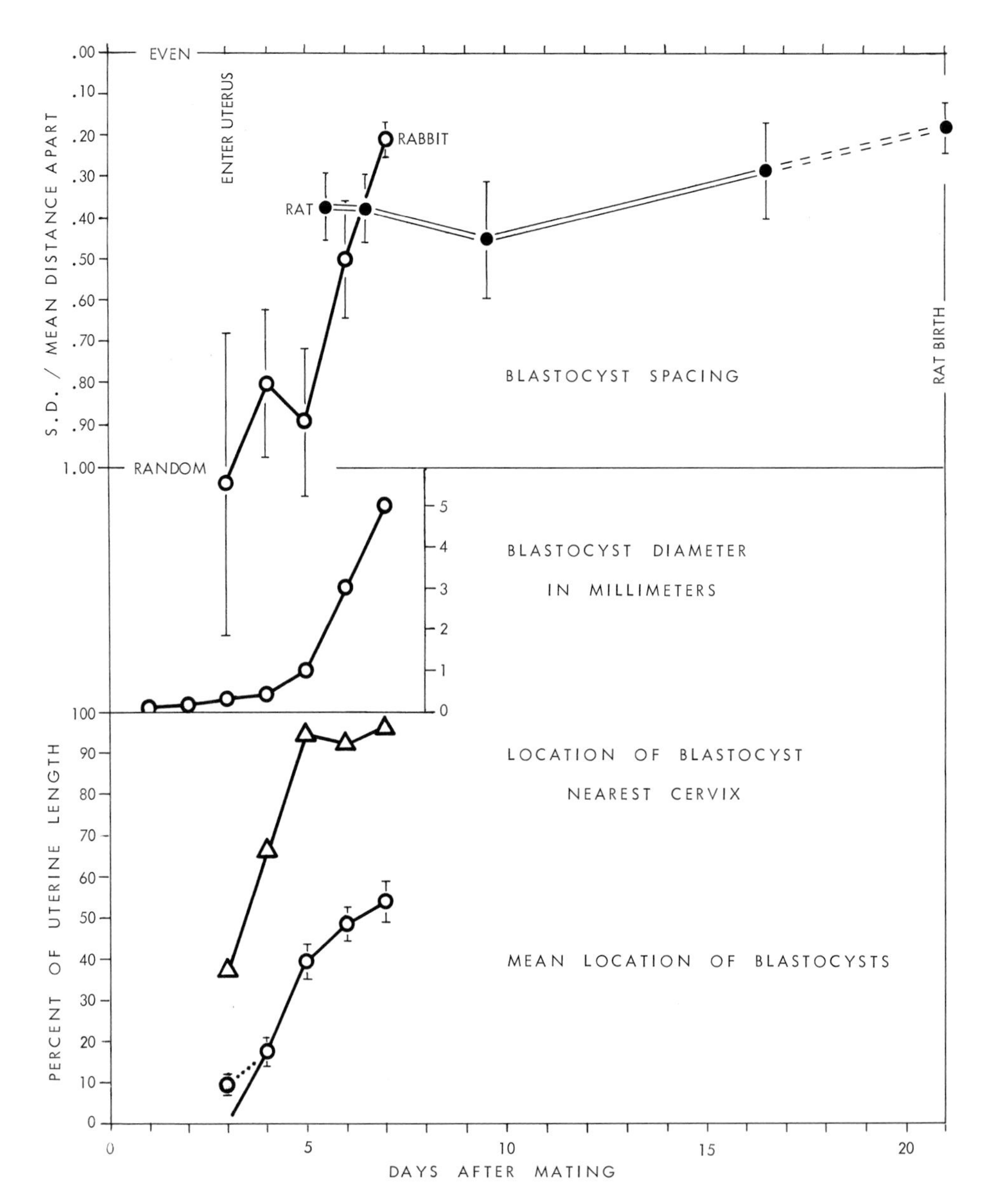

Fig. 2. Rabbit blastocysts (*open symbols*) enter uterus at 3 days after mating, move rapidly but in random arrangement from 3 to 5 days and more slowly but more equidistantly from 5 to 7 days. The most rapidly moving single blastocyst (*triangle*) takes about 2 days to travel its course. Rat conceptuses (*black dots*) are spaced more evenly than random as early as has yet been measured, but at implantation time their spacing is not quite so regular as the rabbit's. The datum graphed at birth is derived from placental scar location actually observed 12 days postpartum by Momberg and Conaway (1956). Circles and dots are means; bars extend 2 standard errors.

statistic is the coefficient of variation: the mean of the distances between blastocysts in a horn divided into the standard deviation (Böving 1956). Perfectly equidistant spacing is reflected by zero; random spacings vary widely but have an average of unity. Combining the information from the estimates of spacing regularity with the estimates of the mean location of blastocysts each day (fig. 2), one finds that rabbit blastocysts have been scattered relatively rapidly and randomly along the uterine horn at 3, 4, and 5 days after mating, but at 6 and 7 days net progress has slowed while spacing has become increasingly and significantly more equidistant than random although never perfectly regular.

The nonrandomness of rabbit blastocyst spacing implies that it has a specific cause and is not just the result of passive scattering. Preformed implantation sites in the uterus occasionally have been spoken of as a possible explanation, although it is inconsistent with the fact that equidistant spacing occurs regardless of the number of blastocysts per horn—and the number varies. Moreover, when there is but a single blastocyst in a horn it lodges near the midpoint rather than near the oviductal end of the horn where one might suppose the first available site to be located. Alternatively blastocysts might be spaced, like beads on a string, by the blastocysts' own space-taking properties. However such spacing by crowding would not explain the midpoint location of single blastocysts and is specifically precluded by the dimensions of rabbit blastocysts (0.05–5 mm in diam.) and uterine horns (200 mm in length); when blastocysts are present in normal numbers (1–10) they do not touch. The fact that equidistant spacing is achieved by blastocysts that do not touch implies that they exert their spacing effects on each other by way of the uterus. Blastocysts are kept away not only from each other but from the ends of the uterine horn. These anatomical results imply that the spacing mechanism involves an impulse that originates at each end of the uterine horn and wherever there is a blastocyst within the horn, that it is conducted along the uterus in both directions and moves neighboring blastocysts away. That tells where to look; the spacing schedule (fig. 2) tells when; and some crude notions about the power requirements and the output of cilia versus muscle suggest what to look for.

Propagated uterine contraction waves were looked for at 6 and 7 days after mating and were recorded by motion pictures taken at 4 frames per sec to give 6-fold acceleration when projected at 24 frames per sec. Recognition of contraction waves was aided by a ring of lights around the lens. The specular reflections are in two parallel rows along the edges of the cylindrical uterus; but wherever there is a circumferential contraction the rows converge or disappear; and wherever there is a bulge, such as around a blastocyst, the rows diverge and usually form a circle. Propagated waves of uterine constriction were observed to arise from each end of the uterine horns and from wherever they contained a blastocyst (fig. 3).

The technique for observation was turned into an experimental method for testing the hypothesis that the stimulus provided by the blastocyst is uterine distention. (The hypothesis was derived from noting that spacing begins at the same time that blastocysts expand rapidly [fig. 2]; support came from discovering that an abnormally small blastocyst had lodged midway but half distance from neighbors of normal size normally separated from each other and other blastocysts of normal size.) Spherical beads were placed in the uterus. Beads the size of unattached blastocysts stimulated waves of contraction like those stimulated by blastocysts. Such waves of contraction were able to move beads the size of unattached blastocysts; they were unable to move beads the size of attached blastocysts. Attached blastocysts, of course, were unmoved by them. The uterine contraction waves do not move the distending object but move away from it equally in both directions and move adjacent objects away from it.

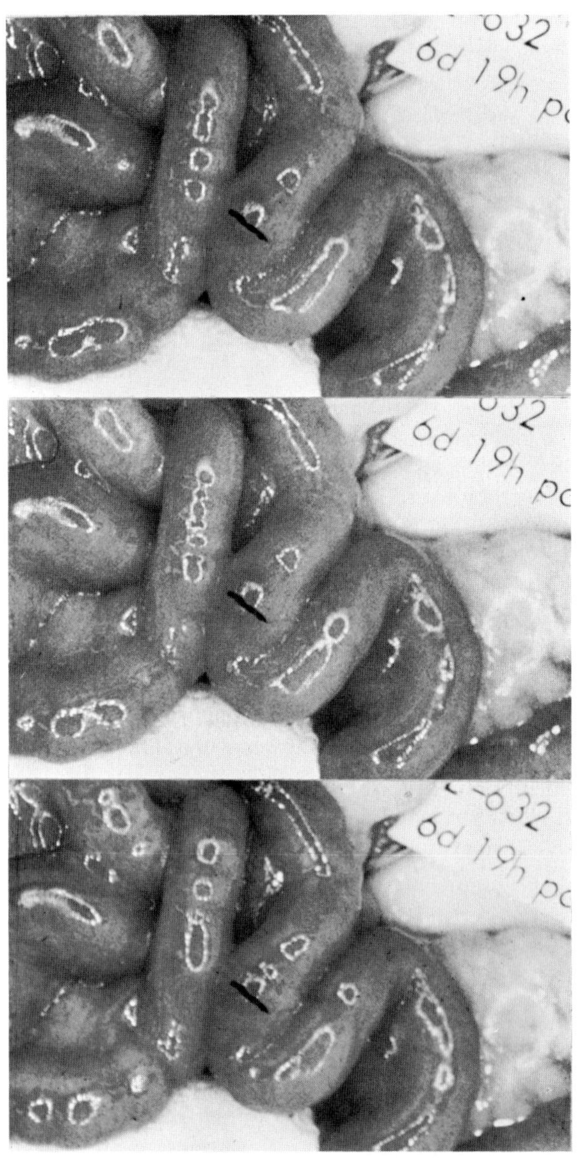

Fig. 3. Rabbit uterus *in vivo*. Dark zones between circles of reflected light mark uterine constrictions. Their progress along a uterine horn was recorded by motion pictures, of which these are cropped frames. The short black thread in the center marks the dome housing the single blastocyst in this uterine horn. Waves travel away from it in both directions.

The elements of the rabbit blastocyst spacing mechanism may be summarized.

Stimulus:	*a*) Spontaneous (at ends of horn)
	b) Uterine distention by expanding blastocyst
Effector:	Uterine muscle
Action:	Propulsive propagated circumferential contractions
Result:	Equidistant spacing

The investigation and this account began with quantitative anatomical analysis of the end result and progressed through simple *in vivo* observations and experiments sufficient only to identify the causative elements. Each should ultimately be explored in all details but for the present can be discussed only in one or another general aspect.

Of the spontaneous stimulus it may be mentioned that electrical activity characteristic of pacemakers has been observed at the ends of the uterine horns but was interpreted as probably able to arise elsewhere along the uterine horns (Kao 1959). (Perhaps the waves of contraction coming from near the oviductal insertion in uteri of *Macaca mulatta* near term [Ivy, Hartman, and Koff 1931] hint at relevance to human conceptus transport and localization, but the hint deserves to be taken with some reservation because the observation is of another species and at the wrong end of pregnancy.)

Concerning the rabbits' distention stimulus, one may ask how the blastocyst expands. It pumps itself up. That conclusion is suggested by the observation that blastocysts cut in half can reform into 2 closed spheres, each of which reexpands (Daniel 1963). (A comparable but spontaneous collapse and reexpansion of mouse blastocysts was described by Cole [1967].) More specifically, the normal 10,000-fold increase of volume between 4 and 8 days *post coitum* is attributable to water intake by energy-coupled flows in the cellular barriers between blood, uterine lumen, and blastocyst cavity. Fluids from the three compartments were collected for freezing point determinations by the Ramsay micro method; from the measurements chemical potentials were calculated with allowance for possible differences in hydrostatic (blood) pressure. At 4 days after mating there was no significant difference between uterine fluid and blastocyst fluid. At 5 and 6 days after mating the potential gradients were such as would make water flow to both blood and blastocyst from the uterine lumen. At 7 and 8 days after mating when uterine fluid no longer separated blastocyst from blood because implantation had begun, the potential gradient was such as to let one expect water to flow from blastocyst to blood. It did not. As a matter of fact the water flow into the blastocyst became most rapid at that time, which moreover was the time when the potential gradient was steepest in the opposed direction. Accordingly the water flow that did occur during rabbit blastocyst expansion must be attributed to something other than a simple osmotic process (Tuft and Böving 1969, 1970).

Concerning the distention stimulus, there is also the question whether it is the only stimulus, since spacing has been said to occur in species whose blastocysts do not expand. The rat and guinea pig are examples. Diagnoses of even or uneven rat spacing have been derived from visual impressions (Krehbiel and Plagge 1962) or by counting (Frazer 1955). Frazer considered spacing to be even when the number of ova in the cervical half of a horn was approximately equal to the number in the tubal half, and by that criterion he found even spacing only if there were 5 or more ova per horn. Krehbiel and Plagge disagreed about the relation of distribution to number. They proposed a hypothesis that even as well as random arrangements stem from random scattering of blastocysts by the uterus. The probability of a random process yielding an even distribution of 5 or more items is very small and becomes smaller the greater the number. I suggest that their hypothesis overlooks the principle that wherever the eye sees a regularity in nature the mind may see an invitation to look for a specific cause. Actually neither method of judging evenness was appropriate. Census or counting methods do not justify conclusions about the equality of distances (Böving 1965*a*), and visual impressions of distance are not precise enough to make what can and should be a quantitative determination.

Measurements of the distances between rat conceptuses made by ruler and calculated as for rabbit blastocyst spacing yield coefficients of variation that indicate that the distances are significantly more nearly equal than random but not perfectly equal (fig. 2). At the time of implantation rat spacing is significantly less equal than rabbit spacing, but it improves in the course of pregnancy as conceptuses grow and impose equal spacing by their own space-taking properties. Spacing by crowding appears to begin sooner when there are many conceptuses per horn, and within a horn it may begin sooner in regions where conceptuses are crowded together (fig. 4). In short, the string-of-beads phenomenon comes into operation sometime after implantation in the rat. But what of the degree of regularity in spacing that has developed already by the time of implantation when conceptuses are too small to touch each other? When does it develop? Presumptive implantation sites as early as 4.5 days after mating were revealed by intravenous injection of Evans Blue approximately according to the method of Psychoyos (1960) (fig. 4). Unfortunately there were a few uncertain spots in the preliminary tests; so histological methods were tried. They too present some uncertainties that have yet to be resolved; there are no data worth reporting. Only after it is ascertained when the regularity of spacing begins to develop can one know when to look for the rat's preimplantation spacing mechanism at work—that is to say when to look for clues to the operation of a spacing mechanism presumably different from that of the rabbit and conceivably more akin to that of the human whose blastocyst expands only slightly and whose massive uterus seems unlikely to be stimulated through distention by so small an object.

Having examined the stimulus of normal and negligible distention, we next turn to excessive distention. Although not relevant to the normal human blastocyst-uterus interaction, excessive distention nevertheless may be relevant to the human IUD-uterus interaction. But first a bit of background. Spherical beads the size and shape of a blastocyst shortly before implantation were inserted, 1 per uterus (with sham operation of the control horn), and left for a few days during the spacing period. Sometimes the expected displacement of conceptuses occurred, sometimes not, and sometimes development was stunted, especially that of the nearest conceptus. Why were the results so inconsistent? The bead dimension no doubt was unnatural insofar as it could match the size of the rapidly expanding blastocyst only briefly. In addition, regardless of how precisely the beads were measured their stimulating action through distention of the uterus no doubt varied according to

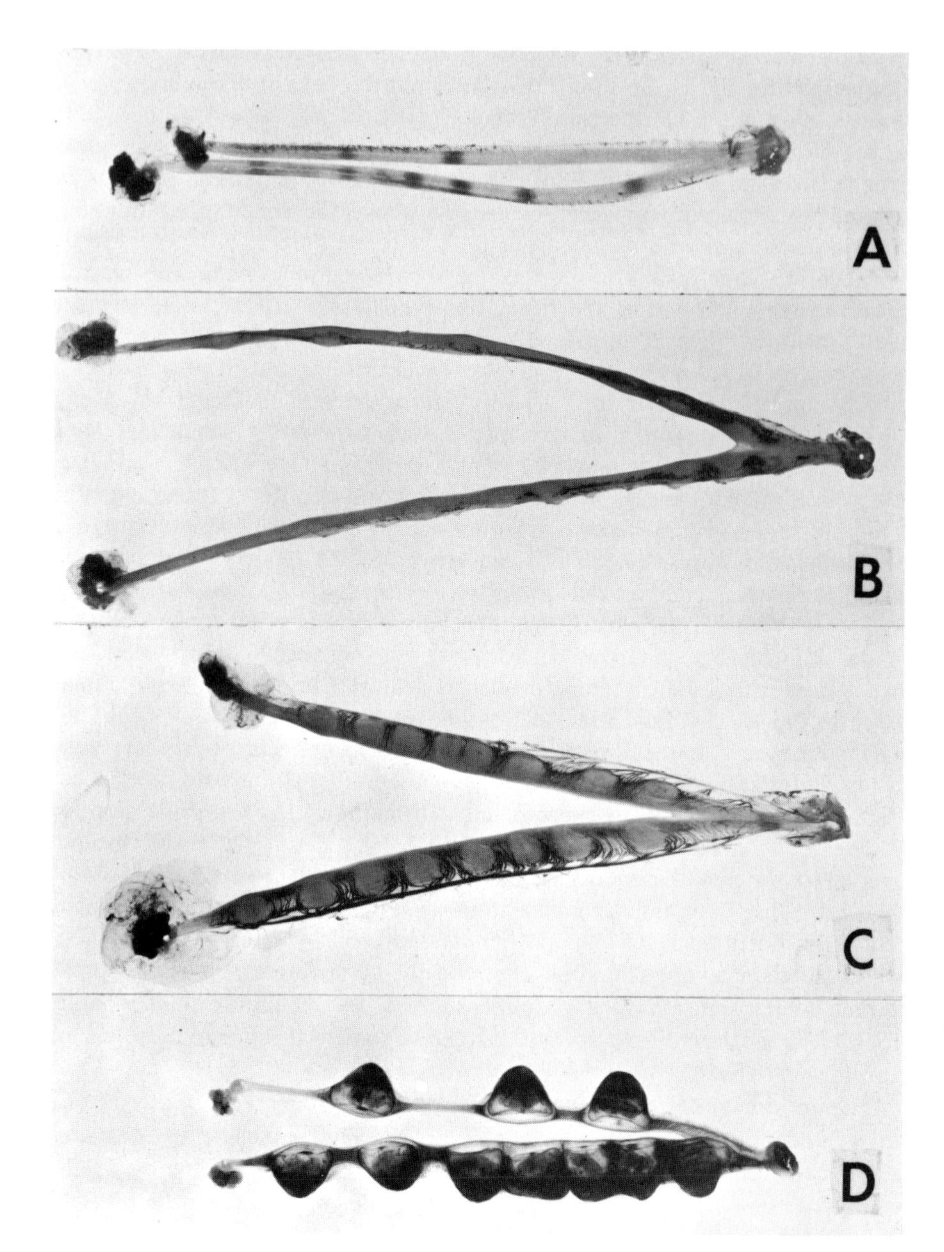

Fig. 4. (*a*) Rat uterus 4.5 days after mating. Intravenously injected Evans Blue reveals presumptive implantation sites.

(*b*) At 5.5 days after mating. Positions of conceptuses are marked by slight swellings, and spacing is more even than random.

(*c*) At 7.5 days after mating. Conceptuses are obvious as swellings and translucent regions. With many conceptuses there may be spacing by crowding even this early.

(*d*) At 13.5 days after mating. Each conceptus causes a pronounced swelling. With few conceptuses (*upper horn*) there may still be space between them; with more conceptuses (*lower horn*) spacing by crowding is most severe. (By courtesy of Carnegie Institution of Washington) whether the uterus was of large or small diameter; and uteri vary considerably. I was concerned too about the suture that anchored the bead in the uterus as well as the sutures closing the uterine slits for bead insertion and the sham operation. To avoid uterine sutures, to block the cervix and keep blastocysts from being expelled into the vagina, and above all to be very sure that a strong distention stimulus really was operating I placed very large beads just above the cervix after dilating the cervix and inserting the beads *per vaginam*. The beads were 9.5–10.5 mm in diameter—much too large to be transported. Surprisingly, instead of finding blastocysts displaced to the tubal end of the horn, I saw no blastocysts at all in the treated horns. A few contralateral effects were noted, including a case of some displacement of conceptuses toward, not away from, the cervix; and there was a case of bilaterally empty horns. The reaction was clearly different in kind from a spacing reaction. In some unrelated experiments in my laboratory a medical student, Thomas Storch, noted that tying off the cervix at 6 days after mating caused a similar total loss in 3 out of 3 cases. Accordingly not only distention but also interference with uterine fluid flow may deserve to be considered a possible consequence of large objects within the uterus, especially since Heap (1962) reported that pronounced changes in chemical composition of rabbit uterine fluid followed uterine ligation. On the other hand, normal blastocysts only slightly later in pregnancy would appear to cause equal distention and blockade with no detrimental effect. Perhaps the timing is important; nevertheless prudence requires that some reservations be attached to the idea of fluid blockade as anticonceptant.

Thoughts turn in the direction of the luteolytic effects of beads placed in guinea pig uteri (Bland and Donovan 1966), but they soon turn away because that influence extends only to corpora lutea in the ovary on the same side as the bead. Moreover, if the unilateral intrauterine bead reduced the endocrine output of one corpus luteum the endocrine effect on implantation would be expected to be bilateral rather than unilateral and no more serious than unilateral ovariectomy. Infection has not been ruled out, but it does not seem very likely to account for the observations. The occasional influence on the contralateral "control" horn then hints that an IUD may have a systemic effect of as yet undetermined nature in addition to its local action by way of chronic endometrial inflammation, chemical, and mechanical effects.

Consider now the effector. What are the relevant peculiarities of uterine muscle under the endocrine conditions presumably present during blastocyst transport and spacing? If there is truth in the popular notion that progesterone-dominated uterine muscle is flaccid, then either it is nonsense to consider blastocysts to be transported and spaced by it, or it is necessary to suppose that the myometrium that transports blastocysts is not then under effective progesterone-domination in spite of the presence of corpora lutea and endometrial histology that testifies to their activity.

The concept of a uterus flaccid under progesterone domination was modified by Csapo (1955). He perceived that the myometrium under progesterone domination is as capable as it was under estrogen domination of exhibiting massive and powerful contractions, provided that all its contractile elements are stimulated simultaneously. (Simultaneous stimulation occurs when an electric current is run from one end to the other of a muscle strip or through the solution containing it.)

Accordingly the significant progesterone effect is an impairment of conduction. Waves of contraction travel the entire length of the uterus under estrogen domination but peter out in shorter and shorter distances as progesterone domination increases until finally muscle elements contract with no coordination and negligible net effect. The early minimum impairment of conduction is consistent with the early rapid transport of small blastocysts randomly along the uterus. The intermediate phase with propulsive force decreasing in proportion to distance from its source might well provide the gradients of repulsion that convey each blastocyst as far as possible from contraction-stimulating neighbors (or an end of a horn) with maximum and consequently equal spacing the result. The late phase with severe loss of contractile coordination exhibits irritability without power. That final "flaccidity" is consistent with the antimesometrial yielding of the uterus to the continually expanding blastocyst whereby the blastocyst becomes lodged in a "dome" and the transport and spacing mechanism is turned off.

Points of agreement, although numerous, are not complete. There is a timing problem. Progesterone secretion begins before ovulation (Forbes 1953) rather than 5 days after mating, when spacing begins. The discrepancy cannot be argued away by claiming that it takes 5 days for the myometrium to react to the increased progesterone in the blood; progesterone effects on muscle are detectable 21 hr after progesterone injection or 24 hr after mating (Schofield 1954, 1955, 1957). The delay could, however, stem not from the effector of spacing but from the stimulus—blastocyst expansion—which, like the muscle behavior, is progesterone-dependent (Corner 1928) but which, like the spacing, begins 4 or 5 days after mating. To explain the delay, one might make the general guess that there is a time-dependent requirement for blastocyst expansion and the specific guess that it concerns development of the water "pump" that expands the rabbit blastocyst.

What is the mechanical nature of the blastocyst propulsion during spacing of rabbit blastocysts? It has been spoken of as "peristalsis" and perhaps properly so. "Peristalsis" really means propulsion of contents through a tube by circular muscle contractions that are propagated along the axis of the tube, but custom has linked the term to intestinal motion. In the intestine, thanks to reflexes in the myenteric plexus, a peristaltic wave begins "upstream" of the distending object that stimulates it. It moves "downstream" carrying the distending object before it. The rabbit uterus has no ganglionated nervous apparatus comparable to the myenteric plexus (Pallie, Corner, and Weddell 1954). The waves of contraction, stimulated by the distention of the uterus brought about by blastocyst or bead, move from the point of distention equally (though not necessarily simultaneously) in both directions and do not propel the distending object but repel its neighbors. Propagated contractions may pass beyond such a neighbor, or contraction waves moving toward each other may meet and cancel out or cross and continue on their way. The uterine motion, although "peristaltic," is distinct in nature from intestinal peristalsis.

How does a propagated wave of circumferential contraction actually carry a blastocyst before it? The facile reply is that the contraction raises the pressure behind the blastocyst higher than the pressure and resistance ahead of it, so it moves ahead, just like the piston in an engine. But what pressure is being raised? One might think uterine fluid pressure. But there is little fluid present in the uterus of a rabbit during the transport and spacing period; when opened, the endometrium is barely moist. A volume of no more than 0.5 ml can be collected even through the artifice of estrogen treatment (Lutwak-Mann 1959), and a volume of 0.1 to 0.5 ml was estimated to be present by a dye dilution method (Kulangara 1960) that not only avoids estrogen, which is inappropriate to the progesterone-dominated spacing period, but seems likely to sample the fluid in the crypts and convolutions of progestational endometrium better than aspiration or blotting methods. If uterine fluid is not the medium of force transfer in a hydraulic or engine piston system, there is an odd alternative. It is conceivable that the soft, highly convoluted endometrium might act as a fluid of sorts, moving a bit along the uterine axis in response to pressure difference, and yet moving back after the pressure wave passes simply

TABLE 1

FORCE NECESSARY TO PROPEL BEADS THROUGH PROGESTERONE-DOMINATED UTERUS

mm Diameter	mm^2 × Sect. Area	gm Force	gm/mm^2 Force/Area	mm Hg Equivalent
2.1	3.5	0.1	0.03	2
2.5	4.9	0.1	0.02	2
3.2	7.8	0.6	0.08	6
3.4	9.1	0.7	0.08	6
3.8	11.4	0.8	0.07	5
4.3	14.1	0.8	0.06	4
4.5	15.8	0.9	0.06	4
5.0	19.6	1.0	0.05	4
5.3	21.9	1.5	0.07	5
6.0	28.3	2.5	0.09	7

NOTE: Beads of various diameters (column 1) were pushed through a uterine horn progesterone-dominated by 12.5 mg Delalutin given 3 days previously. The force required to initiate movement of the bead through the uterine lumen was measured by the bend of a spring wire suitably calibrated; it is recorded in column 3. The pressure calculated to be equivalent to the required force per cross sectional area of each bead is recorded in the last column. The observed pressure across a 4.8 mm bead in the same uterus reached a maximum of about 2 mm Hg.

(Note added March 1970) In pregnant horns 6 and 7 days after mating, studied by an electronic system for force transduction, amplification and continuous recording, resistance forces about 10 times those of the above table have been usual. A fluid pressure difference as high as 5 mm Hg has been observed across a 5 mm bead.

because it is anchored (Böving 1959). The blastocyst could be carried forward by such a mechanism; the unanswered question is whether it might somehow avoid being carried back.

The hydraulic or engine piston hypothesis must be evaluated in a way that dodges the uncertainty about the endometrial behavior. Beads of various sizes, covering the range of blastocysts undergoing spacing and a bit on either side of that range, were placed at the end of thin but stiff wires. The force necessary to push them through a progesterone-dominated uterus was measured by the calibrated bend of a piece of spring wire affixed to a board. The results of this crude dynamometry are presented in table 1. For comparison, pressure determinations were made in front of and behind a bead 4.8 mm in diameter (fig. 5) placed, for want of pregnant animals in the summer, in the same female rabbit treated with 12.5 mg Delalutin 3 days previously. The maximum pressure changes at the rate of the visible contractions were about 2 mm Hg, too low to move beads the size of blastocysts during the 6- to 7-day *post coitum* period of spacing but perhaps barely enough to move spheres of 2.1 to 2.5 mm characteristic of 5.5 days. It would be easy to remark primly that the measuring techniques are too crude and the data too few and close to permit a conclusion at the present time, but one can do a little better than that by recognizing a bias and making some incidental observations.

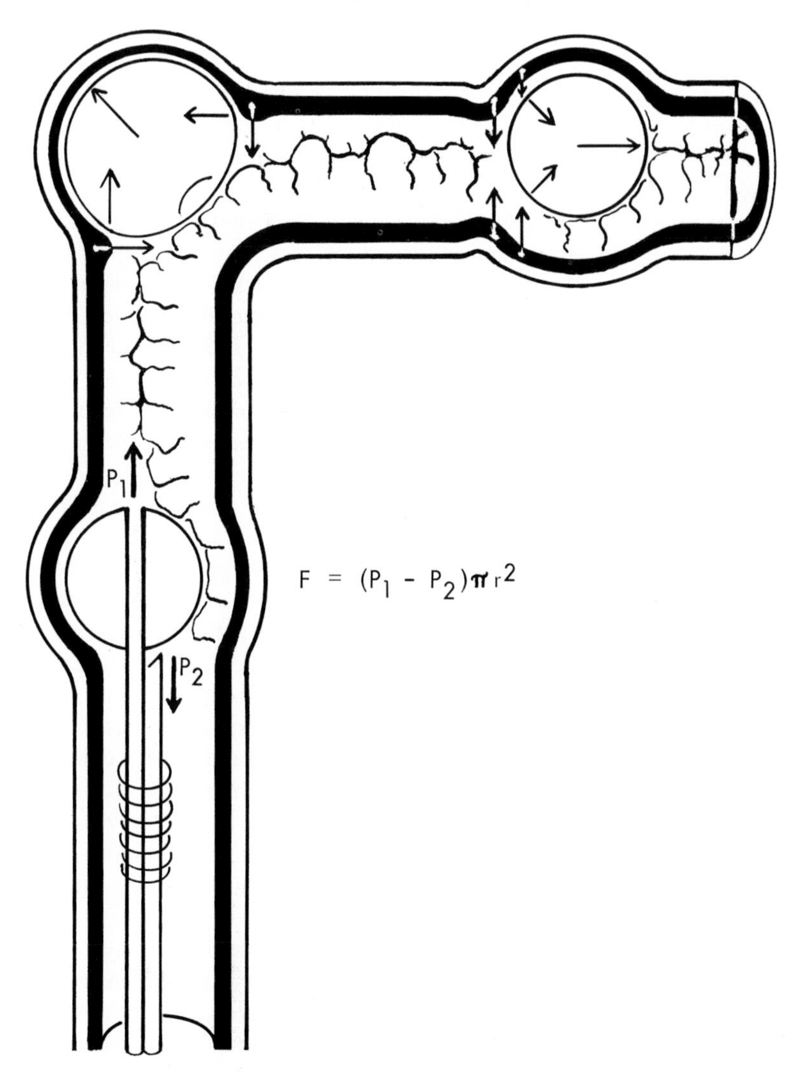

Fig. 5. An intubated bead (*lower left*) permits measuring pressure ahead of and behind a bead the size and shape of a blastocyst. If the propulsive force as well as the pressure difference can be measured accurately enough, a comparison of them could show whether blastocyst propulsion is hydraulic or by squeeze (*upper right*). Subsequently the blastocyst balloons out the anti-mesometrial uterine wall, flexes the uterine horn 90°, and becomes "grasped" in the dome by circular muscle of the adjacent undistended uterus. The "grasp" is a consequence of the 90° flexion placing both arms of the uterus in the mesometrial hemisphere of the dome, so that the resultant of forces from circular muscle contraction can push the blastocyst only upward into the dome (*upper left*).

There is a problem of what the technique does to what it measures. Pressure behind a blastocyst can rise above the pressure ahead only until the pressure difference overcomes the resistance to blastocyst movement and the blastocyst is moved, releasing some of the pressure difference. The same maximum pressure difference would be recorded across the bead if the bead moved as readily as a blastocyst of similar size. The bead probably does not move as readily, because the polyethylene tubes are stiff and the silk wrapping them together presents a certain roughness (fig. 5); actually, the bead was not seen to be moved significantly. Therefore the maximum pressure difference recorded around the bead is not the pressure at which motion begins for a 4.8 mm sphere but the maximum pressure difference that the uterus is able to exert on a sphere of that size. Even that is not enough. Thus, the hydraulic or engine piston hypothesis is even weaker than it appears at first glance.

The possibility of saving the engine piston hypothesis by considering negative pressures was raised by occasional observations of a pressure drop of about 10 mm Hg over 1 to 3 min. However, the pressure drop was usually about the same on both sides of the bead, so no pressure difference and propulsive force could be attributed to that action of a uterus on a single bead. Before the conceivable effect from and on neighboring beads could be tested, it was recognized that the slow pressure drop was an artifact traceable to imprecise leveling of the transducer with respect to the catheter opening and perhaps influenced by catheter occlusion or leakage around the catheter.

The engine piston hypothesis is difficult to rule in or out because it has at least two aspects. The fluid pressures that have been measured are not great enough to produce the required force. The pressure that one might expect to be transmitted to a blastocyst or bead by the soft, viscous, and yet epithelially bounded endometrium is not known, and it is not likely to be adequately reflected by the catheter and transducer system that measures fluid pressure. Might the supposed endometrial force work with the inadequate force from fluid pressure not just additively but cooperatively? Since the endometrium is anchored there is a question whether the blastocyst, if pushed forward, is also pulled back at the end of a contraction. Conceivably the small fluid pressures might prevent the blastocyst from moving back and yet permit stretched endometrium to slip back as the wave of contraction moves by, giving a ratchet mechanism. Whether some combination of hydraulics accomplishes or even contributes to the observed result remains tantalizingly unanswered.

One general point should be made: to understand blastocyst propulsion in hydraulic terms, a pressure difference across the locus and dimension of a blastocyst must be measured. (General pressure measurements in the uterine lumen are like the barometric pressure—they may go up or down according to whether it is a good day for blastocysts to go sailing, but they are no measure of the wind that moves the ship.) If the pressure in the direction of motion had been sufficiently lower than the pressure behind, the engine hypothesis would have been supported. It was not. Suppose the observed pressure difference was not simply inadequate but in the direction opposite to the motion. I have seen that repeatedly, but not consistently. If such pressure were recorded during normal motion (and it is not easy to be sure that the motion of an intubated bead is or even can be normal) then hydraulic pressure change would have to be the result rather than the cause of propulsion, and we could speak of a pump piston hypothesis rather than an engine piston hypothesis. Preliminary differential pressure transducer measurements thus fail to show that propulsion is hydraulic, but one gains a strong impression by just looking at the movies that a constriction wave simply grabs a bead and milks it on down the uterus by the same reverse wedge principle that propels an orange seed when you squeeze it between the fingers (fig. 5, top right). If that were the mechanism the uterine fluid or mucus would be not so much a pressure transmitter as a lubricant, and that role seems more suited to its volume and distribution.

In summary, the exact propulsive mechanism by which rabbit blastocysts are transported and spaced along the uterus is not known, but the questions and principles of exploration that have been outlined can probably yield a decisive answer, given apparatus more refined than my spring wire dynamometer and naturally pregnant animals rather than progesterone-treated ones.

(Subsequent studies have used pregnant animals and have also determined the force exerted on a bead not by the experimenter but by the uterus. Both engine piston and pump piston effects have been recorded, but the hydraulic pressures associated with the forces measured have been too small to account for them and have occasionally been absent or have appeared a second or so later. Accordingly hydraulic effects are now considered secondary to uterine squeezing of the bead and the bead's consequent motion.)

II. Blastocyst Grasp by Uterus

Transport and spacing of blastocysts is terminated by the expansion of blastocysts continuing until the uterus can no longer move them along. Further expansion of the approximately spherical blastocysts within the cylindrical uterus causes the uterine wall to yield not symmetrically but antimesometrially, flexing the uterus about 90° and letting the blastocyst lodge in the bulged-out dome (fig. 5, top left). Since the two arms of the uterus adjacent to each blastocyst are only 90° apart, they both insert into the mesometrial hemisphere of the dome. Constriction of the circular muscles in the undistended part of the uterus next to the blastocyst cannot propel the blastocyst along the uterine axis but must tend to push the blastocyst up into the dome. That action has been called "grasp." The grasp phase of implantation has been studied cinematographically, and the principal evidence for its existence is the observation that a blastocyst lodged in such an antimesometrial dome drops down out of it when the uterine grasp is released by injecting fluid into the uterus until it is distended (Böving 1952, 1960). Unless thus artificially released the grasp holds the rabbit blastocyst antimesometrially and is thought responsible for the consistent antimesometrial location of the initial blastocyst attachment.

III. Lemma Adhesion

The preceding experimental procedure occasionally fails to dislodge a blastocyst, in which case one may see regions of adhesion pulled out, taffylike, until they snap apart. A blastocyst responding so was still surrounded by lemmas. The outer one deposited by the uterus has been called "gloiolemma" in recognition of its sticky property. The stickiness is inducible over the entire surface of the blastocyst when placed in very alkaline solution, whereas acidification hardens the unattached surface or any adhesion that may have become established, such as that to the bottom of a dish. Histological evidence confirms the lemma adhesion as a natural step in implantation (Böving 1963). That the development of adhesion is localized at the abembryonic pole of the blastocyst, as is the natural elicitation of an alkaline reaction, suggests that alkali-induced adhesion may be responsible for the consistent orientation of the blastocyst.

IV. Trophoblast Adhesion

Removal of the lemmas exposes trophoblast on the surface of the blastocyst. The trophoblast experiences similar stickiness and hardness responses to pH. Trophoblast adhesion has not been successfully examined by the *in vivo* chamber, but there is ample histological evidence for its being a natural step in implantation. Trophoblast adhesion very definitely is generated selectively where uterine epithelium has a blood vessel at its base (Böving 1962), and this is probably true even for the earlier lemma adhesion (Böving 1963). Thus adhesions "aim" the subsequent trophoblast penetrations straight for a vessel.

V. Trophoblast Penetration

Invasive attachment of rabbit trophoblast to the uterus has several mechanical aspects. Some of them occur at the same time in different locations of the same uterus, introducing uncertainty as to the sequence of events. I am of the opinion that trophoblast adhesion to epivascular epithelium precedes trophoblast penetration through or between the cells of that epithelium to the vessel below, and I surmise that the penetration mechanism involves a softening of the bonds between epithelial cells and between the superficial or cytotrophoblast cells on the outside of the trophoblast knob. Histologically, syncytiotrophoblast within the knob appears to have been extruded through both the cytotrophoblast and the loosened epithelium, and the probable stimulus for the reduction of cell cohesion, pH rise, experimentally causes both epivascular dehiscence of uterine epithelium and discharge of trophoblast knobs. The fact that knobs pop and extrude their contents indicates that they have an internal pressure; it is presumed to be the force driving syncytiotrophoblast through the uterine epithelium. That inference is also drawn from histological observations. The contours of knobs are spheroidal before attachment and are fanned out or pointed afterward, reflecting tension after shrinkage. Stages intermediate between adhesion (with the trophoblast entirely on the lumenal surface of the uterine epithelium) and complete penetration (with the trophoblast all the way through the epithelium) are very rare, suggesting that the penetration is rapid, as befits an explosive extrusion rather than amoeboid probing or other cellular movements that come to mind (Böving 1962, 1963, 1965*b*, 1966).

VI. Trophoblast Spread

The mean width of penetration in a 7-day-old specimen was 17 μ; the comparable dimension a day later was 71 μ. The change has been referred to as "spread" and has been discussed as if there were a 90° change in the direction of invasion. That is putting it too simply except perhaps for some knobs that look as if they have expanded from a single locus (Böving 1962, fig. 20). Other invasions give the impression that there have been multiple penetrations with subsequent entrapment of epithelial cells and cell debris among trophoblastic protrusions through the epithelium (Böving 1962, figs. 7, 8, 9). A conceivably subsequent stage sees the protrusions merge, leaving epithelial cell debris and occasionally whole but rounded-up epithelial cells engulfed in trophoblastic cytoplasm (Böving 1962, fig. 10).

VII. Other Aspects as Seen by Electronmicrography

Entirely different mechanisms of attachment and orientation of blastocysts are revealed by electronmicroscopy. Enders and Schlafke (1969) find an early attachment between microvilli of trophoblast and uterine epithelium progressing by stages to intimate attachment. That attachment presumably is consummated in the fusion reported by Larsen (1961) in which the cell membranes disappear, leaving trophoblastic and uterine nuclei in a merged but not necessarily mixed cytoplasm. By light microscopy one finds fusion present generally in the early mesometrial placentation of the rabbit, but I have seen it only rarely in antimesometrial attachment. That, however, is partly the fault of my own observation. It was not until after Larsen's publication that with extreme care I looked at the antimesometrial attachment with an oil immersion lens and discovered that what I had stated to be a boundary (Böving 1962, fig. 17) was actually a row of vacuoles. After that educational experience I have wondered whether the microvillar attachment and more intimate but still superficial attachment found by Enders and Schlafke might be the stage that I called "adhesion" or whether the microvillar attachment is so delicate that it was lost in my ordinary histological preparations.

Electronmicroscopy and a different species have contributed another new idea. I had reasoned that the first adhesion between blastocyst and uterus in the rabbit, even just by lemmas, provides an anchor that fixes the orientation of blastocyst with respect to uterus. Kirby, Potts, and Wilson (1967) have shown that orientation of the mouse blastocyst with respect to the uterus is not determined until after the zona pellucida or oolemma is lost. They suggested that the inner cell mass travels around the interior of the trophoblast shell once the latter has become fixed to the uterine lumen, although blastocysts transferred to the anterior chamber of the eye consistently appear to have attached by abembryonic pole first.

VIII. Later Stages

Mechanics of blastocyst transport and spacing have been stressed in this paper, whereas the stages of grasp, lemma adhesion, trophoblast adhesion, penetration, and spread have been treated only superficially. The still later stages of trophoblast growth through the stroma and the final healing of the uterine epithelium characteristic of *Homo* have not been mentioned at all, because there is no comparable situation in the rabbit and it has not been explored mechanically in species where it occurs. Although contraceptive efforts are customarily directed to earlier mechanisms in preference to late ones, the invasion should be studied, if not to back up other methods of contraception then for the interest implicit in an invasive growth by a tissue likely to be fractionally different genetically and metabolically from that of the host, like cancer, but with the crucial difference that its invasion normally becomes arrested—a behavior one would like to understand and imitate for therapeutic purposes. Reduced cohesion between cells mentioned in connection with rabbit trophoblast invasion has been attributed to epivascular pH rise accompanying CO_2 blowoff from bicarbonate accumulated in the blastocyst prior to implantation, and that bicarbonate accumulation has been linked to glycolsis imposed by insufficiently aerobic conditions in the uterus before implantation (Böving 1963). Such ideas of blastocysts shifting from a glycolytic to an aerobic metabolism in connection with implantation remained rather speculative until Yochim and Mitchell (1968) applied the new oxygen microelectrode to rat uteri and discovered that in fact oxygen tension was low before implantation and rose markedly at the time of implantation. With that the discussion has moved from biomechanics to chemical aspects of implantation that either serve or are served by the mechanical adjustments to which our concern has been arbitrarily restricted.

IX. Summary

The principal anatomical stages of implantation and the mechanisms responsible for them have been mentioned. Rabbit blastocyst transport, especially the spacing phase of transport, has been emphasized. Quantitative anatomical analysis, *in vivo* observation, and experimentation identify stimulus, effector, action, and result. The result has been considered in detail elsewhere and is only reviewed here, but the three causative elements of the mechanism are discussed in one or another general aspect.

The stimulus for spacing is blastocyst expansion sufficient to distend the uterus, Freezing-point osmometry of blastocyst fluid, uterine fluid, and maternal plasma shows that the expansion cannot be explained by osmosis and must depend on active pumping of water. Excessive uterine distention or blockade, instead of just influencing spacing, destroys blastocysts. The absence of uterine distention in rats, guinea pigs, and other mammals is associated with a different kind of spacing mechanism.

The effector of rabbit blastocyst spacing, uterine muscle, initiates waves of propagated circumferential contraction at both ends of the horn and wherever stimulated by distention within the horn. The waves move in both directions along the uterine horn.

The action of the system propels beads the diameter of blastocysts in the spacing phase of development and does not transport beads the diameter of attached blastocysts. The propulsion contractions initiated by each distending object do not move it but go off in both directions to move its neighbors. Maximum mutual repulsion is seen as equidistant spacing.

The detailed question of whether the propulsion is by fluid pressure or by mechanical squeezing has been probed, and it appears that squeezing is a better explanation than any of the hydraulic explanations that come to mind. A point is made of the decisive importance of measuring not just general intrauterine pressure but the difference in pressure ahead of and behind an object the size of a blastocyst and relating it to the resistance that must be overcome to move that object.

References

Advisory Committee on Obstetrics and Gynecology, Food and Drug Administration. 1968. *Intrauterine contraceptive devices.* Washington: U.S. Department of Health, Education, and Welfare.

Bland, K. P., and Donovan, B. T. 1966. Neural and humoral stimuli from the uterus and the control of ovarian function. In *Egg implantation,* ed. G. E. W. Wolstenholme and M. O'Connor, p. 29. London: J. & A. Churchill.

Böving, B. G. 1952. Internal observation of rabbit uterus. *Science* 116:21.

———. 1956. Rabbit blastocyst distribution. *Amer J Anat* 98:403.

———. 1959. Implantation. *Ann N Y Acad Sci* 75:700.

———. 1960. L'interaction entre les mécanismes physiologiques intervenant dans l'implantation du blastocysts chez la lapine. In *Les fonctions de nidation utérine et leurs troubles,* p. 103. Paris: Masson et Cie.

———. 1962. Anatomical analysis of rabbit trophoblast invasion. *Contrib Embryol* 37:33.

———. 1963. Implantation mechanisms. In *Mechanisms concerned with conception,* ed. C. G. Hartman, p. 321. New York: Pergamon Press.

———. 1965*a*. Implantation. In *Fetal homeostasis,* ed. R. M. Wynn, 1:38. New York: N.Y. Acad. Sci.

———. 1965*b*. Anatomy of reproduction. In *Obstetrics,* ed. J. P. Greenhill, p. 3. Philadelphia: Saunders.

———. 1966. Some mechanical aspects of rabbit trophoblast penetration of uterine epithelium. In *Egg implantation,* ed. G. E. W. Wolstenholme and M. O'Connor, p. 72. London: J. & A. Churchill.

Cole, R. J. 1967. Cinemicrographic observations on the trophoblast and zona pellucida of the mouse blastocyst. *J Embryol Exp Morph* 17:481.

Corner, G. W. 1928. Physiology of the corpus luteum: I. The effect of early ablation of the corpus luteum upon embryos and uterus. *Amer J Physiol* 86:74.

Csapo, A. 1955. The mechanism of myometrial function and its disorders. In *Modern trends in obstetrics and gynecology,* ed. K. Bowes, 2d Ser., p. 20. London: Butterworth.

Daniel, J. C., Jr. 1963. Some kinetics of blastocyst formation as studied by the process of reconstitution. *J Exp Zool* 154:231.

Enders, A. C., and Schlafke, S. 1969. Cytological aspects of trophoblast-uterine interaction in early implantation. *Amer J Anat* 125:1.

Forbes, T. R. 1953. Pre-ovulatory progesterone in the peripheral blood of the rabbit. *Endocrinology* 53:79.

Frazer, J. F. D. 1955. The site of implantation of ova in the rat. *J Embryol Exp Morph* 3:332.

Heap, R. B. 1962. Some chemical constituents of uterine washings: A method of analysis with results from various species. *J Endocr* 24:367.

Hertig, A. T.; Rock, J.; and Adams, E. C. 1956. A description of 34 human ova within the first 17 days of development. *Amer J Anat* 98:435.

Ivy, A. C.; Hartman, C. G.; and Koff, A. 1931. The contractions of the monkey uterus at term. *Amer J Obstet Gynec* 22:338.

Kao, C. Y. 1959. Long-term observations of spontaneous electrical activity of the uterine smooth muscle. *Amer J Physiol* 196:343.

Kirby, D. R. S.; Potts, D. M.; and Wilson, I. B. 1967. On the orientation of the implanting blastocyst. *J Embryol Exp Morph* 17:527.

Krehbiel, R. H., and Plagge, J. C. 1962. Distribution of ova in the rat uterus. *Anat Rec* 143:239.

Kulangara, A. C. 1960. The macromolecular environment of the rabbit embryo before and during implantation. In *Year book 59,* p. 361. Washington: Carnegie Institution of Washington.

Larsen, J. F. 1961. Electron microscopy of the implantation site in the rabbit. *Amer J Anat* 109:319.

Lutwak-Mann, C. 1959. Biochemical approach to the study of ovum implantation in the rabbit. *Mem Soc Endocrin* 6:35.

Markee, J. E. 1944. Intrauterine distribution of ova in the rabbit. *Anat Rec* 88:329.

Momberg, H., and Conaway, C. 1956. The distribution of placental scars of first and second pregnancies in the rat. *J Embryol Exp Morph* 4:376.

Pallie, W.; Corner, G. W.; and Weddell, G. 1954. Nerve terminations in the myometrium of the rabbit. *Anat Rec* 118:789.

Psychoyos, A. 1960. La réaction déciduale est précédée de modifications précoces de la perméabilité capillaire de l'utérus. *C R Soc Biol* (*Paris*) 154:1384.

Schofield, B. M. 1954. The influence of estrogen and progesterone on the isometric tension of the uterus in the intact rabbit. *Endocrinology* 55:142.

———. 1955. The influence of the ovarian hormones on myometrial behavior in the intact rabbit. *J Physiol* (*London*) 129:289.

———. 1957. The hormonal control of myometrial function during pregnancy. *J Physiol* (*London*) 138:1.

Tuft, P. H., and Böving, B. G. 1969. The uptake of water by the rabbit blastocyst. In *Year book 67,* p. 455. Washington: Carnegie Institution of Washington.

———. 1970. The forces involved in water uptake by the rabbit blastocyst. *J Exp Zool* 174:165.

Yochim, J. M., and Mitchell, J. A. 1968. Intrauterine oxygen tension in the rat during progestation: Its possible relation to carbohydrate metabolism and the regulation of nidation. *Endocrinology* 83:706.

26

The Biophysics of Nidation

John T. Conrad
Department of Physiology and Biophysics
Department of Obstetrics and Gynecology
School of Medicine
University of Washington, Seattle

Nidation is usually defined as "the embedding of the early embryo in the uterine mucosa leading toward the formation of the decidua oapsularis." The definition is rather narrow and does not cover the subject in its entirety. What goes on before and the many factors involved in presenting an embryo to a receptive uterine mucosa are an essential part of the process, and are related to the biophysics of excitable membranes, contraction of smooth muscle, and changes induced in the ovum as a result of activation. All of these are brought together in the process of ovulation, tubal transport, receptivity of the uterus for implantation, and finally nidation itself. Since it is not practical to discuss all of these problems, only a part of the process will be covered.

The road leading to nidation is a fairly straight path and may be compared more to a sports-car rally than to a Sunday drive in the country. By calling it a rally, I imply that not only is the distance traveled important, but time also is of the essence. The time of each event determines whether a particular ovum is released, fertilized, and nidated. The ovum, smooth muscle contraction, and smooth muscle electrophysiology, important in relation to ovulation, tubal transport, and implantation itself, are the topics discussed in this chapter.

I. The Ovum

As is well known, the reproductive structures of vertebrates undergo cyclic variations of both growth and retrenchment. The ovum has been studied as a chemical and as a histological element, but rarely has a study been made of its membrane characteristics. True, the membrane potentials have been measured (Hori 1958; Maeno 1959; Ito 1962; Kanno and Lowenstein 1963; Morrill and Watson 1966; Morrill, Rosenthal, and Watson 1966), but the full impact of this type of experimentation has yet to be felt.

The measurement of membrane potentials and their manipulation by pharmacological and physical methods may provide a great deal of insight into the mechanism and possibly into control of fertilization. Morrill (1965) found that upon ovulation the water and electrolyte shifts in eggs of *Rana pipiens* reflect the action of ovulatory hormones. The evidence that these hormones affect the egg is circumstantial, however, and a direct measure of a change as a result of the release of pituitary hormones has not been accomplished.

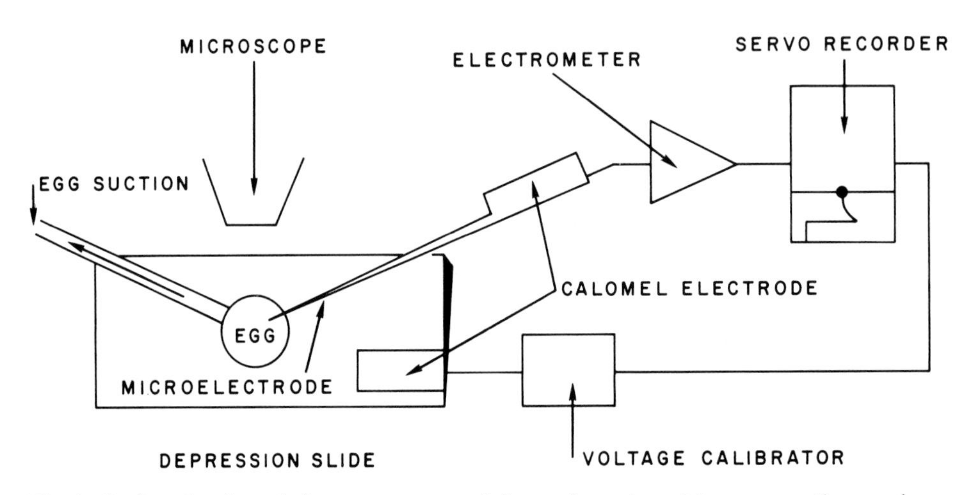

Fig. 1. Outline drawing of the arrangement of the equipment used to measure the membrane potential of frog oocytes.

The membrane potentials may give a degree of insight into the changes in permeability that come about during the reproductive cycle. As an example I would like to present some of the work done in my laboratory (Dawson and Conrad 1969).

This work was an attempt to measure the effect of the pituitary factor luteinizing hormone (LH) and human chorionic gonadotrophin (HCG) on the membrane potential of the unovulated eggs of the frog, *Rana pipiens*. The egg was held by means of a vacuum probe and penetrated by a microelectrode of 5–20 mohm resistance. Figure 1 illustrates in schematic form the general arrangement of the equipment. Upon penetration of the egg membrane by the microelectrode, with the egg in isotonic Ringer's solution, a negative membrane potential developed. This potential varied with the time of year and, if the eggs were retained in the animal, fell to 0 shortly after the normal mating time. The seasonal variations noted in the value of the membrane potential corresponded nicely to the physiological rhythms found in the frog. A dip occurred in January at the height of the winter, possibly indicating hibernation. A late dip in April and May corresponded to the absorption of unshed eggs by the animal. The dip in potential may be prevented by refrigerating the frogs. Seasonal variations in egg potentials may serve to indicate why investigators have obtained variable mean resting potentials, especially if all of the work was confined to a single month.

An experiment with HCG is recorded in figure 2. In this particular case the hormonal material caused a depolarization of the egg membrane. Both materials (LH and HCG) demonstrated a definite effect at all dosage levels.

The purpose of these experiments was to delineate the process by which the

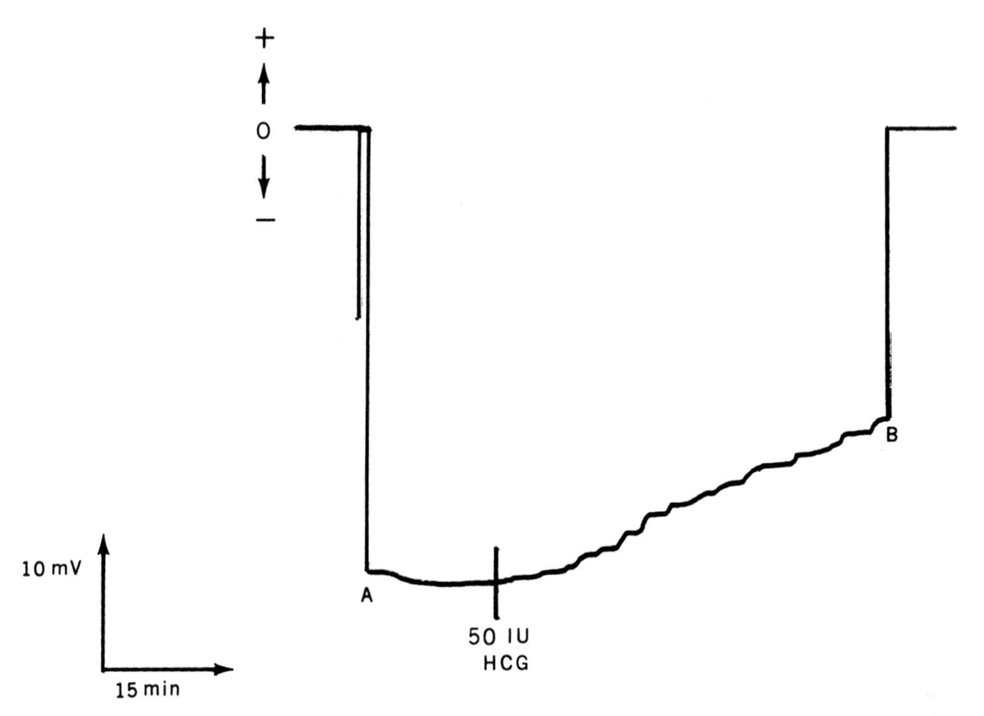

Fig. 2. Representative tracing demonstrating the depolarizing effect of human chorionic gonadotrophin upon the frog oocyte membrane potential. At point *A* the microelectrode penetrated the egg membrane; at point *B* the microelectrode was pulled out of the cell.

quiescence of the maturing oocyte is broken by ovulation. Previous work (Kanno and Lowenstein 1963; Ashman, Kanno, and Lowenstein 1964; Morrill 1965; Morrill and Watson 1966) has been directed at measuring the membrane potential of eggs in the oocyte stage, upon ovulation, following fertilization, and during subsequent cell division. Our particular interest has been the stimulus to egg development about the time of ovulation. It appears that LH (or HCG) is an adequate stimulus to cause changes in the egg membrane permeability as evidenced by the depolarization induced by these hormones.

Although depolarization induced by LH or HCG is only an indication of what is going on inside the cell, it suggests that membrane reactions may control subsequent changes in the cell.

II. Smooth Muscle

Since the ovum is released and transported into the oviduct, muscular contraction becomes of great importance. The contraction cycle may be discussed from various viewpoints; those considered here are bioelectrical, contractile, and viscoelastic.

A. *Bioelectrical Considerations*

Although smooth muscle is considered different from other muscular types, overemphasis has been placed on its unique properties, variability and lability. As Bohr (1964) has pointed out, in overall construction and in basic function smooth muscle resembles skeletal muscle. There are differences as well as similarities, and to these we address ourselves.

The basic features of an idealized smooth muscle structure are contained in figure 3. Here we find that the individual cells may be connected by forking one into another (Rhodin 1962), by fusions of the plasma membranes of adjacent cells, that

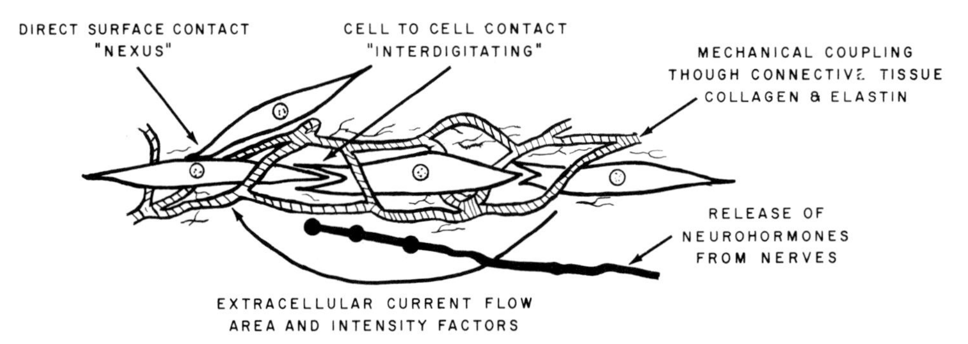

Fig. 3. Idealized smooth muscle structure. Not shown are the small intercellular bridges mentioned in the text.

is the nexus (Dewey and Barr 1962, 1964), or by intercellular bridges (Bergman 1958; Mark 1956; Thaemert 1959). From a functional point of view the ability of smooth muscle structures to conduct an action potential depends, to a great degree, upon the type of connection found between cells. Cell separation by plasma membranes and extracellular space was thought to be a serious impediment to the conduction of the action potential (Thaemert 1959). This may not be true on several counts. Cable properties are found in both "unitary" and "multiunit" smooth muscles (Abe and Tomita 1968). (The classification of "unitary" and "multiunit" is that of Bozler [1948], so named because "unitary" are self-stimulatory and "multiunit" are under nervous control.) The existence of cablelike properties indicates that the individual cells are connected electrically and that the normal parameters of excitation could be followed even without cellular continuity (shown as an extracellular current flow in figure 3).

The importance of the type of connection between cells in smooth muscle structures cannot be overemphasized. On these connections depends the conduction of the action potential from one area to another. If one accepts the concept that these contacts are labile (Lewis 1920; Bergman 1958; Thaemert 1959) and can withdraw upon slight provocation, many of the variations in the contractile pattern during hormone treatment can be explained. Work done in our laboratory (Tahmoush et al. 1968) indicates that conduction velocity in the rat uterus is greatly enhanced during labor. Changes brought about by estrogen injections can increase the velocity of action potential conduction over short distances in the nonpregnant rat uterus, but this increase is not as profound as that induced by labor. Table 1 gives some of the data on these experiments. Questions remain: Are these changes due to hormone induced lability of the smooth muscle interconnections? Or are they membrane effects on the individual cells? Is the increased conduction velocity a measure of the hypertrophy during labor, that is, fewer, but larger, cells between any pair of electrodes?

It may be pointed out that it is difficult to determine the behavior of a whole tissue or organ on the basis of a single cell. Figures 4 and 5 are representative tracings of microelectrode-recorded potentials from human uterine muscle. These tracings are taken from strips of human myometrium of nonpregnant patients; they show

TABLE 1

CONDUCTED ACTION POTENTIALS: *in Vitro* RAT UTERUS
(Extracellular Recording with Multichannel Electrodes)

Sample	Mean cm/sec	St. Dev.
Control		
2 channels, 88 records	2.92	± .757
3 channels, 78 records	3.63	± .536
Primed with Premarin		
2 channels, 182 records	4.75	±3.30
3 channels, 201 records	4.83	±1.52
Pregnant (during labor)		
3 channels, 168 records	7.50	±2.74
4 channels, 420 records	7.95	±2.29
5 channels, 35 records	5.92	±3.56

very little synchrony between the tension and bioelectrical channels. The stimulus for contraction is usually a brief tetanus, involving various areas of the strip. The overall pattern of contraction may be a smooth curve or a series of summed, irregular contractions, depending upon how the bioelectrical tetanus travels down the tissue. Therefore not only are the membrane reactions of the individual cells important, how they are connected is equally important. Sexual steroids may work through the electrical connections between cells of the reproductive smooth muscle by enhancement of labile intercellular bridges, by changes in extracellular resistances, or by modifying the length to width ratio of the cells and thus the cable properties of the tissue.

As mentioned previously, the membrane reactions of the individual cells are influential in controlling smooth muscle responses. One of the more controversial concepts in uterine muscle physiology is that concerning the action of the female sex hormones on excitation. One of the most heavily documented hypotheses is that of Csapo (1955, 1961, 1962). This hypothesis depends upon the alleged ability of progesterone to interfere with excitation by hyperpolarizing the muscle cell membrane in such a manner as to stop contractions. In theory this should make it more difficult to depolarize the myometrial cell to the threshold level for spike generation.

The onset of labor is signified by a lowering of progesterone production by the placenta which leads to a release of the "block." Normal contractions may then begin. A "local" block beneath the placenta is another aspect of this theory. Progesterone produced in the placenta sets up a diffusing gradient which has a strong inhibitory effect in the area. The whole purpose of these inhibitory measures is to prevent contractions which might serve to compromise pregnancy and viability of the fetus.

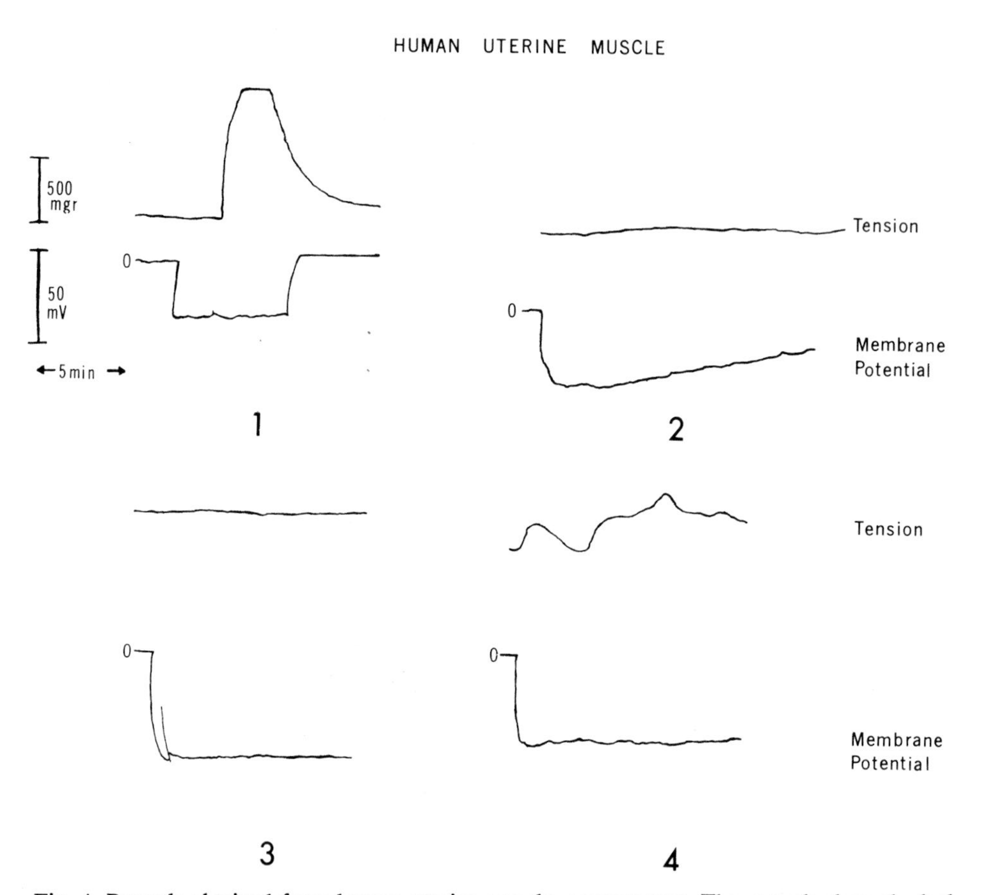

Fig. 4. Records obtained from human uterine muscle, nonpregnant. The records show the lack of agreement between the isometric tension traces and the bioelectrical channel. Note in 3 the appearance of an action potential without any indication of tension development. Conversely, in 1 note the bioelectrical quiet and the development of considerable tension.

In opposition to this hypothesis has been the work of Kao and Nishiyama (1964). Their findings appear to question any increased membrane potential of rabbit uterine muscle in the estrogen-dominated animal as a result of progesterone application. Additional questions are raised by Kao (1967) as to the experimental evidence for this theory.

Conversely, estrogen has been shown to increase the resting potential of the immature myometrium of rabbits and rats (Kao 1967). The most dramatic rises

HUMAN UTERINE MUSCLE

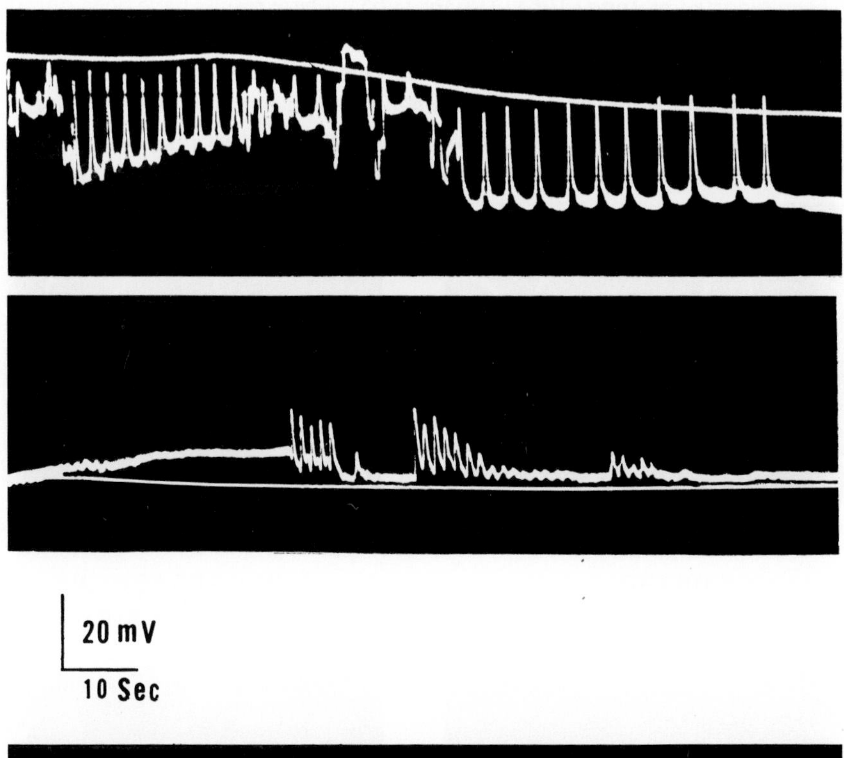

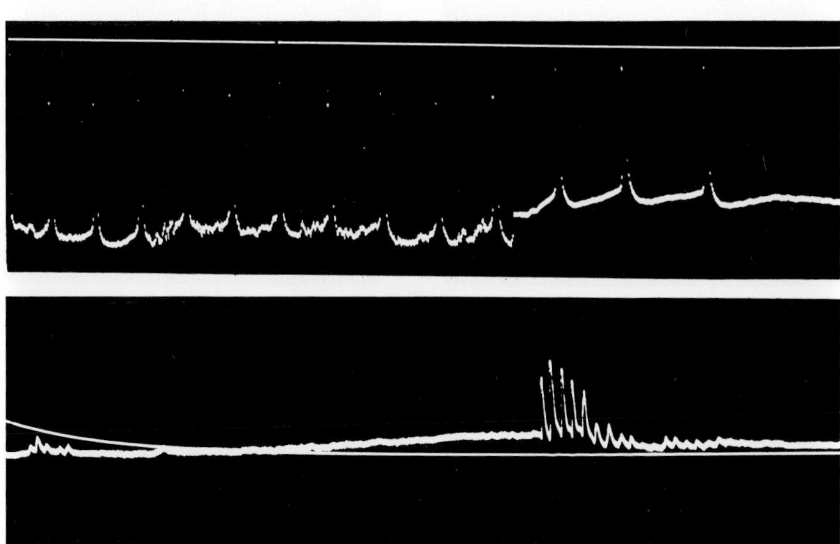

500
mgr

Fig. 5. Examples of the types of records obtained from human uterine muscle. The smooth traces represent tension, the irregular ones membrane potentials. The bottom photographs were obtained when the muscle strip was relaxing, the top two during a contraction. Human uterine muscle, nonpregnant.

have been reported by Jung (1961, 1963*b*, 1965). These results have been questioned by Kao (1967) on the basis of inconsistency with those of other investigators and technical problems inherent in the use of microelectrodes.

It has been stated by Jung (1963*a*) that although estradiol increases resting potential, progesterone blocks muscle activity. The evidence that progesterone caused a

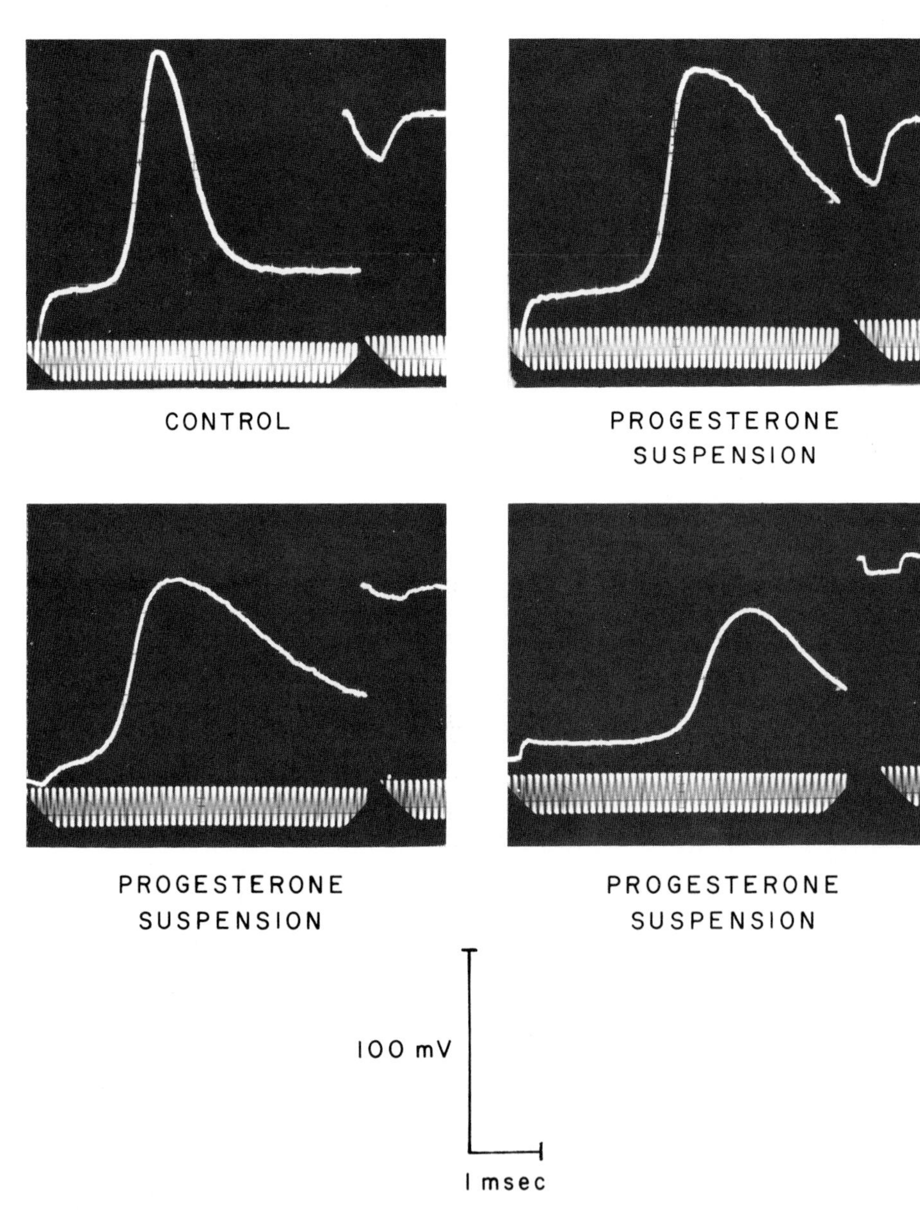

Fig. 6. The results of the application of an aqueous suspension of progesterone to a frog sartorius muscle. There is the development of slowing of the rising phase of the action potential. The zero potential level and the stimulus artifact may be noted by the portion of the oscilloscope trace in the upper right-hand corner of each photograph.

deficiency in the velocity of the rising portion of the action potential led him to believe that the level of the resting potential alone does not cause the block in activity; in addition to the alteration of a membrane potential there is a block in the sodium carrier mechanism.

Our own experimental results appear to support Jung's modification of Csapo's conclusions. Technical problems in working with muscle led us to utilize the muscle cells of the frog sartorius (Conrad, Aoba, and Shimizu 1969) for these experiments.

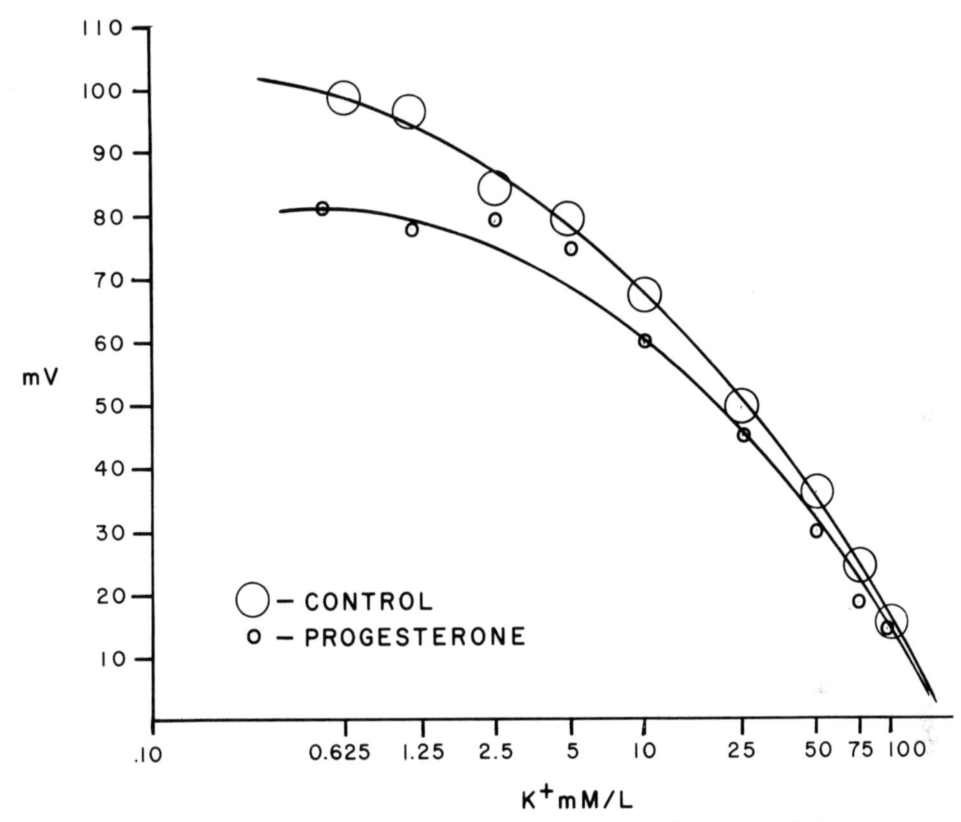

Fig. 7. The effect of an aqueous suspension of progesterone upon the relationship between membrane potentials and the log of the extracellular potassium concentration, frog sartorius muscle.

We administered estradiol monosulfate (E_2S), pregnenolone sulfate, and crystalline progesterone. Pregnenolone and progesterone slowed the rising stroke of the action potential, and decreased the excitability and conduction velocity of the action potential. The effect of the progestins may be noted in figure 6. Estradiol (E_2S) had a dual effect: at low doses (5 μ/cc) we noted multiple action potentials and other indications of hyperexcitability, that is, increased conduction velocity; higher doses (above 100 μ/cc) did not show this behavior.

Progesterone (microcrystals seeded over the entire muscle) likewise affected the relationship of the membrane potential to log concentration of external potassium. Figure 7 summarizes the results of these experiments. Progesterone appears to lower the resting potential in this situation below that of the normal curve, a type of behavior seen when residual sodium leakage across the membrane is increased. Although this is a most extreme method of applying the hormone, that is, over the whole surface, it may indicate the method by which progesterone exerts its action.

It is well known that the sodium permeability of smooth muscle cells is much higher than that of other excitable cells (Casteels 1967). Any treatment that affects sodium permeability would immediately reveal itself as a change in the resting potential. However it might be only secondary to the changes in the action potential.

It is of more than passing interest to note that the effect of Premarin (conjugated steroids from horse urine, Ayerst) and E_2S upon uterine muscle is to block activity in high doses and enhance activity at low doses (Mossman and Conrad 1967).

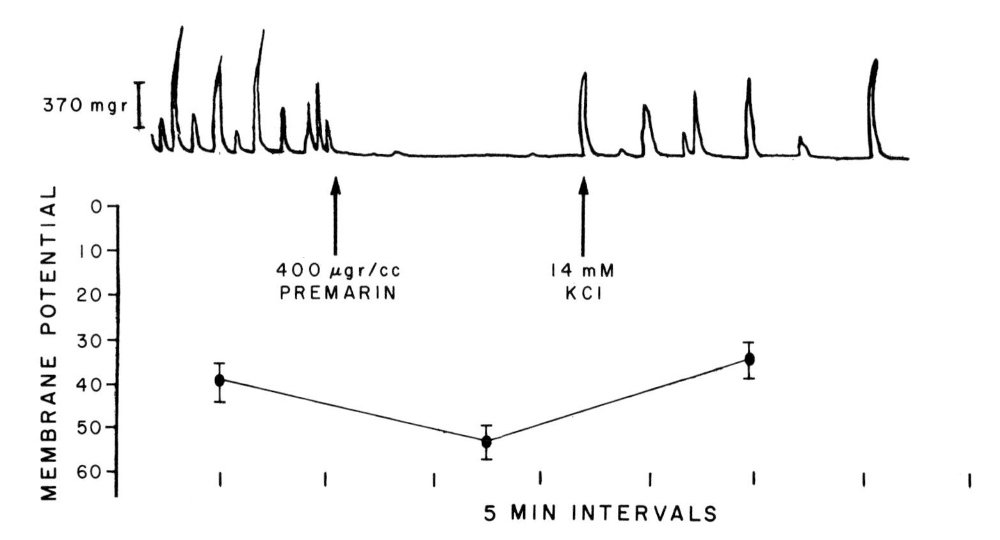

Fig. 8. The hyperpolarizing effect of a mixture of estrogenic materials, Premarin. The vertical bars indicate the standard deviation of the mean values.

Microelectrode studies indicate that during the blockage of activity there is a rise in the resting potential level. Figure 8 shows this effect in a single experiment (Conrad, unpublished observations). Depolarization with KCl lowers the resting potential and contraction begins once again.

The results of these experiments imply that, although both progesterone and estrogen can affect resting potentials, they work by different mechanisms. Progesterone appears to affect the genesis of the action potential, possibly through the availability of sodium carriers. Estrogen, on the other hand, works through the mechanism of the resting potential; potassium permeability may be the system affected in this case, or the pumping rate of sodium extrusion.

Work of this type is subject to one major criticism. The doses of drugs are much larger than those found *in vivo*. It may be that these results are completely meaningless in terms of physiological function. Certainly they should be repeated *in vivo* with physiological amounts of the steroids.

B. *Smooth Muscle Contraction*

Another problem of crucial importance is that of the contractile apparatus of muscle. Through the pioneering work of H. E. Huxley much is known of skeletal muscular contraction. A theory of skeletal muscular contraction was developed by use of the so-called interdigitating filament model (Hanson and Huxley 1955). It states that the contractile material in skeletal muscle is composed of overlapping filaments of actin and myosin arranged in a fashion commensurate with the findings of the electron microscopists and X-ray crystallographers Hanson and Huxley (1955). This array is shown in figure 9. Muscle contraction causes the filaments to slide by one another so that the actin is drawn into the array of myosin molecules.

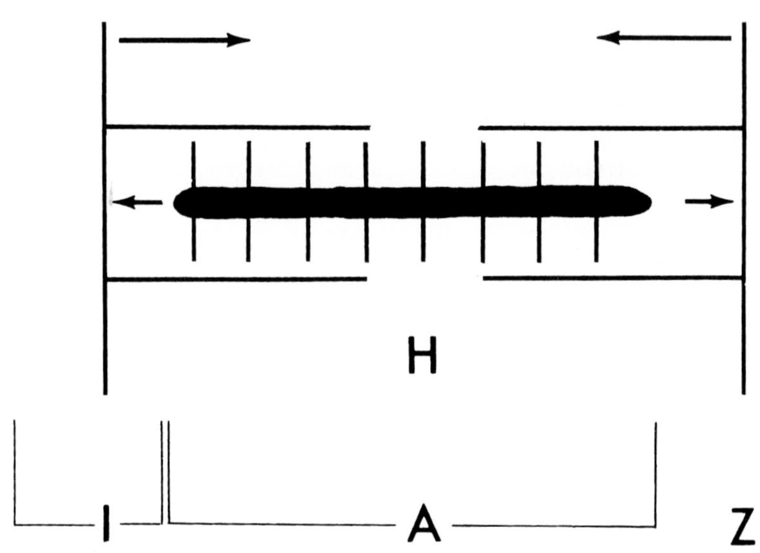

Fig. 9. A drawing showing how the movement of the filaments in skeletal muscle can produce contraction of the muscle. The movement is one of "interdigitation" so that the *Z* lines move closer to one another. The letters *Z, A, H,* and *I* denote these histological features of skeletal muscle.

These molecules do not, of themselves, change in length but they "interdigitate" so that there is an apparent shortening; the z bands approach one another and end-to-end tension is produced (Hanson and Huxley 1955; Huxley 1956, 1957).

Concern with the mechanical coupling of the crossbridges from myosin to actin and how they can produce transilatory movement are the significant points in this theory. These crossbridges serve to link the actin and myosin molecules and to bridge the gap between them. They attained added significance when it was realized that these structures were the only conceivable method by which transilatory movement could be produced (Huxley 1957). Bolstering the crossbridge theory, Szent-Györgyi (1953) found that the crossbridges were the seat of actin-combining and adenosine triphosphatase activity. They however, do not appear to move more than about 100 Å (Huxley 1956). A paradox presented itself: movements larger than 100 Å occurred although crossbridges cannot move more than 100 Å. Two possible explanations were examined: (1) cyclical movement of the crossbridges producing a transilatory movement on one flexure and a detachment, free motion, rebinding on the return stroke and (2) fixed crossbridges and relative movement of the actin along the binding sites of these crossbridges. The available evidence could not resolve this problem.

In a recent paper Huxley (1969) attempted to resolve this problem. By means of X-ray diffraction techniques he constructed a model which may be useful to those interested in both skeletal and smooth muscle contraction. The model shows that the linkage of the crossbridges contained two flexible areas. The light meromyosin (LMM) was bonded to the thick myosin filament, and the linear heavy meromyosin (HMM) tilted out at an angle owing to a flexible linkage. The globular part of the heavy meromyosin, containing the adenosine triphosphatase and actin binding sites, was connected to the tip of the linear heavy meromyosin by a flexible linkage. The two flexible linkages made possible the attachment of the thick filaments to the actin filament over a considerable range of different actin-myosin spacings while retaining their orientation relative to the actin.

It was Huxley's main thesis that the crossbridges, connected by flexible, inelastic linear heavy myosin threads to the myosin backbone, are based *on the actin.* It is this concept to which emphasis should be shifted. Linkage was under tension when contraction occurred, insuring transmission of the generated force. The globular head of the myosin molecule was rigidly attached to the actin filament with a subsequent change in the angle of attachment associated with the splitting of adenosine triphosphate.

These ideas on skeletal muscle may now be applied tentatively to smooth muscle contraction. Csapo (1948, 1950*a, b*) is generally conceded to be the first investigator to study the contractile proteins of the muscle cells of the uterus. He was able to demonstrate the presence of actin and myosin and an increase in actomyosin during pregnancy. He found (Csapo 1950*c*) that ovariectomy decreased actomyosin content and ATPase activity of rabbit uteri and that estrogen injections reversed and even increased the actomyosin and the activity.

It is the general consensus that there is only one actomyosin in the myometrium (Needham and Shoenberg 1967); it differs from skeletal actomyosin. The myosin obtained from uterine actomyosin has a relative viscosity only slightly lower than that of skeletal muscle myosin. This myosin can react with skeletal muscle actin in a quantitatively normal way to give an increase in viscosity that falls when ATP is added (Needham 1962). There is a difference in the amount of globulin proteins obtained from the uterus as compared to skeletal muscle; 1.5–2.5 mg nitrogen per g of wet weight versus 10–11 mg per g of wet weight in skeletal muscle (Needham and Cawkwell 1956; Needham and Williams 1959).

Although as demonstrated above we know the contractile proteins exist in smooth muscle, to prove that they have the properties of an organized array is a more difficult matter. The anatomy of smooth muscle has been studied intensively over the past few years (Shoenberg 1958; Weinstein and Ralph 1951); Caesar Edwards, and Ruska 1957; Mark 1956; Rhodin 1962; Gansler 1956; Nemetschek-Gansler 1967); it is generally concluded that it is very difficult to obtain consistently good images of myofilaments in electron micrographs. In especially well fixed and stained tissues the spindle-shaped muscle cells do show a system of filaments, tightly packed, lying parallel to the long axis of the cells. These filaments are similar in size to the actin filaments in striated muscle (50–80 Å). Great difficulty has been encountered in recognizing filaments which correspond to the myosin of striated muscle; hence it has become a conundrum to piece together the bits of information to form a model of smooth muscle contraction.

There appear to be three explanations for the lack of myosin filaments. The first may be that actin is always in filamentous form, but myosin may be dispersed and form aggregates when properly stimulated (Needham and Shoenberg 1967). The myosin filaments that have been obtained are shorter and thicker than those of skeletal muscle, but they may fit this description only under rather special conditions. There is presumptive evidence for this: an electrical tetanus is usually necessary to begin a contraction cycle in smooth muscle structures; the period the tetanus covers may be the time required for the myosin to aggregate. On the other hand the electrical tetanus may represent only conduction and excitation phenomena in relationship to the recruitment of additional fibers.

The second explanation may lie in syneresis. Syneresis may account for the contraction rather than the sliding filament as put forward by Gansler (1961). A test of this hypothesis might consist of the measurement of both area and volume of cells and the movement of water under contraction. Mullins and Guntheroth (1965) showed by a model, containing a balloon inside a knitted stocking, that the force of expansion may be transferred from the surrounding network to the load. Thus, mechanically, cellular expansion and not end-to-end movement of the cell may be the motive force in smooth muscle contraction. Their theory would be consistent with a model showing syneresis behavior if the syneresis expanded the cross-sectional diameter.

The third possibility is to redefine some of the tenets of the sliding filament hypothesis. The lack of clearly defined myofilaments in an orderly array and points of attachment, assuming they exist, appears to be the vulnerable point in the hypothesis. Rosenbluth (1965) offered evidence that certain myofilaments (paramyosin in the case of Aplysia muscle) and thin filaments in frog jejunum are obliquely located in the muscle. This arrangement of fibrils could account for an ability to generate up to 4 times as much tension as that in a striated series fiber. The important point, however, is that the points of attachment are such that the contractile material exerts force on the plasma membrane at the sides of the fiber.

The concept that the parallel fibrils can generate more tension for a given series arrangement is derived from the fact that forces acting in parallel are additive. Oblique orientation allows a greater spread of lengths without much of a strain on the contractile material. In other words a broader length tension curve, as we will discuss later, is a property of smooth muscle.

The lack in smooth muscle of the orderly array of myofilaments found in skeletal muscle may be only an artifact of the fixation process related to the lability of myosin. However, it may be that there are no thick filaments in smooth muscle. Although myosin is present, it is in the form of crossbridges made of light and heavy meromyosin. Instead of being anchored onto a thick filament of myosin, as in skeletal muscle, it is connected to the membrane on the dense bodies of the smooth muscle cell. The dense bodies may be a special form of myosin (Elliott 1964). The crossbridges have two flexible linkages (Huxley 1969). These linkages remain fixed to the actin and produce longitudinal force in the face of large changes in interfilament separation; they are under tension, thus insuring force transmission during contraction; their flexibility guarantees that the muscle will generate force over a wide range of lengths. A sketch of such a model for smooth muscle may be found in figure 10.

C. *Smooth Muscle Viscoelasticity*

In addition to electrophysiology and contraction of smooth muscle, we must consider its viscosity. Both active contraction and the unequal distribution of forces are required to move the fetus into position for presentation to the birth canal. In

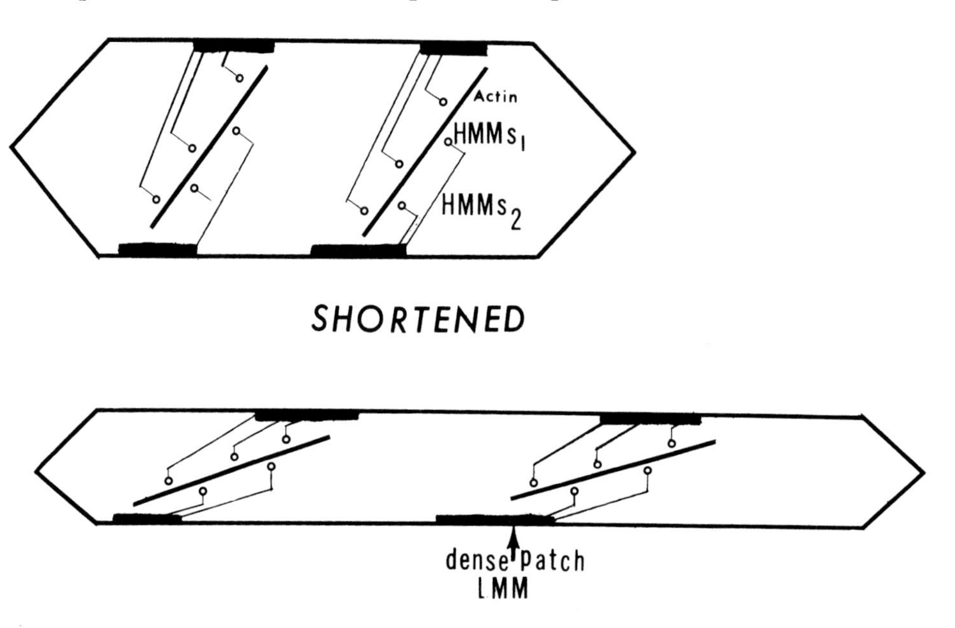

Fig. 10. A sketch of how smooth muscle may contract. $HMMS_1$ refers to heavy meromyosin S_1 and $HMMS_2$ to the heavy meromyosin S_2 fragments. There are linkages at each of these fragments. The dense patches may be made up of material resembling light meromyosin or are just binding points on the cell membrane. The oblique orientation of the actin would allow for tension production over a wide range of lengths, owing to the angular motion of the actin.

order to bring about the coordination of movement preparatory to expulsion of the fetus the lower uterine segment must be stretched out while remaining relatively flaccid and producing only weak contractions.

There are three mechanisms which may account for the ability of the uterus to adapt to volumetric changes without rupturing. The first of these is stress relaxation (Conrad, Kuhn, and Johnson 1966), illustrated in figure 11. Essentially it is a loss of tension over a period of time after tension has been built up by some process, such as passive stretch. This viscoelastic property enables the muscle to remain in a relatively unstressed state in spite of continuous elongation.

The second mechanism is the relatively flat length-tension curve of uterine muscle whereby a considerable increment of length results in but a small production of passive tension (Conrad et al. 1966). This phenomenon is particularly evident in uterine muscle from pregnant patients where the passive length-tension curve is flatter and less steep than the nonpregnant curve. In the human the flatness of the length-tension curve may prevent a buildup of passive tension in the lower uterine segment in addition to the tension relieving property of stress relaxation. Figure 12 is a graphic representation of the results of experiments of Conrad et al. (1966).

The position of the muscle in the active length-tension curve (Conrad and Kuhn 1967) is the third mechanism, the effect of which is pointed out in figure 13.

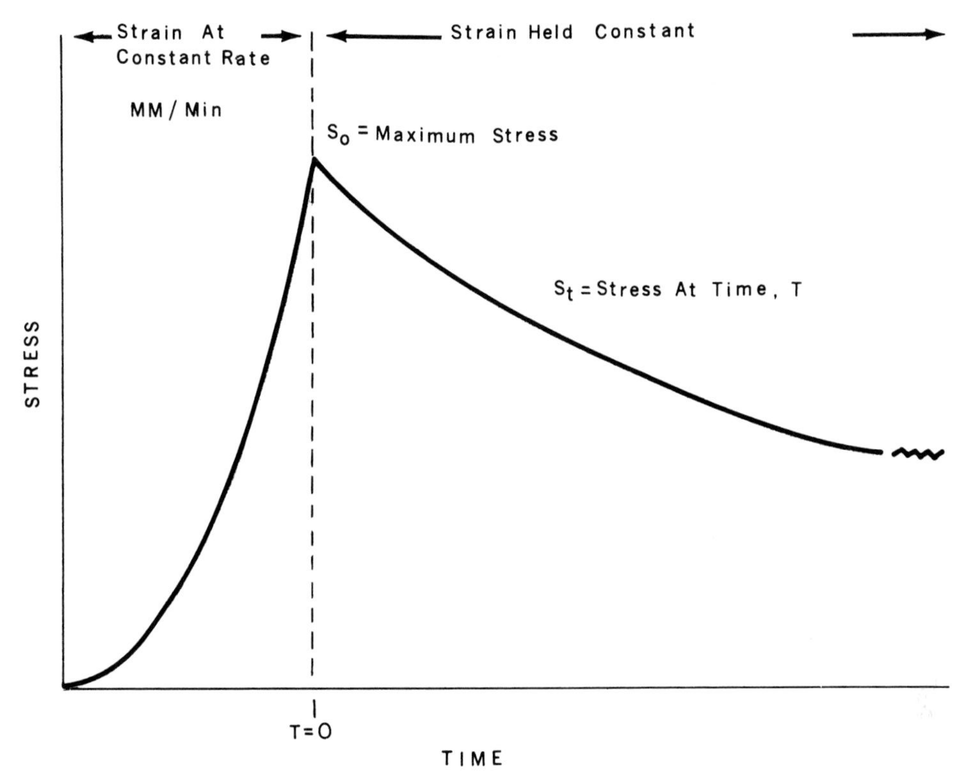

Fig. 11. A diagram illustrating stress relaxation in human uterine muscle. When a strip of muscle is stretched (strained), the stress builds up. If the muscle is then held in a stretched position, the built-up stress declines over a period of time, that is, stress relaxation.

The *in situ* length of this muscle is 30 mm. It should be noted that the muscle is operating below its optimum point for contraction in this patient. That the uterine muscle from pregnant patients is able to contract maximally over a wider range of lengths than that from nonpregnant patients is an interesting observation made in this research and appears to be due to the fact that the development of active tension is related to the passive tension already developed, rather than to length. Therefore with a flatter passive length-tension curve in the pregnant condition there is a greater development of active tension at longer lengths. However if the muscle is overstretched the ability to develop contractile tension declines. The lower uterine

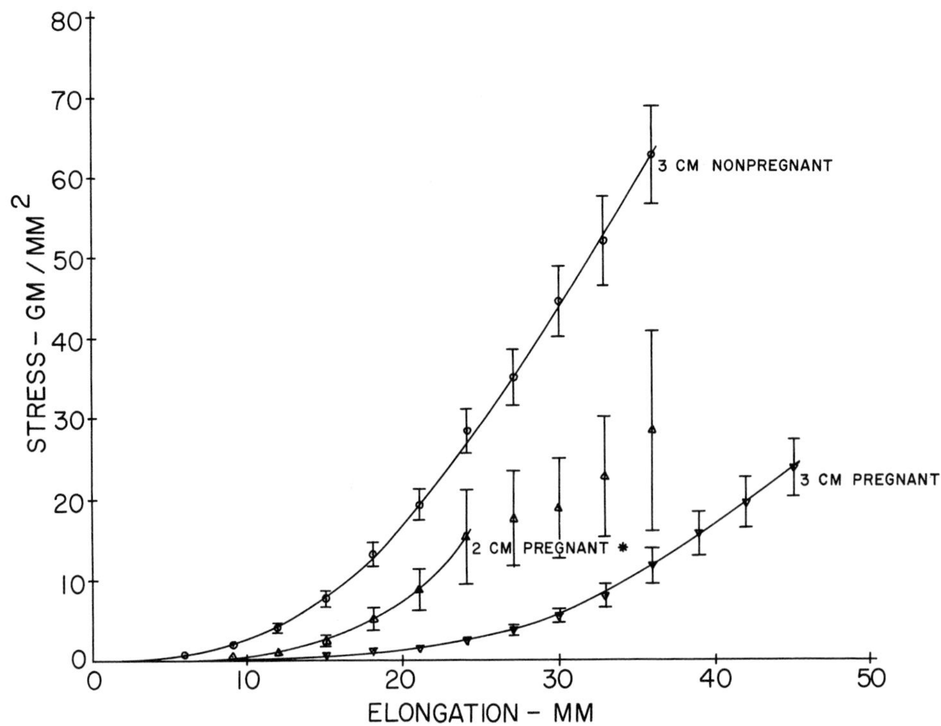

Fig. 12. The passive length-tension curve obtained from strips of human uterine muscle. The 2 cm strips represent an attempt to account for the passive recoil of about 1/3 when the muscle was removed from a pregnant patient. The curve was not continued beyond this point because of yield and breakage. The vertical bars represent the standard error of the mean for each point in question. (Reprinted through the courtesy of the *American Journal of Obstetrics and Gynecology*)

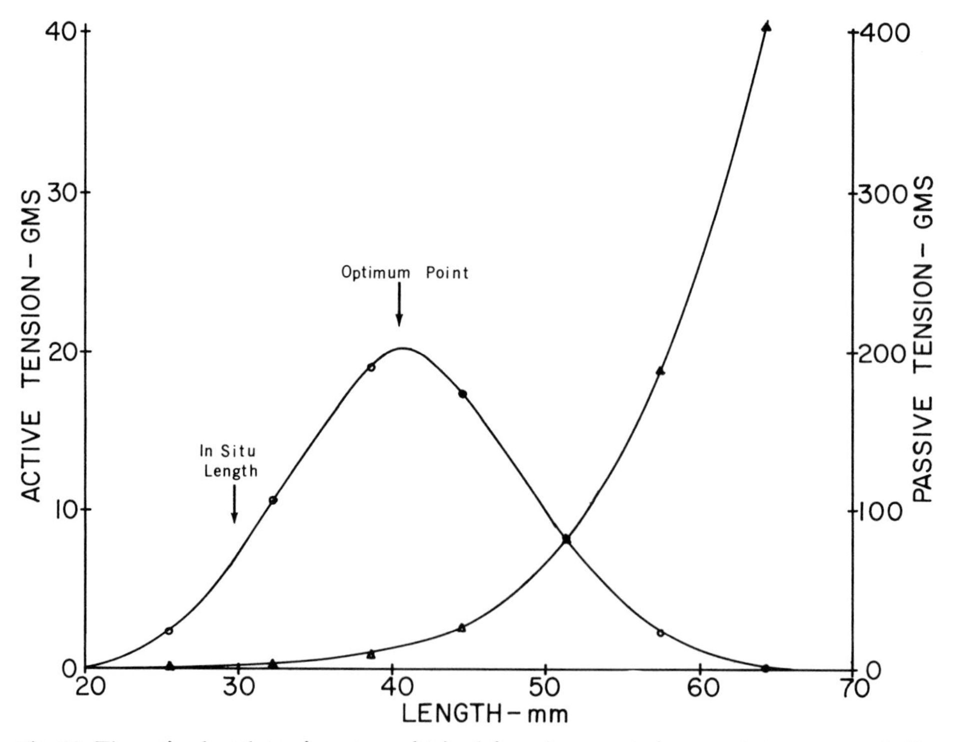

Fig. 13. The active length-tension curve obtained from human uterine muscle, nonpregnant. The *in situ* length is marked on the graph and it can be seen that this point lies below the point of maximum tension production (optimum point). (Redrawn from Conrad and Kuhn 1967)

segment, stretched the most, produces less active tension because it is operating to the right of the "optimal" point.

III. The Problems of Nidation

We have now come full circle, back to nidation. We have attempted to show that the ovum is activated and that this activation can be measured by biophysical means. These measurements are an indirect method of establishing permeability changes, and more work is needed to determine their significance in reproduction.

The problem of the contraction of smooth muscle and its control by electrophysiological means may be closely bound to the problems of embryo orientation, adhesion, penetration, and invasion during implantation. According to Böving (1965) muscular contraction is important in blastocyst transport within the uterus (rabbit). Blastocyst distention of 2–5 mm results in transport with "even" spacing; finally blastocyst distention over 5 mm results in stopping any further transport; implantation then begins.

The biophysical basis for such a mechanism may very well be due to such behavior as demonstrated by the active length-tension curve. Can distention cause a decrease in the active generation of force immediately below the area of implantation? With only a partial elongation the operating position of the muscle on the length-tension curve can shift to the right from a resting position below the optimum point, and more forceful contractions result. This could be the explanation of the situation described by Böving. With greater distention, as the operating point of the muscle on the length-tension curve is shifted beyond the "optimal" point, complete cessation of contraction can result. However, the onset of stress relaxation leads to the development of new contractions. This is prevented by the rapid expansion of the blastocyst, starting again the cycle of stretch and cessation of contractions. Finally the developing placenta begins to secrete progesterone, and this steroid produces quiescence at the site of implantation.

Acknowledgments

This research was supported in part by the Ford Foundation and the Lalor Foundation.

References

Abe, Y. and Tomita, T. 1968. Cable properties of smooth muscle. *J Physiol* (*London*) 196:87.

Ashman, R. F.; Kanno, Y.; and Loewenstein, W. R. 1964. Intercellular electrical coupling at a forming membrane junction in a dividing cell. *Science* 145:604.

Bergman, R. A. 1958. Intercellular bridges in ureteral smooth muscle. *Johns Hopkins Hosp Bull* 102:195.

Bohr, D. F. 1964. Electrolytes and smooth muscle contraction. *Pharmacol Rev* 16:88.

Böving, B. G. 1965. Implantation. In *Fetal homeostasis,* ed. R. M. Wynn, vol. 1. New York: New York Academy of Sciences.

Bozler, E. 1948. Conduction, automaticity and tonus of visceral muscles. *Experientia* 4:213.

Caesar, R.; Edwards, G. A.; and Ruska, H. 1957. Architecture and nerve supply of mammalian smooth muscle tissue. *J Biophys Biochem Cytol* 3:867.

Casteels, R. 1967. The physiology of intestinal smooth muscle: Symposium on gastrointestinal motility. *Amer J Dig Dis* 12:231.

Conrad, J. T.; Aoba, H.; and Shimizu, T. 1969. The effect of estrogenic and progestational steroids upon the transmembrane potentials of cells from the frog sartorius muscle. *Fed Proc* 28:771.

Conrad, J. T.; Johnson, W. L.; Kuhn, W. K.; and Hunter, C. A., Jr. 1966. Passive stretch relationships in human uterine muscle. *Amer J Obstet Gynecol* 96:1055.

Conrad, J. T., and Kuhn, W. 1967. The active length-tension relationship in human uterine muscle. *Amer J Obstet Gynecol* 97:154.

Conrad, J. T.; Kuhn, W.; and Johnson, W. L. 1966. Stress relaxation in human uterine muscle. *Amer J Obstet Gynecol* 95:254.

Csapo, A. I. 1948. Actomyosin content of the uterus. *Nature (London)* 162:218.

———. 1950*a*. Studies on adenosine triphosphatase activity of uterine muscle. *Acta Physiol Scand* 19:100.

———. 1950*b*. Actomyosin of uterus. *Amer J Physiol* 160:46.

———. 1950*c*. Actomyosin formation by estrogen action. *Amer J Physiol* 162:406.

———. 1955. The mechanism of myometrial function and its disorders. In *Modern trends in obstetrics and gynecology,* ed. K. Bowes; 2d series, p. 20. London: Butterworth.

———. 1961. Progesterone and the defense mechanism of pregnancy. In *Defense mechanism of pregnancy,* Ciba Foundation Study Group 9.

———. 1962. Smooth muscle as a contractile unit. Symposium on vascular smooth muscle. *Physiol Rev* 42:7.

Dawson, J. E., and Conrad, J. T. 1969. The effect of human chorionic gonadotropin (HCG) and luteinizing hormone (LH) upon the membrane potential of unovulated frog oocytes. *Physiologist* 12:206.

Dewey, M. M., and Barr, L. 1962. Intracellular connection between smooth muscle cells: The nexus. *Science* 137:670.

———. 1964. A study of the structure and distribution of the nexus. *J Cell Biol* 23:553.

Elliott, G. F. 1964. X-ray diffraction studies on striated and smooth muscles. *Proc Roy Soc Biol* 160:472.

Gansler, H. 1956. Elektronenmikroskopische Untersuchungen am Uterusmuskel der Ratte unter Follikel hormonwirkung. *Virchow Arch Path Anat* 329:235.

———. Struktur und Funktion der glatten Muskulatur: II. Licht- und elektronenmikroskopische Befunde an Hohlorganen von Ratte, Meerschweinchen und Mensch. *Z Zellforsch* 55:724.

Hanson, J., and Huxley, H. E. 1955. The structural basis of contraction in striated muscle. *Sympos Soc Exp Biol* 9:228.

Hori, R. 1958. On the membrane potential of the unfertilized egg of the Medaka, Oryzias latipes and changes accompanying activation. *Embryologia* 4:79.

Huxley, H. E. 1956. Muscular contraction. *Endeavour* 15:177.

———. 1957. The double array of filaments in cross striated muscle. *J Biophys Biochem Cytol* 3:631.

———. 1969. The mechanism of muscular contraction. *Science* 164:1356.

Ito, S. 1962. Resting potential and activation potential of the oryzias egg: II. Changes of membrane potential and resistance during fertilization. *Embryologia* 7:47.

Jung, H. 1961. Zur Erregungsphysiologischen Steuerung des Uterusmuskels durch Oestradiol, Oestrone und Oestriol. *Klin Wschr* 39:1169.

———. 1963*a*. Effects of progesterone on uterine contractility in initiation of labor. In *Proceedings of interdisciplinary conference on the initiation of labor,* ed. J. M. Marshall. Public Health Service Publication #1390. Washington, D.C.: Superintendent of Documents, U.S. Printing Office.

———. 1963*b*. Die Wirkung der ovarial und der placenter-hormone. In *Pharmacology of smooth muscle.* Proc II Int Pharmacol 6:113.

———. 1965. Zur Physiologie und Klinik der hormonalen Uterusregulation. *Fortschr Geburtsh Gynaek* 22:155.

Kanno, Y., and Loewenstein, W. R. 1963. A study of the nucleus and cell membranes of oocytes with an intra-cellular electrode. *Exp Cell Res* 31:149.

Kao, C. Y. 1967. *Ionic basis of electrical activity in uterine smooth muscle in cellular biology of the uterus,* ed. R. M. Wynn. New York: Appleton-Century-Crofts.

Kao, C. Y., and Nishiyama, A. 1964. Ovarian hormones and resting potentials of uterine smooth muscle. *Amer J Physiol* 207:793.

Lewis, M. R. 1920. Muscular contractions in tissue culture. *Carnegie Contrib Embryol* 9:191.

Maeno, T. 1959. Electrical characteristics and activation potential of bufo eggs. *J Gen Physiol* 43:139.

Mark, J. S. T. 1956. An electron microscope study of uterine smooth muscle. *Anat Rec* 125:473.

Morrill, G. A. 1965. Water and electrolyte changes in amphibian eggs at ovulation. *Exp Cell Res* 40:664.

Morrill, G. A.; Rosenthal, J.; and Watson, D. E. 1966. Membrane permeability changes in amphibian eggs at ovulation. *J Cell Physiol* 67:375.

Morrill, G. A., and Watson, D. E. 1966. Transmembrane electropotential changes in amphibian eggs at ovulation, activation and first cleavage. *J Cell Physiol* 67:85.

Mossman, R. G., and Conrad, J. T. 1967. *In vitro* blocking and oxytocic effects of water-soluble estrogens on pregnant human, mouse and rat uteri. *Amer J Obstet Gynec* 99:539.

Mullins, G. L., and Guntheroth, W. G. 1965. A collagen net hypothesis for force transference of smooth muscle. *Nature (London)* 206:592.

Needham, D. M. 1962. Contractile proteins in smooth muscle of the uterus. Symposium on vascular smooth muscle. *Physiol Rev* 42:88.

Needham, D. M., and Cawkwell, J. M. 1956. Some properties of the actomyosin-like protein of the uterus. *Biochem J* 63:337.

Needham, D. M., and Shoenberg, C. F. 1967. The biochemistry of the myometrium. In *Cellular biology of the uterus,* ed. R. M. Wynn, p. 291. New York: Appleton-Century-Crofts.

Needham, D. M., and Williams, J. M. 1959. Some properties of uterus actomyosin and myofilaments. *Biochem J* 73:171.

Nemetschek-Gansler, H. 1967. Ultrastructure of the myometrium. In *Cellular biology of the uterus,* ed. R. M. Wynn, p. 353. New York: Appleton-Century-Crofts.

Rhodin, J. A. G. 1962. Fine structure of vascular walls in mammals. Symposium on vascular smooth muscle. *Physiol Rev* 42:48.

Rosenbluth, J. 1965. Smooth muscle: An ultrastructural basis for the dynamics of its contraction. *Science* 148:1337.

Shoenberg, C. F. 1958. An electron microscope study of smooth muscle in pregnant uterus of the rabbit. *J Biophys Biochem Cytol* 4:609.

Szent-Györgyi, A. G. 1953. Meromyosins, the sub-units of myosin. *Arch Biochem* 42:305.

Tahmoush, A.; Conrad, J. T.; Abrams, R.; and Long, E. 1968. The conduction velocity of action potentials in the *in vitro* rat uterus as modified by estrogens and pregnancy. *Proc Int Union of Physiol Sciences VII.*

Thaemert, J. C. 1959. Intercellular bridges as protoplasmic anastomoses between smooth muscle cells. *J Biophys Biochem Cytol* 6:67.

Weinstein, H. J., and Ralph, P. H. 1951. Myofilament from smooth muscle. *Proc Soc Exp Biol Med* 78:614.

27

Telemetering Systems for Long Term Observations of Uterine Contractility in the Unrestrained Animal

Eli Fromm

Biological Sciences

Biomedical Engineering and Science Program

Drexel University, Philadelphia

Basic smooth muscle physiologic studies to date have for the most part been performed either as *in vitro* experiments on excised tissue in the investigation of the tension-time interrelationships (Csapo 1962; Kuriyama and Csapo 1961; Schofield and Wood 1964) or as *in vivo* experiments with anesthetized or restrained subjects. Although the *in vitro* studies have led to significant basic physiologic findings for the particular environment, *in vivo* studies of smooth muscle contractile activity add an extra dimension by introducing the influences of the normal environment. For information more coincident with normal physiologic function the use of an unanesthetized chronic preparation is suggested, since anesthesia, as well as significant temperature fluctuations and acute surgical procedures, may cause depression of uterine contractile activity (Friedman 1965; Setekleiv 1964). It has been shown that physiologic function pertaining to the reproductive system when investigated on a restrained versus an unrestrained subject exhibits significant variations similar to the responses one might expect from the nervous or cardiovascular systems.

This laboratory has for the past several years attempted to develop a technique for the *in vivo* study of contractile activity of specific small segments of myometrial muscle. The objectives were to realize:

a) a method utilizable in the intact, unanesthetized subject with the uterus in its normal environmental conditions;

b) a method in which the subject was encumbered by neither restraint mechanisms nor transducer-to-data-acquisition system coupling schemes;

c) a method that would not interfere with the normal reproductive functions of the uterus; and

d) a method capable of giving a quantitative measure of contractile activity.

If the subject is to be unencumbered, data must be obtained without physical connection of subject to acquisition station. A radio frequency telemetric system has been devised for this purpose. The earliest of these systems involved a single channel only; more recent developments have been oriented toward multichannel and multilocation concurrent information.

I. History and Status of Uterine Motility Measurements

Several authors have reviewed the earlier *in vivo* methods for the study of motility of both the pregnant and nonpregnant uterus (Hendricks 1964; Karlson 1944; Reynolds, Harris, and Kaiser 1954). The techniques have been used in both human and animal studies and are divided into the two broad categories of internal and external tocodynamometry. The internal type, that in which the force detector is placed within the uterine lumen, is the primary methodology used to date in motility studies of the nonpregnant uterus.

The earliest of experimental studies, in the mid-1800s, of the forces of uterine contractions used obstetrical delivery forceps with a calibrated spring handle to measure the amount of force exerted on the head at the time of delivery. The concept of the use of an intrauterine balloon came about several years later when Schatz (1872) used this technique for determining intrauterine pressures in pregnancy. He used a rather large (20 ml) water-filled rubber bag placed in the uterine cavity, connected to a mercury manometer via water-filled tubing and to a kymograph. This early work had its shortcomings: the bag could not be inserted unless the patient was anesthetized, it created a significant amount of uterine irritation and stretching, and the inertia of both water and mercury systems was found to diminish the accuracy of the results.

However this instrumentality remained in practice into the twentieth century; the systems devised by subsequent investigators used Schatz's technique modified to diminish various sources of error. This technique was used, with slight variations from one investigator to another, to measure not only uterine forces and pressures during pregnancy and labor but also the response of the myometrium to various chemicals and drugs.

The earliest investigation into the use of solid state electronic transducing components for the measurement of varying uterine pressures was carried out by Karlson (1944). He modified a granular carbon microphone and developed very small pressure-sensitive receptors to measure intrauterine pressures. Units placed between the fetal membranes and the uterine wall could determine muscle activity at various locations. Karlson was able to publish recordings made from two separate locations simultaneously. The activity was more of a resultant intracavity pressure than of pressure of specific muscle segments, however.

Alverez and Caldeyro (1950) have done extensive studies of the hydrostatic forces in the pregnant uterus by the transabdominal needle method. The technique requires a fluid coupling from the amnionic sac through the cannula to an appropriate fluid pressure measuring system. As with the intrauterine bag technique the method affords measurement of a total force exerted by the uterus on the fluid-filled amnionic sac in the restrained subject.

Intrauterine catheters have been employed for the measurement of uterine activity. Hendricks (1964) described an open-ended catheter technique in which three individual catheters are positioned at various levels of the uterus and connected by a fluid coupling system to appropriate pressure-sensitive electronic detecting and recording equipment. Simultaneous records of several areas of the uterus may be taken under varying conditions, such as supine, side, and erect positions of the patient. The subject, however, cannot move about freely without interfering with the data acquisition technique. Such techniques as hysterosalpingography and transuterine insufflation are available, and external techniques using either mechanical, electrical, or electromechanical abdominal sensors have been reported on numerous occasions (Karlson 1944; Reynolds, Heard, and Bruns 1947). These devices do not measure uterine pressure alone; the activity and influence of the abdominal muscles are included in the mensuration.

Several recent innovations in the methodology of intrauterine activity detection have been reported. Bangham, Cotes, and Parsons (1966) used a telemetric technique to monitor intrauterine pressures from an unrestrained, unanesthetized animal. The technique records total uterine activity from within the lumen but not segmental information. Simmons, Dracy, and Esler (1965) were able to transmit the uterine electromyogram from the unrestrained animal and measure the electrical activity associated with the segmental mechanical muscle movements. Most recently Behrman and Burchfield (1968), and Behrman, Archie, and O'Brien (1969) have developed an intrauterine device with strain gauges. A number of such gauges may be incorporated in the IUCD to record multisite activity. Thus far it has been connected to data acquisition equipment only by direct wire; it could however be readily adapted to a radio telemetric technique and used as the transducer for the types of systems developed in this laboratory.

II. Status of Implant Biotelemetry

The concept of radio transmission of information from within the body is relatively new. The first generally available paper on the subject of transmitting biologic information via radio signals to a remote receiving station was published in 1957 (Jacobson and Mackay). It dealt with the passage through a human gastrointestinal tract of an internally powered transmitter whose signal contained both pressure and temperature information. The self-powered (active) telemeter had a life of only a few hours. In 1960 Farrar, Zworykin, and Berkley tried a pressure transmission system of the so-called passive type having no self-contained active components and energized totally by external radiation. The extremely close antenna demanded by it limited transmission distance and left the subject virtually bedridden; the data obtained were not much more significant than those acquired from wire.

A number of single channel (Mackay 1968) and multichannel (Freyer, Sandler, and Datnow 1967; Goodman 1967) telemetry systems have been developed, none oriented toward uterine contractile information telemetry.

III. Methods and Procedures

As has been noted this laboratory has pursued the telemetry of uterine muscular activity first via a single channel system and more recently through multichannel development. Both systems will be described.

One cannot completely separate the design of a telemeter from that of the transducer since the output of the telemeter must be capable of modulation by some transducing technique of the variable under investigation. In choosing the appropriate system two objectives have been kept in mind: to facilitate total implantation and to extend the physiologic usefulness of the experiment to 3–5 months, never less than a month. The first was achieved by simplifying the circuitry and minimizing size and weight; the second by a low-drain current to prolong the life of the battery. In reproductive system studies the experiment has to extend over at least several cycles or for the duration of gestation in pregnant animals. For the biotelemetering system to survive the implantation and experimental procedures the individual segments must work concurrently with the biologic interface.

A. *Single Channel Transmitting System*

The telemeter devised for single site transmission is a blocking oscillator modified from a previous design for another variable (Goodman, Gibson, and Marmarou 1964). Its electrical schematic diagram is noted in figure 1 and has a theoretical lifetime of at least 3 months with a small 160-milliampere-hr mercury battery. The variable resistance represents the transducer and in this case has an initial resistance set for approximately 200 ohms.

A slight initial disturbance to the resonant circuit of L_1 and C_1 causes the telemeter to oscillate. To stay below the broadcast band the oscillation frequency of approximately 400 KHz, dependent upon L_1 and C_1 as $\omega = 1/(LC)^{1/2}$ was chosen. The transistor is on with no initial charge on C_2 and allows current flow. As the oscillations continue aided by the feedback loop through L_2 a charge builds up in C_2, causing the transistor's emitter to base voltage to decrease; the charged C_2 turns off the transistor and consequently the battery supply from the circuit. The amount of time necessary for the charge to build up on C_2 is dependent upon the voltage supplied to it through the oscillating circuit, which in turn depends upon the fixed voltage of the battery and the amount of voltage drop across the resistance in the emitter. Thus changes in resistance of the emitter, that is, the transducer, varies the time necessary for the charge to build up on C_2 and consequently the time at which the transistor turns off. The charge on C_2 may leak off through R_2, the fixed resistance between base and collector; its rate of discharge is dependent upon the value of this fixed component. When sufficient charge has leaked off C_2 the transistor is turned on again, since there is sufficient voltage differential between base and emitter, the battery gives energy to the resonant circuit, and the entire process is repeated. Thus with a fixed discharge time the rate at which the transistor is turned on and off is dependent upon the charge rate of C_2 which, as just described, is in turn dependent upon the varying resistance of the transducer in the emitter. It is this changing turn-on-and-off (blocking rate) which is then measured and related to varying forces on the transducer by means of a calibration curve.

B. *Transducer*

Throughout these studies the contractile force transducer as shown in figure 2 was utilized. It was developed specifically to measure the force of contractions and operates on the principle of an electrical resistance change for linear deformation produced by the forces of muscle contraction between the two points of attachment. The material used for the transducing element is Silastic S-2086 (Dow Corning Corporation). By itself it is not considered biologically inert; when fabricated into

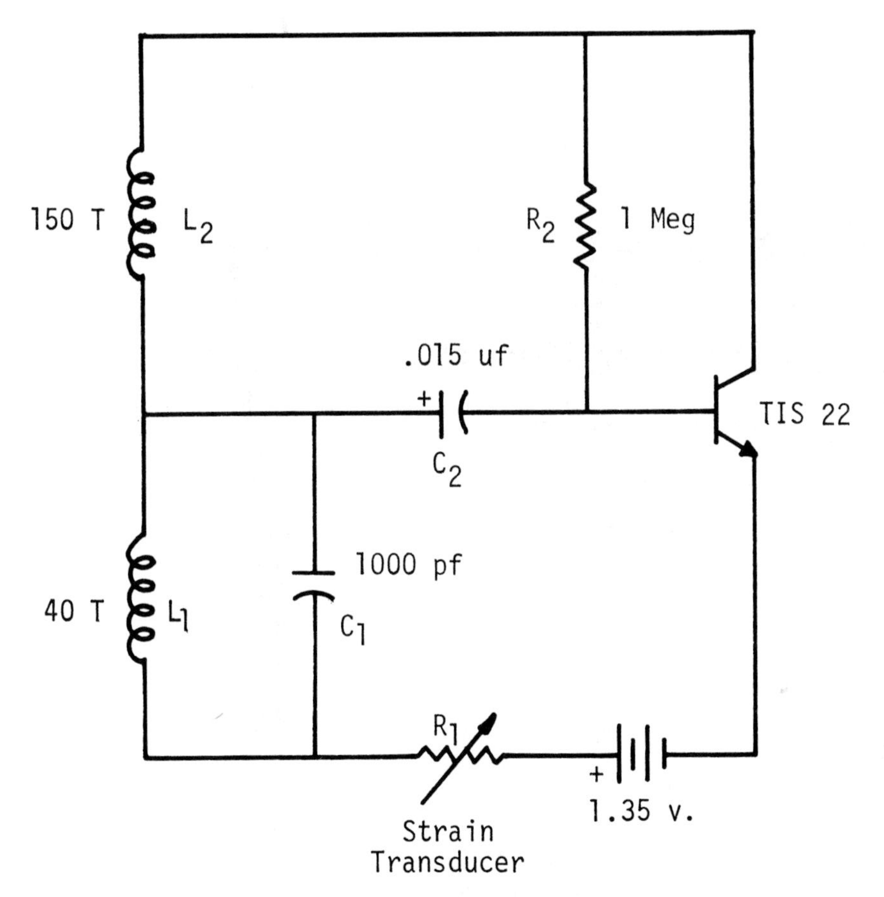

Fig. 1. Schematic drawing of contractile telemeter

the format shown in figure 2, it is capable of chronic implantation, becomes a relatively isotonic device, and has a very linear characteristic curve. During fabrication the lead wires are tied and soldered about the Silastic, then insulated from one another and brought off to whatever distance the particular experiment calls for with respect to transducer-telemeter separation. Only the very small transducer is in contact with the uterus. The separation of lead wire attachments determines the initial resistance of the transducer and may be varied to match the requirements of the telemeter. The outer sheath PE-200 tubing is for the particular diameter transducer shown; it is reduced appropriately for varying diameters of S-2086. The distal end heat seal is employed to prevent body fluid contamination of the system. Attachment to the myometrial muscle is as shown via the two 4–0 sutures centered about the lead wire attachments. These sutures are so placed that they attach to the specific muscle segment under consideration; it is the relationship of the movement at one suture to that at the other which generates the change which is ultimately transmitted.

A characteristic force-resistance curve is shown in figure 3. The initial resistance, 667 ohms in this case, varies as the distance between lead wire attachments and technique varies and must be matched to the telemeter input impedance requirements. The plot and linear regression line are for 500 points of data with 50 readings taken at each value of force. The slope indicates a sensitivity of 1.42 ohms per g of force, and the correlation coefficient for the linear regression equation and points is

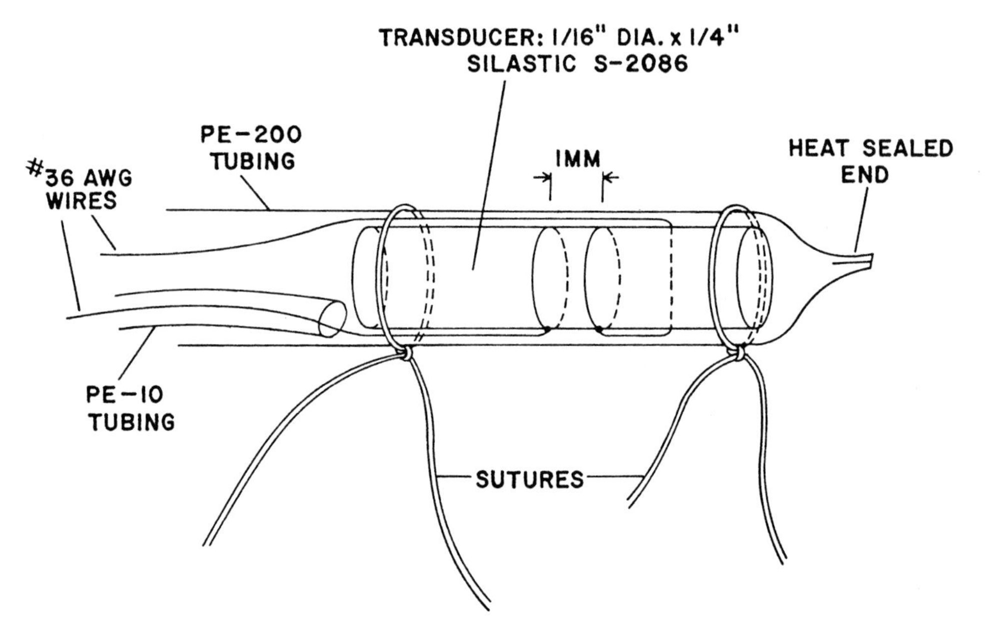

Fig. 2. Schematic drawing of contractile force transducer

0.99. Greater sensitivities could be achieved with more flexible outer tubing for the transducer.

Such a telemetry system as described above has a range limit of approximately 2 ft and is quite directional. To compensate for this, special caging and antenna arrangements were made to allow signal pickup regardless of animal orientation. A special Plexiglas cage was fabricated to avoid the interference of a large metallic interface, and around it were placed three mutually perpendicular antenna for signal pickup in each of the x, y, and z planes. This information was then summed and amplified, and the varying blocking rate, which contained the contractile information, was processed through digital to analog converters and recorded. The details of circuits and construction have been described elsewhere (Fromm 1967) and will not be repeated here.

A total of 11 such single channel telemeters have been implanted in mature female mixed-breed rabbits of good general health weighing between 10 and 12 lb. After anesthetizing the rabbit with sodium pentobarbital (Nembutal) abdominal and transcutaneous dorsal incisions were made through which to insert and attach the device. The transducer was fixed to the myometrium at a point midway between the cervix and tubal isthmus by means of 4–0 silk; the body of the unit was attached to the dorsal muscle subdermally and separated from the transducer by the umbilical cord. Only the transducer, then, was in contact with the uterus. The completed unit prior to implantation is shown in figure 4.

Recordings of contractile information began immediately and continued uninterrupted for the lifetime of the telemeter in each of the 11 implants. The first 6 telemeters were implanted to develop and improve the techniques, materials, and

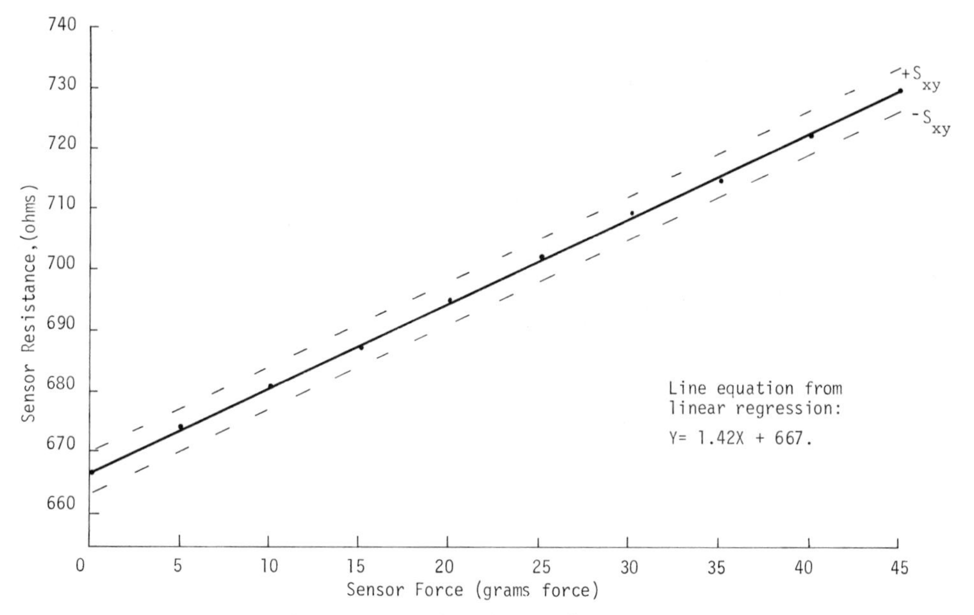

Fig. 3. Characteristic force-resistance curve

methods to make possible prolonged implantation. Pharmacologic agents were not involved. The remaining implants in the series served a dual purpose: the first, of course, was to develop the tool to a more useful, longer life; the second was to determine whether the activity measured was indeed that of uterine contractility. The latter was achieved through the application of pharmacologic agents having specifically demonstrable effects on uterine activity. The drugs were given in the ear vein of the rabbit. In the last 2 rabbits permanent ear vein cannulas were inserted. They remained patent for more than 4 weeks, affording easier intravenous injections.

Epinephrine for its uterine stimulatory and isoxsuprine (Vasodilan) for its uterine relaxant effect (Sollman 1952; Tansy 1961), and ergonovine, histamine, and conjugated estrogens (Premarin) were the agents used. At no time was a second drug given until the activity of the uterus had returned to control level. Where no specific reference for dosage regimen was available a milligram of drug per kg of weight, based on *Physician's Desk Reference* and *Hospital Formulary* recommenda-

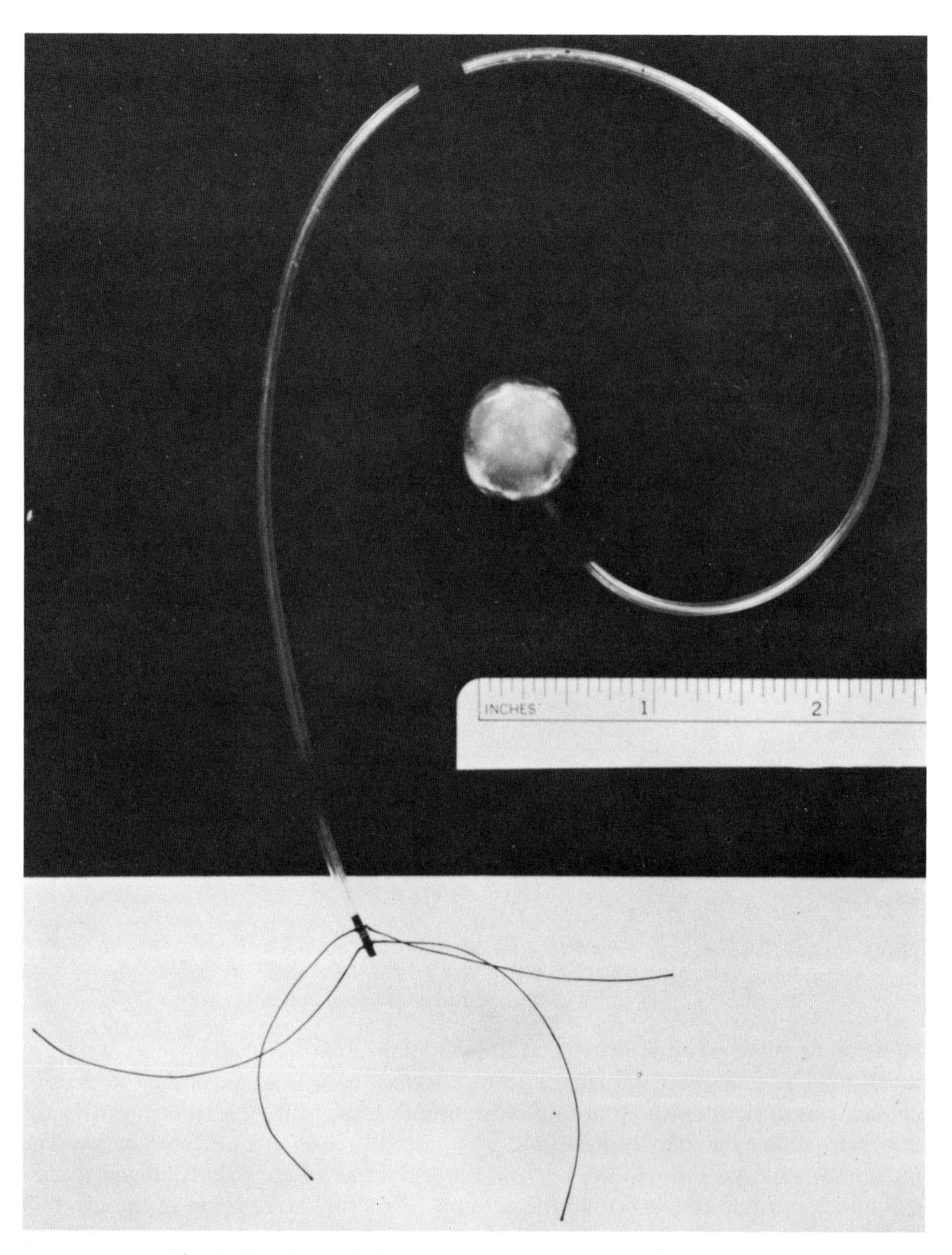

Fig. 4. Completed single channel telemeter prior to implantation

tions, was adopted. Dosage modifications dependent on variations in species differences were made after the initial dosage regimen.

C. *Multichannel Transmitting System*

The capabilities of a single channel of information are limited. Simultaneous information from several sites and perhaps several variables was considered desirable for comparative purposes.

Our laboratory has very recently completed the initial phases of development and construction of a subminiature multichannel contractile force telemetry system. It employs FM–FM multiplexing techniques, is capable of chronic implantation, and offers the versatility of subminiature module construction for easy adaptation to the experimental needs. Several force channel subcarrier oscillators may be used concurrently to modulate the 100 MHz carrier. Each force channel's dimensions are 6.3 mm $\times$ 3.3 mm. The SCO circuit is rather unique in that it uses low resistance (100 to 1000Ω) force sensors while drawing only 50 microamps current from 1.30 vdc Hg cell.

The carrier oscillator measures 6.3 mm $\times$ 3.3 mm $\times$ 3.3 mm. Its current drain can be set from 10 μa to 50 μa depending upon range desired. The circuit is a modified Colpitts using a LID transistor as the active element; variation of the base-collector junction capacitance provides frequency modulation.

All modules are constructed on 10 mil glass epoxy circuit boards. Subminiature discrete components are used throughout and assembly is accomplished under magnification.

All modules are potted in a polymer doped wax. Mechanical support is assured internally for the umbilical cords by mounting them in a circuit board main frame which contains the wax-coated modules; the battery is potted to achieve the same protection as the rest of the system.

As previously indicated these units have not yet undergone *in vivo* evaluation and more specific details must await such future developments. Figure 5 shows one such unit fully encapsulated with its two umbilical cords coiled and the transducers at the distal ends.

D. *Coating Techniques*

One of the difficulties in making a telemeter useful as a biotelemeter arises through contamination of the electronics by body fluids. The problem of tissue reaction to implant materials exists also, of course, but has met with earlier solution owing to earlier recognition (Boone 1965; Braley 1963). The telemeters of Jacobson and Mackay, investigating gastrointestinal activity, were swallowed. Only a short lifetime was required of the devices, hours rather than days. These units therefore needed only coatings which were reasonably effective in preventing fluid seepage; the investigators concerned themselves mainly with the prevention of a gross breach of the encapsulant. When a long term implant is attempted, for days or perhaps months, even the slightest amount of moisture permeability can be disastrous.

Contrary to what had long been an accepted fact, Silastic is not the answer to prevention of contamination in long term implants. In detailed studies (Fromm 1964; Goodman, Gibson, and Marmarou 1964) it has been found that paraffin provided the best fluid and vapor seal for coating a one-piece integral package. It is not practicable, however, in the face of the problems encountered in connecting a remote transducing element to the telemeter body via a flexible umbilical cord; paraffin cracks at the umbilical telemeter body interface and creates a gross breach in the encapsulant. We therefore chose to dope paraffin with a plasticizer which provided limited flexibility to cover the attachment between the polyethylene umbilical cord

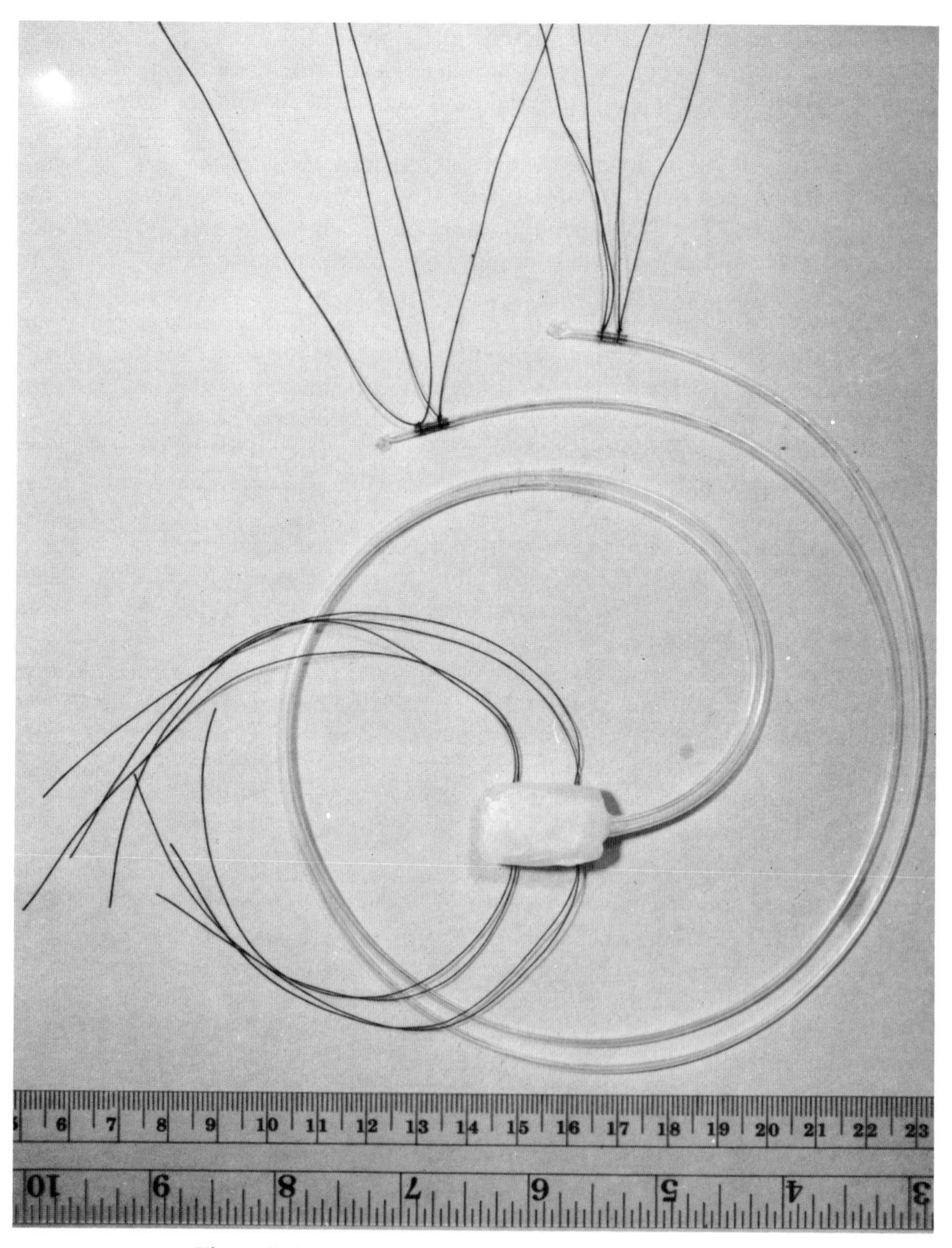

Fig. 5. Fully encapsulated multichannel transmitting unit

and the telemeter body. Such material is commercially available by the trade name of Paraplast; but its cold flow characteristics must be considered. No sutures were fixed to the Paraplast directly; the telemeter body was inserted in a support ring to which the sutures were attached or allowed to float freely.

IV. Results

The single channel system was completed and has transmitted meaningful myometrial contractile information for periods of 6 weeks. The multichannel system is still undergoing engineering evaluation and modifications. It has not yet been used to transmit long term *in vivo* information.

A series of experiments was performed with 5 single channel contractile force telemeter implants to record the effects of pharmacologic agents on uterine activity

TABLE 1

DIFFERENCES BEFORE AND AFTER 2 UNITS PITOCIN

	TOTAL AMPLITUDE PER 5 MIN		AVERAGE AMPLITUDE PER CONTRACTION		NUMBER OF CONTRACTIONS PER 5 MIN		MAXIMUM CONTRACTION	
	Before[a]	After[b]	Before[a]	After[b]	Before[a]	After[b]	Before[a]	After[b]
Tel. #7[c]	1.6	32.7	0.2	3.7	8	9	0.3	5.6
Tel. #8[c]	20.0	34.4	1.3	2.3	15	12.3	3.8	6.0
Tel. #9	6.0	68.2	1.8	4.6	3.5	15.7	2.3	10.9
Tel. #10	0.8	5.3	0.4	1.0	1	5.7	0.5	1.4
Tel. #11	2.5	16.0	1.0	1.8	2.5	9.2	1.2	2.6
	t=2.46 Significant at 5% level		t=2.95 Significant at 2.5% level		t=1.46 Not significant		t=2.53 Significant at 5% level	

NOTE: Differences are multiples of 4 gm units.
[a] One 5-min control period.
[b] An average of 6 successive 5-min intervals immediately following Pitocin.
[c] These figures have accounted for 4 times greater sensitivity.

and to verify directly that the activity recorded could indeed be interpreted as uterine contractions. A study of the actions of these drugs was not the purpose of this series of experiments; therefore the agents chosen were those that could be counted on to stimulate true uterine contractile activity.

A typical response to the administration of 2 units of Pitocin is seen in figure 6. A series of 8 experiments was performed with this dosage in 5 different rabbits utilizing different telemeters. The resultant tracings were evaluated for a 5-min control period prior to the administration of Pitocin and 6 consecutive 5-min intervals after the administration of Pitocin. Table 1 shows for each of these animals a summary of the number of contractions, the average of total amplitude of contraction before and after administration of Pitocin, the average amplitude of contraction, and the maximum contraction for the 5-min period of control before and an average 5 min after administration. A paired t-test was utilized to determine whether the means after drug administration were significantly different from those before the drug administration. The results of these tests are included in the figures mentioned above and indicate that 3 of these 4 parameters show significant differences between the control and experimental results. The differences in average amplitude per contraction in the 5-min interval are significant at the 2.5% level, and the differences of the mean of total amplitude of contractions during a 5-min interval are significant at the 5% level. The differences in the means between control and experimental in the number of contractions per 5-min period appear not to be significant. At times

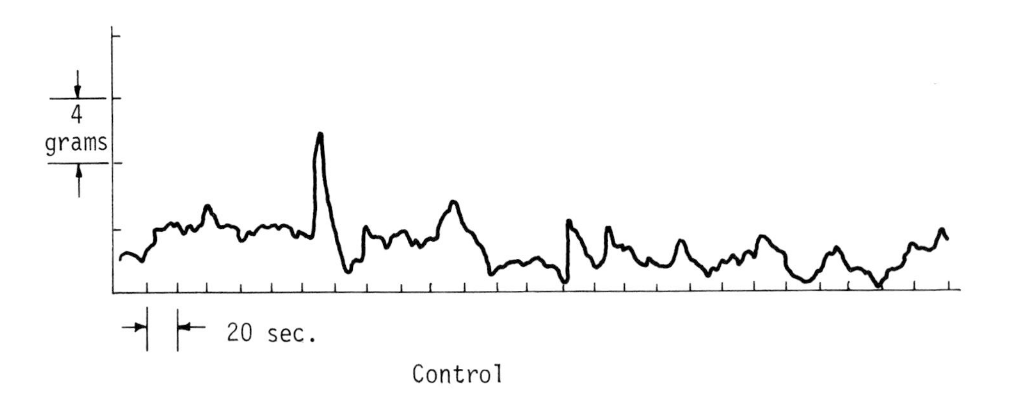

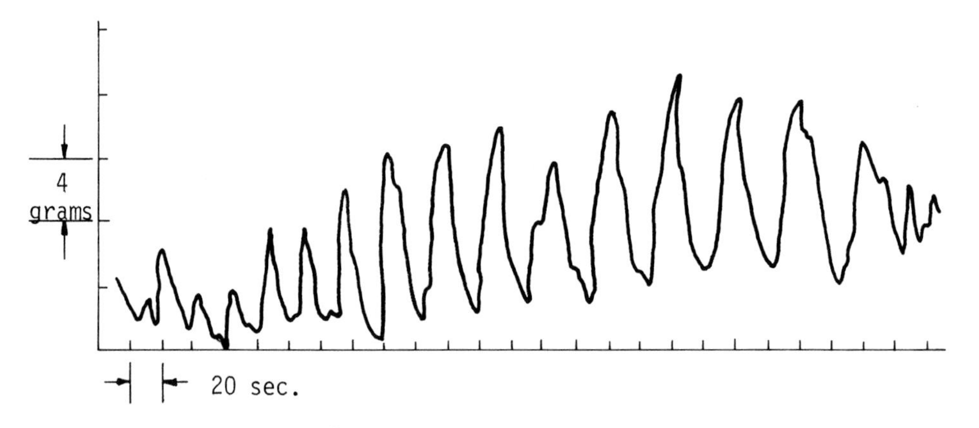

Fig. 6. Typical response to Pitocin (rabbit)

the administration of 2 units of Pitocin resulted in a very noticeable response in periodicity of increased and decreased activity.

Two and 4 milligrams of Premarin were administered intravenously and did not appear to have noticeable effects on the activity measured. When Pitocin was administered ½ hr after the Premarin, activity was considerably increased. Epinephrine was administered as 1 ml of a 1:10,000 solution and in general had no stimulatory effect; occasionally it appeared to cause a moderate decrease in the general tonic state; several times it resulted in one large contraction followed by a lapse into the quiescence characterizing the uterus prior to injection. Figure 7 illustrates this situation. Isoxsuprine was administered as 1 mg intravenously approximately ½ hr after the administration of Pitocin, while the effects of the Pitocin were still noticeable. The injection immediately enhanced the magnitude of the Pitocin induced activity of the uterus; approximately 15 min after administration the activity suddenly ceased. In a later experiment Pitocin was administered approximately ½ hr after the administration of isoxsuprine; the increased activity expected of Pitocin was evident.

Several experiments, which could be carried out only by telemetric technique, were attempted to determine the effects of the traumatic experience of capture and

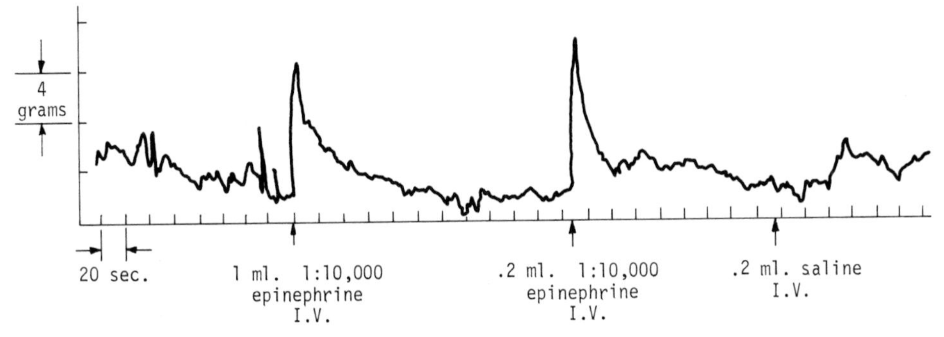

Fig. 7. Response to epinephrine (rabbit)

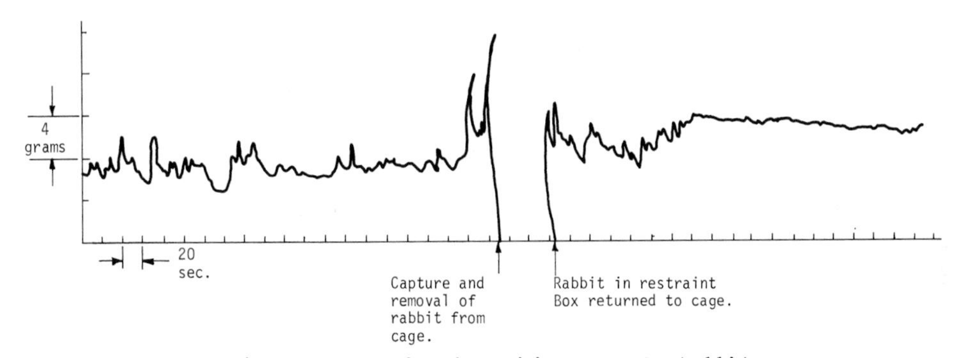

Fig. 8. Response of uterine activity to restraint (rabbit)

restraint on uterine activity. A control tracing of the animal free within the confines of its cage was taken, after which the rabbit was placed in a restraining box. A typical tracing is seen in figure 8. It shows that upon confinement there is a distinct decrease of activity. Normal activity returned within several hours. It is presumed that the animal overcame the traumatic experience as it grew accustomed to its new environment.

V. Discussion

We have shown that transmission of information concerning uterine contractile force in the unrestrained animal is feasible. The usual traumatic reactions experienced by animals undergoing experimental procedures are obviated; the animal is totally unaware of involvement without being anesthetized. It remains intact; complicating effects of anesthetics are disposed of; the uterus is undisturbed in its normal environment, unaltered physiologically or metabolically; normal reproductive function continues unaffected by the implant; segmental information is transmitted entirely separately from activity of the uterus *in toto*.

Extension of our investigations beyond the single channel system has awaited the completion of our multichannel system. With it we expect to study the comparative effects of the IUCD and contractile propagational characteristics in the freely mobile subject among other areas of investigation.

Acknowledgments

Various aspects of the completed and presently continuing efforts have been supported in part by the National Institutes of Health, the Lalor Foundation, and the National Science Foundation.

References

Alverez, H., and Caldeyro, B. R. 1950. Contractility of the human uterus recorded by new methods. *Surg Gynec Obstet* 91:1.

Bangham, D. R.; Cotes, P. M.; and Parsons, J. A. 1966. Measurement of intrauterine pressure changes in the pregnant rhesus monkey by telemetric recording from pressure-sensitive capsules inserted in the uterus. *Mem Soc Endocr* 14:249.

Behrman, S. J.; Archie, J. T.; and O'Brien, O. P. 1969. Myometrial activity and the IUCD. *Amer J Obstet Gynec* 104:1.

Behrman, S. J., and Burchfield, W. 1968. The intrauterine contraceptive device and myometrial activity. *Amer J Obstet Gynec* 100:2.

Boone, J. L. 1965. Silicone rubber insulation for subdermally implanted electronic devices. *Proc XVIII Ann Conf Engin Med Biol* 7:101.

Braley, S. A. 1963. The use of silicones in medical applications. *Proc V Int Conf Med Electron,* p. 41.

Csapo, A. I. 1962. Smooth muscle as a contractile unit. *Physiol Rev* 42: Suppl. 5.

Farrar, J. T.; Zworykin, V. K.; and Berkley, C. 1960. Telemetering of physiologic information from the gastrointestinal tract by an externally energized capsule. *Proc III Int Conf Med Electron,* p. 120.

Freyer, T. B.; Sandler, H.; and Datnow, B. 1967. A multi-channel implantable telemetry system. *Dig VII Int Conf Med Biol Engin.*

Friedman, E. A. 1965. The pharmacology of uterine contractility. *Chicago Med Sch Quart* 25:33.

Fromm, E. 1964. A study of ovarian activity via temperature telemetry. M.S. thesis, Drexel Institute of Technology.

———. 1967. Myometrial contractile force telemetry from the unrestrained rabbit. Ph.D. diss., Jefferson Medical College.

Goodman, R. M. 1967. Simultaneous, multi-parameter data from implantable telemeters. *Dig VII Int Conf Med Biol Engin.*

Goodman, R. M.; Gibson, R. J.; and Marmarou, A. 1964. Instrumentation for study of biological rhythms. Final report #FB2029, vol. 1, 30 Nov. 1964, NASA contract #NASr-146.

Hendricks, C. H. 1964. A new technique for the study of motility in the non-pregnant human uterus. *J Obstet Gynec Brit Comm* 71:712.

Jacobson, B., and Mackay, R. S. 1957. A pH-endoradiosonde. *Lancet* 272:1224.

Karlson, S. 1944. A survey of the *in-vivo* methods used for an objective recording of the motility of the non-pregnant and the pregnant uterus and the efficiency of these methods. *Acta Obstet Gynec Scand* 24: Suppl. 4.

Kuriyama, H., and Csapo, A. 1961. A study of the parturient uterus with the microelectrode technique. *Endocrinology* 68:1010.

Mackay, R. S. 1968. *Bio-medical telemetry*. New York: John Wiley & Sons.

Reynolds, S. R. M.; Harris, J. S.; and Kaiser, I. H. 1954. *Clinical measurement of uterine forces in pregnancy and labor*. Springfield, Ill.: Charles C. Thomas.

Reynolds, S. R. M.; Heard, O. O.; and Bruns, P. 1947. Recording uterine contraction patterns in pregnant women: Application of the strain-gage in a multichannel tokodynamometer. *Science* 106:427.

Schatz, F. 1872. Über die Aetiologie der Lagen der menschlichen Frucht. *Gesellschaft Deutscher Naturforscher und Aerzte. Tageblatt der Versammlung,* p. 175–78.

Schofield, B. M., and Wood, C. 1964. Length-tension relation in rabbit and human myometrium. *J Physiol* (*London*) 175:125.

Setekleiv, J. 1964. Uterine motility of the estrogenized rabbit. *Acta Physiol Scand* 62:68.

Simmons, K. R.; Dracy, A. E.; and Essler, W. O. 1965. Recording uterine activity by radio telemetry techniques. *J Dairy Sci* 48:126.

Sollman, T. 1952. *A manual of pharmacology*. Philadelphia: W. B. Saunders Co.

Tansy, F. M. 1961. *In-vitro* studies on smooth muscle. M.S. thesis, Jefferson Medical College.

28

Implantation as an Immunologic Phenomenon

S. J. Behrman
Center for Research in Reproductive Biology
Department of Obstetrics and Gynecology
Medical Center
University of Michigan, Ann Arbor

Because the blastocyst contains some paternal elements which can be considered foreign to the mother, implantation may be considered a homograft of alien tissue. The highly successful homograft is unusual because of the fine discriminating capacity of the mammalian immune system. This is indicated by its ability to react against a protein which differs from isologous proteins in respect to only one amino acid and the rejection by females in some closely inbred strains of mice of grafts from the male as a result of antigens peculiar to the Y chromosome. That rejection of such grafting is the exception rather than the rule indicates that a unique sequence of events obtains at the time of implantation. This needs to be elaborated upon and understood not only for its intrinsic value but also for its possible contribution to the understanding of the invasiveness of the trophoblast, hybrid vigor, the malignancy of choriocarcinoma, the occurrence of habitual abortion, the toxemias of pregnancy, and possibly even the onset of labor. This human laboratory may contain clues vital to the future of organ transplantation.

It has been suggested that humoral responses to paternal antigens by the female might result in death or immobilization of the sperm (Schwimmer, Ustay, and Behrman 1967); that the humoral and cellular responses could prevent successful implantation of the blastocyst (Brambell 1965; Behrman and Sawada 1966); that it might even result in death of the conceptus after it has implanted (Menge 1969); and that fetal abnormalities could be produced by placental antisera (Brent, Averich, and Drapiewski 1961). Burstein and Blumenthal (1969) presented evidence that various immunologic phenomena occur during pregnancy in the human, both humoral and cellular in origin, in both mother and fetus. Especially interesting is their observation that extensive lymphocytic infiltration occurs in the decidua in certain pregnancies. The literature is replete with speculations, some of which, as Bagshawe (1967) indicates, are contradictory; it is therefore safe to assert that they cannot all be true. With the exception of blood group immunity due to cellular or subcellular antigens which cross the placenta from fetus to mother and then return to damage the fetus, one cannot conclude confidently from the literature that any immune mechanism has yet been firmly established in the fetomaternal relationship or that the mother is effectively immunized against subsequent pregnancies; this is the basis of the discussion in this chapter.

I wish to emphasize the incontestable fact that a normal fetus and its placenta survive in the uterus for a full period of gestation without the development of immunologically typical local cellular response by the mother against the fetus or by the fetus against the mother. Pregnancy presents the biologist with a set of fundamental problems concerning the relationship of pregnancy to immunologic reactions and the fascinating enigma of why the trophoblast resists immunologic attack at the site of implantation.

If we then consider that a fetus has a number of transplantation antigens, differing from those of the mother in whom it is developing, the question that confronts us is how the fetus avoids the destruction that is the almost universal fate of a homograft. Investigations on this problem have been extraordinarily well summarized by Medawar (1953), Billingham (1964), Behrman and Koren (1968), and Kirby (1968*a, b*). The theories are further encapsulated here.

I. Theories

A. *The Fetus Is Antigenically Immature*

As early as 1924 Little proposed the theory that the fetus is antigenically immature. It has now been discarded because of the enormous amount of evidence accumulated against it. Several experiments by Edidin in 1964 showed a rejection of 9-day mouse embryos homotransplanted into preimmunized allogeneic hosts; Simmons and Russell in 1965 and Kirby, Billington, and James in 1966 demonstrated that 4½-day-old mouse eggs failed to develop when transferred to the kidney of specifically immunized hosts; and Dancis, Samuels, and Douglas (1962) injected placental cells from one strain of mice into another and found accelerated rejection of subsequent homologous skin grafts and circulating cytotoxic antibodies—all of which support antigenicity of the placenta and its ability to express antigenicity under certain circumstances. The early embryonic tissue is antigenic and the fetus is capable of responding immunologically as an adult to antigenic stimuli (Uhr et al. 1962). There is little doubt that the embryo has transplantation antigens. The question is how early they develop and why they are not expressed.

B. *The Mother Is Immunologically Inert*

The thymolymphatic system of the body produces the cellular elements (lymphocytes and plasma cells) capable of responding to antigenic stimuli with either soluble antibody production or delayed immune response. The function of the thymolymphatic system is roughly reflected by the production of cellular elements under an antigenic stimulation such as the homograft. This immunologic reactivity is weakened slightly during pregnancy, but only slightly. Estrogen, a known depressant of thymic function in small mammals, might account for this; progesterone and human chorionic gonadotrophin have not been extensively studied for this purpose. Administration of ACTH brings about a transient fall in the number of circulating lymphocytes and involution of lymphoid tissue (Germuth 1956). However at best the increased corticosteroid production associated with pregnancy, to which both the fetus and placenta probably contribute, can be regarded as affording only the weakest ancillary mechanism for preventing development of maternal isoimmunity during pregnancy in a few species. There has been speculation that a high concentration of steroid hormones is produced locally by the invading trophoblast, preventing sensitization. A conjecture of this kind cannot explain the mother's acceptance of the very early trophoblast at a stage when actually little or no placental steroid production occurs.

A telling experiment to demonstrate maternal immunologic competence was performed by Woodruff in 1958. He injected into the flank of the pregnant rat its own fetal tissues and elicited a prompt and vigorous immune reaction against its own tissue.

To evaluate the possibility of the immunologic competence of lymphocytes Comings (1967) tested their response to phytohemagglutinin during and after pregnancy. There appeared to be no significant difference in the placental transformation of lymphocyte culture during the trimesters of pregnancy compared with cultures after delivery.

As the traffic of cellular and subcellular material across the placenta has been well established (fetal red cells by Creger and Steele 1957; placental fragments by Douglas, Thomas, Carr, Cullen, and Morris 1959; and the leukocytes by Payne 1962), and as both the lymphocytes and the mother's thymolymphatic system are competent, the lessening of the general level of immunologic competence of the mother during pregnancy is more than questionable. The idea that maternal immune response to antigens of the fetus is in general depressed cannot be accepted; some other explanation must be sought.

C. *The Uterus as an Immunologically Privileged Site*

In 1966 Kirby, Billington, and James immunized mice against tissues of a donor strain. When fertilized ova were transplanted under the kidney capsule, they failed to develop. Under the same conditions fertilized eggs transplanted to the uterus developed perfectly well. On this basis Kirby and his associates concluded that the uterus was exerting a protective function or acting as a "privileged site" and that the decidual tissue which forms around the implanting embryo had some protective effect in immunity, a position supported by earlier evidence of Lanman (1965) whose work indicated that pregnancy could not be prejudiced by immunization of the mother against paternal transplantation antigens. On the other hand evidence was presented (Schlesinger 1962) that homologous tumors transplanted into the cornua of nonpregnant, pseudopregnant, and pregnant mice and rats resulted in their accelerated rejection in the presensitized animals. Simmons and Russell in 1963 implanted parathyroid tissue into the uterine cavity. It was also rejected in the fashion of a homograft response if the tissue was of genetically different origin; the parathyroid tissue in controls of the same strain functioned normally in the new location. Clinical evidence bolsters the case against this hypothesis: ectopic pregnancy in the Fallopian tube where there might be decidua, in the ovary, on the broad ligament, and in the bowel where there are no decidua. Billingham in 1964 concluded that there was nothing immunologically peculiar or distinctive about the uterine environment. Simmons and Russell (1962) stated that this unique situation indicates that some property of the placental site, or the trophoblast itself, not only suppresses but also is incapable of expressing (or does not express) its antigenicity and is protected from immunologic reaction.

Evidence is not sufficiently convincing to support either hypothesis: (1) that the uterus is an immunologically privileged site or (2) that the decidua acts as an immunologic protective mechanism; evidence is adequate, however, to remove doubt that something peculiar to the site of implantation exists and that the junction of trophoblast and decidua is unique in preventing the trophoblast from immunizing the mother or the maternal antibodies from influencing the trophoblast.

D. *The Presence of a Barrier between Mother and Fetus (Molecular, Physical, or Chemical)*

On the assumption that the trophoblast itself is antigenic but incapable of expressing itself, a fascinating theory of a fibrinoidlike layer was presented as early as 1925 by Grosser. In 1953 Medawar suggested that the fetus might owe its privileged position as a homograft to the presence of a physical barrier interposed between maternal and fetal tissues. Kirby and his co-workers (Kirby et al. 1964; Kirby, Billington, and James 1966; and Bradbury, Billington, and Kirby 1965) argued that the trophoblast functions as an immunologic buffer zone between mother and fetus and that the antigens were, in fact, present but were prevented from escaping by a layer of "fibrinoid" which enveloped each trophoblast cell in the placenta. This layer of fibrinoid material, mucopolysaccharide, rich in hyaluronic acid and sialic acid was conspicuously increased at the maternal-fetal junction in the placenta of a hybrid fetus as compared with the mother and fetus when they were antigenically identical. Billingham and Silvers (1963) demonstrated that the hamster cheek-pouch was an immunologically privileged site because there existed in the pouch skin a matrix, histochemically very similar to the placental fibrinoid, which absorbed or blocked the passage of transplantion antigens. Bradbury et al. (1969) showed by means of histochemistry and electron microscopy that there was a highly sulfated mucoprotein on the cell surfaces of the human uterine endometrium, between them and the ectopic trophoblast, and between them and a hydatidiform mole. The ability of the human uterus to secrete this sulfated mucopolysaccharide was established recently by Bo, Moore, and Ashburn (1969). On the other hand, Wynn (1969) showed that the trophoblast and endometrial cells were separated by a fine fibrin layer in only hemochorial placentas. He believed that its failure to appear in all placentas reasoned against its general immunologic significance. He argued that if the fibrinoid layer does confer protection on the trophoblast its function would have to be twofold: to prevent the escape of transplantation antigens from the trophoblast to the mother and to shield the tissue from immunologic attack in previously sensitized mothers. Burstein and Blumenthal in 1969 questioned whether the fibrinoid commonly seen in normal pregnancy might not be a variety of hypersensitivity and the result of an immune reaction rather than a protective layer, in which case this "barrier" need not totally separate the trophoblast from the decidua. No fibrinoid layer could be demonstrated (Tai and Halasz 1967) on the trophoblastic microvilli of rabbits. Of great interest is the work of Potts (1965), who demonstrated that fibrinoid material, lacking at the site of mouse egg implant, develops as a layer 2 days after implantation, and only from this time on can the trophoblast be transplanted without suffering immunologic harm.

Since it is obvious that some kind of unique sequence of events exists at the interface between trophoblast and decidua, it behooves us to examine in more detail this area of confrontation. The syncytiomembrane and capillary endothelium, basement membrane, and attenuated trophoblastic syncytium are the only systems intervening between the immunologically competent cells of the fetus and those of the mother; therefore, the barrier mechanism must reside within this area.

But the trophoblast cells, a very thin layer intervening between the fetal capillaries and the maternal blood in the intervillous spaces, constitute the only continuous unbroken frontier in mammals with hemochorial placentas. Surrounding the intervillous spaces are the decidual tissue of the mother, about which more must be learned; the trophoblastic layer of the placental projections, fetal and thus paternally derived; and the fetal components of the placenta, endothelium, stroma, lymphocytes, and leukocytes. This "barrier" must successfully protect the pregnancy from presensitization. Somewhere within it must lie the explanation of why the lymphocytes of the mother cannot recognize the trophoblast and damage it, why there is no detectable hemagglutinating circulating antibody from the mother to the trophoblast. It should be noted that this barrier does not deny access of all antibodies to the fetus. Passive immunity in many animals is achieved by the selective passage of antibodies from mother to fetus before birth.

A plausible theory that this coat acts as a mechanical barrier preventing the efflux of trophoblast antigens or the entry or passage of antibodies via its electronegative charge which repels patrolling lymphocytes is presented by Currie (1968). Evidence that lymphocytes originating from the mother cause cytolysis *in vitro* of trypsin-dissociated trophoblast cells denuded of coat material while trophoblast cells enclosed within the coat remained unaffected lends credence to this thinking. Such phenomena point to the presence of lymphocytes in the mother potentially capable of destroying the trophoblast and to the trophoblast as a target for immunologic attack. The trophoblast, therefore, in all probability is antigenic. It appears that the trophoblast antigens which induce immune cellular response could be released readily from the time the trophoblast is freed from the zona pellucida up to and including the point of apposition to the uterine epithelium before the coat material is deposited. This concept, although fascinating, is too simple. Several objections to the electrostatic rejection of the cells exist. Since the fibrinoid layer is not present in the mouse trophoblast at the time of implantation, the fertilized egg should be the victim of a preinduced state of immunity. In fact the trophoblast of the conceptus, transferred 2 days after implantation, suffers no immunologic damage, presumably because of the fibrinoid layer present at this later time. Simmons and Russell (1966) suggested that the zona pellucida, since it is selectively impermeable to most proteins and all cells, is the mechanism for the survival of the fertilized egg prior to the formation of an effective trophoblastic barrier in the female who has been rendered immune to her mate. The trophoblast normally is completely formed inside the zona pellucida prior to zonal rupture. If the zona acts as an immunologic barrier, its persistence in a normal way explains the successful proliferation of eggs transplanted to immune recipients. If the zona is lost prematurely in some ectopic implantation, as it is in delayed intrauterine implantation, the ovum is vulnerable to circulating antibodies and the trophoblast does not develop. Potts (1965) showed that only at implantation, when fetal and maternal cells are not separated by fibrinoid material, is the egg susceptible to immunologic attack—which raises serious questions concerning the fibrinoid sialomucin layer theory.

The obvious conclusion can be drawn that a very clear understanding of the key role of the trophoblastic tissue is mandatory for insight into the potential immunologic problems of pregnancy. This then leads us to two simple questions: Is there any evidence for immunologic phenomena as a result of pregnancy? Is the trophoblast antigenic or not?

II. Evidence for Immunologic Phenomena of Pregnancy

The probability that the fibrinoid layer is absent in inbred matings is of interest when put into context with the observation of Billington (1964) that size of placentas in interstrain matings in mice is roughly proportional to the genetic disparity between mother and fetus. James (1965) noted an effect on placental size when the mother had been immunized against paternal antigens. Mothers mated with strains against which they had been immunized revealed larger placentas than did the nonimmunized mothers. Kaliss (1966) proposed "immunologic enhancement" as a potential mechanism in which the growth of certain tumors is increased by specific immunization. Larger placentas in genetically different strains is of clinical interest, especially as they are present in Rh incompatibility, a phenomenon which led Kirby to state that the genetic disparity may not be disadvantageous, perhaps even beneficial. He questions whether the larger placentas from genetically dissimilar matings are due to an immunologic reaction between mother and fetus, or to the occurrence of similar genotypic complementation or facilitation in which each inbred genotype compensates for the genetic deficiencies of the other (Clarke and Kirby 1967). Mouse hybrid fetuses are significantly heavier than purebred fetuses at a similar age of gestation (Billington 1963; James 1967), and hybrid mice are heavier at birth than inbred mice in the same maternal strain (McLaren 1965), subject to modification by intrauterine environmental factors.

Gates, Doyle, and Noyes (1961) found that mouse F_1 hybrid blastocysts were larger and had more cells at a given postovulatory age than those of either parental inbred strain. This evidence of selective fertilization is supported by the work of McLaren and Michie (1963) and Michie and Anderson (1966).

III. Is the Trophoblast Antigenic?

A problem arises when we come to the area of confrontation between the uterine epithelium and decidual cells on the one hand and the syncytiotrophoblast on the other, with the vascular area between them and the possibility of the appearance of an amorphous mucopolysaccharide or *sialomucin layer:* we must determine the nature of the two interfaces and learn whether any reaction or interaction takes place.

If there is an interaction, is it a chemical, pharmacologic, or immunologic? If it is an immunologic reaction, we must show evidence of antigenicity of the trophoblast and explain (1) why the trophoblast is a selective barrier, (2) why lymphocytes of the mother cannot recognize the trophoblast, and (3) why the lymphocytes cannot damage the trophoblast. The problem would be simple if it were possible to isolate the trophoblast and define the nature of a pure trophoblastic antigen. An antibody would then be available, and passive transfer of this antibody would be a test for immunologic activity. An antibody to a pure antigen having been produced, it would be possible to label it and determine precisely where the antigen was located. There is evidence of antigen-antibody reaction in most species, including the human. Trophoblast antigenicity has been demonstrated by altered challenge graft survival (Hulka and Mohr 1968) wherein adult C57 BL–6J male mice received either a primary ectopic transplant or a primary and a challenge ectopic transplant of trophoblast tissue obtained from the ectoplacental cones of 7½-day-old C3H–HeJ embryos. Gross and histologic examination of these grafts at 5, 7, and 12 days indicated that growth in the challenge grafts was inhibited; few were grossly successful; they had smaller hemorrhagic reactions; there were fewer viable cells at all stages of growth; and there was better host containment of the colony. Such evidence indicates that exposure to pure trophoblast alters an animal's subsequent reaction to grafts syngeneic with the original trophoblast and can be explained as evidence of immunologic sensitization of the host. In an attempt to elaborate further the antigenicity of the trophoblast and the role of the microenvironment at the interface between maternal and trophoblast and the role of the microenvironment at the interface between maternal and trophoblastic cells, a series of experiments were carried out in our laboratory on the mouse.

IV. Experimental Studies on Trophoblastic Antigenicity

When the 4- to 5-day mouse blastocysts are washed out of the cornua and transplanted under the kidney capsules of homologous males or females, the vast majority of trophoblasts proliferate, but the embryos die. The trophoblast is surrounded by a small hemorrhagic layer with virtually no round-cell infiltration. It survives for approximately 12–15 days, the length of time the trophoblast would have survived had it implanted in the cornua. (See fig. 1 for summary of experiments.) It is possible to separate the ectocone (purely trophoblast) and the embryonic plate structure of the 7–9-day blastocyst (at which time the ectocone in the implanted blastocyst is clearly differentiated) and transplant them separately under the kidney capsule (fig. 1 [2 and 3]). The trophoblast survives without any difficulty and lives for another 8–10 days, approximately the normal length of gestation. The embryonic part succumbs immediately, and there is considerable round-cell infiltration characteristic of an immune reaction surrounding the site of implantation. The antigenicity of the embryo, at least, seems clearly defined. If the 4–5-day blastocyst is implanted under the kidney capsule and allowed to proliferate for 8–10 days, it is possible to transplant this trophoblastic nodule to a monolayer tissue culture (Koren and Behrman 1968), at which time the trophoblast proliferates into

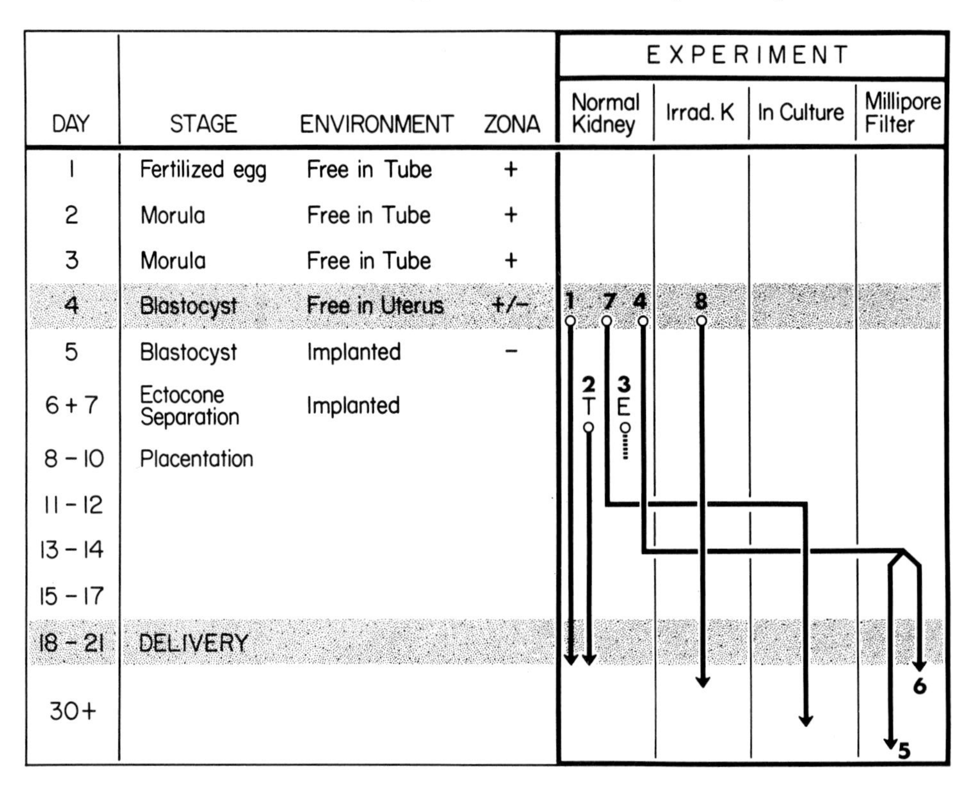

Fig. 1. Composite of stage of growth of blastocyst with environment, state of zona pellucida, and the results of a variety of experiments. (1) Free blastocyst implanted under kidney capsule. (2) Trophoblast ectocone (*T*) implanted under capsule. (3) Embryo (*E*). (4) Free blastocyst implanted under kidney capsule and transplanted in millipore filter (5) with membrane and (6) without membrane into peritoneum; (7) blastocyst under kidney capsule transferred to monolayer culture and (8) blastocyst implanted under kidney capsule of irradiated mouse.

two major cell types: (1) the small trophoblastic cells derived from the primary villous trophoblast, which have glycogen-storing functions, and (2) the larger and flatter cells with long, cytoplasmic processes. As the culture ages, the latter develop into more predominant giant cells with widely spread cytoplasm and notable variation in nuclear morphology. (The ability to culture trophoblast free of enzymatic treatment makes this technique extremely valuable for immunologic study.) The trophoblastic cells in such cultures grow for an additional period of 30 days (fig. 1 [7]), an obvious increase in cell life span to a total of at least 40 days. This indicates that either an interplay between the trophoblastic cell and the kidney at the site of implantation forms a complex interdependent system governing its growth and life span, or that in culture the cells that migrate from the trophoblast grow and survive independently when not influenced by maternal host factors.

Noyes (1959) suggested that the degree of trophoblastic penetration and growth could be determined not only by the erosive and proliferative activity of the trophoblastic cells but also by receptive and defensive mechanisms in the endometrium and in the maternal vascular system. By transplanting fertilized ova from whole-body irradiated inbred C57 BL–6J mice under the kidney capsule (Koren, Abrams, and Behrman 1968*a*), it was possible to show that the transplanted tissue grows rapidly, developing into grossly recognizable nodules of trophoblastic giant cells with a life span of 12–14 days. The trophoblastic tissue in the irradiated animals was surrounded by hematoma 2–5 times larger than in the nonirradiated controls; sections from irradiated recipients contained 2–3 times the number of trophoblastic cells as the nonirradiated controls; the trophoblast was found in irradiated animals as late as 16 days after transplantation when the nodules could no longer be identified in the controls; in the nonirradiated animals the nodules were bordered by a zone of slight fibroblastic proliferation, whereas in the irradiated mice the cellular reaction was definitely diminished. We observe once again that the normal life span of trophoblastic cells in the control animals is limited by the length of gestation, but survival of the trophoblast in the irradiated animals is far longer. This suggests that an intrinsic property of the cells is not exclusively responsible for the trophoblast's limited life span *in vivo;* an additional environmental host factor may operate as well.

Another series of experiments showed that successful homograft is governed (1) directly through physical and cellular contact between two genetically different cell populations and indirectly through humoral antibodies via the extracellular fluid microenvironment wherein these two components become operative (Koren et al. 1970). Blastocysts implanted under the kidney capsule were excised 9–10 days after transfer. Upon careful removal of the kidney capsule the nodules were placed in sterile plastic petri dishes containing Hanks's balanced salt solution at room temperature. After the tissue was gently minced with a scalpel and washed 2 or 3 times, 0.5 mm of tissue was put into a diffusion chamber, consisting of stainless steel screen tubes coated with millipore material (0.45 mμ, pore size), approximately 100 mμ thick (obtained from the Millipore Filter Corporation, Bedford, Massachusetts). By this technique we were able to prevent direct contact between the trophoblastic cells and their environment while allowing the host's extracellular fluids to reach the trophoblastic tissue in the diffusion chamber.

In a second set of diffusion chambers the millipore material was not employed, thus permitting contact between the trophoblast and the host cellular material. Forty-eight mice divided into three groups were studied: (1) donors and recipients of the same inbred strain; (2) donors and recipients from noninbred strains; and (3) donors and recipients of Spartan strain. Each recipient host received two diffusion chambers, one with and one without a membrane. No significant differences in survival of trophoblastic cells were found in the inbred and noninbred strains of mice. At the 8- and 11-day intervals trophoblast was identified in the membrane-covered chambers but not in the open-screen tubes. Once again the trophoblast survived for the duration of normal gestation, but only those trophoblastic cells that were protected from the maternal cellular elements. The results suggest that direct contact of cellular material between maternal and trophoblast cells rather than some simpler humoral interaction is a factor of major importance in the destruction and death, and thus control, of trophoblastic tissues (fig. 1 [5 and 6]).

Fluorescent-tagged globulins from rabbit antimouse placental serum were applied to liver, spleen, uterus implant, and conceptus by both the direct and indirect methods, and to cultures of trophoblast that had been collected and quick frozen at —70° C. The results indicate that by the direct stain all antiplacental sera showed positive staining with pure trophoblastic cell cultures 3–12 days old; negative results were obtained with control normal rabbit or mouse serum. Control cultures of placenta gave positive results also. The indirect stain showed strong positive fluorescence. Passive immunization of pregnant mice by intravenous injection of antiserum obtained from rabbits immunized with placental cell suspensions resulted in indirect positive fluorescence localized only in the placenta and conceptus of these specimens; greatest positive fluorescence was found in the peripheral placenta, and specifically in the trophoblastic cells (Koren, Behrman, and Paine 1969). The control liver, kidney, and spleen preparations were negative throughout the fluorescent staining procedure. Thus, under certain conditions, placental tissue and trophoblast cells, in particular, are capable of expressing antigenic properties as indicated by fluorescent techniques both *in vitro* and *in vivo.*

To prove the antigenicity of the mouse placental tissue by disrupting the placentomaternal relationships through passive immunization, further studies were attempted. Rabbit antimouse trophoblast was produced by immunizing rabbits with placental homogenates and placental suspensions of mice (Koren, Abrams, and Behrman 1968*b*). By hemagglutination, microimmunodiffusion, and immunoelectrophoresis, rabbit antimouse trophoblast was produced, decomplemented, and then absorbed with rabbit red blood cells, serum, and kidney. Antiplacental serum was then injected into the pregnant mice and, by the first or certainly by the second injection, the pregnant mice showed vaginal bleeding never seen in mice given normal rabbit serum. Thirty animals were immunized passively and studied well beyond the expected end of gestation. Of this group only one animal delivered; she had but a single fetus. In the control group injected with normal rabbit serum the pregnancies progressed normally, and every one of the 25 control mice delivered a litter with an average of 6–10 fetuses (see fig. 2).

Brent, Averich, and Drapiewski (1961) found that there was cross-reactivity of antiplacental serum with kidney, and varying doses of this serum resulted in renal damage in all cases and damage to the fetus ranging from abnormalities to a high rate of early abortion. In our animals significant abnormalities were found in the liver and kidneys as well; and the presence of membranous and proliferative glomerular changes suggested the possibility of an immunologically induced glomerular nephritis as well as an impure antiserum. We concluded that the damage to the kidneys was caused by the rabbit globulin improperly removed, rather than the placental or trophoblastic antibody; and since the placental preparations are complex extracts and the resulting antibodies are obviously polyvalent, it was necessary to prove that the abortion was directly due to the antiplacental moiety of the antiserum as administered. Considering the polyvalence and impurity of the antiserum and the renal lesion produced, another experiment was performed wherein a 1-mm piece of trophoblastic nodule homogenate, emulsified with an equal volume of Freund's incomplete adjuvant, was injected into rabbits. The rabbits were bled 4–5 days after the last injection; the serum was pooled and heated at 56° C for 30 min to destroy complement; and the titer for antibodies was determined. To eliminate nonspecific antiserum, mouse red blood cells and lyophilized serum, as well as lyophilized kidney, were used to absorb the antiserum. The antiserum was then put over DEAE cellulose column and then by stepwise elution it was run through paper chromatography. Chromatography showed that the rabbit antimouse trophoblast serum divided into 4 protein fractions: fraction A consisted of approximately 90% pure gamma globulin as shown by agar electrophoresis; fraction B contained beta globulin with some albumin; fraction C, albumin with some contamination of beta and alpha globulin; and fraction D, alpha globulin with minor amounts of albumin and beta globulin. Only fraction A revealed antibodies to

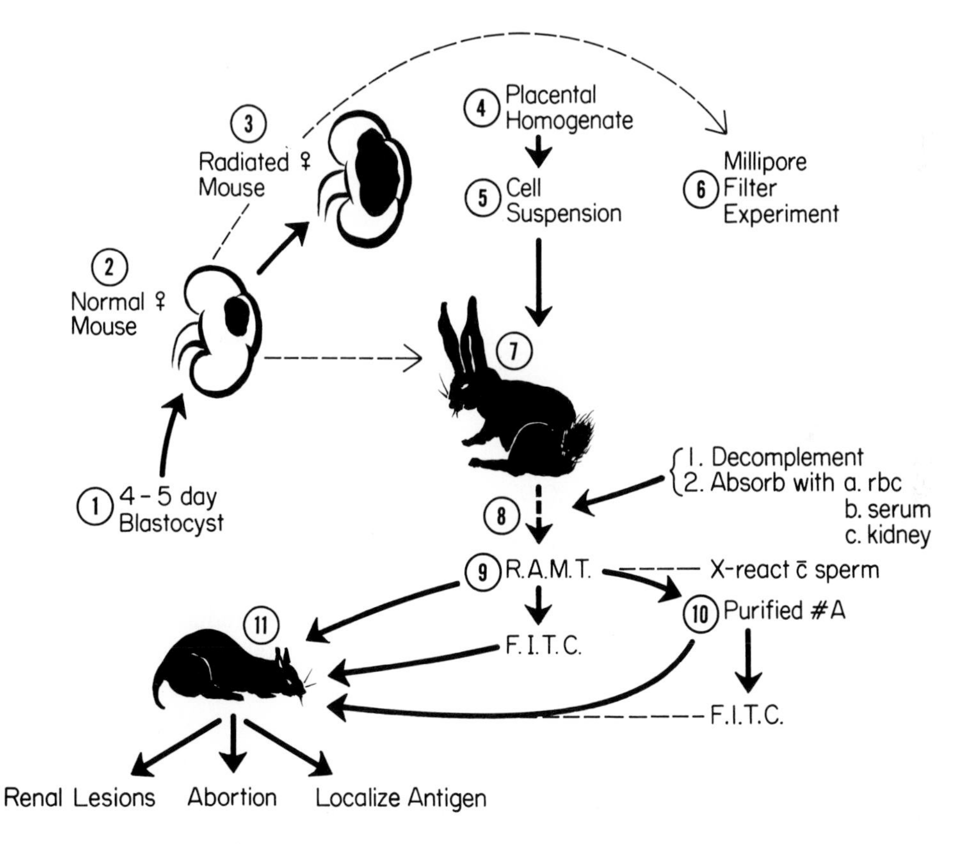

Fig. 2. Diagrammatic representation of experiments performed to support evidence of antigenicity of the placenta. The mouse trophoblast was prepared by (1) implanting 4–5-day blastocysts under (2) the kidney capsule; or from (4) placental homogenate or (5) placental cell suspension. By immunizing the rabbit (9) rabbit antimouse trophoblast (RAMT) was prepared and labeled with fluorescent isothiocyanate. Passive immunization of the pregnant mouse (11) resulted in abortion. Purified fraction A (10) of this antiserum also resulted in abortion.

the trophoblast by hemagglutination, immunodiffusion, and electrophoretic analysis. Thus a reasonable purification of rabbit antimouse trophoblast serum was obtained (see fig. 3).

Twelve pregnant animals at 5, 7, 12, and 16 days of gestation were immunized passively with 2 mg of fraction A of the purified gamma globulin; all of these aborted. Liver, spleen, kidney, uterus, and conceptus of all of these were examined histologically. It is interesting to note that not a single lesion was found in the kidney and there was only an occasional evidence of liver necrosis. Again we found that the antibody given passively acted on the target cells of the trophoblast. By

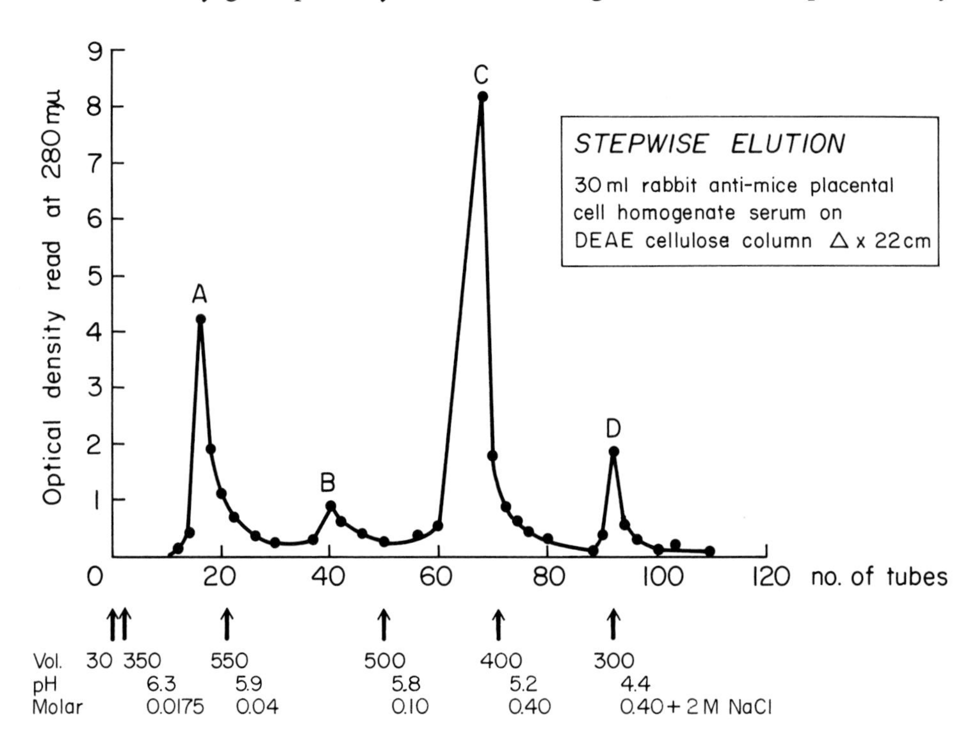

Fig. 3. Stepwise elution; 30 ml rabbit antimouse placental cell homogenate serum on DEAE cellulose column △ × 22 cm.

purifying the antibody, the extent of the renal lesions was reduced considerably. Therefore the renal lesion may not be the result of a placental antibody as suggested by Brent.

V. Summary

It appears that the trophoblast can be antigenic under certain experimental circumstances both *in vivo* and *in vitro,* at the time of implantation, but for some reason as yet undetermined, this antigenicity is not expressed or "turned on." There is evidence that the life span of the trophoblast, but not its invasiveness, is to a great extent governed by the host immunologic reaction and that the important host reaction at the implantation site is cellular rather than humoral. There is evidence of immune reaction but it is attenuated in its severity. One may well ask, May this reaction be protective rather than destructive? Does the phagocytic action of the trophoblast giant cells protect against the antibody? As the trophoblast is antigenic and yet does not create any locally destructive immunologic reaction under normal circumstances, it becomes doubly important to know why the antigenicity is not expressed. Is this a combination of fetal defense by the trophoblast on the one side and suppression of maternal response by hormones on the other side? Are we dealing with immune tolerance or enhancement?

It is conceivable that the trophoblast recognizes its genetic disaffinity with the uterine epithelium through the exchange of information facilitated, in the case of the mouse, by intercellular bridges. RNA transfer of information between the trophoblast and epithelium was proposed by Wilson and Wecker in 1966. Jones and Kemp (1969) suggested that when the trophoblast gives its information to the uterine epithelium, the uterus responds by producing sialomucin which coats the trophoblast, other than the points of adhesion, and isolates it from the epithelium, making it nonadhesive and protecting it from invasion by lymphocytes. On this basis the RNA transfer might be construed as an aid in producing tolerance of the mother to this "foreign protein," or solely as a stimulus for the release of sialomucin, or as a "switch" on the part of the maternal cells to recognize maternal-type genes in the trophoblast cells. That transfer of information from the trophoblast to the mother occurs is evidenced clearly by the following fascinating studies.

Tuffrey, Bishun, and Barnes (1969) studied chimeras. Mouse blastocysts were transferred to uteri of females with distinct marker chromosomes wherein their own blastocysts were growing also. At delivery the spleen, bone marrow, and lymph nodes in nearly all of the offspring from the transferred blastocysts possessed marker cells from the mother or the mother's own fetuses. These colonizing cells from the mother might well have induced tolerance of the fetuses to maternal antigens as maternal skin persists on the fetuses. On the other hand the defensive mechanism of the trophoblast is illustrated clinically by Hulka and Mohr (1968) and Burstein and Blumenthal (1969). They observed maternal serum globulin binding by placental trophoblast of the patient providing the serum and showed that the binding substance in the maternal serum is gamma (G) globulin and possibly also complement. The evidence is strong that the trophoblast prevents antibodies of the maternal serum from reaching the fetus. Whereas this may be true in the first trimester, there is as yet no evidence that such a mechanism exists from the first few days of implantation up to the point of rejection on the 10th day. This may well be a fruitful area of future research.

References

Bagshawe, K. D. 1967. Immunological aspects of trophoblast. *J Obstet Gynaec Brit Comm* 74:829.

Behrman, S. J., and Koren, Z. 1968. Immunology of the conceptus. In *The yearbook of obstetrics and gynecology,* ed. J. P. Greenhill, p. 28. Chicago: Year Book Medical Publishers.

Behrman, S. J. and Sawada, Y. 1966. Heterologous and homologous inseminations with human semen frozen and stored in a liquid-nitrogen refrigerator. *Fertil Steril* 17:457.

Billingham, R. E. 1964. Transplantation immunity and the maternal-fetal relation. *New Eng J Med* 270:667.

Billingham, R. E., and Silvers, W. K. 1963. Sensitivity to homografts of normal tissues and cells. *Ann Rev Microbiol* 17:531.

Billington, W. D. 1963. Studies on the implantation of the mammalian embryo. Ph.D. thesis, University of Wales.

———. 1964. Influence of immunological dissimilarity of mother and foetus on size of placenta in mice. *Nature (London)* 202:317.

Bo, W. J.; Moore, P. J.; and Ashburn, M. J. 1969. The effect of a foreign body on glycogen and sulfomucopolysaccharides of the uterus. *Fertil Steril* 20:351.

Bradbury, S.; Billington, W. D.; and Kirby, D. R. S. 1965. A histochemical and electron microscopical study of the fibrinoid of the mouse placenta. *J Roy Micr Soc* 84:199.

Bradbury, S.; Billington, W. D.; Kirby, D. R. S.; and Williams, E. A. 1969. Surface mucin of human trophoblast. *Amer J Obstet Gynec* 104:416.

Brambell, F. W. R. 1965. In *The early conceptus, normal and abnormal,* ed. W. W. Park. Edinburgh: E. & S. Livingstone.

Brent, R. L.; Averich, E.; and Drapiewski, V. A. 1961. Production of congenital malformations using tissue antibodies: I. Kidney antisera. *Proc Soc Exp Biol Med* 106:523.

Burstein, R. H., and Blumenthal, H. T. 1969. The immune reactions of normal pregnancy. *Amer J Obstet Gynec* 104:671.

Clarke, B., and Kirby, D. R. S. 1967. Some genetic aspects of fertilization, implantation and development. *J Obstet Gynaec Brit Comm* 74:839.

Comings, D. E. 1967. Lymphocyte transformation in response to phytohemagglutinin during and following pregnancy. *Amer J Obstet Gynec* 97:213.

Creger, W. P., and Steele, M. R. 1957. Human fetomaternal passage of erythrocytes. *New Eng J Med* 256:158.

Currie, G. A. 1968. *In-vitro* studies of the foeto-maternal relationship. *J Reprod Fertil* Suppl. 3, *Immunological aspects of pregnancy.* Proc I Sympos Soc Study Fertil. Exeter: Blackwell Sci Publ.

Dancis, J.; Samuels, B. D.; and Douglas, G. W. 1962. Immunological competence of the placenta. *Science* 136:382.

Douglas, G. W.; Thomas, L.; Carr, M.; Cullen, N. M.; and Morris, R. 1959. Trophoblast in circulating blood during pregnancy. *Amer J Obstet Gynec* 78:960.

Edidin, M. 1964. Transplantation antigens in the mouse embryo: Fate of early embryo tissues transplanted to adult hosts. *J Embryol Exp Morph* 12:309.

Gates, A. H.; Doyle, L. L.; and Noyes, R. W. 1961. A physiologic basis for heterosis in hybrid mouse foetuses. *Amer Zool* 1:449.

Germuth, F. G. 1956. The role of adrenocortical steroids in infection, immunity and hypersensitivity. *Pharm Rev* 8:1–24.

Grosser, O. 1925. Ueber Fibrin und Fibrinoid in der Placenta. *Z Anat* 76:304.

Hulka, J. F., and Mohr, K. 1968. Trophoblast antigenicity demonstrated by altered challenge graft survival. *Science* 161:696.

James, D. A. 1965. Effects of antigenic dissimilarity between mother and foetus on placental size in mice. *Nature (London)* 205:613.

———. 1967. Some effects of immunological factors on gestation in mice. *J Reprod Fertil* 14:265.

Jones, B. M., and Kemp, R. B. 1969. Self-isolation of the foetal trophoblast. *Nature* 221:829.

Kaliss, N. 1966. Immunologic enhancement: Conditions for its expression and its relevance for grafts of normal tissues. *Ann N Y Acad Sci.* 129:155.

Kirby, D. R. S. 1968*a*. Transplantation and pregnancy. In *Human transplantation,* ed. F. T. Rapaport and J. Dausset. New York: Grune & Stratton.

———. 1968*b*. Immunological aspects of pregnancy. In *Advances in reproduction,* ed. A. McLaren, 3:33. New York: Academic Press.

Kirby, D. R. S.; Billington, W. D.; Bradbury, S.; and Goldstein, D. J. 1964. Antigen barrier of the mouse placenta. *Nature (London)* 204:548.

Kirby, D. R. S.; Billington, W. D.; and James, D. A. 1966. Transplantation of eggs to the kidney and uterus of immunized mice. *Transplantation* 4:713.

Koren, Z.; Abrams, G.; and Behrman, S. J. 1968*a*. The role of host factors in mouse trophoblastic tissue growth. *Amer J Obstet Gynec* 100:570.

———. 1968*b*. Antigenicity of mouse placental tissue. *Amer J. Obstet Gynec* 102:340.

Koren, Z., and Behrman, S. J. 1968. Organ culture of pure mouse trophoblast. *Amer J Obstet Gynec* 100:576.

Koren, Z.; Behrman, S. J.; and Paine, P. J. 1969. Antigenicity of trophoblastic cells indicated by fluorescein technique. *Amer J Obstet Gynec* 104:50.

Koren, Z.; Srivannaboon, S.; Abrams, G.; and Behrman, S. J. 1970. The role of microenvironment in trophoblastic tissue growth and invasion. In press.

Lanman, J. T. 1965. Transplantation immunity in mammalian pregnancy: Mechanisms of fetal protection against immunologic rejection. *J Pediat* 66:525.

Little, C. C. 1924. Genetics of tissue transplantation in mammals. *J Cancer Res* 8:75.

McLaren, A. 1965. Genetic environmental effects on the fetal and placental growth in mice. *J Reprod Fertil* 9:75.

McLaren, A., and Michie, D. 1963. Nature of the systemic effect of litter size on gestation period in mice. *J Reprod Fertil* 6:139.

Marcuse, P. M. 1954. Pulmonary syncytial giant cells embolism: Report of maternal death. *Obstet Gynec* 3:210.

Medawar, P. B. 1953. Some immunological and endocrinological problems raised by evolution of viviparity in vertebrates. In Soc for Exp Biol. *Evolution VII.* Symposia of the Society for Experimental Biology, no. 7, p. 320. Cambridge: at the University Press.

Menge, A. C. 1969. Early embryo mortality in heifers isoimmunized with semen and conceptus. *J Reprod Fertil* 18:67.

Michie, D., and Anderson, N. F. 1966. A strong selective effect associated with a histocompatibility gene in the rat. *Ann N Y Acad Sci* 129:88.

Noyes, R. W. 1959. Trophoblast: Problems of invasion and transport. *Ann N Y Acad Sci* 80:54.

Payne, R. 1962. The development and persistence of leukoagglutinins in parous women. *Blood* 19:411.

Potts, D. M. 1965. Implantation: An electronmicroscopical study with special reference to the mouse. Ph.D. thesis, Cambridge University.

Schlesinger, M. 1962. Uterus of rodents as site for manifestation of transplantation immunity against transplantable tumors. *J Nat Cancer Inst* 28:927.

Schwimmer, W. B.; Ustay, K. A.; and Behrman, S. J. 1967. An evaluation of immunologic factors of infertility. *Fertil Steril* 18:167.

Simmons, R. L., and Russell, P. S. 1962. Antigenicity of mouse trophoblast. *Ann N Y Acad Sci* 99:717.

———. 1963. Potential immunologic interactions of the placental site. Presented at II Rochester Trophoblast Conf.

———. 1965. Histocompatibility antigens in transplanted mouse eggs. *Nature (London)* 208:698.

———. 1966. In *Abstract VII Internat Transplantation Conference.*

Tai, Chiaki, and Halasz, N. A. 1967. Histocompatibility antigen transfer *in utero:* Tolerance in progeny and sensitization in mother. *Science* 158:125.

Tuffrey, M.; Bishun, N. P.; and Barnes, R. D. 1969. Do maternal cells enter the foetus? *Nature (London)* 221:1029.

Uhr, J. W.; Dancis, J.; Franklin, E. C.; Finkelstein, M. S., and Lewis, E. W. 1962. The antibody response to bacteriophase phi-X 174 in newborn premature infants. *J Clin Invest* 41:1509.

Wilson, D. B., and Wecker, E. E. 1966. Quantitative studies on the behavior of sensitized lymphoid cells *in vitro:* III. Conversion of "normal" lymphoid cells to an immunologically active status with RNA derived from isologous lymphoid tissues of specifically immunized rats. *J Immunol* 97:512.

Woodruff, M. F. A. 1958. Transplantation immunity and the immunological problem of pregnancy. *Proc Roy Soc B* 148:68.

Wynn, R. M. 1969. Noncellular components of the placenta. *Amer J Obstet Gynec* 103:723.

29

Immunological Implications of Comparative Placental Ultrastructure

Ralph M. Wynn

Department of Obstetrics and Gynecology
University of Illinois at the Medical Center
Chicago

The enigma of immunological protection of the placenta is thoroughly discussed in Billingham's review (1964) and in numerous subsequent reports for which it provided the impetus. It now appears that, of the four major hypotheses he suggested, three have been all but discarded. Fetal somatic tissues have been shown to be antigenic; immunologic reactivity has been demonstrated in the mother; and the uterus has not proved to be an immunologically privileged site in the customary sense. There is now a virtual consensus that the immunologic barrier is anatomic, namely the trophoblast and its pericellular deposits.

In assessing the role of extracellular components as immunological barriers it is important to distinguish several possibly related, but morphologically and histogenetically discrete structures. The ensuing paragraphs will define and describe decidual capsular material, histologically demonstrable fibrinoid, hypertrophied basal laminae, trophoblastic terminal webs (which may be confused with extracellular deposits), and ultrastructurally detectable sialomucins. Secondary effects of cellular interaction of trophoblast and endometrium will be differentiated from primary immunological factors in protection of the trophoblast.

I. Ultrastructure of the Placental Membrane

Comparative electron microscopy of the placenta has supported the concept of the prime role of the trophoblast in maintaining the immunological barrier (Enders 1965; Mossman 1967; Wynn 1967*b,* 1968, 1969). In all placentas examined by electron microscopy at least 1 layer of trophoblast has been found to persist. Many of the rodents and lagomorphs, formerly thought to have hemoendothelial placentas (Mossman 1937), retain 1 or more layers of trophoblast to the end of gestation. In the rabbit 2 layers of trophoblast persist, forming the so-called hemodichorial placenta (Enders 1965; Wynn and Davies 1965) (fig. 1). Enders (1965), Mossman (1967), and Wynn (1968), among others, have illustrated a variety of ultrastructural modifications of the trophoblast of the rodents. The attenuated barrier of the mouse, rat, and hamster (fig. 2) consists of 3 layers of trophoblast, the hemotrichorial condition. It appears that the placentas of essentially all rodents comprise some type of hemochorial membrane.

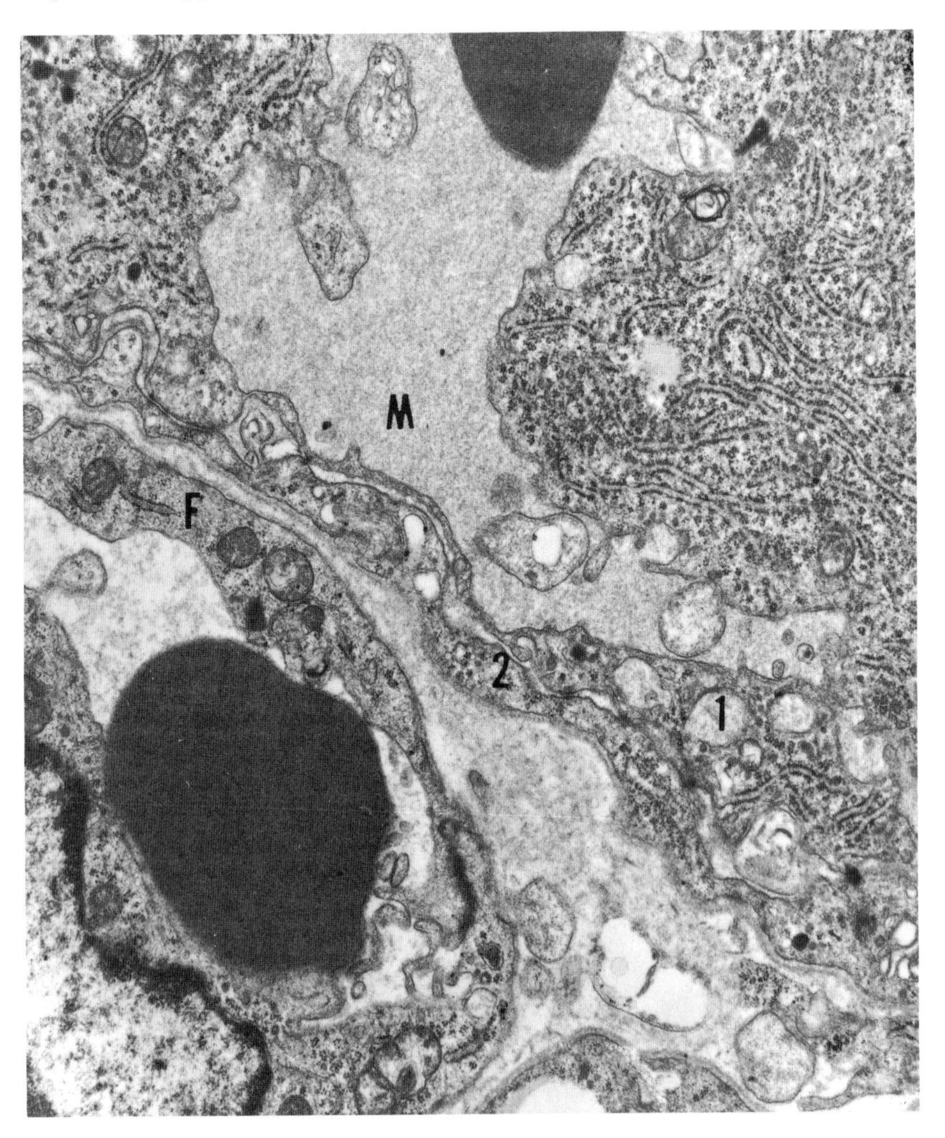

Fig. 1. Hemodichorial placental membrane of rabbit showing 2-layered trophoblast (1, 2), fetal capillary (*F*), and maternal blood space (*M*). Note the well-developed endoplasmic reticulum of the trophoblast. ×15,000.

Two genera, *Castor,* the beaver, and *Pedetes,* the spring hare, formerly thought to retain maternal endothelium, have recently been reclassified as hemodichorial (Fischer and Luckett 1969). One possible exception is the kangaroo rat, *Dipodomys,* which was recently reported to have an endotheliochorial placenta (King and Tibbitts 1969). It is possible, however, that the cellular layer lining the maternal blood spaces in this geomyid rodent has displaced the maternal endothelium at an earlier stage of development to form a hemochorial placenta as in other members of the order Rodentia. In that case the placenta of *Dipodomys* would be hemodi-

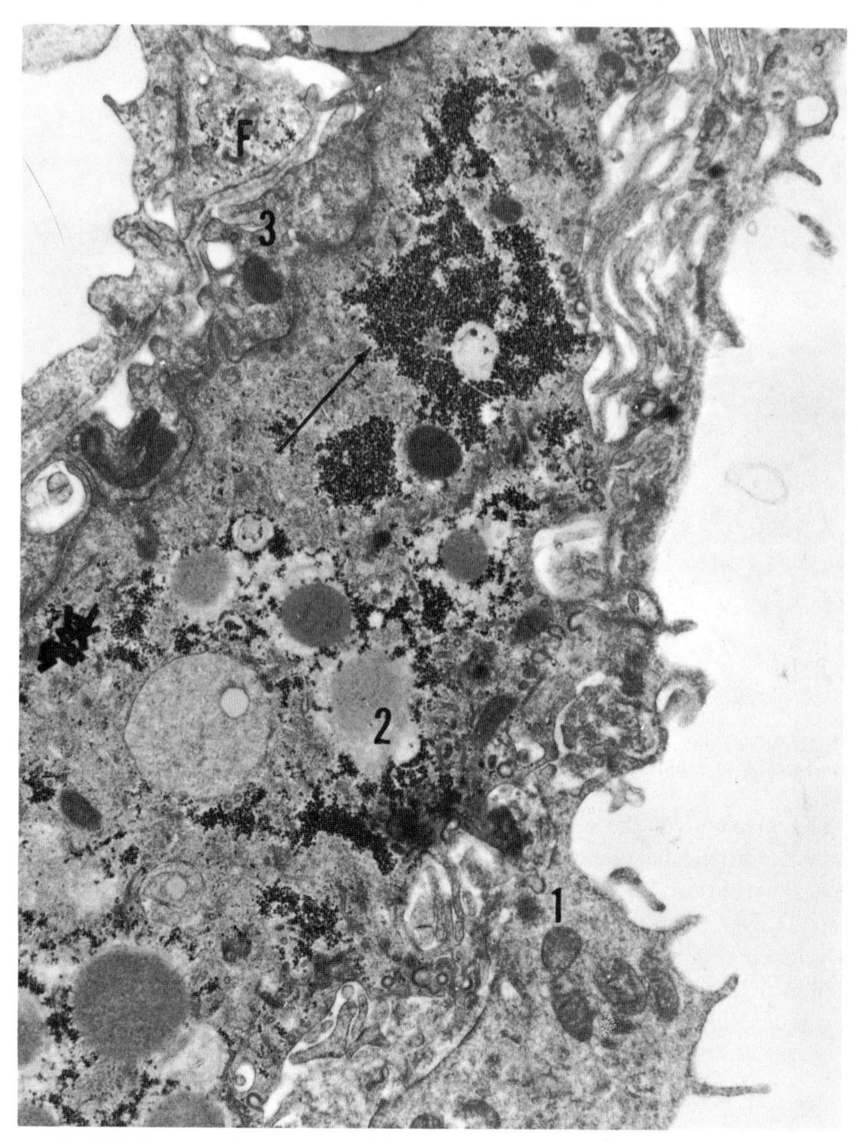

Fig. 2. Hemotrichorial membrane of golden hamster showing 3-layered trophoblast (1, 2, 3), the outer layer of which is cellular. Note fetal capillary endothelium (*F*) and large deposits of glycogen in the middle layer of the trophoblast (*arrow*). $\times$15,000.

chorial also. The acellular structure beneath the layer lining the maternal blood spaces could be the retained basal lamina of the original maternal endothelium as in the placentas of several members of the order Chiroptera (Enders and Wimsatt 1968). The placentas of these bats and of *Dipodomys* differ from the typically endotheliochorial organs of most carnivores in that the maternal endothelium in the cat (Wynn and Björkman 1968) and dog (Wynn and Corbett 1969), for example, consists of hypertrophied cells that rest on a greatly thickened basal lamina or intermediate layer.

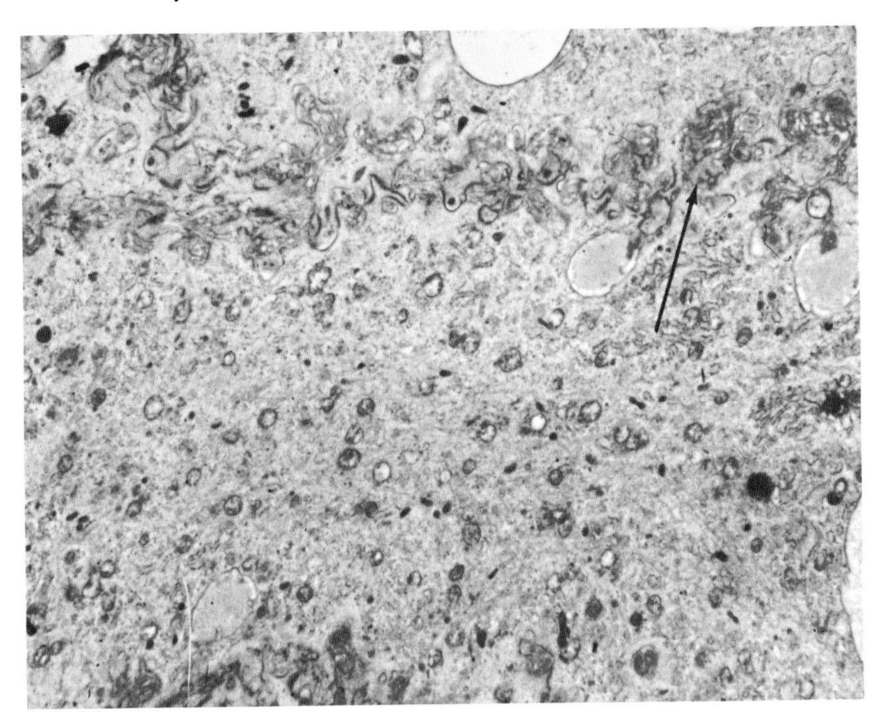

Fig. 3. Basal decidua of guinea pig showing pericellular capsular material (*arrow*) similar in electron-density to a basal lamina. ×11,000.

The placenta of the shrew *Sorex* is still tentatively classified as endothelio-endothelial on the basis of light-microscopic examination (Mossman and Owers 1963). It is reasonable to predict, however, that electron microscopy will reveal at least 1 layer of trophoblast or an endotheliochorial labyrinth. If trophoblast is indeed lacking in this genus, immunological protection of its placenta must be attributed to another tissue. In that case the noncellular components may assume even greater significance in the immunology of placentation.

II. Decidual Capsular Material

Some 6 years ago we began a comparative ultrastructural examination of the decidua. In the guinea pig (Wynn 1964), the first animal studied, we described the pericellular capsular material (fig. 3), which was most prominent in the basal decidua, and attempted to assess the role of this noncellular barrier in the limitation of trophoblastic invasion and, possibly, in the immunological regulation of normal pregnancy. In the guinea pig's deciduotrophoblastic junctional zone, tissues of fetal and maternal origin are intimately juxtaposed. The decidual pericellular deposit corresponds to the amylase-resistant, periodic acid-Schiff (PAS)-positive beaded capsules noted with light microscopy. Although complex invaginations of the decidual plasma membranes characterize both basal and parietal deciduae, the endometrium closest to the trophoblast exhibits the most prominent accumulations of homogeneous material of moderate electron density within the convolutions. Near the surface of the basal decidua many granules appear to empty their contents into recesses formed by inflections of the plasma membranes into the decidual cytoplasm. The homogeneity and electron density of the material in some of the cytoplasmic granules near the surface of these cells are identical with those of the deposits outside the cell and in the subendothelial spaces around the larger vessels of the decidua.

In the guinea pig the invasive trophoblast is separated from viable decidua by a zone of necrotic tissue. At the junction of viable and necrotic decidua the apparently normal tissue contains large numbers of Golgi complexes and lysosome-like dense bodies. At this point leukocytic infiltration ceases and trophoblastic invasion is apparently halted. Although trophoblast may be surrounded by a similar amorphous deposit, much of the pericellular material in the guinea pig is elaborated by the decidual cytoplasm.

The highly invasive trophoblast of the guinea pig is purely syncytial, forming a hemomonochorial labyrinth (Enders 1965; Wynn 1968). In contrast, in the mouse the invasive trophoblast of the early placenta is cellular (Kirby and Malhotra 1964). The persistence of a cellular layer of trophoblast has been demonstrated in the definitive placentas of several other rodents (Enders 1965; Mossman 1967; Wynn 1968). Syncytial transformation is thus prerequisite to neither trophoblastic invasiveness nor the formation of noncellular barriers.

Although no satisfactory evidence has been produced to show that the uterus is a privileged site in the same sense as the anterior chamber of the eye, Billington (1967) has shown that mouse eggs transferred to hyperimmune hosts grow better in the uterus than in the kidney. He concluded that the uterus provides some protection against the "immunological" reaction that prevents the development of eggs transferred to the kidney. The success of ectopic pregnancies in the human and of normal gestation in species with adeciduate placentation, however, casts doubt on the prime importance of the decidua and its capsular material in immunological protection, although their mechanical function in stemming trophoblastic invasion cannot be denied.

To assess the role of the trophoblast in inducing the formation of this capsular material we attempted to compare the decidual reactions of the hamster in true pregnancy (fig. 4) and in air-induced deciduomas (fig. 5) (Orsini et al. 1970). Although the ultrastructural features of both deciduae appeared similar during the first few days, the question of the noncellular deposits could not be answered by this technique since the deciduoma began to regress before the time at which the pericellular capsules were well developed in true pregnancy.

III. Pathologic Placentation

In a recent study of invasive hydatidiform mole (Wynn and Harris 1967) we attempted to compare the cellular relations of trophoblast and decidua with those in normal human placentation. The trophoblast of invasive mole (*chorioadenoma destruens*) maintains the ultrastructural features of its counterparts in normal placenta, benign hydatidiform mole, and choriocarcinoma. The endometrium in closest relation to the penetrating trophoblast, however, is ultrastructurally less complex than normal decidua basalis (Wynn 1967*a*), more nearly resembling the parietal decidua. There is relatively little necrosis of contiguous trophoblast and decidua, and less noncellular material is deposited between antigenically dissimilar cells (fig. 6). The decrease in necrosis and fibrinoid suggests a less extensive interaction of fetal and maternal cells in invasive mole.

In our comparative studies of normal placentation (Wynn 1967*b*) we found

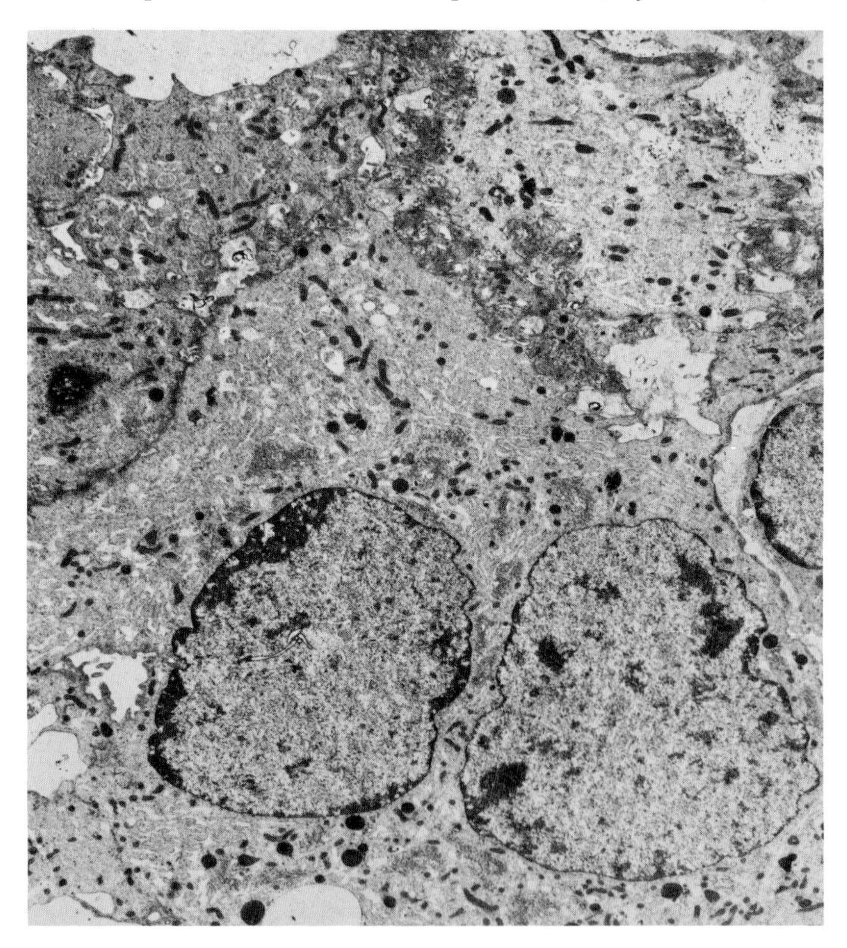

Fig. 4. Decidua of pregnant hamster 7½ days after ovulation. Relatively little intercellular amorphous material has been deposited between the well-developed decidual elements. Compare with figure 5. ×4,500.

that in species with the most penetrative trophoblast there was greatest deposition of fibrinoid and most extensive necrosis of fetal and maternal cells. The findings in invasive mole thus at first seem paradoxical in that its highly penetrating trophoblast creates less fibrinoid in its wake. We suggested, therefore, that in this trophoblastic growth the abnormally poor ultrastructural development of the decidua is correlated with formation of less fibrinoid, which we regarded as one manifestation of tissue reaction equivalent to that found at the site of a homograft. Having escaped from endometrial control, the trophoblast may cause necrosis and local hemorrhage in the myometrium. Whether the invasive molar trophoblast is not recognized as foreign tissue or whether the primary defect lies in the blunted capacity of the decidua to limit trophoblast by formation of pericellular barriers cannot be ascertained by electron microscopy alone. If even normal trophoblast is not recognized as foreign, the difference between normal placentation and chorionic neoplasia is inherent in some property of trophoblastic invasiveness rather than

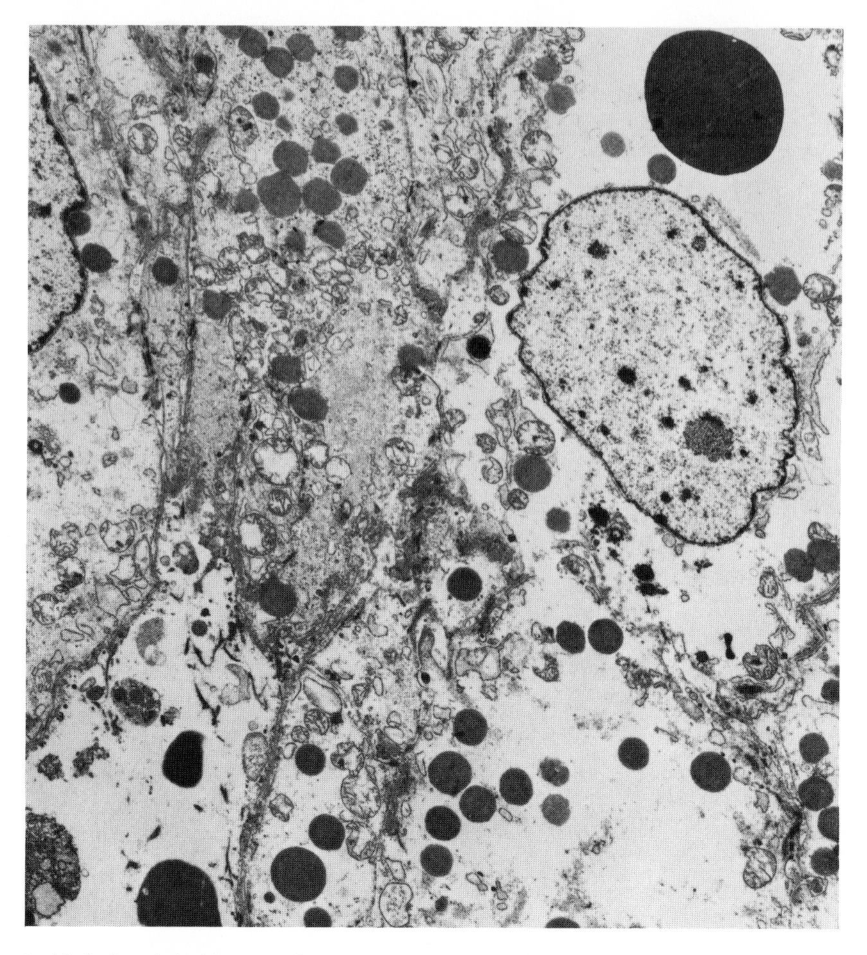

Fig. 5. Air-induced deciduoma of pseudopregnant hamster at postovulatory stage equivalent to that of true pregnancy shown in figure 4. Decidual tissue has begun to regress before formation of capsular material. ×4,500.

in primarily immunological factors. In placenta accreta, which is characterized by invasion of the myometrium by trophoblast and absence of well-developed capsular material and fibrinoid, the primary defect may be a poor decidual reaction.

IV. Hypertrophic Basal Laminae

To extend our observations of hemochorial placentas, in which trophoblastic invasiveness seemed positively correlated with ultrastructural complexity of the decidua and formation of histologically demonstrable noncellular barriers, we

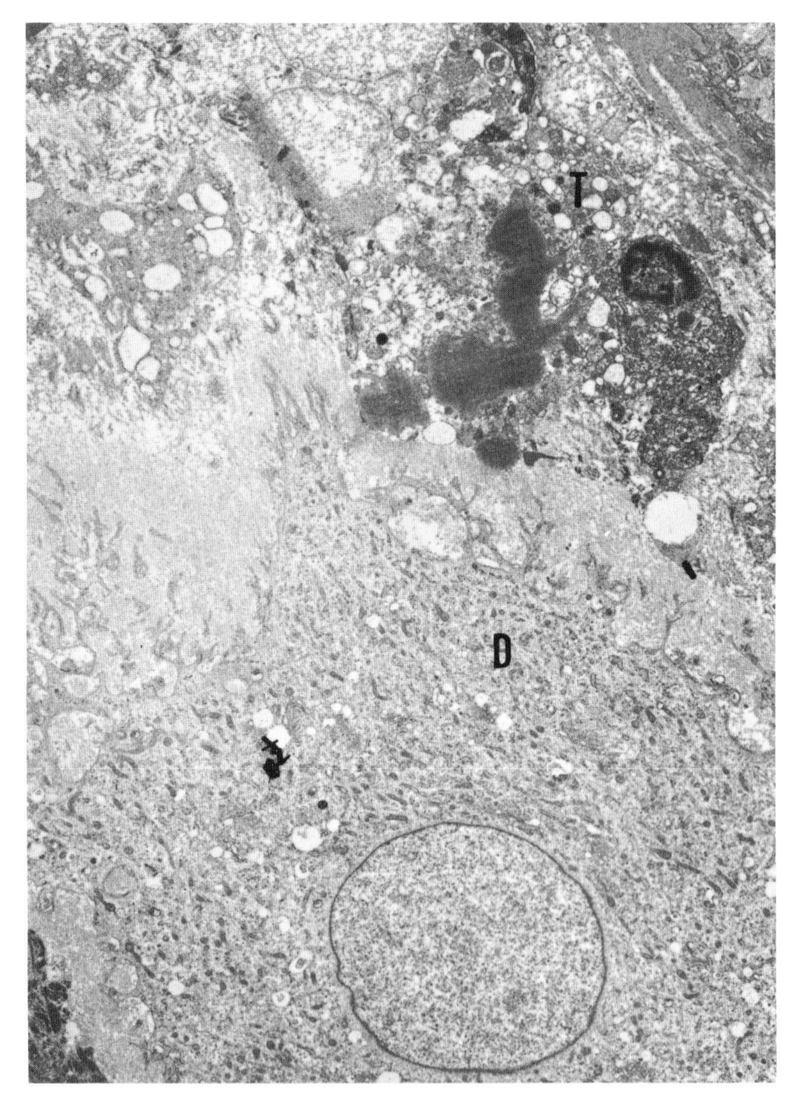

Fig. 6. Intimate relation of decidua (*D*) and trophoblast (*T*) in invasive hydatidiform mole. Note the fairly well-preserved decidua in proximity to trophoblast and the relatively little fibrinoid. ×4,600.

examined the placentas of several carnivores. In the hyena (Wynn and Davies 1965), which is unique among carnivores in possessing a partially villous hemochorial placenta, no extracellular deposit is associated with the syncytial trophoblast exposed only to circulating maternal blood. In the junctional zone, however, the basal endometrium is contacted or penetrated by tips of chorionic processes that are covered by either syncytial or cellular trophoblast (Wynn 1967*b*). As in the endotheliochorial labyrinth of typical carnivores there is no true decidua; instead there is a transformed gestational endometrium that consists of glandular debris, giant cells, and blood. Both syncytial and cellular forms of trophoblast elicit similar reactions in the endometrium and form the same amount of amorphous material. The deposition of fibrinoid and related extracellular material depends not on the cytologic type of trophoblast but on its inherent invasiveness, as reflected perhaps in the histologic classification of the definitive placenta.

Well-defined noncellular barriers around the maternal capillaries have been found in all endotheliochorial placentas examined by electron microscopy. The prominence of some of these structures led Wislocki (1954) to suggest the term "vasochorial" in preference to endotheliochorial. It is unlikely, however, that this material is a remnant of the original endometrial connective tissue, as its attempted entry into the Grosser classification might imply. It is more likely a product of the maternal endothelium that is continually eroded by the trophoblast.

Our own studies of the feline and canine placentas indicate that the prominent extracellular barriers in these species are best described as hypertrophic basal laminae. In the cat's placenta (Wynn and Björkman 1968), an extracellular deposit of this kind usually, but not always, separates trophoblast from maternal capillaries and giant cells. Although the endometrial giant cells are characteristic of the feline membrane in particular, similar well-defined "basement membranelike" structures may be found in the placentas of all the carnivores examined by electron microscopy (Wynn 1968). In the cat there is, in addition, a prominent trophoblastic terminal web, which under the light microscope may be mistaken for fibrinoid or a thick "basement membrane" (fig. 7). In certain areas the maternal surface of the trophoblast may contact the endometrial capillary with only a typical thin endothelial basal lamina intervening. The approximation of apparently viable fetal and maternal elements is greater in endotheliochorial placentas than in hemochorial, but less than in epitheliochorial forms. Although noncellular material is usually obvious between trophoblast and maternal capillary, the trophoblast and the genetically dissimilar giant cells may be closely juxtaposed in an essentially syndesmochorial relation. These noncellular basal laminae in the endotheliochorial labyrinth may provide mechanical protection of the uterus similar to that afforded by the capsular material in typical deciduate hemochorial placentation. They do not necessarily imply immunological protection of the trophoblast, however.

The dog's placenta (Wynn and Corbett 1969) resembles that of the cat except for the much less prominent terminal web and the absence of endometrial stromal cells within the canine lamellae. The barrier between the maternal capillary endothelium and the trophoblast in the dog's placenta appears ultrastructurally to be a hypertrophied basal lamina that is even more prominent than that in the cat (fig. 8). No such thick noncellular barrier lies between the genetically identical trophoblast and fetal capillary endothelium; they are separated only by typical basal laminae. The homologous thick basal lamina in the ferret's placenta has been termed an "intermediate layer" (Lawn and Chiquoine 1965). A similar, but thinner, "interstitial membrane" was described in the labyrinth of certain bats that were originally thought to have endotheliochorial placentas. Recent electron microscopic studies have reclassified these labyrinths as hemodichorial (Enders and Wimsatt 1968).

The endotheliochorial placenta has trophoblast of intermediate invasiveness and only moderate development of noncellular barriers between antigenically dis-

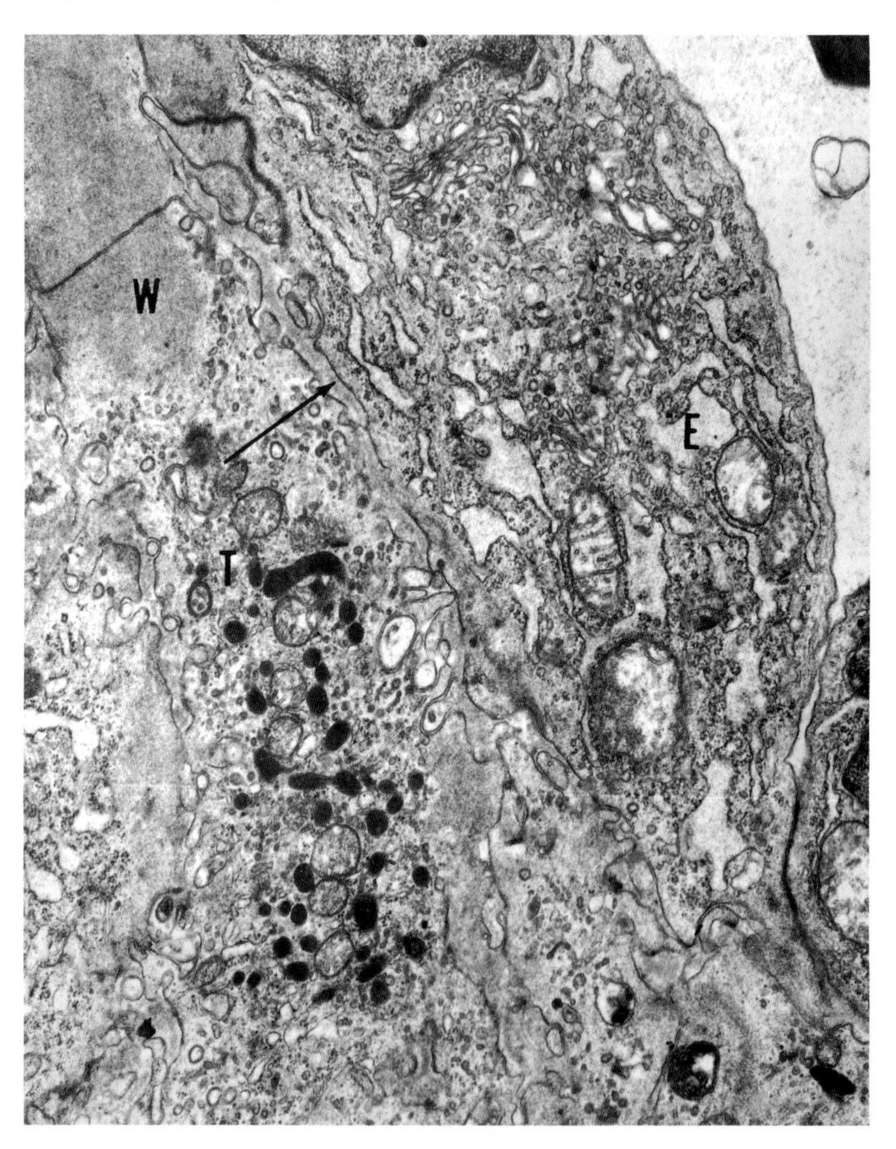

Fig. 7. Placental membrane of cat showing maternal endothelium (*E*) with its basal lamina (*arrow*) and trophoblast (*T*) with terminal web (*W*). ×15,000.

similar tissues. The histologically demonstrable decidual capsular material and fibrinoid and the hypertrophied basal laminae may thus reflect the interplay of trophoblastic invasion and endometrial defense that characterizes successful gestation.

V. Placental Fibrinoid

The suggestion by Bardawil and Toy (1959) that the fibrinoid in the human placenta creates an immunological no-man's-land led to a renewal of interest in this material. Much of the recent work on placental fibrinoid was stimulated by the hypothesis of Kirby et al. (1964) that the deposition of this material may be a gen-

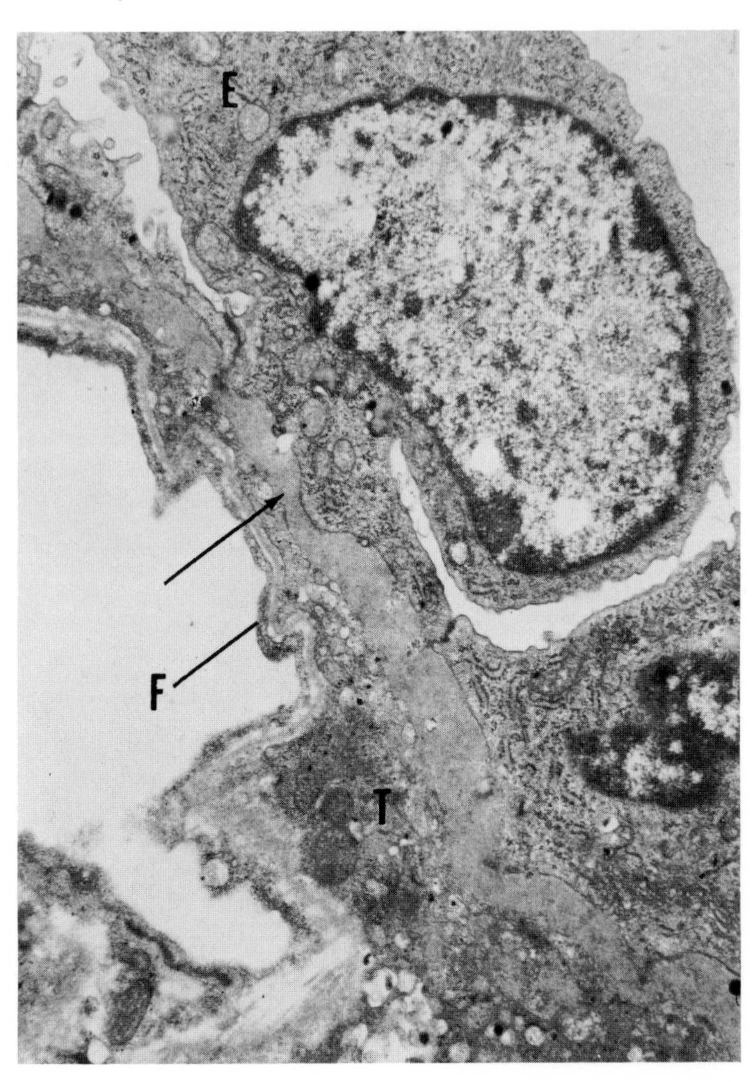

Fig. 8. Placental membrane of dog showing maternal endothelium (*E*) with hypertrophic basal lamina (*arrow*), trophoblast (*T*), and fetal capillary (*F*). ×15,000.

eral phenomenon of mammalian placentation, serving mechanically to prevent the passage of fetal transplantation antigens to the mother. Detailed reviews of the histochemistry of placental fibrinoid are found in several current reports (Wynn 1968, 1969). Our electron microscopic studies have not provided criteria for the reliable histogenetic identification of fibrinoid and related amorphous materials in the placenta. We have used the term in the restricted conventional sense of the histopathologist to refer to a group of substances recognizable with the light microscope.

In the human placenta (fig. 9) (Wynn 1967*a*), as in that of the rat (fig. 10) (Kirby et al. 1964; Schiebler and Knoop 1959), the basal region is characterized by varying proportions of maternal and fetal cells, both viable and degenerating,

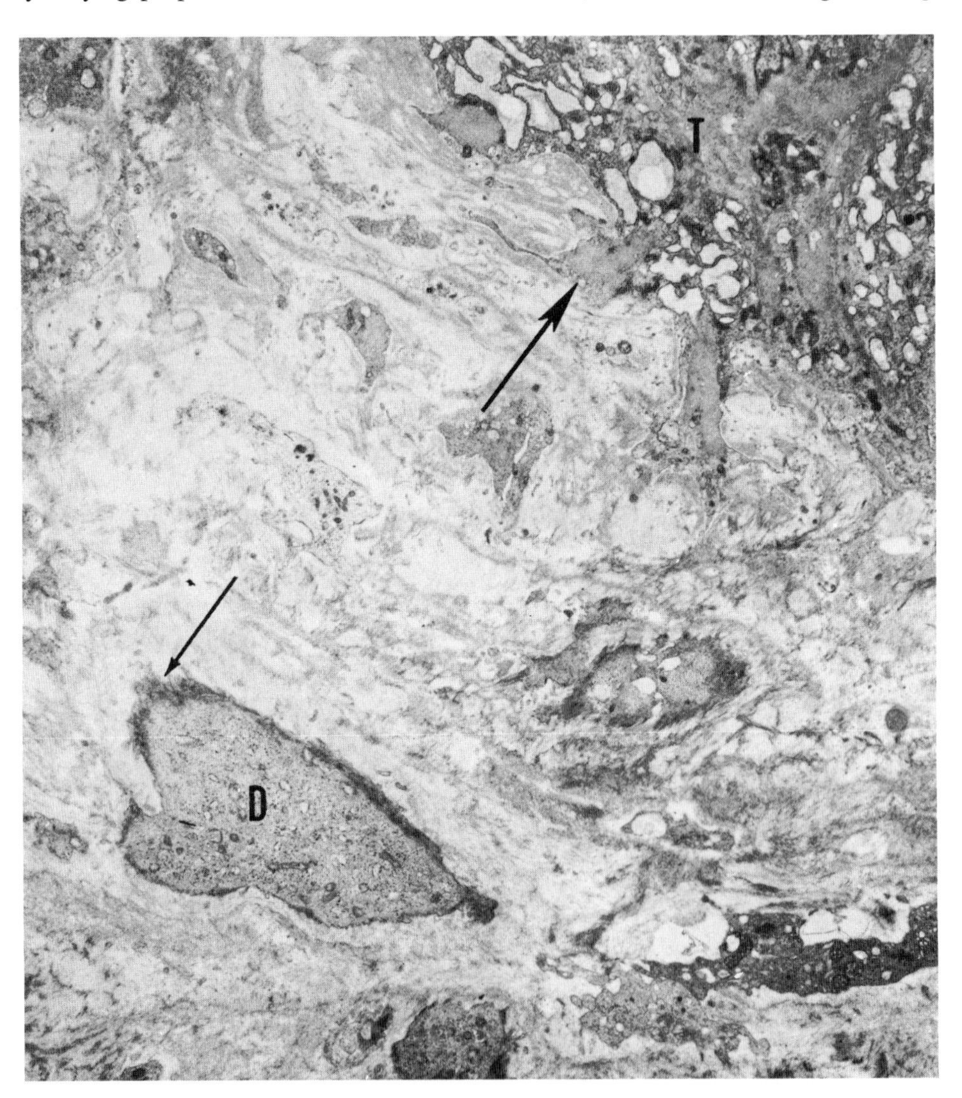

Fig. 9. Human basal plate showing decidua (*D*) surrounded by amorphous material (*thin arrow*) and degenerating trophoblast (*T*) encased in fibrinoid (*heavy arrow*). ×4,600.

and fibrinoid. In the mature placenta ultrastructurally well-preserved areas of trophoblast and decidua remain separated by regressing tissues (decidua and trophoblast) or fibrinoid, or both. The fibrinoid appears to arise from decidual or trophoblastic cytoplasm or through transformation of intercellular collagenous fibers. The material may appear ultrastructurally identical with that elaborated by the viable decidua to form its pericellular capsules. Some trophoblastic elements are completely buried in fibrinoid; others are surrounded by more fibrillar material that closely follows the cytoplasmic outlines and appears to be intracellular. The decidual elements may be surrounded also by capsules of fine fibrillar or amorphous

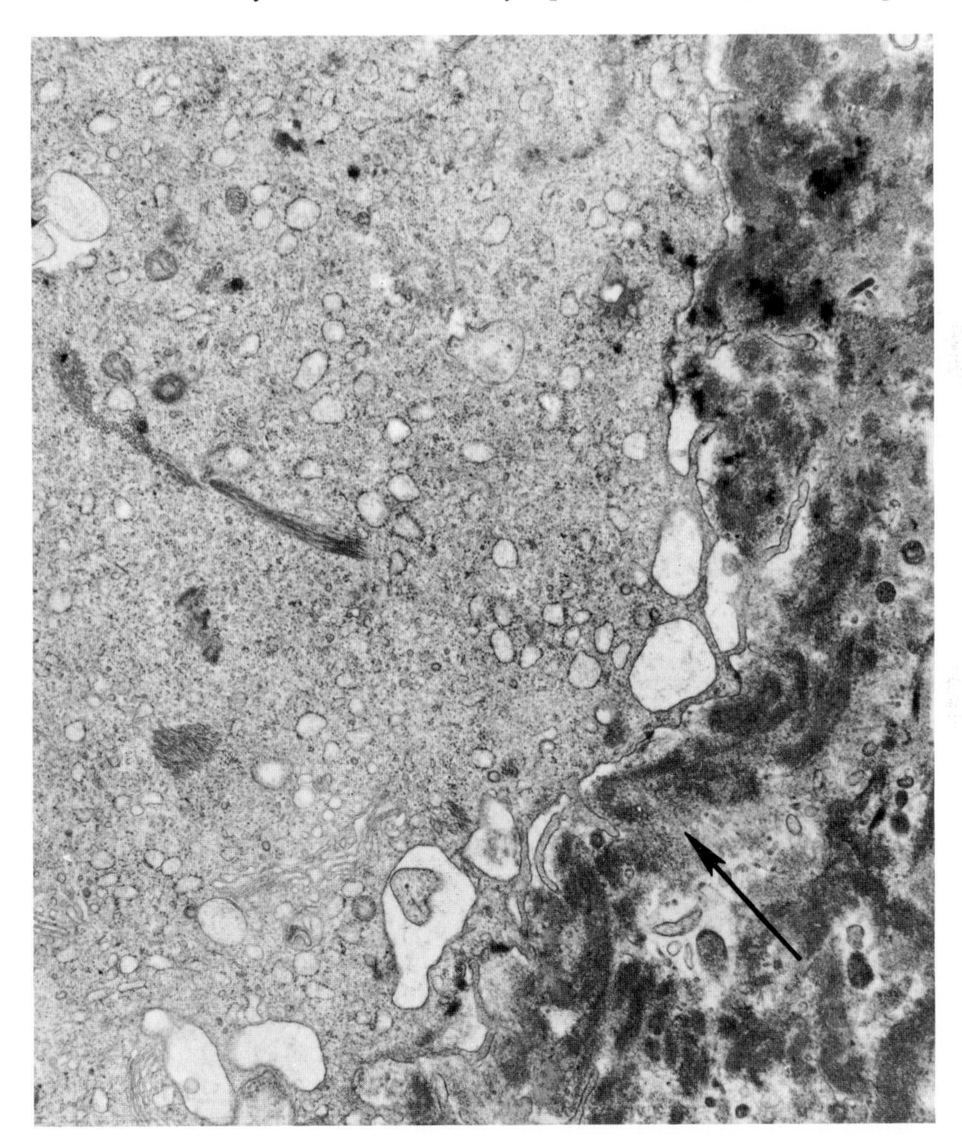

Fig. 10. Trophoblastic giant cell in base of rat's placenta surrounded by fibrinoid (*arrow*). ×15,800.

material that appear to persist after destruction of the remainder of the cell. Necrosis of decidua thus may precede or follow its encasement in fibrinoid. Where fibrinoid forms an almost complete demarcation between maternal and fetal tissues, the so-called layer of Nitabuch appears. In the macaque's basal plate (Wynn 1967*b*) there is less intimate intermingling of trophoblastic and endometrial cells, and consequently less necrosis and fibrinoid. These features of the macaque's basal plate may reflect the inherently lower degree of invasiveness of the monkey's trophoblast compared with that of man.

Our observations support the conclusion of many other workers that the term fibrinoid is merely descriptive of a large group of substances more or less closely related to fibrin. We have demonstrated the apparent conversion of fibrillar to amorphous material and even the transformation of typical collagen fibers into similarly appearing masses. In most hemochorial placentas such histologically demonstrable fibrinoids are readily revealed in areas in which trophoblast and endometrium are in intimate contact, but not around trophoblast exposed directly to maternal blood. Were all labyrinthine or villous trophoblast surrounded by this deposit, placental transfer would be abolished. In the definitive labyrinth of the endotheliochorial placenta little or no noncellular material, other than a hypertrophic basal lamina, separates maternal and fetal tissues. Neither deposition of fibrinoid nor extensive necrosis occurs in epitheliochorial forms (fig. 11), even in regions of direct apposition of fetal and maternal tissues (Wynn 1967*b*, 1968, 1969). The epitheliochorial condition may perhaps reflect inherently decreased cytolytic activity of the trophoblast or increased resistance of the endometrium. If so, the absence of fibrinoid may be the result of less necrosis of the simply apposed chorion and endometrium.

In hemochorial placentas no fibrinoid layer is noted in association with the very early trophoblast. At this stage the invading trophoblast may depend on some other means of protection, perhaps enzymatic or phagocytic. Simmons and his coworkers examined the ectopic implants of mouse ova transferred under the renal capsule of male recipients and found no electron-dense fibrinoid barrier between the trophoblastic elements of either isogeneic or allogeneic transplants (Simmons, Cruse, and McKay 1967). We have demonstrated intimately related trophoblast and decidua without interposed fibrinoid in chorionic neoplasms (Wynn and Harris 1967). Although transplanted and neoplastic trophoblast may not provide valid models, we nevertheless find histologically demonstrable fibrinoid consistently only in those regions of the placenta where antigenically mature fetal cells and immunologically competent maternal tissues other than blood are in direct contact.

VI. Sialomucins

To detect noncellular mucopolysaccharide barriers at the ultrastructural level we examined the trophoblastic microvilli of placentas in a variety of animals by means of a colloidal iron reaction modified for electron microscopy (Wynn 1969). Although in epitheliochorial placentas, such as those of the mare, sow, cow, and deer, microvilli of chorionic and endometrial epithelia interdigitate without extensive necrosis or formation of fibrinoid, we have observed an extracellular coat of mucopolysaccharides on the interlocking plasma membranes. The reaction products do not differ morphologically, however, from those produced by similar treatment of epithelial microvilli of a variety of known antigenic tissues (fig. 12). The unique feature of the trophoblast may be its coating on all sides by this sialomucin rather than any particular histochemical property. Although the colloidal iron deposited around the trophoblast as shown by the Hale reaction may represent some of the material demonstrated histochemically by the PAS stain, we hesitate to designate the sialomucin coating of the trophoblastic plasma membranes as fibrinoid (fig. 13).

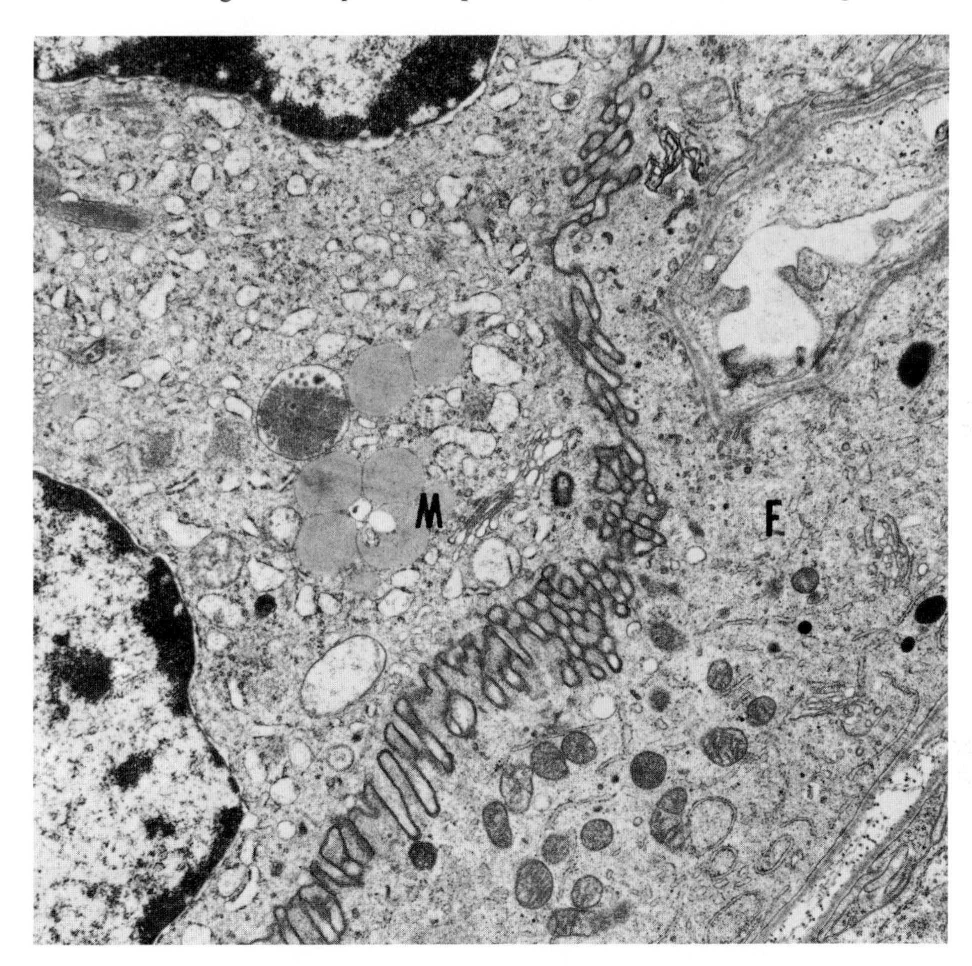

Fig. 11. Junction of fetal (*F*) and maternal (*M*) epithelia in sheep's placenta. Note absence of fibrinoid and necrosis. ×15,000.

Even though rigid criteria for separation of fibrin, fibrinoids, and related extracellular components remain difficult or impossible to formulate, it is desirable to avoid further confusion by restricting the term "fibrinoid" to histologically demonstrable deposits, excluding such ultrastructural features as the glycocalyx. In most hemochorial forms the trophoblastic plasma membranes demonstrate the Hale reaction, but at the tight junctions observed in areas of fetomaternal cellular contact in the placentas of the sheep (fig. 14) and cow, for example, the epithelia may be di-

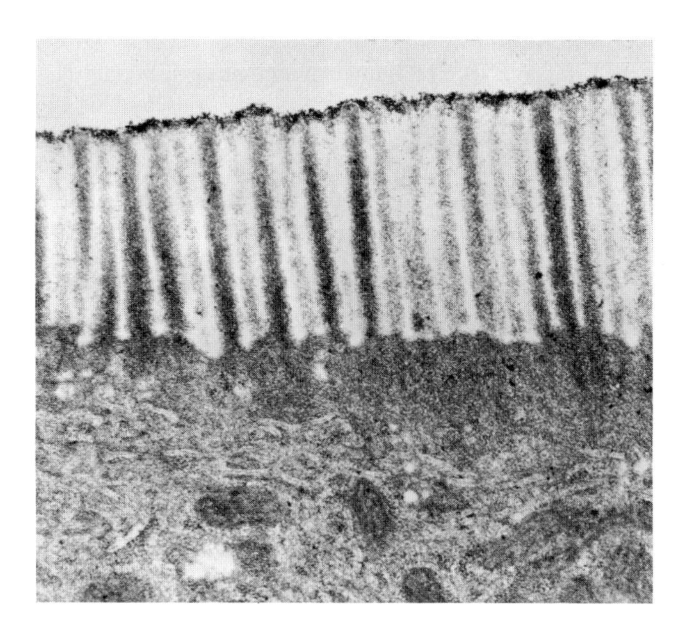

Fig. 12. Hamster gut showing deposit of colloidal iron on microvilli. Hale stain. ×32,000

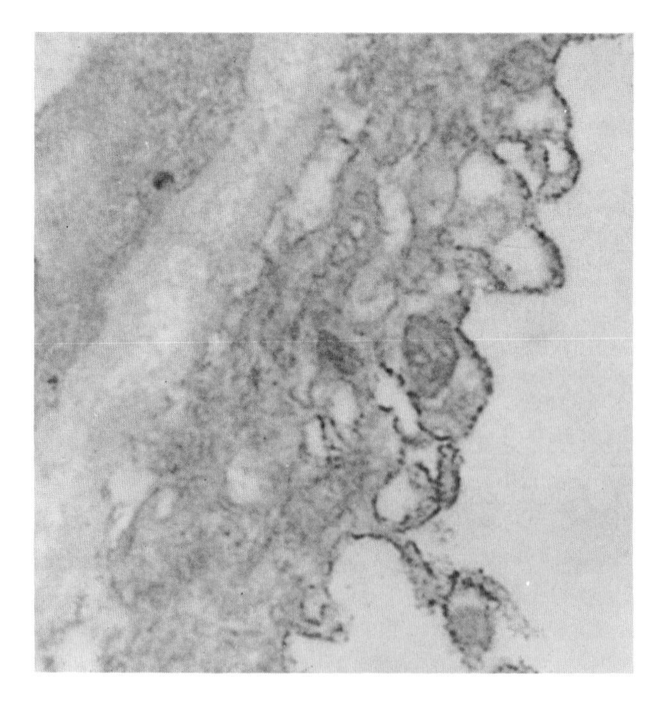

Fig. 13. Human syncytiotrophoblast showing colloidal iron on surface of microvilli. Hale stain. Compare with figure 12. ×20,000.

rectly apposed with no intervening demonstrable mucopolysaccharide. It is possible, of course, that the trophoblast in these thick membranes may need no protection, since it is not normally exposed to immunologically competent lymphocytes.

The interesting hypothesis of Currie and Bagshawe (1967) holds that pericellular sialomucins present a chemical barrier to immunologically competent cells. They reported that maternal lymphocytes respond *in vitro* to trophoblast as they do to other allogeneic tissues and originally suggested that peritrophoblastic sialomucin confers a high electronegative surface charge. Since lymphocytes have a negative charge also, they reasoned that the trophoblast escapes attack *in vivo* by electro-

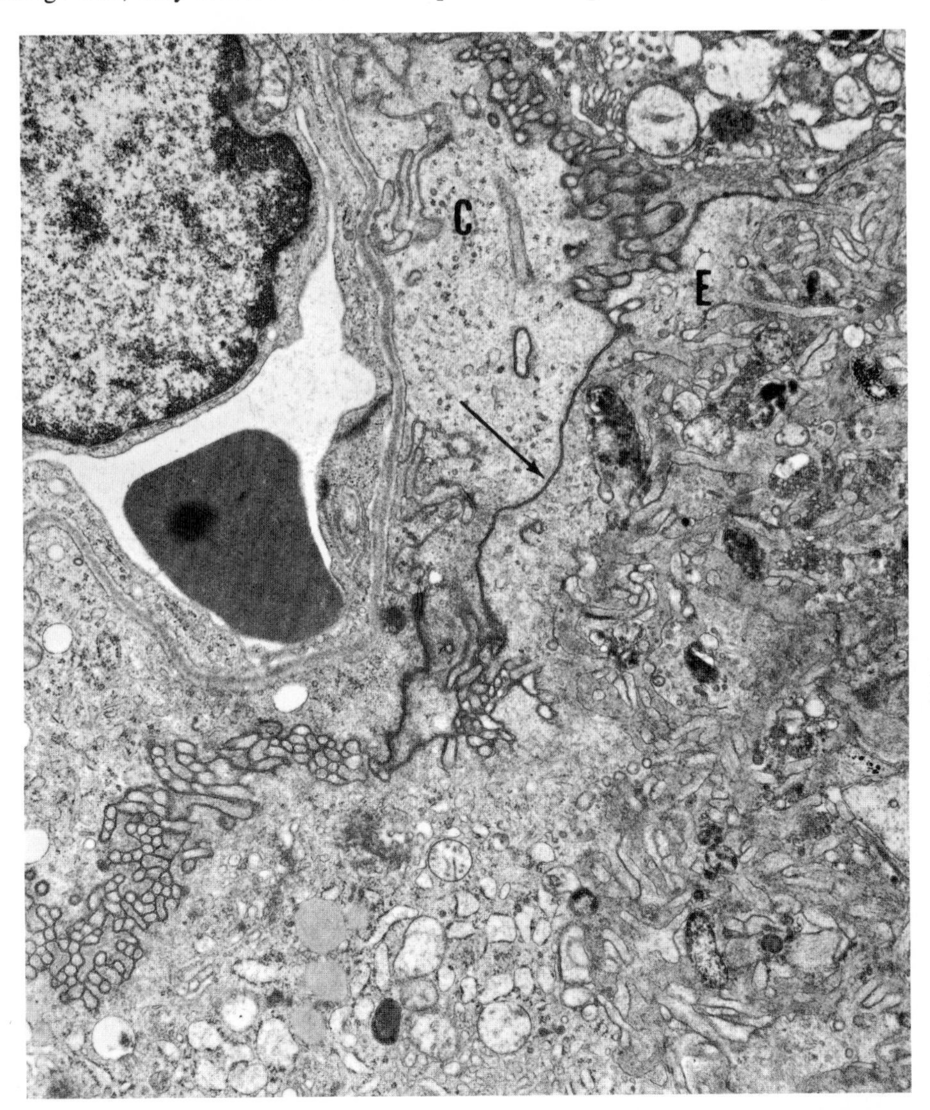

Fig. 14. Fetomaternal junction in sheep's placenta showing chorionic (*C*) and endometrial (*E*) epithelia connected by tight junction (*arrow*) without necrosis of either tissue or demonstrable sialomucin deposit. ×15,000.

chemical repulsion of maternal lymphocytes. They showed that trophoblast grown *in vitro* undergoes gross cytolysis in the presence of maternal lymphocytes and concluded that this allogeneic inhibition indicates that trophoblast expresses antigenicity that is detectable by maternal cells.

More recently Currie and his co-workers performed a key experiment that adds considerable weight to the argument favoring an immunological role for the sialomucins (Currie, Van Doorninck, and Bagshawe 1968). They showed that attacking the mucopolysaccharide coat with neuraminidase results in the ability of the treated trophoblastic cells to express strong histocompatibility antigens on transplantation. This observation is consistent with the hypothesis that mucopolysaccharides of the trophoblastic surface mask histocompatibility antigens. Neuraminidase is said to disrupt the O-glycoside link between sialic acid and its amino sugar, thus removing these terminal acid groups from sialomucins of the plasma membrane. Although they have not abandoned the idea that electrostatic effects of the carboxyl groups of sialic acid may be important, they appear to favor the idea that the sialic acids may act by steric hindrance to prevent contact with underlying antigenic groups.

It remains to be learned whether the immunological behavior of the ectoplacental cone is identical with that of the trophoblast in the definitive placental membrane or whether there is some maturation that may involve loss of histocompatibility antigens. It is important also to prove that these divested cells are viable. We are presently attempting to learn, by means of a combination of histochemical and ultrastructural techniques, whether the sialomucins of trophoblastic microvilli differ from those on other epithelia or whether their anatomic distribution around the entire surface of the trophoblast is a unique feature of this tissue.

VII. Conclusion

Little doubt remains that noncellular components play an important role in placentation. It seems not unreasonable to construct an anatomic classification of placentas, analogous to Grosser's scheme, based on the origin and prominence of these noncellular deposits. The various pericellular structures must be differentiated, however, according to origin, magnitude, and function. The decidual capsules and the hypertrophic basal laminae in the hemochorial and endotheliochorial placentas, respectively, are very likely products of maternal tissues that are partially eroded by the trophoblast and that serve a mechanical role in restricting trophoblastic invasion. The fibrinoids that are recognized by conventional histochemical techniques may arise from a secretion or degeneration of the fetal and maternal tissues and represent the cellular interplay of trophoblast and endometrium. The ultrastructurally demonstrable sialomucins, which correspond to the widely distributed glycocalyces of plasma membranes in general, may play an important role as antigenic barriers. Their unique properties require detailed study by ultracytochemical methods. If, in fact, they are crucial to the immunological protection of the trophoblast, they probably owe their effects to anatomic rather than chemical factors in that the trophoblast, in contrast to other tissues, may be covered on all sides by sialomucins in areas in which antigenically mature fetal tissues and immunologically competent maternal cells are in direct contact. Finally, it is most significant that in every placenta examined by electron microscopy at least one layer of trophoblast persists throughout gestation. It is reasonable to conclude that immunological protection of the placental homograft is afforded by the trophoblast and the pericellular sialomucins on its plasma membranes.

References

Bardawil, W. A., and Toy, B. L. 1959. The natural history of choriocarcinoma: Problems of immunity and spontaneous regression. *Ann N Y Acad Sci* 80:197.

Billingham, R. E. 1964. Transplantation immunity and the maternal-fetal relation. *New Eng J Med* 270:667.

Billington, W. D. 1967. Transplantation immunity and the placenta. *J Obstet Gynaec Brit Comm* 74:834.

Currie, G. A., and Bagshawe, K. D. 1967. The masking of antigens on trophoblast and cancer cells. *Lancet* 1:708.

Currie, G. A.; Van Doorninck, W.; and Bagshawe, K. D. 1968. Effect of neuraminidase on the immunogenicity of early mouse trophoblast. *Nature* (*London*) 219: 191.

Enders, A. C. 1965. A comparative study of the fine structure of the trophoblast in several hemochorial placentas. *Amer J Anat* 116:29.

Enders, A. C., and Wimsatt, W. A. 1968. Formation and structure of the hemodichorial chorio-allantoic placenta of the bat (*Myotis lucifugus*). *Amer J Anat* 122: 453.

Fischer, T. V., and Luckett, W. P. 1969. The haemodichorial placenta of the beaver, *Castor canadensis*. *J Reprod Fertil* 18:166.

King, B. F., and Tibbitts, F. D. 1969. The ultrastructure of the placental labyrinth in the kangaroo rat, *Dipodomys*. *Anat Rec* 163:543.

Kirby, D. R. S.; Billington, W. D.; Bradbury, S.; and Goldstein, D. J. 1964. Antigen barrier of the mouse placenta. *Nature* (*London*) 204:548.

Kirby, D. R. S., and Malhotra, S. K. 1964. Cellular nature of the invasive trophoblast. *Nature* (*London*) 201:520.

Lawn, A. M., and Chiquoine, A. D. 1965. The ultrastructure of the placental labyrinth of the ferret (*Mustela putorius furo*). *J Anat* 99:47.

Mossman, H. W. 1937. Comparative morphogenesis of the fetal membranes and accessory uterine structures. *Contrib Embryol Carnegie Inst Wash* 26:129.

———. 1967. Comparative biology of the placenta and fetal membranes. In *Fetal homeostasis,* ed. R. M. Wynn, 2:13. New York: New York Academy of Sciences.

Mossman, H. W., and Owers, N. 1963. The shrew placenta: Evidence that it is endothelio-endothelial in type. *Amer J Anat* 113:245.

Orsini, M. W.; Wynn, R. M.; Harris, J. A.; and Bulmash, J. M. 1970. Comparative ultrastructure of the decidua in pregnancy and pseudopregnancy. *Amer J Obstet Gynec* 106:14.

Schiebler, T. H., and Knoop, A. 1959. Histochemische und electronenmikroskopische Untersuchungen an der Rattenplazenta. *Z Zellforsch* 50:494.

Simmons, R. L.; Cruse, V.; and McKay, D. G. 1967. The immunologic problem of pregnancy: II. Ultrastructure of isogeneic and allogeneic trophoblastic transplants. *Amer J Obstet Gynec* 97:218.

Wislocki, G. B. 1954. In discussion following chapter by E. C. Amoroso, in *Gestation,* ed. L. B. Flexner, 1:221. New York: Josiah Macy Jr. Foundation.

Wynn, R. M. 1964. Ultrastructure of the deciduotrophoblastic junction of the guinea pig. *Amer J Obstet Gynec* 90:690.

———. 1967*a*. Fetomaternal cellular relations in the human basal plate: An ultrastructural study of the placenta. *Amer J Obstet Gynec* 97:832.

———. 1967*b*. Comparative electron microscopy of the placental junctional zone. *Obstet Gynec* 29:644.

———. 1968. Morphology of the placenta. In *Biology of gestation,* ed. N. S. Assali, 1:93. New York: Academic Press.

———. 1969. Noncellular components of the placenta. *Amer J Obstet Gynec* 103: 723.

Wynn, R. M., and Björkman, N. 1968. Ultrastructure of the feline placental membrane. *Amer J Obstet Gynec* 102:34.

Wynn R. M., and Corbett, J. R. 1969. Ultrastructure of the canine placenta and amnion. *Amer J Obstet Gynec* 103:878.

Wynn, R. M., and Davies, J. 1965. Comparative electron microscopy of the hemochorial villous placenta. *Amer J Obstet Gynec* 91:533.

Wynn, R. M., and Harris, J. A. 1967. Ultrastructure of trophoblast and endometrium in invasive hydatidiform mole (chorioadenoma destruens). *Amer J Obstet Gynec* 99:1125.

30

Studies of the Hormonal Barrier to Trophoblast Rejection

J. F. Hulka

Department of Obstetrics and Gynecology
School of Medicine
University of North Carolina, Chapel Hill

The successful implantation, nutrition, and nonrejection of genetically foreign tissue such as the blastocyst has generated many hypotheses regarding the mechanisms involved. In this symposium the data regarding the need for an immunological barrier at all (i.e., the evidence that the blastocyst and its trophoblast is antigenic) has been reviewed by Behrman (chapter 28). Assuming that sufficient data are available to make trophoblastic antigenicity a viable hypothesis, various possible barriers to trophoblastic rejection have been examined by Wynn (chapter 29) (noncellular barriers) and Kirby (chapter 23) (fibrinoid barrier). The purpose of this report is to evaluate the data pertinent to the hypothesis that a hormonal or humoral agent peculiar to human pregnancy is responsible for this "barrier" to rejection. Since the endocrine changes in pregnancy are dramatic, workers in this field have studied the relationship between the known hormones of pregnancy and the immunologic phenomena. Since the trophoblast cell is directly exposed to the maternal host at implantation as well as throughout human pregnancy, the biology of its survival has received most study. The data from these studies will be the subject of this review.

I. Ovarian Hormones

An obvious hypothesis to explore is that the hormones of pregnancy produced by the ovary induce changes in the immunologic reactivity of the maternal host sufficient to allow the implanting trophoblast to survive. This possibility gained strength when Billingham, Krohn, and Medawar (1951) demonstrated that the steroid hormone cortisone had a profoundly inhibitory effect on primary immune responses in skin transplantation studies. Aside from the obvious histological effects on the reproductive system, the steroid hormones of the corpus luteum estrogens and progesterone, appear to be fair game with which to explore this hypothesis. Early studies of estradiol and progesterone, however (Krohn 1954; Medawar and Sparrow 1956), failed to reveal any inhibiting effect on skin graft rejection. Additional negative observations with progesterone have been reported recently (Hulka, Mohr, and Lieberman 1965; Simmons, Price, and Ozirkis 1968). The synthetic progestins, norethynodrel and norethindrone, do prolong skin graft survival (Hulka, Mohr, and Lieberman 1965), but such graft prolongation was characteristic only of synthetic steroid agents. These agents (as well as natural progesterone) significantly inhibited any graft-prolongation effect of cortisone (Hulka and Mohr 1967) in rabbit skin grafts, which appears to negate any contribution that progesterone could make to the concept that corticosteroid levels in pregnancy help graft prolongation. Thus it is unlikely that the progesterone produced by the corpus luteum affects trophoblast survival through a systemic immunologic mechanism.

Estrogens have been observed to prolong skin grafts in rat endometrium (Zipper et al. 1966) and trophoblast transplants in the human (Lajos et al. 1964); this circumstance may be explained by mechanisms which are not immunologic. Simmons, Price, and Ozirkis (1968) observed prolongation of skin graft survival in mice receiving very high doses of estradiol, and noted that estrogens, concentrated as they are locally in the uterus (Roy, Mahesh, and Greenblatt 1964), could be a factor in trophoblast survival. However, they appeared to favor local secretion of estrogen (to be discussed below) by the trophoblast as the more plausible explanation of trophoblast survival. The bulk of the data (Krohn 1954; Medawar and Sparrow 1956; Hulka, Mohr, and Lieberman 1965) does not reveal any skin graft prolongation evidence by estrogen levels compatible with pregnancy. Thus current data do not support the hypothesis that the ovarian hormones prolong trophoblast survival by a systemic immunologic effect.

II. Trophoblastic Hormones: Systemic Effects

That the trophoblast secretes a substance (other than estrogen or progesterone reviewed above) which inhibits its immunologic rejection by a systemic effect is another hypothesis. That this is biologically feasible is evidenced by the common though remarkable occurrence of the elaboration of chorionic gonadotrophin by the 21- or 22-day-old human ovum (Hertig 1964) in sufficient quantities to sustain the corpus luteum. Perhaps the trophoblast is capable also of altering the systemic immune response with a humoral agent.

The hypothesis is considerably weakened as a biologic principle by the failure of pregnancy of any stage to alter skin graft survival in species such as mice or cattle (Billingham and Lampkin 1957; Medawar and Sparrow 1956). Nevertheless the demonstration of prolongation of skin graft survival in human pregnancy (Andersen and Monroe 1962) makes the study of the systemic immunologic effects of nonsteroid trophoblastic hormones reasonable.

A. *Chorionic Gonadotrophin*

Pearse and Kaiman (1967) noted a prolongation of skin graft survival with high systemic doses of HCG in mice, but on repetition of the study Pearse and Curtis

(1969) observed no such survival with HCG alone. HCG did appear to be synergistic with cortisone, however. Younger (1969) perceived an immunosuppressive effect of HCG on systemic antibody production in mice. The possibility that HCG was working through its action on the steroid secretions of the intact mouse gonads or adrenals rather than directly on the immune mechanism in these studies remains to be explored.

In our laboratories we have failed to demonstrate altered skin graft rejection rates in castrated rabbits treated with HCG (Hulka and Mohr 1969). The available data make HCG an unlikely agent of the trophoblast in its struggle for immunologic survival at implantation.

B. *Placental Lactogen*

The recent description of human placental lactogen (HPL) as a hormone produced by the placenta in voluminous quantities, plus the wry fact that a function for the abundant hormone in pregnancy has not been established, makes HPL a suitable candidate as an agent of immunosuppression. Again, as we will report below, studies in our laboratory failed to demonstrate altered skin graft survival in castrated rabbits receiving this hormone.

C. *Other Agents*

Thoroughness and cleanliness compel the investigator to evaluate the known and purified products of the trophoblast in seeking an immunosuppressive agent to support the hypothesis of this presentation. However, there is no logical reason to assume that the known steroid or protein hormones, all of which have essentially endocrine end-points for assay, are involved in immunosuppression with graft prolongation as the goal. Studies of the systemic administration of crude human placental extracts (Marino and Bevain 1961; Erlik and Barzilai 1968; Pearse and Curtis 1969) reported in the literature to date have indicated slight but tantalizing prolongation of graft survival in the various experimental models used. Although these observations were not confirmed in our laboratories (see below), the hypothesis of a systemic immunosuppressant produced by the trophoblast (or more correctly, by the placenta in these studies), other than the known protein and steroid hormones, seems worthy of further exploration before it is completely rejected.

III. Trophoblastic Hormones: Local Effect

We turn now to the hypothesis that the trophoblast is capable of secreting a substance that abrogates immune responses at a local rather than a systemic level. Such a hypothesis is suggested by the frequency with which both syngeneic and allogeneic ectopic transplants of endocrine glands have been used as a technique in rat endocrine studies. The testes and the ovary, both steroid-secreting glands, survive allogeneic grafting over weak histocompatibility differences (Linder 1961; Cock 1962). Strong histocompatibility differences can be overcome in ovarian transplants by using a corneal capsule to protect the foreign tissue during its initial month or two of residence. Such grafts have been shown to survive for months in monkeys and rabbits (Castellanos and Sturgis 1961; Sturgis and Castellanos 1962) and to become vascularized and accepted by the host (Shaffer and Hulka 1969). Human studies with ovarian tissue encapsulated in corneal tissue have been performed by Sturgis, Charles, Castellanos, and Hulka (personal communications), and have also supported graft survival as measured by estrogen production. These data suggest that a steroid-producing graft (such as an ovary) has a better chance of surviving than a non-hormone-producing graft. Thus endocrine activity of the trophoblast may be a factor in the local survival at the implantation site. These considerations led to the following survey of the local and systemic influence of trophoblast hormones on graft rejection.

Adult male New Zealand white rabbits weighing 2.5–3 kg were used throughout these studies. All animals receiving hormones systemically were castrated a week prior to the injection. Animals were divided into groups of 5; after the initiation of a hormone schedule they underwent 1 $\times$ 1 cm allograft exchanges of ear skin with autograft controls as described previously (Hulka, Mohr, and Lieberman 1965).

TABLE 1

DOSAGE SCHEDULE FOR LOCAL AND SYSTEMIC HORMONE ADMINISTRATION

	Local (Daily)	Systemic
Estrogen	Premarin—4 mgm	Estradiol valerate—10 mgm weekly
Progesterone	Proluton—3.5 mgm	Progesterone—20 mgm thrice weekly
Chorionic gonadotrophin	1,600 i.u.	2,000 i.u. daily
Placental lactogen	0.1 mgm	0.1 mgm daily
Cortisone	1.66 mgm	5 mgm twice weekly
Crude placental extract	0.5 cc	1 cc daily

These exchanges were performed under general intravenous Surital anesthesia. For local administration of hormone a chamber made of plastic was devised, filled with surgical gauze soaked in solution containing the appropriate hormone. This chamber was airtight and recharged daily to maintain saturation. The amount needed to maintain the chamber on a daily basis was rather constant from animal to animal; calculation of the amount absorbed led to the local dosage schedule reported below. Figure 1 illustrates the chamber with a graft on the day of rejection to show the end point used.

The hormones chosen for administration were estrogen, progesterone, chorionic gonadotrophin, and placental lactogen. Controls included crude placental extract and cortisone; vehicles used to dissolve the hormones (saline, and sesame oil) were given locally and systemically. Because placental extracts were stored and administered with antibiotics, an additional control group received topical antibiotics on the graft. Table 1 indicates the dosage schedule for the various hormones.

After grafting the animals were maintained on either local or systemic administration, and the grafts were inspected daily for signs of rejection. Since rejection occurred over several days, both the day of the first sign of rejection and the day of complete graft discoloration are depicted in the results in section IV.

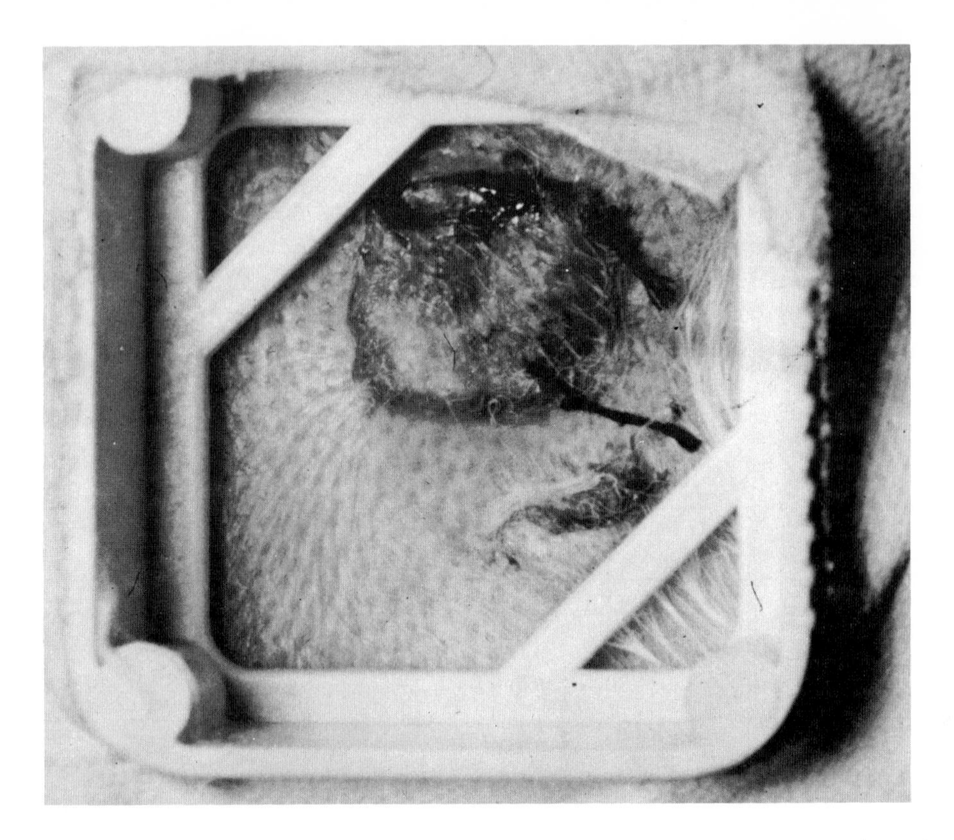

Fig. 1. Plastic chamber applied to the skin for local administration of hormone; a graft on the day of rejection shows the end point used.

TREATMENT	SYSTEMIC DAY OF REJECTION	LOCAL DAY OF REJECTION
CONTROLS	6–8	7–8
	8–10	7–9
	8–10	7–9
	8–10	8–9
		8–10
CHORIONIC GONADOTROPIN	7–9	1-5–6
	7–9	6–7
	7–10	6–7
	7–10	6–7
	8–9	7–8
PLACENTAL LACTOGEN	6–8	6–9
	7–9	6–8
	7–9	7–8
	8–10	8–10
	8–10	8–10

Fig. 2. Graft rejection with placental protein hormones

IV. Results

Figures 2 and 3 present the observation of these allograft rejections under both local and systemic hormonal influence. The significant prolongation of allografts under both local and systematic cortisone administration serves as confirmation of previous reports (Billingham, Krohn, and Medawar 1951) and a positive control of technique in the present study. There were no deaths or technical failures in this series. There is a difference in the day of allograft rejection in animals exposed to local chorionic gonadotrophin and the group exposed to local crude placental extract, but no statistically significant differences from the vehicle controls were evident.

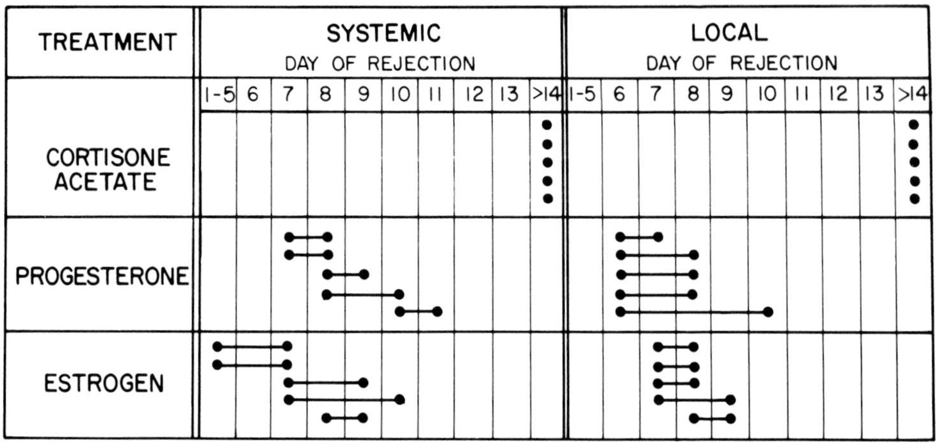

Fig. 3. Graft rejection with placental steroid hormones

V. Conclusions

The failure to date to demonstrate immunosuppressive activities of the hormones of pregnancy in prolonging experimental graft survival could lead one to conclude that the hormones do not play a role in prolonging trophoblast survival. Other conclusions are also possible. Perhaps the experimental designs to date have been too insensitive to be pertinent to the hypothesis. In most of the studies reported, including our own, skin grafts were used as end points; possibly skin tissue richly endowed with histocompatibility antigen is not an appropriate model for the less antigenically endowed implanting trophoblast. Perhaps the hormones studied are not the humoral agent active in such suppression, as suggested by occasional reports of some prolongation of graft survival with whole placental extract. These possibilities should be explored further before abandoning the hypothesis that the trophoblast secretes an immunosuppressive factor essential for its survival in a foreign host.

References

Andersen, R. H., and Monroe, L. W. 1962. Experimental study of the behavior of adult human skin homografts during pregnancy. *Amer J Obstet Gynec* 84:1096.

Billingham, R. E.; Krohn, P. L.; and Medawar, P. B. 1951. Effect of cortisone on survival of skin homografts in rabbits. *Brit Med J* 1:1157.

Billingham, R. E., and Lampkin, G. H. 1957. Further studies of tissue homotransplantation in cattle. *J Embryol Exp Morph* 5:351.

Castellanos, H., and Sturgis, S. H. 1961. Function of ovarian homografts in corneal chambers. *Surg Forum* 12:426.

Cock, A. G. 1962. The survival of testis homografts in fowls, and their effect on subsequent skin grafts. *Transplantation Bull* 29:96.

Erlik, D., and Barzilai, A. 1968. Does the placenta contain an antirejection factor? *Israel J Med Sci* 4:310.

Hertig, A. T. 1964. Gestational hyperplasia of endometrium. *Lab Invest* 13:1153.

Hulka, J. F., and Mohr, K. 1967. Interference of cortisone-induced homograft survival by progestins. *Amer J Obstet Gynec* 97:407.

———. 1969. Placental hormones and graft rejection. *Amer J Obstet Gynec* 104:889.

Hulka, J. F.; Mohr, K.; and Lieberman, M. W. 1965. Effect of synthetic progestational agents on allograft rejection and circulating antibody production. *Endocrinology* 77:897.

Krohn, P. L. 1954. The effects of pregnancy on the survival of skin homografts with rabbits. *J Endocr* 11:78.

Lajos, L.; Gorcs, J.; Szekely, J.; Csaba, I.; and Domany, S. 1964. The immunologic and endocrinologic basis of successful transplantation by human trophoblast. *Amer J Obstet Gynec* 85:595.

Linder, O. E. A. 1961. Comparisons between survival of grafted skin ovaries and tumors in mice across histocompatibility barriers of different strength. *J Nat Cancer Inst* 27:351.

Marino, H., and Bevain, F. 1961. In *Proc Int Sympos Tissue Transplantation,* University of Chile, Santiago.

Medawar, P. B., and Sparrow, E. 1956. The effects of adrenocortical hormones, adrenocorticotrophic hormone and pregnancy on skin transplantation immunity in mice. *J Endocr* 14:240.

Pearse, W. H., and Curtis, G. L. 1969. Skin allograft survival and lowered maternal immunologic activity. *Amer J Obstet Gynec* 104:419.

Pearse, W. H., and Kaiman, H. 1967. Human chorionic gonadotropin and skin homograft survival. *Amer J Obstet Gynec* 98:572.

Roy, S.; Mahesh, V. B.; and Greenblatt, R. B. 1964. Effects of clomiphene on the physiology of reproduction in rats. *Acta Endocrin* 47:669.

Shaffer, C. F., and Hulka, J. F. 1969. Ovarian transplantation. *Amer J Obstet Gynec* 103:78.

Simmons, R. L.; Price, A. L.; and Ozirkis, A. J. 1968. The immunologic problem of pregnancy. *Amer J Obstet Gynec* 100:908.

Sturgis, S. H., and Castellanos, H. 1962. Ovarian homografts in organic filter chambers. *Ann Surg* 156:367.

Younger, J. B. 1969. Effect of human chorionic gonadotropin on antibody production. *Amer J Obstet Gynec* 105:9.

Zipper, J.; Ferrando, G.; Saez, G.; and Tchernitchin, A. 1966. Intrauterine grafting in rats of autologous and homologous adult rat skin. *Amer J Obstet Gynec* 94:1056.

31

Pregnancy as an Experimental System for the Study of Immunologic Phenomena

R. S. Weiser

Department of Microbiology
School of Medicine
University of Washington, Seattle

I. The Fetus as an Allograft

A. *Introduction*

Most investigators who have worked on the immunologic aspects of pregnancy have been primarily concerned with employing knowledge and tools in immunology to gain a better understanding of pregnancy. In this presentation I will engage in some reverse thinking and speculate about ways in which study of the various immunologic aspects of pregnancy promises to add to an understanding of immunologic phenomena including tolerance, enhancement, immune deviation, allograft rejection, tumor resistance, and autoimmune disease.

The membranes which invest the developing embryo in various viviparous animals differ both anatomically and functionally even among the mammals. For example the placentas of horses and cattle do not allow the passage of antibody to the fetus, whereas the placenta of man does. Such differences may, of course, profoundly influence immunologic events attending pregnancy. Unless indicated otherwise the present discussion will center on man and other animals possessing placentas that allow the passage of antibody to the fetus.

The central immunologic problem of pregnancy within species, that of the fetus as a natural allograft, has held the attention of many investigators for more than a decade. Major unanswered questons are: Why is the conceptus neither harmed nor rejected as an allograft? and Why does the fetus, which is not totally immunologically inert, fail to react against the mother?

Medawar (1953) was the first to point out that, except in highly inbred populations, the intraspecies mammalian conceptus represents an allograft by virtue of paternally determined antigens lacking in the mother. (Henceforth the term "paternal antigens" will be used to designate those paternally determined transplantation antigens of the fetus which are not shared by the mother.) In contrast to surgically placed grafts, which are almost invariably rejected by the recipient, the conceptus normally persists throughout pregnancy without evidence of allograft rejection, despite the fact that trophoblast cells are directly exposed to the maternal circulation. This phenomenon cannot be explained fully on the basis of existing knowledge and remains one of the most fascinating enigmas in nature.

It is possible that rejection of the conceptus as an allograft occurs more often than has been recognized. Rejection of the conceptus due to an immunologic reaction against nontransplantation antigens such as the red cell antigens of the Rh system is known to occur, and it is possible that rejection may sometimes occur owing to soluble antigens, such as those of the gamma globulin allotypes alleged to cause fetal death in mice (Lieberman and Dray 1964).

Over the years many theories have been advanced to explain the persistence of the fetus as an allograft. Most of them have little or no validity. Since much of the literature has been covered in recent articles (Woodruff 1957; Billingham 1964; Currie 1968; Simmons 1969; Lanman 1965), only selected references will be cited in this presentation.

The various theories which have been advanced include the following: (1) the uterus is a privileged site against immunologic attack (if the uterus serves as a privileged site it is for only a brief period of a few days at the time of blastocyst implantation before trophoblast cells become coated with sialomucin [Kirby, Billington, and James 1966]); (2) local steroid production by the placenta protects it against immunologic attack; (3) a general nonspecific suppression of maternal immunity occurs during pregnancy; (4) specific immunologic tolerance develops in the mother; (5) trophoblast cells have a unique capacity to destroy maternal immune cells; (6) the fetus is antigenically immature and does not possess histocompatibility antigens in amounts sufficient either to stimulate an effective immune response in the mother or to render fetal tissues susceptible as immunologic targets; and (7) trophoblast, which in hemochorial placentas presents a continuous physical barrier separating the circulations of mother and fetus, both resists maternal immunologic attack and protects other fetal tissues against such attack.

There is now strong support for the elimination of all except the last theoretical mechanism as neither sufficient nor necessary to the prevention of rejection of the fetus. At best all of the other mechanisms proposed must be regarded as of secondary importance.

The mechanism which protects the fetus against immune forces of graft rejection must be strong indeed, for in rabbits experimental hyperimmunization of foster-mothers before pregnancy against unshared parental histocompatibility antigens does not usually influence the allogeneic fetus (Lanman, Dinerstein, and Fikrig 1962). The strength of the protection afforded the fetus is even more impressive when one considers the frequent success of interspecies hybridization in which genetic and antigenic disparity between mother and fetus is profound and in which maternal immune responses should be correspondingly great.

That an immunologic barrier resides in the placenta was made evident by the finding that when tissues of the fetus are separated from the placenta and transplanted to ectopic sites in the mother they are promptly rejected (Woodruff 1957). The most apparent anatomical structure in the placenta which could serve as a barrier to rejection of the fetus is trophoblast, which consists of a continuous layer of fetal cells separating other fetal elements from maternal tissues (in the hemochorial placenta trophoblast is directly exposed to maternal blood). That the immunologic barrier in the placenta is trophoblast was clearly indicated by the observation of Simmons and Russell (1962) that when the F_1 conceptus is divided and its components are grafted to an ectopic site in immune female allogeneic parental pure strain hosts the fetus is destroyed but trophoblast cells survive and proliferate.

The ability of trophoblast to function as a barrier evidently results from its capacity to resist immune attack by either alloantibody or alloimmune cells and to prevent immunocompetent cells from reaching and harming the fetus. This apparently results from an abundance of surface sialoglycoprotein (sialomucin) which is thought to protect trophoblast cells against immune attack, presumably by masking surface antigens (Kirby et al. 1964). When one considers that in certain areas the fetal and maternal circulations are separated only by a thin layer of syncytial trophoblast and a thin plate of endothelial cytoplasm it appears likely that trophoblast plays a major role in regulating the passage of many materials between mother and fetus, including cells and macromolecular substances such as immunoglobulins.

The observation that, in certain species at least, maternal cytotoxic antibodies specific for transplantation antigens of target cells of the fetus can pass the placenta but fail to produce serious injury to the fetus (Lanman and Herod 1965) clearly implies that the important and necessary functon of trophoblast in protecting the fetus against immunologic attack is not that of shielding the fetus against humoral alloantibodies but rather that of limiting the passage of maternal immunocompetent cells to the fetus. It is conceivable that trophoblast cells may block the passage of immune cells to the fetus more effectively than they block passage of uncommitted immunocompetent cells. Although support for this concept is lacking it is possible that immune cells could be trapped preferentially by trophoblast either specifically or nonspecifically. For example it is known that immune lymphocytes have greater nonspecific stickiness for glass than nonimmune lymphocytes (Salerno and Pontieri 1969).

The common failure of humoral alloantibodies against transplantation antigens to cause serious injury to the fetus, despite their demonstrable cytotoxicity for target cells *in vitro,* is in keeping with the generally accepted view that such antibodies also play a lesser role than immune cells in the rejection of surgical allografts and in tissue injury associated with autoimmune disease. To illustrate, it is common experience that both allograft immunity and autoimmune disease can be transferred readily with immune cells but not with immune serum. In the case of the fetus it is only when alloantibody is directed against circulating elements such as red cells that life-threatening injury results, for example, Rh disease.

The possible mechanisms responsible for the limited toxicity of alloimmune sera against transplantation antigens are especially deserving of attention. Comparative studies of antisera of xenogeneic and allogeneic origin have provided some clues for explaining the phenomenon. Whereas alloimmune sera do not commonly produce serious *in vivo* toxicity, xenoimmune sera are commonly toxic. The difference is evidently due to the greater antigenic diversity between animals in xenogeneic systems. The antibodies in alloimmune sera are directed against a limited number of the antigens of the target cell but the antibodies of xenoimmune sera are directed against essentially all antigens of the target species. Thus the chance that potentially injurious antigen-antibody systems capable of activating complement will be involved is profoundly greater in the case of xenoimmune sera than alloimmune sera. Support for this concept is provided by the work of Möller and Möller (1966). The cytotoxic activity of an antiserum evidently depends on its capacity to fix and activate complement in critical locations on the cell surface; this rests not only on the concentration and nature of antibodies in the serum but upon the concentration, nature, and geographic positioning of antigens exposed on the cell surface. In general, antisera, such as those of xenogeneic origin, containing an abundance of antibodies of different specificities directed against a large number of different antigens on the cell surface have the greatest complement fixing and cytotoxic potential. However there are additional considerations; antibodies, particularly alloantibodies and autoantibodies, are often less effective in producing cell lysis *in vivo* than *in vitro;* this may result from an unexplained coating of cells with denatured complement (Evans, Turner, and Bingham 1967) or complementlike components which takes place *in vivo.* It is possible that this mechanism augments the effector aspects of enhancement produced by enhancing antibodies, and that it may thus aid in sparing the fetus and in lessening immunologic injury in autoimmune disease.

B. *The Trophoblast Controversy*

Although it is generally held that trophoblast is the decisive barrier to rejection of the fetus as an allograft there is no consensus regarding the manner by which it exercises this function. Questions which are still disputed or which invite exploration are:

1. Does trophoblast contain paternal histocompatibility antigens?
2. Does the barrier function of trophoblast cells depend on a surface coating of sialomucin?
3. What is the basis of the "allograft tolerance" developed by the mother and fetus during pregnancy and what factors contribute to its development?
4. Does maternal immunologic tolerance, enhancement, or immune deviation contribute significantly to retention of the fetus as an allograft?
5. What is the significance of trophoblast and graft tolerance between fetus and mother with respect to the evolution of species and the economy of the individual?
6. What clues may studies on the immunology of pregnancy provide toward a better understanding of basic immunologic phenomena and immunologic diseases?
7. Why do maternal lymphocytes fail to produce graft-versus-host disease in the fetus?

Certain of these questions will now be considered.

It is generally accepted that histocompatibility antigens remain weak in the developing embryo. Nevertheless they are present in early blastocysts (Simmons and Russell 1966) including a sufficient quantity of paternal histocompatibility antigen to render blastocysts vulnerable to immune attack, as indicated by the observation that they are readily rejected when placed in extrauterine positions in alloimmune hosts but survive when placed in extrauterine positions in syngeneic hosts (Kirby, Billington, and James 1966).

One theory that has been entertained is that the early blastocyst contains paternal histocompatibility antigens but that its trophoblastic elements do not contain, or at least do not manifest, such antigens. Of great interest are the recent reports indicating that trophoblast cells contain paternal histocompatibility antigens but that their immunogenicity is masked by surface sialomucin (Kirby et al. 1964; Bradbury, Billington, and Kirby 1965; Currie, van Doorninck, and Bagshawe 1968).

The truth of this concept has been shown most convincingly by Currie (1968), who demonstrated that trophoblast cells divested of sialomucin by neuraminadase treatment induced strong immunity to skin allografts, whereas untreated trophoblast cells did not. Currie and Bagshawe (1967) also showed that trophoblast cells divested of sialomucin with trypsin are killed on *in vitro* exposure to maternal lymphocytes. They favor the concept that sialomucin acts by steric hindrance to prevent the intimate contact between immune cells and target cells needed for target cell destruction.

Trophoblast is evidently not impregnable to all sorts of immune attack, for Simmons and Russell (1967) have found that mouse trophoblast cells are destroyed when transplanted to immunized rats. This result in a xenogeneic system suggests the possibility that alloimmune forces might also cause at least limited injury to trophoblast cells. Perhaps the enlarged placentas observed in allogeneic as compared to syngeneic matings are due to such injury. This idea is in accord with the observation that the enlargement of placentas accompanying allogeneic pregnancies can be abolished by inducing specific immunologic tolerance in the mothers before mating or accentuated by immunizing mothers with paternal antigens before mating (James 1965). The finding by Breyere and Sprenger (1969) that hyperimmunization of mothers before pregnancy with paternal antigens can reduce the litter size of mice is also compatible with this concept. Thus in allogeneic systems it is probable that trophoblast cells do not completely protect the conceptus against immune attack, albeit in most instances the fetus remains ostensibly healthy.

As I will emphasize later, trophoblast may, in some species at least, be aided in its protective role by events of pregnancy that oppose allograft rejection: classical immunologic tolerance, enhancement, and immune deviation. Nevertheless the capacity of trophoblast to oppose rejection of the fetus is evidently tremendous and in most allogeneic matings is probably fully sufficient to protect the fetus. Impressive evidence of this has been provided by Lanman, Dinerstein, and Fikrig (1962), who transplanted fertilized rabbit ova into allogeneic foster-mothers preimmunized with the antigens of both true parents to insure the development of "double-dose" maternal immunity against the fetus. Pregnancy was normal and the offspring were healthy even though the mother promptly rejected paternal skin grafts placed during midpregnancy or within weeks after onset of pregnancy! These results ruled out the possibility that during pregnancy a developing maternal tolerance might abolish an effective existing immunity, and provide strong support for the view that in the rabbit it is necessary to consider only one role for trophoblast as a barrier, that of blocking the effector limb of the immune response. Trophoblast is presumed to accomplish this by its unique capacity to resist attack by immune cells and to prevent such cells from reaching the fetus. Prior to the discovery that the fetus can develop normally in hyperimmunized mothers it was thought that any hypothetical placental barrier of fetal origin such as trophoblast must of necessity prevent immunization of the mother by paternal antigens present in the barrier itself or in other fetal components. This concept is still deserving of attention, especially in man and animals in which the placenta allows free passage of antibodies of the class IgG. In the course of evolution it is conceivable that with the advent of free placental passage of "immune" antibodies counterforces arose which promoted graft tolerance and thus provided additional protection to the fetus. With respect to trophoblast, it is now obvious that its lack of immunogenicity or its possible unique capacity to induce tolerance, need not be invoked to explain either retention of the fetus or the development of graft tolerance. Indeed, an alternative and equally likely source of antigen for inducing the afferent responses of both maternal and fetal tolerance and immunity is leukocytes which probably cross the placenta in greater numbers than was formerly suspected. Pulvertaft and Pulvertaft (1966) reported that lymphocyte transformation can be demonstrated with blood taken from the umbilical cord vein at delivery, presumably because such blood contains an admixture of maternal and fetal lymphocytes. Whether this represents a terminal event accompanying delivery is uncertain.

Allogeneic differences involve many antigens; evidently in pregnancy some succeed in stimulating forces of maternal immunity (Soren 1967); others incite tolerance (Breyere and Burhoe 1963). In theory the net result of maternal responses depends on the relative forces of tolerance engendered versus those of immunity. In man and most of the laboratory animals studied the fetomaternal responses during pregnancy have been commonly those of graft tolerance rather than immunity. The investigations of Avery and Hunt (1968) in mice indicate that the development of maternal tolerance is favored by repeated exposure to small doses of paternal antigen rather than exposure to large doses of antigen.

The nature of the placenta and its permeability to antigens and antibodies undoubtedly influences the net immune response. For example in cattle, in which the placenta does not permit the passage of antibody, it has been reported that maternal immunity to the fetus, but not tolerance, develops (Billingham and Lampkin 1957). The role of placental permeability in determining the success of interspecies hybridization is not clear.

Even though the maternal afferent responses of immunity and tolerance may commonly be of only secondary significance with respect to the fate of the fetus as an allograft they are important events of basic biologic interest.

II. Immunologic Events Associated with Pregnancy

In view of the great difficulty of inducing tolerance in adult animals it is a most singular finding that, in some species at least, both mothers and their offspring commonly develop strong fetomaternal graft tolerance, as evidenced by prolonged retention of exchange grafts of skin placed after birth (Anderson 1969). Whether such graft tolerance depends on classical immunologic tolerance, enhancement, or immune deviation is uncertain. It is of singular interest, however, that with respect to lymphocyte blastogenesis induced by the mixed leukocyte technique the lymphocytes of most pregnant women show a depressed specific responsiveness to their husbands' leukocytes as compared to those of unrelated males (Lewis et al. 1966).

Early evidence of mutual fetomaternal graft tolerance in man was presented by Peer, Bernhard, and Walker (1958). The full significance of their important findings remains to be realized. In a series of well-conducted experiments they demonstrated tolerance to grafts of full-thickness skin exchanged between mother and child but were unable to find such tolerance between father and child. They also demonstrated that mothers were more tolerant of skin grafts from their children than children were of skin grafts from their mothers. This was surprising in view of the greater ease with which artificial tolerance is produced in the fetus and newborn than in the adult. Other noteworthy findings were that tolerance was exhibited to grafts placed as long as 1 to 2 months after parturition and that occasional grafts were still viable 8 months or more after grafting. This is a most remarkable finding in view of the fact that skin grafts are among the most easily rejected of all grafts. One case of singular interest was that of switch grafts between a 12-year-old boy and his mother; both remained viable for at least 120 days. Her tolerance after such a prolonged period could conceivably have been due to a subsequent unrecognized pregnancy; but what could his prolonged acceptance of her graft have been due to? Was this a rare fortuitous case of close histocompatibility between donor and recipient? Perhaps the donors of the grafts which survived the longest possessed the closest histocompatibility with their respective recipients. For interpreting the results it would be advantageous to know the degree of histocompatibility that existed in the various mother-child pairs and, since classical tolerance is presumed to persist only as long as antigen persists, how long foreign leukocytes may have persisted in their circulations or colonized in their tissues. Possibly maternal antibodies against allotype globulins of the child reached and suppressed certain of his lymphocytes *in utero* (Marcuson and Roitt 1969). Suppression of the production of specific allotype globulin by antiserum is known to persist for months to years. In any event it is reasonably certain that the most important mechanism contributing to persistence of the grafts was some form of immune response. Indeed it is probable that in man histocompatibility antigens transgress the placenta in both directions and often establish graft tolerance in mothers and their offspring which is stronger and more long-lasting than has been generally suspected.

The mutual transplantation tolerance developed during pregnancy by the human mother and fetus has been shown to occur in several other species, including the armadillo, rat, dog, and sheep (Anderson 1969) but not in the rabbit (Lanman, Dinerstein, and Fikrig 1963). Among rats, tolerance in mothers was always found to be greater than in their offspring and waned gradually over a period of weeks to months. This difference suggests that the bases of the "tolerance" developed by the mother and by the fetus are different. Indeed there is evidence to support the concept that allograft tolerance in mothers may be the net result of several immunologic mechanisms.

Kaliss and Dagg (1964) suggested that immunologic enhancement may be responsible for the maternal tolerance to paternal antigens which develops in mice as the result of multiparity "parity tolerance." They presented good arguments favoring this view, but failed in their attempts to demonstrate enhancement by passive transfer of parity tolerance of mice with the serum of multiparous animals. However the possible role of enhancement in parity tolerance will probably not be ruled out until enhancing antibody can be isolated and used in pure form for passive transfer. This may soon be possible; recent evidence presented by Tokuda and McEntee (1967) indicates that enhancing antibody of the mouse belongs to the gamma 2 subclass of IgG. Recent evidence that parity tolerance is due to enhancing antibody has been obtained by use of the colony inhibition test (K. Hellstrom and I. Hellstrom, personal communication).

An alternative and equally probable basis of "parity tolerance" is "immune deviation." This is a poorly understood immunologic phenomenon in which an induced immune response which normally proceeds to delayed hypersensitivity and cellular immunity is "deviated" in the direction of humoral antibody production and immediate hypersensitivity. Its demonstration is accomplished by administering soluble antigen in saline to an animal before injecting the antigen-adjuvant preparation which would otherwise produce delayed sensitivity and cellular immunity. A good model for this purpose is experimental autoallergic encephalomyelitis. The disease, which is produced by injecting brain antigen incorporated in complete Freund's adjuvant, and which involves the development of cellular immunity and delayed hypersensitivity to brain antigen, can be prevented by prior injection of brain antigen in saline to "deviate" the immune response. Immune deviation has not been passively transferred with serum and hence appears to be distinct from immunologic enhancement.

The source and physical state of antigen responsible for natural tolerance associated with pregnancy and the manner of its induction in the fetus and mother are matters of great interest to the immunologist because of meager knowledge of the mechanisms of induction and maintenance of tolerance, either artificial or natural. The problem of induction of immunologic tolerance is extremely complicated. Recent evidence indicates that several types of cells including macrophages are involved in the various responses to antigen. The nature of the antigen as well as the dosage and route by which antigen reaches immunocompetent cells are important determinants of the immune response; tolerance can be induced by either high or low doses of antigen and more readily with certain antigens than others. In some instances at least, physical aggregation of antigen favors antibody production, whereas disaggregation favors the induction of tolerance.

Is it possible that pregnancy in some way promotes the induction of classical immunologic tolerance, immune deviation or enhancing antibody in the mother or that maternal cells which reach the fetus find an environment peculiarly conducive to the induction of these events?

Could the trophoblast itself possess unique tolerogenic properties, possibly related to the nature or weakness of its paternal antigens or the manner of their release into the maternal circulation? This concept is compatible with the recent findings of Kirby (1969) that extrauterine grafts of allogeneic blastocysts, which give rise to predominately trophoblastic growth, induce specific immunologic tolerance in the recipient. It is notable that in man large numbers of trophoblast cells enter the maternal circulation, principally during the third trimester of pregnancy (Ober 1959), and lodge in the lung, where they undergo prompt lysis for unknown reasons.

In the case of the fetus it is possible that environmental conditions independent of immunologic immaturity render the fetus uniquely capable of developing immunologic tolerance. For example, could the immaturity of the reticuloendothelial system of the fetus, with possible associated deficiency of macrophages to process antigen, have any bearing on the development of tolerance by the fetus?

I am not aware that any work has been done to answer these obvious questions.

With respect to the source of tolerogens there is now abundant evidence that in man and certain other species blood cells can pass in appreciable numbers between mother and fetus. Consequently it is not necessary to limit consideration of maternal immune responses and transplantation tolerance in pregnancy solely to antigens derived from trophoblast, since leukocytes and platelets could contribute equally well. A point of interest in this regard is the finding that the injection of hyaluronidase during pregnancy, to increase placental permeability, greatly increases the tolerance of both newborn rabbits and their mothers to skin grafts (Najarian and Dixon 1963).

The recent work of Tuffrey, Bishun, and Barnes (1969) indicates that in mice there is ample opportunity for tolerance induction in both mother and fetus by leukocytes which pass the placental barrier. By the use of CBA/T6T6 mice carrying embryos developed from implanted CFW blastocysts they showed that as many as ⅓ of the cells of certain tissues of the offspring (spleen, lymph nodes, thymus, bone marrow, and liver) were of maternal origin.

Since parity tolerance increases in mice with succeeding interstrain pregnancies (Breyere and Barrett 1960), it seems possible that the key to the development of graft tolerance in the mother lies in some type of response which is favored by repeated exposure to small amounts of antigen over extended periods of time rather than to antigen with special tolerogenic properties. The idea that graft tolerance is induced by small repeated doses of antigen is more compatible with the development of enhancing antibody or immune deviation than with the development of classical immunologic tolerance.

Results of work on experimental parabiosis of animals may provide clues to events which take place during pregnancy. For example Strober and Murray (1967) reported that plasma taken from interstrain mouse parabionts is capable of specifically prolonging the survival of grafts exchanged between mice of the parabiont strains, presumably because of enhancing antibodies. It is of interest that the plasma lacked hemagglutinating and cytotoxic antibodies. Work on graft versus host disease also suggests mechanisms by which injury of the fetus by maternal lymphocytes may be avoided. Field, Cauchi, and Gibbs (1967) have observed that F_1 hybrid rats which recover from allogeneic graft-versus-host disease produced by injecting parental lymphoid cells often show resistance to rechallenge. Such resistance can be passively transferred by cross-circulation with extracorporeally irradiated blood and, presumably, with serum, and depends on enhancing antibodies produced by the grafted cells!

It is virtually certain that the phenomena of immunologic enhancement, classical immunologic tolerance, and immune deviation are not laboratory artifacts lacking biological significance but instead represent evolutionary mechanisms with benefits to the host, possibly including those of protecting the fetus and avoiding life-threatening injury due to allergy and autoimmune disease.

The idea of exploring these phenomena by studying them in the natural setting of pregnancy is appealing. Obviously many directions could be taken. Species with different types of placentas should be compared and studies on interspecies hybridization should be conducted.

In the limited time remaining I will now give attention to a related area, that of choriocarcinoma.

III. Gestational Choriocarcinoma

Gestational choriocarcinoma (CHO), a maternal tumor derived from trophoblast cells, has long been of interest to oncologists and immunologists because of the potential immunogenicity it should have for its host owing to the paternal transplantation antigens and tumor specific antigens it is presumed to possess. Numerous attempts have been made to immunize CHO patients with their own tumors or with skin grafts from their husbands (Robinson et al. 1967). However, it is doubtful whether any of these attempts have been successful.

The progress of CHO is rapid and death usually results within 6 to 12 months. In light of recent reports that trophoblast cells are endowed with a heavy coat of sialomucin it is logical to ask whether CHO retains this coat during tumor progression and whether sialomucin serves to protect the tumor against rejection. Unfortunately CHO has not been encountered in animals other than man and there has been little opportunity for experimental study of the tumor. Consequently no evidence is available regarding the possible relation of surface sialomucin to the progress of CHO, despite the fact that Currie and his associates (see review Currie 1968) have assembled much data to indicate that the malignancy of transplantable animal tumors is favored by their coating of sialomucin.

There is little information at hand relative to the question whether choriocarcinomas possess paternal transplantation antigens. However, the observation of Robinson et al. (1967) that patients with CHO often show tolerance to skin grafts of their husbands favors this view. Currie and Bagshawe (1967) have reported that choriocarcinoma cells divested of sialomucin with trypsin are destroyed by allogeneic and host lymphocytes *in vitro*. Although the considerable number of cases of spontaneous regression of CHO of maternal origin (Park and Lees 1950) could be due to either paternal transplantation antigen or tumor specific antigen or both, the lack of reports of spontaneous regression of CHO of testicular origin favors the view that regression of CHO in females is due to paternal transplantation antigens.

There is the question whether strong histocompatibility differences between the patient and the tumor exist and whether they disfavor tumor progression; if this could be shown, it would constitute evidence that these tumors possess effective paternal transplantion antigens. There is also the question of whether "tolerance" develops and favors tumor progression. In a recent study involving family histocompatibility typing of antigens determined by the HL-A locus, Mogenson, Kissmeyer-Nielsen, and Hauge (1969) obtained some evidence indicating that the progression of choriocarcinoma may be inversely related to the extent of HL-A incompatibility between husband and wife. However, in view of the recent finding that enhancing antibody is frequently present in the serum of animals and patients with progressing tumors of various types (Hellstrom et al. 1969) it is reasonable to anticipate that enhancement may play the key role in the progression of choriocarcinoma by opposing the forces of cellular immunity; indeed the studies of Robinson et al. (1967) support the concept that some form of immunologic response leading to graft tolerance is involved in progression of this tumor. If enhancing antibodies occur they could render attempts to correlate tumor progression with maternal-paternal histoincompatibility difficult.

References

Anderson, J. M. 1969. The "immunological inertia" of viviparity. *Transplantation Proc* 1:67.

Avery, G. B., and Hunt, C. V. 1968. Maternal sensitization after pregnancy in mice. *Fertil Steril* 19:826.

Billingham, R. E. 1964. Transplantation immunity and the maternal-fetal relation. *New Eng J Med* 270:667.

Billingham, R. E., and Lampkin, G. H. 1957. Further studies in tissue homotransplantation in cattle. *J Embryol Exp Morph* 5:351.

Bradbury, S.; Billington, W. D.; and Kirby, D. R. S. 1965. A histochemical and electron microscopical study of the fibrinoid of the mouse placenta. *J Roy Micr Soc* 84:199.

Breyere, E. J., and Barrett, M. K. 1960. Prolonged survival of skin homografts in parous female mice. *J Nat Cancer Inst* 25:1405.

Breyere, E. J., and Burhoe, S. O. 1963. The nature of the "partial" tolerance induced by parity. *J Nat Cancer Inst* 31:179.

Breyere, E. J., and Sprenger, W. W. 1969. Evidence of allograft rejection of the conceptus. *Transplantation Proc* 1:71.

Currie. G. A. 1968. Immunology of pregnancy: The feto-maternal barrier. *Proc Roy Soc Med* 61:1206.

Currie, G. A., and Bagshawe, K. D. 1967. The masking of antigens on trophoblast and cancer cells. *Lancet* 1:708.

Currie, G. A.; van Doorninck, W.; and Bagshawe, K. D. 1968. Effect of neuraminidase on the immunogenicity of early mouse trophoblast. *Nature* (*London*) 219:191.

Evans, R. S.; Turner, E.; and Bingham, M. 1967. Chronic hemolytic anemia due to cold agglutinins: The mechanism of resistance of red cells to C′ hemolysis by cold agglutinins. *J Clin Invest* 46:1461.

Field, E. O.; Cauchi, M. N.; and Gibbs, J. E. 1967. The transfer of refractoriness to G-V-H disease in F_1 hybrid rats. *Transplantation* 5:241.

Hellstrom, I.; Hellstrom, K. E.; Evans, C. A.; Heppner, G. H.; Pierce, G. E.; and Yang, J. P. 1969. Serum-mediated protection of neoplastic cells from inhibition by lymphocytes immune to their tumor-specific antigens. *Proc Nat Acad Sci USA* 62:362.

James, D. A. 1965. Effects of antigenic dissimilarity between mother and foetus on placental size in mice. *Nature (London)* 205:613.

Kaliss, N., and Dagg, M. K. 1964. Immune response engendered in mice by multiparity. *Transplantation* 2:416.

Kirby, D. R. S. 1969. Is the trophoblast antigenic? *Transplantation Proc* 1:53.

Kirby, D. R. S.; Billington, W. D.; Bradbury, S.; and Goldstein, D. J. 1964. Antigen barrier of the mouse placenta. *Nature (London)* 204:548.

Kirby, D. R. S.; Billington, W. D.; and James, D. A. 1966. Transplantation of eggs to the kidney and uterus of immunized mice. *Transplantation Proc* 4:713.

Lanman, J. T. 1965. Transplantation immunity in mammalian pregnancy: Mechanisms of fetal protection against immunologic rejection. *J Pediat* 66:525.

Lanman, J. T.; Dinerstein, J.; and Fikrig, S. 1962. Homograft immunity in pregnancy: Lack of harm to the fetus from sensitization of the mother. *Ann N Y Acad Sci* 99:706.

———. 1963. The survival time of skin homografts exchanged between mother and offspring in rabbits. *Transplantation Proc* 1:509.

Lanman, J. T., and Herod, L. 1965. Homograft immunity in pregnancy: The placental transfer of cytotoxic antibody in rabbits. *J Exp Med* 122:579.

Lewis, J., Jr.; Whang, J.; Nagel, B.; Oppenheim, J. J.; and Perry, S. 1966. Lymphocyte transformation in mixed leukocyte cultures in women with normal pregnancy or tumors of placental origin. *Amer J Obstet Gynec* 96:287.

Lieberman, R., and Dray, S. 1964. Maternal-fetal mortality in mice with isoantibodies to parental γ-globulin allotypes. *Proc Soc Exp Biol Med* 116:1069.

Marcuson, E. C., and Roitt, I. M. 1969. Transformation of rabbit lymphocytes by anti-allotype serum: Ultrastructure of transformed cells and suppression of responding cells by fetal exposure to anti-allotype serum. *Immunology* 16:791.

Medawar, P. B. 1953. Some immunological and endocrinological problems raised by evolution of viviparity in vertebrates. In *Sympos Soc Exp Biol,* ed. J. F. Danielli and R. Brown, 7:320. New York: Academic Press.

Mogenson, B.; Kissmeyer-Nielsen, F.; and Hauge, M. 1969. Histocompatibility antigens on the HL-A locus in gestational choriocarcinoma. *Transplantation Proc* 1:76.

Möller, G., and Möller, E. 1966. Immune cytotoxicity and immunological enhancement in tissue transplantion. In *Antibodies to biologically active molecules.* Proc II Meeting Fed Europ Biochem Soc Vienna, 1:349. Oxford and New York: Pergamon Press.

Najarian, J. S., and Dixon, F. J. 1963. Induction of tolerance to skin homografts in rabbits by alterations of placental permeability. *Proc Soc Exp Biol Med* 112:136.

Ober, W. B. 1959. Historical perspectives on trophoblast and its tumors. *Ann N Y Acad Sci* 80:3.

Park, W. W., and Lees, J. C. 1950. Choriocarcinoma: A general review with an analysis of five hundred and sixteen cases. *AMA Arch Path* 49:205.

Peer, L. A.; Bernhard, W.; and Walker, J. C., Jr. 1958. Full-thickness skin exchanges between parents and their children. *Amer J Surg* 95:239.

Pulvertaft, R. J. V., and Pulvertaft, I. 1966. Spontaneous "transformation" of lymphocytes from the umbilical-cord vein. *Lancet* 2:892.

Robinson, E.; Ben-Hur, N.; Zuckerman, H.; and Neuman, Z. 1967. Further immunologic studies in patients with choriocarcinoma and hydatidiform mole. *Cancer Res* 27:1202.

Salerno, A., and Pontieri, G. M. 1969. Heterogeneity of populations of antibody-forming cells fractionated by glass bead columns. *Clin Exp Immun* 5:209.

Simmons, R. L. 1969. Histoincompatibility and the survival of the fetus: Current controversies. *Transplantation Proc* 1:47.

Simmons, R. L., and Russell, P. S. 1962. The antigenicity of mouse trophoblast. *Ann N Y Acad Sci* 99:717.

———. 1966. The histocompatibility antigens of fertilized mouse eggs and trophoblast. *Ann N Y Acad Sci* 129:35.

———. 1967. Xenogenic antigens in mouse trophoblast. *Transplantation* 5:85.

Soren, L. 1967. Immunological reactivity of lymphocytes in multiparous females after strain specific matings. *Nature* (*London*) 213:621.

Strober, S., and Murray, J. E. 1967. Studies of the enhancing properties of plasma of murine parabionts. *Transplantation Proc* 5:1371.

Tokuda, S., and McEntee, P. F. 1967. Immunologic enhancement of Sarcoma I by mouse γ-globulin fractions. *Transplantation Proc* 5:606.

Tuffrey, M.; Bishun, N. P.; and Barnes, R. D. 1969. Porosity of the mouse placenta to maternal cells. *Nature* (*London*) 221:1029.

Woodruff, M. F. A. 1957. Transplantation immunity and the immunological problem of pregnancy. *Proc Roy Soc* [*Biol*] 148:68.

Contributors

S. J. BEHRMAN
Center for Research in Reproductive Biology
Department of Obstetrics and Gynecology
Medical Center, University of Michigan
Ann Arbor, Michigan 48104

J. D. BIGGERS
Division of Population Dynamics
School of Hygiene and Public Health
The Johns Hopkins University
Baltimore, Maryland 21205

R. J. BLANDAU
Department of Biological Structure
School of Medicine
University of Washington
Seattle, Washington 98105

C. W. BODEMER
Department of Biomedical History
School of Medicine
University of Washington
Seattle, Washington 98105

B. G. BÖVING
Departments of Gynecology-Obstetrics and Anatomy
School of Medicine
Wayne State University
Detroit, Michigan 48207

B. G. BRACKETT
Division of Reproductive Biology
Department of Obstetrics and Gynecology
School of Medicine

Department of Animal Biology
School of Veterinary Medicine
University of Pennsylvania
Philadelphia, Pennsylvania 19104

R. L. BRINSTER
Laboratory of Reproductive Physiology
School of Veterinary Medicine
University of Pennsylvania
Philadelphia, Pennsylvania 19104

D. H. CARR
Department of Anatomy
McMaster University
Hamilton, Canada

C. A. B. CLEMETSON
Research Laboratory
Department of Obstetrics and Gynecology
Methodist Hospital of Brooklyn
Brooklyn, New York 11215

J. T. CONRAD
Department of Physiology and Biophysics
Department of Obstetrics and Gynecology
School of Medicine
University of Washington
Seattle, Washington 98105

J. DAVIES
Department of Anatomy
School of Medicine
Vanderbilt University
Nashville, Tennessee 37203

A. C. ENDERS
Department of Anatomy
School of Medicine
Washington University
Saint Louis, Missouri 63110

V. H. FERM
Department of Anatomy and Cytology
Dartmouth Medical School
Hanover, New Hampshire 03755

E. FROMM
Biological Sciences
Biomedical Engineering and Science Program
Drexel University
Philadelphia, Pennsylvania 19104

B. J. GULYAS
Department of Anatomy
School of Medicine
Georgetown University
Washington, D.C. 20007

E. S. E. HAFEZ
Reproduction Laboratory
Department of Gynecology-Obstetrics
School of Medicine
Wayne State University
Detroit, Michigan 48207

R. HERTZ
The Population Council
Rockefeller University
New York, New York 10021

H. HESSELDAHL
Department of Anatomy
School of Medicine
Vanderbilt University
Nashville, Tennessee 37203

J. F. HULKA
Department of Obstetrics and Gynecology
School of Medicine
University of North Carolina
Chapel Hill, North Carolina 27514

R. W. JEANLOZ
Department of Biological Chemistry
Harvard University Medical School

Laboratory for Carbohydrate Research
Massachusetts General Hospital
Boston, Massachusetts 02114

D. R. S. KIRBY (*deceased*)
Department of Zoology
University of Oxford
Oxford, England

R. S. KRISHNAN
Department of Biochemistry
Colorado State University
Fort Collins, Colorado 80521

C. LUTWAK-MANN
Unit of Reproductive Physiology and Biochemistry
(Agricultural Research Council)
University of Cambridge
Cambridge, England

V. R. MALLIKARJUNESWARA
Research Laboratory
Department of Obstetrics and Gynecology
Methodist Hospital of Brooklyn
Brooklyn, New York 11215

M. M. MOSHFEGHI
Research Laboratory
Department of Obstetrics and Gynecology
Methodist Hospital of Brooklyn
Brooklyn, New York 11215

H. W. MOSSMAN
Department of Anatomy
University of Wisconsin
Madison, Wisconsin 53706

A. V. NALBANDOV
Department of Animal Science
University of Illinois
Urbana, Illinois 61801

N. O. OWERS
Department of Anatomy
Medical College of Virginia
Virginia Commonwealth University
Richmond, Virginia 23219

R. E. RUMERY
Department of Biological Structure
School of Medicine
University of Washington
Seattle, Washington 98105

D. SZOLLOSI
Department of Biological Structure
School of Medicine
University of Washington
Seattle, Washington 98105

R. S. WEISER
Department of Microbiology
School of Medicine
University of Washington
Seattle, Washington 98105

H. M. WEITLAUF
Department of Anatomy
Department of Obstetrics and Gynecology
University of Kansas Medical Center
Kansas City, Kansas 66103

R. M. WYNN
Department of Obstetrics and Gynecology
University of Illinois at the Medical Center
Chicago, Illinois 60680

J. M. YOCHIM
Department of Physiology and Cell Biology
University of Kansas
Lawrence, Kansas 66044

Author Index

Subject Index